Multimodal Optical Diagnostics of Cancer

Valery V. Tuchin • Jürgen Popp
Valery Zakharov
Editors

Multimodal Optical Diagnostics of Cancer

Springer

Editors
Valery V. Tuchin
Department of Optics and Biophotonics
Saratov State University
Saratov, Russia

Valery Zakharov
Department of Laser and Biotechnical Systems
Samara National Research University
Samara, Russia

Jürgen Popp
Leibniz-Institute of Photonic Technology e.V. and Institute of Physical Chemistry and Abbe Center of Photonics
University of Jena
Jena, Thüringen, Germany

ISBN 978-3-030-44593-5 ISBN 978-3-030-44594-2 (eBook)
https://doi.org/10.1007/978-3-030-44594-2

This Springer imprint is published by the registered company Springer Nature Switzerland AG
The registered company address is: Gewerbestrasse 11, 6330 Cham, Switzerland

Preface

The purpose of this book is to outline the state-of-the-art optical approaches for the enhancement of medical diagnostics and earlier cancer detection and classification. The book brings together a description of the wide variety of optical techniques that have recently emerged for the specific study of the tumors of different organs and tissues. The appearance of this book was stimulated by the recent rapid progress in novel photonics technologies, the development of robust and cost-effective low-noise detectors and lasers, and fiber-optics devices. The multimodal approach provides a unique combination of structural, morphological, molecular, and metabolic information. The complexities of different methods of integration are compensated by the potential increase in both sensitivity and specificity of cancer diagnosis, screening, treatment monitoring, or image-guided intervention.

The diagnosis of cancer is a complex process and requires a number of diagnostic studies. But standard imaging modalities such as computed tomography, magnetic resonance imaging, and positron emission tomography require significant financial resources and infrastructure, which limits access to these modalities and excludes their usage for screening. In contrast, optical imaging strategies, with the potential for reduced cost and enhanced portability, are emerging as additional tools to facilitate the early detection and diagnosis of cancer. Many spectroscopic methodologies and optical imaging technologies have been established for nearly all kinds of cancer. However, the current gold-standard in cancer diagnosis is the examination of a neoplasm by the trained eye of a physician followed by histological examination of an invasive excisional biopsy of the tumor tissue specimen. Pathologists rely on the microscopic analysis of tissue samples or additionally employ immune histochemical or molecular pathological analyses for a more precise diagnosis, classification, and prognosis regarding the cancerous tumor. In the majority of cases, the pathologist works with fixated and embedded tissue samples or, at least, with frozen sections. Intrinsically, the histological process is time-consuming. It is obvious that new methods and approaches are required for quick and reliable *in vivo* production of diagnostically relevant additional information. The optical methods meet these challenges through the implementation of molecular sensitive spectroscopic methods, including linear and nonlinear Raman spectroscopy and its combination with other spectroscopic/optical modalities to a multimodal imaging approach for precise surgical guidance and intraoperative histopathologic examination of tissue.

However, due to the rather long acquisition times of spectroscopic tissue examination, the scanning of larger regions is quite challenging. This problem can be circumvented by combining detailed tumor spectral study with a fast imaging technique providing a quick selection of the region of interest (ROI). At the same time, the diagnostic accuracy depends as on the sensitivity of cancer detection as on the precision of correct ROI selection. Therefore, each optical imaging method also needs to be optimized for cancer detection and a multimodal approach should provide a synchronous sequential multilevel refinement of the diagnosis, both by spatial localization and cancer type detection. Hyperspectral imaging (HIS), optical coherence tomography (OCT), or fluorescence lifetime imaging (FLIM) offers a great potential for such a combination with the possibility of fiber-optics implementation suitable for clinical conditions.

Expanding the role of spectral analysis and optical imaging became a reality in global cancer management, including screening, early detection at the point-of-care, biopsy guidance, and real-time histology. In combination with online data analysis and multivariate statistics, multimodal spectral diagnostics and optical imaging have the potential to aid in the high-sensitive detection and management of precancer and early cancer by combining the tissue topology features and chemical composition.

This book is divided into four parts: Part I *Tumor Tissue Optics and Multimodal Microscopy* (Chaps. 1–4); Part II *Diffuse Spectroscopy and Fluorescence Analysis for Cancer Detection* (Chaps. 5–7); Part III *Raman Spectroscopy and Cancer Diagnostics* (Chaps. 8–10); and Part IV *Multimodal Cancer Imaging* (Chaps. 11–16).

Part I describes the tissue optical properties and their alteration with cancer progression. The malignant tissue alteration results in specific changes in nucleic acid, protein, lipid, and carbohydrate quantities of neoplastic cells. It is the basis for any optical cancer diagnostic applications, which might distinguish the tissue strongly associated with cancer development. The morphologic and biochemical changes that occur with malignant tissue are numerous and in many cases depend on the specific type and location of the cancer.

Chapter 1 reviews the light scattering spectroscopic methods and different approaches for tumor tissue optical properties and spectral features characterization. The inverse Monte Carlo simulation and diffuse approximations are used for modeling of light scattering in tumor and normal tissues. The physiological and optical properties of various types of lung, breast, colorectal, prostate, cervical, bladder, stomach, liver, kidney, skin, oral, and brain cancer are considered. The biochemical diagnostic models have been discussed as an instrument that helps to derive the morphological and biochemical composition, the functional state of cellular metabolism, and, as a result, determine the key features for the pathologist diagnostic decisions for appropriate cancer treatment.

In Chap. 2, the authors discuss the enhancement of optical diagnostics of tumors by using optical clearing (OC) methods. The OC-assisted optical imaging techniques allow for acquiring high-resolution structural and functional images of neoplasms and their microvasculature. The monitoring of OC agents diffusion in tissues with high temporal and depth resolution allows one to differentiate healthy from malignant tissues.

The application of fluorescence lifetime imaging (FLIM) for recognition of changes in structural information accompanied by certain biochemical processes in living organisms is explored in Chap. 3. This chapter describes three main techniques on the basis of fluorescence time-resolved measurements—microscopic, spectroscopic, and macroscopic—and provides multiple examples of their applications in probing metabolism in cancer.

Chapter 4 describes the imaging of extracellular vesicles, including exosomes, for diagnosis, monitoring, and prediction of cancer. Different methods of light scattering, fluorescence, and single- and two-photon resonance absorption for exosome morphology and size measurements are presented.

In Part II, Chap. 5, the authors review the application of functional near-infrared spectroscopy to cancer diagnostics and therapy monitoring. It is discussed the capabilities of tissue differentiation, imaging of microvasculature and blood oxygenation concentrations, the possibility of brain cancer detection, and enhancement of noninvasive diagnostics in combination with other imaging modalities.

Chapter 6 describes the methods of volatile organic compounds measurements and details the spectral analysis of breath air samples for understanding specific biochemical processes and as a diagnostic tool for lung cancer diagnosis. High accuracy of multiclass classification of patients with several pulmonary diseases was demonstrated using a set of binary support vector machine classifiers.

A combined light-induced autofluorescence analysis and diffuse reflectance spectroscopy for the diagnosis of skin tumors are investigated in Chap. 7. It was demonstrated the preference of autofluorescence methods for detection and discrimination of nonpigmented skin tumors, but the whole set of benign, dysplastic, and malignant cutaneous lesions can be diagnosed with high sensitivity and specificity only with a multimodal approach, which serves as an "optical biopsy" tool.

Part III presents Raman spectroscopy as an especially potent technology with respect to molecular sensitivity providing detection of subtle changes in morphology and the distribution of endogenous molecular markers connected to disease initiation and disease progression in a label-free and noninvasive manner. Chapter 8 discusses the application of Raman spectroscopy in combination with advanced multivariate statistical analysis as a universal and cost-effective method of early cancer diagnosis and presents the benefits and the risks of the methodology.

The authors of Chap. 9 discuss the methods of Raman signal enhancement effects such as surface-enhanced Raman scattering, coherent anti-Stokes Raman scattering, and stimulated Raman scattering. Applications of Raman-based approaches for cancer diagnostics are summarized. Complementary optical methods including autofluorescence, optical coherence tomography, second harmonic generation, and two-photon excited fluorescence in combination with Raman scattering as multicontrast modalities are presented for pathological screening of cancer cells and tissues under *ex vivo* and *in vivo* conditions.

Chapter 10 compares the discussed Raman spectroscopic methods for optical analysis and screening of malignant and benign skin tissues. The clinical applications of skin cancer screening using a multimodal approach and *in vivo* Raman spectroscopy are presented.

Part IV describes different aspects of multimodal cancer imaging. In Chap. 11, the optical instruments for minimally invasive interventions in abdominal surgery are considered. Fine-needle optical probes with immediate comparison of tumor and healthy tissue optical properties are presented for the determination of the tissue state, in particular, for the diagnosis of hepatocellular carcinoma. The integration of multimodal fiber-optical probes with standard semiautomatic biopsy systems is discussed.

Chapter 12 describes the applications of multimodal optical coherence tomography (OCT) for characterizing tumorous tissues, including polarization-sensitive OCT providing complementary images in the initial and cross-polarization channels (CP-OCT), label-free angiography (OCA), and compressional variant of OCT-based elastography (OCE). The principles of realization and examples of various biomedical applications are considered for each modality. The authors present the results of using CP-OCT for detection of tumorous and non-tumorous regions in brain and breast tissues, application of OCA for monitoring the results of photodynamic therapy of murine model tumor CT26 and patients' basal cell carcinoma, utilization of OCA for studying mucositis on patients during radiation therapy, as well as monitoring of tumor response to chemotherapy with OCE.

Different statistical, frequency, and stochastic methods of texture analysis of dermatoscopic and OCT images for differentiation of various malignant and benign tumors are discussed in Chap. 13. The recognition of various tumors contemporaneously with a high-score identification of a tumor type is demonstrated in real clinical conditions by calculating comparative personal textural descriptors and implementation of multi-texture analysis when used texture features (Haralick, Tamura, Gabor, Markov Random Field, Complex Directional Field, Fractal Dimensions) complement each other.

Chapter 14 is dedicated to the specification of hyperspectral imaging devices based on acousto-optical tunable filters and their application for the detection and recognition of skin cancer. The basic concepts related to this technology are detailed. The applicability of hyperspectral imagers for endoscopic, microscopic, and macroscopic studies, as well as image reconstruction is demonstrated for different HSI schemes. The details of skin hyperspectral image analysis are presented along with the accuracy of a skin neoplasm detection.

Chapter 15 focuses on innovations using inelastic Raman scattering analysis for molecular fingerprinting of tissue combined with morphological information obtained from OCT. The combination of Raman and OCT, particularly recording co-localized three-dimensional information, could lead to a new form of diagnostic or screening tool for cancer studies.

A description of Terahertz (THz) spectroscopy and imaging for label-free diagnosis of malignancies with different nosology and localization is presented in Chap. 16. The brief introduction to THz technology, peculiarities of THz-wave–tissue interactions, and THz pulsed spectroscopy are provided. The authors discuss THz imaging of brain tumors *ex vivo* and *in vivo* for mice and rat glioma models, human brain gliomas of different grades, as well as the pilot measurement of human brain meningioma.

This book is aimed at researchers, postgraduate and undergraduate students, laser engineers, biomedical engineers, and physicians who are interested in designing and applying laser and optical methods and instruments for cancer diagnostics and treatment, general application of optical methods in medicine and the medical device industry. Because of the large amount of fundamental and basic research on optical methods presented in this book, it should be useful for a broad audience including students and physicians. Physicians and biomedical engineers will be particularly interested in the chapters covering clinical applications and instrumentation. Optical engineers will also find many critical applications to stimulate novel ideas of laser and optical design.

Finally, the editors would like to thank all the authors who devoted their precious time to contribute very interesting and knowledgeable chapters, all who helped us in the preparation of the book, authors and publishers for their permission of reproducing their figures in this book, and the editorial staff of the publisher.

Saratov, Russia — Valery V. Tuchin
Jena, Germany — Jürgen Popp
Samara, Russia — Valery Zakharov

Contents

Contributors

Dmitry N. Artemyev Department of Laser and Biotechnical Systems, Samara National Research University, Samara, Russian Federation

Alexey N. Bashkatov Department of Optics and Biophotonics, Saratov State University, Saratov, Russian Federation

Laboratory of Biophotonics, Tomsk State University, Tomsk, Russian Federation

Wolfgang Becker Becker & Hickl GmbH, Berlin, Germany

Sheyh-Islyam T. Beshplav Burdenko Neurosurgery Institute, Moscow, Russian Federation

Alexey V. Borisov Department of Physics with a Course of Higher Mathematics, Siberian State Medical University, Tomsk, Russian Federation

Department of General and Experimental Physics, National Research Tomsk State University, Tomsk, Russian Federation

Ekaterina G. Borisova Institute of Electronics, Bulgarian Academy of Sciences, Sofia, Bulgaria

Saratov State University, Saratov, Russian Federation

Ivan A. Bratchenko Department of Laser and Biotechnical Systems, Samara National Research University, Samara, Russian Federation

Lyudmila A. Bratchenko Department of Laser and Biotechnical Systems, Samara National Research University, Samara, Russian Federation

Alla B. Bucharskaya Saratov State Medical University, Saratov, Russian Federation

Mingzhou Chen SUPA, School of Physics and Astronomy, University of St Andrew, St Andrews, Great Britain

Kishan Dholakia SUPA, School of Physics and Astronomy, University of St Andrew, St Andrews, Great Britain

Irina N. Dolganova Bauman Moscow State Technical University, Moscow, Russian Federation

Institute of Solid State Physics of the Russian Academy of Sciences, Moscow, Russian Federation

Viktor Dremin Orel State University, Orel, Russian Federation

University of Oulu, Oulu, Finland

Andrey Dunaev Orel State University, Orel, Russian Federation

Vadim Elagin Privolzhsky Research Medical University, Nizhny Novgorod, Russian Federation

Grigory Gelikonov Institute of Applied Physics RAS, Nizhny Novgorod, Russian Federation

Valentin Gelikonov Institute of Applied Physics RAS, Nizhny Novgorod, Russian Federation

Elina A. Genina Department of Optics and Biophotonics, Saratov State University, Saratov, Russian Federation

Laboratory of Biophotonics, Tomsk State University, Tomsk, Russian Federation

Natalia Gladkova Privolzhsky Research Medical University, Nizhny Novgorod, Russian Federation

Ekaterina Gubarkova Privolzhsky Research Medical University, Nizhny Novgorod, Russian Federation

Valery E. Karasik Bauman Moscow State Technical University, Moscow, Russian Federation

Irina Kasatkina Institute of Applied Physics RAS, Nizhny Novgorod, Russian Federation

Alexander G. Khramov Department of Technical Cybernetics, Samara National Research University, Samara, Russian Federation

Image Processing Systems Institute of RAS - Branch of the FSRC "Crystallography and Photonics" RAS, Samara, Russian Federation

Yulia A. Khristoforova Department of Laser and Biotechnical Systems, Samara National Research University, Samara, Russian Federation

Elena Kiseleva Privolzhsky Research Medical University, Nizhny Novgorod, Russian Federation

Lobachevskiy State University of Nizhny Novgorod, Nizhny Novgorod, Russia Federation

Yury V. Kistenev Laboratory of Biophotonics, National Research Tomsk State University, Tomsk, Russian Federation

Department of Physics with a Course of Higher Mathematics, Siberian State Medical University, Tomsk, Russian Federation

Semyon G. Konovalov Department of Laser and Biotechnical Systems, Samara National Research University, Samara, Russian Federation

Vesa Korhonen Oulu University Hospital, Oulu, Finland

Sergey V. Kozlov Department of Oncology, Samara State Medical University, Samara, Russian Federation

Samara Regional Clinical Oncology Dispensary, Samara, Russian Federation

Christoph Krafft Leibniz Institute of Photonic Technologies, Jena, Germany

Sergey Ksenofontov Institute of Applied Physics RAS, Nizhny Novgorod, Russian Federation

Vladimir N. Kurlov Institute of Solid State Physics of the Russian Academy of Sciences, Moscow, Russian Federation

Sergey Kuznetsov Privolzhsky Research Medical University, Nizhny Novgorod, Russian Federation

Igor K. Lednev Department of Chemistry, University at Albany, State University of New York, Albany, NY, USA

Maria M. Lukina Privolzhsky Research Medical University (PRMU), Nizhny Novgorod, Russian Federation

Alexander S. Machikhin Scientific and Technological Center of Unique Instrumentation of the Russian Academy of Sciences, Moscow, Russian Federation

Andrian Mamoshin Orel State University, Orel, Russian Federation

Orel Regional Clinical Hospital, Orel, Russian Federation

Anna Maslennikova Privolzhsky Research Medical University, Nizhny Novgorod, Russian Federation

Lev Matveev Institute of Applied Physics RAS, Nizhny Novgorod, Russian Federation

Alexander Matveyev Institute of Applied Physics RAS, Nizhny Novgorod, Russian Federation

Alexander Moiseev Institute of Applied Physics RAS, Nizhny Novgorod, Russian Federation

Alexander A. Moryatov Department of Oncology, Samara State Medical University, Samara, Russian Federation

Samara Regional Clinical Oncology Dispensary, Samara, Russian Federation

Oleg O. Myakinin Department of Laser and Biotechnical Systems, Samara National Research University, Samara, Russian Federation

Teemu Myllylä University of Oulu, Oulu, Finland

Luís M. C. Oliveira School of Engineering, Physics Department, Polytechnic Institute of Porto, Porto, Portugal

Centre of Innovation in Engineering and Industrial Technology, ISEP, Porto, Portugal

Andrey E. Orlov Samara Regional Clinical Oncology Dispensary, Samara, Russian Federation

Anton Plekhanov Privolzhsky Research Medical University, Nizhny Novgorod, Russian Federation

Jürgen Popp Leibniz Institute of Photonic Technologies, Jena, Germany

Institute of Physical Chemistry and Abbe Center of Photonics, Jena, Germany

Elena Potapova Orel State University, Orel, Russian Federation

Alexander A. Potapov Burdenko Neurosurgery Institute, Moscow, Russian Federation

Vitold E. Pozhar Scientific and Technological Center of Unique Instrumentation of the Russian Academy of Sciences, Moscow, Russian Federation

Nicole M. Ralbovsky Department of Chemistry, University at Albany, State University of New York, Albany, NY, USA

Dmitry S. Raupov Department of Laser and Biotechnical Systems, Samara National Research University, Samara, Russian Federation

Igor V. Reshetov First Moscow State Medical University (Sechenov University), Moscow, Russian Federation

Elena Sedova Privolzhsky Research Medical University, Nizhny Novgorod, Russian Federation

Vladislav I. Shcheslavskiy Becker & Hickl GmbH, Berlin, Germany

Pavel Shilyagin Institute of Applied Physics RAS, Nizhny Novgorod, Russian Federation

Marina V. Shirmanova Privolzhsky Research Medical University (PRMU), Nizhny Novgorod, Russian Federation

Marina Sirotkina Privolzhsky Research Medical University, Nizhny Novgorod, Russian Federation

Alexander Sovetsky Institute of Applied Physics RAS, Nizhny Novgorod, Russian Federation

Dmitriy Terpelov Institute of Applied Physics RAS, Nizhny Novgorod, Russian Federation

Petranka Troyanova University Hospital "Tsaritsa Yoanna – ISUL", Sofia, Bulgaria

Valery V. Tuchin Department of Optics and Biophotonics, Saratov State University, Saratov, Russian Federation

Laboratory of Biophotonics, Tomsk State University, Tomsk, Russian Federation

Laboratory of Laser Diagnostics of Technical and Living Systems, Institute of Precision Mechanics and Control of the Russian Academy of Sciences, Saratov, Russian Federation

Laboratory of Molecular Imaging, Bach Institute of Biochemistry, Research Center of Biotechnology of the Russian Academy of Sciences, Moscow, Russian Federation

Laboratory of Femtomedicine, ITMO University, St. Petersburg, Russian Federation

Alex Vitkin University of Toronto and University Health Network, Toronto, ON, Canada

Dmitry Vorontsov Privolzhsky Research Medical University, Nizhny Novgorod, Russian Federation

Denis A. Vrazhnov Department of General and Experimental Physics, National Research Tomsk State University, Tomsk, Russian Federation

Laboratory of Molecular Imaging and Photoacoustics, Institute of Strength Physics and Materials Science of Siberian Branch Russian Academy of Sciences, Tomsk, Russian Federation

Konstantin Yashin Privolzhsky Research Medical University, Nizhny Novgorod, Russian Federation

Natalia V. Yunusova Tomsk National Research Medical Center of the Russian Academy of Sciences, Tomsk, Russian Federation

Department of Biochemistry and Molecular Biology with a Course of Clinical Laboratory Diagnostics, Siberian State Medical University, Tomsk, Russian Federation

Elena V. Zagaynova Privolzhsky Research Medical University (PRMU), Nizhny Novgorod, Russian Federation

Lobachevskiy State University of Nizhny Novgorod, Nizhny, Novgorod, Russian Federation

Vladimir Zaitsev Institute of Applied Physics RAS, Nizhny Novgorod, Russian Federation

Valery P. Zakharov Department of Laser and Biotechnical Systems, Samara National Research University, Samara, Russian Federation

Kirill I. Zaytsev Prokhorov General Physics Institute of the Russian Academy of Sciences, Moscow, Russian Federation

Bauman Moscow State Technical University, Moscow, Russian Federation

Evgeny Zherebtsov Orel State University, Orel, Russian Federation

University of Oulu, Oulu, Finland

Part I
Tumor Tissue Optics and Multimodal Microscopy

Chapter 1
Malignant Tissue Optical Properties

Alexey N. Bashkatov, Valery P. Zakharov, Alla B. Bucharskaya, Ekaterina G. Borisova, Yulia A. Khristoforova, Elina A. Genina, and Valery V. Tuchin

1.1 Introduction

Despite the development of medicine, cancer remains one of the most dangerous diseases nowadays. World Health Organization (WHO) has reported 18.1 million new cancer cases and 9.6 million cancer deaths in 2018 [1]. Therefore, the detection and treatment of cancer is one of the most challenges for medicine in the twenty-first century. An effective solution of the problem is the use of modern

A. N. Bashkatov (✉) E. A. Genina
Department of Optics and Biophotonics, Saratov State University, Saratov, Russian Federation

Laboratory of Biophotonics, Tomsk State University, Tomsk, Russian Federation
e-mail: bashkatovan@sgu.ru; geninaea@sgu.ru

V. P. Zakharov · Y. A. Khristoforova
Department of Laser and Biotechnical Systems, Samara National Research University, Samara, Russian Federation
e-mail: zakharov@ssau.ru

A. B. Bucharskaya
Saratov State Medical University, Saratov, Russian Federation
e-mail: Ckp@sgmu.ru

E. G. Borisova
Institute of Electronics, Bulgarian Academy of Sciences, Sofia, Bulgaria
e-mail: borisova@ie.bas.bg

V. V. Tuchin
Department of Optics and Biophotonics, Saratov State University, Saratov, Russian

Laboratory of Biophotonics, Tomsk State University, Tomsk, Russian

Laboratory of Laser Diagnostics of Technical and Living Systems, Institute of Precision Mechanics and Control of the Russian Academy of Sciences, Saratov, Russian

Laboratory of Molecular Imaging, Bach Institute of Biochemistry, Research Center of Biotechnology of the Russian Academy of Sciences, Moscow, Russian
e-mail: tuchinvv@mail.ru

V. V. Tuchin et al. (eds.), *Multimodal Optical Diagnostics of Cancer*,
https://doi.org/10.1007/978-3-030-44594-2_1

interdisciplinary technologies. Most often, if the tumor is diagnosed earlier and treated, the patient will have a better prognosis and much greater opportunities for complete recovery. Many recent technological innovations have used physics principles, such as optics and coherent photonics, to improve early diagnostic and therapeutic procedures to reduce cancer incidence and mortality.

The development of optical methods in modern medicine in the field of diagnostics, surgery, and therapy has stimulated the study of the optical properties of human and animal tissues, since the efficiency of optical sensing of tissues depends on the photon propagation and fluence rate distribution [2].

Examples of diagnostic use are the following: monitoring of blood oxygenation and tissue metabolism, analysis of main tissue components, detection of malignant neoplasms, and recently proposed various techniques for optical imaging. The latter is particularly interesting for virtual optical biopsy and the precise determination of tumor boundaries during surgical operations. Therapeutic usage mostly includes applications in photodynamic therapy. For all these applications, knowledge of the optical properties of tissues is of great importance for the interpretation and quantification of diagnostic data and for predicting the distribution of light and absorbed energy for therapeutic and surgical use.

In this chapter, we provide an overview of the optical properties of benign and malignant tumors measured over a wide wavelength range and discuss the main cancer markers for various types of tumors.

1.2 Tumor Optical Properties Measurements: A Brief Description

Among the numerous methods for measuring the optical properties of tissue, the most widely used are integrating spheres spectroscopy, reflectance spectroscopy as well as Raman and fluorescence spectroscopy.

Iterative methods for processing experimental data, as a rule, take into account discrepancies between the refractive indices at the boundaries of the sample as well as the multilayer nature of the sample. The following factors are responsible for the errors in the estimated values of the optical coefficients and need to be borne in mind in a comparative analysis of the optical parameters obtained in various experiments [3]:

- The physiological conditions of tissues (the degree of hydration, homogeneity, species-specific variability, frozen/thawed or fixed/unfixed state, *in vitro*/*in vivo* measurements, smooth/rough surface);
- The geometry of irradiation;
- The matching/mismatching interface refractive indices;
- The numerical aperture of photodetectors;
- The separation of radiation experiencing forward scattering from unscattered radiation;
- The theory used to solve the inverse problem.

To analyze the propagation of light under multiple scattering conditions, it is assumed that absorbing, fluorescence, and scattering centers are uniformly distributed across the tissue. UV-A, visible, or NIR radiation is usually subjected to anisotropic scattering characterized by a clearly apparent direction of photons undergoing single scattering, which may be due to the presence of large cellular organelles (mitochondria, lysosomes, Golgi apparatus, etc.) [3–5].

When the scattering medium is illuminated by unpolarized light and/or only the intensity of multiply scattered light needs to be computed, a sufficiently strict mathematical description of continuous wave (CW) light propagation in a medium is possible in the framework of the scalar stationary radiation transfer theory (RTT) [3–6]. This theory is valid for an ensemble of scatterers located far from each other and has been successfully used to develop some practical aspects of tissue optics. The main stationary equation of RTT for average spectral power flux density $I_\lambda(\vec{r},\vec{s})$ (in W/cm² sr) for wavelength λ at point $\vec{r}$ in the given direction $\vec{s}$ and monochromatic irradiation has the form

$$\frac{\partial I_\lambda(\vec{r},\vec{s})}{\partial s} = -\mu_t I_\lambda(\vec{r},\vec{s}) + \frac{\mu_s}{4\pi}\int_{4\pi} I_\lambda(\vec{r},\vec{s})\,p(\vec{s},\vec{s}')\,d\Omega' + \varepsilon(\vec{s},\vec{s}'), \quad (1.1)$$

where $p(\vec{s},\vec{s'})$ is the scattering phase function, 1/sr; $d\Omega'$ is the unit solid angle about the direction $\vec{s}'$, sr; $\mu_t = \mu_a + \mu_s$ is the total attenuation coefficient, 1/cm; μ_a is the absorption coefficient, 1/cm; μ_s is the scattering coefficient, 1/cm; and $\varepsilon(\vec{s},\vec{s}')$ is the internal light source, which accumulates the effects of fluorescence and Raman spectroscopy. However, in most practically interesting cases, the measurement of the absorption and scattering coefficients of tissues can be performed neglecting the effects of fluorescence and Raman scattering, since their quantum efficiency is relatively small. It is equivalent to Eq. (1.1) in the absence of the internal radiation sources.

The scalar approximation of the radiative transfer equation (RTE) gives poor accuracy when the size of the scattering particles is much smaller than the wavelength, but provides acceptable results for particles comparable to and exceeding the wavelength [7].

The phase function $p(\vec{s},\vec{s'})$ describes the scattering properties of the medium and is actually the probability density function for scattering in the direction $\vec{s}'$ of a photon traveling in the direction $\vec{s}$; in other words, it characterizes an elementary scattering event. If scattering is symmetric relative to the direction of the incident wave, then the phase function depends only on the scattering angle θ (angle between directions $\vec{s}$ and $\vec{s}'$), i.e., $p(\vec{s},\vec{s'}) = p(\theta)$. The assumption of random distribution of scatterers in a medium (i.e., the absence of spatial correlation in the tissue structure) leads to normalization:

$$\int_0^\pi p(\theta)\,2\pi\sin\theta\,d\theta = 1.$$

In practice, the phase function is usually well approximated with the aid of the postulated Henyey–Greenstein function [2–6, 8]:

$$p(\theta) = \frac{1}{4\pi} \frac{1 - g^2}{\left(1 + g^2 - 2g\cos\theta\right)^{3/2}}, \tag{1.2}$$

where g is the scattering anisotropy parameter (mean cosine of the scattering angle θ):

$$g = <\cos\theta> = \int_0^{\pi} p(\theta)\cos\theta 2\pi \sin\theta d\theta.$$

The value of g varies in the range from −1 to 1; $g = 0$ corresponds to isotropic (Rayleigh) scattering, $g = 1$ to total forward scattering (Mie scattering at large particles), and $g = -1$ to total backward scattering [3–9].

Other phase functions commonly used to analyze the propagation of light in turbid media, including tissue, are the small-angle scattering phase function [10, 11], the Mie phase function [12–15], the δ-Eddington phase function [16, 17], the Reynolds–McCormick phase function [18–20], the Gegenbauer kernel phase function [14, 15, 21, 22], and their modifications [23–25].

1.2.1 Integrating Sphere Spectroscopy

Integrating sphere spectroscopy (ISS) is commonly used as an optical calibration and measurement tool and, in particular, it is successfully used to measure optical properties of tissues [2, 3, 5]. A detailed theory of the integrating sphere spectroscopy is presented in [26–32]. The inner surface of an integrating sphere is uniformly coated with highly reflective diffuse materials (exceeding 0.98) to achieve homogenous distributions of light radiation at the sphere's inner wall. A light beam falling on the inner surface of an integrating sphere is evenly scattered to all directions (Lambertian reflections) and the light fluxes are evenly distributed (spatially integrated) on the homogenous inner surface of the sphere after multiple Lambertian reflections. A standard integrating sphere usually has three ports: an input port, an output port, and a third port for the detector. In certain applications, the fourth port is also used so that the specular reflection beam can go out from the sphere in a light trap. However, for real integrating spheres, the surfaces do not have perfect Lambertian reflection. To prevent measurement errors by specular reflection, baffle(s) coated with a highly reflective material is often placed inside the sphere to further diffuse the specular reflection and avoid the direct reflection from reaching the detector.

There are several advantages of using spectroscopy with integrating sphere for measuring the spectral reflectance and transmittance of tissue samples, in comparison with direct measurement of the samples by a spectrometer. First, in a regular spectrometer measurement the incident light directly illuminates the sample surface, and the detected reflectance often has a dependency on the angle and distance between the incident beam and the detector. When an integrating sphere is used, all backreflected fluxes are captured and normalized by the sphere. Therefore, the angular dependency is no longer an issue. Second, the detector-object distance is often fixed in the integrating sphere measurement. Even if there is a small change in the sample-sphere distance, it will not affect the results of the measurements as long as all reflected light bounces back into the sphere. Additionally, by using integrating spheres, the spectral measurements are less dependent on the shape of the light beam and the homogeneity of the sample, since both incident light beam and the reflected/scattered light will be normalized on the inner surface of the sphere before being captured by the detector.

The optical parameters of tissue samples (namely the absorption coefficient μ_a, the scattering coefficient μ_s, and the anisotropy factor of scattering g) could be measured by various methods. The single- or double-integrating sphere method combined with collimated transmittance measurement (see Fig. 1.1) is the most often used for *in vitro* tissue studies. Briefly, this approach implies either sequential or simultaneous determination of three parameters: the total transmittance $T_t = T_c + T_d$ (T_d is the diffuse transmittance), the diffuse reflectance R_d, and the collimated transmittance $T_c = I_d/I_0$ (I_d is the intensity of transmitted light measured using a distant photodetector with a small aperture, and I_0 is the intensity of incident radiation).

Any three measurements from the following five are sufficient for the evaluation of all three optical parameters [3]:

1. Total (or diffuse) transmittance for collimated or diffuse radiation;
2. Total (or diffuse) reflectance for collimated or diffuse radiation;
3. Absorption by a sample placed inside an integrating sphere;
4. Collimated transmittance;
5. Angular distribution of radiation scattered by the sample.

The optical parameters of the tissue are deduced from these measurements using different theoretical expressions or numerical simulations: the inverse Monte Carlo (IMC) [33–41] or inverse adding-doubling (IAD) [42–51] methods, or methods based on the diffusion approximation of the transfer equation [52–56]. However, the diffusion approximation has limitations, including describing tissue with a low albedo and accurate consideration of boundary conditions. To overcome these shortcomings other techniques such as the IAD and the IMC are the most commonly used.

The adding-doubling technique is a numerical method for solving the 1D transport equation in slab geometry. It can be used for tissue with an arbitrary phase function, arbitrary angular distribution of the spatially uniform incident radiation, and infinite beam size as lateral light losses cannot be taken into account. The angular distribution of the reflected radiance (normalized to an incident diffuse flux) is given by Prahl et al. [42]:

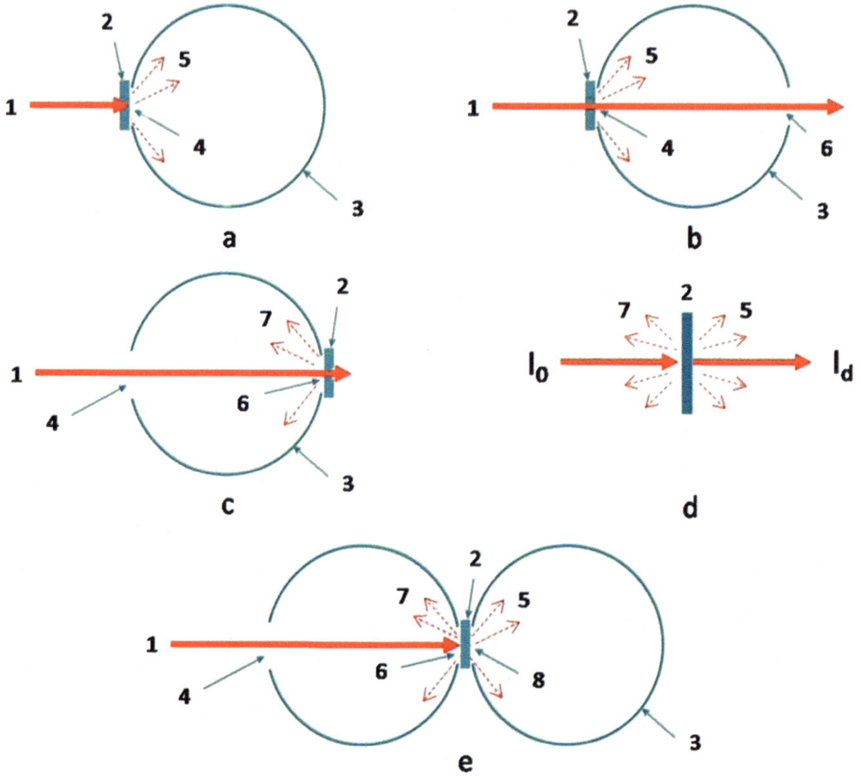

Fig. 1.1 Measurement of tissue optical properties using an integrating sphere. (**a**) Total transmittance mode, (**b**) diffuse transmittance mode, (**c**) diffuse reflectance mode, (**d**) collimated transmittance mode, (**e**) double-integrating sphere. 1 is the incident beam; 2 is the tissue sample; 3 is the integrating sphere; 4 is the entrance port; 5 is the transmitted radiation; 6 is the exit port; 7 is the diffuse reflected radiation

$$I_{\mathrm{ref}}\left(\eta_c\right)=\int_0^1 I_{\mathrm{in}}\left(\eta_c'\right)R\left(\eta_c',\eta_c\right)2\eta_c' d\eta_c', \tag{1.3}$$

where $I_{\mathrm{in}}(\eta_c)$ is an arbitrary incident radiance angular distribution, η_c is the cosine of the polar angle, and $R\left(\eta_c',\eta_c\right)$ is the reflection redistribution function determined by the optical properties of the slab.

The distribution of the transmitted radiance can be expressed in a similar manner, with obvious substitution of the transmission redistribution function $T\left(\eta_c',\eta_c\right)$. If M quadrature points are selected to span over the interval (0, 1), the respective matrices can approximate the reflection and transmission redistribution functions:

$$R\left(\eta_{\mathrm{ci}}',\eta_{\mathrm{cj}}\right)\rightarrow R_{\mathrm{ij}};\ T\left(\eta_{\mathrm{ci}}',\eta_{\mathrm{cj}}\right)\rightarrow T_{\mathrm{ij}}. \tag{1.4}$$

These matrices are referred to as the reflection and transmission operators, respectively. If a slab with boundaries indexed as 0 and 2 is comprised of two layers, (01) and (12), with an internal interface 1 between the layers, the reflection and transmission operators for the whole slab (02) can be expressed as:

$$\boldsymbol{T}^{02} = \boldsymbol{T}^{12}\left(\boldsymbol{E} - \boldsymbol{R}^{10}\boldsymbol{R}^{12}\right)^{-1}\boldsymbol{T}^{01},$$

$$\boldsymbol{R}^{20} = \boldsymbol{T}^{12}\left(\boldsymbol{E} - \boldsymbol{R}^{10}\boldsymbol{R}^{12}\right)^{-1}\boldsymbol{R}^{01}\boldsymbol{T}^{12} + \boldsymbol{R}^{21},$$

$$\boldsymbol{T}^{20} = \boldsymbol{T}^{10}\left(\boldsymbol{E} - \boldsymbol{R}^{12}\boldsymbol{R}^{10}\right)^{-1}\boldsymbol{T}^{21},$$

$$\boldsymbol{R}^{02} = \boldsymbol{T}^{10}\left(\boldsymbol{E} - \boldsymbol{R}^{12}\boldsymbol{R}^{10}\right)^{-1}\boldsymbol{R}^{12}\boldsymbol{T}^{01} + \boldsymbol{R}^{10}, \quad (1.5)$$

where $\boldsymbol{E}$ is the identity matrix defined in this case as:

$$E_{\mathrm{ij}} = \frac{1}{2\eta_{\mathrm{ci}} w_i}\delta_{\mathrm{ij}}, \quad (1.6)$$

where w_i is the weight assigned to the i-th quadrature point and δ_{ij} is a Kronecker delta symbol, $\delta_{\mathrm{ij}} = 1$ if $i = j$, and $\delta_{\mathrm{ij}} = 0$ if $i \neq j$.

The definition of the matrix multiplication also slightly differs from the standard. Specifically,

$$\left(\boldsymbol{AB}\right)_{\mathrm{jk}} \equiv \sum_{j=1}^{M} A_{\mathrm{ij}} 2\eta_{\mathrm{cj}} w_j B_{\mathrm{jk}}. \quad (1.7)$$

Equations (1.5) allow one to calculate the reflection and transmission operators of a slab when those of the comprising layers are known. The idea of the method is to start with a thin layer for which the RTE can be rather easily simplified and solved, producing the reflection and transmission operators, and then proceed by doubling the thickness of the layer until the thickness of the whole slab is reached. Several techniques exist for layer initialization. The single-scattering equations for reflection and transmission for the Henyey–Greenstein function are given by van de Hulst [57] and Prahl [58]. The refractive index mismatch can be taken into account by adding effective boundary layers of zero thickness and having the reflection and transmission operators determined by Fresnel's formulas. Both total transmittance and reflectance of the slab are obtained by straightforward integration of Eq. (1.3). Different methods of performing the integration and the IAD program provided by S. A. Prahl [42, 58] allow one to obtain the absorption and the scattering coefficients from the measured diffuse reflectance R_{d} and total transmittance T_t of the tissue slab. This program is the numerical solution to the steady-state RTE (Eq. (1.1)) realizing an iterative process, which estimates the reflectance and transmittance from a set of optical parameters until the calculated reflectance and transmittance match the measured values. Values for the anisotropy factor$g$ and the refractive index$n$ must be provided to the program as input parameters.

It was shown that using only four quadrature points, the IAD method provides optical parameters that are accurate to within 2–3% [42]. Higher accuracy, however, can be obtained by using more quadrature points, but it would require increased computation time. Another valuable feature of the IAD method is its validity for the study of samples with comparable absorption and scattering coefficients [42], since other methods based on only diffusion approximation are inadequate. Furthermore, since both anisotropic phase function and Fresnel reflection at boundaries are accurately approximated, the IAD technique is well suited to optical measurements of biological tissues and blood held between two glass slides. The adding-doubling method provides accurate results in cases when the side losses are not significant, but it is less flexible than the Monte Carlo (MC) technique.

Both the real geometry of the experiment and the tissue structure may be complicated. Therefore, inverse Monte Carlo method has to be used if reliable estimates are to be obtained. A number of algorithms to use the IMC method are available now in the literature [5, 15, 19, 33, 37–39, 59–61]. Many researches use the Monte Carlo (MC) simulation algorithm and program provided by S. L. Jacques, and L. Wang et al. [35, 62, 63]. The MC technique is employed as a method to solve the forward problem in the inverse algorithm for the determination of the optical properties of tissues and blood. The MC method is based on the formalism of the RTT, where the absorption coefficient is defined as a probability of a photon to be absorbed per unit length, and the scattering coefficient is defined as the probability of a photon to be scattered per unit length. The effects of fluorescence and Raman scattering may be also taken into account in a similar way by introducing the probability of generating new photons with different frequencies for the correspondingly absorbed or scattered initial photons. Using these probabilities, a random sampling of photon trajectories is generated. Among the firstly designed IMC algorithms, similar algorithms for determining all three optical parameters of the tissue (μ_a, μ_s, and g) based on the *in vitro* evaluation of the total transmittance, diffuse reflectance, and collimated transmittance using a spectrophotometer with integrating spheres can be also mentioned [5, 15, 33, 37, 38, 40, 41, 44, 50, 60, 61, 64]. The initial approximation (to speed up the procedure) is achieved with the help of the Kubelka–Munk theory, specifically its four-flux variant [3, 5, 33, 37, 38, 65–67]. The algorithms take into consideration the sideways loss of photons, which becomes essential in sufficiently thick samples. Similar results have been obtained using the condensed IMC method [5, 60, 61, 68–73]. Figure 1.2 demonstrates the typical flowchart of the IMC method [41].

In the basic MC algorithm a photon described by three spatial coordinates and two angles (x, y, z, θ, ϕ) is assigned its weight $W = W_0$ and placed in its initial position, depending on the source characteristics. The step size s of the photon is determined as $s = -\ln(\xi)/\mu_t$, where ξ is the random number between (0, 1). The direction of the photon's next movement is determined by the scattering phase function substituted as the probability density distribution. Several approximations for the scattering phase function of tissue and blood have been used in MC simulations. They

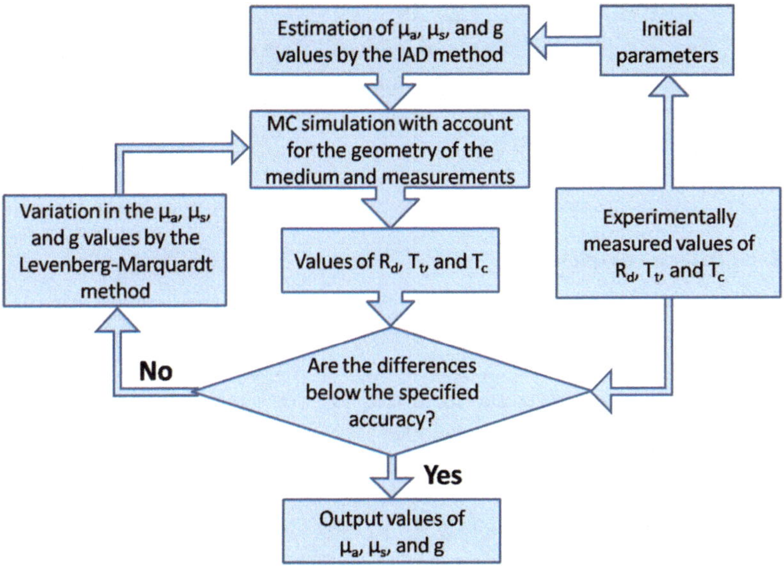

Fig. 1.2 The typical flowchart of the IMC method [41]

include two empirical phase functions widely used to approximate the scattering phase function of tissue and blood, Henyey–Greenstein phase function (HGPF) (see Eq. (1.2)), the Gegenbauer kernel phase function (GKPF), and Mie phase function.

In most cases, azimuthal symmetry is assumed. This leads to $p(\phi) = 1/2\pi$ and, consequently, $\phi_{\text{rnd}} = 2\pi\xi$. At each step, the photon loses part of its weight due to absorption: $W = W(1 - \Lambda)$, where $\Lambda = \mu_s/\mu_t$ is the albedo of the medium.

When the photon reaches the boundary, part of its weight is transmitted according to the Fresnel equations. The amount transmitted through the boundary is added to the reflectance or transmittance. Since the refraction angle is determined by the Snell's law, the angular distribution of the out-going light can be calculated. The photon with the remaining part of the weight is specularly reflected and continues its random walk.

When the photon's weight becomes lower than a predetermined minimal value, the photon can be terminated using "Russian roulette" procedure [35, 62, 63]. This procedure saves time, since it does not make sense to continue the random walk of the photon, which will not essentially contribute to the measured signal. On the other hand, it ensures that the energy balance is maintained throughout the simulation process.

The MC method has several advantages over the other methods because it may take into account mismatched medium-glass and glass-air interfaces, losses of light at the edges of the sample, any phase function of the medium, and the finite size and arbitrary angular distribution of the incident beam. The only disadvantage of this method is the long time needed to ensure good statistical convergence, since it is a statistical approach. The standard deviation of a quantity (diffuse reflectance, transmittance, etc.) approximated by MC technique decreases proportionally to $1/\sqrt{N}$, where N is the total number of launched photons. It is worthy of note that stable operation of the algorithm is maintained by generation of from 10^5 to 5×10^5 photons per iteration. Two to five iterations are usually necessary to estimate the optical parameters with approximately 2% accuracy.

1.2.2 Diffuse Backscattered Reflectance Spectroscopy

Diffuse backscattered reflectance spectroscopy (BS) [5, 72–87] is well suited for use in biomedical applications due to its low instrumentation cost, easy implementation, and non-destructive measurement setup. Hence, many different BS measurement configurations have been developed. Optical fiber arrays and non-contact reflectance imagery are two typical sensing configurations in BS measurement, which can be implemented with fiber-optic probe (FOP), monochromatic imaging (MCI), and hyperspectral imaging (HSI). In the FOP measurement, a single spectrometer, multiple spectrometers, or a spectrograph-camera combination coupled with multiple detection fibers can be used to measure diffuse reflectance at different distances from the light incident point. Moreover, it is also desirable to measure a tissue sample at a greater depth. To overcome the shortcomings of a rigid FOP, a flexible FOP with numerous optical fibers covering a spatial distance range of 0–30 mm can be used for measuring the tissue optical properties. Optical fibers have to be coupled to a multichannel hyperspectral imaging system, which allows simultaneous acquisition of reflectance spectra from the sample. The use of several different sizes of fibers for the probe also expands effectively the dynamic range of the camera, allowing acquiring spectra at greater depth of the sample.

As a non-contact method, MCI is more suitable for measuring optical properties of tissues for monochromatic irradiation. A laser diode or a combination of a supercontinuum laser and a monochromator can be used to illuminate a sample at a specific wavelength. The diffuse reflectance is acquired with a CCD camera. This BS configuration is simple and relatively easy to implement. The acquired 2D scattering images are reduced to 1D scattering profiles by radial averaging when the scattering images are axisymmetric with respect to the laser incident point. However, this assumption is not satisfied for anisotropic tissues where the light is guided by the tissue fibers. For example, in the case of bovine muscle tissue, the effect of the fibers resulted in scatter spots with a rhombus shape. Measurement at multiple wavelengths requires sequential wavelength scanning. In addition, a substantial portion of the signal of each pixel comes from the surrounding areas, which may affect the accuracy of the measurement. Therefore, the characterization of the point-spread function (PSF) is necessary in order to minimize errors in the obtained intensity values for the image data interpretation.

In the hyperspectral imaging, spectral and spatial information is acquired simultaneously and, therefore, it has advantageous for measuring diffuse reflectance profiles over a broad spectral range. As a rule typical hyperspectral imaging-based BS system in line scan mode has high spatial resolution and mainly consists of a high-performance CCD camera, an imaging spectrograph, a zoom or prime lens, a light source, and an optical fiber coupled with a focusing lens for delivering a broadband beam to the sample.

As an indirect method for optical property measurement, computation of the optical parameters from the BS measurements usually requires sophisticated modeling based on the diffusion approximation of radiative transfer theory or MC simulation, coupled with appropriate inverse algorithms. Numerical methods are generally required for solving the radiative transfer equation or using inverse MC simulation. These methods are flexible and allow possibility for modeling of different geometries of experimental setups but they may be subjected to statistical uncertainties during the estimation of the reflectance. Moreover, one of the major drawbacks with the numerical methods is that they require substantial computational time. To overcome the shortcomings the condensed IMC method can be used, that is, a library of MC simulated BS profiles for a grid of μ_s, μ_a and g values can be calculated, and then the library can be used either as a look-up table or for training a neural network.

Another way to reconstruct the tissue optical parameters (such as μ_a and reduced scattering coefficient $\mu_s' = \mu_s(1-g)$) has been proposed by Zonios et al. [88–91]. Their approach is based on diffusion approximation and assumes that $R(\lambda) = \frac{\mu_s'(\lambda)}{k_1 + k_2\mu_a(\lambda)}$. Here $R(\lambda)$ is the diffuse reflectance, λ is the wavelength, k_1 and k_2 are constants that depend on the probe geometry. The optical coefficients μ_a and μ_s' can be related to the absorption and scattering properties of the tissue through Eqs. (1.8) and (1.9) (for example):

$$\begin{aligned}\mu_a(\lambda) = C_{\mathrm{Hb}}\left[\alpha \cdot \varepsilon_{\mathrm{HbO}}(\lambda) + (1-\alpha)\varepsilon_{\mathrm{Hb}}(\lambda)\right] + C_w \cdot \varepsilon_w(\lambda) \\ + C_{\mathrm{mel}} \cdot \varepsilon_{\mathrm{mel}}(\lambda) + C_{\mathrm{col}} \cdot \varepsilon_{\mathrm{col}}(\lambda) + \ldots,\end{aligned} \tag{1.8}$$

where C_{Hb} is the total concentration of hemoglobin, α is the oxygen saturation of hemoglobin, C_w is the concentration of water, C_{mel} is the concentration of melanin, C_{col} is the concentration of collagen, and $\varepsilon_{\mathrm{HbO}}$, $\varepsilon_{\mathrm{Hb}}$, ε_w, $\varepsilon_{\mathrm{mel}}$, $\varepsilon_{\mathrm{col}}$ are the absorption coefficients of oxyhemoglobin, deoxyhemoglobin, water, melanin, and collagen, respectively.

$$\mu_s'(\lambda) = \frac{A}{\lambda^w}, \tag{1.9}$$

where parameter A is defined by the concentration of scattering particles in the tissue, and the wavelength exponent w is independent of the particles concentration, characterizes the mean size of the particles, and defines the spectral behavior of the scattering coefficient [92].

Accurate estimation of optical parameters by inverse algorithms is not an easy task due to the complexity of analytical solutions and potential experimental errors in the measurement of diffuse reflectance from a medium. Moreover, for many biological materials, the values of the absorption coefficient over a specific spectral region (especially in the region from 700 to 900 nm) are rather small that makes it more difficult to obtain an accurate estimation of the optical parameters. For these reasons, it is generally considered acceptable or accurate when errors for measuring μ_a and $\mu_s^{'}$ are within 10%. In general, the estimation of optical parameters can be defined as the nonlinear least-squares optimization problem with several important assumptions, that is, constant variance errors, uncorrelated errors, and a Gaussian error distribution. The results will not be valid if these assumptions are violated. In addition, for estimating the optical parameters of layered media, the increased number of free parameters can dramatically increase the computational time, further exacerbating the estimation of optical parameters, and/or causing ill-posed problems. Different strategies such as a multi-step method, sensitivity analysis, statistical evaluation, etc. [93] have been proposed to optimize the inverse algorithms and improve the estimation accuracies.

1.2.3 Raman Spectroscopy

Raman Scattering Neoplastic cells are characterized by increased nuclear material, an increased nuclear-to-cytoplasmic ratio, increased mitotic activity, abnormal chromatin distribution, and decreased differentiation [94, 95]. There is a progressive loss of cell maturation, and proliferation of these undifferentiated cells results in increased metabolic activity. The morphologic and biochemical changes that occur with malignant tissue are numerous and in many cases depend on the specific type and location of the cancer. Biochemical tumor markers include cell surface antigens, cytoplasmic proteins, enzymes, and hormones. These general features of neoplastic cells result in specific changes in nucleic acid, protein, lipid, and carbohydrate quantities and/or conformations [95]. There are multiple molecular markers, located in the membrane, the cytoplasm, the nucleus, and the extracellular space that may be indicative of neoplasia. As most biological molecules are Raman active, with distinctive spectra in the fingerprint region (500–1800 cm^{-1}), vibrational spectroscopy is a desirable tool for cancer detection.

Raman spectroscopy is based on the inelastic scattering of photons by molecular bond vibrations. Therefore the alteration of molecular signatures in a cell or tissue undergone cancer transformation can be detected by noninvasive Raman scattering without labeling.

In general, the majority of scattered photons have the same frequency as incident photons when light passes through the tissue (Fig. 1.3). This is known as Rayleigh or elastic scattering. However, a very small portion of photons alters the energy after collision with molecular due to inelastic of Raman scattering. The energy difference between the incident and scattered photons (Raman shift, measured by wavenumber

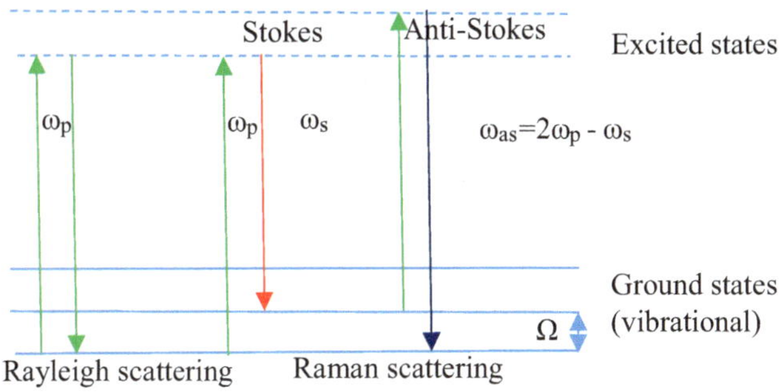

Fig. 1.3 Energy level diagram for elastic (Rayleigh) and inelastic spontaneous Raman scattering

in cm^{-1}) corresponds to the vibrational energy of the specific molecular bond interrogated [96–98].

The ground state vibrational frequencies and energies vary depending on the strengths of bonds and masses of atoms involved in the normal mode motion. The greatest varieties of vibrational transitions in biological molecules occur in the fingerprint range (500–1800) cm^{-1}. Signatures in the higher wavenumber (HW) range (2800–3500) cm^{-1} arise from transitions between states of modes involving symmetric or asymmetric stretching of C–H bonds. The intensity of the Raman peaks for a particular molecule is directly proportional to the concentration of that molecule within a sample so the resulting spectrum is a superposition of Raman response of all the Raman active molecules from within a sample. Therefore, a Raman spectrum is an intrinsic molecular fingerprint of the sample, revealing detailed information about DNA, protein, and lipid content as well as macromolecular conformations, which can be extracted from the measured spectra. The spectral capacity of encoding chemical information can be estimated as the maximum number of distinct spectral states one can discriminate and include up to 50 spectral peaks in the entire Raman spectrum [99]. The original analyses for Raman signals are based on differences in intensity, shape, and location of the various Raman bands between normal and cancerous cells and tissues. These characteristic Raman bands elucidate not only information about biological components of the cell but also their quality, quantity, symmetry, and orientation. They can be used for understanding the spectral signature as it pertains to the disease process. However, it should be taken into account that high sensitivity to small biochemical changes is accompanied by weak Raman signal (inelastic scattering cross section is ~10^{-30} cm^2/molecular) often in the presence of high background. Therefore, significant problems exist for acquiring viable Raman signatures inherent to the chemically complex and widely varying biological tissue. The primary challenge for obtaining Raman spectra from biological materials is the intrinsic fluorescence, which is ubiquitously presented in almost all tissues and in several orders of magnitude intense than Raman signal.

Typical Raman setup is shown in Fig. 1.4 and consists of three primary components—laser source (1), sample light delivery and collection module (2), and spectrometer with CCD detector (3). The diagnostic effectiveness of Raman system is

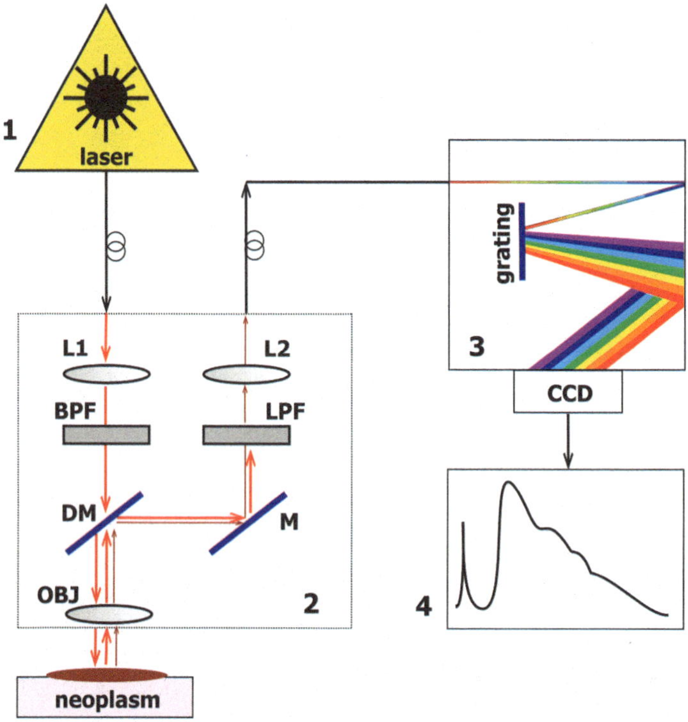

Fig. 1.4 The typical Raman setup: 1—laser; 2—sample light delivery and collection module; 3—spectrometer with CCD detector; 4—PC. *L1, L2* fiber-coupling lens, *OBJ* objective lens, *DM* dichroic mirror, *M* mirror, *BPF* bandpass filter, *LPF* longpass filter

tightly bound by the instrumentation parameters, which have to be chosen very carefully to measure the weak Raman signals. Generally, the choice of instrumentation is always a compromise between different factors driven by tissue under study and the pathophysiological processes. For example, the laser power is limited by signal-to-noise ratio (SNR) and maximum permissible exposure.

The key component of a Raman system is the detector, which in most cases is a charged coupled device (CCD). Several important factors have to be considered when choosing the appropriate CCD array for any Raman spectroscopy application. Specifically, the noise level and the quantum efficiency (QE) are of great importance. A typical CCD camera used in spectroscopy consists of a rectangular chip wherein the horizontal axis corresponds to the wavelength/wavenumber axis and the vertical axis is used to stack multiple fibers for increased throughput, which can subsequently be binned for improved SNR. While different types of chips are commercially available for different applications, a back-illuminated, deep-depletion CCD provides the highest QE in the NIR region. Most CCDs use a thermoelectric (TE) multistage Peltier system to actively cool the camera down to at least −70 °C in order to realize excellent dark noise performance. In fact, current Raman systems for most biomedical applications are only limited by shot noise. Selection of appropriate wavelengths for excitation is often governed by the reduction of fluorescence and scattering background, which decreases with wavelength increasing. However, due to $1/\lambda^4$ dependency the Raman intensity also reduces with wavelength increasing and quantum efficiency of silicon-based CCD detector falls rapidly for wavelengths over 1000 nm.

Overall, researchers in this field tend to prefer 785 nm and 830 nm excitation as a reasonable compromise for most tissues. A comprehensive overview of different Raman-instrumentation schemes and various probe designs is given in [100–102].

Data Processing and Analysis The direct background subtraction from raw signal may be achieved by excitation wavelength shifting within a few nanometers with following differentiation of acquiring signals, but such hardware technique requires specific design considerations including the use of tunable stabilized lasers [103, 104]. The other common methods for fluorescence elimination use software-based mathematical techniques like frequency-domain filtering [105], wavelet transformation [106], polynomial fitting [107, 108]. The polynomial curve fitting has an advantage over other fluorescence reduction techniques due to its inherent ability to retain the spectral contours and intensities of the input Raman spectra and minimal presence of artificial peaks in low SNR spectra [100, 108].

As Raman scattering intensity is extremely weak the measured Raman spectra require significant noise smoothing and binning for extraction of the underlying Raman bands, including median filter, the moving average window filter, the Gaussian filter, the Savitzky–Golay filter of various orders [109–111], and multivariate statistical approaches to remove the higher order components and noise [112].

Raman spectra are complex in nature as tissue contains a diverse set of small and large biomolecules. The vibrational frequencies associated with different functional groups and backbone chains, for example, in proteins, saccharides, and nucleic acids often overlap, thus, making it difficult to assign a specific observed band in the Raman spectrum to a specific functional group of a particular molecule in the tissue [113]. Moreover, while the peak location of an isolated functional group of atoms is typically known, the actual peak location of a functional group in a molecule may slightly differ from the isolated case because of interactions and bonding with its neighbors. Nevertheless, functional groups associated with specific molecules often give rise to relatively narrow and well-resolved bands in the Raman spectra. Table 1.1 summaries the major Raman spectral modes, where spectral differences have been found for normal and cancerous tissues [94, 95, 98, 114–121]. A detailed description of Raman spectral modes for malignant tissues may be found in Refs. [94, 115]. Characteristic Raman peaks arise from nucleic acids, lipids (C–C, C–O stretching), proteins (C–C, C–N stretching), and C–O stretching of carbohydrates in the region between 800 and 1200 cm^{-1}; C–N stretching and N–H bending (amide III band) with contributions from proteins (CH_3CH_2 wagging, twisting, bending), polysaccharides, lipids (CH_3CH_2 twisting, wagging, bending), and nucleic acids in the region between 1200 and 1400 cm^{-1}; C–H, CH_2, and CH_3 vibrations in the region between 1400 and 1500 cm^{-1}; C=O stretching vibrations (amide I band), proteins

Table 1.1 Major molecular vibrational modes and biochemical assignments observed for normal and malignant tissues [94, 95, 98, 114–121]

Vibrational modes	Band frequency[a] (cm^{-1})	Chemical
C–C stretching mode	855	Protein (collagen)
C–C stretching mode	935	Proline and valine, protein (α-helix)/ glycogen
C–C symmetric ring breathing mode	1004	Phenylalanine (collagen)
C–H bending mode	1031	Phenylalanine
PO_2^- symmetric stretching mode	1080	Phospholipids
C–C stretching mode	1158	Carotenoids
PO_2^- asymmetric stretching mode	1259	Phospholipids
CH_2 twisting mode	1302	Proteins (collagen)/phospholipids
CH_2CH_3 wagging mode	1335	DNA/protein (collagen)
CH_2 scissoring mode	1444	Lipids
C=C stretching mode	1518	Carotenoids
C=O C–C stretching mode	1600–1700	Amide I, α-helix conformation
C=C stretching mode	1618	Porphyrin, protein (tryptophan)
OH bending mode	1642	Water
C=C stretching mode	1660	Unsaturated bonds of lipids
C=O stretching mode	1750	Lipids
CH_2 symmetric stretching mode	2854	Saturated bonds of lipids
(CH_2)C–H asymmetric stretching mode	2888	Saturated bonds of lipids
CH_2 asymmetric stretching mode	2884	
CH_2 asymmetric stretching mode	2926	Saturated bonds of lipids
CH_3 symmetric stretching mode	2930	proteins
H–C=C stretching mode	3009	Unsaturated bonds of lipids
OH stretching mode	3258	Water
OH stretching mode	3410	Water
OH stretching mode	3450	Water

[a]Please note that Raman vibrations may shift in wavenumber depending on the sample and that these values should not be regarded as absolute

(C=C), nucleic acids, and lipids (C=C stretch) in the region between 1500 and 1760 cm^{-1}; CH_2 symmetric and asymmetric stretching modes of lipids and proteins in the region between 2850 and 3000 cm^{-1}; OH stretching modes of water in the region from 3100 to 3500 cm^{-1}. Despite there being clear Raman bands in malignant tissue that probably connected to the abundance of different biomolecules, there were no unique peaks that could be assigned to any type of cancer alone.

The prognostic value of Raman diagnostics follows from its inherent chemical specificity, which makes it possible to determine changes in the content of the tumor compared with the surrounding tissue. Observed content alteration includes several biomarkers, such as relative abundance of DNA [122–124], changes in structural and hydrogen bonding information for lipid, protein, and nucleic acids [122, 125, 126], variation of collagen and elastin context [122, 123, 127–130], increase or decrease in chemical components content like tryptophan [113, 123, 128, 129], keratin [113], carotenoids [129, 131], glycogen [123, 128, 131], cholesterol ester [122, 131], tyrosine and proline [123, 128]. Most markers have several peaks, facilitating increased robustness in detection. In some cases it is possible to determine identity of diagnostically relevant species by a few factors. For example, Haka et al. [122] have shown that relative abundances in calcium hydroxyapatite and calcium oxalate dehydrate correlate with malignancies in breast cancer. Peak pairs also can provide information on protein-to-DNA and protein-to-lipid ratios. It has been shown by several research groups that the ratio of intensities at 1455 cm^{-1} and 1655 cm^{-1} may be used for classification of tumor *vs.* normal tissue in the lung, brain, breast, colon, and cervix [109, 124, 129], since the 1655 cm^{-1} band corresponds to the C=O stretching of collagen and elastin, and the 1445 cm^{-1} band (CH_2 scissoring) varies with the lipid-to-protein ratio. But in most cases spectral changes between healthy and diseased tissue appear in the context of entire highly complex spectra from the tissue, and diagnostic information may be derived only with a help of spectral pattern recognition approaches. The Raman spectra also contain hidden links between different bands of the spectrum due to the contribution of the same chemical components. This leads to the emergence of multiple correlations. Consequently, multivariate statistical techniques have become the accepted practice for the development of discrimination and classification algorithms for diagnostic applications. Chemometrics is one of the powerful tools that are able to identify variations that lead to accurate and reliable separation of malignant and normal tissue. In the past few years discrimination techniques such as linear and nonlinear regression [132–134], principal component analysis (PCA) for data compression [112, 114] as well as classification techniques such as support vector machines (SVMs) [135], neural networks [125, 136], classification trees [137, 138], partial least-squares discriminant analysis (PLS-DA) [139, 140] have been employed. One of the perceived advantages of PLS-DA is that it has the ability to analyze highly collinear and noisy data. As a result, a combination of Raman spectral data and chemometrics is capable of differentiation between cancer and normal tissues as surveyed from the publications reviewed in Table 1.2.

Tissue Analysis To assess the applicability of Raman spectroscopy for the clinical diagnosis of cancer, numerous studies have been conducted with extracted tissues that have been frozen (with liquid nitrogen or dry ice) at the time of collection and thawed for study or fixed in formalin to prevent deterioration. The fixation process chemically alters the tissue, primarily cross-linking the collagen proteins, and thus, affects the Raman spectral signature of the tissue. Although some differences are observed in the Raman spectra of fresh and fixed tissues, the variation appears to be small and does not fundamentally affect the potential diagnostic capability of the spectrum [95].

Lyng et al. [141] have examined formalin fixed paraffin preserved specimens of benign lesions (fibrocystic, fibroadenoma, intraductal papilloma) and cancer

Table 1.2 General overview of Raman spectroscopy cancer studies

Relevant cancer	Raman method	Sampling type	Sample number	Spectral range (cm^{-1})	Laser (nm)	Data analysis	Sensitivity (%)	Specificity (%)	Year	Ref.
Lung cancer	Raman probe	*Ex vivo* (tissue)	28	700–1800	785	Ratio 1444/1655	94	92	2003	[129]
Lung cancer	Raman probe	*In vivo*	80	2775–3040	785	PC-GDA; PLS-DA, LOOCV	90	65	2016	[156]
Lung cancer	Raman probe	*In vivo*	26	700–1700; 1600–3500	785	PCA-LDA, LOOCV	90	90	2008	[155]
Lung cancer	Raman probe + OCT	*Ex vivo* (tissue)	22	800–1800	785	LDA, QDA	91	79	2015	[111]
Breast cancer	Micro-spectrometer	*In vitro* (cell lines)	60	500–1800; 2800–3050	532	PCA; LDA	100	91	2015	[114]
Breast cancer	Micro-spectrometer	*Ex vivo* (tissue)	150	500–3500	532	LRA-marker, PCA	83	77	2011	[157]
Breast cancer	Micro-spectrometer	*Ex vivo* (tissue)	146	500–3500	514, 532	LRA-marker, PCA?	83	77	2012	[158]
Breast cancer	Micro-spectrometer	*Ex vivo* (tissue)	58	800–1800	830	PhysM	94	96	2005	[122]
Breast cancer	Raman probe	*Ex vivo* (tissue)	?	600–1800	830	PhysM	?	?	2002	[98]
Breast cancer	Raman probe	*Ex vivo* (tissue)	21	800–1800	830	PhysM	83	93	2009	[117]
Breast cancer	Raman probe	*In vivo* (biopsy)	33	800–1800	830	PhysM; LR, LOOCV	84	86	2011	[116]

Breast cancer	Raman probe	*In vivo* (biopsy)	33	800–1800	830	SVM, LOOCV	64.3	100	2013	[159]
Breast cancer	Raman probe	*Ex vivo* (tissue)	90	800–1800	830	PCA	88	93	2002	[160]
Breast cancer	Micro-spectrometer	*Ex vivo* (tissue)	44	200–3600	514	PCA-PLS	72	83	2008	[161]
Breast cancer	Raman probe	*In vivo*	133	800–1800	785	SVM, AWKH	94.9	93.8	2014	[162]
Colorectal cancer	Micro-spectrometer	*Ex vivo* (tissue)	88	610–1700	785	SVM	92.3	94.2	2014	[163]
Colorectal cancer	Micro-spectrometer	*Ex vivo*	88	1000–4000		EWHK	81.38	92.69	2017	[164]
Colorectal cancer	Micro-spectrometer	*In vitro* (cell lines)		300–1800; 1800–3200	532	PCA-LDA; SVM; PLS	100	98	208	[153]
Colorectal cancer	Raman probe	*Ex vivo* *In vivo*	54	800–1800	785	PCA-LDA	91 (adenomatous polyps)	95	2003	[154]
Colorectal cancer	Micro-spectrometer	*Ex vivo*	81	400–4000	532	PCA-LDA PLS-DA	72.8 77.7	85.9 91	2016	[165]
Colorectal cancer	Raman probe	*Ex vivo*	242	220–3310	785	SVM, LOPOCV	79	74	2017	[166]
GI cancer	Raman probe	*In vivo*	107	800–1800	785	PhysM; PLS-DA, LOOCV	90.9	93.9	2011	[167]

(continued)

Table 1.2 (continued)

Relevant cancer	Raman method	Sampling type	Sample number	Spectral range (cm^{-1})	Laser (nm)	Data analysis	Sensitivity (%)	Specificity (%)	Year	Ref.
GI cancer	Raman probe	*In vivo*	305	800–1800	785	PLS-DA	90	73.3	2012	[168]
GI cancer	Raman probe	*In vivo*	450	800–1800	785	PLS-DA	81.3	88.3	2014	[316]
GI cancer (stomach)	Raman probe	*In vivo*	67	800–1800	785		94	93.4	2010	[169]
Gastric dysplasia	Raman probe	*In vivo*	157	400–3600	785	PCA-LDA	89.3 (FP+HW) 89.3 (FP) 75 (HW)	92.2 (FP+HW) 84.4 (FP) 82 (HW)	2016	[170]
Gastric dysplasia	Raman probe	*In vivo*	191	800–1800; 2800–3600	785	PLS-DA, LOPCV	90.9	95.9	2016	[171]
Esophageal cancer	Raman probe	*In vivo*	92	800–1800	785	PLS-LDA, LOOCV	81	94	2016	[172]
Gastric cancer	Raman probe	*In vivo*	29	800–1800	785	Ratio 1600/1300	91.7	97.7	2019	[173]
Gastric dysplasia	Raman probe	*In vivo*	66	800–1800	785	PLS-DA, LOPCV; PhysM	92.5	93.1	2015	[174]
Esophageal SCC	Raman probe	*In vivo*	48	800–1800; 2800–3600	785	PLS-DA, LOPCV	92.7	93.6	2015	[175]
Gastric cancer	Raman probe	*In vivo*	67	800–1800	785	ACO-LDA, LOOCV	89.3	97.8	2010	[176]
Gastric cancer	Raman probe	*In vivo*	83	800–1800	785	PLS-DA	83.33 (dysplasia) 84.91 (adenocarcinoma)	95.8 (dysplasia) 95.57 (adenocarcinoma)	2013	[177]

Gastric cancer	Raman probe + NIR AF	*In vivo*	81	800–1800	785	PCA-LDA	97.9	91.5	2011	[178]
Stomach ulcers	Raman probe	*In vivo*	71	800–1800	785	PLS-DA, LOOCV	90.8(mucosa) 84.7 (benign ulcers) 82.1 (malignant ulcers)	93.8 (mucosa) 94.5 (benign ulcers) 95.3 (malignant ulcers)	2010	[179]
Barrett's esophagus	Raman probe	*In vivo*	373	800–1800	785	PLS-DA	87	84.7	2014	[180]
Gastric cancer	Raman probe	*In vivo*	73	800–1800	785	CART	88.9	92.9	2008	[181]
Gastric dysplasia	Raman probe	*In vivo*	30	800–1800	785	PCA-LDA	94.4	96.3	2010	[182]
Gastric cancer	Raman probe	*In vivo*	62	800–1800	785	PCA = MLR	78 (intestinal adenocarcinoma) 75 (diffuse adenocarcinoma)	95 (intestinal adenocarcinoma) 96 (diffuse adenocarcinoma)	2010	[183]
Gastric dysplasia	Raman probe	*In vivo*	44	800–1800	785	PCA-LDA, LOOCV	95.2	90.9	2008	[184]
Barrett's esophagus	Raman probe	*Ex vivo*	673	400–1850	830	PCA	86	88	2014	[185]
Liver	Micro-spectrometer	Cell lines		500–3100	488	PCA; SVM	100	98	2014	[186]
Liver	Micro-spectrometer	*Ex vivo* (tissue)	15	800–1800	785, 1064	SMLR, LOOCV	100	89	2015	[132]

(continued)

Table 1.2 (continued)

Relevant cancer	Raman method	Sampling type	Sample number	Spectral range (cm^{-1})	Laser (nm)	Data analysis	Sensitivity (%)	Specificity (%)	Year	Ref.
Bladder cancer	Raman probe	*Ex vivo* (tissue)	29	800–1800	785	PCA-LDA	89 (benign vs normal) 79 (malignant vs normal)	79 (benign vs normal) 89 (malignant vs normal)	2005	[187]
Bladder cancer	Raman probe + OCT	*Ex vivo* (tissue)	8	300–3200	785	PCA, kNN	80 (OCT) 99 (RS-OCT)	60 (OCT) 87 (RS-OCT)	2018	[188]
Bladder cancer	Raman probe	*Ex vivo* (tissue)	54	400–1800	785	PCA-LDA	78.5	78.9	2009	[189]
Bladder cancer	Raman probe	*In vivo*	38	400–1800	785	PCA-LDA	85	79	2010	[190]
Bladder cancer	Raman probe	*Ex vivo* (tissue)	32	400–3200	785	PCA; ANN	90.3 (low grade) 97.5 (high grade)	98 (low grade) 96.4 (high grade)	2018	[191]
Prostate cancer	Raman probe	*Ex vivo* (tissue)	18	800–1800; 2800–3550	671, 785	SVM, LOOCV	85	89	2018	[192]
Prostate cancer	Micro-spectrometer	*In vitro* (cell lines)	200	400–1800	830	PCA, LDA	98	99	2005	[128]
Prostate cancer	Raman probe	*Ex vivo* (tissue)	38	800–1800	785	PCA-LDA	87 (benign vs norm) 84 (malignant vs norm)	84(benign vs norm) 87 (malign vs norm)	2005	[187]
Prostate cancer	Raman probe	*Ex vivo* (tissue)	32	500–1800	785	NNA, LOOCV	87	86	2018	[193]
Prostate cancer	Micro-spectrometer	*Ex vivo* (tissue)	50	700–1800	632.8	PCA; SVM, LOOCV	85.7	88.9	2013	[194]

Prostate cancer	Micro-spectrometer	*Ex vivo* (tissue)	74	700–1800	1064	SVM	89	98	2018	[195]
Prostate cancer	Micro-spectrometer	*Ex vivo* (tissue)	366	600–1800	785	PCA	89	98	2010	[196]
Cervical precancer	Raman probe	*In vivo*	93	800–1800	785	PC-LDA, LOOCV	100	93	2014	[197]
Cervical precancer	Raman probe	*In vivo*	29	800–1800	785	PLS-DA, dCV	72.5	89.2	2011	[198]
Cervical precancer	Raman probe	*In vivo*	84	800–1800	785	PC-LDA	81	87.1	2013	[199]
Cervical precancer	Raman probe	*In vivo*	44	800–1800; 2800–3800	785	PLS-DA	85	81.7	2012	[200]
Cervical precancer	Raman probe	*In vivo*	46	2800–3700	785	PCA-LDA	93.5	97.8	2009	[201]
Cervical precancer	Raman probe	*In vivo*	172	1000–1800	785	SMRL	96.5 (dysplasia)	97.8	2011	[202]
Cervical precancer	Raman probe	*In vivo*	145	800–1800	785	SMRL	96.2	90.7	2009	[203]
Cervical precancer	Raman probe	*In vivo*	79	800–1800	785	LR	89 (high grade)	81 (benign)	2007	[204]
Cervical precancer	Micro-spectrometer	*Ex vivo* (tissue)	35	400–1800	532	PCA-LDA PLS-DA	74.9 95.5	89.9 92.7	2018	[205]

(continued)

Table 1.2 (continued)

Relevant cancer	Raman method	Sampling type	Sample number	Spectral range (cm^{-1})	Laser (nm)	Data analysis	Sensitivity (%)	Specificity (%)	Year	Ref.
Ovarian cancer	Micro-spectrometer	*In vitro* (cell lines)		700–1600	785	PCA-LDA, LOOCV	81.3	83.8	2017	[206]
Ovarian cancer	Micro-spectrometer	*Ex vivo*	15	800–1800	785	PCA	87.5	83.8	2008	[207]
Oral cancer	Raman probe	*Ex vivo* (tissue)	12	1000–1800	783,785	PCA, LDA	86.1	94.4	2014	[104]
Nasopharyngeal carcinoma	Raman probe	*In vivo*	95	800–1800; 2800–3600		PCA-LDA, LOOCV	93.6 (FP+HW) 89.2 (FP) 89.7 (HW)	92.6 (FP+HW) 89.9 (FP) 89 (HW)	2017	[208]
Oral SCC	Raman probe	*In vivo*	24	300–2000	488	MCR-ALS	92	100	2016	[209]
Oral SCC	Raman probe	*In vivo*	50	1200–1800	785	PCA-LDA	96	83	2012	[210]
Oral cancer	Raman probe	*In vivo*	515	900–1750	785	SMRL	96 (SCC vs norm) 99 (submucous fibrosis vs norm)	99 (SCC vs norm) 98 (submucous fibrosis vs norm)	2014	[211]
Oral SCC	Raman probe	*In vivo*	57	800–1800	785	PC-LDA, LOOCV	80	29.7	2017	[212]
Laryngeal cancer	Raman probe	*Ex vivo*	47	950–1650	785	PCA	69	94	2005	[213]
Laryngeal cancer	Micro-spectrometer	*Ex vivo*	19	400–1800	830	PCA-DA	76 (dysplasia) 92 (carcinoma)	91 (dysplasia) 90 (carcinoma)	2000	[124]
Nasopharyngeal cancer	Raman probe	*In vivo*	79	800–3600	785	PLS-DA	91	95	2017	[214]
Thyroid cancer	Micro-spectrometer	Cell lines	116	600–1800	783	NNA	95	92	2009	[215]

Thyroid cancer	Micro-spectrometer	Cell lines		400–1800	532	PC-LDA	94 (FTC vs benign) 63 (PTC vs benign) 79 (UTC vs benign) 100 (MTC vs benign)	92 (FTC) 91 (PTC) 96 (UTC) 96 (MTC)	2018	[216]
Thyroid cancer	Micro-spectrometer	*Ex vivo*	38	200–3400	532	PCA-LDA, LOOCV	93	100	2016	[217]
Brain cancer	Needle Raman probe	*In vivo* (biopsy)	19	2600–3800	671	SVM, LOOCv	80	90	2018	[218]
Brain cancer	Raman probe	*In vivo*	6	400–1800	785	ANN	93	91	2015	[219]
Brain cancer	Raman probe	*In vivo* (biopsy)	10	400–1800	785	BT, LOOCV	84	89	2015	[220]
Brain cancer	Raman probe + BS + AFS	*In vivo* (biopsy)	15		785	SVM, LOPCV	100 (RS + IFS) 92 (RS)	9 (RS + IFS)3 90 (RS)	2017	[221]
Skin (melanoma)	Micro-spectrometer	*Ex vivo* (tissue)	19	800–1800	1064	LDA	93	99	2010	[120]
Skin (MM, BCC, SCC)	Micro-spectrometer	*In vivo*	190	800–1800	830	Biomodel	??	???	2017	[222]
Skin (MM, BCC, SCC, DN, AK)	Raman probe	*In vivo*	65	800–1800	830	PhysM PCA-LDA, LOOCV	95 95	71 (MM vs norm) 18 (BCC, SCC vs norm) 64 (MM vs norm) 10 (BCC, SCC vs norm)	2018	[121]

(continued)

Table 1.2 (continued)

Relevant cancer	Raman method	Sampling type	Sample number	Spectral range (cm^{-1})	Laser (nm)	Data analysis	Sensitivity (%)	Specificity (%)	Year	Ref.
Skin (MM, BCC, SCC, PL, AK)	Raman probe + LIFS	*In vivo*	76	400–1800	830	PCA	90 (BCC/SCC vs norm)	85 (BCC/SCC vs norm)	2014	[223]
							100 (MM vs PL)	100 (MM vs PL)		
Skin (MM, BCC)	Raman probe	*Ex vivo* (tissue)	47	400–1800	830	PhysM; LLS	98.2	93.3	2012	[224]
Skin	Raman probe	*In vivo*	645	500–1800	785	PC-LDA; PLS, LOOCV	95	54	2015	[225]
Skin	Raman probe	*In vivo*	30	500–1800	830	PhysM	90	92	2019	[226]
Skin (MM, BCC, SK, PL, AK)	Micro-spectrometer	*Ex vivo*	134	800–1800; 2500–3500	1064	NNA	85	99	2004	[125]
Skin cancer	Raman probe	*Ex vivo*	23	800–1800	785	LDA, QDA	89 (MM vs non-MM) 78 (BCC vs non-BCC, RS) 89 (BCC vs non-BCC, RS-OCT)	88 (RS), 93 (RS-OCT) 85 (RS) 96 (RS-OCT)	2015	[111]
Skin cancer	Raman probe	*In vivo*	453	500–1800	785	PC-LDA; PLS, LOOCV	90 (cancer vs benign) 95 (MM vs benign)	64 38	2012	[227]

Skin cancer	Raman probe + autofluorescence	*Ex vivo*	79	800–1800	785	PCA-DA	89.7 (RS) 94.6 (RS + NIR AF) 100 (RS + NIR + UV AF)	67 (RS) 92.5 (RA+NIR AF) 96.3 (RS + NIR + UV AF)	2017	[228]
Skin cancer	Raman probe + autofluorescence	*In vivo*	56	800–1800	785	PLS-DA	100 (MM vs benign) 73.7 (MM vs other) 94.7 (BCC vs other)	100 (MM vs benign) 97.4 (MM vs other) 89.2 (BCC vs other)	2018	[140]
Skin cancer	Raman probe	*In vivo*	104	300–1750	785	PCA; PLS-DA	87 (MM vs PN) 63 (BCC vs norm) 80 (SCC vs norm)	94 (MM vs PN) 83 (BCC vs norm) 89 (BCC vs norm)	2015	[229]
Skin cancer	Raman probe	*In vivo*	289	600–1800	785	PLS-DA	91 (cancer vs benign) 97 (MM vs benign)	75 (cancer vs benign) 78 (MM vs benign)	2008	[230]

GI gastrointestinal, *BCC* basal cell carcinoma, *SCC* squamous cell carcinoma, *MM* malignant melanoma, *AK* actinic keratosis, *PL* melanoma pigmented lesion, *PN* pigmented nevi, *SK* seborrheic keratosis, *DN* dysplastic nevi, *FTC* follicular thyroid carcinoma, *PTC* papillary thyroid carcinoma, *UTC* undifferentiated thyroid carcinoma, *MTC* medullary thyroid carcinoma, *PhysM* physiological (biochemical) model, *BS* diffuse backscattered reflectance spectroscopy, *AFS* autofluorescence spectroscopy, *LIFS* laser-induced fluorescence spectroscopy, *PCA* principal component analysis, *LDA* linear discriminant analysis, *SVM* support vector machine, *LRA-marker* lipids structure and relative abundance, *ACO* ant colony optimization, *MCR* multivariate curve resolution analysis, *EWHK* entropy weight local-hyperplane k-nearest neighbor, *LLS* linear least-squares fitting, *PLS-DA* partial least-squares discriminant analysis, *LOOCV* leave-one-out cross-validation, *LOPOCV* leave-one-patient-out cross-validation, *PC-LDA* principal components-logistic regression analysis, *MLR* multinomial logistic regression, *q SMLR* sparse multinomial logistic regression, *LP-ratio* lipid-to-protein ratio (1444/1665), *NNA* neural network analysis, *dCV* double cross validation, *PC-GDA* principal components with generalized discriminant analysis, *QDA* quadratic discriminant analysis, *LR* logistic regression analysis, *AWKH* adaptive weight k-local hyperplane algorithm, *kNN* k-nearest neighbor, *ANN* artificial neural network, *CART* classification and regression tree, *BT* boosted trees (invasive ductal carcinoma and lobular carcinoma) aiming an aid to histopathological diagnosis of breast cancer. Several modes of vibration have been found to be significantly different between the benign and malignant tissues. The band at 1662 cm^{-1} is assigned to the amide I mode originating mainly from proteins and nucleic acids. The two weak bands at 1610 and 1585 cm^{-1} observed in the breast tissue are due to the ν(C=C) modes of aromatic amino acids (phenylalanine, tyrosine, and tryptophan). The band at 1448 cm^{-1} is assigned to the $\nu(CH_2/CH_3)$ modes from a combination of lipoproteins from the cell membrane, adipose tissue, and nucleic acids. The amide III bands are observed in the region of 1295–1200 cm^{-1}, which are attributed to a combination of ν(CN) and ν(NH) modes of the peptide bond ν(–CONH). The bands at 936 and 856 cm^{-1} are assigned to the ν(C–C) modes of proline and valine and the ν(C–CH) modes of proline and tyrosine, respectively. The spectrum exhibits three major characteristic bands in this region including those due to the ν(C=C) mode at 1515 cm^{-1}, the ν(C–C) mode at 1156 cm^{-1}, and the ring breathing mode at 1004 cm^{-1}. The performance of the different algorithms PCA-LDA, PCA-QDA, PLS-DA, linear c-SVC, linear nu-SVC, RBF c-SVC, and RBF nu-SVC has been evaluated using sensitivity and specificity calculated based on the results from the Raman data and histopathology as the gold standard. PCA-LDA, PCA-QDA, and PLS-DA models have achieved similar sensitivity and specificity of 80%. SVM models have achieved sensitivity and specificity exceeding 90%, but required more processing time than other models.

Cell Lines The complexity of tissue structure and environment makes the interpretation of tissue Raman spectra difficult. An understanding of the molecular, microscopic, and macroscopic origin of observed tissue Raman signals may be achieved by *in vitro* study of Raman spectra of biologically important molecules in solution, in single living cells, in cell cultures prepared from surgically removed human tissues [142] or established with cancer cell lines [114]. Cell lines are widely used in many aspects of laboratory research and particularly as *in vitro* models in cancer research. They have a number of advantages, for example, they are easy to handle and represent an unlimited self-replicating source that can be grown in almost infinite quantities. In addition, they exhibit a relatively high degree of homogeneity and ease of handling [143]. Raman spectra not only reveal differences in biological composition between cell lines but also represent the combined effect of these parameters in order to study various aspects of elementary biological processes such as the cell cycle, cell differentiation, and apoptosis.

The majority of researchers have primarily been focused on spectral differences in the fingerprint range 600–1800 cm^{-1} as it includes peaks that can be assigned to different biochemical compounds, such as lipids, proteins, or nucleic acids. The lipid content and the chemical structure of these compounds, for instance, can be evaluated using peak frequencies of 1754 cm^{-1} (C=O), 1656 cm^{-1} (C=C), 1440 cm^{-1} (CH_2 bend), and 1300 cm^{-1} (CH_2 twist). Specification of the protein content of biological samples can also be understood from 1656 cm^{-1} (amide I), 1450 cm^{-1} (CH_2 bend), 1100–1375 cm^{-1} (amide III), and 1004 cm^{-1} (phenylalanine) [99, 115, 127, 142–149].

Oshima et al. [150] have demonstrated differences among cultures of normal and cancerous lung cell lines, namely adenocarcinoma and squamous cell carcinoma with low to medium and high malignancy. Single-cell Raman spectra have been obtained by using 532-nm excitation wavelength. Strong bands at 747, 1127, and 1583 cm^{-1} have been assigned to cytochrome *c* (cyt-c) indicating resonance near 550 nm with excitation light. Peaks at 1449, 1257, 1003, and 936 cm^{-1} have been assigned to the CH_2 deformation, amide III, the symmetric ring breathing bands of phenylalanine of the protein, and C–C stretching, respectively. The bands at 720, 785, 830, 1086, 1340, 1421, and 1577 cm^{-1} have been assigned to nucleic acids (DNA and RNA). The overlapping modes of the amide I band of protein and the C=C stretching band of lipids form strong Raman peak at 1659 cm^{-1}. PCA has successfully applied and 80% accuracy has been achieved in discrimination between four cancer cell lines.

Guo et al. [151] have reported that Raman spectroscopy can be used to differentiate malignant hepatocytes from normal liver cells. It has been shown that the strong bands at 1447 and 1656 cm^{-1} can be attributed to the CH_2 deformation mode and the C=C stretching mode of the lipids and proteins, respectively. The band originating at 786 cm^{-1} can be assigned to the O–P–O stretching mode of DNA. The bands appearing at 1004 and 1032 cm^{-1} can be assigned to the symmetric ring breathing mode and the C–H in-plane bending mode of phenylalanine, respectively. Statistical methods such as *t* test, PCA, and LDA have been used to analyze the Raman spectra of both cell lines. The results of *t* test have confirmed that the intensities of these bands are considerably different between two cell lines, except for the 1585 and 1625–1720 cm^{-1} bands.

Crow et al. [128] have studied different prostatic adenocarcinoma cell lines and have found that principal components allow identification of molecular species from their Raman peaks and provide an understanding of the origins of the statistical variations. PC1 represents increased concentrations of nuclear acids (721, 783, 1305, 1450, and 1577 cm^{-1}), DNA backbone (O–P–O) (827 and 1096 cm^{-1}), and unordered proteins (1250 and 1658 cm^{-1}). PC2 represents decreased concentrations of α-helix proteins (935, 1263, and 1657 cm^{-1}) and phospholipids (719, 1094, 1125, and 1317 cm^{-1}). PC3 represents decreased concentrations of lipids (1090, 1302, and 1373 cm^{-1}), glycogen (484 cm^{-1}), and nucleic acids (786, 1381, and 1576 cm^{-1}). The PCA/LDA algorithm has achieved near perfect identification of each cell line, with sensitivities ranging from 96 to 100% and specificities all 99% or higher.

Krishna et al. [152] have used micro-Raman spectroscopy to investigate randomly mixed cancer cell populations, including human promyelocytic leukemia, human breast cancer, and human uterine sarcoma, as well as their respective pure cell lines. According to the results, cells from different origins can display variances in their spectral signatures and the technique can be used to identify a cell type in a mixed cell population via its spectral signatures.

Recent attention has been directed towards the use of high-wavenumber range (2800–3600 cm^{-1}), as the HW spectral range exhibits stronger tissue Raman signals with less autofluorescence interference. In this spectral region most of the spectral features obtained from tissue are overlapping symmetric and asymmetric stretching of CH_2 and CH_3 vibrations of phospholipids and proteins with four main peaks, located at 2850 cm^{-1}, 2880 cm^{-1}, 2920 cm^{-1}, and 2960 cm^{-1} [144, 147, 148]. There are also minor peaks of SH-stretching vibrations 2500–2600 cm^{-1} [115, 145] and broad band of OH-stretching vibrations (primarily due to water) in the spectral interval 3100–3500 cm^{-1} [145–147]. The CH stretch vibrations are sensitive to their environment by direct coupling and through Fermi resonances with C–H bending modes near 1500 cm^{-1}. Together, these influences can introduce significant shifts and broadening of the CH stretch peaks [99, 127].

For example, Telari et al. [114] have studied Raman spectra of normal (MCF-10A) and two breast cancerous cell lines with different concentrations of nucleic acid (MDA-MB-436 and MCF-7) in fingerprint and HW ranges using noninvasive dispersive micro-Raman system equipped with a 532-nm laser. Peak intensities have shown clear differences among three cell lines in lipids (2934 cm^{-1}), amide I (1658 cm^{-1}), and amide III (1244 cm^{-1}) ranges. PCA with the whole spectral range has shown good overall separation between the three cell lines, but it has not formed separate clusters representing "normal" and "cancerous/diseased" classes [114]. This suggests a very large biochemical variation even between the two breast cancerous cell lines. The MCF-7 cell line appears to be much higher in lipids compared to MDA-MB 436 and MCF 10A, and PCA works well to single out this cell line in view of the high-wavenumber region, which includes major peaks of symmetric and asymmetric stretching CH_2 vibrations of lipids at 2882 cm^{-1}, C–H, CH_2 symmetric vibrations in lipids and proteins (2940 cm^{-1}, 2921 cm^{-1}, and 2948 cm^{-1}). Although MCF-7 and MDA-MB-436 are both breast cancer subtypes, the MDA-MB-436 does not appear to contain lipids at a concentration vastly different to those found in the normal MCF-10A cell line. Instead, the difference lies more in the relative protein and amino acid concentrations, which may be identified in fingerprint region for adenine and guanine (1337 cm^{-1}), CH_2 deformation of lipids, adenine, and cytosine (1258, 1299, and 1304 cm^{-1}), and methylene twisting vibrations (1294 cm^{-1}) and different conformations in C=O stretching of proteins (1687 cm^{-1}), anti-parallel ß-sheets of amide I (1670 cm^{-1}), tryptophan or ß-sheet of protein (1621 cm^{-1}), C=C of phenylalanine ring vibration, tyrosine (1607 cm^{-1}), and tryptophan (1548 cm^{-1}).

Gala de Pablo et al. [153] have studied Raman spectra distinction (Fig. 1.5) between primary (SW480) and secondary (SW620) tumor cells, derived from a primary Duke's stage B adenocarcinoma and secondary tumor in a lymph node from the same patient. The CH_2 and CH_3 stretching contributions in the region of 2800–3200 cm^{-1} have shown higher overall intensity for primary tumor cells and a greater CH_2:CH_3 ratio for secondary cells, indicating differences in lipid composition between the two cell lines with higher lipid content for the larger size of primary cells.

When normalizing to the amide I band, secondary tumor cells (SW620) show a larger contribution of α-helix proteins, saccharides, nucleic acids, and double bonds related bands, whereas primary tumor cells (SW480) show larger contribution of lipids, β-sheet, and disordered structure proteins. Principal component analysis with linear discriminant analysis yields the best classification between the SW620/SW480 cell lines, with an accuracy of 98.7 ± 0.3% (standard error).

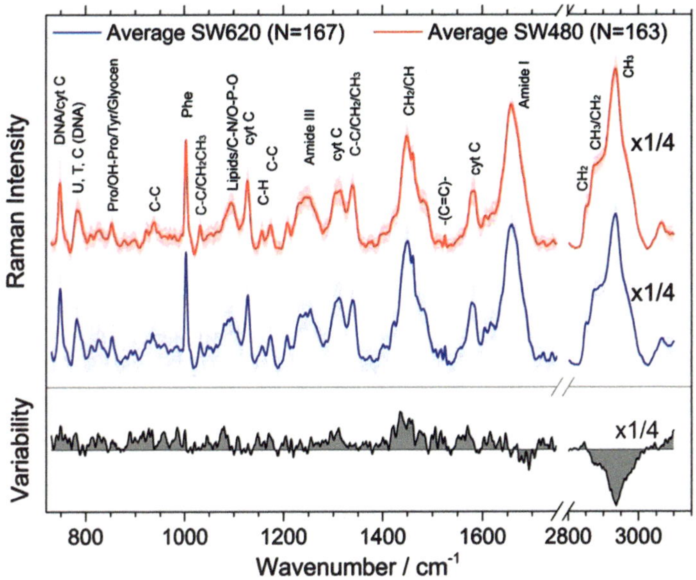

Fig. 1.5 Average single-cell spectra and variability spectrum, for primary (SW480) and secondary (SW620) tumor cells, normalized to the amide I peak. The error around the average shows one standard deviation. The region around 2900 cm^{-1} is shown reduced by a factor of 4 to enhance the details in the fingerprint region. Adapted with permission from [153]

Laser-Trapped Single-Cell Diagnostics The combination of laser tweezers and Raman detection is a very attractive application for the identification of malignant cells in cytological diagnosis systems. Chen et al. [142] have employed PCA analysis of Raman spectra from laser trapping of single cell of colorectal epithelial cells solution to differentiate cancerous and normal epithelial cells. The higher concentrations of nuclear acids and proteins in cancerous cells are reflected in major variations and an increase in Raman intensities at 788 cm^{-1} (DNA backbone O–P–O stretching), 853 cm^{-1} (ring breathing mode of tyrosine and C–C stretching of proline ring), 938 cm^{-1} (C–C backbone stretching of protein α-helix), 1004 cm^{-1} (symmetric ring breathing of phenylalanine), 1095 cm^{-1} (DNA PO_2^- symmetric stretching), 1257 cm^{-1} (amide III β-sheet), 1304 cm^{-1} (lipids CH_2 twist), 1446 cm^{-1} (CH_2 deformation of all components in cell), and 1657 cm^{-1} (C=O stretching of amide I α-helix). The PCA scores have been fed into logistic regression algorithm to determine the parameter equation that best differentiates the cancer cells from the normal ones, obtaining an overall sensitivity of 82.5% and specificity of 92.5%.

The extensive *ex vivo* studies have helped to form a reliable and detailed database of accurate Raman peak definitions and have given the knowledge about differences in spectral features of normal, benign, and malignant tissues (see references in Table 1.1). However, the real benefit of the method can only be explored through *in vivo* applications, which has become possible due to the advantages in the detector technology and progress in the development of miniature Raman fiber-optic probes. As such, it has been a significant movement from *ex vivo* to *in vivo* studies in recent years. A partial list of different Raman clinical applications for cancer diagnostics can be found in Table 1.2. Strong efforts have been made towards transfer of *ex vivo* tissue statistical models and classifiers to an *in vivo* clinical situation. For example, Molchovsky et al. [154] have found that the *ex vivo* classifier has not performed well; indeed, the PCA analyses of *ex vivo* and *in vivo* tissue are different. Therefore, the designed models need to be adapted to *in vivo* applications. *In vivo* studies are focused on three major clinical targets: early cancer diagnosis, biopsy guidance, and oncologic surgery guidance. As it may be seen in Table 1.2 the average sensitivity and specificity obtained using Raman spectroscopy for different cancer types vary from 83 up to 96% and from 77 up to 94%, respectively. It is interesting to point out that multimodal approaches, combining different modalities (OCT, fluorescence and Raman spectroscopy), improve the sensitivity of *in vivo* Raman diagnostic system by 5–8% and allow the more accurate diagnosis of premalignant lesions. Implementation of biophysical models together with cross-validation algorithms allows obtaining a statistical predictor for cancer diagnostics with biochemical semi-quantitative justification.

1.2.4 Fluorescence Spectroscopy

Light-induced autofluorescence spectroscopy is a very attractive tool for early diagnosis of cancer due to its high sensitivity, easy-to-use methodology for measurements, lack of need for an exogenous contrast agents' application, possibilities for real-time measurements, and noninvasive character of the detection technique in general, which allows one to work *in vivo* without pre-preparation of the samples [5, 79, 80, 83–85]. Highly-sensitive cameras and narrow-band filters application nowadays allow obtaining fluorescent maps of the tissues investigated in 2-D image modality, which support the exact tumor borders and safety margins determination, which is required and very useful information in the following therapeutical planning. Fluorescence spectroscopy is a very sensitive tool with broad applicability for tumor detection. Its diagnostic sensitivity depends on many factors related to the lesions investigated: their biochemical content, metabolic state, morphological structure, localization and stage of tumor development.

Internal chemical compounds, which can fluoresce after irradiation with a light, are called endogenous fluorophores. Investigation of such chemicals' fluorescent emission properties can give information about their concentration, distribution into the different tissue structures and layers, as well about alterations in microenvironment, related to disease progress, including changes in pH, temperature, or chemical transformations or reactions, preceded in these fluorophores. Typical endogenous fluorophores used for evaluation of the tissue state are divided into several groups depending on their chemical nature, including amino acids, proteins, co-enzymes, vitamins, lipids, and porphyrins. Protein cross-links, being overmolecular structures, which are related to the tissues' extracellular matrices, add their fluorescent signals as well, to enrich the picture of emission properties that can be used for tissues' biomedical diagnostics. Both the endogenous fluorophores concentration and distribution into the tissues depend on the metabolic and structural peculiarities of the tissue investigated. Some of them during alteration of microenvironment of the tissue or cells, where they are situated, go through chemical transformations as well, which can alter their emission properties and also can be used for evaluation of the processes of tumor growth and metabolic activity in the lesions investigated.

The compounds that absorb the light without re-emission in normal conditions in the form of fluorescence signal are known as endogenous chromophores and also influent the emission response, when fluorescent spectroscopy technique is used for analysis and can significantly alter the emission detected from the tissues investigated. In the ultraviolet (UV) spectral range, most of the biologically important molecules including amino acids, DNA, RNA, structural proteins, co-enzymes, and lipids absorb light. Typical endogenous chromophores with absorption bands in visible and near-infrared range, where the tissue endogenous fluorescence is observed typically, are melanin (pheo- and eumelanin, the pigments typical for mammal skin and eye tissues), pigment in the red blood cells—hemoglobin, in its oxidized and reduced form (oxy- and deoxyhemoglobin), and bilirubin (yellow pigment, product of catabolism of heme in hemoglobin). These absorbers can have significant influence on the emission signal from the tissue investigated due to filtering effect, when they directly absorb excitation light leading to decreased effective absorption in the fluorophores and lower yield of emitted photons as a result, and due to lower levels of their excitation, as well indirectly, when they reabsorb the resultant emission from the fluorophores. Their absorption bands are observed in the reported emission spectra for different types of tumors and localizations. The cancerous tissues are characterized by different content and distribution of such chromophores in the tissue volume. Therefore, their influence on the emission spectra is significant index to the process of malignization being non-specific but diagnostically-important additive indicator for tumor development process in the tissue investigated.

Fluorescent properties investigated for the tissue cancer diagnostics needs are based on the steady-state or time-resolved measurements of excitation and emission spectral and fluorescent decay properties, respectively.

Steady-state fluorescence spectroscopy technique is based on the detection of fluorescence intensity as a function of the registered wavelength (energy and frequency) for fixed excitation wavelength. Each fluorophore is characterized with specific pair of excitation (1) and emission (2) maxima—(1) wavelength with light absorption maximal efficiency, which is transformed to a fluorescent signal and (2) wavelength, where the fluorescent intensity observed is maximal by its absolute value in comparison with all others into the emission range for a given compound. Such pair of excitation and emission wavelengths is unique for each fluorophore appeared in the biological tissue and can be used as indicator of the presence of this compound. If multiple excitation wavelengths are used for consequent detection of such fluorescent intensities functions of registered wavelength, the so-called excitation–emission matrix can be developed, which allows to address whole set of endogenous fluorophores in a complex sample, such as biological tissues that are consisted from a mixture of several different fluorescent compounds. Excitation–emission matrices developed in such a way consist of specific islands with high fluorescent emission detected that correspond to the specific pairs of excitation and emission wavelengths. In ideal case, the number of such "islands" in excitation–emission matrix corresponds to the number of endogenous fluorophores existing in the tissue investigated. The fluorescent emission intensity corresponds to the number of excited molecules of given type of fluorophore, which re-emit light, that correlate directly to the quantity of this compound in the sample investigated. In such a way, the steady-state fluorescence intensity measurements allow the determination of the fluorophores' presence and concentration inside of the object investigated, and they are broadly used in experimental studies of neoplasia due to simple approaches needed for spectral data detection, processing, and analysis.

Time-resolved fluorescence spectroscopy technique is not so popular for biological tissue investigations due to the required very sensitive and fast detection equipment, which lead to higher costs for the last. Time-resolved fluorescence allows finding the values of the fluorescence decay time of the endogenous fluorophores after irradiation with short pulse of excitation light. Fluorescence decay time, also called fluorescence lifetime, occurs as emissive decays from the excited to ground singlet – state energy levels of the endogenous fluorophore molecule. The typical decay time for diagnostically important fluorophores lies in the region from picoseconds to nanoseconds. This parameter is specific for a given chemical compound by its value, but also can vary due to strong sensitivity to the small perturbations in the microenvironment around such fluorophore molecule. Information about the fluorescent decay time and its deviations allows to obtain knowledge about the interaction with surrounding molecules for a given fluorophore and for the microenvironment conditions for the molecular ensemble in general.

In the process of malignancy development prominent alterations in biochemical and morphological properties of the biological tissues are observed. They can lead to significant differences in the fluorescent spectra of normal and abnormal biological tissues, which can be detected and used as diagnostic indicators and/or as predictors of tumor lesion development.

Table 1.3 presents the typical endogenous fluorophores and chromophores, the dynamics of their fluorescent properties, which are indicative of malignant alterations in the biological tissues. Reasons for these changes are also briefly indicated, according to the investigations of research groups referred.

The most often alterations observed and discussed in the literature are related to the changes in the ratio of NADH/NAD^+ that lead to changes in the level of the autofluorescent intensity—reduced form of the coenzyme NAD^+ is not fluorescent, but its concentration increases in tumor cells due to alteration in their metabolism related to hypoxic environment in the tumor, which leads to general decrease of the tumor fluorescent intensity in 420–460 nm spectral region. Fluorescence intensity decrease in the region of 470–500 nm is observed as well due to the tissue partial destruction in the process of tumor lesion growth and changes in the extracellular matrix and decrease or even partial demolition in the structural protein content in the area of tumor. That extracellular matrix damages affect the signals coming from

Table 1.3 Endogenous fluorophores—excitation and emission maxima, the dynamics of their fluorescence intensity from normal to cancerous tissue development and origins of the observed alterations in cancer tissues' fluorescent properties

Endogenous fluorophore	Excitation (nm)	Emission (nm)	Origins of fluorescent changes observed in cancerous tissues
Amino acids			
Phenylalanine	260	280	General increase in the signal from amino acids due to increased metabolic activity in tumor tissues [231, 232]
Tyrosine	275	300	
Tryptophan	280	350	
Structural proteins			
Collagen	280, 320–350, 390	370–440	General decrease in the collagen and elastin fluorescent signals is usually observed. Protein cross-links fluorescence signal is lower in tumor vs. normal tissues due to destruction of extracellular matrix (ECM) during tumor lesion growth. ECM degradation is associated with the secretion of proteolytic enzymes such as metalloproteinases [233] and changes in morphology caused by the expression of tumor cell matrix proteins, such as laminins [234]
Elastin	290–325	340, 400	
Collagen cross-links	380–420	440–500	
Elastin cross-links	320–360, 400	480–520	
Keratin	380–400, 450–470	500–550	For skin tumors an increase in keratin content is usually observed and in non-pigmented tumors the fluorescence signal is higher than in normal tissue [231]
Enzymes and co-enzymes			
NADH	290, 350–370	440, 460	NAD^+/NADH increasing ratio is observed in tumor tissues, which leads to general decrease in the fluorescence signal, due to non-fluorescent form NAD^+ [235]
NADPH	340	460	
FAD, Flavins	420–450	520–550	The excess of tyrosine and tryptophan residues resulting from their intensive metabolism in tumor cells is a prerequisite for a lower intensity of FAD fluorescence from tumor relative to healthy tissues due to quenching the fluorescence of FAD [236]. Hypoxia also leads to a decrease in flavin content in tumor cells [237]
FMN	420–500	520–570	
Vitamins			
Vitamin A	327	510	Fluorescence of vitamins is reported in tumor tissues by different research groups, especially the fat-soluble vitamins, but no specific tendencies related to their appearance and peculiar dynamics of fluorescent properties are reported in literature [238]
Vitamin K	335	480	
Vitamin D	290, 350–390	400–480	
Vitamin B6 compounds	315, 330, 340	385, 400, 425	
Vitamin B12	275	305	
Lipids			
Phospholipids	430	500, 540	Lamina propria of GIT tract (as an example) is saturated with lipofuscin granules and as this part of the mucosal layer progressively diminishes with the progress of tumors, a decrease in the intensity of lipofuscin fluorescence is observed [239]
Lipofuscin	340–395	430–460, 540	
Ceroid	340–395	430–460, 540	

(continued)

Table 1.3 (continued)

Endogenous fluorophore	Excitation (nm)	Emission (nm)	Origins of fluorescent changes observed in cancerous tissues
Porphyrins	390–450, 630	635–690, 704	Increased fluorescence in tumors due to ferrochelatase enzyme blocking in tumor cells, leading to abnormal increase in porphyrins' content in tumor cells vs. normal ones. Level of porphyrins' fluorescence emission correlates with the stage of tumor growth [231, 240, 241]

collagen and elastin, the main structural proteins, as well as from the cross-link protein structures. In some specific cases the opposite tendency is observed, where the tumor reveals increased metabolic activity, fast growth, and low pigmentation, such as for cutaneous squamous cell carcinoma (SCC) lesions. There, the autofluorescence intensity can be higher than that of surrounding normal skin, and in advanced stages of SCC flavin green fluorescence can be detected and easily observed even with naked eye.

Red fluorescence signals *in vivo* are also observed and reported in the literature. Hypothesis related to the origin of this signal is related to accumulation of endogenous porphyrins in the tumor cells of various types of tumors. The specific signature of fluorescent emission with bright maximum at 635 nm and less pronounced 704 nm fluorescence peaks related to the endogenous porphyrins can be observed in advanced stages of tumor growth (grade III and IV), which make it specific but not optimistic index of lesion development. However, usually the fluorescent maxima at 635 and 704 nm on the initial stage of lesion development are with low intensity or even absent and not typically observed for lesions on grade I or II of their development. Porphyrins' fluorescent signal can be increased using exogenous delta-aminolevulinic acid application, which is precursor of protoporphyrin IX. After accumulation of 5-ALA in the cells it transforms to heme of hemoglobin. In normal cells for few hours all chain of heme synthesis is accomplished, but in tumor ones, due to blockage of enzyme ferrochelatase the iron ion cannot be added to the protoporphyrin IX molecule, which will transform it to heme molecule, and the concentration of PpIX is rapidly increased in the cancerous area. In many clinical applications exogenous fluorophores from the family of porphyrins photosensitizers are applied as exogenous fluorescent contrast agents.

1.3 Optical and Physiological Properties of Malignant Tissues

1.3.1 Lung cancer

In both sexes combined, lung cancer is the most commonly diagnosed cancer (11.6% of the total cases) and the leading cause of cancer death (18.4% of the total cancer deaths) [1]. Reasons for the high mortality rate are the fact that patients tend to be diagnosed at an advanced stage and a lack of effective treatments. Part of the diagnostic process is white light or fluorescence bronchoscopy combined with tissue biopsy for definitive pathology. A problem with this technique is that it suffers from either low sensitivity or specificity and it is difficult to ensure the representativeness and quality of the biopsies during the procedure [242].

In early study Huang et al. [129] demonstrated the potential of Raman spectroscopy to differentiate accurately normal bronchial tissue specimens, squamous cell carcinoma, and adenocarcinoma. The Raman spectra of malignant tumor tissue were characterized by higher intensity bands corresponding to nucleic acids (PO_2^- asymmetric stretching 1223 cm^{-1} and CH_3CH_2 wagging 1335 cm^{-1}), tryptophan (752, 1208, 1552, and 1618 cm^{-1}), and phenylalanine (1004, 1582, and 1602 cm^{-1}) and lower signals for phospholipids (CH_2CH_3 bending modes 1302 and 1445 cm^{-1}) and proline (855 cm^{-1}), compared to normal tissue. The peak at 1078 cm^{-1} in normal tissue due to the C–C or C–O stretching mode of phospholipids was shifted to 1088 cm^{-1} in tumor tissue and had lower normalized percentage signals, reflecting a decreased vibrational stability of lipid chains in tumors. The authors found that the ratio of the Raman band intensity at 1445 cm^{-1} (CH_2 scissoring) and 1655 cm^{-1} (C=O stretching of collagen and elastin) had high discrimination power between normal and tumor tissues with sensitivity and specificity of 94% and 92%, respectively. Zakharov et al. [111, 243] used three ratios of maximum scattering intensities in the 1300–1340 cm^{-1} bands, in the 1640–1680 cm^{-1} bands, and in the 1440–1460 cm^{-1} bands to separate lung tumor from healthy tissue with following differentiation adenocarcinoma and squamous cell carcinoma by ratios contrast with surrounding normal tissue. It was achieved sensitivity and specificity of 91% and 79%, respectively. However, the diagnostically useful information contained not only in a few peaks, the entire spectral information could be important for the accuracy of tissue classification and cancer detection.

Similar spectral features were obtained by Magee et al. [244] using shifted subtracted Raman spectroscopy for reduction of the fluorescence from the lung tissue and principal component with a leave-one-out analysis for accurate tissues classification. The first *in vivo* study was conducted in 2008 by Short et al. [148]. The authors fail to obtain precise Raman spectra in fingerprint range due to high fluorescence background, which were explained by high levels of hemoglobin close to the tissue surface. Clear Raman peaks were registered only in HW range, where the intensity ratio of extracted Raman peaks to the fluorescence was six times greater for the most intense Raman peaks compared to those in the fingerprint range. Preliminary research on 26 patients demonstrated that the combination of Raman spectroscopy with white light bronchoscopy and autofluorescence bronchoscopy could reduce the number of unnecessary biopsies and achieve the sensitivity and specificity above 90% for detection of lung cancer and high-grade dysplasia lesions [159].

Recently, McGregor et al. [156] used the bronchoscopic Raman spectroscopy *in vivo* in 80 patients. The authors acquired Raman spectra from the high-wavenumber region (from 2775 to 3040 cm^{-1}) with an acquisition time of 1 s. Major Raman peaks were observed for CH_2 symmetric stretching modes of fatty acids and lipids at 2850 cm^{-1}; CH_3 symmetric stretching modes at 2885 cm^{-1}; overlapping CH vibrations in proteins and CH_3 asymmetric stretching modes of lipids and nucleic acids at 2940 cm^{-1}; in-plane and out-of-plane asymmetric CH_3 stretching in lipid and fatty acid molecule at 2965 cm^{-1} and 2990 cm^{-1}. It was found that spectra with malignant lesions presented a distinctive loss in lipid at 2850 cm^{-1}. The intensity of the inflammation group was relatively higher than all other categories between 2850 cm^{-1} and 2900 cm^{-1}. To extract a more reliable correlation of spectra with pathology, principal components with generalized discriminant analysis and PLS-DA with leave-one-out cross-validation (LOOCV) were used for spectral classification. The detection of high-grade dysplasia and malignant lung lesions resulted in a reported sensitivity of 90% at a specificity of 65%. In 2018 same group developed novel miniature Raman probe (1.35 mm in diameter) capable of navigating the peripheral lung architecture [245]. The *in vivo* collected spectra showed lipid, protein, and deoxyhemoglobin signatures in fingerprint (1350–1800 cm^{-1}) and HW (2300–2800 cm^{-1}) ranges that might be useful for classifying pathology.

It is known that repeated exposure to carcinogens, in particular, cigarette smoke, leads to lung epithelium dysplasia. Further, it leads to genetic mutations and affects protein synthesis and can disrupt the cell cycle and promote carcinogenesis. The most common genetic mutations responsible for lung cancer development are MYC, BCL2, and p53 for small cell lung cancer (SCLC) and EGFR, KRAS, and p16 for non-small cell lung cancer (NSCLC) [246–248]. The broad divisions of small cell lung cancer (SCLC) and non-small cell lung cancer (NSCLC) represent more than 95% of all lung cancers.

Small Cell Lung Cancer Histologically, SCLC is characterized by small cells with scant cytoplasm and no distinct nucleoli. The WHO classifies SCLC into three cell subtypes: oat cell, intermediate cell, and combined cell (SCLC with NSCLC component, squamous, or adenocarcinoma). SCLC is almost usually with smoking. It has a higher doubling time and metastasizes early; therefore, it is always considered a systemic disease on diagnosis. The central nervous system, liver, and bone are the most common sites. Certain tumor markers help differentiate SCLC from NSCLC. The most commonly tested tumor markers are thyroid transcription factor-1, CD56, synaptophysin, and chromogranin. Characteristically, NSCLC is associated with a paraneoplastic syndrome which can be the presenting feature of the disease.

Non-small Cell Lung Cancer Five types of NSCLC are distinguished: squamous cell carcinoma, adenocarcinoma, adenosquamous carcinoma, large cell carcinoma, and carcinoid tumors. Squamous cell carcinoma is characterized by the presence of intercellular bridges and keratinization. These NSCLCs are associated with smoking and occur predominantly in men. Squamous cell cancers can present as Pancoast tumor and hypercalcemia. Pancoast tumor is the tumor in the superior sulcus of the lung. The brain is the most common site of recurrence postsurgery in cases of Pancoast tumor.

Adenocarcinoma is the most common histologic subtype of NSCLC. It is also the most common cancer in women and non-smokers. Classic histochemical markers include napsin A, cytokeratin-7, and thyroid transcription factor-1. Lung adenocarcinoma is further subdivided into acinar, papillary, and mixed subtypes.

Adenosquamous carcinoma comprises 0.4–4% of diagnosed NSCLC. It is defined as having more than 10% mixed glandular and squamous components. It has a poorer prognosis than either squamous and adenocarcinomas. Molecular testing is recommended for these cancers.

Large cell carcinoma lacks the differentiation of a small cell and glandular or squamous cells [249].

Optical and physiological properties of the lung tumor tissue were investigated in [36, 250–252]. Earlier Qu et al. [250] investigated the optical properties of 10 human lung tumor samples (without indicating the king of the tumors) using integrating sphere technique and IAD method in the spectral range from 400 to 700 nm. The result of the measurements is presented in Table 1.4. Fishkin et al. [251] measured the optical properties of human large-cell primary lung carcinoma using multi-wavelength frequency-domain photon migration instrument and found significant absorption differences between normal and tumor tissue at all wavelengths. Scattering changes were less significant, but exhibited consistent wavelength-dependent behavior. Lower tumor scattering parameters (versus normal tissue) could be due to a loss of cellularity and increased water content in necrotic zones [251]. The authors demonstrated that total hemoglobin content varied from 29.2 ± 2.4 to 42.9 ± 2.9 μM for normal tissue and from 85.1 ± 8.2 to 102 ± 10 μM for tumor tissue. Deoxyhemoglobin content varied from 6.22 ± 0.64 to 9.68 ± 1.04 μM for normal tissue and from 15.9 ± 3.2 to 20.2 ± 5.2 μM for tumor tissue. Oxyhemoglobin content is varied from 23.0 ± 2.1 to 33.2 ± 2.7 μM for normal tissue and from 66.0 ± 7.4 to 86.0 ± 9.6 μM for tumor tissue. In turn oxygenation degree varied from 77.4 ± 8.2 to 82.2 ± 8.3 (%) for normal tissue and from 77 ± 18 to 84 ± 13 (%) for tumor tissue. Water content varied from 3.95 ± 1.94 to 5.87 ± 1.31 M for normal tissue and 20.1 ± 10.8 M for tumor tissue [251]. Similar results were obtained by Fawzy et al. [252]. In the study, the author measured *in vivo* 100 reflectance spectra of normal tissue, benign and malignant lesions (small cell lung cancer, combined squamous cell carcinoma and non-small cell lung cancer, non-small cell lung cancer, and adenocarcinoma) in 22 patients. As follows from their analysis, the mean value of the blood volume fraction was higher for malignant lesions (0.065 ± 0.03) compared to the benign lesions (0.032 ± 0.02). The mean value of the oxygen saturation parameter was reduced from 0.90 ± 0.11 for benign lesions to 0.78 ± 0.13 for malignant lesions [252]. The significant increasing in the volume fraction of blood in malignant tissue related to the overgrowth of the tumor microvasculature [253]. A significant decrease in blood oxygenation in malignant lesions was consistent with hypoxia-related changes during the development of cancer [254], which could be related to the increase in tissue metabolism and a high proliferation rate of the cancerous cells [252].

Table 1.4 The optical properties of lung tumor tissues measured *in vitro* and *in vivo* [36, 250, 251]

Tissue	λ (nm)	μ_a (cm^{-1})	μ_s (cm^{-1})	g	μ_s' (cm^{-1})	Remarks
In vitro human lung tumor ($n = 10$)	400	20.56	299.2	0.918	24.42	ISS, IAD; data from [250] $\mu_s = 5.209 \times 10^3\lambda^{-0.476}$ $\mu_{s'} = 4.638 \times 10^5\lambda^{-1.639}$ $g = 0.921 + 0.061(1-\exp(-(\lambda^{-419.2})/324.91))$, [λ] in nm
	450	6.27	286.2	0.9266	21.02	
	500	2.97	271.4	0.934	17.79	
	550	2.18	257.9	0.942	14.89	
	600	1.38	245.3	0.947	12.97	
	650	1.13	238.5	0.952	11.39	
	700	1.02	236.2	0.958	10.01	
In vitro Lewis lung carcinoma ($n = 2$). Mouse	350	7.598	–	–	14.58	ISS, IMC; data from [36] $\mu_s' = 4.238 \times 10^4\lambda^{-1.355}$, [λ] in nm
	400	8.883	–	–	12.63	
	450	8.294	–	–	10.81	
	500	4.289	–	–	9.55	
	550	7.274	–	–	8.32	
	600	3.716	–	–	7.40	
	650	1.787	–	–	6.57	
	700	1.495	–	–	5.88	
	750	1.344	–	–	5.34	
	800	1.155	–	–	4.88	
	850	1.071	–	–	4.49	
	900	1.077	–	–	4.21	
	950	1.158	–	–	3.91	
	1000	1.158	–	–	3.69	
Normal tissue/human large-cell primary lung carcinoma ($n = 1$). *In vivo* measurement	674	0.0883 ± 0.0061/ 0.174 ± 0.022	–	–	10.7 ± 0.4/ 10.4 ± 0.9	Frequency-domain photon migration spectroscopy, *in vivo*, data from [251]
	811	0.0892 ± 0.0050/ 0.177 ± 0.013	–	–	9.99 ± 0.27/ 9.23 ± 0.5	
	849	0.0915 ± 0.0030/ 0.190 ± 0.010	–	–	9.65 ± 0.15/ 9.20 ± 0.3	
	956	0.127 ± 0.0305/ 0.186 ± 0.058	–	–	6.3 ± 0.95/ 4.7 ± 2.65	

1.3.2 Breast Cancer

The second leading cause of cancer-related deaths in women worldwide is breast cancer with a worldwide incidence rate of more than 2,088,849 (11.6% of the total cases) and a mortality rate of 522,000 (6.6% of the total cancer deaths) [1]. A low-dose X-ray mammogram is the most common technique used for screening of microcalcifications in breast cancers. Mammography is not effective in dense female breasts and does not discriminate whether a lesion is benign or malignant. Therefore, it is always followed by either surgical excision biopsy or needle biopsy and only 36.5% of found microcalcifications are identified as malignant tumor [255]. The intra-surgical assessment of the tumor margins is often quite challenging and requires an objective and rapid guidance to eliminate the risk of additional resections, whose rate varies between 7 and 73% as reported by different institutions [256].

Frank et al. [257] demonstrated in 1995 the possibility of Raman biopsy by fiber-optic sampling through a hypodermic needle. It was remotely shown that differences between benign lesion (fibrocystic) and infiltrating ductal carcinoma were smaller than those between normal and malignant specimens. Rehman et al. [126] reported about spectral differences between the nuclear grades of ductal carcinoma *in situ* and invasive ductal carcinoma of the breast. It was confirmed the increase of protein content and relative decrease in the lipids/acylglyceride content in the cancerous tissues. The intensity at 1662 cm^{-1} (amide I group of proteins) varied with the degree of fatty acid unsaturation and it depended mainly on the lipid-to-protein ratio. The normal tissue showed weaker intensity at 1442 cm^{-1}, which represented CH_2 scissoring and CH_3 bending in lipids and proteins, and increased with the increase in nuclear grades. The same trend was observed for the OH–NH–CH peaks in the 2700–3500 cm^{-1} region, indicating varying concentrations of fatty acyl chains, phospholipids, cholesterol, creatine, proteins, and nucleic acids.

Haka et al. [160] showed that the types of microcalcifications could be easily distinguished based on the presence or absence of vibrational bands characteristic of calcium oxalate dihydrate at 912 cm^{-1} and 1477 cm^{-1} and calcium hydroxyapatite at 960 cm^{-1}, and that their relative abundances correlated with malignancies in breast cancer [99]. Further Saha et al. [116] presented *ex vivo* studies for real-time identification of microcalcifications in stereotactic core needle breast biopsy specimens collected from freshly excised tissue from 33 patients. The authors employed ordinary least-squares fitting to approximate the acquired spectra with a breast model, including fit coefficients for total calcium, collagen, and fat. Further the same group demonstrated the utility of Raman spectroscopy as a guidance tool for mastectomy procedures [159]. The modified classification model used SVM algorithm for the diagnosis of lesions irrespective of microcalcification status followed by logistic regression algorithm for detection of microcalcifications. The accuracy obtained for differentiation between normal, fibrocystic change, fibroadenoma, and breast cancer was 82.2% [159]. Haka et al. [117] reported on the feasibility to use Raman spectroscopy for *in vivo* diagnostics in an operation surgery environment.

With their classification model, they reached an overall accuracy of 93% (28 of 30) [258]. The highest sensitivity and specificity (94.9% and 93.8%, respectively) were achieved by Li et al. [162] with proposed adaptive weight k-local hyperplane (AWKH) algorithm, which extended K-local hyperplane distance nearest-neighbor algorithm of breast cancer classification.

Brozek-Pluska et al. [161] showed clear differences in carotenoids and fatty acid composition and products of their metabolism between cancerous tissue and surrounding noncancerous. The most pronounced differences were observed in the region of the bands at 1158 and 1518 cm^{-1} assigned to the C–C and C=C stretching modes of carotenoids, the symmetric and asymmetric C–H vibrations of lipids at 2850 and 2940 cm^{-1}, and the region of the OH stretching mode of water around 3300 cm^{-1}. The Raman intensities of lipid peaks were significantly smaller in the cancerous tissue than in the noncancerous tissue as in fingerprint region (854, 1444, 1660, 1750 cm^{-1}) as in HW spectral range (2888, 2926 cm^{-1}). It was reported to effectively diagnose early-stage breast cancer with a sensitivity of 72% for malignant tissue and 62% for benign tissue and a specificity for normal tissue of 83% [161]. It was also observed by Abramczyk et al. [157, 158] after testing the same Raman system on 150 patients. The group found that the fatty acid composition and products of their metabolism in cancerous breast tissue had an increased content of 20-carbon essential fatty acid, whereas surrounding noncancerous tissue was almost identical to monounsaturated oleic acid. This study suggested that carotenoids and lipids can be used as Raman biomarkers in breast cancer pathology.

Optical properties of breast tumors were investigated in [45, 259–264], the data partially summarized in [265] and presented in Table 1.5. Zhang et al. [45] compared the optical properties of normal breast tissue, benign and malignant neoplasm using integrating sphere technique and IAD method in the spectral range from 400 to 2200 nm. The authors observed an increase in water concentration and a decrease in lipid content in malignant tissue compared with normal tissue. Moreover, as can be seen in Table 1.5, spectral behavior of the scattering properties of tumor tissue is determined primarily by relative small (so-name Rayleigh) scatterers in comparison with normal tissue. Earlier similar results were obtained in [259, 263, 264] for *in vitro* and *ex vivo* experiments in visible and near-infrared spectral range. *In vivo* studies of normal and malignant breast tissues were performed by Fantini et al. [260], Grosenick et al. [261], and Cerussi et al. [262] using frequency- and time-domain techniques. The authors observed an increase in both the absorption and scattering coefficients for the tumor tissue in comparison with normal tissue. Moreover, the authors found a significant increase in total hemoglobin concentration: from 17.3 ± 6.2 μmol/L for normal tissue to 53 ± 32 μmol/L for tumor tissue [261] or from 17.5 ± 7.5 μM (normal tissue) to 24.7 ± 9.8 μM (tumor tissue) [262]. At the same time, the blood oxygen saturation was not changed: 74 ± 7% for normal tissue and 72 ± 14% for tumor tissue [261] or 67.7 ± 9.3% for normal tissue and 67.5 ± 8.4% for tumor tissue [262]. Lipid content in the tumor tissue decreased: from 66.1 ± 10.3% (normal tissue) to 58.5 ± 14.8%, and on the contrary the water content increased: from 18.7 ± 10.3% (normal tissue) to 25.9 ± 13.5% for tumor [262]. Similar results were presented in [266–268].

Table 1.5 The optical properties of breast tumor tissues measured in *in vitro* and *in vivo* [45, 259–264]

Tissue	λ (nm)	μ_a (cm^{-1})	μ_s (cm^{-1})	g	μ_s' (cm^{-1})	Remarks
Normal breast tissue (n = 16)/ benign tumor tissue (n = 5)/ malignant tumor (n = 11)	400	9.57 ± 5.61/5.54 ± 1.91/9.29 ± 5.57	–	–	20.4 ± 8.8/40.9 ± 4.0/40.9 ± 4.0	ISS, IAD, *in vitro*; data from [45]: $\mu_{s'} = 2.072 \times 10^9\lambda^{-3.173} + 19.876\lambda^{-0.103}$ (normal tissue), $\mu_{s'} = 7.115 \times 10^8\lambda^{-2.8} + 30.69\lambda^{-0.187}$ (benign and malignant tumor tissues), [λ] in nm
	500	3.23 ± 1.78/1.79 ± 0.28/3.16 ± 1.69	–	–	16.2 ± 6.0/30.3 ± 4.4/30.3 ± 4.4	
	600	1.61 ± 0.96/0.86 ± 0.01/1.69 ± 0.91	–	–	13.9 ± 4.6/22.6 ± 3.3/22.6 ± 3.3	
	700	1.16 ± 0.45/0.65 ± 0.01/0.79 ± 0.24	–	–	12.2 ± 4.0/18.0 ± 2.7/18.0 ± 2.7	
	800	1.13 ± 0.49/0.63 ± 0.03/0.75 ± 0.19	–	–	11.3 ± 3.9/15.2 ± 2.8/15.2 ± 2.8	
	900	1.13 ± 0.48/0.65 ± 0.28/0.65 ± 0.01	–	–	10.8 ± 3.7/13.0 ± 2.9/13.0 ± 2.9	
	1000	0.99 ± 0.25/0.71 ± 0.12/0.84 ± 0.03	–	–	10.4 ± 3.6/11.4 ± 2.9/11.4 ± 2.9	
	1100	1.20 ± 0.46/0.68 ± 0.05/1.01 ± 0.15	–	–	10.0 ± 3.6/10.2 ± 2.6/10.2 ± 2.6	
	1200	2.56 ± 0.22/1.90 ± 0.17/1.85 ± 0.11	–	–	9.6 ± 3.6/9.0 ± 2.7/9.0 ± 2.7	
	1300	1.55 ± 0.06/1.56 ± 0.16/1.79 ± 0.24	–	–	9.6 ± 3.9/8.3 ± 2.3/8.3 ± 2.3	
	1400	4.92 ± 0.64/9.78 ± 0.51/12.30 ± 3.01	–	–	9.5 ± 3.7/8.6 ± 2.0/8.6 ± 2.0	
	1500	4.65 ± 1.02/10.05 ± 1.19/13.23 ± 4.08	–	–	9.4 ± 3.5/9.0 ± 1.7/9.0 ± 1.7	
	1600	2.58 ± 0.06/4.81 ± 0.61/6.21 ± 1.46	–	–	9.1 ± 3.9/7.7 ± 1.7/7.7 ± 1.7	
	1700	6.63 ± 0.22/5.43 ± 1.21/5.39 ± 1.23	–	–	9.1 ± 3.8/7.1 ± 1.9/7.1 ± 1.9	
	1800	5.46 ± 0.51/6.77 ± 0.74/7.45 ± 1.82	–	–	9.3 ± 4.0/7.0 ± 1.8/7.0 ± 1.8	
	1900	14.29 ± 3.02/31.60 ± 4.99/36.19 ± 8.98	–	–	9.5 ± 4.6/8.5 ± 2.1/8.5 ± 2.1	
	2000	12.48 ± 4.80/33.31 ± 8.72/42.20 ± 14.94	–	–	9.3 ± 4.4/8.8 ± 2.3/8.8 ± 2.3	
	2100	8.02 ± 2.76/18.94 ± 3.57/22.88 ± 7.46	–	–	9.2 ± 4.2/8.0 ± 2.0/8.0 ± 2.0	
	2200	7.10 ± 2.30/13.71 ± 2.17/16.19 ± 4.93	–	–	9.6 ± 4.5/7.5 ± 1.4/7.5 ± 1.4	

(continued)

Table 1.5 (continued)

Tissue	λ (nm)	μ_a (cm^{-1})	μ_s (cm^{-1})	g	μ_s' (cm^{-1})	Remarks
Human breast (normal glandular tissue) ($n = 3$)/human breast (normal adipose tissue) ($n = 7$)/human breast (fibrocystic tissue) ($n = 8$)/ human breast (fibroadenoma) ($n = 6$)/human breast (ductal carcinoma) ($n = 9$)	500	3.42/2.73/2.28/ 4.02/2.60	461.8/313.8/ 879.2/ 447.7/426.4	0.947/0.971/ 0.980/ 0.970/0.954	24.3/9.1/18.0/ 13.4/19.4	ISS, IMC, *in vitro*, data from [259] for normal glandular tissue $\mu_s' = 2.802 \times 10^9\lambda^{-3.029} + 18.615\lambda^{-0.146}$ and $g = 0.737 + 0.229(1-\exp(-(\lambda^{-282.7})/93.5))$; for normal adipose tissue $\mu_s' = 5.717 \times 10^8\lambda^{-3.227} + 17.657\lambda^{-0.13}$ and $g = 0.741 + 0.236(1-\exp(-(\lambda+23.3)/148))$; for fibrocystic tissue: $\mu_s' = 1.556 \times 10^9\lambda^{-3.012} + 22.187\lambda^{-0.157}$ and $g = 0.749 + 0.234(1-\exp(-(\lambda+15.1)/177.8))$; for fibroadenoma $\mu_s' = 1.625 \times 10^9\lambda^{-3.029} + 17.803\lambda^{-0.229}$ and $g = 0.749 + 0.235(1-\exp(-(\lambda^{-255.9})/89.4))$; for ductal carcinoma $\mu_s' = 1.951 \times 10^9\lambda^{-3.016} + 19.858\lambda^{-0.186}$ and $g = 0.727 + 0.236(1-\exp(-(\lambda)/156.5))$; [λ] in nm
	600	0.92/0.99/0.47/ 1.76/1.61	431.5/294.1/ 623.5/ 492.6/337.5	0.959/0.972/ 0.977/ 0.979/0.958	17.6/8.1/14.6/ 10.3/14.3	
	700	0.48/0.81/0.24/ 0.53/0.44	409.1/306.3/ 568.0/ 438.4/277.4	0.965/0.975/ 0.978/ 0.982/0.961	14.1/7.7/12.3/ 7.9/10.9	
	800	0.55/0.82/0.28/ 0.34/0.34	332.7/313.8/ 548.8/ 384.0/233.0	0.965/0.976/ 0.981/ 0.983/0.962	11.7/7.4/10.5/ 6.5/8.9	
	900	0.67/0.84/0.39/ 0.79/0.45	275.1/306.1/ 536.5/ 327.3/181.2	0.965/0.976/ 0.981/ 0.983/0.957	9.7/7.4/9.8/ 5.5/7.8	
	1000	0.90/0.90/0.63/ 1.57/0.64	213.7/306.2/ 485.6/ 269.1/143.7	0.957/0.976/ 0.982/ 0.982/0.950	9.1/7.4/8.8/ 5.0/7.2	
	1100	0.82/1.14/0.84/ 1.48/0.52	200.2/332.2/ 465.9/ 209.2/123.0	0.961/0.977/ 0.983/ 0.979/0.946	7.8/7.8/7.8/ 4.4/6.7	
Human papillary cancer *in vivo* ($n = 1$)	690	0.084 ± 0.014	–	–	15.0 ± 0.3	Frequency-domain mammography. *In vivo* [260]
	825	0.085 ± 0.017	–	–	12.7 ± 0.3	
Human healthy breast tissue/ carcinomas ($n = 87$)	670	0.036 ± 0.008/0.110 ± 0.066	–	–	10.5 ± 1.3/13.5 ± 4.7	Time-domain optical mammography. *In vivo* [261]
	785	0.039 ± 0.011/0.100 ± 0.060	–	–	9.5 ± 1.4/11.6 ± 3.9	
	843	0.036 ± 0.005/0.118 ± 0.096	–	–	8.4 ± 0.4/12.2 ± 1.7	
	884	0.059 ± 0.016/0.124 ± 0.089	–	–	8.0 ± 1.0/9.1 ± 1.9	

(continued)

Table 1.5 (continued)

Tissue	λ (nm)	μ_a (cm^{-1})	μ_s (cm^{-1})	g	μ_s' (cm^{-1})	Remarks
Human adenocarcinoma (n = 1)	650	0.069	–	–	8.44	Diffuse optical spectroscopy *in vivo*. Data from [262]: $\mu_{s'} = 571.326\lambda^{-0.65}$; [λ] in nm
	700	0.044	–	–	8.06	
	750	0.053	–	–	7.70	
	800	0.044	–	–	7.39	
	850	0.055	–	–	7.10	
	900	0.093	–	–	6.84	
	950	0.125	–	–	6.60	
	1000	0.131	–	–	6.40	
Normal adipose tissue (n = 39)/ invasive ductal carcinoma (n = 28)	400	16.191/9.049	–	–	13.7/23.9	Reflectance spectroscopy *ex vivo*. IMC. Data from [263]: $\mu_{s'} = 1.168 \times 10^9\lambda^{-3.26} + 3.26\lambda^{-0.129}$ (for normal adipose tissue) and $\mu_{s'} = 1.004 \times 10^9\lambda^{-3.186} + 23.151\lambda^{-0.037}$ (for invasive ductal carcinoma); [λ] in nm
	450	21.796/2.293	–	–	12.9/21.8	
	500	5.556/0.590	–	–	11.6/21.3	
	550	1.209/1.209	–	–	11.1/20.5	
	600	0.155/0.144	–	–	11.0/18.7	
Normal adipose tissue (n = 38)/ fibrous/benign tissue (n = 8)/ malignant tissue (invasive ductal carcinoma) (n = 37)	350	2.80/4.45/5.78	–	–	14.2/28.3/33.5	Reflectance spectroscopy *ex vivo*. IMC. Data from [264]: $\mu_s' = 4.094 \times 10^8\lambda^{-3.168} + 33.493\lambda^{-0.192}$ (for normal adipose tissue); $\mu_s' = 6.721 \times 10^8\lambda^{-3.085} + 30.765\lambda^{-0.075}$ (for fibrous/ benign tissue); $\mu_s' = 6.98 \times 10^8\lambda^{-3.055} + 33.39\lambda^{-0.063}$ (for malignant tissue); [λ] in nm
	400	7.49/9.33/8.50	–	–	12.9/25.9/30.5	
	450	8.76/6.90/5.30	–	–	12.6/25.2/29.9	
	500	3.25/2.95/2.12	–	–	11.7/23.1/27.5	
	550	0.97/1.64/1.62			10.7//20.9/24.6	
	600	0.74/1.37/1.38			10.1/19.5/22.4	

1.3.3 Skin Cancer

Malignant melanoma (MM) is the most aggressive form of skin cancer with a worldwide incidence of 287,723 (1.6% of the total cases) and mortality rate as high as 60,712 (0.6% of the total cancer deaths) [1]. Although nonmelanoma skin cancers (NMSC), such as the basal cell carcinoma (BCC) and squamous cell carcinoma (SCC), are associated with low mortality rate deaths, they are particularly common in fair-skinned populations of European descent (5.8% of the total cases), with high incidence rates found in Australia/New Zealand, North America, and Northern Europe [1, 269]. Standard type of treatment of skin cancer is complete removal of the lesion with a high cure rate without reducing life expectancy (5-year relative survival rate for melanoma is approximately 92% [269]). Therefore, investigations of Raman spectroscopy for skin cancer primarily focused on early detection and discrimination of skin tumor types.

In early *ex vivo* studies Gniadecka et al. [125, 270] revealed clear-cut changes in skin tumor tissues allowing to differentiate MM from normal skin, seborrheic keratosis, and BCC. The major spectra alteration was found in the region 1200–1750 cm^{-1}. An increase in the intensity of Raman bands was observed for pigmented tumors in the region 2500–3500 cm^{-1}. To demonstrate spectral changes for proteins, the ratio between the amide I band and $\delta(CH_2)(CH_3)$ in proteins and lipids (I_{1650}/I_{1450}) and the ratio between the amide III and lipids at around 1320 cm^{-1} were calculated (I_{1270}/I_{1320}). Neural network analysis of Raman spectra in a range 200–3500 cm^{-1} achieved a diagnostic sensitivity of 85% and specificity of 99% for the diagnosis of MM.

In 2008 Zhao et al. [230] reported 289 *in vivo* measurements of Raman spectra from nine different types of lesions. The authors stated a sensitivity of 91% and a specificity of 75% in differentiating malignant lesions from benign lesions. Further, the same group utilized a Raman probe to study MM, BCC, SCC, actinic keratosis, atypical nevi, melanocytic nevi, blue nevi, and seborrheic keratosis from 453 patients in 2012 [227] and from 645 patients in 2015 [225]. The collected single-point spectra were acquired in 1 s and subjected to principal component with generalized discriminant analysis (PC-GDA) and PLS for statistical data evaluation. The sensitivity to differentiate skin cancer versus benign lesions ranged between 95 and 99%, with a related specificity of 15–54%. Similar results were achieved by Silveira et al. [224] for a set of 145 spectra from biopsy fragments of normal ($n = 30$), BCC ($n = 96$), and MM ($n = 19$) skin tissues. The authors applied the best-fitting model to the spectra of biochemicals and verified that actin, collagen, elastin, and triolein were the most important biochemicals representing the spectral features of skin tissues.

Typical *in vivo* Raman spectra for normal skin tissue, melanoma, and BCC are depicted in Fig. 1.6. Overall all skin lesions appear to share similar major Raman peaks and bands in fingerprint region. There are no distinctive Raman peaks or bands that can be uniquely assigned to specific skin cancers by visual inspection alone. The strongest Raman peak is located around 1445 cm^{-1} with other major Raman bands centered at 855, 936, 1002, 1271, 1302, 1655, and 1745 cm^{-1}. The

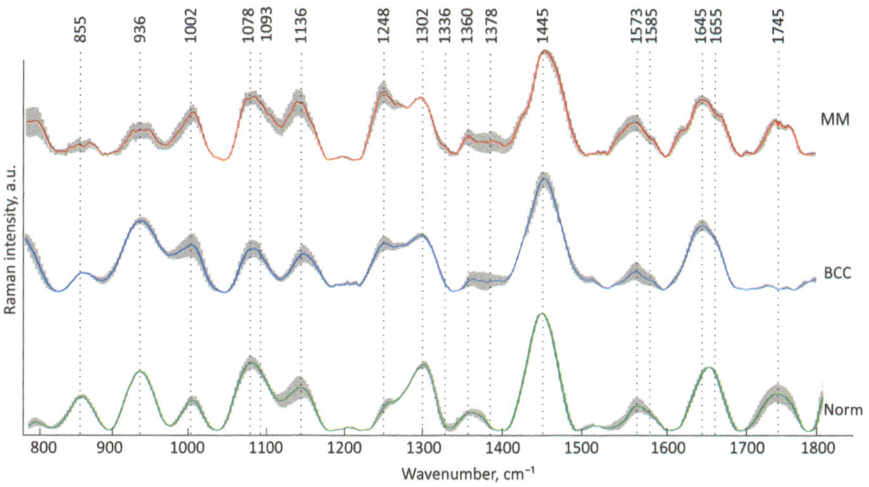

Fig. 1.6 Average Raman spectra and standard deviation of melanoma (MM), basal cell carcinoma (BCC), and normal skin (NORM). Each spectrum was acquired by Raman setup [140] and preprocessed with baseline removal, smoothing by the Savitzky–Golay method, data normalization, and centering

development of the malignant skin disease increases the content of metabolic products in the pathological areas of the skin, changes the concentration of proteins and lipids. Proteins predominantly contribute to the appearance of bands in the spectral range 1240–1270, 1340, 1440–1460, and 1665 cm^{-1}, the spectral features arising from the contribution of lipids, predominantly of triolein, are observed in the 1271–1301, 1440, 1650–1660 cm^{-1} bands [224]. One of the significant differences between malignant and benign formations is the process of metabolism and destruction of collagen. Cells of malignant tumors form fast-growing, low-differentiated structures, and the development of such structures is accompanied by the increased activity of collagenase [95]. Collagenase destroys the molecular bonds of collagen fibers, and changes in Raman spectra of skin tissue can be observed in 1248, 1454, and 1665 cm^{-1} bands associated with peaks of collagen [121].

Schleusener et al. [229] performed *in vivo* measurements on 104 subjects with lesions using a multi-fiber Raman probe, which was optimized for collecting scattered light from within the epidermal layer's depth down to the basal membrane, where early stages of skin cancer developed. NMSC were discriminated from normal skin with a balanced accuracy of 73% (BCC) and 85% (SCC) using partial least-squares discriminant analysis (PLS-DA). Discriminating MM and pigmented nevi (PN) resulted in a balanced accuracy of 91%.

Lim et al. [223] employed fiber-optic Raman probe in combination with fluorescence and diffuse backscattered reflectance techniques in order to improve diagnostic outcomes. Raman, fluorescence, and reflectance spectra were acquired from 137 lesions in 76 patients. Raman spectroscopy alone demonstrated to achieve 100% sensitivity and specificity for discriminating melanoma from benign pigmented lesions, but only 68% sensitivity and 55% specificity for distinguishing NMSC from normal tissues. However, for multimodal approach NMSC were classified with a sensitivity of 95% and specificity of 71%. To support the data analysis, the group also analyzed the spectral contributions of individual skin components such as collagen, elastin, triolein, nuclei, keratin, ceramide, melanin, and water, by fitting spectra obtained *in vitro* using Raman microscopy [222]. It was demonstrated that the biophysical model had consistent diagnostic capability similar to statistical PCA-LDA model with leave-one-lesion-out cross-validation. More importantly, the biophysical model captured the relevant biophysical changes accounting for the diagnosis. In particular, the authors found that collagen and triolein were the most important biomarkers in discriminating MM from benign pigmented lesions, and BCC had a significantly different concentration of nucleus, keratin, collagen, triolein, and ceramide compared to surrounding healthy skin. Recently, the group demonstrated the capability to use developed biophysical model for skin cancer margin assessment in BCC surgery resection [226].

In 2017 Bratchenko et al. [228] tested combined Raman and autofluorescence *ex vivo* diagnostics of MM ($n = 39$) and BCC ($n = 40$) in near-infrared and visible regions. The authors stated the accuracy of 97.3% for discriminating each of skin cancers in multimodal approach, whereas the determined accuracy for each modality separately was 79%. Further, the same group has performed *in vivo* measurements on 17 MM, 18 BBC, and 19 various types of benign neoplasms with portable Raman system and confirmed the higher accuracy of multimodal approach [140]. The diagnostic efficiency of portable system was defined by PLS-DA analysis of entire spectra, taking into account each feature of the spectra in a range from 300 to 1800 cm^{-1} including maxima intensities and exact bands position estimation. For combined Raman and autofluorescence diagnostics, the authors reported the accuracy of 89.5% and 91.1% for classifying MM vs other neoplasms and BCC vs other neoplasms, respectively.

The other multimodal approach includes combination of optical coherence tomography (OCT) and Raman spectroscopy (RS) for noninvasive characterization of skin lesions based on either morphological or biochemical features of disease [271]. Although the OCT not clearly defines features associated with malignancy, it provides a morphological context to guide placement of the RS acquisition axes for specific biochemical analysis of the tissue. In 2015, Zakharov et al. [111] reported the increase in average accuracy of *in vivo* diagnosis of skin tumors (9 MM, 9 BCC, and various benign tumors) by multimodal RS-BS-OCT system. It was shown that these methods were complementary and increased the diagnostic specificity for a variety of tumor types by 5–11%. In 2018, Varkentin et al. [272] presented trimodal RS-OA-OCT with optoacoustic (OA) modality, which provided precise tumor depth determination due to potentially deeper penetration compared to OCT. The Raman signal was collected via the OCT scanning lens to maximize the signal-to-noise ratio of the measured signal while keeping radiation levels below maximum permissible exposure limits. The preliminary results of first RS-OA-OCT clinical trials showed good agreement with histology results and distinctive differences in Raman data between normal skin and different areas of melanocytic lesions.

The optical properties of skin tumor tissue are presented in Table 1.6, which demonstrates qualitatively similar behavior for all investigated nonmelanoma types of skin tumors [39, 273]. Scattering coefficient gradually decreases with the increasing wavelength. Quantitatively, infiltrative BCC is characterized by a higher scattering coefficient in comparison with the scattering of nodular BCC and SCC. The higher scattering coefficient of infiltrative BCC may be explained by its structural characteristics. Typically, these tumors have thin strands or cords of tumor cells extending into the surrounding highly scattering dermis. The value of μ_s' for SCC is consistently lower than for both types of BCC in the entire wavelength range [39].

Similar result was obtained by Garcia-Uribe et al. [273] (see Table 1.6). Higher light scattering in cancerous tissue can be explained by the larger average effective size of the scattering centers. SCC *in situ* has not yet penetrated through the basement membrane of the dermoepidermal junction. SCCs typically appear as scaling plaques with sharply defined red color. Histologically, all epidermal layers may contain atypical keratinocytes. The larger amount of atypical keratinocytes in SCCs can increase the light scattering in this type of skin lesion and significantly affect its contribution to diffusely reflected light on the surface. SCCs can penetrate the basement membrante to become invasive [273].

Actinic keratosis can appear rough and scaly and can develop into SCCs. Histologically, actinic keratosis is recognized by the presence of atypical keratinocytes in the deeper parts of the epidermis. Defective maturation of the superficial epidermal layers results in parakeratosis, alternating with hyperkeratosis [274, 275]. The amounts of atypical keratinocytes and collagen are factors related to the amount of light scattering in the lesion [273].

BCC is derived from the basal layer of keratinocytes, the deepest cell layer of the epidermis. BCCs can present nodular aggregates of basal cells in the dermis and exhibit peripheral palisading and retraction artifacts. Melanin can also be present in the tumor and in the surrounding stroma, as observed in pigmented BCCs. The aggregation of basal cells can increase the light scattering in these types of malignant lesions. The progression of seborrheic keratosis into BCC and SCC is rare [276, 277]. Seborrheic keratosis, composed of basaloid cells admixed with some squamous cells, can be pigmented when some cells contain melanin transferred from neighboring melanocytes [273].

Cugmas et al. [81] investigated *ex vivo* optical properties of canine skin and *ex vivo* and *in vivo* subcutaneous tumors and found that average water volume fraction in the skin samples was 81.4%. Darkly pigmented skin contained almost 10 times more melanin (2.22 mmol/L) than lightly pigmented skin (0.26 mmol/L). The authors estimated melanin concentrations of 115.0 and 443.5 mg/L for the lightly and darkly pigmented human skin, respectively. The average hemoglobin mass concentration was 1.07 g/L and saturation was 46%. For tumors the water volume fraction was around 82%, saturation was slightly above 50%. However, benign tumors contained 0.62 g/L of hemoglobin and malignant tumors contained 7.93 g/L of hemoglobin [81].

Table 1.6 The optical properties of skin tumor tissues measured in *in vitro* and *in vivo* [39, 273]

Tissue	λ (nm)	μ_a (cm^{-1})	μ_s' (cm^{-1})	Remarks
Infiltrative basal cell carcinoma (n = 6)/nodular basal cell carcinoma (n = 5)/squamous cell carcinoma (n = 8)	370	5.19/7.19/7.70	67.0/47.7/43.9	ISS, IMC, *in vitro*; data from [39]; spectral range 400–1350 nm $\mu_s' = 8.608 \times 10^8\lambda^{-2.767} + 109.912\lambda^{-0.335}$ (infiltrative basal cell carcinoma), $\mu_s' = 7.641 \times 10^8\lambda^{-2.791} + 67.29\lambda^{-0.284}$ (nodular basal cell carcinoma), $\mu_s' = 8.389 \times 10^8\lambda^{-2.811} + 42.079\lambda^{-0.263}$ (squamous cell carcinoma), [λ] in nm
	400	5.20/7.04/8.68	65.7/46.2/43.5	
	450	3.99/4.72/7.54	57.2/42.7/38.9	
	500	2.06/2.52/2.46	46.5/35.9/30.4	
	550	1.56/1.67/2.43	37.7/30.2/25.5	
	600	1.20/1.11/1.47	31.9/26.2/21.4	
	650	0.94/0.75/0.83	27.9/22.6/18.1	
	700	0.87/0.55/0.65	25.3/19.8/16.1	
	750	0.77/0.37/0.52	22.8/17.9/14.2	
	800	0.64/0.19/0.37	20.5/15.8/12.9	
	850	0.52/0.09/0.28	19.0/14.7/12.0	
	900	0.40/0.05/0.22	17.4/13.6/11.2	
	950	0.44/0.05/0.24	15.9/12.9/10.5	
	1000	0.49/0.03/0.25	14.4/12.3/9.8	
	1050	0.48/0.01/0.27	12.9/11.8/8.9	
	1100	0.53/0.01/0.26	12.2/11.3/8.6	
	1150	0.89/0.05/0.54	11.6/10.8/8.4	
	1200	1.14/0.11/0.79	11.1/10.4/8.2	
	1250	1.15/0.08/0.73	10.7/10.2/7.9	
	1300	1.53/0.23/1.00	10.4/9.9/7.6	
	1350	3.70/1.81/2.47	10.6/10.0/8.1	
	1400	13.79/11.19/10.68	13.1/13.1/11.3	
	1450	24.06/17.01/21.21	17.1/16.3/14.6	
	1500	17.98/10.45/13.81	14.7/14.3/12.7	
	1550	10.38/5.71/8.63	11.5/11.5/9.8	
	1600	5.95/3.43/5.33	11.6/10.3/8.9	

(continued)

Table 1.6 (continued)

Tissue	λ (nm)	μ_a (cm^{-1})	μ_s' (cm^{-1})	Remarks
Skin type I/skin type II/skin type III/ common nevi/ dysplastic nevi/ melanoma/actinic keratosis/ seborrheic keratosis/ squamous cell carcinomas/ basal cell carcinomas	455	0.60/0.74/0.83/ 1.20/1.40/1.78/0.79/ 0.96/1.17/1.28	12.8/13.5/14.1/ 15.4/19.8/22.2/22.9 /20.7/27.0/22.4	Oblique incidence diffuse reflectance, *in vivo*; data from [273]; for skin type I: $\mu_{s'} = 1.095 \times 10^9\lambda^{-3.256} + 20.658\lambda^{-0.111}$; for skin type II: $\mu_{s'} = 1.212 \times 10^9\lambda^{-3.236} + 21.316\lambda^{-0.114}$; for skin type III: $\mu_{s'} = 9.575 \times 10^8\lambda^{-3.2} + 23.143\lambda^{-0.117}$; for common nevi: $\mu_{s'} = 5.818 \times 10^8\lambda^{-3.177} + 22.682\lambda^{-0.084}$; for dysplastic nevi: $\mu_{s'} = 7.607 \times 10^8\lambda^{-3.111} + 24.242\lambda^{-0.07}$; for melanoma: $\mu_{s'} = 8.341 \times 10^8\lambda^{-3.149} + 22.058\lambda^{-0.031}$; for actinic keratosis: $\mu_{s'} = 1.067 \times 10^9\lambda^{-3.123} + 23.134\lambda^{-0.042}$; for seborrheic keratosis: $\mu_s' = 1.248 \times 10^9\lambda^{-3.114} + 23.535\lambda^{-0.08}$; for squamous cell carcinomas: $\mu_{s'} = 1.619 \times 10^9\lambda^{-3.104} + 21.012\lambda^{-0.024}$; for basal cell carcinomas: $\mu_{s'} = 9.903 \times 10^8\lambda^{-3.131} + 22.773\lambda^{-0.04}$; [λ] in nm
	500	0.46/0.57/0.63/ 1.08/1.24/1.59/0.70/ 0.85/0.97/1.05	12.2/12.7/13.4/ 15.0/18.7/20.9/21.8 /19.2/24.8/21.2	
	550	0.55/0.63/0.68/ 1.12/1.29/1.67/0.78/ 0.96/1.01/1.15	11.6/12.1/12.8/ 14.6/17.9/20.1/20.9 /18.1/23.2/20.4	
	600	0.29/0.36/0.37/ 0.68/0.76/0.97/0.42/ 0.52/0.55/0.63	11.1/11.5/12.2/ 14.2/17.3/19.5/20.0 /17.0/21.9/19.6	
	650	0.21/0.28/0.31/ 0.50/0.57/0.72/0.33/ 0.39/0.42/0.48	10.8/11.1/11.8/ 13.8/16.8/19.2/19.5 /16.2/21.0/19.1	
	700	0.19/0.24/0.27/ 0.39/0.44/0.55/0.28/ 0.30/0.33/0.40	10.6/10.8/11.5/ 13.5/16.4/19.1/18.9 /15.7/20.3/18.7	
	750	0.16/0.21/0.24/ 0.30/0.33/0.43/0.25/ 0.27/0.27/0.36	10.4/10.5/11.3/ 13.4/16.1/19.0/18.5 /15.1/19.7/18.4	
	765	0.18/0.21/0.24/ 0.30/0.34/0.42/0.24/ 0.26/0.26/0.37	10.4/10.5/11.2/ 13.3/16.0/19.1/18.4 /15.0/19.6/18.4	

1.3.4 Colorectal Cancer

Over 1.8 million new colorectal cancer cases and 881,000 deaths are estimated to occur in a year, accounting for about 1 in 10 cancer cases and deaths [1]. Overall, colorectal cancer ranks third in terms of incidence but second in terms of mortality. Surgery is the only curative modality for localized colon cancer. Colonoscopy is the most sensitive instrument for screening and detection of early-stage malignancies and premalignant polyps (adenomas). However, colonoscopy miss rates are about 20% for adenomas [277].

Li et al. [163] identified the differences between colorectal normal and cancer tissues in five spectral bands in the regions around 815–830, 935–945, 1131–1141, 1447–1457, and 1665–1675 cm^{-1}. The strongest signals were observed at 1004 cm^{-1} (C–C stretching ring breathing of phenylalanine), 1323 cm^{-1} (CH_3CH_2 twisting of proteins and nucleic acids), 1450 cm^{-1} ($\delta(CH_2)$ of phospholipids and collagen), and 1665 cm^{-1} (C=O stretching mode of amide I and lipids). It was shown that normalized intensities of Raman bands in the ranges of 800–860 and 1580–1660 cm^{-1} were greater in normal tissue than in cancer tissue, while Raman signals at 1210–1400 cm^{-1} increased in cancer tissue, which correlated with dysplasia progression. Raman peaks 1323 cm^{-1} became widened and intense in cancer tissue than in normal colorectal tissue, revealing the increase of nucleic acid contents in tumor cells. The Raman bands 1665–1675 cm^{-1} which are attributed to the amide I bands of protein in the α-helix conformation were increased in malignant tissue, suggesting that malignancy may be associated with an increase in the relative amounts of protein in the β-pleated sheet or random coil conformation. The authors stated the increase in the intensity of amide I band 1665–1675 cm^{-1} in malignant tissue is associated with the increase in relative amount of protein in the β-pleated sheet and significant decrease in the intensity of band 1131–1141 cm^{-1} (C–N stretching mode of proteins, lipids) indicating relative reduction of lipid content in cancer tissue in accordance with early micro-Raman investigations [278]. The diagnostic statistical model was built with the help of the ant colony optimization, and support vector machine provided a diagnostic accuracy of 93.2% for identifying colorectal cancer from normal tissue.

In a similar study Widjaja et al. [279] investigated *ex vivo* 105 colonic tissue specimens from 59 patients (41 normal, 18 hyperplastic polyps, and 46 adenocarcinomas) using PCS-SVM diagnostic algorithm that utilized the entire Raman spectrum from 800 to 1800 cm^{-1}. The Gaussian radial basis function kernel SVM algorithm was proven to be the best classifier for providing the highest diagnostic specificity 98.1–99.7% and 100% sensitivity for multiclass classification. Wood et al. [280] measured Raman spectra from a total of 356 colon biopsies (81 of normal colon mucosa, 79 of hyperplastic polyps, 92 of adenomatous polyps, 64 of adenocarcinoma, and 40 of ulcerative colitis) from 177 patients. Spectral classification accuracies comparing pathology pairs ranged from 72.1 to 95.9% for 10-s acquisitions and from 61.5 to 95.1% for 1-s acquisitions, reflecting the improved signal-to-noise ratio with longer spectral acquisition times.

Short et al. [155] and Li et al. [164] separately analyzed Raman spectra in regions from 1000 to 1800 cm^{-1} and from 2800 to 3800 cm^{-1} for *ex vivo* samples collected during endoscopic biopsy. It was shown that the peak intensity of C–H stretching vibration bands relating to the lipids (near 2958 cm^{-1}, 2924 cm^{-1}, and 2858 cm^{-1}) decreased and even disappeared in the spectra of malignant tissues due to essential consumption of fat in carcinoma development. The entropy weight local-hyperplane k-nearest-neighbor classifier provided a sensitivity of 81.38% and a specificity of 92.69% for differentiating cancer from colitis samples. Petersen et al. [166] performed Raman fiber-optical measurements of 242 colon biopsy samples. The authors stated that better accuracy was achieved for leave-one-patient-out cross-validation in comparison with leave-one-spectrum-out cross-validation schemes due to minimization of systematic errors. Cancer was differentiated from normal tissue with a sensitivity of 79%, specificity of 83%, and an accuracy of 81%. PCA-LDA and PLS-DA discrimination models were compared on the same dataset of Raman spectra acquired in normal ($n = 78$) and cancerous ($n = 81$) colorectal tissues resulting in the preference of PLS-DA algorithm with LOOCV [170]. PLS-DA modeling yielded a diagnostic accuracy of 84.3% for colorectal cancer detection, while the accuracy of PCA-LDA classification was 79.2%.

Bergholt et al. [119] demonstrated that simultaneous Raman endoscopy in fingerprint and high-wavenumber regions provided a diagnostic sensitivity of 90.9% and specificity of 83.3% for differentiating colorectal adenoma from hyperplastic polyps, which was superior to considering either region alone. It was found that adenomas were associated with significantly reduced Raman peak intensities at 1078 cm^{-1} (C=C stretching), 1425 cm^{-1} ($\delta(CH_2)$ scissoring), 2850 and 2885 cm^{-1} (symmetric and asymmetric CH_2 stretching), and 3009 cm^{-1} compared to hyperplastic polyps pointing to a relative reduction in lipid content. An up-regulated protein content was largely indicated by the biomarkers at 1004 cm^{-1} (symmetric C–C stretching, ring breathing of phenylalanine) and band broadening of the 1655 cm^{-1} (amide I C=O stretching mode of proteins). The peak-ratio of the asymmetric to symmetric OH stretching (defined as mean intensity ratio I_{3250}/I_{3400}) showed significant differences, which represented the evidence of re-arrangements in hydrogen-bonded networks in epithelial cells caused by local interactions with macromolecules such as proteins [281]. The fingerprint range contains highly specific information about proteins, lipids, and DNA conformations. On the other hand, the HW technique contains information related to the CH_2/CH_3 stretching of lipids/proteins, as well as intense water bands reflecting the local conformation of water that are not contained in the FP range. The complementary properties of the FP and HW Raman spectral modalities for enhancing tissue diagnosis can partially be explained by back-tracking the misclassified spectra of each Raman modality.

Physiological differences between normal and tumorous tissues were investigated in [282, 283] and colon tumor optical properties were investigated in [284, 285]. In *in vivo* studies, Zonios et al. [282] showed that normal colorectal mucosa had a total hemoglobin concentration of 13.6 ± 8.8 mg/dL, whereas the corresponding value for the adenomatous polyp was approximately 72.0 ± 29.2 mg/dL. The hemoglobin oxygen saturation was found to be 0.59 ± 0.08 and 0.63 ± 0.1,

Table 1.7 The optical properties of colon tumor tissues measured in *in vitro* [284, 285]

Tissue	λ (nm)	μ_a (cm^{-1})	μ_s (cm^{-1})	g	μ_s' (cm^{-1})	Remarks
Healthy colon tissue (n = 30)/colon carcinoma (n = 30)	850	0.8 ± 0.1/ 0.4 ± 0.1	93.9 ± 22.5/ 71.4 ± 12.2	0.93 ± 0.01/ 0.92 ± 0.01	–	ISS, IMC, *in vitro*; data from [284]
	980	0.7 ± 0.1/ 0.5 ± 0.1	86.4 ± 21.1/ 63.3 ± 12.0	0.92 ± 0.01/ 0.92 ± 0.01	–	
	1060	0.3 ± 0.1/ 0.2 ± 0.1	82.1 ± 20.7/ 59.7 ± 12.5	0.92 ± 0.01/ 0.92 ± 0.02	–	
Colon adenocarcinoma CC 531 (n = 14)	632.8	1.4 ± 0.2	280 ± 20	0.946 ± 0.004	15.0 ± 2.0	ISS, IAD, *in vitro*; data from [285]
	1064	2.5 ± 0.5	180 ± 10	0.952 ± 0.005	9.0 ± 1.0	

respectively. The effective scattering size was found to be 0.94 ± 0.44 μm for polyps and 0.56 ± 0.18 μm for normal mucosa. It was found that the values for scatterer density are (9.2 ± 7.5) × 10^8 mm^{-3} and (3.5 ± 4.0) × 10^8 mm^{-3} for the normal mucosa and the adenomatous polyp, respectively. The mean oxygen saturation was determined by Knoefel et al. [283] as 37 ± 19, 46 ± 13, 45 ± 10, and 49 ± 15% for adenocarcinomas, adenomatous polyps, hyperplastic polyps, and normal mucosa, respectively. Colon tissue properties are presented in Table 1.7.

1.3.5 *Cervical Cancer*

With an estimated 570,000 cases and 311,000 deaths in 2018 worldwide, this disease ranks as the fourth most frequently diagnosed cancer and the fourth leading cause of cancer death in women [1]. Cervical cancer ranks second in incidence and mortality behind breast cancer; however, it is the most commonly diagnosed cancer in 28 countries and the leading cause of cancer death in 42 countries, the vast majority of which are in Sub-Saharan Africa and South-Eastern Asia [1]. In 2001, Utzinger et al. [109] first demonstrated *in vivo* detection of squamous dysplasia, a precursor of cervical cancer, using Raman fiber-optic probe in 24 measurements in 13 patients. It was introduced simple algorithm of classification involving two intensity ratios I_{1454}/I_{1556} and I_{1330}/I_{1454}, which were correspondently greater and lower for samples with squamous dysplasia than all other tissue types. By integrating analytical algorithms with data collection, diagnostic accuracies as high as 88% were achieved [203]. Observed strong peaks at 1660 (amide I), 1450 (δ(CH_2) deformation), and 1340 cm^{-1} (DNA) of the Raman spectra were characteristic of a cervical tumor, which indicated increased DNA and protein while decreased peaks at 1280 and 1240 cm^{-1} indicated collagenous proteins [197]. By considering the variations in the Raman spectra of normal cervix due to the hormonal or menopausal status of women, the diagnostic accuracy was improved from 88 to 94% [118, 286]. To further increase the diagnostic accuracy, the authors also incorporated spectral variations linked to confounding factors, such as age, race, smoking habits, body mass index, and menopausal status in cervical Raman spectra [287].

Duraipandian et al. [198] reported an *in vivo* investigation on cervical precancer detection based on the measurement of 105 near-infrared Raman spectra from 57 sites *in vivo* of 29 patients. The authors employed a genetic algorithm partial least-squares discriminant analysis (GA-PLS-DA-dCV) to identify seven significant bands associated with lipids, proteins, and nucleic acids in tissue 925–935 cm^{-1} (CCH deformation mode of glycogen, C–C stretching mode of protein and collagen), 979–999 cm^{-1} (phospholipids), 1080–1090 cm^{-1} (PO_2^- symmetric stretching mode of nucleic acids and C–C stretching mode of phospholipids), 1240–1260 cm^{-1} (amide III), 1320–1340 cm^{-1} (CH_3CH_2 wagging of nucleic acids and proteins), 1400–1420 cm^{-1} (CH_3 bending vibration of proteins), and 1625–1645 cm^{-1} (C=O stretching mode amide I, a-helix). It was achieved a diagnostic accuracy of 82.9% for differentiation of low- and high-grade precancerous lesions. The potential of high-wavenumber (2800–3700 cm^{-1}) Raman spectroscopy for *in vivo* detection of cervical precancer has been investigated by Mo et al. [201]. Significant differences in CH_2 stretching bands of lipids at 2850 and 2885 cm^{-1}, CH_3 stretching bands of proteins at 2940 cm^{-1}, and the broad Raman band of water at 3400 cm^{-1} were observed in normal and dysplastic cervical tissue. A follow-up study by Duraipandian et al. [288] explored the advantages of using both the low- and high-wavenumber regions for *in vivo* detection of cervical precancer, acquiring 473 Raman spectra from 35 patients. Raman spectral differences between normal and dysplastic cervical tissue were observed at 854, 937, 1001, 1095, 1253, 1313, 1445, 1654, 2946, and 3400 cm^{-1}, mainly related to proteins, lipids, glycogen, nucleic acids, and water content in the tissue. PLS-DA together with LOPOCV yielded the diagnostic sensitivities of 84.2%, 76.7%, and 85.0%, respectively; specificities of 78.9%, 73.3%, and 81.7%, respectively; and overall diagnostic accuracies of 80.3%, 74.2%, and 82.6%, respectively, using FP, HW, and integrated FP/HW Raman spectroscopic techniques for *in vivo* diagnosis of cervical precancer.

1.3.6 Prostate Cancer

Prostate cancer is among the most common cancers in men worldwide, with an incidence of 1,276,106 (7.1% of the total cases) and mortality rate of 358,989 (3.8% of the total cancer deaths) [1]. The diagnosis of prostate cancer is often made through transrectal ultrasound guided prostatic biopsy. Ten to twelve core biopsies are recommended [289]. When the diagnosis of prostate cancer is made, grading is used to describe the histologic appearance of the tumor cells. Prostate cancer is most commonly graded using the Gleason score (GS), which categorizes the degree of cancerous tissue as compared to normal prostate tissue.

As Raman probes are suitable for use during endoscopic, laparoscopic, or open procedures they can be used for screening, biopsy, margin assessment, and monitoring of prostate cancer treatment efficacy [290]. Patel and Martin [291] used Raman spectroscopy to characterize the transitional, central, and peripheral zones of normal prostates, revealing larger concentrations of DNA and RNA in the peripheral zone, as well as differences in the relative concentration of lipids and proteins between the three zones. Crow et al. [187, 292] and Stone et al. [123] showed the ability to differentiate prostate cancer into three categories (GS < 7, GS $= 7$, and GS > 7) and from benign prostatic hyperplasia with 89% accuracy from *in vitro* analysis of frozen biopsies. They found that the nucleus/cytoplasm (actin) ratio increased with malignancy, with malignant Raman spectra showing increased DNA concentration. An increase in the relative concentration of choline and cholesterol was shown to be associated with malignancy, potentially representing increased cell membrane synthesis from increased proliferation and increased necrosis, respectively.

Devpura et al. [196] detected benign epithelia and adenocarcinoma, distinguishing GS of 6, 7, and 8 in deparaffinized bulk tissues. The intensity of 782 cm^{-1}, associated with DNA bases, increased with malignancy, as did the ratio of I_{726}/I_{634}. An intensity of 726 cm^{-1} correlated with DNA, and the intensity of 634 cm^{-1} was constant across all tissues, which again demonstrated a relative increase in DNA content with malignancy. Spectral variation by Gleason score was observed in the ranges 900–1000 and 1292–1352 cm^{-1}. Adenocarcinoma was identified using PCA with 94% sensitivity and 82% specificity, and Gleason scores of 6, 7, and 8 were distinguished with 81% accuracy. PCA and SVM were used by Wang et al. [194] to classify the spectra of 50 patients into two groups according to their GS (≤ 7 and >7), achieving 88% sensitivity, specificity, and accuracy. In the experiments performed with the 1064-nm laser, significant differences were found predominantly in the 1000–1450 cm^{-1} range [195]. Using SVM the prostate samples were classified into malignant and benign with 96% accuracy and prediction of their GSs with 95% accuracy. The specificity of the method was consistently high, with an average of 98%. The sensitivity varied from 67 to 100%, with an average of 89%.

In 2018, Aubertin et al. [193] demonstrated availability of accurate diagnosis and grading of prostate cancer using handheld contact Raman fiber probe with the 785 nm excitation laser. It was shown that the sensitivity and specificity of differentiation of benign and cancerous tissues vary depending on GS from 76 to 90% and from 73 to 89%, respectively. Later, the same group enhanced the accuracy of prostate cancer determination up to 91% by duel wavelength excitation (785 nm and 671 nm) with simultaneous Raman spectra measurements in fingerprint and high-wavenumber regions and SVM classification model with leave-one-patient-out cross-validation procedure.

Optical properties of experimental prostate tumors *in vivo* were investigated in [293]. The authors measured absorption and reduced scattering coefficients of R3327-AT and R3327-H prostate tumors at 630 and 789 nm and found that the absorption coefficient of the tumors was 0.9 ± 0.4 (630 nm) and 0.4 ± 0.2 (789 nm) for R3327-AT tumor, and 0.9 ± 0.4 (630 nm) and 0.5 ± 0.3 (789 nm) for R3327-H tumor. The reduced scattering coefficient of the tumors was 10.1 ± 3.5 (630 nm) and 5.3 ± 1.4 (789 nm) for R3327-AT tumor, and 12.3 ± 3.2 (630 nm) and 6.7 ± 1.7 (789 nm) for R3327-H tumor.

1.3.7 Bladder Cancer

Tumors of the genitourinary system account for almost one-fourth of malignancies [294]. Bladder cancer is the sixth most commonly occurring cancer in men and the 17th most commonly occurring cancer in women. There were almost 550,000 new cases in 2018 and mortality was 199,922 (2.1% of the total cancer deaths) [1].

According to the clinical course, there are muscle-noninvasive (TIS, Ta, T1), muscle-invasive (T2–T4), and metastatic bladder cancer. Superficial and muscle-invasive tumors of the bladder in 90–95% are represented by urothelial carcinoma, but differ in molecular genetic, morphological, and immuno-histochemical characteristics. Muscle-invasive bladder cancer (MIBC) is a potentially fatal disease, as patients die within 24 months without treatment. In 50% of patients with magnetic resonance imaging operated radically, relapse develops, which is associated with the morphological stage of development of the primary tumor and the state of the regional lymph nodes. The most common localization of metastases of urothelial cancer is regional lymph nodes (78%), liver (38%), lungs (36%), bones (27%), adrenal glands (21%), and intestines (13%), less often (1–8%) metastases develop in the heart, brain, kidneys, spleen, pancreas, meninges, uterus, ovaries, prostate [295, 296].

To reduce the mortality of patients with MIBC, the early detection of relapses and metastases in the pre- and postoperative periods is of leading importance. Metastasis in patients with bladder cancer most often affects the regional lymph nodes of the pelvis, the bifurcation area of the common iliac arteries, distant metastases capture the bones, liver, and lungs [297]. After radical treatment in the first year, the probability of tumor recurrence reaches 10–67%, progression over 5 years—0–55% [298].

The WHO 1973 grading system proposed by Mostofi et al. [299] differentiates papillary urothelial lesions into three grades: G1, G2, and G3 [300]. Tumors are graded according to the degree of cellular and architectural atypia. The lowest grade (G1) displays nearly no atypia, while the highest grade (G3) displays major atypia with major architectural disorders, such as loss of polarity or pseudostratification.

In 1997, a new multidisciplinary consensus meeting was held to revise terminology and provide updated recommendations to the WHO on the pathology of urothelial carcinomas. The WHO/International Society of Urological Pathology (ISUP) classification of 1998 distinguishes papilloma, papillary urothelial neoplasm of low malignant potential (PUNLMP), and low-grade (LG) and high-grade (HG) carcinomas.

The WHO 2016 system is based on the WHO/ISUP 1998 classification and the WHO 2004 classification, which refined the criteria of WHO/ISUP 1998. According to the WHO 2016 system, pTa and pT1 tumors are graded into LG and HG and all detrusor muscle-invasive urothelial carcinomas are considered to be HG tumors. pTa tumors do not invade the lamina propria (no lymphovascular invasion and distant metastasis). However, pT1 tumors grow under the basement membrane into the lamina propria, and lymphovascular invasion and metastasis can be seen in these cases. In many instances pathologists identify pT1 tumors as HG tumors, independently of their atypia [301, 302].

There are no reliable screening tests available for detecting bladder cancer; hence, the diagnosis is usually made based on clinical signs and symptoms. Microscopic or gross painless hematuria is the most common presentation and a hematuria investigation in an otherwise asymptomatic patient detects bladder neoplasm in roughly 20% of gross and 5% of microscopic cases [303, 304].

Currently, the definition of generally accepted criteria for the stage of bladder cancer, such as the depth of tumor invasion, the degree of differentiation of cells, the defeat of regional lymph nodes for prediction, does not always lead to a positive treatment outcome. This is also confirmed by the analysis of the long-term results of treatment of patients with the same diagnosed stage of bladder cancer. Some patients after organ-preserving surgery have a favorable outcome, while others relapse and tumor progression quickly develops [305].

Cystoscopy is an essential procedure for the diagnosis and treatment of bladder cancer, allowing for direct access to a tumor for biopsy, fulguration, and/or resection. Low-grade (LG), papillary (Ta) tumors can be reliably eradicated with one treatment but more advanced disease (high grade and/or T1) often requires repeat resection for complete eradication. Following an initial diagnosis of HG Ta or T1 between 40% and 78% of re-TUR specimens can contain residual disease, with muscle invasion presented in 2% and 14%, respectively [306–309].

Carcinoma *in situ* (CIS) of the urinary bladder is extremely hard to diagnose. The symptoms are highly unspecific and the small, flat CIS lesions can easily be missed, thus remaining unseen in standard white light cystoscopy. Photodynamic diagnosis (PDD) is recommended by the European Association of Urology (EAU) as a diagnostic procedure in cases of suspected CIS [310]. PDD represents a great enhancement in the urological diagnosis of CIS of the urinary bladder and is a superior method to standard white light cystoscopy in cases, where CIS is suspected [311].

The standard for diagnosis is cystoscopy, with biopsies of suspicious lesions, and transurethral resection to confirm the diagnosis [302]. Early *in vitro* and *ex vivo* studies demonstrated the potential of Raman spectroscopy to detect bladder tumors by increased DNA and cholesterol content and decreased collagen content [123, 312, 313]. In 2005, Crow et al. [187] were the first research group to integrate a fiber-optic probe into Raman spectroscopy to differentiate between benign and malignant bladder tissue *in vitro*. In 2009, Grimbergen and colleagues [189] also investigated the potential of using Raman spectroscopy for bladder tissue diagnosis during cystoscopy by examining 107 bladder tissue biopsies *ex vivo* using an endoscopic probe. Raman spectral measurements were obtained from fresh tissue samples immediately after surgery with an integration time of 2 s. Developed PCA-LDA model with a leave-one-out cross-validation distinguishes normal and malignant tissue with sensitivity and specificity of 78.5% and 78.9%, respectively.

In 2010, Draga et al. [190] were the first research group to investigate the use of Raman fiber-optic probe [189] for the diagnosis of bladder cancer *in vivo*. Raman spectra were obtained during transurethral resection of bladder tumor procedures on 38 patients. Spectra were measured with a previously reported high-volume Raman probe [194] with a penetration depth of 2 mm for integration times of 1–5 s. The authors found a significant peak at 875 cm^{-1} in the spectra of normal bladder tissues, which could be assigned to hydroxyproline, an important molecular component of collagen. In addition, the cancer spectra also showed significant elevated peaks at the wavenumbers 1003, 1208, 1580, and 1601 cm^{-1} and 1208, 1548, and 1617 cm^{-1}, which might be attributed to the amino acidsphenylalanine and tryptophan, respectively. Bladder cancer spectra were significantly expressed by elevated intensities of the wavenumbers 680, 789, 1180, 1580, and 1610 cm^{-1} that were most likely due to nucleotide chains. However, the relative increase of lipid content in malignant mucosa *in vivo* was not ascertain in contradiction to the results of [192, 200], which could be explained by the influence of Raman spectra collected from deeper layers due to the use of the high-volume Raman probe. PCA-LDA and leave-one-out cross-validation were used to distinguish cancer from normal tissue, achieving a sensitivity of 85% and specificity of 79%.

And most recently, in 2012, Barman et al. [314] proposed the use of a confocal fiber-optic Raman probe to increase the specificity (in terms of tissue depth discrimination) for bladder cancer diagnosis. The confocal probe was designed by placing a pinhole aperture into the high-volume probe to decrease the depth of field to 280 μm, thus suppressing the spectral information from surrounding regions and from deeper tissue layers beyond the region of interest. All spectra were preprocessed and diagnostic algorithms were developed using PCA and logistic regression analysis along with a leave-one-out cross-validation. The high-volume probe produced a sensitivity of 85.7% and specificity of 85.7%, whereas the confocal probe had a sensitivity of 85.7% and specificity of 100%. The significant increase in specificity values of the confocal probe in comparison to the high-volume probe was associated with the smaller depth of field values, giving this particular device an advantage in the application of Raman probes for real-time *in vivo* diagnosis of bladder pathology.

In 2018, Chen et al. [191] implemented a low-resolution fiber-optic Raman sensing system for different bladder pathologies discrimination. With the help of a specially trained and cross-validated PCA-ANN classification model, an overall diagnostic accuracy of 93.1% was obtained for the determination of normal, low-grade, and high-grade bladder tissues.

Bovenkamp et al. [188] demonstrated the applicability of RS-OCT for improved diagnosis, effective staging, and grading of bladder cancer by linking the complementary information provided by either modality. OCT well discriminated urothelium, lamina propria, and muscularis layers, which specifically identify the pathological degeneration of the tissue. Raman spectroscopy determines the molecular characteristics via point measurements at suspicious sites. It was shown that OCT differentiated healthy and malignant tissues with an accuracy of 71% in tumor staging and Raman spectroscopy yielded an accuracy of 93% in discriminating low-grade from high-grade lesions.

1.3.8 Stomach and Esophageal Cancer

Stomach cancer (cardia and noncardia gastric cancer combined) remains an important cancer worldwide and is responsible for over 1,000,000 new cases in 2018 and an estimated 783,000 deaths (equating to 1 in every 12 deaths globally), making it the fifth most frequently diagnosed cancer and the third leading cause of cancer death. Esophageal cancer is the eighth most frequent cancer with a worldwide incidence rate of more than 572,034 (3.2% of the total cases) and a mortality rate of 508,585 (5.3% of the total cancer deaths) [1]. By adapting Raman fiber-optic probe designs for endoscopic compatibility, *in situ* measurements of these disease targets have been enabled by many research groups (Table 1.3). In 2008, Teh et al. [181, 184] studied 73 gastric tissue samples from 53 patients and found that Raman peaks at 875 and 1745 cm^{-1} to be two of the most significant features to discriminate gastric cancer from normal tissue. A sensitivity of 90% and specificity of 95% between cancerous and healthy tissue were reported. In a follow-up studies, it was demonstrated sensitivity of 94% and specificity of 96.3% for distinction of gastric dysplastic tissue with a help of narrow-band image-guided Raman endoscopy associated with PCA-LDA model [182]; predictive accuracies were evaluated as 88, 92, and 94% for normal stomach and intestinal- and diffuse-type gastric adenocarcinomas, respectively [182]. Bergholt with co-authors performed *in vivo* studies for diagnostic gastric dysplasia in Barrett's esophagus [315, 316], premalignant and malignant lesions in the upper gastrointestinal tract [176], ulcers in the stomach [179], gastric dysplasia and neoplasia [167], intestinal metaplasia, Helicobacter pylori infection, and adenocarcinoma [177].

Gastric tissue Raman spectra contain a large contribution from triglyceride (major peaks at 1078, 1302, 1445, 1652, and 1745 cm^{-1}) that reflects the interrogation of subcutaneous fat in the gastric wall [167, 179]. Remarkable Raman spectral alterations are observed in the Raman peaks 875, 936, 1004, 1078, 1265, 1302, 1335, 1445, 1618, 1652, and 1745 cm^{-1} between different tissue pathologies due to major pathological features such as upregulation of mitotic and proteomic activity, increase in DNA contents and relative reduction in lipid as well as the onset of angiogenesis leading to neovascularization in the tissue [169, 178]. It has been demonstrated that the diagnostic capabilities can be optimized through the combination of near-infrared autofluorescence with Raman spectroscopy [178]. A total of 1098 normal tissue samples and 140 cancer gastric tissue samples from 81 patients were measured with a spectral acquisition time of 0.5 s. The differentiation between gastric cancer and normal tissue was achieved with a sensitivity of 97.9% and specificity of 91.5%. It was proposed and tested several probabilistic models for online *in vivo* diagnostics and pathology prediction: PLS-DA [168], PCA-LDA [178, 184], ACO-LDA [176], CART [181]. The PLS-DA modeling provided the predictive accuracy of 80.0% [168], suggesting that Raman endoscopy with the integration of online diagnostic framework could be a diagnostic screening tool for real-time *in vivo* gastric cancer identification.

It was demonstrated that the acquisition of both the fingerprint and the high-wavenumber regions of a Raman spectrum meaningfully enhanced the detection of esophageal neoplasia [175] and gastric intestinal metaplasia [176] *in vivo* in comparison with each region alone. A total of 157 patients were included. PCA-LDA model with the leave-one tissue site-out, cross-validation on *in vivo* tissue Raman spectra yielded the diagnostic sensitivities of 89.3%, 89.3%, and 75.0% and specificities of 92.2%, 84.4%, and 82.0%, respectively, by using the integrated FP/HW, FP, and HW Raman techniques for identifying intestinal metaplasia from normal gastric tissue. Wang et al. [174] achieved a diagnostic accuracy of 93.0% (sensitivity of 92.5%; specificity of 93.1%) for differentiating gastric dysplasia from normal gastric tissue by using the beveled fiber-optic Raman probe, which was superior to the diagnostic performance (accuracy of 88.4%; sensitivity of 85.8%; specificity of 88.6%) by using the volume Raman probe.

Optical properties of stomach and esophagus, measured in the spectral range from 300 to 1140 nm in [317], are presented in Table 1.8. It is clearly visible that the

Table 1.8 The optical properties of stomach/esophageal tumor tissues measured *in vitro* [317]

Tissue	λ (nm)	μ_a (cm^{-1})	μ_s (cm^{-1})	g	μ_s' (cm^{-1})	Remarks
Stomach (healthy)/ adenocarcinoma ($n = 21$)	300	44.14/33.10	233.4/231.7	0.752/0.807	57.9/44.7	ISS; Kubelka–Munk; *in vitro*; data from [317]: for healthy stomach tissue $\mu_s = 2.632 \times 10^4 \lambda^{-0.829}$ $\mu_s' = 2.925 \times 10^9 \lambda^{-3.135} + 16.336\lambda^{-0.278}$ and $g = 0.801 + 0.147 (1 - \exp(-(\lambda^{-330.724})/106.543))$; for adenocarcinoma $\mu_s = 1.481 \times 10^5 \lambda^{-1.106}$ $\mu_s' = 1.503 \times 10^9 \lambda^{-3.044} + 17.073\lambda^{-0.273}$ and $g = 0.841 + 0.106 (1 - \exp(-(\lambda^{-331.335})/112.427))$; [λ] in nm
	350	40.26/14.90	196.1/213.8	0.825/0.857	34.3/30.5	
	400	72.37/19.72	192.6/186.6	0.871/0.889	24.8/20.6	
	450	39.56/14.40	152.1/163.8	0.900/0.910	15.2/14.7	
	500	9.13/3.89	155.7/156.2	0.918/0.923	12.8/11.9	
	550	17.60/4.35	139.3/142.9	0.929/0.932	9.9/9.7	
	600	6.18/2.63	138.9/134.4	0.936/0.937	8.9/8.4	
	650	3.10/1.50	139.5/126.0	0.941/0.941	8.2/7.5	
	700	2.46/0.91	134.8/118.9	0.943/0.943	7.6/6.8	
	750	1.52/0.21	122.7/107.6	0.945/0.944	6.7/5.9	
	800	0.98/0.19	113.4/96.8	0.946/0.945	6.1/5.3	
	850	0.49/0.18	104.2/88.9	0.947/0.946	5.5/4.8	
	900	0.51/0.15	81.9/69.1	0.947/0.946	4.3/3.7	
	950	0.37/0.17	84.6/69.2	0.948/0.947	4.4/3.7	
	1000	0.27/0.10	82.3/66.0	0.948/0.947	4.3/3.5	
	1050	0.22/0.20	80.1/62.8	0.948/0.947	4.2/3.3	
	1100	0.19/0.27	76.8/60.3	0.948/0.947	4.0/3.2	
	1140	0.15/0.21	75.3/89.2	0.948/0.947	3.9/4.7	

(continued)

Table 1.8 (continued)

Esophagus (healthy)/ squamous cell carcinoma ($n = 15$)	300	37.22/36.53	224.5/229.3	0.811/0.827	42.5/39.7	ISS. Kubelka–Munk. *In vitro*; data from [317]: for healthy esophagus tissue $\mu_s = 8.131 \times 10^4\lambda^{-1.021}$ $\mu_s' = 2.63 \times 10^9\lambda^{-3.178} + 13.711\lambda^{-0.214}$ and $g = 0.849 + 0.094 (1-\exp(-(\lambda^{-328.74})/84.581))$; for squamous cell carcinoma $\mu_s = 5.229 \times 10^5\lambda^{-1.347}$ $\mu_s' = 1.521 \times 10^9\lambda^{-3.098} + 15.494\lambda^{-0.229}$ and $g = 0.86 + 0.061(1-\exp(-(\lambda^{-329.851})/69.083))$; [λ] in nm
	350	28.35/15.67	192.5/189.3	0.870/0.875	25.0/23.6	
	400	43.18/22.41	165.9/166.3	0.903/0.899	16.2/16.8	
	450	34.27/16.38	151.6/137.6	0.921/0.910	12.0/12.3	
	500	5.49/2.77	147.2/129.6	0.931/0.916	10.2/10.9	
	550	11.61/5.80	131.1/109.7	0.936/0.918	8.4/8.9	
	600	4.69/2.82	127.6/97.9	0.939/0.920	7.8/7.9	
	650	2.05/1.41	122.0/89.4	0.941/0.920	7.2/7.1	
	700	1.97/1.04	114.7/80.0	0.942/0.921	6.7/6.3	
	750	0.93/1.04	103.8/73.3	0.942/0.921	6.0/5.8	
	800	0.94/1.04	94.4/66.5	0.943/0.921	5.4/5.3	
	850	1.14/0.85	86.4/60.5	0.943/0.921	4.9/4.8	
	900	1.14/0.83	69.8/50.6	0.943/0.921	4.0/4.0	
	950	1.14/1.03	68.7/48.5	0.943/0.921	3.9/3.8	
	1000	0.86/1.04	66.1/46.6	0.943/0.921	3.8/3.7	
	1050	0.93/0.83	63.6/43.2	0.943/0.921	3.6/3.4	
	1100	1.14/1.04	60.5/40.4	0.943/0.921	3.4/3.2	
	1140	1.15/0.83	58.9/40.2	0.943/0.921	3.4/3.2	

absorption of light in healthy tissues (both the stomach and the esophagus) predominates over absorption in tumor tissues (adenocarcinoma and squamous cell carcinoma). At the same time, scattering properties of these tissues are comparable across the entire wavelength range.

1.3.9 Oral Cancer

The global incidence rate for oral cancer is 354,864 (2.0% of the total cases) and mortality is 177,384 (1.9% of the total cancer deaths) [1]. These are mostly associated with tobacco and alcohol use, which affect the entire upper aerodigestive tract mucosa resulting in molecular changes that can progress further into carcinomas. The diagnoses of oral cancers are typically performed using a biopsy and histopathology of the tissue [318]. The first *in vivo* study of site wise variations in the human oral cavity was carried out by Guze et al. [319] in high-wavenumber region and by Bergholt et al. [320] in the fingerprint region. It was found that the Raman signal was not influenced by gender or ethnicity [319]; however, the inter-anatomical variability is significant and should be considered as an important parameter in the interpretation and rendering of Raman diagnostic algorithms for oral tissue diagnosis and characterization [320]. Singh et al. [321, 322] reported the discrimination of normal control, premalignant, and cancerous sites with prediction accuracies ranging from 75 to 98% depending on oral cancer location in smoking and non-smoking population. Krishna et al. [211] investigated the classification between spectra acquired from multiple normal sites of 28 healthy volunteers and 171 patients with oral lesions. Using probability-based multiclass diagnostic algorithm, each oral tissue type (squamous cell carcinoma, submucosa fibrosis, leukoplakia, and normal mucosa) was correctly classified in 89%, 85%, 82%, and 85% of the cases, respectively.

Sahu et al. [323] investigated the influence of anatomical differences between subsites on healthy vs pathological classification by examining Raman spectra acquired from 85 oral cancer and 72 healthy subjects. Mean spectra indicated predominance of lipids in healthy buccal mucosa, contribution of both lipids and proteins in lip, while major dominance of protein was found in tongue spectra. From healthy to tumor, changes in protein secondary-structure, DNA, and heme-related features were observed. PC-LDA followed by LOOCV yielded an overall classification of 98%, 54%, 29%, and 67% for healthy, contralateral normal, premalignant, and malignant conditions.

Amelink et al. [324] examined the physiological differences between normal and tumorous (squamous cell carcinoma) oral mucosa and stated that oxygen saturation was $(95 \pm 5)\%$ for normal tissue and $(81 \pm 21)\%$ for SCC; the vessel diameter was 24 ± 14 μm for normal tissue and 25 ± 12 μm for SCC, and the blood volume fraction was equal $(1.0 \pm 0.9)\%$ for normal tissue and $(2.2 \pm 2.3)\%$ for SCC.

1.3.10 Liver Cancer

The global incidence rate for liver cancer is 841,080 (4.7% of the total cases) and mortality is 781,631 (8.2% of the total cancer deaths) [1]. Surgical intervention is often indicated as a potential treatment for liver cancers identified at early stages. The liver's highly specialized tissues regulate a wide variety of high-volume biochemical reactions and are characterized by strong background autofluorescence that overwhelms the Raman scattered signal in fingerprint region. For Raman excitation laser with wavelength 785 nm the feasibility of recovering the spectral signature from bulk liver specimens with sufficient signal-to-noise ratios for interpretation was achieved only from high-wavenumber regions [325]. Therefore, most of the investigations with liver malignancy were conducted in cell lines and thin slices of tissue, where the use of confocal collection geometry reduced autofluorescence [186, 326]. For example, Tolstic et al. [186] reported the precise multivariate PCA-SVM separation of two types of hepatocellular carcinoma cells by the recognition of spectral pattern with peak intensities at 2900–2850, 1655, 1440, 1304, 1266, and 1060 cm^{-1}. The results confirmed that a lot of molecular differences were hidden in lipids and associated with specific wavenumbers of unsaturated fatty acids.

Pence at al. [132] reported the use of a dispersive 1064 nm Raman system using a low-noise indium-gallium-arsenide (InGaAs) array to discriminate highly autofluorescent bulk tissue *ex vivo* specimens from healthy liver, adenocarcinoma, and hepatocellular carcinoma. The resulting spectra were combined with a multivariate discrimination algorithm, sparse multinomial logistic regression (SMLR), to predict class membership of healthy and diseased tissues, and spectral bands selected for robust classification were extracted. These spectral bands included retinol, heme, biliverdin, or quinones (1595 cm^{-1}); lactic acid (838 cm^{-1}); collagen (873 cm^{-1}); and nucleic acids (1485 cm^{-1}). It was achieved 100% sensitivity and 89% specificity for normal versus tumor classification.

Cholangiocarcinoma (CC) is a group of malignant tumors originating from bile duct epithelium. According to WHO classification [327] the term cholangiocarcinoma is reserved for carcinomas arising in the intrahepatic bile ducts. The prognosis of this malignancy is dismal owing to its silent clinical character, difficulties in early diagnosis, and limited therapeutic approaches.

CC is the most common malignant tumor of the biliary tract found in the bile duct epithelial cells and the second most common primary tumor of the liver [328]. Depending on anatomical localization, CC is classified as intrahepatic CC or extrahepatic CC, including perihilar CC and distal CC. The Bismuth–Corlette classification provides preoperative assessment of local spread. The anatomic margins for distinguishing intra- and extrahepatic CCs are the second-order bile ducts [329].

The Liver Cancer Study Group of Japan proposed in 2000 a new classification based on growth (morphologic) characteristics being identified as mass forming, periductal-infiltrating, and intraductal-growing types [330].

Intrahepatic CC is a primary liver malignancy arising from the epithelial cells of the distal branch intrahepatic bile duct [331]. The incidence of intrahepatic CC exhibits wide geographical variation and generally accounts for between 5 and 30% of primary liver cancers [332–334]. Approximately 67% of CCs are perihilar.

Distal CCs are those that arise in the mid or distal bile duct. They are potentially amenable to pancreaticoduodenectomy.

The Classification of Malignant Tumors (TNM) of the American Joint Committee on Cancer and the International Union Against Cancer applies to all primary carcinomas of the liver, including hepatocellular carcinomas, intrahepatic bile duct carcinomas, and mixed tumors [335].

Hilar CC arises from the extrahepatic bile ducts (right and left hepatic ducts at or near their junction) and is considered an extrahepatic carcinoma [336].

In most peripheral CCs, hard, compact, and grayish-white massive or nodular lesions are found in the liver. They may grow inside the dilated bile duct lumen or show an infiltrative growth along the portal pedicle. Usually the tumors are not big compared to the whole liver. Hemorrhage and necrosis are infrequent, and the association with cirrhosis is only occasional. Tumor located just beneath the capsule of the liver shows umbilication, as in metastatic liver cancer.

In most hilar CCs, the tumor infiltrates and proliferates along the extrahepatic bile duct, which is thickened in most cases. Mass formation can be minimal and there can be thickening and enlargement of the portal region. The infiltration in the liver has an arborescent appearance. Extensive parenchymal infiltration is also observed in most cases [337].

Currently, surgical resection remains the most effective treatment for intrahepatic CC [338]. The prognosis for patients with this disease remains disappointing despite advances in the operative and nonoperative management [339]. A positive bile duct resection margin is correlated with higher local recurrence rate and poor prognosis and its role is similar to a positive lymph node [340].

However, because of vague symptomatic presentation, most patients are at an advanced stage by the time of diagnosis, and only nearly one-third of patients are eligible for surgical resection [341]. As a result, the overall outcome of intrahepatic CC remains extremely poor, in which patients who are unable to undergo surgical resection have a less than 10% survival rate at 5 years. Moreover, the reported outcome after hepatic resection is also not optimistic, with a 5-year survival rate of 30–35% [342].

The principal reason for the dismal outcome of surgical treatment is the high incidence of postoperative intrahepatic CC recurrence, in which more than 60% of patients can subsequently develop cancer recurrence after hepatic resection.

The earliest description of CC of bile ducts without palpable surface liver mass on laparotomy was described by Sanford in 1952 [343]. Tsushimi et al. hypothesized that CC arising within bile ducts was from ectopic liver tissue [344].

1.3.11 Thyroid Cancer

Thyroid cancer is responsible for 567,000 cases worldwide, ranking in ninth place for incidence. The global incidence rate in women of 10.2 per 100,000 is three times higher than in men; the disease represents 5.1% of the total estimated female cancer burden, or 1 in 20 cancer diagnoses in 2018 [1]. The diagnosis is commonly based on clinical perceptions and ultrasonography-guided fine needle aspiration, often presenting inconclusive results, which can indicate surgery as the main treatment for these cases. Therefore, the need to biochemically characterize the thyroid gland during surgery is extremely important. Raman spectroscopy may greatly speed up the diagnostic process whether pre-operatively or in the theater setting. The applicability of Raman spectroscopy for thyroid cancer diagnostics was confirmed in cell line studies [215, 216, 345]. In 2009, Harris et al. [215] reported the accuracy of 95% for identification of cancerous cell lines using neural network analysis. O'Dea et al. [216] demonstrated the possibility to correctly classify cell lines representing benign thyroid cells and various subtypes of thyroid cancer. Spectral differences were consistently observed between the benign and cancerous cell lines with the strongest signals occurring at ~470, ~780, 855, 941, ~1230, 1278, 1343, 1402, 1436, 1456, 1571, 1650, 1690, and 1677 cm^{-1}, representing significant differences in the molecular composition of carbohydrates, nucleic acids, lipids, protein structures, and amides. A PC-LDA model was applied to examine the possibility of correct classification of various subtypes of thyroid cancer. The well-differentiated papillary and follicular thyroid carcinoma cell lines were detected with sensitivities >90% and specificities >80%, although the model yielded lower performance scores for identifying the undifferentiated thyroid carcinoma cell lines (sensitivities of 77% and specificities of 73%). Rau et al. [217] demonstrated *ex vivo* the significant presence of carotenoids in papillary thyroid carcinoma with respect to the healthy tissue. The authors stated the sensitivity of 93% and specificity 100% in discrimination of papillary and follicular thyroid carcinoma using combined fingerprint and high-wavenumber regions of Raman spectra and PC-LDA statistical model with leave-one-out cross-validation. In 2019, Medeiros-Neto et al. [346] compared *in vivo* and *ex vivo* spectra of papillary carcinomas, confirming the efficacy of the technique in the biochemical identification of the analyzed tissue. The intense peaks related to an increased amount of DNA were registered at 1017, 991, 829, and 810 cm^{-1} (*in vivo* samples) and at 1421, 1324, 828, 810 cm^{-1} (*ex vivo* samples). The amino acid tyrosine, a very important metabolite for the proper functioning of the thyroid gland, was evidenced by the peaks at 1205, 863, 854 cm^{-1} (*in vivo*) and at 1605, 1206, 863, 853, and 828 cm^{-1} (*ex vivo*). The phenylalanine, produced from the hydroxylation process and essential for the thyroid, was observed at the peak situated at 1174 cm^{-1} (*in vivo*), and at 1174 and 1103 cm^{-1} (*ex vivo*), proving the increase in protein concentration in carcinogenic tissues. Another important observed amino acid was tryptophan, at 1366 and 877 cm^{-1} (*in vivo*), and at 1556, 1360, and 878 cm^{-1} (*ex vivo*). This amino acid is important because it participates in the production of the hormones serotonin and melatonin and of the enzyme tryptamine, which are tumor growth inhibitory substances.

1.3.12 Brain Cancer

The global incidence rate for brain cancer is 296,851 (1.6% of the total cases) and mortality is 241,037 (2.5% of the total cancer deaths) [1]. Raman spectroscopy is a potential modality that can identify the margins of the tumor intraoperatively.

For example, Kalkanis et al. [347, 348] demonstrated identification of normal gray matter and white matter from pathologic glioblastoma and necrosis in frozen brain tissue sections by imaging of relative concentrations of 1004, 1300, 1344, and 1660 cm^{-1}, which correspond primarily to protein and lipid content. Leslie et al. [349] investigate the application of Raman spectroscopy to diagnose pediatric brain tumors acquiring Raman spectra from fresh tissue samples. Support vector machine analysis was used to classify spectra using the pathology diagnosis as a gold standard. Normal brain (321 spectra), glioma (246 spectra), and medulloblastoma (82 spectra) were identified with 96.9, 96.7, and 93.9% accuracy, respectively. Jermyn et al. [350] demonstrated that a handheld Raman probe could detect cancer cells intraoperatively that could not be detected by T1-contrast-enhanced and T2-weighted MRI. The gliomas were detected with 93% sensitivity and 91% specificity using supervised machine learning boosted-trees classification algorithm that utilized all spectral data. Recently, the same research group demonstrated the increase in accuracy of brain cancer detection by multimodal optical system from 91% for standalone RS to 97% when combined with fluorescence analysis [319].

Desrochers et al. [218] tested the use of high-wavenumber Raman spectroscopy in a practical fiber-optic probe that satisfied the stringent miniaturization constraints required for direct integration with a commercial brain biopsy needle. As it was expected, in comparison with gray matter, the white matter spectrum demonstrated larger contributions from lipids (2845 cm^{-1}) and a lower contribution from proteins and nucleic acids (2930 cm^{-1}) in comparison with gray matter. These data demonstrated an increase in the protein/lipid ratio for dense cancer compared to normal brain samples, consistent with findings [351] in brain *ex vivo* samples. However, when compared with that of normal brain, the protein/lipid ratio in infiltrated samples did not show significant differences. The authors showed that HW Raman spectroscopy could detect human dense cancer with >60% cancer cells *in situ* during surgery with a sensitivity and specificity of 80% and 90%, respectively.

Optical properties of brain tumors were studied in [41, 43, 77, 352–355] and summary of those investigations is presented in Table 1.9. Genina et al. [41] measured absorption, scattering, reduced scattering coefficients, and scattering anisotropy factor of brain tissues in a wide spectral range from 350 to 1800 nm (see Table 1.9) for healthy rats and rats with model C6 glioblastoma. Glioblastoma multiforme (GBM) is the most common and aggressive form among all brain tumors (grade IV WHO). The development of the glioblastoma is accompanied by the following main symptoms: headaches, dysfunction of memory and general brain function, visual impairment, poor speech, impaired sensitivity and motor activity, pathological changes in behavior, loss of appetite, etc.

Gebhart et al. [43] investigated human glioma optical properties using integrating sphere technique and inverse adding-doubling method in the spectral range 400–1300 nm. Astrocytoma of optic nerve and medulloblastoma were investigated

Table 1.9 The optical properties of brain tumor tissues measured in *in vitro* [41, 43, 77, 352–355]

Tissue	λ (nm)	μ_a (cm^{-1})	μ_s (cm^{-1})	g	μ_s' (cm^{-1})	Remarks
Healthy brain tissue (n = 12)	350	1.95 ± 0.93	413.1 ± 69.6	0.767 ± 0.038	95.8 ± 18.8	ISS; IMC; *in vitro*; data from [41] $\mu_s = 1.165 \times 10^5\lambda^{-1.178} + 345.365\lambda^{-0.024}$ $\mu_{s'} = 1.177 \times 10^5\lambda^{-1.147} + 48.742\lambda^{-0.308}$ and $g = 0.739+0.215 (1-\exp(-(\lambda^{-505.34})/705.497))$; [λ] in nm
	400	1.80 ± 0.85	400.5 ± 67.3	0.771 ± 0.038	91.6 ± 17.6	
	450	1.31 ± 0.62	384.9 ± 64.6	0.765 ± 0.038	90.4 ± 17.8	
	500	0.80 ± 0.38	371.7 ± 61.4	0.737 ± 0.036	97.4 ± 19.4	
	550	0.93 ± 0.44	366.0 ± 60.9	0.766 ± 0.038	85.4 ± 17.0	
	600	0.78 ± 0.37	357.5 ± 58.9	0.766 ± 0.038	83.3 ± 16.5	
	650	0.69 ± 0.33	351.1 ± 57.2	0.770 ± 0.038	80.4 ± 15.9	
	700	0.70 ± 0.33	346.0 ± 55.5	0.784 ± 0.039	74.5 ± 14.7	
	750	0.73 ± 0.34	341.7 ± 54.1	0.797 ± 0.039	69.3 ± 13.7	
	800	0.75 ± 0.35	337.9 ± 52.9	0.810 ± 0.040	64.1 ± 12.6	
	850	0.76 ± 0.36	335.4 ± 52.4	0.819 ± 0.041	60.4 ± 11.8	
	900	0.77 ± 0.36	331.5 ± 52.1	0.829 ± 0.041	56.5 ± 10.9	
	950	0.79 ± 0.37	328.4 ± 51.5	0.840 ± 0.042	52.2 ± 10.0	
	1000	0.78 ± 0.37	326.2 ± 51.4	0.850 ± 0.042	48.8 ± 9.3	
	1050	0.76 ± 0.36	324.1 ± 51.3	0.856 ± 0.042	46.5 ± 8.8	
	1100	0.75 ± 0.35	322.1 ± 51.3	0.863 ± 0.043	44.0 ± 8.3	
	1150	0.80 ± 0.38	320.0 ± 51.2	0.871 ± 0.043	41.0 ± 7.7	
	1200	0.83 ± 0.39	318.1 ± 51.2	0.878 ± 0.043	38.6 ± 7.3	
	1250	0.79 ± 0.37	316.5 ± 51.1	0.882 ± 0.044	37.1 ± 7.0	
	1300	0.79 ± 0.37	315.0 ± 51.2	0.887 ± 0.044	35.4 ± 6.7	
	1350	0.91 ± 0.43	313.1 ± 51.1	0.893 ± 0.044	33.2 ± 6.3	
	1400	1.63 ± 0.83	313.3 ± 51.4	0.891 ± 0.044	33.9 ± 6.8	
	1450	2.44 ± 1.32	314.6 ± 51.9	0.881 ± 0.044	37.1 ± 7.8	
	1500	1.96 ± 1.03	311.3 ± 51.5	0.892 ± 0.044	33.4 ± 6.8	
	1550	1.41 ± 0.70	308.8 ± 51.2	0.903 ± 0.045	29.7 ± 5.8	
	1600	1.13 ± 0.55	307.6 ± 51.3	0.908 ± 0.045	28.1 ± 5.5	
	1650	1.01 ± 0.49	306.9 ± 51.5	0.911 ± 0.045	27.1 ± 5.2	
	1700	1.03 ± 0.50	306.4 ± 51.5	0.913 ± 0.045	26.4 ± 5.1	
	1750	1.13 ± 0.56	305.7 ± 51.5	0.914 ± 0.045	26.1 ± 5.1	
	1800	1.17 ± 0.8	305.4 ± 51.5	0.914 ± 0.045	26.2 ± 5.1	

(continued)

Table 1.9 (continued)

Tissue	λ (nm)	μ_a (cm^{-1})	μ_s (cm^{-1})	g	μ_s' (cm^{-1})	Remarks
7-day C6-glioma ($n = 6$)	350	5.58 ± 1.58	164.3 ± 61.2	0.524 ± 0.120	78.2 ± 15.3	ISS; IMC; *in vitro*; data from [41] $\mu_s = 2.235 \times 10^5\lambda^{-1.16} + 41.77\lambda^{-0.209}$ $\mu_s' = 7.449 \times 10^5\lambda^{-1.59} + 185.056\lambda^{-0.446}$ and $g = 0.521+0.255 (1-\exp(-(\lambda^{-368.293})/167.548))$; [λ] in nm
	400	5.21 ± 1.59	157.6 ± 53.2	0.571 ± 0.104	67.5 ± 14.8	
	450	4.17 ± 1.31	154.3 ± 46.2	0.610 ± 0.098	60.1 ± 14.6	
	500	3.08 ± 1.03	152.5 ± 39.2	0.638 ± 0.096	55.1 ± 14.7	
	550	3.18 ± 1.08	148.3 ± 37.1	0.686 ± 0.087	46.5 ± 12.9	
	600	2.88 ± 0.99	142.2 ± 32.9	0.709 ± 0.080	41.3 ± 11.6	
	650	2.68 ± 0.94	135.4 ± 28.2	0.729 ± 0.072	36.7 ± 10.4	
	700	2.62 ± 0.92	129.5 ± 23.5	0.748 ± 0.062	32.6 ± 9.1	
	750	2.61 ± 0.90	123.4 ± 19.6	0.762 ± 0.055	29.2 ± 8.0	
	800	2.63 ± 0.88	117.4 ± 16.5	0.773 ± 0.049	26.5 ± 7.0	
	850	2.63 ± 0.87	109.1 ± 13.6	0.777 ± 0.045	24.2 ± 6.3	
	900	2.57 ± 0.81	101.5 ± 12.8	0.779 ± 0.043	22.3 ± 5.5	
	950	2.53 ± 0.76	94.6 ± 11.6	0.781 ± 0.040	20.6 ± 4.9	
	1000	2.49 ± 0.75	87.9 ± 23.8	0.777 ± 0.040	19.5 ± 4.5	
	1050	2.39 ± 0.73	81.9 ± 21.9	0.774 ± 0.040	18.4 ± 4.1	
	1100	2.33 ± 0.72	76.7 ± 20.4	0.772 ± 0.040	17.4 ± 3.9	
	1150	2.33 ± 0.71	72.0 ± 18.9	0.767 ± 0.040	16.7 ± 3.6	
	1200	2.29 ± 0.71	67.9 ± 17.6	0.764 ± 0.039	16.0 ± 3.3	
	1250	2.19 ± 0.69	64.0 ± 16.4	0.761 ± 0.039	15.2 ± 3.1	
	1300	2.13 ± 0.68	60.5 ± 15.4	0.757 ± 0.039	14.6 ± 3.0	
	1350	2.25 ± 0.69	57.6 ± 14.6	0.750 ± 0.039	14.3 ± 2.7	
	1400	3.55 ± 0.69	58.4 ± 14.0	0.703 ± 0.041	17.3 ± 2.7	
	1450	5.04 ± 0.76	60.1 ± 14.6	0.649 ± 0.043	21.0 ± 2.8	
	1500	4.08 ± 0.71	54.8 ± 13.4	0.675 ± 0.042	17.8 ± 2.5	
	1550	2.97 ± 0.67	50.2 ± 12.2	0.707 ± 0.040	14.6 ± 2.3	
	1600	2.41 ± 0.67	47.2 ± 11.7	0.719 ± 0.041	13.2 ± 2.2	
	1650	2.16 ± 0.65	44.9 ± 11.4	0.720 ± 0.041	12.5 ± 2.2	
	1700	2.15 ± 0.64	43.8 ± 11.4	0.717 ± 0.042	12.3 ± 2.1	
	1750	2.30 ± 0.64	43.3 ± 11.3	0.709 ± 0.041	12.5 ± 2.0	
	1800	2.40 ± 0.60	42.6 ± 11.1	0.701 ± 0.040	12.7 ± 1.8	

(continued)

Table 1.9 (continued)

Tissue	λ (nm)	μ_a (cm^{-1})	μ_s (cm^{-1})	g	μ_s' (cm^{-1})	Remarks
10-day C6-glioma ($n = 5$)	350	8.76 ± 1.94	253.5 ± 21.8	0.522 ± 0.080	121.0 ± 18.0	ISS; IMC; *in vitro*; data from [41] $\mu_s = 9267 \times 10^3\lambda^{-0.702} + 246.785\lambda^{-0.09}$ $\mu_s' = 3.328 \times 10^5\lambda^{-1.349} + 60\lambda^{-0.385}$ and $g = 0.514+0.371(1-\exp(-(\lambda^{-355.468})/289.003))$; [λ] in nm
	400	8.22 ± 2.04	250.3 ± 23.0	0.566 ± 0.083	108.4 ± 18.0	
	450	6.55 ± 1.26	249.4 ± 24.4	0.617 ± 0.074	95.3 ± 18.3	
	500	4.85 ± 0.63	253.0 ± 26.1	0.656 ± 0.067	86.8 ± 19.7	
	550	5.15 ± 0.74	246.8 ± 25.8	0.698 ± 0.058	74.4 ± 17.2	
	600	4.61 ± 0.56	240.6 ± 27.6	0.723 ± 0.055	66.5 ± 16.3	
	650	4.27 ± 0.44	235.6 ± 29.1	0.748 ± 0.053	59.2 ± 14.9	
	700	4.19 ± 0.44	231.8 ± 30.6	0.772 ± 0.053	52.7 ± 13.3	
	750	4.19 ± 0.44	228.1 ± 32.3	0.792 ± 0.052	47.4 ± 11.8	
	800	4.22 ± 0.45	224.4 ± 32.7	0.807 ± 0.053	43.1 ± 10.5	
	850	4.22 ± 0.50	220.5 ± 31.0	0.820 ± 0.053	39.5 ± 9.2	
	900	4.19 ± 0.47	215.0 ± 29.6	0.829 ± 0.055	36.6 ± 8.1	
	950	4.17 ± 0.48	210.3 ± 28.8	0.837 ± 0.056	34.1 ± 7.0	
	1000	4.11 ± 0.50	206.1 ± 27.9	0.843 ± 0.056	32.1 ± 6.5	
	1050	4.00 ± 0.50	202.2 ± 27.0	0.849 ± 0.058	30.4 ± 6.1	
	1100	3.93 ± 0.50	198.7 ± 26.0	0.854 ± 0.059	28.8 ± 5.7	
	1150	3.96 ± 0.53	195.4 ± 24.9	0.858 ± 0.060	27.6 ± 5.4	
	1200	3.93 ± 0.54	192.5 ± 24.0	0.862 ± 0.061	26.4 ± 5.1	
	1250	3.78 ± 0.54	189.8 ± 23.1	0.867 ± 0.062	25.2 ± 4.8	
	1300	3.71 ± 0.54	187.4 ± 22.2	0.870 ± 0.062	24.2 ± 4.5	
	1350	3.92 ± 0.61	185.0 ± 21.7	0.871 ± 0.063	23.7 ± 4.3	
	1400	5.78 ± 0.96	185.5 ± 20.6	0.850 ± 0.062	27.7 ± 4.8	
	1450	7.96 ± 1.38	186.9 ± 18.6	0.821 ± 0.061	33.3 ± 5.7	
	1500	6.55 ± 1.12	182.7 ± 18.1	0.843 ± 0.063	28.6 ± 4.9	
	1550	4.97 ± 0.86	179.3 ± 18.5	0.866 ± 0.064	23.9 ± 4.1	
	1600	4.19 ± 0.72	177.3 ± 18.3	0.877 ± 0.064	21.7 ± 3.7	
	1650	3.83 ± 0.68	175.8 ± 17.8	0.882 ± 0.065	20.6 ± 3.4	
	1700	3.89 ± 0.67	175.1 ± 17.4	0.883 ± 0.066	20.4 ± 3.5	
	1750	4.08 ± 0.74	174.5 ± 17.3	0.881 ± 0.065	20.7 ± 3.4	
	1800	4.16 ± 0.76	174.0 ± 16.8	0.880 ± 0.065	20.8 ± 3.3	

(continued)

Table 1.9 (continued)

Tissue	λ (nm)	μ_a (cm^{-1})	μ_s (cm^{-1})	g	μ_s' (cm^{-1})	Remarks
30-day C6-glioma ($n = 1$)	350	6.21	278.3	0.607	82.5	ISS; IMC; *in vitro*; data from [41] $\mu_s = 2.413 \times 10^5\lambda^{-1.104} + 27.561\lambda^{-0.038}$ $\mu_s' = 3.82 \times 10^6\lambda^{-1.89} + 251.221\lambda^{-0.438}$ and $g = 0.596+0.231(1-\exp(-(\lambda^{-362.975})/128.855))$; [λ] in nm
	400	5.60	249.4	0.643	67.2	
	450	4.71	241.3	0.694	55.7	
	500	3.67	243.2	0.739	47.8	
	550	3.67	238.6	0.776	40.3	
	600	3.41	232.6	0.801	34.9	
	650	3.24	221.1	0.814	30.9	
	700	3.18	206.9	0.822	27.7	
	750	3.17	192.7	0.825	25.3	
	800	3.20	178.6	0.825	23.5	
	850	3.19	168.1	0.825	22.1	
	900	3.10	157.2	0.826	20.6	
	950	3.05	149.0	0.826	19.5	
	1000	2.96	141.7	0.825	18.7	
	1050	2.83	133.3	0.823	17.7	
	1100	2.71	127.3	0.822	17.1	
	1150	2.70	121.3	0.820	16.5	
	1200	2.65	116.4	0.818	16.0	
	1250	2.51	110.4	0.816	15.2	
	1300	2.41	106.0	0.814	14.8	
	1350	2.58	102.6	0.798	14.8	
	1400	4.49	104.8	0.754	19.2	
	1450	6.65	109.1	0.719	25.3	
	1500	5.23	101.2	0.729	20.7	
	1550	3.56	93.1	0.762	16.4	
	1600	2.83	88.3	0.783	14.2	
	1650	2.46	85.6	0.791	13.4	
	1700	2.40	84.4	0.792	13.2	
	1750	2.60	83.3	0.790	13.5	
	1800	2.76	82.5	0.789	13.9	
Glioma ($n = 39$)	400	21.50	–	–	38.0	ISS; IAD; *in vitro*; data from [43] $\mu_s' = 2.25 \times 10^7\lambda^{-2.279} + 266.6\lambda^{-0.495}$; [λ] in nm
	500	4.08	–	–	28.1	
	600	2.35	–	–	22.7	
	700	1.42	–	–	18.3	
	800	1.34	–	–	15.5	
	900	1.39	–	–	14.4	
	1000	1.88	–	–	11.2	
	1100	1.70	–	–	10.0	
	1200	2.85	–	–	8.9	
	1300	2.62	–	–	8.8	

(continued)

Table 1.9 (continued)

Tissue	λ (nm)	μ_a (cm^{-1})	μ_s (cm^{-1})	g	μ_s' (cm^{-1})	Remarks
Astrocytoma of optic nerve (n = 1)	674	1.4 ± 0.3	–	–	12.5 ± 1.0	Spatially resolved diffuse reflectance; *in vivo*; human; data from [77]
	811	1.2 ± 0.3	–	–	9.5 ± 1.0	
	849	0.9 ± 0.3	–	–	7.6 ± 1.0	
	956	1.5 ± 0.3	–	–	7.3 ± 1.0	
Medulloblastoma (n = 1)	674	2.6 ± 0.5	–	–	14.0 ± 1.0	Spatially resolved diffuse reflectance; *in vivo*; human; data from [77]
	849	1.0 ± 0.2	–	–	10.7 ± 1.0	
	956	0.75 ± 0.2	–	–	4.0 ± 1.0	
Human glioblastoma (n = 14)	350	12.25	–	–	54.9	IS; IMC; *in vitro*; μ_s' = 1.514 × 104λ-0.948; [λ] in nm; data from [352]
	400	13.68	–	–	49.4	
	450	7.63	–	–	44.9	
	500	4.10	–	–	42.3	
	550	4.49	–	–	38.9	
	600	2.80	–	–	36.9	
	650	1.79	–	–	35.5	
	700	1.46	–	–	32.6	
	750	1.20	–	–	29.2	
	800	1.02	–	–	26.4	
	850	0.95	–	–	24.5	
	900	0.86	–	–	23.1	
	950	0.94	–	–	21.9	
	1000	0.97	–	–	21.1	
Human meningioma (n = 9)	350	11.17	–	–	43.9	ISS; IMC; *in vitro* $\mu_{s'} = 2.098 \times 10^5\lambda^{-1.448}$; [λ] in nm; data from [352]
	400	12.18	–	–	36.9	
	450	8.15	–	–	30.7	
	500	4.38	–	–	25.4	
	550	4.93	–	–	22.0	
	600	2.95	–	–	20.1	
	650	2.09	–	–	17.2	
	700	1.76	–	–	15.9	
	750	1.42	–	–	13.6	
	800	1.20	–	–	12.8	
	850	1.06	–	–	12.3	
	900	0.86	–	–	11.1	
	950	0.93	–	–	10.8	
	1000	0.95	–	–	9.7	

(continued)

Table 1.9 (continued)

Tissue	λ (nm)	μ_a (cm^{-1})	μ_s (cm^{-1})	g	μ_s' (cm^{-1})	Remarks
Human oligodendroglioma ($n = 3$)	350	11.31	–	–	23.1	ISS; IMC; *in vitro*; data from [352] $\mu_s' = 7.147 \times 10^3\lambda^{-0.981}$; [λ] in nm
	400	16.17	–	–	18.9	
	450	9.93	–	–	17.9	
	500	5.48	–	–	16.1	
	550	8.21	–	–	14.7	
	600	2.68	–	–	14.1	
	650	1.85	–	–	13.2	
	700	1.63	–	–	11.8	
	750	1.43	–	–	10.7	
	800	1.37	–	–	9.9	
	850	1.34	–	–	9.7	
	900	1.26	–	–	9.0	
	950	1.38	–	–	8.2-	
	1000	1.38	–	–	8.2	
Human metastasis ($n = 6$)	350	22.99	–	–	76.9	ISS; IMC; *in vitro*; data from [352] $\mu_s' = 8.367 \times 10^3\lambda^{-0.79}$; [λ] in nm
	400	25.40	–	–	71.3	
	450	18.46	–	–	63.9	
	500	12.23	–	–	60.1	
	550	12.83	–	–	60.8	
	600	7.43	–	–	57.4	
	650	3.81	–	–	53.1	
	700	2.86	–	–	49.6	
	750	2.52	–	–	45.0	
	800	2.15	–	–	41.9	
	850	2.08	–	–	39.2	
	900	2.03	–	–	38.2	
	950	2.11	–	–	36.1	
	1000	1.94	–	–	33.6	

(continued)

Table 1.9 (continued)

Tissue	λ (nm)	μ_a (cm^{-1})	μ_s (cm^{-1})	g	μ_s' (cm^{-1})	Remarks
Human low-grade glioma (astrocytoma WHO grade II)	350	17.25	230.9	0.679	74.1	ISS; IMC; *in vitro*; data from [353] $\mu_s = 5.337 \times 10^4\lambda^{-0.931}$ $\mu_s' = 3.204 \times 10^{14}\lambda^{-5} + 10.894\lambda^{-0.258}$; $g = 0.719+0.247(1-\exp(-(\lambda^{-353.984})/67.287))$; [λ] in nm
	400	15.31	204.6	0.867	27.3	
	450	6.85	182.8	0.909	16.7	
	500	3.86	164.1	0.932	11.1	
	550	4.12	149.4	0.938	9.2	
	600	3.52	140.1	0.958	5.8	
	650	2.37	123.8	0.958	5.2	
	700	2.01	120.8	0.964	4.3	
	750	2.07	110.5	0.964	3.9	
	800	1.72	102.3	0.968	3.2	
	850	1.88	100.2	0.968	3.2	
	900	1.90	96.5	0.967	3.1	
	950	2.06	90.1	0.966	3.1	
	1000	1.98	89.3	0.966	2.9	
	1050	1.85	82.0	0.967	2.6	
	1100	1.96	82.3	0.971	2.4	
Human high-grade glioma (WHO grade III)	400	11.14	88.7	0.892	9.5	ISS; IMC; *in vitro*; data from [353] $\mu_s = 3.397 \times 10^3\lambda^{-0.608}$; $\mu_{s'} = 4.107 \times 10^{13}\lambda^{-5.049} + 21.957\lambda^{-0.187}$; $g = 0.891 + 5.608 \times 10^{-3}(1-\exp(-(\lambda^{-409.326})/23.73))$; [λ] in nm;
	450	8.19	82.7	0.894	8.7	
	500	7.16	76.3	0.897	7.8	
	550	5.52	71.9	0.891	7.8	
	600	5.20	71.5	0.902	7.0	
	650	4.62	65.6	0.895	6.9	
	700	4.39	61.2	0.896	6.3	
	750	3.80	61.0	0.898	6.2	
	800	3.14	60.2	0.897	6.2	
Human glioma (n = 1)	400	6.39	–	–	8.3	ISS; Kubelka–Munk; *in vitro*; data from [354] $\mu_s' = 8.928 \times 10^{10}\lambda^{-3.975} + 16.391\lambda^{-0.209}$; [λ] in nm
	450	16.47	–	–	7.5	
	500	12.91	–	–	5.4	
	550	9.93	–	–	4.5	
	600	7.44	–	–	4.5	
	650	2.62	–	–	4.5	
	700	1.80	–	–	4.3	
	750	1.35	–	–	4.4	
	800	0.84	–	–	4.4	
	850	0.93	–	–	4.2	
	900	0.94	–	–	4.2	
	950	1.12	–	–	4.4	
	1000	1.22	–	–	4.5	
	1050	0.91	–	–	4.4	
	1100	0.51	–	–	4.4	

(continued)

Table 1.9 (continued)

Tissue	λ (nm)	μ_a (cm^{-1})	μ_s (cm^{-1})	g	μ_s' (cm^{-1})	Remarks
Human meningioma (n = 6)/human astrocytoma (WHO grade II) (n = 4)	360	3.14/11.78	194.7/218.5	0.818/0.710	35.4/63.3	ISS; IMC; *in vitro*; data from [355] For human meningioma $\mu_s = 7.137 \times 10^3\lambda^{-0.59}$ $\mu_s' = 7.358 \times 10^{10}\lambda^{-3.673} + 18.164\lambda^{-0.229}$; $g = 0.825+0.134 (1-\exp(-(\lambda^{-368.148})/78.556))$ For human astrocytoma $\mu_s = 9.254 \times 10^4\lambda^{-1.025}$; $\mu_s' = 6.181 \times 10^{14}\lambda^{-5.146} + 8.463\lambda^{-0.215}$; $g = 0.903+0.06 (1-\exp(-(\lambda^{-410.5})/33.7))$ [λ] in nm
	400	3.75/15.98	196.1/200.3	0.872/0.889	25.1/22.3	
	450	2.09/3.13	191.2/177.3	0.912/0.943	16.7/10.2	
	500	1.07/2.03	183.4/155.4	0.932/0.958	12.4/6.5	
	550	0.81/2.37	182.2/145.0	0.943/0.961	10.3/5.6	
	600	0.67/1.19	174.9/130.3	0.953/0.962	8.2/4.9	
	650	0.39/0.76	163.2/119.4	0.956/0.958	7.2/5.0	
	700	0.29/0.41	154.9/111.9	0.956/0.959	6.9/4.5	
	750	0.22/0.51	146.7/102.6	0.959/0.963	6.0/3.8	
	800	0.22/0.49	138.8/96.9	0.959/0.967	5.7/3.2	
	850	0.27/0.39	131.1/90.6	0.958/0.965	5.6/3.2	
	900	0.21/0.31	125.9/86.6	0.958/0.963	5.3/3.2	
	950	0.20/0.48	121.8/81.9	0.958/0.959	5.1/3.3	
	1000	0.36/0.44	116.8/78.8	0.956/0.961	5.1/3.1	
	1050	0.42/0.37	115.0/74.5	0.961/0.963	4.5/2.8	
	1100	0.64/0.48	115.2/72.2	0.964/0.968	4.1/2.3	

in vivo by Bevilacqua et al. [77] using spatially resolved diffuse reflectance. Absorption and reduced scattering coefficients of different human tumor tissues (glioblastoma, meningioma, oligodendroglioma, and metastasis) were investigated by Honda et al. [352] by double-integrating sphere and IMC technique. Schwarzmaier et al. [353] investigated optical properties of human low (astrocytoma WHO grade II) and high (WHO grade III) grade glioma with integrating sphere technique and IMC method. Optical properties of glioma were investigated by Sterenborg et al. [354]. Yaroslavsky et al. [355] studied optical properties of human meningioma and astrocytoma (WHO grade II).

1.3.13 Kidney Cancer

The global incidence of renal cell cancer is increasing annually and the causes are multifactorial. It ranks the second most common neoplasm found in the urinary system [356].

Renal cell carcinoma (RCC) is the commonest solid lesion within the kidney and accounts for approximately 90% of all kidney malignancies and 2–3% of all cancers, with the highest incidence occurring in Western countries [357]. The proportion of small and incidental renal tumors has significantly increased owing to the widespread use of abdominal imaging. Consequently, more than 50% of RCCs are currently detected incidentally [358]. Diagnosis and subtyping of RCC can usually be accomplished through a thorough morphologic investigation of the resected tumor, which in itself offers valuable prognostic information [301]. The main subtypes of RCC are clear cell, papillary, chromophobe, collecting duct, and unclassified [359].

The most frequent histological type of RCC is clear cell renal cell carcinoma, it occurs in 75% of all primary kidney cancers. Papillary and chromophobe RCC are less common subtypes [360, 361].

The classic clear cell renal cell carcinoma has a yellow-brown cut surface and it is inhomogeneous due to hemorrhage and necrosis. Macroscopically, it is relatively well separated from the normal renal tissue, but there may be a risk to form microscopic tumor satellites. The tumor cells are derived from the proximal convoluted tubule. The rich content of glycogen and fat in the cytoplasm of the cells produces a clear appearance in conventional staining. But there are also eosinophilic, sarcomatoid, and mixed patterns of differentiation.

The distinction of clear cell RCC from papillary renal cell carcinomas is not particularly difficult, but the distinction between RCC and other neoplasms in some cases is problematic and requires additional research methods, such as immunohistochemistry. Occasionally, the tumor cells harbor granular to pink eosinophilic cytoplasm and can resemble chromophobe RCC, which more typically contains polygonal cells with transparent to reticulated cytoplasm rimmed by thickened cell membranes [362].

The most effective treatment of RCC remains the surgical resection of the tumor mass by partial or total nephrectomy [363]. The functional benefits of nephron-sparing procedures have driven the indication of partial nephrectomy, which is recommended as the standard treatment in patients with T1a tumors [364]. Adjuvant therapy after nephrectomy has not been proven to prolong survival or to have any significant patient benefit [365].

The less invasive approaches include percutaneous radiofrequency ablation and laparoscopically assisted cryoablation. Indications for thermal ablations are usually small renal masses in elderly more comorbid patients unable to undergo surgical intervention and patients with bilateral tumors or solitary kidney [366].

Optical properties of kidney tumor transplanted in rat were investigated in [48] (see Table 1.10).

Table 1.10 The optical properties of kidney tumor tissues measured in *in vitro* [48]

Tissue	λ (nm)	μ_a (cm^{-1})	μ_s (cm^{-1})	g	μ_s' (cm^{-1})	Remarks
Tumor capsule (n = 10)	350	20.56 ± 12.12	152.0 ± 74.9	0.563 ± 0.091	58.8 ± 35.3	ISS; IMC; *in vitro*; in the spectral range 350–1850 nm $\mu_s = 1.968 \times 10^8\lambda^{-2.44} + 59.956\lambda^{-0.023}$; $\mu_s' = 4.293 \times 109\lambda^{-3.116} + 83.768\lambda^{-0.298}$ and in the spectral range 350–1350 nm: $g = 0.544 + 0.27 (1–\exp(–(\lambda^{-364.598})/126.05))$; [λ] in nm
	400	24.06 ± 14.72	143.9 ± 69.7	0.587 ± 0.085	53.5 ± 33.4	
	500	7.19 ± 3.33	101.3 ± 40.1	0.745 ± 0.072	24.6 ± 10.6	
	600	5.86 ± 2.66	86.3 ± 30.6	0.779 ± 0.070	18.9 ± 7.7	
	700	4.27 ± 2.27	76.1 ± 24.8	0.798 ± 0.073	15.8 ± 6.3	
	800	4.35 ± 2.25	69.9 ± 21.5	0.805 ± 0.066	13.8 ± 5.5	
	900	4.53 ± 2.07	64.8 ± 19.5	0.806 ± 0.064	12.5 ± 5.2	
	1000	4.34 ± 1.79	61.1 ± 18.3	0.809 ± 0.063	11.5 ± 4.9	
	1100	3.95 ± 1.61	57.7 ± 17.4	0.812 ± 0.061	10.6 ± 4.6	
	1200	4.10 ± 1.42	55.5 ± 16.9	0.811 ± 0.059	10.2 ± 4.5	
	1300	3.81 ± 1.28	53.1 ± 16.3	0.812 ± 0.058	9.7 ± 4.3	
	1400	8.45 ± 1.29	56.9 ± 17.2	0.756 ± 0.037	13.5 ± 4.5	
	1500	10.97 ± 1.45	57.2 ± 17.7	0.721 ± 0.029	15.9 ± 4.4	
	1600	5.51 ± 1.28	49.8 ± 16.1	0.787 ± 0.046	10.3 ± 4.0	
	1700	4.80 ± 1.38	47.9 ± 15.7	0.792 ± 0.049	9.6 ± 3.9	
	1800	5.79 ± 1.33	48.3 ± 16.1	0.776 ± 0.043	10.5 ± 4.0	
	1900	21.99 ± 6.04	76.9 ± 30.6	0.583 ± 0.084	31.1 ± 9.6	
	2000	24.88 ± 4.74	69.4 ± 28.7	0.526 ± 0.190	34.5 ± 7.5	
	2100	13.98 ± 1.32	54.4 ± 19.1	0.651 ± 0.036	19.2 ± 4.9	
	2200	11.05 ± 1.42	50.8 ± 17.8	0.671 ± 0.036	15.9 ± 5.3	
	2300	13.91 ± 1.96	52.7 ± 18.4	0.628 ± 0.034	18.9 ± 5.9	
	2400	19.38 ± 3.99	61.2 ± 22.3	0.560 ± 0.054	26.7 ± 7.1	
	2500	28.69 ± 11.64	75.3 ± 37.2	0.423 ± 0.312	42.5 ± 16.5	

(continued)

Table 1.10 (continued)

Tissue	λ (nm)	μ_a (cm^{-1})	μ_s (cm^{-1})	g	μ_s' (cm^{-1})	Remarks
Tumor tissue on the periphery (n = 10)	350	26.52 ± 4.13	209.4 ± 40.4	0.697 ± 0.085	53.6 ± 7.1	ISS; IMC; *in vitro*; in the spectral range 350–1850 nm $\mu_s = 4.105 \times 10^8\lambda^{-2.519} + 102.777\lambda^{-0.03}$ $\mu_s' = 4.085 \times 10^9\lambda^{-3.149} + 84.286\lambda^{-0.238}$ and in the spectral range 350–1350 nm: g = 0.691+0.145 (1–exp(–(λ–361.102)/195.813)); [λ] in nm
	400	30.51 ± 4.41	199.4 ± 25.5	0.714 ± 0.063	51.4 ± 6.1	
	500	11.08 ± 3.41	151.7 ± 24.8	0.768 ± 0.056	27.6 ± 3.9	
	600	9.39 ± 2.98	133.9 ± 17.1	0.796 ± 0.041	22.7 ± 2.6	
	700	8.09 ± 2.63	121.1 ± 14.3	0.809 ± 0.035	19.8 ± 1.9	
	800	8.67 ± 2.65	112.2 ± 10.1	0.818 ± 0.029	18.9 ± 1.6	
	900	8.91 ± 2.14	104.7 ± 8.5	0.826 ± 0.026	18.0 ± 1.2	
	1000	8.48 ± 1.62	98.8 ± 7.9	0.831 ± 0.024	17.1 ± 1.1	
	1100	7.84 ± 1.41	93.2 ± 7.4	0.834 ± 0.023	16.2 ± 0.8	
	1200	7.61 ± 1.43	89.3 ± 7.1	0.833 ± 0.023	15.8 ± 0.7	
	1300	7.04 ± 1.45	85.3 ± 6.8	0.835 ± 0.023	15.1 ± 0.7	
	1400	11.57 ± 2.16	88.1 ± 5.1	0.788 ± 0.027	18.8 ± 1.3	
	1500	13.36 ± 2.61	87.1 ± 4.4	0.759 ± 0.033	20.2 ± 1.9	
	1600	7.55 ± 2.05	78.3 ± 5.8	0.814 ± 0.029	15.1 ± 1.1	
	1700	6.69 ± 2.02	75.5 ± 6.1	0.819 ± 0.029	14.4 ± 1.0	
	1800	7.61 ± 2.15	75.3 ± 5.7	0.805 ± 0.030	15.1 ± 1.2	
	1900	28.74 ± 3.25	109.7 ± 1.5	0.639 ± 0.038	39.2 ± 4.8	
	2000	29.32 ± 3.66	100.8 ± 0.5	0.610 ± 0.057	39.3 ± 5.9	
	2100	15.75 ± 2.60	81.4 ± 3.7	0.721 ± 0.043	22.5 ± 2.5	
	2200	12.64 ± 2.20	77.4 ± 5.2	0.749 ± 0.041	19.2 ± 1.8	
	2300	16.01 ± 2.88	79.3 ± 5.7	0.722 ± 0.042	21.9 ± 1.7	
	2400	22.76 ± 2.57	89.9 ± 2.3	0.654 ± 0.061	30.8 ± 4.7	
	2500	35.99 ± 2.01	110.8 ± 2.5	0.536 ± 0.093	51.8 ± 11.5	

(continued)

Table 1.10 (continued)

Tissue	λ (nm)	μ_a (cm^{-1})	μ_s (cm^{-1})	g	μ_s' (cm^{-1})	Remarks
Tumor tissue in the center (n = 10)	350	33.79 ± 12.92	140.6 ± 71.6	0.497 ± 0.144	66.8 ± 25.9	ISS; IMC; *in vitro*; in the spectral range 350–1850 nm $\mu_s = 4.89 \times 10^5\lambda^{-1.47} + 95.529\lambda^{-0.062}$ $\mu_s' = 3.002 \times 10^9\lambda^{-3.012} + 72.049\lambda^{-0.251}$ and in the spectral range 350–1350 nm: $g = 0.484 + 0.34 (1-\exp(-(\lambda-382.365)/158.714))$; [λ] in nm
	400	43.09 ± 16.82	139.9 ± 75.0	0.483 ± 0.199	68.5 ± 30.7	
	500	11.03 ± 3.70	115.3 ± 53.5	0.703 ± 0.103	31.1 ± 9.3	
	600	9.37 ± 3.06	106.1 ± 47.1	0.745 ± 0.091	24.5 ± 6.8	
	700	6.59 ± 1.98	98.0 ± 42.4	0.778 ± 0.081	19.8 ± 5.4	
	800	6.65 ± 1.97	92.9 ± 39.1	0.795 ± 0.066	17.6 ± 4.8	
	900	6.54 ± 1.82	87.8 ± 36.6	0.805 ± 0.056	15.9 ± 4.5	
	1000	6.14 ± 1.59	83.3 ± 34.4	0.814 ± 0.049	14.5 ± 4.1	
	1100	5.59 ± 1.44	79.0 ± 32.3	0.821 ± 0.045	13.4 ± 3.9	
	1200	5.55 ± 1.25	75.8 ± 30.8	0.822 ± 0.042	12.7 ± 3.8	
	1300	5.14 ± 1.11	72.4 ± 29.3	0.825 ± 0.041	12.0 ± 3.5	
	1400	10.03 ± 1.23	75.1 ± 29.4	0.775 ± 0.055	15.9 ± 4.0	
	1500	12.27 ± 1.33	74.2 ± 28.8	0.743 ± 0.067	17.7 ± 4.0	
	1600	6.52 ± 0.63	66.3 ± 26.5	0.802 ± 0.051	12.3 ± 3.3	
	1700	5.72 ± 0.61	63.7 ± 25.4	0.807 ± 0.048	11.6 ± 3.2	
	1800	6.67 ± 0.53	63.3 ± 25.3	0.792 ± 0.056	12.3 ± 3.2	
	1900	26.09 ± 4.16	83.9 ± 37.7	0.526 ± 0.128	36.1 ± 10.5	
	2000	28.29 ± 4.14	78.3 ± 36.1	0.438 ± 0.180	38.8 ± 10.1	
	2100	15.29 ± 2.25	66.4 ± 27.7	0.652 ± ± 0.122	20.6 ± 4.1	
	2200	12.27 ± 1.78	63.4 ± 25.8	0.700 ± 0.102	17.2 ± 3.7	
	2300	15.08 ± 2.58	64.8 ± 26.7	0.661 ± 0.112	19.8 ± 4.3	
	2400	22.52 ± 4.69	71.1 ± 30.5	0.555 ± 0.129	28.5 ± 6.7	
	2500	36.68 ± 8.44	80.1 ± 36.7	0.378 ± 0.180	45.6 ± 14.2	

1.3.14 Pancreatic Cancer

Pancreatic adenocarcinoma (AC), the 4th leading cause of cancer death in the USA with a 5-year survival rate of less than 6%, is often detected at late stages of development when treatment is ineffective. Intraductal papillary mucinous neoplasm (IPMN) is a precursor lesion of pancreatic cancer, characterized by an intraductal proliferation of neoplastic cells with mucin production [367].

Lee et al. [367] measured human pancreatic malignant precursor, IPMN, using methods of reflectance and fluorescence spectroscopy. They found morphological property differences between normal tissue, IPMN, and AC (see Table 1.11).

Optical properties of normal and cancerous pancreas were investigated in [368, 369] (see Table 1.12) using integrating sphere technique.

Table 1.11 Morphological properties of normal tissue, IPMN, and AC

Tissue	Nuclear diameter (μm)	Refractive index	Cell density (cm^{-1})	Collagen density (cm^{-1})
Normal	8.89 ± 0.13	1.372 ± 0.002	(8.07 ± 0.12) × 10^7	(1.28 ± 0.16) × 10^6
IPMN	11.50 ± 0.88	1.394 ± 0.004	(7.15 ± 0.09) × 10^7	(6.15 ± 1.56) × 10^6
AC	11.64 ± 0.37	1.396 ± 0.002	(7.21 ± 0.09) × 10^7	(8.58 ± 0.62) × 10^6

Table 1.12 The optical properties of pancreas tumor tissues measured *in vitro* [368, 369]

Tissue	λ (nm)	μ_a (cm^{-1})	μ_s (cm^{-1})	g	μ_s' (cm^{-1})	Remarks
Normal tissue/ malignant pancreas neoplasm ($n = 8$)	500	29.84/12.07	–	–	27.3/46.0	ISS; IAD; *in vitro*; data from [368]
	550	11.38/6.25	–	–	27.0/20.6	
	600	11.58/5.38	–	–	28.4/19.7	
	650	22.17/18.1	–	–	17.1/28.3	
Pancreatic neuroendocrine tumor Freezing/ Paraffin embedding	1064	0.9 ± 0.1/ 56 ± 3	130 ± 1/ 539 ± 4	0.82 ± 0.01/ 0.96 ± 0.01	–	ISS, IMC; *in vitro*; data from [369]

1.4 Biochemical Cancer Model

To date, most Raman studies of cancerous tissues have used multivariate statistical algorithms to describe the spectral differences of spectral data, such as PCA-LDA or PLS-DA. However, the principal components and loading vectors are difficult to relate to the biophysical origin of the disease, such as the microstructural organization of proteins and lipids and the functional state of cellular metabolism, which are the key features for the pathologist diagnostic decisions for appropriate cancer treatment. Therefore, several research groups have proposed biochemical diagnostic models extracting physiologically relevant markers from Raman spectra of tissues (Table 1.13). The biochemical model derives the morphological and biochemical composition of the modeling tissue from its Raman spectrum. The building blocks of the model are Raman active components either measured directly from synthetic/purified chemicals [123, 167, 169, 177, 224, 313, 316, 370–373] or morphologically extracted from tissue sections *in situ* [98, 122, 222]. In the last case, a Raman spectrometer is coupled to a microscope and is scanned across the tissue section to obtain Raman images that can then be correlated with serial hematoxylin–eosin stained sections to identify relevant morphologic components and their Raman

Table 1.13 Raman biochemical diagnostics models of tumor tissues

Cancer	Model	Method	Components	Research Group	Ref.
Breast	Morphologically derived components	NNLS	Cell cytoplasm, cell nucleus, collagen, fat, cholesterol-like, β-carotene, calcium hydroxyapatite, calcium oxalate dehydrate, water	Shafer-Peltier et al. (2002)	[98]
Breast	Morphologically derived components	NNLS	Cell cytoplasm, cell nucleus, collagen, fat, cholesterol-like, β-carotene, calcium hydroxyapatite, calcium oxalate dehydrate, and water	Haka et al. (2005)	[122]
Brain	Synthetic/ purified chemicals	OLS	Cholesterol, cholesterol ester, lipids, proteins, nucleic acids, carotene, phosphate buffer solution	Bergner et al. (2013)	[370]
Oral	Synthetic/ purified chemicals	NNCLS	Hydroxyapatite, oleic acid, DNA, collagen, and keratin	Bergholt et al. (2012)	[320]
Cervix	Synthetic/ purified chemicals	NNLS	Actin, DNA, histone, collagen, glycogen, protein—human serum albumin, glycerol triolate, cholesterol, and β-carotene	Daniel et al. (2016)	[371]
Stomach	Synthetic/ purified chemicals	NNCLS	Actin, albumin, collagen type I, DNA, histones, triolein, pepsinogen, and phosphatidylcholine	Huang et al. (2010)	[169]
GI	Synthetic/ purified chemicals	NNCLS	Actin, albumin, pepsin, pepsinogen, B-NADH, RNA, DNA, myosin, hemoglobin, collagen I, collagen II, collagen V, mucin 1, mucin 2, mucin 3, flavins, elastin, phosphatidylcholine, cholesterol, glucose, glycogen, triolein, histones, beta-carotene	Bergholt et al. (2011)	[167]
GI	Synthetic/ purified chemicals	NNLS	Actin, collagen, DNA, histones, triolein, glycogen	Bergholt et al. (2011)	[316]
GI	Synthetic/ purified chemicals	NNLS	Glycogen, DNA, oleic acid, collagen I, choline, actin, triolein	Shetti et al. (2006)	[372]
Bladder; prostate	Synthetic/ purified chemicals	OLS	Actin, β-carotene type 1, cholesterol, choline, collagen type 1, collagen type 3, collagen type 4, DNA, glycogen, lycopene, oleic acid, PSA, triolein	Stone et al. (2006)	[123]
Bladder	Synthetic/ purified chemicals	OLS	Collagen I and III, actin, albumin, DNA, RNA, cholesterol, β-carotene, glycogen, gangliosides, phosphatidylcholine, phosphatidylethanolamine, arachidonic acid, palmitic acid, palmitoleic acid, stearic acid, oleic acid methyl ester, linoleic acid	de Jong et al. (2006)	[313]

(continued)

Table 1.13 (continued)

Cancer	Model	Method	Components	Research Group	Ref.
Skin	Synthetic/purified chemicals	OLS	Oleic acid, palmitic acid, collagen I, keratin, hemoglobin	Zhao et al. (2008)	[374]
Skin	Synthetic/purified chemicals	OLS	Triolein, cholesterol, actin, collagen I, collagen III, elastin, keratin, DNA, phenylalanine, carotene, melanin, squalene	Silveira et al. (2012)	[224]
Skin	Morphologically derived components	NNLS	Collagen, elastin, triolein, cell nucleus, keratin, ceramide, melanin, water	Feng et al. (2017)	[222]

GI gastrointestinal, *OLS* ordinary least-squares fitting algorithm, *NNLS* non-negativity-constrained least-squares minimization, *NNCLS* non-negativity-constrained least-squares minimization

signature. *In situ* constituents better represent the milieu of biological tissues that cannot be recapitulated in a synthetic environment. For example, collagen can be presented in human tissue in many different forms, each one having a slightly different Raman spectrum. However, if both of them are included in the model, it may lead to overfitting and unstable results [123, 372]. In addition, skin constituents synthesized in the lab or from commercial sources are not in their natural state. Nevertheless, the advantage of using synthetic/purified chemicals as model components is that they can be easily measured using the same Raman instrument as used to measure the biological specimen, providing evaluation of more specific molecular constituent changes.

Construction of a biochemical model of the tissue relies on three assumptions: first, that the Raman spectrum of a mixture equals to the weighted linear sum of the individual components of the mixture; second, that the biological morphological features, such as cells, have the same Raman spectrum from one patient to another; and third, that the basis spectra included in the model are sufficiently distinct to enable their differentiation based on their Raman spectrum [98]. In such approach the fitting of the vector normalized constituents to the mean spectra of the different pathologies can be performed by linear least-squares analysis with a nonnegative constraint for model fitting (NNLS), according to the following equation:

$$X = cS + E, \tag{1.10}$$

where X is the measured spectra of the tissue, c is the matrix of concentrations to be predicted, and S is the matrix of spectral components. This can be used to provide a linear "best fit" of the spectral components with minimum residuals. Here E gives the error or residual, which can be mainly attributed to the noise in the measured signal. The NNLS model presumes that the Raman spectra measured from the tissue is a linear combination of its biochemical components' spectra and the signal intensity scales linearly with the relative concentration of biochemical components in the tissue. The biochemical model determines the relative concentration profiles of major tissue biochemical constituents responsible for prominent tissue Raman spectral features and its changes associated with disease progression. If components are not included the omitted variable bias can introduce some errors in the fit [375]. Observation of the residual E enables the quality of the fit to be observed and any remaining features of the spectra to be included in the next iteration of the model.

One important factor that may influence the performance of the model is a collinearity of the basis spectra. Collinearity is a common issue in linear regression that can lead to an unstable result [123]. Any collinearity in the components selected will skew the fit. An example being amino acids and the proteins containing them being used in the same model. Hence, the collinearity coefficients of the basis components must be calculated:

$$R = \frac{x^T y}{\left(x^T x\right)\left(y^T y\right)}, \tag{1.11}$$

where x and y are any two component spectra and T indicates the transpose of the respective spectra. The orthogonality matrix represents the degree of orthogonality between the components. If the orthogonality value is zero, then the two components are orthogonal, and if the orthogonality value is one, then the two components are identical. For instance, DNA and RNA have an orthogonality value close to 1. As usual, this equation is used for the initial evaluation of the model components.

The choices of biochemical substances used in the model are mainly based on their known presence in the correspondent tissue, and the contributions they would give to the observed tissue spectrum. For example, Stone et al. [123] diagnosed bladder and prostate cancer by quantifying differences in actin, collagen, choline, triolein, oleic acid, cholesterol, and DNA, assessing the gross biochemical changes in each pathology. De Jong et al. [313] demonstrated that spectra from normal bladder tissue showed a higher collagen content, while spectra from tumor tissue were characterized by higher lipid, nucleic acid, protein, and glycogen. Haka et al. [122] used a biochemical model to correlate changes in the amounts of fat (adipocytes), collagen, cholesterol, and calcium oxalate in the cell nucleus and cytoplasm, aiming at breast cancer diagnosis *in vivo*.

Huang et al. [169] developed biochemical model for effective gastric cancer diagnosis, including eight reference tissue constituents (actin, albumin, collagen type I, DNA, histones, triolein, pepsinogen, and phosphatidylcholine). The authors showed that albumin, nucleic acid, phospholipids, and histones were found to be the most significant features for diagnosing the epithelial neoplasia of the stomach, giving rise to an overall accuracy of 93.7%.

In 2011, Bergholt et al. [167] based on NNLS analyses of over 35 basis reference Raman spectra obtained from different biomolecules associated with GI tissue (e.g., actin, albumin, pepsin, pepsinogen, B-NADH, RNA, DNA, myosin, hemoglobin, collagen I, collagen II, collagen V, mucin 1, mucin 2, mucin 3, flavins, elastin, phosphatidylcholine, cholesterol, glucose, glycogen, triolein, histones, beta-carotene, etc.) showed that the following five biochemicals, i.e., actin, histones, collagen type I, DNA, and triolein were the most significant Raman-active biochemical constituents that could effectively characterize gastric and esophageal tissue with very small fit-residuals. For instance, DNA represented nucleic acids within the cell nucleus; triolein represented typical lipid signals; actin and histones resembled proteins of different conformations and were the major components of the cytoskeleton and chromatin, respectively, whereas collagen type I was a substantial part of the extracellular matrix [95, 123, 169]. In a follow-up study [307] the authors added glycogen to the model, which was present in the squamous epithelium.

The comparisons of the mean *in vivo* measured Raman spectra and the reconstructed Raman spectra of different normal and cancerous tissues are shown in Fig. 1.7 and indicate good fitting (with residual less than 10%) [167]. The diagnostic sensitivity of GI cancer was 97.0%, and the specificity was 95.2% [177, 316].

The six-component biochemical model showed that neoplastic tissue was mainly associated with a decrease in actin, collagen, lipids, and glycogen, while an increase in DNA and histones concentration [316]. More specifically, it was found a significant increased fit coefficient of DNA (highly related to the Raman peak at 1335 cm^{-1}) and histones (associated with amide III at 1265 cm^{-1} and amide I (C=O) stretching vibration at 1655 cm^{-1}), the relative reduction in actin, which represented a major part of the Raman signal originating from cell cytoplasm. This reflected the increase in the nuclear-to-cytoplasm ratio of neoplastic cells, which was a well-established qualitative indicator of malignancies used by pathologists. The fit coefficient of collagen representing the extracellular matrix was noted to be lower for cancerous as compared to normal tissue. The model also revealed a considerable decrease in fit coefficients of lipids (associated with Raman peaks at 1078, 1302, 1445, 1745 cm^{-1}) and a reduction in glycogen in cancer tissue related to the abnormal glucose metabolism in cancer cells [372]. Incorporation of significant biochemical fit coefficients in LDA-DA statistical analysis provided an inherent separation of different tissue types based on the biomolecular information.

Feng et al. [222] designed skin biochemical model with eight primary model components: collagen, elastin, triolein, nucleus, keratin, ceramide, melanin, and water, which were collected from human skin *in situ* and were averaged over multiple patients. Those components contained both biochemical and structural

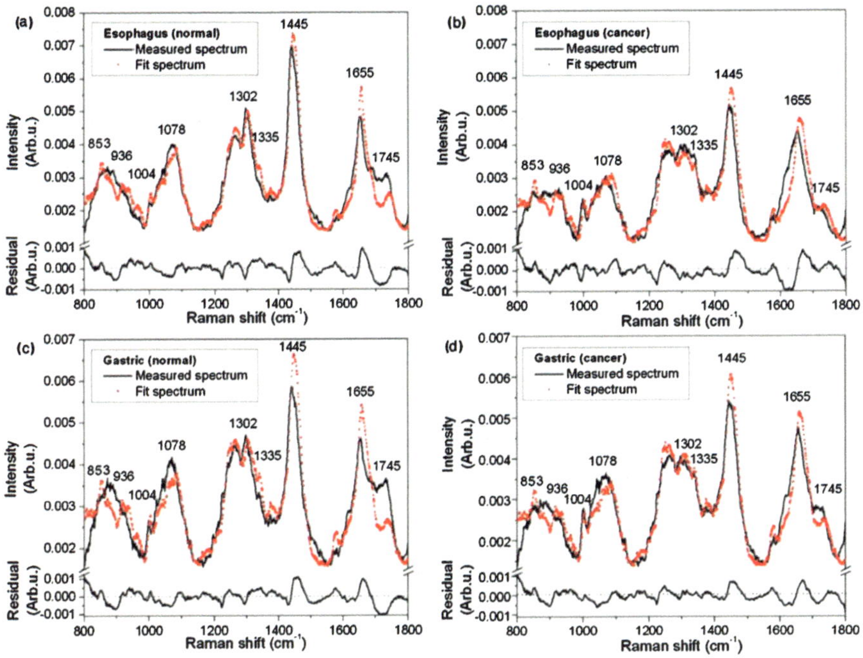

Fig. 1.7 Comparison of the *in vivo* Raman spectra measured with the reconstructed tissue Raman spectra through the employment of the five basis reference Raman spectra: (**a**) normal esophagus, (**b**) esophageal cancer, (**c**) normal gastric, (**d**) gastric cancer. Residuals (measured spectrum minus fit spectrum) are also shown in each plot. Reprinted with permission from [167]

information. For instance, nucleus referred to the nuclear material in the cell. Collagen and elastin referred to dermal extracellular matrix. Keratin represented epidermal extracellular matrix. Triolein mainly represented subcutaneous fat. The fit coefficients provided the relative concentration of those components and were used as the input variables of the discriminant analysis. The authors showed that a biophysical model could achieve consistent diagnostic performance with the statistical model while simultaneously extracting the relevant biomarkers accounting for the diagnosis [121, 226].

In general, Raman spectral biochemical modeling in conjunction with linear discriminant analysis shows good classification results and also provides new insights into biochemical origins of Raman spectroscopy for tissue diagnosis and characterization.

1.5 Summary

Numerous experimental studies have shown the capability of Raman spectroscopy for malignant lesion detection and grading based on objective and quantifiable molecular information. Consequently, the use of this technique can reduce the number of unnecessary biopsies and guide the precise tumor margin detection improving the surgical outcome of the patients.

Despite the advantages that the Raman spectroscopy can offer, there are some challenges existing in cancer diagnosis with Raman spectroscopy. First, it has low measurement speed. One way to overcome this is to complement Raman spectroscopy with other techniques, such as autofluorescence imaging and OCT. Image-guided Raman spectroscopy substantially reduces the time spent on redundant or non-relevant Raman measurements.

The translation for clinical use involves the development of comprehensive spectral databases and tissue classification methodologies, which can draw effective diagnostic information from usually overlapping Raman spectra with subtle spectral differences between neoplastic and normal tissues. Validation studies need to be performed to confirm that classification diagnostic algorithms developed on *ex vivo* specimens are applicable to *in vivo* tissues. Deep learning training, using large numbers of spectra, can also discriminate different cancer types and become predictors of the aggressiveness of the cancer. Combination of statistical and biochemical models can lead to new classification methodologies that can be comparable with current gold standards.

Recently optical properties of many kinds of tumor have been investigated spectroscopically using integrating sphere spectroscopy and reflectance spectroscopy methods. From the analysis of the above presented tables we can conclude that during tumor development both absorption and scattering properties of tumor tissues increase (as a rule) in comparison with normal tissue. Hemoglobin and water content increase, whereas oxygenation degree does not change or changes insignificantly. The presented data can be used for the development of novel methods of cancer diagnostics and treatment.

Acknowledgments The work was supported by the Russian Foundation of Basic Research grant # 18-52-16025 (A. N. Bashkatov and E. A. Genina) and grant # 19-52-18001 Bolg_a (V. P. Zakharov), by Bulgarian Science Fund grant # KP06-Russia/19/28.09.2019 (E.G. Borisova) and by the Government of the Russian Federation grant # 14.W03.31.0023 (V. V. Tuchin).

Disclosure The chapter was written through the contribution of all the authors. All the authors declare that there is no conflict of interest for this chapter.

References

1. Bray, F., Ferlay, J., Soerjomataram, I., et al.: Global cancer statistics 2018: GLOBOCAN estimates of incidence and mortality worldwide for 36 cancers in 185 countries. CA Cancer J. Clin. **68**, 394–424 (2018)
2. Bashkatov, A.N., Genina, E.A., Tuchin, V.V.: Tissue optical properties. In: Boas, D.A., Pitris, C., Ramanujam, N. (eds.) Handbook of Biomedical Optics, pp. 67–100. Taylor & Francis Group LLC/CRC Press Inc, Boca Raton (2011)
3. Cheong, W.F., Prahl, S.A., Welch, A.J.: A review of the optical properties of biological tissue. IEEE J. Quantum Electron. **26**(12), 2166–2185 (1990)
4. Vo-Dinh, T. (ed.): Biomedical Photonics Handbook: Biomedical Diagnostics, 2nd edn. CRC Press, Boca Raton (2019)
5. Tuchin, V.V.: Tissue Optics: Light Scattering Methods and Instruments for Medical Diagnosis, 3rd edn. SPIE Press, Washington (2015)
6. Welch, A.J., van Gemert, M.J.C. (eds.): Optical-Thermal Response of Laser-Irradiated Tissue, 2nd edn. Springer, New York (2011)
7. Mishchenko, M.I., Travis, L.D., Lacis, A.A.: Scattering, Absorption, and Emission of Light by Small Particles. Cambridge University Press, Cambridge (2002)
8. Henyey, L.G., Greenstein, J.L.: Diffuse radiation in the galaxy. Astrophys. J. **93**, 70–83 (1941)
9. Ishimaru, A.: Electromagnetic Wave Propagation, Radiation, and Scattering. From Fundamentals to Applications, 2nd edn. Wiley, Hoboken (2017)
10. Borovoi, A.G., Naats, E.I., Oppel, U.G.: Scattering of light by a red blood cell. J. Biomed. Opt. **3**(3), 364–372 (1998)
11. Turcu, I.: Effective phase function for light scattered by disperse systems – the small-angle approximation. J. Opt. A Pure Appl. Opt. **6**, 537–543 (2004)
12. Bohren, C.F., Huffman, D.R.: Absorption and Scattering of Light by Small Particles. Wiley, New York (1998)
13. Canpolat, M., Mourant, J.R.: High-angle scattering events strongly affect light collection in clinically relevant measurement geometries for light transport through tissue. Phys. Med. Biol. **45**, 1127–1140 (2000)
14. Vaudelle, F., L'Huillier, J.P., Askoura, M.L.: Light source distribution and scattering phase function influence light transport in diffuse multi-layered media. Opt. Commun. **392**, 268–281 (2017)
15. Yaroslavsky, A.N., Yaroslavsky, I.V., Goldbach, T., et al.: Influence of the scattering phase function approximation on the optical properties of blood determined from the integrating sphere measurements. J. Biomed. Opt. **4**(1), 47–53 (1999)
16. Chai, C., Chen, Y., Li, P., et al.: Improved steady-state diffusion approximation with an anisotropic point source and the δ-Eddington phase function. Appl. Opt. **46**(21), 4843–4851 (2007)
17. Cong, W., Shen, H., Cong, A.X., et al.: Integral equations of the photon fluence rate and flux based on a generalized Delta-Eddington phase function. J. Biomed. Opt. **13**(2), 024016 (2008)
18. Reynolds, L.O., McCormick, N.J.: Approximate two-parameter phase function for light scattering. J. Opt. Soc. Am. **70**(10), 1206–1212 (1980)
19. Friebel, M., Roggan, A., Muller, G., et al.: Determination of optical properties of human blood in the spectral range 250 to 1100 nm using Monte Carlo simulations with hematocrit-dependent effective scattering phase functions. J. Biomed. Opt. **11**(3), 034021 (2006)
20. Kienle, A., Patterson, M.S., Ott, L., et al.: Determination of the scattering coefficient and the anisotropy factor from laser Doppler spectra of liquids including blood. Appl. Opt. **35**(19), 3404–3412 (1996)

21. Hammer, M., Yaroslavsky, A.N., Schweitzer, D.: A scattering phase function for blood with physiological haematocrit. Phys. Med. Biol. **46**, N65–N69 (2001)
22. Sharma, S.K., Banerjee, S.: Role of approximate phase functions in Monte Carlo simulation of light propagation in tissues. J. Opt. A Pure Appl. Opt. **5**, 294–302 (2003)
23. Chamot, S., Migacheva, E., Seydoux, O., et al.: Physical interpretation of the phase function related parameter γ studied with a fractal distribution of spherical scatterers. Opt. Express. **18**(23), 23664–23675 (2010)
24. Calabro, K.W., Bigio, I.J.: Influence of the phase function in generalized diffuse reflectance models: review of current formalisms and novel observations. J. Biomed. Opt. **19**(7), 075005 (2014)
25. Kortun, C., Hijazi, Y.R., Arifler, D.: Combined Monte Carlo and finite-difference time-domain modeling for biophotonic analysis: implications on reflectance-based diagnosis of epithelial precancer. J. Biomed. Opt. **13**(3), 034014 (2008)
26. Clare, J.F.: Comparison of four analytic methods for the calculation of irradiance in integrating spheres. J. Opt. Soc. Am. A. **15**(12), 3086–3096 (1998)
27. Hanssen, L.: Integrating-sphere system and method for absolute measurement of transmittance, reflectance, and absorbance of specular samples. Appl. Opt. **40**(19), 3196–3204 (2001)
28. Hisdal, B.: Correction of directional reflectance measured in the integrating sphere. Opt. Acta. **15**(2), 139–151 (1968)
29. Jacquez, J.A., Kuppenheim, H.F.: Theory of the integrating sphere. J. Opt. Soc. Am. **45**(6), 460–470 (1955)
30. Jernshoj, K.D., Hassing, S.: Analysis of reflectance and transmittance measurements on absorbing and scattering small samples using a modified integrating sphere setup. Appl. Spectrosc. **63**(8), 879–888 (2009)
31. Pickering, J.W., Moes, C.J.M., Sterenborg, H.J.C.M., et al.: Two integrating spheres with an intervening scattering sample. J. Opt. Soc. Am. A. **9**(4), 621–631 (1992)
32. Pickering, J.W., Prahl, S.A., van Wieringen, N., et al.: Double-integrating-sphere system for measuring the optical properties of tissue. Appl. Opt. **32**(4), 399–410 (1993)
33. Yaroslavsky, I.V., Yaroslavsky, A.N., Goldbach, T., et al.: Inverse hybrid technique for determining the optical properties of turbid media from integrating-sphere measurements. Appl. Opt. **35**(34), 6797–6809 (1996)
34. Sassaroli, A., Martelli, F.: Equivalence of four Monte Carlo methods for photon migration in turbid media. J. Opt. Soc. Am. A. **29**(10), 2110–2117 (2012)
35. Jacques SL (2011) Monte Carlo modeling of light transport in tissue (steady state and time of flight). In: Welch AJ, van Gemert MJC (eds) Optical-Thermal Response of Laser-Irradiated Tissue, 2nd edn. Springer, New York p 109-144
36. Honda, N., Ishii, K., Terada, T., et al.: Determination of the tumor tissue optical properties during and after photodynamic therapy using inverse Monte Carlo method and double integrating sphere between 350 and 1000 nm. J. Biomed. Opt. **16**(5), 058003 (2011)
37. Bashkatov, A.N., Genina, E.A., Kochubey, V.I., et al.: Optical properties of human colon tissues in the 350-2500 nm spectral range. Quantum Electr. **44**(8), 779–784 (2014)
38. Bashkatov, A.N., Genina, E.A., Kozintseva, M.D., et al.: Optical properties of peritoneal biological tissues in the spectral range of 350-2500 nm. Opt. Spectrosc. **120**(1), 1–8 (2016)
39. Salomatina, E., Jiang, B., Novak, J., et al.: Optical properties of normal and cancerous human skin in the visible and near-infrared spectral range. J. Biomed. Opt. **11**(6), 064026 (2006)
40. Bashkatov, A.N., Genina, E.A., Tuchin, V.V.: Optical properties of skin, subcutaneous, and muscle tissues: a review. J. Innov. Opt. Health Sci. **4**(1), 9–38 (2011)
41. Genina, E.A., Bashkatov, A.N., Tuchina, D.K., et al.: Optical properties of brain tissues at the different stages of glioma development in rats: pilot study. Biomed. Opt. Express. **10**(10), 5182–5197 (2019)
42. Prahl, S.A., van Gemert, M.J.C., Welch, A.J.: Determining the optical properties of turbid media by using the adding-doubling method. Appl. Opt. **32**(4), 559–568 (1993)

43. Gebhart, S.C., Lin, W.-C., Mahadevan-Jansen, A.: *In vitro* determination of normal and neoplastic human brain tissue optical properties using inverse adding-doubling. Phys. Med. Biol. **51**, 2011–2027 (2006)
44. Sardar, D.K., Yust, B.G., Barrera, F.J., et al.: Optical absorption and scattering of bovine cornea, lens and retina in the visible region. Lasers Med. Sci. **24**(6), 839–847 (2009)
45. Zhang, Y., Chen, Y., Yu, Y., et al.: Visible and near-infrared spectroscopy for distinguishing malignant tumor tissue from benign tumor and normal breast tissues *in vitro*. J. Biomed. Opt. **18**(7), 077003 (2013)
46. Belikov, A.V., Zagorul'ko, A.M., Smirnov, S.N., et al.: Optical properties of human eye cataractous lens *in vitro* in the visible and near-IR ranges of the spectrum. Opt. Spectrosc. **126**(5), 574–579 (2019)
47. Carneiro, I., Carvalho, S., Henrique, R., et al.: Measuring optical properties of human liver between 400 and 1000 nm. Quantum Electr. **49**(1), 13–19 (2019)
48. Genin, V.D., Genina, E.A., Bucharskaya, A.B., et al.: Investigation of the change of tumor optical properties after laser-induced plasmon-resonant photothermal treatment of transplanted tumors in rats. Proc. SPIE. **10716**, 107160 (2018)
49. Soleimanzad, H., Gurden, H., Pain, F.: Optical properties of mice skull bone in the 455- to 705-nm range. J. Biomed. Opt. **22**(1), 010503 (2017)
50. Yust, B.G., Minum, L.C., Sardar, D.K.: Optical absorption and scattering of bovine cornea, lens, and retina in the near-infrared region. Lasers Med. Sci. **27**(2), 413–422 (2012)
51. Zamora-Rojas, E., Aernouts, B., Garrido-Varo, A., et al.: Optical properties of pig skin epidermis and dermis estimated with double integrating spheres measurements. Innov. Food Sci. Emerg. Technol. **20**, 343–349 (2013)
52. Prahl, S.A.: The diffusion approximation in three dimensions. In: Welch, A.J., van Gemert, M.J.C. (eds.) Optical-Thermal Response of Laser-Irradiated Tissue, pp. 207–231. Plenum Press, New York (1995)
53. Cilesiz, I.F., Welch, A.J.: Light dosimetry: effects of dehydration and thermal damage on the optical properties of the human aorta. Appl. Opt. **32**(4), 477–487 (1993)
54. Maitland, D.J., Walsh Jr., J.T., Prystowsky, J.B.: Optical properties of human gallbladder tissue and bile. Appl. Opt. **32**(4), 586–591 (1993)
55. Marchesini, R., Clemente, C., Pignoli, E., et al.: Optical properties of *in vitro* epidermis and their possible relationship with optical properties of *in vivo* skin. J. Photochem. Photobiol. B. **16**(2), 127–140 (1992)
56. Marchesini, R., Pignoli, E., Tomatis, S., et al.: *Ex vivo* optical properties of human colon tissue. Lasers Surg. Med. **15**, 351–357 (1994)
57. van de Hulst, H.C.: Multiple Light Scattering. Tables, Formulas and Applications. Academic Press, New York (1980)
58. Prahl, S.A.: Light transport in tissue. Ph.D. Thesis, University of Texas, Austin (1988)
59. Bashkatov, A.N., Genina, E.A., Kochubey, V.I., et al.: Optical properties of human stomach mucosa in the spectral range from 400 to 2000 nm: prognosis for gastroenterology. Med. Laser Appl. **22**, 95–104 (2007)
60. Dam, J.S., Dalgaard, T., Fabricius, P.E., et al.: Multiple polynomial regression method for determination of biomedical optical properties from integrating sphere measurements. Appl. Opt. **39**, 1202–1209 (2000)
61. Graaff, R., Koelink, M.H., de Mul, F.F.M., et al.: Condensed Monte Carlo simulations for the description of light transport. Appl. Opt. **32**, 426–434 (1993)
62. Wang, L., Jacques, S.L., Zheng, L.: MCML - Monte Carlo modeling of light transport in multi-layered tissues. Comp. Methods Progr. Biomed. **47**, 131–146 (1995)
63. Jacques, S.L., Wang, L.: Monte Carlo modeling of light transport in tissue. In: Welch, A.J., van Gemert, M.J.C. (eds.) Optical-Thermal Response of Laser-Irradiated Tissue, pp. 73–100. Plenum Press, New York (1995)

64. Ugryumova, N., Matcher, S.L., Attenburrow, D.P.: Measurement of bone mineral density via light scattering. Phys. Med. Biol. **49**, 469–483 (2004)
65. Hammer, M., Roggan, A., Schweitzer, D., et al.: Optical properties of ocular fundus tissues - an *in vitro* study using the double-integrating-sphere technique and inverse Monte Carlo simulation. Phys. Med. Biol. **40**, 963–978 (1995)
66. Friebel, M., Helfmann, J., Netz, U., et al.: Influence of oxygen saturation on the optical scattering properties of human red blood cells in the spectral range 250 to 2000 nm. J. Biomed. Opt. **14**, 034001 (2009)
67. van Gemert, M.J.C., Jacques, S.L., Sterenborg, H.J.C.M., et al.: Skin optics. IEEE Trans. Biomed. Eng. **36**, 1146–1154 (1989)
68. Einstein, G., Aruna, P.R., Ganesan, S.: Monte Carlo based model for diffuse reflectance from turbid media for the diagnosis of epithelial dysplasia. Optik. **181**, 828–835 (2019)
69. Hennessy, R., Lim, S.L., Markey, M.M., et al.: Monte Carlo lookup table-based inverse model for extracting optical properties from tissue-simulating phantoms using diffuse reflectance spectroscopy. J. Biomed. Opt. **18**(3), 0370033 (2013)
70. Hourdakis, C.J., Perris, A.: A Monte Carlo estimation of tissue optical properties for use in laser dosimetry. Phys. Med. Biol. **40**, 351–364 (1995)
71. Wang, Q., Le, D., Ramella-Roman, J., et al.: Broadband ultraviolet-visible optical property measurement in layered turbid media. Biomed. Opt. Express. **3**(6), 1226–1240 (2012)
72. Palmer, G.M., Ramanujam, N.: Monte Carlo-based inverse model for calculating tissue optical properties. Part I: theory and validation on synthetic phantoms. Appl. Opt. **45**(5), 1062–1071 (2006)
73. Palmer, G.M., Ramanujam, N.: Monte Carlo-based inverse model for calculating tissue optical properties. Part II: application to breast cancer diagnosis. Appl. Opt. **45**(5), 1072–1078 (2006)
74. Amouroux, M., Diaz-Ayil, G., Blondel, W.C.P.M., et al.: Classification of ultraviolet irradiated mouse skin histological stages by bimodal spectroscopy: multiple excitation autofluorescence and diffuse reflectance. J. Biomed. Opt. **14**(1), 014011 (2009)
75. Lu, R., van Beers, R., Sayes, W., et al.: Measurement of optical properties of fruits and vegetables: a review. Postharvest Biol. Technol. **159**, 111003 (2020)
76. Baltussen, E.J.M., Snaebjornsson, P., Konig, S.G.B., et al.: Diffuse reflectance spectroscopy as a tool for real-time tissue assessment during colorectal cancer surgery. J. Biomed. Opt. **22**(10), 106014 (2017)
77. Bevilacqua, F., Piguet, D., Marquet, P., et al.: *In vivo* local determination of tissue optical properties: applications to human brain. Appl. Opt. **38**(22), 4939–4950 (1999)
78. Bigio, I.J., Bown, S.G., Briggs, G., et al.: Diagnosis of breast cancer using elastic-scattering spectroscopy: preliminary clinical results. J. Biomed. Opt. **5**(2), 221–228 (2000)
79. Borisova, E., Trojanova, P., Avramov, L.: Reflectance measurements of skin lesions – non-invasive method for diagnostic evaluation of pigmented neoplasia. Proc. SPIE-OSA. **5862**, 58620A (2005)
80. Borisova, E., Genova-Hristova, T.S., Troyanova, P., et al.: Optical UV-VIS-NIR spectroscopy of benign, dysplastic and malignant cutaneous lesions *ex vivo*. Proc. SPIE. **10685**, 106853T (2018)
81. Cugmas, B., Plavec, T., Bregar, M., et al.: Detection of canine skin and subcutaneous tumors by visible and near-infrared diffuse reflectance spectroscopy. J. Biomed. Opt. **20**(3), 037003 (2015)
82. Dhar, S., Lo, J.Y., Palmer, G.M., et al.: A diffuse reflectance spectral imaging system for tumor margin assessment using custom annular photodiode arrays. Biomed. Opt. Express. **3**(12), 3211–3222 (2012)
83. Hamblin, M.R., Avci, P., Gupta, G.K. (eds.): Imaging in Dermatology. Elsevier Inc., London (2016)
84. Boas, D.A., Pitris, C., Ramanujan, N. (eds.): Handbook of Biomedical Optics. CRC Press, Boca Raton (2011)

85. Tuchin, V.V. (ed.): Handbook of Optical Biomedical Diagnostics, 2nd edn. SPIE Press, Bellingham (2016)
86. Wang, L.V., Wu, H.: Biomedical Optics: Principles and Imaging. Wiley, Hoboken (2007)
87. Zakharov, V.P., Timchenko, P.E., Timchenko, E.V., et al.: Backscattering spectroscopy for assessing skin tumor. J. Biomed. Photon Eng. **1**(2), 164–168 (2015)
88. Zonios, G., Dimou, A.: Modeling diffuse reflectance from semi-infinite turbid media: application to the study of skin optical properties. Opt. Express. **14**(19), 8661–8674 (2006)
89. Zonios, G., Dimou, A., Bassukas, I., et al.: Melanin absorption spectroscopy: new method for noninvasive skin investigation and melanoma detection. J. Biomed. Opt. **13**(1), 014017 (2008)
90. Zonios, G., Dimou, A., Carrara, M., et al.: *In vivo* optical properties of melanocytic skin lesions: common nevi, dysplastic and malignant melanoma. Photochem. Photobiol. **86**, 236–240 (2010)
91. Zonios, G.: Investigation of reflectance sampling depth in biological tissues for various common illumination/collection configurations. J. Biomed. Opt. **19**(9), 097001 (2014)
92. Shchyogolev, S.Y.: Inverse problems of spectroturbidimetry of biological disperse systems: an overview. J. Biomed. Opt. **4**(4), 490–503 (1999)
93. Press, W.H., Tuekolsky, S.A., Vettering, W.T., et al.: Numerical Recipes in C: The Art of Scientific Computing. Cambridge University Press, Cambridge (1992)
94. Movasaghi, Z., Rehman, S., Rehman, I.U.: Raman spectroscopy of biological tissues. Appl. Spectrosc. Rev. **42**(5), 493–541 (2007)
95. Mahadevan-Jansen, A., Richards-Kortum, R.R.: Raman spectroscopy for the detection of cancers and precancers. J. Biomed. Opt. **1**(1), 31–70 (1996)
96. Pacia, M.Z., Mateuszuk, L., Buczek, E., et al.: Rapid biochemical profiling of endothelial dysfunction in diabetes, hypertension and cancer metastasis by hierarchical cluster analysis of Raman spectra. J. Raman Spectrosc. **47**, 1310–1317 (2016)
97. Chan, J.W., Taylor, D.S., Zwerdling, T., et al.: Micro-Raman spectroscopy detects individual neoplastic and normal hematopoietic cells. Biophys. J. **90**(2), 648–656 (2005)
98. Shafer-Peltier, K.E., Haka, A.S., Fitzmaurice, M., et al.: Raman microspectroscopic model of human breast tissue: implications for breast cancer diagnosis *in vivo*. J. Raman Spectrosc. **33**(7), 552–563 (2002)
99. Cicerone, M.T., Camp, C.H.: Histological coherent Raman imaging: a prognostic review. Analyst. **143**(1), 33–59 (2017)
100. Pence, I.J., Mahadevan-Jansen, A.: Clinical instrumentation and applications of Raman spectroscopy. Chem. Soc. Rev. **45**(7), 1958–1979 (2016)
101. Cordero, E., Latka, I., Matthäus, C., et al.: *In-vivo* Raman spectroscopy: from basics to applications. J. Biomed. Opt. **23**(7), 071210 (2018)
102. Krafft, C., Schie, I.W., Meyer, T., et al.: Developments in spontaneous and coherent Raman scattering microscopic imaging for biomedical applications. Chem. Soc. Rev. **45**(7), 1819–1849 (2016)
103. Kendall, C., Day, J., Hutchings, J., et al.: Evaluation of Raman probe for oesophageal cancer diagnostics. Analyst. **135**(12), 3038–3041 (2010)
104. Knipfer, C., Motz, J., Adler, W., et al.: Raman difference spectroscopy: a non-invasive method for identification of oral squamous cell carcinoma. Biomed. Opt. Express. **5**(9), 3252–3265 (2014)
105. Mosier-Boss, P.A., Lieberman, S.H., Newbery, R.: Fluorescence rejection in Raman spectroscopy by shifted-spectra, edge detection, and FFT filtering techniques. Appl. Spectrosc. **49**, 630–638 (1995)
106. Cai, T.T., Zhang, D., Ben-Amotz, D.: Enhanced chemical classification of Raman images using multiresolution wavelet transformation. Appl. Spectrosc. **55**, 1124–1130 (2001)
107. Lieber, C.A., Mahadevan-Jansen, A.: Automated method for subtraction of fluorescence from biological Raman spectra. Appl. Spectrosc. **57**, 1363–1367 (2003)

108. Zhao, J., Lui, H., McLean, D.I., et al.: Automated autofluorescence background subtraction algorithm for biomedical Raman spectroscopy. Appl. Spectrosc. **61**(11), 1225–1232 (2007)
109. Utzinger, U., Heintzelman, D.L., Mahadevan-Jansen, A., et al.: Near-infrared Raman spectroscopy for *in vivo* detection of cervical precancers. Appl. Spectrosc. **55**, 955–959 (2001)
110. Edgell, W.F., Schmidlin, E., Balk, M.W.: A computer-spectrometer interactive system for infrared spectroscopy. Appl. Spectrosc. **34**(4), 420–434 (1980)
111. Zakharov, V.P., Bratchenko, I.A., Artemyev, D.N., et al.: Comparative analysis of combined spectral and optical tomography methods for detection of skin and lung cancers. J. Biomed. Opt. **20**(2), 025003 (2015)
112. Song L-M, W.K., Molckovsky, A., Wang, K.K., et al.: Diagnostic potential of Raman spectroscopy in Barrett's esophagus. Proc. SPIE. **5692**, 140–146 (2005)
113. Devpura, S., Thakur, J.S., Sethi, S., et al.: Diagnosis of head and neck squamous cell carcinoma using Raman spectroscopy: tongue tissues. J. Raman Spectrosc. **43**, 490–496 (2012)
114. Talari, A.C.S., Evans, C.A., Holen, I., et al.: Raman spectroscopic analysis differentiates between breast cancer cell lines. J. Raman Spectrosc. **46**, 421–427 (2015)
115. Talari, A.C.S., Movasaghi, Z., Rehman, S., et al.: Raman spectroscopy of biological tissues. Appl. Spectrosc. Rev. **50**(1), 46–111 (2015)
116. Saha, A., Barman, I., Dingari, N.C., et al.: Raman spectroscopy: a real-time tool for identifying microcalcifications during stereotactic breast core needle biopsies. Biomed. Opt. Express. **2**, 2792–2803 (2011)
117. Haka, A.S., Volynskaya, Z., Gardecki, J.A., et al.: Diagnosing breast cancer using Raman spectroscopy: prospective analysis. J. Biomed. Opt. **14**(5), 054023 (2009)
118. Duraipandian, S., Zheng, W., Ng, J., et al.: Effect of hormonal variation on *in vivo* high wavenumber Raman spectra improves cervical precancer detection. Proc. SPIE. **8214**, 82140A (2012)
119. Bergholt, M.S., Lin, K., Wang, J., et al.: Simultaneous fingerprint and high-wavenumber fiber-optic Raman spectroscopy enhances real-time *in vivo* diagnosis of adenomatous polyps during colonoscopy. J. Biophotonics. **9**, 333–342 (2016)
120. Oliveira, A.F., Santos, I.D.A.O., Cartaxo, S.B., et al.: Differential diagnosis in primary and metastatic cutaneous melanoma by FT-Raman spectroscopy. Acta Cir. Bras. **25**(5), 434–439 (2010)
121. Feng, X., Moy, A.J., Nguyen, H.T.M., et al.: Raman biophysical markers in skin cancer diagnosis. J. Biomed. Opt. **23**(5), 057002 (2018)
122. Haka, A.S., Shafer-Peltier, K.E., Fitzmaurice, M., et al.: Diagnosing breast cancer by using Raman spectroscopy. Proc. Natl. Acad. Sci. **102**(35), 12371–12376 (2005)
123. Stone, N., Prieto, M.C.H., Crow, P., et al.: The use of Raman spectroscopy to provide an estimation of the gross biochemistry associated with urological pathologies. Anal. Bioanal. Chem. **387**(5), 1657–1668 (2007)
124. Stone, N., Stavroulaki, P., Kendall, C., et al.: Raman spectroscopy for early detection of laryngeal malignancy: preliminary results. Laryngoscope. **110**, 1756–1763 (2000)
125. Gniadecka, M., Philipsen, P.A., Wessel, S., et al.: Melanoma diagnosis by Raman spectroscopy and neural networks: structure alterations in proteins and lipids in intact cancer tissue. J. Invest. Dermatol. **122**(2), 443–449 (2004)
126. Rehman, S., Movasaghi, Z., Tucker, A.T., et al.: Raman spectroscopic analysis of breast cancer tissues: identifying differences between normal, invasive ductal carcinoma and ductal carcinoma *in situ* of the breast tissue. J. Raman Spectrosc. **38**, 1345–1351 (2007)
127. Nijssen, A., Maquelin, K., Santos, L.F., et al.: Discriminating basal cell carcinoma from perilesional skin using high wave-number Raman spectroscopy. J. Biomed. Opt. **12**(3), 034004 (2007)
128. Crow, P., Barrass, B., Kendall, C., et al.: The use of Raman spectroscopy to differentiate between different prostatic adenocarcinoma cell lines. Br. J. Cancer. **92**(12), 2166–2170 (2005)

129. Huang, Z., McWilliams, A., Lui, H., et al.: Near-infrared Raman spectroscopy for optical diagnosis of lung cancer. Int. J. Cancer. **107**, 1047–1052 (2003)
130. Kaminaka, S., Ito, T., Yamazaki, H., et al.: Near-infrared multichannel Raman spectroscopy toward real-time *in vivo* cancer diagnosis. J. Raman Spectrosc. **33**, 498–502 (2002)
131. Koljenović, S., Choo-Smith, L.P., Schut, T.C.B., et al.: Discriminating vital tumor from necrotic tissue in human glioblastoma tissue samples by Raman spectroscopy. Lab. Investig. **82**, 1265–1277 (2002)
132. Pence, I.J., Patil, C.A., Lieber, C.A., et al.: Discrimination of liver malignancies with 1064 nm dispersive Raman spectroscopy. Biomed. Opt. Express. **6**(8), 2724–2737 (2015)
133. Krishnapuram, B., Carin, L., Figueiredo, M.A.T., et al.: Sparse multinomial logistic regression: fast algorithms and generalization bounds. IEEE Trans. Pattern Anal. Mach. Intell. **27**(6), 957–968 (2005)
134. Farrés, M., Platikanov, S., Tsakovski, S., et al.: Comparison of the variable importance in projection (VIP) and of the selectivity ratio (SR) methods for variable selection and interpretation. J. Chemom. **29**, 528–536 (2015)
135. Brereton, R.G., Lloyd, G.R.: Support vector machines for classification and regression. Analyst. **135**, 230–267 (2010)
136. González-Solís, J.: Discrimination of different cancer types clustering Raman spectra by a super paramagnetic stochastic network approach. PLoS ONE. **14**(3), e0213621 (2019)
137. Hess, K.R., Abbruzzese, M.C., Lenzi, R., et al.: Classification and regression tree analysis of 1000 consecutive patients with unknown primary carcinoma. Clin. Cancer Res. **5**, 3403–3410 (1999)
138. Garzotto, M., Beer, T.M., Hudson, R.G., et al.: Improved detection of prostate cancer using classification and regression tree analysis. J. Clin. Oncol. **23**(19), 4322–4329 (2005)
139. Gromski, P.S., Muhamadali, H., Ellis, D.I., et al.: A tutorial review: Metabolomics and partial least squares-discriminant analysis – a marriage of convenience or a shotgun wedding. Anal. Chim. Acta. **879**, 10–23 (2015)
140. Khristoforova, Y.A., Bratchenko, I.A., Myakinin, O.O., et al.: Portable spectroscopic system for *in vivo* skin neoplasms diagnostics by Raman and autofluorescence analysis. J. Biophotonics. **12**(4), e201800400 (2019)
141. Lyng, F.M., Traynor, D., Nguyen, T.N.Q., et al.: Discrimination of breast cancer from benign tumours using Raman spectroscopy. PLoS ONE. **14**(2), e0212376 (2019)
142. Chen, K., Qin, Y., Zheng, F., et al.: Diagnosis of colorectal cancer using Raman spectroscopy of laser-trapped single living epithelial cells. Opt. Lett. **31**, 2015–2017 (2006)
143. Burdall, S.E., Hanby, A.M., Lansdown, M.R., et al.: Breast cancer cell lines: friend or foe? Breast Cancer Res. **5**(2), 89–95 (2003)
144. Koljenović, S., Schut, T.B., Vincent, A., et al.: Detection of meningioma in dura mater by Raman spectroscopy. Anal. Chem. **77**(24), 7958–7965 (2005)
145. Gniadecka, M., Nielsen, O.F., Christensen, D.H., et al.: Structure of water, proteins, and lipids in intact human skin, hair, and nail. J. Invest. Dermatol. **110**(4), 393–398 (1998)
146. Kline, N.J., Treado, P.J.: Raman chemical imaging of breast tissue. J. Raman Spectrosc. **28**, 119–124 (1997)
147. Sato, H., Yamamoto, Y.S., Maruyama, A., et al.: Raman study of brain functions in live mice and rats: a pilot study. Vib. Spectrosc. **50**(1), 125–130 (2009)
148. Short, M.A., Lam, S., McWilliams, A., et al.: Development and preliminary results of an endoscopic Raman probe for potential *in vivo* diagnosis of lung cancers. Opt. Lett. **33**, 711–713 (2008)
149. Cheng, W., Liu, M., Liu, H., et al.: Micro-Raman spectroscopy used to identify and grade human skin pilomatrixoma. Microsc. Res. Tech. **68**, 75–79 (2005)
150. Oshima, Y., Shinzawa, H., Takenaka, T., et al.: Discrimination analysis of human lung cancer cells associated with histological type and malignancy using Raman spectroscopy. J. Biomed. Opt. **15**(1), 017009 (2010)

151. Guo, J., Du, B., Qian, M., et al.: Raman spectroscopic identification of normal and malignant hepatocytes. Chin. Opt. Lett. **7**, 60–63 (2009)
152. Krishna, C.M., Sockalingum, G.D., Kegelaer, G., et al.: Micro-Raman spectroscopy of mixed cancer cell populations. Vib. Spectrosc. **38**, 95–100 (2005)
153. Gala de Pablo, J., Armistead, F.J., Peyman, S.A., et al.: Biochemical fingerprint of colorectal cancer cell lines using label-free live single-cell Raman spectroscopy. J. Raman Spectrosc. **49**, 1323–1332 (2018)
154. Molckovsky, A., Song, L.M., Shim, M.G., et al.: Diagnostic potential of near-infrared Raman spectroscopy in the colon: differentiating adenomatous from hyperplastic polyps. Gastrointest. Endosc. **57**(3), 396–402 (2003)
155. Short, M.A., Lam, S., McWilliams, A.M., et al.: Using laser Raman spectroscopy to reduce false positives of autofluorescence bronchoscopies: a pilot study. J. Thorac. Oncol. **6**(7), 1206–1214 (2011)
156. McGregor, H.C., Short, M.A., McWilliams, A., et al.: Real-time endoscopic Raman spectroscopy for *in vivo* early lung cancer detection. J. Biophotonics. **10**, 98–110 (2016)
157. Abramczyk, H., Brozek-Pluska, B., Surmacki, J., et al.: The label-free Raman imaging of human breast cancer. J. Mol. Liq. **164**(1–2), 123–131 (2011)
158. Abramczyk, H., Brozek-Pluska, B., Surmacki, J., et al.: Raman "optical biopsy" of human breast cancer. Prog. Biophys. Mol. Biol. **108**(1–2), 74–81 (2012)
159. Barman, I., Dingari, N.C., Saha, A., et al.: Application of Raman spectroscopy to identify microcalcifications and underlying breast lesions at stereotactic core needle biopsy. Cancer Res. **73**, 3206–3215 (2013)
160. Haka, A.S., Shafer-Peltier, K.E., Fitzmaurice, M., et al.: Identifying microcalcifications in benign and malignant breast lesions by probing differences in their chemical composition using Raman spectroscopy. Cancer Res. **62**, 5375–5380 (2002)
161. Brożek-Płuska, B., Placek, I., Kurczewski, K., et al.: Breast cancer diagnostics by Raman spectroscopy. J. Mol. Liq. **141**(3), 145–148 (2008)
162. Li, Q., Gao, Q., Zhang, G.: Classification for breast cancer diagnosis with Raman spectroscopy. Biomed. Opt. Express. **5**(7), 2435–2445 (2014)
163. Li, S., Chen, G., Zhang, Y., et al.: Identification and characterization of colorectal cancer using Raman spectroscopy and feature selection techniques. Opt. Express. **22**, 25895–25908 (2014)
164. Li, Q., Hao, C., Kang, X., et al.: Colorectal cancer and colitis diagnosis using Fourier transform infrared spectroscopy and an improved K-nearest-neighbour classifier. Sensors. **17**(12), 2739 (2017)
165. Liu, W., Sun, Z., Chen, J., et al.: Raman spectroscopy in colorectal cancer diagnostics: comparison of PCA-LDA and PLS-DA models. J. Spectrosc. **2016**, 1603609 (2016)
166. Petersen, D., Naveed, P., Ragheb, A., et al.: Raman fiber-optical method for colon cancer detection: cross-validation and outlier identification approach. Spectrochim. Acta Pt. A: Mol. Biomol. Spectrosc. **181**, 270–275 (2017)
167. Bergholt, M.S., Zheng, W., Lin, K., et al.: Characterizing variability in *in vivo* Raman spectra of different anatomical locations in the upper gastrointestinal tract toward cancer detection. J. Biomed. Opt. **16**(3), 037003 (2011)
168. Duraipandian, S., Bergholt, M.S., Zheng, W., et al.: Real-time Raman spectroscopy for *in vivo*, online gastric cancer diagnosis during clinical endoscopic examination. J. Biomed. Opt. **17**(8), 081418 (2012)
169. Huang, Z., Teh, S.K., Zheng, W., et al.: *In vivo* detection of epithelial neoplasia in the stomach using image-guided Raman endoscopy. Biosens. Bioelectron. **26**(2), 383–389 (2010)
170. Lin, K., Wang, J., Zheng, W., et al.: Rapid fiber-optic Raman spectroscopy for real-time *in vivo* detection of gastric intestinal metaplasia during clinical gastroscopy. Cancer Prev. Res. **9**(6), 476–483 (2016)
171. Wang, J., Lin, K., Zheng, W., et al.: Fiber-optic Raman spectroscopy for *in vivo* diagnosis of gastric dysplasia. Faraday Discuss. **187**, 377–392 (2016)

172. Ishigaki, M., Maeda, Y., Taketani, A., et al.: Diagnosis of early-stage esophageal cancer by Raman spectroscopy and chemometric techniques. Analyst. **141**, 1027–1033 (2016)
173. Lijian, Y., Yunjiang, R., Jianhua, D., et al.: A feasibility study of using fiber-optic Raman spectrum system for fast diagnosis of gastric cancer. Opto-Electron. Eng. **46**(4), 180645 (2019)
174. Wang, J., Lin, K., Zheng, W., et al.: Comparative study of the endoscope-based bevelled and volume fiber-optic Raman probes for optical diagnosis of gastric dysplasia *in vivo* at endoscopy. Anal. Bioanal. Chem. **407**, 8303–8310 (2015)
175. Wang, J., Lin, K., Zheng, W., et al.: Simultaneous fingerprint and high-wavenumber fiberoptic Raman spectroscopy improves *in vivo* diagnosis of esophageal squamous cell carcinoma at endoscopy. Sci. Rep. **5**(1), 12957 (2015)
176. Bergholt, M.S., Zheng, W., Lin, K., et al.: *In vivo* diagnosis of gastric cancer using Raman endoscopy and ant colony optimization techniques. Int. J. Cancer. **128**, 2673–2680 (2011)
177. Bergholt, M.S., Zheng, W., Ho, K.Y., et al.: Fiber-optic Raman spectroscopy probes gastric carcinogenesis *in vivo* at endoscopy. J. Biophotonics. **6**, 49–59 (2013)
178. Bergholt, M.S., Zheng, W., Lin, K., et al.: Combining near-infrared-excited autofluorescence and Raman spectroscopy improves *in vivo* diagnosis of gastric cancer. Biosens. Bioelectron. **26**(10), 4104–4110 (2011)
179. Bergholt, M.S., Zheng, W., Lin, K., et al.: Raman endoscopy for *in vivo* differentiation between benign and malignant ulcers in the stomach. Analyst. **135**, 3162–3168 (2010)
180. Bergholt, M.S., Zheng, W., Ho, K.Y., et al.: Fiber-optic confocal Raman spectroscopy for real-time *in vivo* diagnosis of dysplasia in Barrett's esophagus. Gastroenterology. **146**, 27–32 (2014)
181. Teh, S.K., Zheng, W., Ho, K.Y., et al.: Diagnosis of gastric cancer using near-infrared Raman spectroscopy and classification and regression tree techniques. J. Biomed. Opt. **13**(3), 034013 (2008)
182. Huang, Z., Bergholt, M.S., Zheng, W., et al.: *In vivo* early diagnosis of gastric dysplasia using narrow-band image-guided Raman endoscopy. J. Biomed. Opt. **15**(3), 037017 (2010)
183. Teh, S.K., Zheng, W., Ho, K.Y., et al.: Near-infrared Raman spectroscopy for early diagnosis and typing of adenocarcinoma in the stomach. Br. J. Surg. **97**, 550–557 (2010)
184. Teh, S.K., Zheng, W., Ho, K.Y., et al.: Diagnostic potential of near-infrared Raman spectroscopy in the stomach: differentiating dysplasia from normal tissue. Br. J. Cancer. **98**(2), 457–465 (2008)
185. Almond, L.M., Hutchings, J., Lloyd, G., et al.: Endoscopic Raman spectroscopy enables objective diagnosis of dysplasia in Barrett's esophagus. Gastrointest. Endosc. **79**(1), 37–45 (2014)
186. Tolstik, T., Marquardt, C., Matthäus, C., et al.: Discrimination and classification of liver cancer cells and proliferation states by Raman spectroscopic imaging. Analyst. **139**, 6036–6043 (2014)
187. Crow, P., Molckovsky, A., Stone, N., et al.: Assessment of fiberoptic near-infrared Raman spectroscopy for diagnosis of bladder and prostate cancer. Urology. **65**(6), 1126–1130 (2005)
188. Bovenkamp, D., Sentosa, R., Rank, E., et al.: Combination of high-resolution optical coherence tomography and Raman spectroscopy for improved staging and grading in bladder cancer. Appl. Sci. ***8***, 2371 (2018)
189. Grimbergen, M.C.M., van Swol, C.F.P., Draga, R.O.P., et al.: Bladder cancer diagnosis during cystoscopy using Raman spectroscopy. Proc. SPIE. **7161**, 716114 (2009)
190. Draga, R.O.P., Grimbergen, M.C.M., Vijverberg, P.L.M., et al.: *In vivo* bladder cancer diagnosis by high-volume Raman spectroscopy. Anal. Chem. **82**(14), 5993–5999 (2010)
191. Chen, H., Li, X., Broderick, N., et al.: Identification and characterization of bladder cancer by low-resolution fiber-optic Raman spectroscopy. J. Biophotonics. **11**, e201800016 (2018)

192. Aubertin, K., Desroches, J., Jermyn, M., et al.: Combining high wavenumber and fingerprint Raman spectroscopy for the detection of prostate cancer during radical prostatectomy. Biomed. Opt. Express. **9**(9), 4294–4305 (2018)
193. Aubertin, K., Trinh, V.Q., Jermyn, M., et al.: Mesoscopic characterization of prostate cancer using Raman spectroscopy: potential for diagnostics and therapeutics. BJU Int. **122**(2), 326–336 (2018)
194. Wang, L., He, D., Zeng, J., et al.: Raman spectroscopy, a potential tool in diagnosis and prognosis of castration-resistant prostate cancer. J. Biomed. Opt. **18**(8), 087001 (2013)
195. Magalhães, F.L., Machado, A.M.C., Paulino, E., et al.: Raman spectroscopy with a 1064-nm wavelength laser as a potential molecular tool for prostate cancer diagnosis: a pilot study. J. Biomed. Opt. **23**(12), 121613 (2018)
196. Devpura, S., Thakur, J.S., Sarkar, F.H., et al.: Detection of benign epithelia, prostatic intraepithelial neoplasia, and cancer regions in radical prostatectomy tissues using Raman spectroscopy. Vib. Spectrosc. ***53***(2), 227–232 (2010)
197. Shaikh, R.S., Dora, T.K., Chopra, S., et al.: *In vivo* Raman spectroscopy of human uterine cervix: exploring the utility of vagina as an internal control. J. Biomed. Opt. **19**(8), 087001 (2014)
198. Duraipandian, S., Zheng, W., Ng, J., et al.: *In vivo* diagnosis of cervical precancer using Raman spectroscopy and genetic algorithm techniques. Analyst. **136**(20), 4328–4336 (2011)
199. Duraipandian, S., Zheng, W., Ng, J., et al.: Near-infrared-excited confocal Raman spectroscopy advances *in vivo* diagnosis of cervical precancer. J. Biomed. Opt. **18**(6), 067007 (2013)
200. Duraipandian, S., Zheng, W., Ng, J., et al.: Simultaneous fingerprint and high-wavenumber confocal Raman spectroscopy enhances early detection of cervical precancer *in vivo*. Anal. Chem. ***84***(14), 5913–5919 (**2012**)
201. Mo, J., Zheng, W., Low, J.J.H., et al.: High wavenumber Raman spectroscopy for *in vivo* detection of cervical dysplasia. Anal. Chem. **81**(21), 8908–8915 (2009)
202. Vargis, E., Kanter, E.M., Majumder, S.K., et al.: Effect of normal variations on disease classification of Raman spectra from cervical tissue. Analyst. **136**, 2981–2987 (2011)
203. Kanter, E.M., Vargis, E., Majumder, S., et al.: Application of Raman spectroscopy for cervical dysplasia diagnosis. J. Biophotonics. **2**, 81–90 (2009)
204. Robichaux-Viehoever, A., Kanter, E.M., Shappell, H., et al.: Characterization of Raman spectra measured *in vivo* for the detection of cervical dysplasia. Appl. Spectrosc. **61**, 986–993 (2007)
205. Duraipandian, S., Traynor, D., Kearney, P., et al.: Raman spectroscopic detection of high-grade cervical cytology: using morphologically normal appearing cells. Sci. Rep. **8**, 15048 (2018)
206. Moradi, H., Ahmad, A., Shepherdson, D., et al.: Raman micro-spectroscopy applied to treatment resistant and sensitive human ovarian cancer cells. J. Biophotonics. **10**, 1327–1334 (2017)
207. Maheedhar, K., Brat, R.A., Malini, R., et al.: Diagnosis of ovarian cancer by Raman spectroscopy: a pilot study. Photomed. Laser Surg. ***26***(2), 83–90 (2008)
208. Lin, K., Zheng, W., Lim, C.M., et al.: Real-time *in vivo* diagnosis of nasopharyngeal carcinoma using rapid fiber-optic Raman spectroscopy. Theranostics. **7**(14), 3517–3526 (2017)
209. Chen, P.H., Shimada, R., Yabumoto, S., et al.: Automatic and objective oral cancer diagnosis by Raman spectroscopic detection of keratin with multivariate curve resolution analysis. Sci. Rep. **6**, 20097 (2016)
210. Singh, S.P., Deshmukh, A., Chaturvedi, P., et al.: *In vivo* Raman spectroscopy for oral cancers diagnosis. Proc. SPIE. **8219**, 82190K (2012)
211. Krishna, H., Majumder, S.K., Chaturvedi, P., et al.: *In vivo* Raman spectroscopy for detection of oral neoplasia: a pilot clinical study. J. Biophotonics. **7**, 690–702 (2014)

212. Malik, A., Sahu, A., Singh, S.P., et al.: *In vivo* Raman spectroscopy–assisted early identification of potential second primary/recurrences in oral cancers: an exploratory study. Head Neck. **39**, 2216–2223 (2017)
213. Lau, D.P., Huang, Z., Lui, H., et al.: Raman spectroscopy for optical diagnosis in the larynx: preliminary findings. Lasers Surg. Med. **37**, 192–200 (2005)
214. Ming, L.C., Gangodu, N.R., Loh, T., et al.: Real time near-infrared Raman spectroscopy for the diagnosis of nasopharyngeal cancer. Oncotarget. **8**(30), 49443–49450 (2017)
215. Harris, A.T., Garg, M., Yang, X.B., et al.: Raman spectroscopy and advanced mathematical modelling in the discrimination of human thyroid cell lines. Head Neck Oncol. **1**, 38 (2009)
216. O'Dea, D., Bongiovanni, M., Sykiotis, G.P., et al.: Raman spectroscopy for the preoperative diagnosis of thyroid cancer and its subtypes: an *in vitro* proof-of-concept study. Cytopathology. **30**, 51–60 (2019)
217. Rau, J.V., Graziani, V., Fosca, M., et al.: Raman spectroscopy imaging improves the diagnosis of papillary thyroid carcinoma. Sci. Rep. **6**, 35117 (2016)
218. Desroches, J., Jermyn, M., Pinto, M., et al.: A new method using Raman spectroscopy for *in vivo* targeted brain cancer tissue biopsy. Sci. Rep. **8**, 1792 (2018)
219. Jermyn, M., Mok, K., Mercier, J., et al.: Intraoperative brain cancer detection with Raman spectroscopy in humans. Sci. Transl. Med. **7**(274), 274ra19 (2015)
220. Desroches, J., Jermyn, M., Mok, K., et al.: Characterization of a Raman spectroscopy probe system for intraoperative brain tissue classification. Biomed. Opt. Express. **6**, 2380–2397 (2015)
221. Jermyn, M., Mercier, J., Aubertin, K., et al.: Highly accurate detection of cancer *in situ* with intraoperative, label-free, multimodal optical spectroscopy. Cancer Res. **77**(14), 3942–3950 (2017)
222. Feng, X., Moy, A.J., Nguyen, H., et al.: Raman active components of skin cancer. Biomed. Opt. Express. **8**(6), 2835–2850 (2017)
223. Lim, L., Nichols, B.S., Migden, M.R., et al.: Clinical study of noninvasive *in vivo* melanoma and nonmelanoma skin cancers using multimodal spectral diagnosis. J. Biomed. Opt. **19**(11), 117003 (2014)
224. Silveira, L., Silveira, F.L., Zângaro, R.A., et al.: Discriminating model for diagnosis of basal cell carcinoma and melanoma *in vitro* based on the Raman spectra of selected biochemicals. J. Biomed. Opt. **17**(7), 077003 (2012)
225. Zhao, J., Lui, H., Kalia, S., et al.: Real-time Raman spectroscopy for automatic *in vivo* skin cancer detection: an independent validation. Anal. Bioanal. Chem. **407**, 8373 (2015)
226. Feng, X., Fox, M.C., Reichenberg, J.S., et al.: Biophysical basis of skin cancer margin assessment using Raman spectroscopy. Biomed. Opt. Express. **10**, 104–118 (2019)
227. Lui, H., Zhao, J., McLean, D., et al.: Real-time Raman spectroscopy for *in vivo* skin cancer diagnosis. Cancer Res. **72**(10), 2491–2500 (2012)
228. Bratchenko, I.A., Artemyev, D.N., Myakinin, O.O., et al.: Combined Raman and autofluorescence *ex vivo* diagnostics of skin cancer in near-infrared and visible regions. J. Biomed. Opt. **22**(2), 027005 (2017)
229. Schleusener, J., Gluszczynska, P., Reble, C., et al.: *In vivo* study for the discrimination of cancerous and normal skin using fibre probe-based Raman spectroscopy. Exp. Dermatol. **24**, 767–772 (2015)
230. Zhao, J., Lui, H., McLean, D.I., et al.: Real-time Raman spectroscopy for non-invasive skin cancer detection - preliminary results. Conf. Proc. IEEE Eng. Med. Biol. Soc. **2008**, 3107–3109 (2008)
231. Li, B., Xie, S.: Autofluorescence excitation-emission matrices for diagnosis of colonic cancer. World J. Gastroenterol. **11**, 3931–3934 (2005)
232. Bobone, S., van de Weert, M., Stella, L.: A reassessment of synchronous fluorescence in the separation of TRP and TYR contributions in protein emission and in the determination of conformational changes. J. Mol. Struct. **1077**, 68–76 (2014)

233. Zeng, Z., Cohen, A., Guillem, J.: Loss of basement membrane type IV collagen is associated with increased expression of metalloproteinases 2 and 9 (MMP-2 and MMP-9) during human colorectal tumorigenesis. Carcinogenesis. **20**(5), 749–755 (1999)
234. Alexandrova, A.: Evolution of cell interactions with extracellular matrix during carcinogenesis. Biochemistry. **73**(7), 733–741 (2008)
235. Liu, Z., Pouli, D., Alonzo, C., et al.: Mapping metabolic changes by noninvasive, multiparametric, high-resolution imaging using endogenous contrast. Sci. Adv. **4**, eaap9302 (2018)
236. Kolenc, O.I., Quinn, K.P.: Evaluating cell metabolism through autofluorescence imaging of NAD (P)H and FAD. Antioxid. Redox Signal. **30**(6), 875–889 (2019)
237. Yang, S., Park, Y., Cho, J., et al.: Regulation of hypoxia responses by flavin adenine dinucleotide-dependent modulation of hif-1α protein stability. EMBO J. **36**, 1011–1028 (2017)
238. Richards-Kortum, R., Rava, R., Petras, R., et al.: Spectroscopic diagnosis of colonic dysplasia. Photochem. Photobiol. **53**(6), 777–786 (1991)
239. DaCosta, R., Kost, J., Lothar, L.: Confocal fluorescence microscopy/macroscopy and microspectrofluorimetry analysis of human colorectal tissues. J. Analyt. Morphol.: Cell Vision. **4**, 24–29 (1997)
240. DaCosta, P., Wilson, Б., Marcon, H.: Photodiagnostic techniques for the endoscopic detection of premalignant gastrointestinal lesions. Dig. Endosc. **15**(3), 153–173 (2003)
241. Genova, T.S., Borisova, E., Zhelyazkova, A., et al.: Colorectal cancer stage evaluation using synchronous fluorescence spectroscopy technique. Opt. Quant. Electron. **48**(8), 378 (2016)
242. Sun, J., Garfield, D.H., Lam, B., et al.: The value of autofluorescence bronchoscopy combined with white light bronchoscopy compared with white light alone in the diagnosis of intraepithelial neoplasia and invasive lung cancer: a metaanalysis. J. Thorac. Oncol. **6**(8), 1336–1344 (2011)
243. Zakharov, V.P., Bratchenko, I.A., Artemyev, D.N., et al.: Lung neoplasm diagnostics using Raman spectroscopy and autofluorescence analysis. J. Biomed. Photon Eng. **1**(1), 70–76 (2015)
244. Magee, N.D., Villaumie, J.S., Marple, E.T., et al.: *Ex vivo* diagnosis of lung cancer using a Raman miniprobe. J. Phys. Chem. B. **113**(23), 8137–8141 (2009)
245. McGregor, H.C., Short, M.A., Lam, S., et al.: Development and *in vivo* test of a miniature Raman probe for early cancer detection in the peripheral lung. J. Biophotonics. **11**(11), e201800055 (2018)
246. Burns, D.M.: Primary prevention, smoking, and smoking cessation: implications for future trends in lung cancer prevention. Cancer. **89**(11 Suppl), 2506–2509 (2000)
247. Lindeman, N.I., Cagle, P.T., Aisner, D.L., et al.: Updated molecular testing guideline for the selection of lung cancer patients for treatment with targeted tyrosine kinase inhibitors: guideline from the college of American pathologists. The International Association for the Study of Lung Cancer, and the Association for Molecular Pathology. J. Mol. Diagn. **20**(2), 129–159 (2018)
248. Cagle, P.T., Allen, T.C., Olsen, R.J.: Lung cancer biomarkers: present status and future developments. Arch. Pathol. Lab. Med. **137**(9), 1191–1198 (2013)
249. Rajdev, K., Siddiqui, A.H., Ibrahim, U., et al.: An unusually aggressive large cell carcinoma of the lung: undiagnosed until autopsy. Cureus. **10**(2), e2202 (2018)
250. Qu, J., MacAulay, C., Lam, S., et al.: Optical properties of normal and carcinomatous bronchial tissue. Appl. Opt. **33**(31), 7397–7405 (1994)
251. Fishkin, J.B., Coquoz, O., Anderson, E.R., et al.: Frequency-domain photon migration measurements of normal and malignant tissue optical properties in a human subject. Appl. Opt. **36**(1), 10–20 (1997)
252. Fawzy, Y.S., Petek, M., Tercelj, M., et al.: *In vivo* assessment and evaluation of lung tissue morphologic and physiological changes from non-contact endoscopic reflectance spectroscopy for improving lung cancer detection. J. Biomed. Opt. **11**(4), 044003 (2006)
253. Jain, R.K.: Determinants of tumor blood flow: a review. Cancer Res. **48**(10), 2641–2658 (1988)

254. Höckel, M., Vaupel, P.: Tumor hypoxia: definitions and current clinical biologic, and molecular aspects. J. Natl. Cancer Inst. **93**(4), 266–276 (2001)
255. Johnson, J.M., Dalton, R.R., Wester, S.M., et al.: Histological correlation of microcalcifications in breast biopsy specimens. Arch. Surg. **134**(7), 712–716 (1999)
256. Vos, E.L., Siesling, S., Baaijens, M.H.A., et al.: Omitting re-excision for focally positive margins after breast-conserving surgery does not impair disease-free and overall survival. Breast Cancer Res. Treat. **164**(1), 157–167 (2017)
257. Frank, C.J., McCreery, R.L., Redd, D.C.B.: Raman spectroscopy of normal and diseased human breast tissues. Anal. Chem. **67**(5), 777–783 (1995)
258. Haka, A.S., Volynskaya, Z., Gardecki, J.A., et al.: *In vivo* margin assessment during partial mastectomy breast surgery using Raman spectroscopy. Cancer Res. **66**(6), 3317–3322 (2006)
259. Peters, V.G., Wyman, D.R., Patterson, M.S., et al.: Optical properties of normal and diseased human breast tissues in the visible and near infrared. Phys. Med. Biol. **35**(9), 1317–1334 (1990)
260. Fantini, S., Walker, S.A., Franceschini, M.A., et al.: Assessment of the size, position, and optical properties of breast tumors *in vivo* by noninvasive optical methods. Appl. Opt. **37**(10), 1982–1989 (1998)
261. Grosenick, D., Wabnitz, H., Moesta, K.T., et al.: Time-domain scanning optical mammography: II. Optical properties and tissue parameters of 87 carcinomas. Phys. Med. Biol. **50**, 2451–2468 (2005)
262. Cerussi, A., Shah, N., Hsiang, D., et al.: *In vivo* absorption, scattering, and physiologic properties of 58 malignant breast tumors determined by broadband diffuse optical spectroscopy. J. Biomed. Opt. **11**(4), 044005 (2006)
263. Zhu, C., Palmer, G.M., Breslin, T.M., et al.: Diagnosis of breast cancer using diffuse reflectance spectroscopy: comparison of a Monte Carlo versus partial least squares analysis based feature extraction technique. Lasers Surg. Med. **38**, 714–724 (2006)
264. Zhu, C., Palmer, G.M., Breslin, T.M., et al.: Diagnosis of breast cancer using fluorescence and diffuse reflectance spectroscopy: a Monte-Carlo-model-based approach. J. Biomed. Opt. **13**(3), 034015 (2008)
265. Konovalov, A.B., Genina, E.A., Bashkatov, A.N.: Diffuse optical mammotomography: state-of-the-art and prospects. J. Biomed. Photon Eng. **2**(2), 020202 (2016)
266. Tromberg, B.J., Shah, N., Lanning, R., et al.: Non-invasive *in vivo* characterization of breast tumors using photon migration spectroscopy. Neoplasia. **2**(1-2), 26–40 (2000)
267. Nachabe, R., Evers, D.J., Hendriks, B.H.W., et al.: Diagnosis of breast cancer using diffuse optical spectroscopy from 500 to 1600 nm: comparison of classification methods. J. Biomed. Opt. **16**(8), 087010 (2011)
268. Grosenick, D., Rinneberg, H., Cubeddu, R., et al.: Review of optical breast imaging and spectroscopy. J. Biomed. Opt. **21**(9), 091311 (2016)
269. Siegel, R.L., Miller, K.D., Jemal, A.: Cancer statistics. 2019. CA Cancer J. Clin. **69**, 7–34 (2019)
270. Gniadecka, M., Wulf, H.C., Mortensen, N.N., et al.: Diagnosis of basal cell carcinoma by Raman spectroscopy. J. Raman Spectrosc. **28**(2-3), 125–129 (1997)
271. Patil, C.A., Kirshnamoorthi, H., Ellis, D.L., et al.: A clinical instrument for combined Raman spectroscopy - optical coherence tomography of skin cancers. Lasers Surg. Med. **43**, 143–151 (2011)
272. Varkentin, A., Mazurenka, M., Blumenröther, E., et al.: Trimodal system for *in vivo* skin cancer screening with combined optical coherence tomography - Raman and colocalized optoacoustic measurements. J. Biophotonics. **11**, e201700288 (2018)
273. Garcia-Uribe, A., Zou, J., Duvic, M., et al.: *In vivo* diagnosis of melanoma and nonmelanoma skin cancer using oblique incidence diffuse reflectance spectrometry. Cancer Res. **72**(11), 2738–2745 (2012)

274. Anwar, J., Wrone, D.A., Kimyai-Asadi, A., et al.: The development of actinic keratosis into invasive squamous cell carcinoma: evidence and evolving classification schemes. Clin. Dermatol. **22**, 189–196 (2004)
275. Goldberg, L.H., Joseph, A.K., Tschen, J.A.: Proliferative actinic keratosis. Int. J. Dermatol. **33**, 341–345 (1994)
276. Mikhail, G.R., Mehregan, A.H.: Basal cell carcinoma in seborrheic keratosis. J. Am. Acad. Dermatol. **6**, 500–506 (1982)
277. Wallace, M.B., Keisslich, R.: Advances in endoscopic imaging of colorectal neoplasia. Gastroenterology. **138**, 2140–2150 (2010)
278. Gao, Z., Hu, B., Ding, C., et al.: Micro Raman spectra for lipids in colorectal tissue. Spectrosc. Spectr. Anal. **30**, 692–696 (2010)
279. Widjaja, E., Zheng, W., Huang, Z.: Classification of colonic tissues using near-infrared Raman spectroscopy and support vector machines. Int. J. Oncol. **32**, 653–662 (2008)
280. Wood, J.J., Kendall, C., Hutchings, J., et al.: Evaluation of a confocal Raman probe for pathological diagnosis during colonoscopy. Color. Dis. **16**, 732–738 (2014)
281. Leikin, S., Parsegian, V.A., Yang, W., et al.: Raman spectral evidence for hydration forces between collagen triple helices. Proc. Natl. Acad. Sci. U. S. A. **94**(21), 11312–11317 (1997)
282. Zonios, G., Perelman, L.T., Backman, V., et al.: Diffuse reflectance spectroscopy of human adenomatous colon polyps *in vivo*. Appl. Opt. **38**(31), 6628–6637 (1999)
283. Knoefel, W.T., Kollias, N., Rattner, D.W., et al.: Reflectance spectroscopy of pancreatic microcirculation. J. Appl. Physiol. **80**(1), 116–123 (1996)
284. Holmer, C., Lehmann, K.S., Risk, J., et al.: Colorectal tumors and hepatic metastases differ in their optical properties – relevance for dosimetry in laser-induced interstitial thermotherapy. Lasers Surg. Med. **38**, 296–304 (2006)
285. Beek, J.F., Blokland, P., Posthumus, P., et al.: *In vitro* double-integrating-sphere optical properties of tissues between 630 and 1064 nm. Phys. Med. Biol. **42**, 2255–2261 (1997)
286. Kanter, E.M., Majumder, S., Kanter, G.J., et al.: Effect of hormonal variation on Raman spectra for cervical disease detection. Am. J. Obstet. Gynecol. **200**(512), e1–e5 (2009)
287. Vargis, E., Mahadevan-Jansen, A., Byrd, T., et al.: Sensitivity of Raman spectroscopy to normal patient variability. J. Biomed. Opt. **16**(11), 117004 (2011)
288. Duraipandian, S., Zheng, W., Ng, J., et al.: Integrated fingerprint and high wavenumber confocal Raman spectroscopy for *in vivo* diagnosis of cervical precancer. Proc. SPIE. **8572**, 85720Z (2013)
289. Mottet, N., Bellmunt, J., Bolla, M., et al.: AU-ESTRO-SIOG guidelines on prostate cancer. Part 1: screening, diagnosis, and local treatment with curative intent. Eur. Urol. **71**(4), 618–629 (2017)
290. Kast, R.E., Tucker, S.C., Killian, K., et al.: Emerging technology: applications of Raman spectroscopy for prostate cancer. Cancer Metastasis Rev. **33**, 673–693 (2014)
291. Patel, I.I., Martin, F.L.: Discrimination of zone-specific spectral signatures in normal human prostate using Raman spectroscopy. Analyst. **135**(12), 3060–3069 (2010)
292. Crow, P., Stone, N., Kendall, C.A., et al.: The use of Raman spectroscopy to identify and grade prostatic adenocarcinoma *in vitro*. Br. J. Cancer. **89**, 106–108 (2003)
293. Arnfield, M.R., Chapman, J.D., Tulip, J., et al.: Optical properties of experimental prostate tumors *in vivo*. Photochem. Photobiol. **57**(2), 306–311 (1993)
294. Schiff, D., Trump, D.L., Wen, P.Y.: Neurologic complications of genitourinary malignancies. In: Schiff, D., Wen, P.Y. (eds.) Cancer Neurology in Clinical Practice, pp. 327–338. Humana Press Inc., Totowa (2003)
295. Stein, J.P., Lieskovsky, G., Cote, R., et al.: Radical cystectomy in the treatment of invasive bladder cancer: long-term results in 1054 patients. J. Clin. Oncol. **19**(3), 666–675 (2001)
296. García, J., Santomé, L., Anido, U., et al.: Current metastatic bladder cancer: second-line treatment and recommendations of the genitourinary tumor division of the Galician oncologic society (SOG-GU). Oncol. Rep. **18**, 72 (2016)

297. Liedberg, F., Mansson, W.: Lymph node metastasis in bladder cancer. Eur. Urol. **49**, 13–21 (2006)
298. Babjuk, M., Burger, M., Zigeuner, R., et al.: EAU guidelines on non-muscle-invasive urothelial carcinoma of the bladder: update 2013. Eur. Urol. **64**(4), 639–653 (2013)
299. Mostofi, F.K.S.L., Torloni, H.: Histologic Typing of Urinary Bladder Tumors: International Histological Classification of Tumors. World Health Organization, Geneva (1973)
300. Epstein, J.I., Amin, M.B., Reuter, V.R., et al.: The World Health Organization/International Society of Urological Pathology consensus classification of urothelial (transitional cell) neoplasms of the urinary bladder. Bladder Consensus Conference Committee. Am. J. Surg. Pathol. **22**, 1435–1448 (1998)
301. Eble, J.N., Sauter, G., Epstein, J.I., et al.: World Health Organization Classification of Tumours. Pathology and Genetics of Tumours of the Urinary System and Male Genital Organs. IARC Press, Lyon (2004)
302. Comperat, E.M., Burger, M., Gontero, P., et al.: Grading of urothelial carcinoma and the new "World Health Organisation classification of tumours of the urinary system and male genital organs 2016". Eur. Urol. Focus. **5**(3), 457–466 (2019)
303. Khadra, M.H., Pickard, R.S., Charlton, M., et al.: A prospective analysis of 1,930 patients with hematuria to evaluate current diagnostic practice. J. Urol. **163**(2), 524–527 (2000)
304. Mishriki, S.F., Nabi, G., Cohen, N.P.: Diagnosis of urologic malignancies in patients with asymptomatic dipstick hematuria: prospective study with 13 years' follow-up. Urology. **71**(1), 13–16 (2008)
305. Ma, B., Li, H., Zhang, C., et al.: Lymphovascular invasion, ureteral reimplantation and prior history of urothelial carcinoma are associated with poor prognosis after partial cystectomy for muscle-invasive bladder cancer with negative pelvic lymph nodes. Eur. J. Surg. Oncol. **39**(10), 1150–1156 (2013)
306. Mano, R., Shoshany, O., Baniel, J., et al.: Resection of ureteral orifice during transurethral resection of bladder tumor: functional and oncologic implications. J. Urol. **188**(6), 2129–2133 (2012)
307. Naselli, A., Puppo, P.: En bloc transurethral resection of bladder tumors: a new standard? J. Endourol. **31**(S1), S20 (2017)
308. Bolat, D., Gunlusoy, B., Aydogdu, O., et al.: Comparing the short-term outcomes and complications of monopolar and bipolar transurethral resection of bladder tumors in patients with coronary artery disease: a prospective, randomized, controlled study. Int. Braz. J. Urol. **44**(4), 717–725 (2018)
309. Cumberbatch, M.G.K., Foerster, B., Catto, J.W.F., et al.: Repeat transurethral resection in non-muscle-invasive bladder cancer: a systematic review. Eur. Urol. **73**(6), 925–933 (2018)
310. Babjuk, M., Böhle, A., Burger, M., et al.: Guidelines on non-muscle invasive bladder cancer (Tar. T1 and Cis). Eur. Assoc. Urol. **13**, 913 (2015)
311. Lipiński, M.I., Różański, W., Markowski, M.P.: Photodynamic diagnosis – current tool in diagnosis of carcinoma *in situ* of the urinary bladder. Contemp. Oncol. **19**(4), 341–342 (2015)
312. Santos, I.P., Barroso, E.M., Schut, T.C.B., et al.: Raman spectroscopy for cancer detection and cancer surgery guidance: translation to the clinics. Analyst. **142**(17), 3025–3047 (2017)
313. De Jong, B.W.D., Schut, T.C.B., Maquelin, K., et al.: Discrimination between nontumor bladder tissue and tumor by Raman spectroscopy. Anal. Chem. **78**(22), 7761–7769 (2006)
314. Barman, I., Dingari, N.C., Singh, G.P., et al.: Selective sampling using confocal Raman spectroscopy provides enhanced specificity for urinary bladder cancer diagnosis. Anal. Bioanal. Chem. **404**, 3091 (2012)
315. Bergholt, M.S., Zheng, W., Ho, K.Y., et al.: Real-time depth-resolved fiber optic Raman endoscopy for *in vivo* diagnosis of gastric precancer. Proc. SPIE. **8939**, 89390M (2014)
316. Bergholt, M.S., Zheng, W., Lin, K., et al.: *In vivo* diagnosis of esophageal cancer using image-guided Raman endoscopy and biomolecular modeling. Technol. Cancer Res. Treat. **10**(2), 103–112 (2011)

317. Holmer, C., Lehmann, K.S., Wanken, J., et al.: Optical properties of adenocarcinoma and squamous cell carcinoma of the gastroesophageal junction. J. Biomed. Opt. **12**(1), 014025 (2007)
318. Sahu, A., Krishna, C.M.: Optical diagnostics in oral cancer: an update on Raman spectroscopic applications. J. Can. Res. Ther. **13**, 908–915 (2017)
319. Guze, K., Short, M.A., Sonis, S., et al.: Parameters defining the potential applicability of Raman spectroscopy as a diagnostic tool for oral disease. J. Biomed. Opt. **14**(1), 014016 (2009)
320. Bergholt, M.S., Zheng, W., Huang, Z.: Characterizing variability in *in vivo* Raman spectroscopic properties of different anatomical sites of normal tissue in the oral cavity. J. Raman Spectrosc. **43**, 255–262 (2012)
321. Singh, S.P., Deshmukh, A., Chaturvedi, P., et al.: *In vivo* Raman spectroscopic identification of premalignant lesions in oral buccal mucosa. J. Biomed. Opt. **17**(10), 105002 (2012)
322. Singh, S.P., Sahu, A., Deshmukh, A., et al.: *In vivo* Raman spectroscopy of oral buccal mucosa: a study on malignancy associated changes (MAC)/cancer field effects (CFE). Analyst. **138**, 4175–4182 (2013)
323. Sahu, A., Deshmukh, A., Hole, A.R., et al.: *In vivo* subsite classification and diagnosis of oral cancers using Raman spectroscopy. J. Innov. Opt. Health Sci. **9**(5), 1650017 (2016)
324. Amelink, A., Kaspers, O.P., Sterenborg, H.J.C.M., et al.: Non-invasive measurement of the morphology and physiology of oral mucosa by use of optical spectroscopy. Oral Oncol. **44**, 65–71 (2008)
325. Huang, N., Short, M., Zhao, J., et al.: Full range characterization of the Raman spectra of organs in a murine model. Opt. Express. **19**(23), 22892–22909 (2011)
326. Hawi, S.R., Campbell, W.B., Kajdacsy-Balla, A., et al.: Characterization of normal and malignant human hepatocytes by Raman microspectroscopy. Cancer Lett. **110**, 35–40 (1996)
327. Hamilton, S.R., Aaltonen, L.A. (eds.): World Health Organization Classification of Tumours. Pathology and Genetics of Tumours of the Digestive System, p. 173. IARC Press, Lyon (2000)
328. Blechacz, B., Gores, G.J.: Cholangiocarcinoma: advances in pathogenesis, diagnosis, and treatment. Hepatology. **48**(1), 308–321 (2008)
329. Bismuth, H., Corlette, M.B.: Intrahepatic cholangioenteric anastomosis in carcinoma of the hilus of the liver. Surg Gynecol Obstet. **140**(2), 170–178 (1975)
330. Kudo, M., Kitano, M., Sakurai, T., et al.: General rules for the clinical and pathological study of primary liver cancer, nationwide follow-up survey and clinical practice guidelines: the outstanding achievements of the Liver Cancer Study Group of Japan. Dig. Dis. **33**(6), 765–770 (2015)
331. Ishak, K.G., Goodman, Z.D., Stocker, J.T.: Tumours of the liver and intrahepatic bile ducts. In: Atlas of Tumour Pathology, 3rd Series, Fascicle 31. Armed Forces Institute of Pathology, Washington, DC (2001)
332. Shaib, Y., El-Serag, H.B.: The epidemiology of cholangiocarcinoma. Semin. Liver Dis. **24**, 115–125 (2004)
333. Jan, Y.Y., Yeh, C.N., Yeh, T.S., et al.: Clinicopathological factors predicting long-term overall survival after hepatectomy for peripheral cholangiocarcinoma. World J. Surg. **29**, 894–898 (2005)
334. Razumilava, N., Gores, G.J.: Cholangiocarcinoma. Lancet. **383**, 2168–2179 (2014)
335. Greene, F.L., Page, D.L., Fleming, I.D., et al. (eds.): AJCC Cancer Staging Manual, 6th edn. Springer, New York (2003)
336. Albores-Saavedra, J., Henson, D.E., Klimstra, D.S.: Tumours of the Gallbladder, Extrahepatic Bile Ducts, and Ampulla of Vater. Atlas of Tumour Pathology. 3rd-Series, Fascicle 27. Armed Forces Institute of Pathology, Washington DC (2000)
337. Guglielmi, A., Ruzzenente, A., Iacono, C.: Reporting cholangiocarcinoma: pathological aspects. In: Guglielmi, A., Ruzzenente, A., Iacono, C. (eds.) Surgical Treatment of Hilar and Intrahepatic Cholangiocarcinoma, pp. 3–15. Springer, Milan (2007)

338. Patel, T.: Increasing incidence and mortality of primary intrahepatic cholangiocarcinoma in the United States. Hepatology. **33**, 1353–1357 (2001)
339. McMasters, K.M., Curley, S.A.: Treatment of cholangiocarcinoma. In: Curley, S.A. (ed.) Liver Cancer, pp. 95–116. Springer, New York (1998)
340. Sasaki, R., Takeda, Y., Funato, O., et al.: Significance of ductal margin status in patients undergoing surgical resection for extrahepatic cholangiocarcinoma. World J. Surg. **31**(9), 1788–1796 (2007)
341. Moeini, A., Sia, D., Bardeesy, N., et al.: Molecular pathogenesis and targeted therapies for intrahepatic cholangiocarcinoma. Clin. Cancer Res. **22**, 291–300 (2016)
342. de Jong, M.C., Nathan, H., Sotiropoulos, G.C., et al.: Intrahepatic cholangiocarcinoma: an international multi-institutional analysis of prognostic factors and lymph node assessment. J. Clin. Oncol. **29**, 3140–3145 (2011)
343. Sanford, C.H.: Primary malignant disease of the liver. Ann. Intern. Med. **37**, 304–312 (1952)
344. Tsushimi, T., Enoki, T., Harada, E., et al.: Ectopic hepatocellular carcinoma arising in the bile duct. J. Hepato-Biliary-Pancreat. Surg. **12**, 266–268 (2005)
345. Taylor, J.N., Mochizuki, K., Hashimoto, K., et al.: High-resolution Raman microscopic detection of follicular thyroid cancer cells with unsupervised machine learning. J. Phys. Chem. B. **123**(20), 4358–4372 (2019)
346. Medeiros-Neto, L.P., Soto, C.A.T., Chagas, M.J., et al.: *In vivo* Raman spectroscopic characterization of papillary thyroid carcinoma. Vib. Spectrosc. **101**, 1–9 (2019)
347. Kalkanis, S.N., Kast, R.E., Rosenblum, M.L., et al.: Raman spectroscopy to distinguish grey matter, necrosis, and glioblastoma multiforme in frozen tissue sections. J. Neuro-Oncol. **116**, 477 (2014)
348. Kast, R., Auner, G., Yurgelevic, S., et al.: Identification of regions of normal grey matter and white matter from pathologic glioblastoma and necrosis in frozen sections using Raman imaging. J. Neuro-Oncol. **125**, 287 (2015)
349. Leslie, D.G., Kast, R.E., Poulik, J.M., et al.: Identification of pediatric brain neoplasms using Raman spectroscopy. Pediatr. Neurosurg. **48**, 109–117 (2012)
350. Jermyn, M., Desroches, J., Mercier, J., et al.: Raman spectroscopy detects distant invasive brain cancer cells centimeters beyond MRI capability in humans. Biomed. Opt. Express. **7**, 5129–5137 (2016)
351. Ji, M., Orringer, D.A., Freudiger, C.W., et al.: Rapid, label-free detection of brain tumors with stimulated Raman scattering microscopy. Sci. Transl. Med. **5**(201), 201ra119 (2013)
352. Honda, N., Ishii, K., Kajimoto, Y., et al.: Determination of optical properties of human brain tumor tissues from 350 to 1000 nm to investigate the cause of false negatives in fluorescence-guided resection with 5-aminolevulinic acid. J. Biomed. Opt. **23**(7), 075006 (2018)
353. Schwarzmaier, H.-J., Eickmeyer, F., Fiedler, V.U., et al.: Basic principles of laser induced interstitial thermotherapy in brain tumors. Med. Laser Appl. **17**, 147–158 (2002)
354. Sterenborg, H.J.C.M., van Gemert, M.J.C., Kamphorst, W., et al.: The spectral dependence of the optical properties of human brain. Lasers Med. Sci. **4**, 221–227 (1989)
355. Yaroslavsky, A.N., Schulze, P.C., Yaroslavsky, I.V., et al.: Optical properties of selected native and coagulated human brain tissues *in vitro* in the visible and near infrared spectral range. Phys. Med. Biol. **47**, 2059–2073 (2002)
356. Wu, X., Shu, X.: Epidemiology of renal cell carcinoma. In: Oya, M. (ed.) Renal Cell Carcinoma. Springer, Tokyo (2017)
357. Ferlay, J., Soerjomataram, I., Dikshit, R., et al.: Cancer incidence and mortality worldwide: sources, methods and major patterns in GLOBOCAN 2012. Int. J. Cancer. **136**, E359–E386 (2015)

358. Novara, G., Ficarra, V., Antonelli, A., et al.: Validation of the 2009 TNM version in a large multi-institutional cohort of patients treated for renal cell carcinoma: are further improvements needed? Eur. Urol. **58**, 588–595 (2010)
359. Tan, P.H., Cheng, L., Rioux-Leclercq, N., et al.: ISUP Renal Tumor Panel. Renal tumors: diagnostic and prognostic biomarkers. Am. J. Surg. Pathol. **37**(10), 1518–1531 (2013)
360. Chow, W.H., Dong, L.M., Devesa, S.S.: Epidemiology and risk factors for kidney cancer. Nat. Rev. Urol. **7**(5), 245–257 (2010)
361. Petejova, N., Martinek, A.: Renal cell carcinoma: review of etiology, pathophysiology and risk factors. Biomed. Pap. **160**(2), 183–194 (2016)
362. Abrahams, N.A., MacLennan, G.T., Khoury, J.D., et al.: Chromophobe renal cell carcinoma: a comparative study of histological, immunohistochemical and ultrastructural features using high throughput tissue microarray. Histopathology. **45**, 593–602 (2004)
363. Ljungberg, B., Bensalah, K., Bex, A., et al.: Guidelines on renal cell carcinoma. Eur. Urol. **67**, 913–992 (2013)
364. Bamias, A., Escudier, B., Sternberg, C.N., et al.: Current clinical practice guidelines for the treatment of renal cell carcinoma: a systematic review and critical evaluation. Oncologist. **22**(6), 667–679 (2017)
365. Aitchison, M., Bray, C.A., Van Poppel, H., et al.: Adjuvant 5-flurouracil, alpha-interferon and interleukin-2 versus observation in patients at high risk of recurrence after nephrectomy for renal cell carcinoma: results of a phase III randomised European Organisation for Research and Treatment of Cancer (Genito-Urinary Cancers Group)/National Cancer Research Institute trial. Eur. J. Cancer. **50**, 70–77 (2014)
366. Ljungberg, B., Bensalah, K., Canfield, S., et al.: EAU guidelines on renal cell carcinoma: 2014 update. Eur. Urol. **67**(5), 913–924 (2015)
367. Lee, S.Y., Lloyd, W.R., Chandra, M., et al.: Characterizing human pancreatic cancer precursor using quantitative tissue optical spectroscopy. Biomed. Opt. Express. **4**(12), 2828–2834 (2013)
368. Kiris, T., Akbulut, S., Kiris, A., et al.: Optical characterization of pancreatic normal and tumor tissues with double integrating sphere system. Proc. SPIE. **9321**, 932116 (2015)
369. Saccomandi, P., Larocca, E.S., Rendina, V., et al.: Estimation of optical properties of neuroendocrine pancreas tumor with double-integrating-sphere system and inverse Monte Carlo model. Lasers Med. Sci. **31**, 1041–1050 (2016)
370. Bergner, N., Medyukhina, A., Geiger, K.D., et al.: Hyperspectral unmixing of Raman microimages for assessment of morphological and chemical parameters in non-dried brain tumor specimens. Anal. Bioanal. Chem. **405**(27), 8719–8728 (2013)
371. Daniel, A., Aruna, P., Ganesan, S., et al.: Biochemical assessment of human uterine cervix by micro-Raman mapping. Photodiagn. Photodyn. Ther. **17**, 65–74 (2017)
372. Shetty, G., Kendall, C., Shepherd, N., et al.: Raman spectroscopy: elucidation of biochemical changes in carcinogenesis of oesophagus. Br. J. Cancer. **94**, 1460–1464 (2006)
373. Ding, H., Dupont, A.W., Singhal, S., et al.: Effect of physiological factors on the biochemical properties of colon tissue – an *in vivo* Raman spectroscopy study. J. Raman Spectrosc. **48**, 902–909 (2017)
374. Zhao, J., Lui, H., McLean, D.I., et al.: Integrated real-time Raman system for clinical *in vivo* skin analysis. Skin Res. Technol. **14**, 484–492 (2008)
375. Sowa, M.G., Smith, M.S.D., Kendall, C., et al.: Semi-parametric estimation in the compositional modeling of multicomponent systems from Raman spectroscopic data. Appl. Spectrosc. **60**(8), 877–883 (2006)

Chapter 2
Optical Clearing of Biological Tissues: Prospects of Application for Multimodal Malignancy Diagnostics

Elina A. Genina, Luís M. C. Oliveira, Alexey N. Bashkatov, and Valery V. Tuchin

2.1 Introduction

The reduction of tissue light scattering gives improvement of image quality and precision of spectroscopic information, decreasing of irradiating light beam distortion and its sharp focusing. Various physical and chemical actions such as compression, dehydration, and impregnation by biocompatible chemical agents are widely described in literature as tools for controlling of tissue optical properties. All approaches have benefits in specific applications. Special interest is related to optical differentiation of pathology modified tissues. For example, compression of soft tissues as mucosa and skin are developed in optical coherence tomography (OCT)

E. A. Genina (✉) · A. N. Bashkatov
Department of Optics and Biophotonics, Saratov State University, Saratov, Russian Federation

Laboratory of Biophotonics, Tomsk State University, Tomsk, Russian Federation
e-mail: geninaea@sgu.ru

L. M. C. Oliveira
School of Engineering, Physics Department, Polytechnic Institute of Porto, Porto, Portugal

Centre of Innovation in Engineering and Industrial Technology, ISEP, Porto, Portugal
e-mail: lmo@isep.ipp.pt

V. V. Tuchin
Department of Optics and Biophotonics, Saratov State University, Saratov, Russian Federation

Laboratory of Biophotonics, Tomsk State University, Tomsk, Russian Federation

Laboratory of Laser Diagnostics of Technical and Living Systems, Institute of Precision Mechanics and Control of the Russian Academy of Sciences, Saratov, Russian Federation

Laboratory of Molecular Imaging, Bach Institute of Biochemistry, Research Center of Biotechnology of the Russian Academy of Sciences, Moscow, Russian Federation

Laboratory of FemtoMedicine, ITMO University, St. Petersburg, Russian Federation

V. V. Tuchin et al. (eds.), *Multimodal Optical Diagnostics of Cancer*,
https://doi.org/10.1007/978-3-030-44594-2_2

for increasing in the brightness of the OCT-image and, thus, contrast of tumor visualization inside tissues [1, 2]. Dehydration of upper layers of tissues under action of hyperosmotic optical clearing agents (OCAs) allows for increasing of optical probing depth of spectral methods in VIS-NIR [3, 4] and THz ranges [5, 6]. The immersion optical clearing (OC) is based on the impregnation (immersion) of the tissue by a biocompatible agent, possessing sufficiently high refractive index (RI) to match the RIs of the scatterers and the surrounding medium, penetrating into the interstitial space of the tissue. This approach is widely used in optical imaging techniques, such as laser speckle contrast imaging [7–9], OCT [10–13], Raman spectroscopy [14, 15], multiphoton microscopy [16, 17], etc., and demonstrates high potentiality of their mutual use not only for getting high-resolution structural and functional tissue images *in vitro*, but also *in vivo*. The study of scattering kinetics during penetration of OCAs in tissues allows one to evaluate the OCA diffusion and permeability coefficients [10, 18–21]. Using these techniques, based mostly on collimated light transmittance measurements for *in vitro* studies and OCT for *in vivo*, the diffusion rates of glucose, some drugs, and OCAs were determined in eye tissues [22–25], muscle [26], skin [20, 27], *dura mater* [28], arterial [29], and lung [30] tissues. The monitoring of OCA diffusion with high temporal and depth resolution allows one, in turn, to differentiate healthy from malignant tissues [31–36].

In the chapter, we discuss the possibilities of enhancement of optical diagnostics of tumors using immersion OC approach.

2.2 Mechanisms of Immersion Optical Clearing

The OCAs can be roughly divided into polyatomic alcohols (glycerol, polyethylene glycol (PEG), polypropylene glycol, combined mixtures on the base of polypropylene glycols and polyethylene glycols, mannitol, sorbitol, xylitol, etc.) [5, 6, 26, 28, 32, 34, 37–40], solutions of sugars (glucose, dextrose, fructose, ribose, sucrose, etc.) [18, 19, 25, 27–30, 33, 36, 41], organic acids (oleic and linoleic acids) [42, 43], organic solvents (dimethyl sulphoxide (DMSO) [44, 45], and x-ray contrast agents (verografin, trasograf, and omnipaque) [46–48]. For increasing effectiveness of the multicomponent OCAs, organic solvents (DMSO, thiazone, azone, ethanol, salicylic acid) [11, 13, 40, 42, 43, 49–56], propylene glycol [56, 57], and oleic acid [43, 58–60] are included to their composition. Recently, specially optimized multicomponent solutions as FocusClear™ [61–64], BABB [64–66], Sca*l*e A2, U2, and S [64, 66, 67], CLARITY [16, 66], FASTClear [68], Clear^T [66], SeeDB [64, 69], CUBIC [64, 66, 70], PACT [64, 71], PARS [64, 71], 3DISCO [72], iDISCO [64, 73], uDISCO [64, 74], and others are developed for application *in vitro* at studies with modern microscopic techniques.

Such properties as hydrophilicity, lipophilicity, hyperosmoticity, ability to dissolve lipids and others determine mechanisms of optical clearing. The most of OCAs are hydrophilic ones and include hyperosmotic components such as polyatomic alcohols and sugars. For them several physical and chemical mechanisms of the light scattering reduction are proposed and described as follows: (1) dehydration of tissue components, (2) partial replacement of interstitial fluid with the immersion agent, and (3) structure modification or dissociation of tissue proteins [4, 5, 18, 28, 37, 39, 42, 46, 51, 75–86]. The contact of hyperosmotic OCA with the tissue surface causes the water diffusion from the tissue. These processes produce fast and considerable clearing effect, since, first, tissue becomes denser, and the ordering of scattering components increases, and, second, the concentration of salts and proteins dissolved in the interstitial fluid increases and, therefore, the RI of the interstitial fluid becomes closer to that of the scattering fibrils. It has been shown that hyperosmotic OCAs cause decreasing in weight and volume of tissue samples [37, 85, 86]. The measurement of the tissue spectrum at the wavelengths corresponded to the water absorption band in the NIR and THz ranges has allowed assess the water content change in the tissue during OC [4–6, 26, 84]. In Ref. [87] a quantitative assessment of the dehydration of skin and tumor tissue when using hyperosmotic OCAs by the method of multi-wavelength refractometry for the visible and NIR spectral regions has been presented. The resulting decrease in the RI of the OC solution made it possible to estimate the volume of the fluid extracted from the tissue [37]. However, the complete dehydration of tissue does not occur. When hygroscopic OCAs penetrate tissue the reverse process of rehydration after dehydration can be observed, for example, glycerol, ethylene glycol, glucose solutions, etc., induce rehydration due to the bounding of water molecules inside tissues [37, 86].

For fibrous tissues, such as sclera, *dura mater*, dermis, muscle, etc., both processes, namely, the water loss and the diffusion of the hyperosmotic OCAs into the tissues, occur simultaneously, since molecular size of OCAs is much smaller than the mean separation between fibrils [46]. Rehydration can induce a decrease in tissue transparency after achievement of the maximal value by some percents [80, 86].

Low-molecular OCAs (for example, sugars and polyatomic alcohols) can induce a reversible solubility of tissue protein and lead to the additional reduction of tissue scattering due to decrease of the basic scatterer size [40, 81–83, 86]. The collagen fibers have complex self-organization structure and are the main scattering centers in many fibrous tissues. The measured forces between the triple helices of type I collagen at different temperatures, pH, and solute concentrations [88], together with studies on the interaction of low-molecular agents with collagen leading to its dissolution related to violation of the water network of hydrogen bonds [89], have formed the basis for a theoretical description of this process [40, 81–83]. A molecular mechanism involving a partial substitution of water molecules connected with collagen by molecules of the OCAs, subsequent reversible collagen dissolution, and adjustment of the RI to the index of interfibrillar medium characterizes the post diffusion stage of tissue OC.

The main mechanism of tissue OC with lipophilic OCAs as, for example, oils, includes only penetration of OCAs into intercellular space and matching effect between the RIs of the interstitial fluid and scatterers. However, diffusion of oils into water-containing interstitial space of fibrous tissues is hindered [42]. As the oils are lipophilic, they interact with the lipophilic components, i.e. intercellular lipids (ICL), which consists of free fatty acid, ceramides, and cholesterol with an equimolar ratio [90, 91]. Thus, the use of oils is effective for the optical clearing of such tissues as, for example, stratum corneum (SC). The ICL matrix in the SC plays the key role in maintaining the skin barrier function, the trans-epidermal water loss regulation, penetration of xenobiotics, and drug delivery [91–93]. The skin barrier function is determined by the lateral organization of ICL, which can have three main forms: the orthorhombic structure, which has less permeable and highly ordered states, the hexagonal structure, which is medium permeable and more disordered than orthorhombic, and the fluid state, which is the most permeable and disordered [91, 94–96]. In the mammalian SC, all types of ICL structures are present and vary depth-dependently [96–98].

In vivo studies have shown that most oils are distributed in the uppermost layers of SC [99, 100]. Some oils cause swelling of the skin [99]. In other studies it has been demonstrated that the oils are able to permeate the SC at the application of chemicals (e.g., ethanol) disrupting the SC's barrier integrity [101] or sonophoresis [43]. Thus, oils have a tendency to disorder ICL, disrupt the skin barrier function [102], and enhance the permeation of other substances including hydrophilic OCAs [43, 58–60].

The study of the OC mechanisms in both normal and pathological tissues can provide differentiated results that allow for discriminating pathology. Generally, water content is higher in tumor tissue because angiogenesis occurs to compensate for the lack of oxygen and nutrients. Additionally, the number of cell nuclei (per unit area) is greater in tumor tissue than in normal tissue because the rapid proliferation of tumor cells results in a higher number of cells in the tumor region. An approximately 5% increase of water content in tumor tissues and a greater than 15% increase in percentage of cell nuclei per unit area in tumor tissues were reported [21, 103]. High cell density resulted in higher refractive index by approximately 0.05 in tumor tissues [103]. Since pathological tissues present higher sized structures, have higher mobile water content and higher protein concentrations, the water and OCA diffusivity will be different between normal and pathological tissues. Besides, pathological tissues have higher protein concentrations [104]; evaluation of protein dissociation during OC treatments may also allow for discrimination between normal and pathological tissues.

2.3 Evaluation of Water and OCA Diffusivity in Normal and Pathological Tissues

The OC method can be used to retrieve differentiated information from normal and malignant tissues. When some pathologies like cancer start developing in tissue locations, high-sized structures are built and proteins aggregate in those locations at greater concentrations than in other locations where the tissue is normal [104]. Considering the water content in tissues, similar values have been observed for the total water in normal and pathological mucosa tissues from the human colorectal wall [105], while for the mobile water, pathological tissues present about 5% more than normal tissues [21]. Such difference indicates that when pathology develops, some of the bound water becomes available to transport nutrients and can eventually flow out, if stimulated.

To evaluate differentiated diffusivity of water and OCA in normal and pathological tissues, we need to model such diffusivity and apply such model to experimental results obtained from both tissues. Considering that in most biological tissues, water and OCA diffusion is not made through membranes, we can assume that such diffusion obeys the laws of free diffusion [106]. This means that Fick's law for diffusion can be used to describe the OCA flux in the tissue [18, 46, 106–108].

$$\frac{\partial C_a(x,t)}{\partial t} = D_a \frac{\partial^2 C_a(x,t)}{\partial t^2}. \tag{2.1}$$

The time dependence of the OCA concentration C_a at any unidirectional position x between the two surfaces of the tissue slab is characterized by Eq. (2.1), where the diffusion coefficient of the OCA in the slab is expressed as D_a [18, 26]. The diffusion time τ of the OCA can be calculated from the sample thickness d, and D_a through Eq. (2.2) [18, 26, 46, 107–109].

$$\tau = \frac{d^2}{D_a \times \pi^2}. \tag{2.2}$$

Using treating solutions with a volume 10x higher than the slab's volume to guarantee a continuous OCA flux into the tissue, the amount of dissolved matter m_t in the tissue at an instant t relative to its equilibrium value m_∞ can be determined by Eq. (2.3) [26, 106, 107].

$$\frac{m_t}{m_\infty} = \frac{\int_0^d C_a(x,t)dx}{C_{a0} \times d} = 1 - \frac{8}{\pi^2}\left[\exp\left(-\frac{t}{\tau}\right) + \frac{1}{9}\exp\left(-\frac{9t}{\tau}\right) + \frac{1}{25}\exp\left(-\frac{25t}{\tau}\right) + \ldots\right]. \tag{2.3}$$

The volume averaged concentration of the OCA $C_a(t)$ that is located inside the tissue slab at a particular time of treatment t is represented by the ratio in Eq. (2.3). Using a first order approximation, the solution of Eq. (2.3) is given by [18, 26, 46, 108].

$$C_a(t) = \frac{1}{d}\int_0^d C_a(x,t)dx \cong C_{a0}\left[1 - \exp\left(-\frac{t}{\tau}\right)\right]. \tag{2.4}$$

A relation between the time dependence of OCA concentration within the slab and its characteristic diffusion time in the tissue (τ) is established in Eq. (2.4) [26, 109].

We should consider that Eqs. (2.1)–(2.4) are valid for the description of free diffusion of one type of molecules in a medium. If two fluxes are induced in the system, such as the water flux out and the OCA flux into the tissue, these equations can also be applicable, but the estimated diffusion coefficient D_a or the diffusion time τ in that case will characterize the diffusivity of the global mixed flux [26].

The partial replacement of water by the OCA in the tissue will provide changes both in tissue thickness and in the RI of the interstitial locations and/or inside tissue cells. By measuring tissue thickness and its collimated transmittance (T_c) during treatment, we should be able to evaluate those changes, as it has already been demonstrated [21, 26, 105, 109–113].

If we normalize $C_a(t)$ in Eq. (2.4) to the agent concentration in the treating solution C_{a0}, we obtain a time dependence relation that can be used to fit the T_c increase observed during treatments [21, 26, 105, 109, 111–113].

$$T_c(\lambda,t) = \frac{C_a(t)}{C_{a0}} \cong 1 - \exp\left(-\frac{t}{\tau}\right). \tag{2.5}$$

Some studies have been published where Eqs. (2.2) and (2.5) were used to evaluate the diffusion properties of water and OCAs, such as glucose, glycerol, or ethylene glycol in various tissues [21, 26, 109, 114]. One of those studies was made using both normal and pathological mucosa from the human colorectal wall and allowed to obtain different free water content and different diffusion properties for glucose in normal and pathological tissues [21, 114].

For the case of colorectal carcinoma, adenomatous polyps begin their development in the innermost layer of the colorectal wall—the mucosa, progressing from polyps into invasive adenocarcinoma, which sequentially infiltrates the outer layers of the wall, first the submucosa and then the *muscularis propria*—see Fig. 2.1 [114]. The risk of infiltrating adjacent organs is high if early detection is not made.

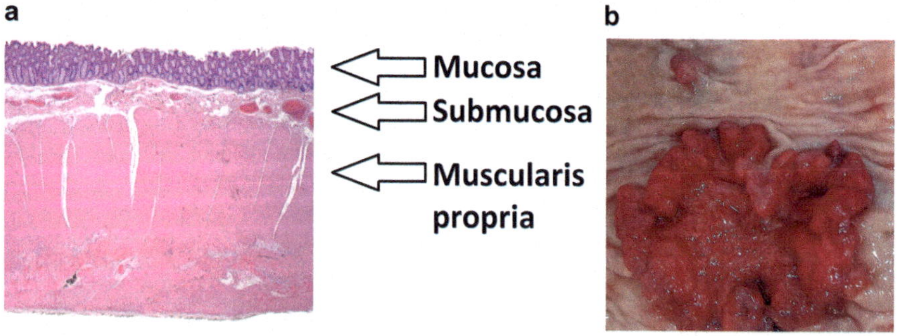

Fig. 2.1 (**a**) Human colorectal histology section showing the distinct layers that compose its wall and (**b**) The gross features of an invasive colorectal adenocarcinoma viewed in a surgical specimen. Images collected at the Portuguese Oncology Institute of Porto, Portugal (reprinted with permission from Ref. [114]

With the objective of obtaining discriminating data between normal and pathological mucosa from the human colorectal wall that can be useful for future diagnostic procedures, both tissues were submitted to OC treatments with glucose solutions. Glucose was diluted in distilled water in different concentrations between 10 and 54% to obtain different osmolarities. After collection from surgical specimens, all fresh normal and pathological samples were prepared with a cryostat microtome with approximated circular slab-form, having a diameter of ~1 cm and a fixed thickness of $d = 0.5$ mm [21]. Considering a particular treatment with a specific glucose osmolarity, a tissue sample was immersed in the solution for 30 min to acquire the collimated transmittance (T_c) spectra during the treatment. The solution had a volume of 10× the sample volume. Three normal mucosa samples were treated with each of the solutions and such procedure was repeated for pathological samples [21, 114]. Since both tissues and glucose do not show significant absorption bands between 600 and 800 nm, wavelengths within this range were selected to evaluate T_c time dependence. Figure 2.2 shows the mean T_c time dependencies at those wavelengths for some of the treatments of normal and pathological mucosa.

According to graphs (a) and (b) of Fig. 2.2, solutions containing low glucose concentrations have more water than the mobile water in the tissues and consequently the water flux out and OCA flux in are limited to the first 5–6 min of treatment. This means that the data after the beginning of saturation must be neglected for the following calculations [21, 114]. Similar behavior is presented in graphs (e) and (f) for treatments with highly concentrated solutions. The water balance between the treating solution and the mobile water in the tissues has been found for solutions containing 40% of glucose for normal mucosa and containing 35% of glucose for pathological mucosa. In these cases (represented in graphs (c) and (d) of Fig. 2.2), a smooth behavior is obtained during the entire treatment.

With the exception of the data in graphs (c) and (d) of Fig. 2.2, the data from all other treatments needed to be trimmed to the first minutes, before saturation is reached. The following step consisted of displacing each of the T_c time dependencies to have $T_c = 0$ at $t = 0$ (untreated tissue) and then normalizing to the highest value in the dataset. After this rearrangement of T_c data, each of the datasets in each treatment has been fitted with a curve described by Eq. (2.5) to estimate τ. By averaging the obtained τ values for each particular treatment, a representation of the mean diffusion time as a function of glucose concentration in the treating solution can be made. Such representation is made in Fig. 2.3 for normal and pathological mucosa of the human colorectal wall [21, 114]. The correspondent numerical values are presented in Table 2.1.

The estimated diffusion time values in Fig. 2.3 have been connected using a smooth spline to evaluate their dependence on the glucose concentration in the treating solution. We see from Fig. 2.3 that the diffusion time is maximum for a glucose concentration of 35.7% in the treating solution for the case of pathological mucosa and for a concentration of 40.6% in the case of normal mucosa. This means that pathological mucosa has a higher mobile water content (64.3%) than normal mucosa (59.4%). From Fig. 2.3, we can also retrieve the τ values for unique glucose (or water) diffusion in both tissues. For glucose, τ has a mean value of 302.4 s in

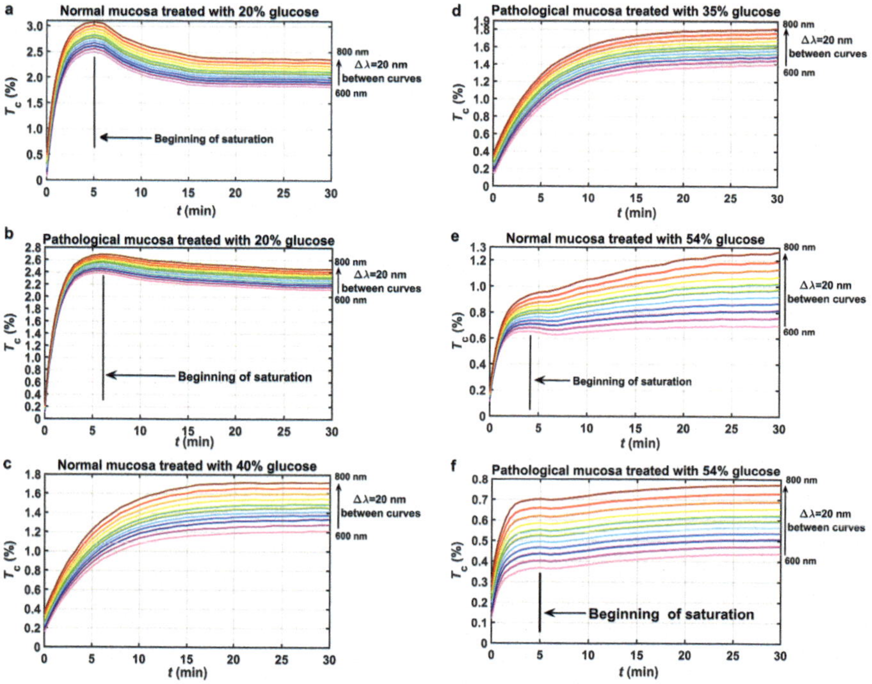

Fig. 2.2 T_c time dependencies of normal mucosa treated with: (**a**) 20% glucose, (**c**) 40% glucose, and (**e**) 54% glucose. T_c time dependencies of pathological mucosa treated with: (**b**) 20% glucose, (**d**) 35% glucose, and (**f**) 54% glucose (reprinted with permission from Ref. [114])

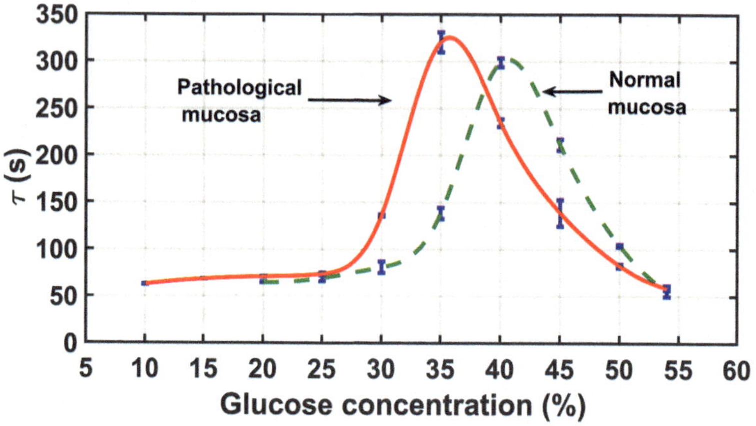

Fig. 2.3 Mean diffusion time as a function of the glucose concentration in the treating solution (reprinted with permission from Ref. [114])

Table 2.1 Mean and standard deviation values for the glucose diffusion time in normal and pathological colorectal mucosa

Tissue type	Normal mucosa									
Glucose concentration (%)	10	15	20	25	30	35	40	45	50	54
Mean diffusion time ($\bar{\tau}$) (s)	—	—	65.1	69.4	81.1	138.4	299.2	211.5	104.3	55.7
Standard deviation	—	—	0.2	3.2	6.1	5.9	4.7	6.1	1.3	5.9
	Pathological mucosa									
Mean diffusion time ($\bar{\tau}$) (s)	62.9	68.6	71.1	73.9	136.1	320.6	234.9	139.0	82.7	58.4
Standard deviation	0.5	0.2	0.5	1.5	1.1	10.6	4.1	14.0	2.0	1.7

normal mucosa and a mean value of 325.1 s in pathological mucosa, showing that it takes more time for glucose to diffuse in pathological tissue. For water, τ presents similar mean values for both tissues: 55.7 s in normal mucosa and 58.4 s in pathological mucosa [21, 114].

Using these values in Eq. (2.2) with the corresponding sample thickness d retrieved from the graphs in Fig. 2.4, the diffusion coefficients for glucose and water can be calculated for normal and pathological mucosa. These values are: $D_{glucose} = 5.8 \times 10^{-7}$ cm^2/s and $D_{H_2O} = 3.3 \times 10^{-6}$ cm^2/s in normal mucosa and $D_{glucose} = 4.4 \times 10^{-7}$ cm^2/s and $D_{H_2O} = 2.4 \times 10^{-6}$ cm^2/s in pathological mucosa [114].

Each of the datasets in graphs of Fig. 2.4 represents the mean of three thickness variation studies, where all natural samples (at $t = 0$) had 0.5 mm thickness. Considering the initial thickness decrease that is associated with tissue dehydration and occurs within the first 2 min, the standard deviation (SD) for the three studies begins with high values, typically near 10% of the mean thickness values represented. SD decreases during the dehydration of tissues, reaching values between 1.5 and 2% of the mean thickness near 2 min of treatment. After 2 min, as a result of glucose inclusion, thickness begins to increase again until it stabilizes between 10 and 15 min. From 2 min to thickness stabilization, the SD decreases to zero.

This study showed that glucose diffusion properties are different for normal and pathological colorectal mucosa. It has also demonstrated the existence of different mobile water content in both tissues. Similar studies can be performed for other tissues and using different OCAs. Although this method can be applied only for *ex vivo* tissues, an *in vivo* methodology can also be applied using a diffuse reflectance (R_d) setup [110] to acquire spectral measurements and OCT imaging to evaluate sample thickness variations during treatment [25].

Combination OCT with OC has a great prospect as a routine research tool for analysis of identifying the boundaries between normal and diseased breast tissue *in vitro* and *in vivo*. Ref. [32] shows that the OCT imaging depth and imaging contrast of breast tissues have been improved after application of 60% glycerol in the 2-D OCT images. Besides, the permeability coefficient of a hyperosmotic agent in

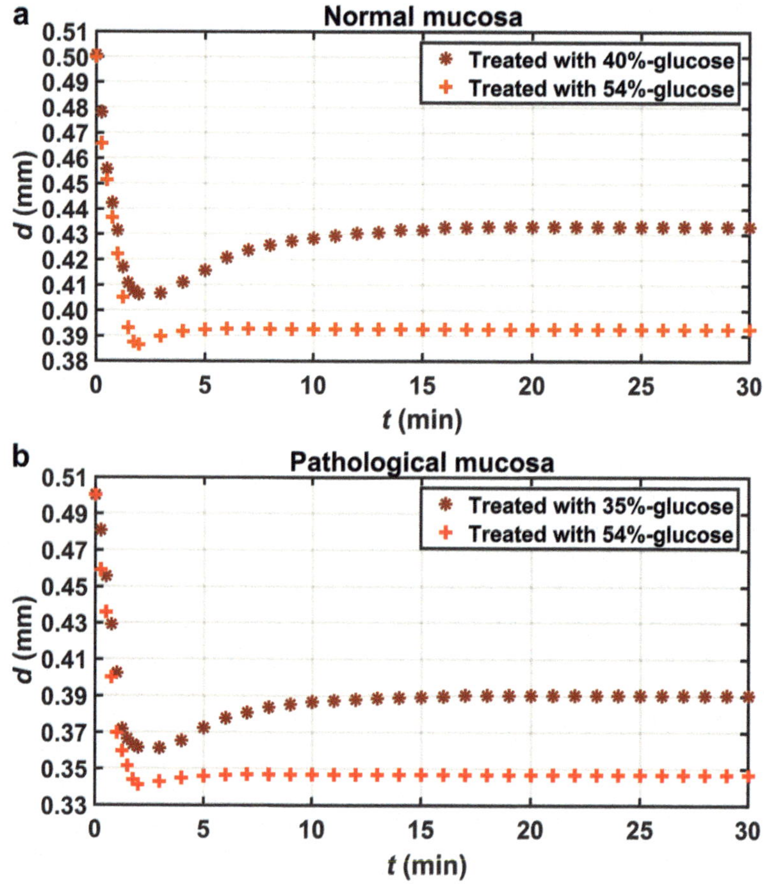

Fig. 2.4 Thickness variations for (**a**) normal and (**b**) pathological mucosa under treatment with glucose solutions (reprinted with permission from Ref. [114])

a specific region of tissues *ex vivo* can be calculated by the OCT signal slope (OCTSS) method, which is computed by analyzing the slope changes in the OCT signal caused by the hyperosmotic agent diffusion [24, 25, 27, 32]. Due to the OCA diffusion in tissues, the light scattering coefficient decreases in the specific depth region, which can be clearly seen from the OCTSS graphs (see Fig. 2.5) [32].

It is clearly seen that the slope of OCT signals for breast tissues decreases with the diffusion of glycerol into the tissues, therefore, the water and intercellular fluids are drawn out from tissues. Then the reverse process starts due to water is drawn back into the cells as a result of its affinity for water.

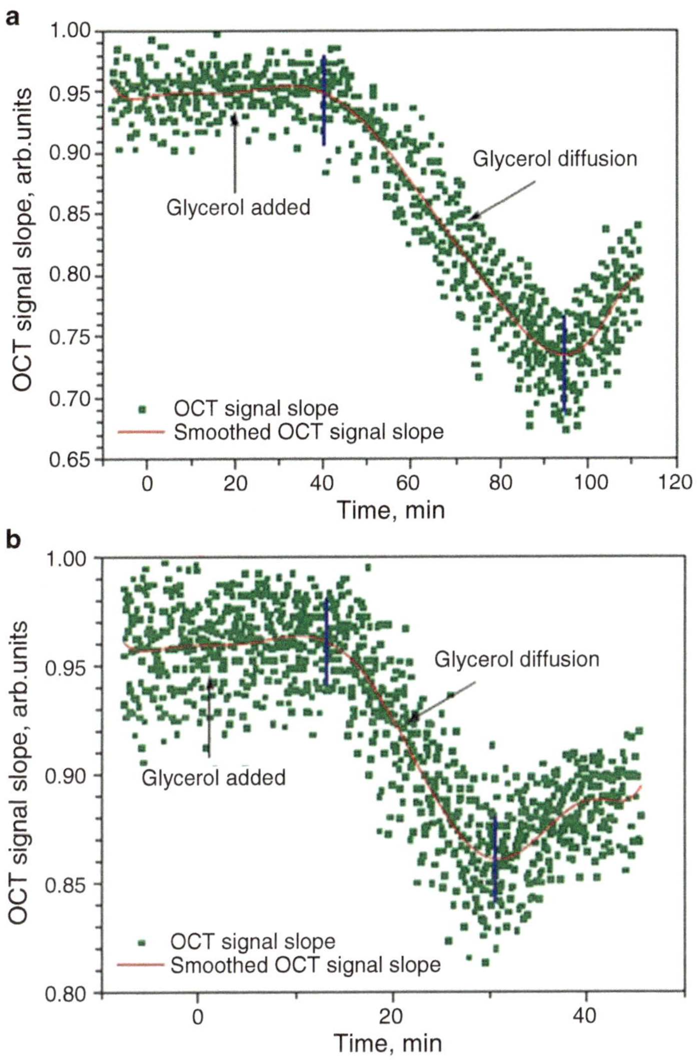

Fig. 2.5 OCTSS graphs as a function of time from normal breast tissues treatment with 60% glycerol (**a**), breast cancer tissues treatment with 60% glycerol (**b**). The monitored region was about 290 μm in thickness and 300 μm away from the surface. The tick marks indicate the interval of glycerol diffusion (reprinted with permission from Ref. [32])

A linear region with minimal fluctuations from the 1-D OCT signal is selected to analyse the changes of the sample optical properties as a function of time after treatment. The permeability coefficients of the 60% glycerol of breast tissues were calculated using the following equation:

$$P = \frac{z_{\text{region}}}{t_{\text{region}}}, \quad (2.6)$$

where P is the permeability coefficients, the z_{region} is the thickness of the selection region, and t_{region} is the time for diffusing through that region [25].

The permeability coefficient of 60% glycerol is $(3.14 \pm 0.07) \times 10^{-5}$ cm/s in breast cancer tissues, and $(0.89 \pm 0.02) \times 10^{-5}$ cm/s in normal breast tissues, respectively. Thus, the permeability coefficient of glycerol in cancer tissues is 3.54-fold than that in normal tissues.

The same approach has been used for the study of diffusion of 20% aqueous solution of glucose in normal stomach and gastric tumor tissues [115]. The permeability coefficients in a model gastric cancer and stomach tissue have been evaluated as $(5.32 \pm 0.17) \times 10^{-5}$ cm/s and $(0.94 \pm 0.04) \times 10^{-5}$ cm/s, respectively, that also demonstrates a higher permeability coefficient of tumor tissues compared to normal tissues quantified from OCT images.

It has been shown that terahertz radiation is promising for establishing early stages of skin cancer [116, 117]. As compared with healthy tissues, cancer cells contain a larger amount of water, which intensively absorbs radiation in the THz. The use of THz spectroscopy (from 16 to 30 THz) has allowed for estimating dehydration degree of skin and model tumor of liver cancer in rat under action of the following OCAs: 40% aqueous solution of glucose, propylene glycol (99.9%), polyethylene glycol 600, and glycerol (99.9%). Significant increase in transmission for the transplanted-tumor tissue has been observed in this spectral range due to hyperosmotic impact of the OCAs that can be an addition criterion of tissue malignancy [5].

2.4 Discrimination of Pathological from Normal Tissues Using Protein Dissociation

The third OC mechanism, protein dissociation, is also important and it allows discriminating pathological from normal tissues. According to Ref. [104], pathological tissues have higher protein concentrations than normal tissues. It has also been demonstrated that during the application of OC treatments to the skin, proteins dissociate under the action of agents like glycerol or ethylene glycol [118, 119], and such dissociation is reversible after treatment application [118]. Proteins present a significant absorption band at 200 nm [120], meaning that if we can measure at that wavelength, we can possibly obtain differentiated data for normal and pathological tissues regarding protein dissociation. Using both normal and pathological colorectal mucosa samples, T_c spectra between 200 and 1000 nm from these samples under treatment with glycerol (93% pure) were measured. The samples were prepared in the same way as the samples used in the study to evaluate glucose diffusion properties presented above. As we can see from Fig. 2.6, the T_c levels at deep UV are very low for both tissues. For visible—NIR wavelengths, T_c increases fast at the beginning and then saturates. For deep UV, near 200 nm, we do not see variations due to the low levels of T_c.

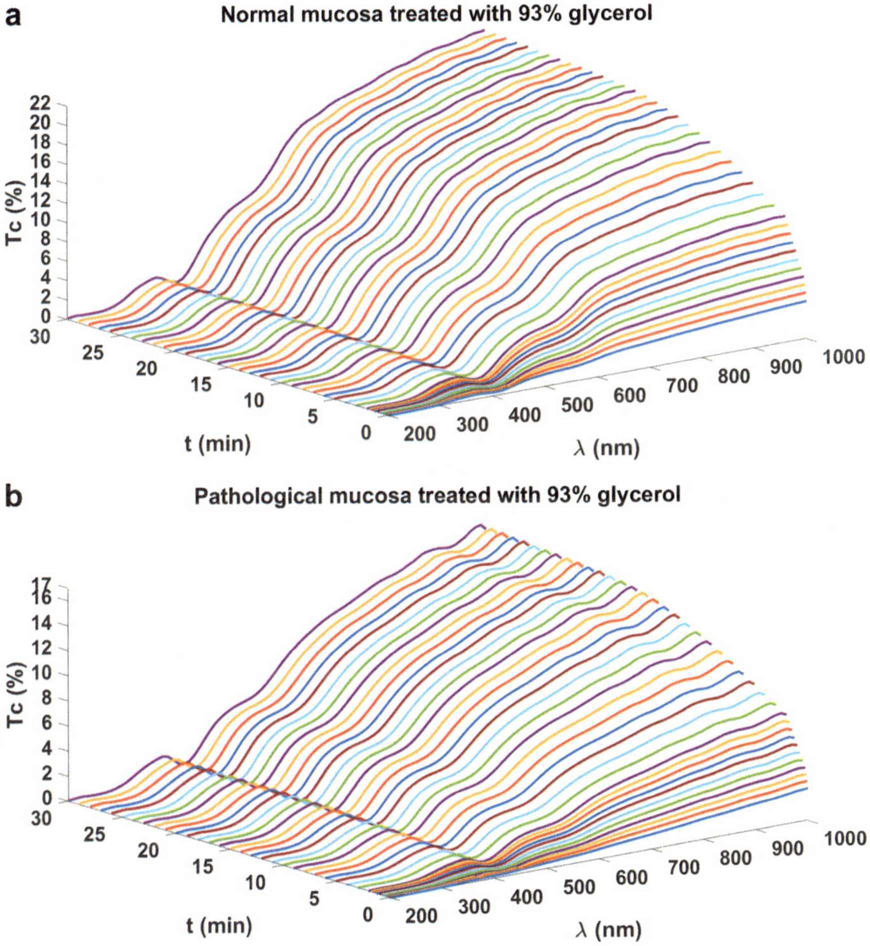

Fig. 2.6 Time dependence for tissue T_c spectra during treatment with 93% glycerol: (**a**) normal mucosa and (**b**) pathological mucosa

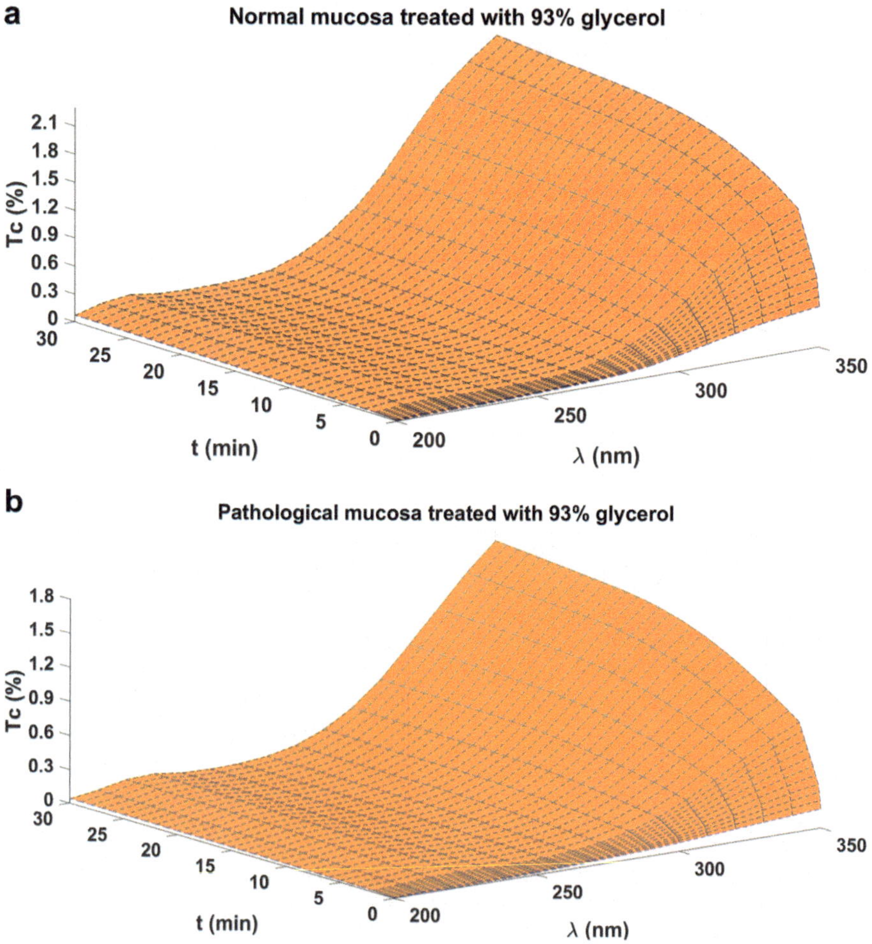

Fig. 2.7 T_c increase at low wavelengths during treatment with 93% glycerol: (**a**) normal mucosa and (**b**) pathological mucosa

If we zoom the above graphs between 200 and 350 nm, we can see that OC also creates a T_c increase on both sides of the DNA absorption band (located at 260 nm [120])—see Fig. 2.7.

From the graphs in Fig. 2.7 we already see that T_c also increases above and below 260 nm, but the increase observed at 200 nm seems negligible. For better visualization of such increase at short wavelengths and comparison of it with the increase seen for visible-NIR wavelengths, it is need to calculate the percent increase of Tc for the entire wavelength range, using Eq. (2.7):

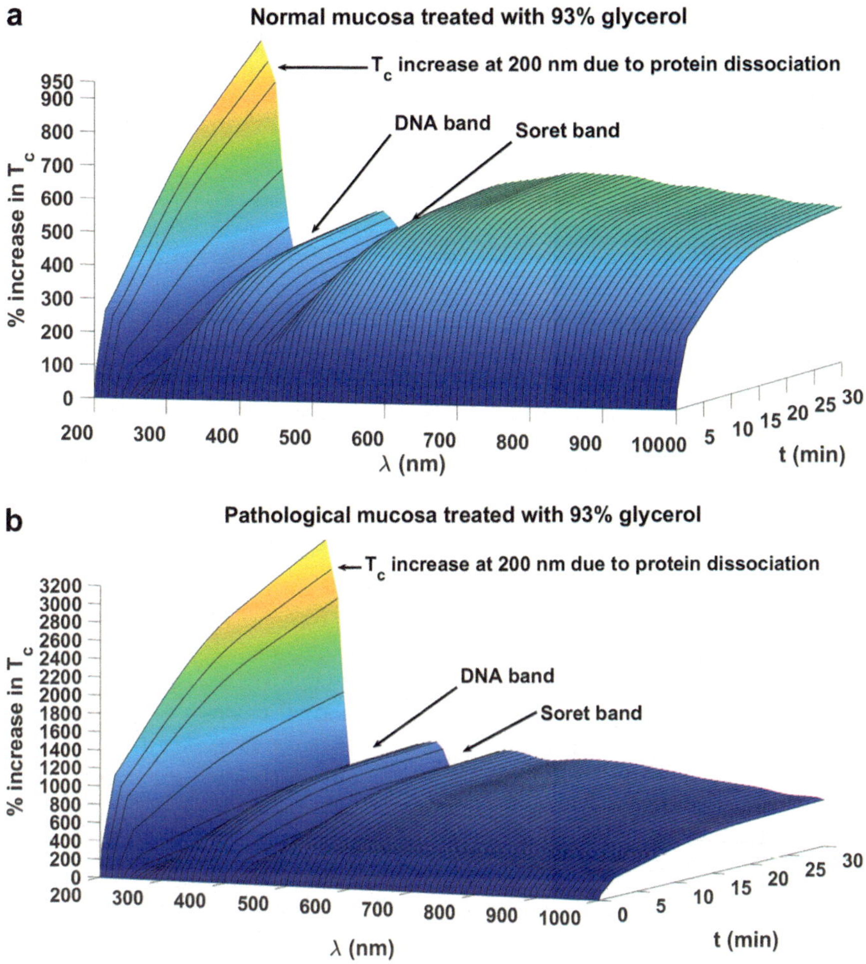

Fig. 2.8 Percent increase of T_c during treatment with 93% glycerol: (**a**) normal mucosa and (**b**) pathological mucosa

$$\% \uparrow T_c(\lambda, t) = \frac{T_c(\lambda, t) - T_c(\lambda, t=0)}{T_c(\lambda, t=0)} \times 100\%. \tag{2.7}$$

The resulting graphs from this calculation are presented in Fig. 2.8.

The graphs in Fig. 2.8 provide more information than the previous. First, we see that above 260 nm, T_c increases fast at the beginning and then saturates. This behavior is observed between 260 nm (DNA absorption band [120]) and 420 nm (Soret band) and above 420 nm. In these ranges, the T_c time dependencies can be fitted

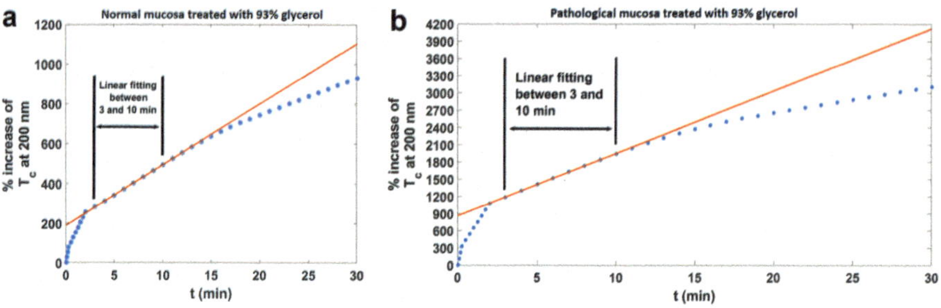

Fig. 2.9 Linear fitting of the percent increase of T_c at 200 nm for: (**a**) normal mucosa and (**b**) pathological mucosa

with a curve described by Eq. (2.5), as we have done for the determination of glucose diffusion properties above. For wavelengths shorter than 260 nm, we see a different type of increase in T_c. Such anomalous increase is created by protein dissociation, and different magnitudes for this increase between normal and pathological tissues are observed. Considering the 30 min treatment, we see that for normal mucosa, the percent increase reaches approximately 950%, while for the pathological mucosa, a percent increase of about 3200% is observed. The fact that a higher percent increase is observed for pathological mucosa proves the concept that these tissues have higher protein concentrations than normal mucosa. Considering the time dependencies for both tissues at 200 nm, the data cannot be fitted by Eq. (2.5). Since for future diagnostic procedures it is desirable to use a fast method, we have considered only the data between 3 and 10 min. Such data can be fitted for both tissues by a straight line, as presented in graphs of Fig. 2.9.

Since the data between 3 and 10 min has linear behavior, the slope of the calculated lines can be found and shown that it is bigger for pathological mucosa. The slope obtained for the case of normal mucosa is 27.4. while for the case of pathological mucosa the slope is 79.1.

Although these measurements have been obtained from *ex vivo* tissue samples, same methodology can be implemented using an R_d setup for *in vivo* measurements.

2.5 Optical Clearing Assisted Multimodal Tumor Imaging

The OC technique allows greatly enhancing the resolution, contrast, and visualization depth of optical methods for tumor diagnostics. In particular, OC-assisted microscopy has opened up new avenues to visualize the entire central nervous system, blood microvessel system, whole organs, and whole body with subcellular resolution in health and tumors, allowed for single-cell studies and biomedical engineering [121]. For example, Cui et al. [122] used OCA in Hyper Spectral

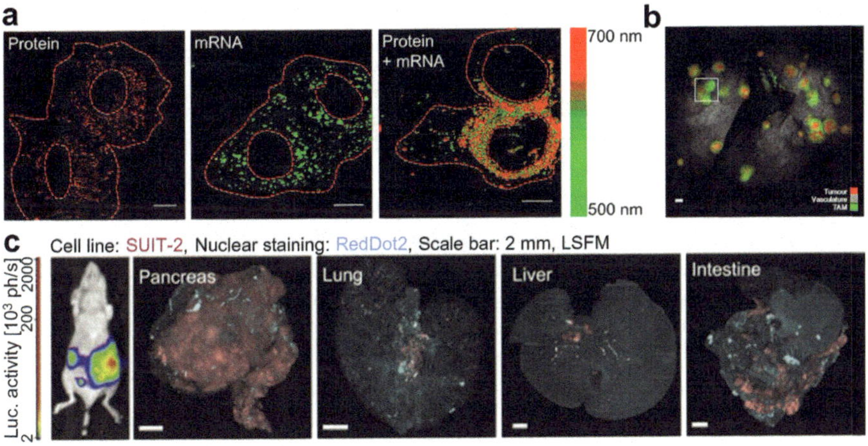

Fig. 2.10 (**a**) Multiplex detection of HER2 protein and mRNA in the same cancer cell. (Scale bars: 10 μm) (reprinted from Ref. [122] with permission), (**b**) Visualization of tumor cells (red) and tumor-associated macrophages (green) in lung (reprinted from Ref. [133] with permission), (**c**) Simultaneous visualization of cancer metastases within various organs at the whole-body clearing (reprinted with permission from Ref. [135])

Dark-Field Microscopy (HSDFM) that allowed quantitative mapping of the dimerization-activated receptor kinase HER2 (Human Epidermal Growth Factor Receptor 2) in a single cancer cell by non-fluorescent approach. *In situ* multiplex detection of mRNA and HER2protein was made possible by HSDFM in a dual-labeled cell (see Fig. 2.10a). The authors also showed high spatiotemporal resolution at the characterization of intracellularly grown gold particulates achieved owing to optical clearing [122].

Using OC-assisted label-free stimulated Raman scattering microscopy Wei et al. [123] could reveal significant differences in lipid-to-protein ratios between healthy brain and glioblastoma tissues and interestingly, also between glioblastoma tumor cells itself. Fluorescent confocal imaging, two-photon imaging, photoacoustic imaging, and image reconstruction of optically cleared brain tissue (using CLARITY and iDISCO) allowed quantification of the 3D microvascular characteristics in healthy mouse brains and in tissues with diffuse, infiltrative growing GBM8 brain tumors [124].

Treatment of human prostate biopsy specimens [125] and bladder tumors [126] with either CUBIC or PACT resulted in GFP (Green Fluorescent Protein) fluorescence preservation, minimal tissue deformation, and feasibility of whole-mount two-photon imaging. van Royen et al. [127] showed that Murray's Clear combined with immunochemistry allowed to shed light on architectural differences between grades of prostate cancer. Using iDISCO protocols the authors characterized 3D architecture of both benign and pre-cancerous prostate lesions [128] and described two distinct growth patterns of prostate cancer in patients [129].

Hume et al. [130] proved CUBIC to be an effective clearing agent for human breast tumor biopsy and further applied this method in conjunction with immunostaining to monitor structural integrity of engineered 3D *in vitro* adipose tissue model recapitulating breast tumor microenvironment.

Pioneering lung BABB clearing was reported by Scott et al. [131], who presented the first detailed view on the relationship between nerves, vessels, and airways architecture, detected novel patterns of pleural innervations, and connectivity patterns of pleural nerves. Solvent-based clearing proved also beneficial for the detection and analysis of lung tumor metastases [132]. OC of lungs was also essential for studies analyzing tumor-immune microenvironment, for example, 3D pattern of tumor-associated macrophages distribution (see Fig. 2.10b) [133, 134].

Kubota et al. [135] demonstrated the possibilities of whole-body clearing for the monitoring of tumor metastases in the variety of cancer cell lines using CUBIC (see Fig. 2.10c).

OC was successfully used to monitor the hidden tumors: for example, in laser speckle contrast imaging of microvessels [136] and prostate tumor [8]; in optical transmission and emission tomography of blood vessels stained with light-absorbing inks [137, 138]; in Raman spectroscopy of subcutaneous tumor [139].

Pires et al. [140] used white-light diffuse reflectance spectroscopy and near-infrared optical coherence tomography to evaluate the effect of a topically applied OCA in melanoma *in vivo* and to image the microvascular network. Improved contrast resolution of the microvasculature in the pigmented melanoma tissue was achieved with optical clearing to a depth of ~750 μm.

2.6 Conclusion

We have reviewed the specific features and methods of optical clearing related to visualization of tumors and differentiation of healthy and malignant tissues. The impact of the OCA on a tissue allows for efficient control of the optical properties, particularly, the reduction of tissue scattering coefficient, which facilitates the increase of efficiency of optical imaging (optical biopsy) in cancer diagnostics. Differences in structure and composition of normal and pathological tissue cause differences in OCA diffusivity and protein dissociation during OC treatments that may also allow discrimination between them.

Acknowledgments The work was supported by the Portuguese grant FCT-UID/EQU/04730/2019 and Russian Federation grants RFBR 17-02-00358 and 18-52-16025 NTSNIL_a, and Russian Federation Governmental grant 14.W03.31.0023. The authors are thankful to the colleagues of the Portuguese Oncology Institute of Porto for collecting, preparing and photographing the biological tissues in some of the research presented in this chapter.

References

1. Agrba, P.D., Kirillin, M.Y., Abelevich, A.I., et al.: Compression as a method for increasing the informativity of optical coherence tomography of biotissues. Opt. Spectrosc. **107**(6), 853–858 (2009)
2. Drew, C., Milner, T.E., Rylander, C.G.: Mechanical tissue optical clearing devices: evaluation of enhanced light penetration in skin using optical coherence tomography. J. Biomed. Opt. **14**(6), 064019 (2009)
3. Oliveira, L., Lage, A., Clemente, M.P., Tuchin, V.: Rat muscle opacity decrease due to the osmosis of a simple mixture. J. Biomed. Opt. **15**(5), 055004 (2010)
4. Yu, T., Wen, X., Tuchin, V.V., et al.: Quantitative analysis of dehydration in porcine skin for assessing mechanism of optical clearing. J. Biomed. Opt. **16**(9), 095002 (2011)
5. Kolesnikov, A.S., Kolesnikova, E.A., Kolesnikova, K.N., et al.: THz monitoring of the dehydration of biological tissues affected by hyperosmotic agents. Phys. Wave Phenom. **22**(3), 169–176 (2014)
6. Smolyanskaya, O.A., Schelkanova, I.J., Kulya, M.S., et al.: Glycerol dehydration of native and diabetic animal tissues studied by THz-TDS and NMR methods. Biomed. Opt. Express. **10**(3), 1198–1215 (2018)
7. Wang, J., Zhang, Y., Xu, T., et al.: An innovative transparent cranial window based on skull optical clearing. Laser Phys. Lett. **9**, 469–473 (2012)
8. Abookasis, D., Moshe, T.: Reconstruction enhancement of hidden objects using multiple speckle contrast projections and optical clearing agents. Opt. Commun. **300**, 58–64 (2013)
9. Feng, W., Shi, R., Zhang, C., et al.: Vizualization of skin microvascular dysfunction of type I diabetic mice using *in vivo* skin optical clearing method. J. Biomed. Opt. **24**(3), 031003 (2019)
10. Larin, K.V., Ghosn, M.G., Bashkatov, A.N., et al.: Optical clearing for OCT image enhancement and in-depth monitoring of molecular diffusion. IEEE J. Sel. Top Quantum Electron. **18**, 1244–1259 (2012)
11. Wen, X., Jacques, S.L., Tuchin, V.V., Zhu, D.: Enhanced optical clearing of skin *in vivo* and optical coherence tomography in-depth imaging. J. Biomed. Opt. **17**, 066022 (2012)
12. Choi, W.J., Wang, R.K.: Volumetric cutaneous microangiography of human skin *in vivo* by VCSEL swept-source optical coherence tomography. Quantum. Electron. **44**(8), 740–745 (2014)
13. Zhao, Q., Dai, C., Fan, S., et al.: Synergistic efficacy of salicylic acid with a penetration enhancer on human skin monitored by OCT and diffuse reflectance spectroscopy. Sci. Rep. **6**, 34954 (2016). https://doi.org/10.1038/srep34954
14. Huang, D., Zhang, W., Zhong, H., et al.: Optical clearing of porcine skin tissue *in vitro* studied by Raman microspectroscopy. J. Biomed. Opt. **17**, 015004 (2012)
15. Zhang, Y., Liu, H., Tang, J., et al.: Noninvasively imaging subcutaneous tumor xenograft by handheld Raman detector with assistance of optical clearing agent. ACS Appl. Mater. Interfaces. **9**, 17769–17776 (2017)
16. Chang, E.H., Argyelan, M., Aggarwal, M., et al.: The role of myelination in measures of white matter integrity: combination of diffusion tensor imaging and two-photon microscopy of CLARITY intact brains. NeuroImage. **147**, 253–261 (2017)
17. Richardson, L., Vargas, G., Brown, T., et al.: Redefining 3Dimensional placental membrane microarchitecture using multiphoton microscopy and optical clearing. Placenta. **53**, 66–75 (2017)
18. Bashkatov, A.N., Genina, E.A., Tuchin, V.V.: Measurement of glucose diffusion coefficients in human tissues. In: Tuchin, V.V. (ed.) Handbook of Optical Sensing of Glucose in Biological Fluids and Tissues, pp. 587–621. CRC Press, Boca Raton (2009)
19. Ullah, H., Hussain, F., Ikram, M.: Optical coherence tomography for glucose monitoring in blood. Appl. Phys. B Lasers Opt. **120**, 355–366 (2015)

20. Guo, X., Guo, Z., Wei, H., et al.: *In vivo* comparison of the optical clearing efficacy of optical clearing agents in human skin by quantifying permeability using optical coherence tomography. Photochem. Photobiol. **87**, 734–740 (2011)
21. Carneiro, I., Carvalho, S., Henrique, R., et al.: A robust *ex vivo* method to evaluate the diffusion properties of agents in biological tissues. J. Biophotonics. **12**, e201800333 (2019)
22. Genina, E.A., Bashkatov, A.N., Zubkova, E.A., et al.: Measurements of Retinalamin diffusion coefficient in human sclera by optical spectroscopy. Opt. Lasers Eng. **46**, 915–920 (2008)
23. Genina, E.A., Bashkatov, A.N., Tuchin, V.V., et al.: Cortexin diffusion in human eye sclera. Quantum Electron. **41**(5), 407–413 (2011)
24. Ghosn, M.G., Tuchin, V.V., Larin, K.V., et al.: Nondestructive quantification of analyte diffusion in cornea and sclera using optical coherence tomography. Invest. Ophthalmol. Vis. Sci. **48**, 2726–2733 (2007)
25. Ghosn, M.G., Carbajal, E.F., Befrui, N., et al.: Differential permeability rate and percent clearing of glucose in different regions in rabbit sclera. J. Biomed. Opt. **13**, 021110 (2008)
26. Oliveira, L.M., Carvalho, M.I., Nogueira, E.M., Tuchin, V.V.: Diffusion characteristics of ethylene glycol in skeletal muscle. J. Biomed. Opt. **20**, 051019 (2015)
27. Ghosn, M.G., Sudheendran, N., Wendt, M., et al.: Monitoring of glucose permeability in monkey skin *in vivo* using optical coherence tomography. J. Biophotonics. **3**, 25–33 (2010)
28. Bashkatov, A.N., Genina, E.A., Sinichkin, Y.P., et al.: Glucose and mannitol diffusion in human *dura mater*. Biophys. J. **85**(5), 3310–3318 (2003)
29. Larin, K.V., Ghosn, M.G., Ivers, S.N., et al.: Quantification of glucose diffusion in arterial tissues by using optical coherence tomography. Laser Phys. Lett. **4**(4), 312–317 (2007)
30. Guo, X., Wu, G., Wei, H., et al.: Quantification of glucose diffusion in human lung tissues by using Fourier domain optical coherence tomography. Photochem. Photobiol. **88**, 311–316 (2012)
31. Ghosn, M.G., Carbajal, E.F., Befrui, N.A., et al.: Permeability of hyperosmotic agent in normal and atherosclerotic vascular tissues. J. Biomed. Opt. **13**(1), 010505 (2008)
32. Zhong, H.Q., Guo, Z.Y., Wei, H.J., et al.: Quantification of glycerol diffusion in human normal and cancer breast tissues *in vitro* with optical coherence tomography. Laser Phys. Lett. **7**, 315–320 (2010)
33. Tuchina, D.K., Shi, R., Bashkatov, A.N., et al.: *Ex vivo* optical measurements of glucose diffusion kinetics in native and diabetic mouse skin. J. Biophotonics. **8**(4), 332–346 (2015)
34. Tuchina, D.K., Bashkatov, A.N., Bucharskaya, A.B., et al.: Study of glycerol diffusion in skin and myocardium *ex vivo* under the conditions of developing alloxan-induced diabetes. J Biomed Photon Eng. **3**(2), 020302 (2017)
35. Tuchina, D.K., Tuchin, V.V.: Optical and structural properties of biological tissues under diabetes mellitus. J Biomed Photon Eng. **4**(2), 020201 (2018)
36. Carvalho, S., Gueiral, N., Nogueira, E., et al.: Glucose diffusion in colorectal mucosa - a comparative study between normal and cancer tissues. J. Biomed. Opt. **22**(9), 091506 (2017)
37. Genina, E.A., Bashkatov, A.N., Korobko, A.A., et al.: Optical clearing of human skin: comparative study of permeability and dehydration of intact and photothermally perforated skin. J. Biomed. Opt. **13**, 021102 (2008)
38. Genina, E.A., Bashkatov, A.N., Tuchin, V.V.: Optical clearing of cranial bone. Adv Opt Technol. **2008**, 267867 (2008)
39. Wen, X., Mao, Z., Han, Z., et al.: *In vivo* skin optical clearing by glycerol solutions: mechanism. J. Biophotonics. **3**(1–2), 44–52 (2010)
40. Zhu, Z., Wu, G., Wei, H., et al.: Investigation of the permeability and optical clearing ability of different analytes in human normal and cancerous breast tissues by spectral domain OCT. J. Biophotonics. **5**, 1–8 (2012)
41. Feng, W., Shi, R., Ma, N., et al.: Skin optical clearing potential of disaccharides. J. Biomed. Opt. **21**(8), 081207 (2016)
42. Choi, B., Tsu, L., Chen, E., et al.: Determination of chemical agent optical clearing potential using *in vitro* human skin. Lasers Surg. Med. **36**(2), 72–75 (2005)

43. Zaitsev, S.V., Svenskaya, Y.I., Lengert, E.V., et al.: Optimized skin optical clearing for optical coherence tomography monitoring of encapsulated drug delivery through the hair follicles. J. Biophotonics. (2019). https://doi.org/10.1002/jbio.201960020
44. Zimmerley, M., McClure, R.A., Choi, B., Potma, E.O.: Following dimethyl sulfoxide skin optical clearing dynamics with quantitative nonlinear multimodal microscopy. Appl. Opt. **48**(10), D79–D87 (2009)
45. Andanson, J.M., Chan, K.L.A., Kazarian, S.G.: High-throughput spectroscopic imaging applied to permeation through the skin. Appl. Spectrosc. **63**(5), 512–517 (2009)
46. Tuchin, V.V., Maksimova, I.L., Zimnyakov, D.A., et al.: Light propagation in tissues with controlled optical properties. J. Biomed. Opt. **2**, 401–417 (1997)
47. Sdobnov, A.Y., Darvin, M.E., Lademann, J., Tuchin, V.V.: A comparative study of *ex vivo* skin optical clearing using two-photon microscopy. J. Biophotonics. **10**(9), 1115–1123 (2017)
48. Sdobnov, A.Y., Darvin, M.E., Schleusener, J., et al.: Hydrogen bound water profiles in the skin influenced by optical clearing molecular agents - quantitative analysis using confocal Raman microscopy. J. Biophotonics. **12**(5), e201800283 (2019)
49. Masoumi, S., Ansari, M.A., Mohajerani, E., et al.: Combination of analytical and experimental optical clearing of rodent specimen for detecting beta-carotene: phantom study. J. Biomed. Opt. **23**(9), 095002 (2018). https://doi.org/10.1117/1.JBO.23.9.095002
50. Chang, C.-H., Myers, E.M., Kennelly, M.J., Fried, N.M.: Optical clearing of vaginal tissues, *ex vivo*, for minimally invasive laser treatment of female stress urinary incontinence. J. Biomed. Opt. **22**(1), 018002 (2017)
51. Genina, E.A., Bashkatov, A.N., Kolesnikova, E.A., et al.: Optical coherence tomography monitoring of enhanced skin optical clearing in rats *in vivo*. J. Biomed. Opt. **19**(2), 021109 (2014)
52. Gallwas, J., Stanchi, A., Ditsch, N., et al.: Effect of optical clearing agents on optical coherence tomography images of cervical epithelium. Lasers Med. Sci. **30**, 517–525 (2015)
53. Hatanaka, T., Shimoyama, M., Sugibayashi, K., Morimoto, Y.: Effect of vehicle on the skin permeability of drugs: polyethylene glycol 400-water and ethanol-water binary solvents. J. Control. Release. **23**, 247–260 (1993)
54. Inamori, T., Ghanem, A.-H., Higuchi, W.I., Srinivasan, V.: Macromolecule transport in and effective pore size of ethanol pretreated human epidermal membrane. Int. J. Pharm. **105**, 113–123 (1994)
55. Lane, M.E.: Skin penetration enhancers. Int. J. Pharm. **447**, 12–21 (2013)
56. Levang, A.K., Zhao, K., Singh, J.: Effect of ethanol/propylene glycol on the *in vitro* percutaneous absorption of aspirin, biophysical changes and macroscopic barrier properties of the skin. Int. J. Pharm. **181**, 255–263 (1999)
57. Zhi, Z., Han, Z., Luo, Q., Zhu, D.: Improve optical clearing of skin *in vitro* with propylene glycol as a penetration enhancer. J Innov Opt Health Sci. **2**(3), 269–278 (2009)
58. Jiang, J., Wang, R.K.: How different molarities of oleic acid as enhancer exert its effect on optical clearing of skin tissue *in vitro*. J. Xray Sci. Technol. **13**, 149–159 (2005)
59. Ding, Y., Wang, J., Fan, Z., et al.: Signal and depth enhancement for *in vivo* flow cytometer measurement of ear skin by optical clearing agents. Biomed. Opt. Express. **4**(11), 2518–2526 (2013)
60. Liu, Y., Yang, X., Zhu, D., Luo, Q.: Optical clearing agents improve photoacoustic imaging in the optical diffusive regime. Opt. Lett. **38**(20), 4236–4239 (2013)
61. Chiang, A.S., Lin, W.Y., Liu, H.P., et al.: Insect NMDA receptors mediate juvenile hormone biosynthesis. Proc. Natl. Acad. Sci. U. S. A. **99**, 37–42 (2002)
62. Fu, Y.-Y., Lin, C.-W., Enikolopov, G., et al.: A microtome-free 3-dimensional confocal imaging method for visualization of mouse intestine with subcellular-level resolution. Gastroenterology. **137**(2), 453–465 (2009)
63. Moy, A.J., Capulong, B.V., Saager, R.B., et al.: Optical properties of mouse brain tissue after optical clearing with FocusClear™. J. Biomed. Opt. **20**(9), 095010 (2015)

64. Richardson, D.S., Lichtman, J.W.: Clarifying tissue clearing. Cell. **162**, 246–257 (2015)
65. Dodt, H.U., Leischner, U., Schierloh, A., et al.: Ultramicroscopy: three-dimensional visualization of neuronal networks in the whole mouse brain. Nat. Methods. **4**, 331–336 (2007)
66. Seo, J., Choe, M., Kim, S.-Y.: Clearing and labeling techniques for large-scale biological tissues. Mol. Cells. **39**(6), 439–446 (2016)
67. Hama, H., Kurokawa, H., Kawano, H., et al.: Sca*l*e: a chemical approach for fluorescence imaging and reconstruction of transparent mouse brain. Nat. Neurosci. **14**, 1–10 (2011)
68. Perbellini, F., Liu, A.K.L., Watson, S.A., et al.: Free-of-acrylamide SDS-based tissue clearing (FASTClear) for three dimensional visualization of myocardial tissue. Sci. Rep. **7**, 5188 (2017). https://doi.org/10.1038/s41598-017-05406-w
69. Ke, M.-T., Nakai, Y., Fujimoto, S., et al.: Super-resolution mapping of neuronal circuitry with index-optimized clearing agent. Cell Rep. **14**, 2718–2732 (2016)
70. Susaki, E.A., Tainaka, K., Perrin, D., et al.: Whole-brain imaging with single-cell resolution using chemical cocktails and computational analysis. Cell. **157**, 726–739 (2014)
71. Yang, B., Treweek, J.B., Kulkami, R.P., et al.: Single-cell phenotyping within transparent intact tissue through whole-body clearing. Cell. **158**, 945–958 (2014)
72. Erturk, A., Becker, K., Jahrling, N., et al.: Three-dimensional imaging of solvent-cleared organs using 3DISCO. Nat. Protoc. **7**(11), 1983–1995 (2012)
73. Renier, N., Wu, Z., Simon, D.J., et al.: iDISCO: a simple, rapid method to immunolabel large tissue samples for volume imaging. Cell. **159**, 896–910 (2014)
74. Pan, C., Cai, R., Quacquarelli, F.P., et al.: Shrinkage-mediated imaging of entire organs and organisms using uDISCO. Nat. Methods. **13**(10), 859–867 (2016)
75. Rylander, C.G., Stumpp, O.F., Milner, T.E., et al.: Dehydration mechanism of optical clearing in tissue. J. Biomed. Opt. **11**, 041117 (2006)
76. Genina, E.A., Bashkatov, A.N., Tuchin, V.V.: Tissue optical immersion clearing. Expert Rev. Med. Devices. **7**, 825–842 (2010)
77. LaComb, R., Nadiarnykh, O., Carey, S., Campagnola, P.J.: Quantitative second harmonic generation imaging and modeling of the optical clearing mechanism in striated muscle and tendon. J. Biomed. Opt. **13**, 021109 (2008)
78. Hovhannisyan, V., Hu, P.-S., Chen, S.-J., et al.: Elucidation of the mechanisms of optical clearing in collagen tissue with multiphoton imaging. J. Biomed. Opt. **18**(4), 046004 (2013)
79. Mao, Z., Zhu, D., Hu, Y., et al.: Influence of alcohols on the optical clearing effect of skin *in vitro*. J. Biomed. Opt. **13**, 021104 (2008)
80. Genina, E.A., Bashkatov, A.N., Sinichkin, Y.P., Tuchin, V.V.: Optical clearing of skin under action of glycerol: *ex vivo* and *in vivo* investigations. Opt. Spectrosc. **109**(2), 225–231 (2010)
81. Hirshburg, J.M., Ravikumar, K.M., Hwang, W., Yeh, A.T.: Molecular basis for optical clearing of collagenous tissues. J. Biomed. Opt. **15**(5), 055002 (2010)
82. Wang, J., Ma, N., Shi, R., et al.: Sugar-induced skin optical clearing: from molecular dynamics simulation to experimental demonstration. IEEE J. Sel. Top. Quantum Electron. **20**(2), 7101007 (2014)
83. Berezin, K.V., Dvoretskiy, K.N., Chernavina, M.L., et al.: Molecular modeling of immersion optical clearing of biological tissues. J. Mol. Model. **24**(2), 45 (2018). https://doi.org/10.1007/s00894-018-3584-0
84. Xu, X., Wang, R.K.: The role of water desorption on optical clearing of biotissue: studied with near infrared reflectance spectroscopy. Med. Phys. **30**, 1246–1253 (2003)
85. Tuchina, D.K., Genin, V.D., Bashkatov, A.N., et al.: Optical clearing of skin tissue *ex vivo* with polyethylene glycol. Opt. Spectrosc. **120**(1), 28–37 (2016)
86. Oliveira, L., Carvalho, M.I., Nogueira, E., Tuchin, V.V.: Optical clearing mechanisms characterization in muscle. J Innov Opt Health Sci. **9**, 1650035 (2016)
87. Lazareva, E.N., Dyachenko, P.A., Bucharskaya, A.B., et al.: Estimation of dehydration of skin by refractometric method using optical clearing agents. J Biomed Photon Eng. **5**(2), 020305 (2019)

88. Leikin, S., Rau, D.C., Parsegian, V.A.: Temperature-favoured assembly of collagen is driven by hydrophilic not hydrophobic interactions. Nat. Struct. Biol. **2**(3), 205–210 (1995)
89. Kuznetsova, N., Chi, S.L., Leikin, S.: Sugars and polyols inhibit fibrillogenesis of type I collagen by disrupting hydrogen-bonded water bridges between the helices. Biochemistry. **37**(34), 11888–11895 (1998)
90. Weerheim, A., Ponec, M.: Determination of stratum corneum lipid profile by tape stripping in combination with high-performance thin-layer chromatography. Arch. Dermatol. Res. **293**(4), 191–199 (2001)
91. Schaefer, H., Redelmeier, T.E.: Skin Barrier. Karger, Basel (1996)
92. Michaels, A.S., Chandrasekaran, S.K., Shaw, J.E.: Drug permeation through human skin: theory and *in vitro* experimental measurement. AICHE J. **21**(5), 985–996 (1975)
93. Bolzinger, M.A., Briancon, S., Pelletier, J., Chevalier, Y.: Penetration of drugs through skin, a complex rate-controlling membrane. Curr. Opin. Colloid Interface Sci. **17**(3), 156–165 (2012)
94. Feingold, K.R., Elias, P.M.: Role of lipids in the formation and maintenance of the cutaneous permeability barrier. Biochim. Biophys. Acta. **1841**(3), 280–294 (2014)
95. Damien, F., Boncheva, M.: The extent of orthorhombic lipid phases in the stratum corneum determines the barrier efficiency of human skin *in vivo*. J. Invest. Dermatol. **130**(2), 611–614 (2010)
96. Choe, C., Schleusener, J., Lademann, J., Darvin, M.E.: *In vivo* confocal Raman microscopic determination of depth profiles of the stratum corneum lipid organization influenced by application of various oils. J. Dermatol. Sci. **87**, 183–191 (2017)
97. Doucet, J., Potter, A., Baltenneck, C., Domanov, Y.A.: Micron-scale assessment of molecular lipid organization in human stratum corneum using microprobe X-ray diffraction. J. Lipid Res. **55**(11), 2380–2388 (2014)
98. Choe, C., Lademann, J., Darvin, M.E.: A depth-dependent profile of the lipid conformation and lateral packing order of the stratum corneum *in vivo* measured using Raman microscopy. Analyst. **141**(6), 1981–1987 (2016)
99. Choe, C., Lademann, J., Darvin, M.E.: Confocal Raman microscopy for investigating the penetration of various oils into the human skin *in vivo*. J. Dermatol. Sci. **79**, 171–178 (2015)
100. Patzelt, A., Lademann, J., Richter, H., et al.: *In vivo* investigations on the penetration of various oils and their influence on the skin barrier. Skin Res. Technol. **18**(3), 364–369 (2012)
101. Horita, D., Hatta, I., Yoshimoto, M., et al.: Molecular mechanisms of action of different concentrations of ethanol in water on ordered structures of intercellular lipids and soft keratin in the stratum corneum. Biochim. Biophys. Acta. **1848**(5), 1196–1202 (2015)
102. Correa, M.C.M., Mao, G.R., Saad, P., et al.: Molecular interactions of plant oil components with stratum corneum lipids correlate with clinical measures of skin barrier function. Exp. Dermatol. **23**(1), 39–44 (2014)
103. Yamaguchi, S., Fukushi, Y., Kubota, O., et al.: Origin and quantification of differences between normal and tumor tissues observed by terahertz spectroscopy. Phys. Med. Biol. **61**, 6808–6820 (2016)
104. Peña-Llopis, S., Brugarolas, J.: Simultaneous isolation of high-quality DNA, RNA, miRNA and proteins from tissues for genomic applications. Nat. Protoc. **8**, 2240–2255 (2013)
105. Carneiro, I., Carvalho, S., Silva, V., et al.: Kinetics of optical properties of human colorectal tissues during optical clearing: a comparative study between normal and pathological tissues. J. Biomed. Opt. **23**, 121620 (2018)
106. Tuchin, V.V.: Optical Clearing of Tissues and Blood, pp. 12–15. SPIE Press, Bellingham (2006)
107. Kotyk, A., Janacek, K.: Membrane Transport: An Interdisciplinary Approach. Plenum Press, New York (1977)
108. Kreft, M., Lukšič, M., Zorec, T.M., et al.: Diffusion of D-glucose measured in the cytosol of a single astrocyte. Cell. Mol. Life Sci. **70**, 1483–1492 (2013)

109. Oliveira, L., Carvalho, M.I., Nogueira, E.M., Tuchin, V.V.: The characteristic time of glucose diffusion measured for skeletal muscle tissue at optical clearing. Laser Phys. **23**, 075606 (2013)
110. Tuchina, D.K., Timoshina, P.A., Tuchin, V.V., et al.: Kinetics of rat skin optical clearing at topical application of 40% glucose: *ex vivo* and *in vivo* studies. IEEE J. Sel. Top. Quantum Electron. **25**, 7200508 (2019)
111. Carneiro, I., Carvalho, S., Henrique, R., et al.: Simple multimodal technique for evaluation of free/bound water and dispersion of human liver tissue. J. Biomed. Opt. **22**, 125002 (2017)
112. Carneiro, I., Carvalho, S., Henrique, R., et al.: Water content and scatterers dispersion evaluation in colorectal tissues. J Biomed Phothon Eng. **3**, 040301 (2017)
113. Carneiro, I., Carvalho, S., Henrique, R., et al.: Kinetics of optical properties of colorectal muscle during optical clearing. IEEE J. Sel. Top. Quantum Electron. **25**, 7200608 (2019)
114. Carvalho, S., Gueiral, N., Nogueira, E., et al.: Glucose diffusion in colorectal mucosa – a comparative study between normal and cancer tissues. J. Biomed. Opt. **22**, 091506 (2017)
115. Xiong, H., Guo, Z., Zeng, C., et al.: Application of hyperosmotic agent to determine gastric cancer with optical coherence tomography *ex vivo* in mice. J. Biomed. Opt. **14**(2), 024029 (2009)
116. Wallace, V.P., Fitzgerald, A.J., Pickwell, E., et al.: Terahertz pulsed spectroscopy of human basal cell carcinoma. Appl. Spectrosc. **60**(10), 1127–1133 (2006)
117. Jung, E.A., Lim, M., Moon, K., et al.: Terahertz pulse imaging of micro-metastatic lymph nodes in early-stage cervical cancer patients. J Opt Soc Korea. **15**(2), 155–160 (2011)
118. Yeh, A.T., Choi, B., Nelson, J.S., Tromberg, B.J.: Reversible dissociation of collagen in tissues. J. Invest. Dermatol. **121**, 1332–1335 (2003)
119. Hirshburg, J., Choi, B., Nelson, J.S., Yeh, A.T.: Collagen solubility correlates with skin optical clearing. J. Biomed. Opt. **11**, 040501 (2006)
120. Cheung, M.C., Evans, J.G., McKenna, B., Ehrlich, D.J.: Deep ultraviolet mapping of intracellular protein and nucleic acid in femtograms per pixel. Cytometry A. **79**, 920–932 (2012)
121. Matryba, P., Kaczmarek, L., Gołąb, J.: Advances in ex situ tissue optical clearing. Laser Photon Rev. **13**, 1800292 (2019). https://doi.org/10.1002/lpor.201800292
122. Cui, Y., Wang, X., Ren, W., et al.: Optical clearing delivers ultrasensitive hyperspectral dark-field imaging for single-cell evaluation. ACS Nano. **10**(3), 3132–3143 (2016)
123. Wei, M., Shi, L., Shen, Y., et al.: Volumetric chemical imaging by clearing-enhanced stimulated Raman scattering microscopy. Proc. Natl. Acad. Sci. U. S. A. **116**(14), 6608–6617 (2019). https://doi.org/10.1073/pnas.1813044116
124. Lagerweij, T., Dusoswa, S.A., Negrean, A., et al.: Optical clearing and fluorescence deep-tissue imaging for 3D quantitative analysis of the brain tumor microenvironment. Angiogenesis. **20**, 533–546 (2017)
125. Glaser, A.K., Reder, N.P., Chen, Y., et al.: Light-sheet microscopy for slide-free non-destructive pathology of large clinical specimens. Nat. Biomed. Eng. **1**(7), 0084 (2017). https://doi.org/10.1038/s41551-017-0084
126. Tanaka, N., Kaczynska, D., Kanatani, S., et al.: Mapping of the three-dimensional lymphatic microvasculature in bladder tumours using light-sheet microscopy. Br. J. Cancer. **118**(7), 995–999 (2018). https://doi.org/10.1038/s41416-018-0016-y
127. van Royen, M.E., Verhoef, E.I., Kweldam, C.F., et al.: Three-dimensional microscopic analysis of clinical prostate specimens. Histopathology. **69**(6), 985–992 (2016). https://doi.org/10.1111/his.13022
128. Verhoef, E.I., van Cappellen, W.A., Slotman, J.A., et al.: Three-dimensional architecture of common benign and precancerous prostate epithelial lesions. Histopathology. **74**(7), 1036–1044 (2019). https://doi.org/10.1111/his.13848
129. Verhoef, E.I., van Cappellen, W.A., Slotman, J.A., et al.: Three-dimensional analysis reveals two major architectural subgroups of prostate cancer growth patterns. Mod. Pathol. **32**, 1032–1041 (2019). https://doi.org/10.1038/s41379-019-0221-0

130. Hume, R.D., Berry, L., Reichelt, S., et al.: An engineered human adipose/collagen model for *in vitro* breast cancer cell migration studies. Tissue Eng. Part A. **24**(17–18), 1309–1319 (2018). https://doi.org/10.1089/ten.TEA.2017.0509
131. Scott, G.D., Blum, E.D., Fryer, A.D., Jacoby, D.B.: Tissue optical clearing, three-dimensional imaging, and computer morphometry in whole mouse lungs and human airways. Am. J. Respir. Cell Mol. Biol. **51**(1), 43–55 (2014). https://doi.org/10.1165/rcmb.2013-0284OC
132. von Neubeck, B., Gondi, G., Riganti, C., et al.: An inhibitory antibody targeting carbonic anhydrase XII abrogates chemoresistance and significantly reduces lung metastases in an orthotopic breast cancer model *in vivo*. Int. J. Cancer. **143**(8), 2065–2075 (2018). https://doi.org/10.1002/ijc.31607
133. Cuccarese, M.F., Dubach, J.M., Pfirschke, C., et al.: Heterogeneity of macrophage infiltration and therapeutic response in lung carcinoma revealed by 3D organ imaging. Nat. Commun. **8**, 14293 (2017). https://doi.org/10.1038/ncomms14293
134. Joshi, N.S., Akama-Garren, E.H., Lu, Y., et al.: Regulatory T cells in tumor-associated tertiary lymphoid structures suppress anti-tumor T cell responses. Immunity. **43**(3), 579–590 (2015). https://doi.org/10.1016/j.immuni.2015.08.006
135. Kubota, S.I., Takahashi, K., Nishida, J., et al.: Whole-body profiling of cancer metastasis with single-cell resolution. Cell Rep. **20**(1), 236–250 (2017). https://doi.org/10.1016/j.celrep.2017.06.010
136. Zhu, D., Lu, W., Weng, Y., et al.: Monitoring thermal-induced changes in tumor blood flow and microvessels with laser speckle contrast imaging. Appl. Opt. **46**(10), 1911–1917 (2007)
137. Oldham, M., Sakhalkar, H., Oliver, T., et al.: Three-dimensional imaging of xenograft tumors using optical computed and emission tomography. Med. Phys. **33**(9), 3193–3202 (2006)
138. Oldham, M., Sakhalkar, H., Oliver, T., et al.: Optical clearing of unsectioned specimens for three-dimensional imaging via optical transmission and emission tomography. J. Biomed. Opt. **13**(2), 021113 (2008)
139. Zhang, Y., Liu, H., Tang, J., et al.: Non-invasively imaging subcutaneous tumor xenograft by handheld Raman detector, with assistance of optical clearing agent. ACS Appl. Mater. Interfaces. **9**(21), 17769–17776 (2017)
140. Pires, L., Demidov, V., Vitkin, I.A., et al.: Optical clearing of melanoma *in vivo*: characterization by diffuse reflectance spectroscopy and optical coherence tomography. J. Biomed. Opt. **21**(8), 081210 (2016)

Chapter 3
Exploring Tumor Metabolism with Time-Resolved Fluorescence Methods: from Single Cells to a Whole Tumor

Marina V. Shirmanova, Vladislav I. Shcheslavskiy, Maria M. Lukina, Wolfgang Becker, and Elena V. Zagaynova

3.1 Introduction

Multiple alterations in cancer metabolism are now evident. The key underlying reasons for metabolic changes accompanying neoplastic transformation are the high bioenergetic and biosynthetic demands of growing and proliferating cells and adaptation to limited nutrient and oxygen supply and to an acidic environment.

One of the primary metabolic features of cancer is enhanced rate of glycolysis. While it is natural to switch to glycolysis in hypoxic conditions, which occur in many advanced tumors, cancer cells actively use glycolysis even in the presence of oxygen—a phenomenon known as the Warburg effect—or aerobic glycolysis [1]. Aerobic glycolysis makes cancer cells insensitive to fluctuations of oxygen in cellular environment and gives them several other important advantages. Specifically, it provides a constant supply of metabolic intermediates to support tumor growth, reduces level of reactive oxygen species (ROS), thus, protecting cells against ROS-mediated cell death, generates adenosine triphosphate (ATP) more rapidly than oxidative phosphorylation (OXPHOS), and produces lactate that has an important functional role in promoting tumorigenesis [2, 3].

Marina V. Shirmanova and Vladislav I. Shcheslavskiy contributed equally to this chapter.

M. V. Shirmanova (✉) · M. M. Lukina
Privolzhsky Research Medical University (PRMU), Nizhny Novgorod, Russian Federation
e-mail: niibmt@pimunn.ru

V. I. Shcheslavskiy (✉) · W. Becker
Becker & Hickl GmbH, Berlin, Germany
e-mail: vis@becker-hickl.de

E. V. Zagaynova
Privolzhsky Research Medical University (PRMU), Nizhny Novgorod, Russian Federation
Lobachevskiy State University of Nizhny Novgorod, Nizhny Novgorod, Russian Federation

V. V. Tuchin et al. (eds.), *Multimodal Optical Diagnostics of Cancer*,
https://doi.org/10.1007/978-3-030-44594-2_3

Mitochondrial respiration is frequently impaired in cancer cells due to dysfunctional mitochondria. Meanwhile, there is a growing body of evidence that some cancer cells preserve mitochondrial function and rely mainly on OXPHOS for energy production [4, 5].

The fate of lactate secreted by glycolytic cancer cells remains to be clarified. Recent findings suggest that adjacent oxidative cancer cells can consume and use it as a source for the tricarboxylic acid (TCA) cycle intermediates and for ATP production [6]. Moreover, cancer-associated fibroblasts, being in a metabolic crosstalk with cancer cells, also can extrude lactate to directly feed cancer cells [7].

Glucose is a major, but not the only fuel source for cancer cells. Many cancer cells have been shown to rely heavily upon an exogenous supply of glutamine [8]. Glutamine is converted to glutamate, which is further metabolized to α-ketoglutarate for utilization through the TCA cycle. Another important energy source in cancer cells is fatty acids, catabolized by the fatty acid oxidation (FAO; also known as β-oxidation) pathway [9]. FAO produces a pool of the reduced electron carriers NADH and FADH2, which are oxidized in the mitochondrial electron transport chain for ATP production, and acetyl CoA, which together with oxaloacetate is transformed in the TCA cycle into citrate, which on export to the cytoplasm, can enter NADPH-producing reactions.

There are number of factors that collectively determine metabolic phenotype of a tumor, including the tissue of origin, the underlying cancer genetics and subtype, the tumor microenvironment, and other variables such as diet and host physiology, which lead to a vast intertumor heterogeneity in metabolism. In addition, individual cells within a tumor can explore distinct metabolic programs. The emerging view on cancer metabolism is that malignant cells have an ability to switch between different biochemical pathways in order to satisfy their immediate biosynthetic and energetic demands and to adopt to varying microenvironment [10]. However, the mechanisms by which cancer cells coordinate their metabolic activities and the conditions upon which cancer cells upregulate specific metabolic pathways are poorly defined.

Multiple studies suggest that cellular metabolism has an impact not only on biological behavior of tumors, but also on treatment response. The links between metabolic processes and response to cancer therapeutics are complex and not fully understood. On the one side, metabolism and its derangement are involved in the initiation and execution of cell death at multiple levels and in regulation of cell survival [11, 12]. A hallmark of cancer cells is the ability to evade apoptosis. On the other side, increased aerobic glycolysis and glutamine metabolism under suppression of mitochondrial respiration in cancer cells contribute to resistance to various therapies by providing higher intracellular ATP and NADPH levels and extracellular acidification, to name a few proposed mechanisms [13]. Significant intertumor and intratumor heterogeneity and metabolic plasticity of tumors impose additional difficulties for cancer treatment.

To reach a deeper understanding of cancer metabolism, highly sensitive, quantitative, high-resolution metabolic assays are critically needed. Established methods to characterize cellular metabolic activity include both solely laboratory techniques, e.g. mass spectrometry, routine analytical methods, molecular and genetic analyses, optical biosensors, and clinical imaging, e.g. positron emission tomography (PET) with ^{18}F-labeled fluoro-2-deoxyglucose, ^{13}C magnetic resonance spectroscopy (MRS). All these methods, however, either destructive or require the use of exogenous labeling of molecules of interests. Time-resolved optical imaging of endogenous fluorophores has emerged as a powerful approach for non-invasive, label-free assessment of metabolic state of living cells *in vitro* and tissues *in vivo*.

The book chapter is organized as following. Section 3.2 gives biochemical basis of optical metabolic imaging by introducing fluorescent metabolic cofactors. In Sect. 3.3 we address theoretical background for time-correlated single photon counting (TSCPC) technique as all the described systems are based on it. Fluorescence lifetime imaging on a microscopic scale together with the examples of metabolic imaging in cancer cells, spheroids, and tumors is presented in Sect. 3.4. Section 3.5 reports about the developed system for fiber-based fluorescence lifetime measurements and its applications for interrogation of metabolic state of a tumor *in vivo*. Section 3.6 describes fluorescence lifetime imaging on a macroscale and its applications in cancer studies. Finally, the conclusions of the book chapter are presented in the Summary section.

3.2 Fluorescence Lifetime of Metabolic Cofactors as an Indicator of a Metabolic State

There are two groups of fluorescent molecules involved in various metabolic pathways in cells (1) reduced nicotinamide adenine dinucleotide NADH and its phosphorylated derivative NADPH (collectively denoted NAD(P)H) and (2) oxidized flavin adenine dinucleotide FAD and flavin mononucleotide FMN. These molecules function as coenzymes, being linked with various dehydrogenases. There is a constant interchange between the reduced and the oxidized states of these molecules, which jointly constitute a redox couple (NAD+/NADH, NADP+/NADPH, FAD/FADH2, FMN/FMNH2). The primary role of these cofactors, except NADPH, is the transfer of electrons across the respiratory chain of mitochondria. Reduced electron carriers NADH and FADH2 are produced in catabolic pathways by transferring electrons from the substrate to NAD+ and FAD. Subsequently, NADH feeds electrons into complex I and FADH2—into complex II of the mitochondrial electron transport chain (ETC). ETC and oxidative phosphorylation (OXPHOS) reoxidize NADH and FADH2 and trap the released energy in the form of adenosine triphosphate (ATP). FMN, a prosthetic group of NADH-ubiquinone oxidoreductase (complex I), accepts electrons from NADH and passes to FeS centers (iron-sulfur clusters). NADPH acts as a reducing agent during the synthesis of nucleic acids and lipids and plays a central role in defense against oxidative damage and detoxification of reactive oxygen species.

NAD(P) H, FAD and FMN possess intrinsic fluorescence, detection of which is the basis for a group of methods referred as "optical metabolic imaging." NADH and NADPH are spectrally identical—excitation peaks at 260 nm and 340 nm, emission peak at 460 nm. Flavin cofactors absorb light maximally at ~450 nm and emit at ~520 nm, with minor variations in spectra of FAD and FMN.

While fluorescence of NAD(P)H is derived from both mitochondria and cytosol (mitochondrial NAD(P)H fluorescence is dominant), flavin fluorescence originates predominantly from mitochondrial flavoproteins [14], although additional subcellular localizations (e.g., nucleus) cannot be ruled out.

The first approach to probe metabolism using fluorescence of the metabolic cofactors was developed in 1970th by B. Chance, who measured oxidation-reduction states of mitochondria taking the ratio of the fluorescence intensities of oxidized flavoproteins to reduced NADH [15, 16].

The "optical redox ratio" has been widely used as a simple, label-free metric of cellular redox states [16–18]. In a general case, the redox ratio is sensitive to the changes in the rates of glycolysis and mitochondrial respiration. NADH is generated during glycolysis, therefore enhanced glycolysis causes decrease in FAD/NADH ratio, which is often observed in cancer. Activation of oxidative metabolism increases the portion of oxidized and decreases the portion of reduced cofactors, which results in increased FAD/NADH ratio. However, the intensity-based redox ratio imaging and interpretation face some challenges associated with the fact that emission intensity is influenced by many factors besides concentration of the fluorophores, e.g. conformation or binding to proteins, light scattering and absorption—especially in tissues, excitation power, and participation of the same cofactors in various metabolic pathways.

Fluorescence lifetimes of NAD(P)H and flavoproteins provide an additional dimension to probing the cellular metabolic status.

For NAD(P)H, the fluorescence lifetime largely depends on binding to enzyme. This observation was first reported by Lakowicz et al. in 1992 [19]. Freely diffusing NAD(P)H has distinctly shorter fluorescence lifetime (~0.3–0.5 ns) due to the large rate of non-radiative relaxation mediated by small scale motion of the nicotinamide ring [20]. The fluorescence lifetime of protein-bound NADH is, at least, threefold higher and strongly vary—on average, in the range 1.2–2.5 ns in solutions [21, 22], when bound to different enzymes, which is presumably attributed to different levels of conformational restriction upon binding. Moreover, the fluorescence lifetime of enzyme-bound NADH may increase upon formation of a ternary complex with the specific substrate [23]. Protein-bound NADPH possesses a significantly larger fluorescence lifetime (~4.4 ns) than protein-bound NADH within the cellular environment [24]. Therefore, fluorescence lifetime of NAD(P)H depends on the composition and activity of NADH-binding enzymes and the amount of the phosphorylated form of NADH.

Currently, nonlinear curve-fitting analysis is the most common approach to analyze fluorescence lifetimes. With a large range of different enzymes to which NAD(P)H can bind, it is natural that its fluorescence is characterized by a complex multi-exponential decay. However, limited photon counts, acquisition and computational times, especially when measurements are made in live cells or tissues, allow to resolve most effectively only two lifetime components. When using a bi-exponential function for fitting NAD(P)H fluorescence decay, the first (short, t1) component is attributed to its free, and the second (long, t2) component to its protein-bound state, and their relative amplitudes (a1 ~ 85% and a2 ~ 15%) reflect the relative amounts of free and protein-bound NAD(P)H. The three-exponential model for fitting NAD(P)H fluorescence decay is appropriate for estimation of the contribution of protein-bound NADPH, as was suggested by Blacker et al. [24]. Three-exponential fitting requires larger photon numbers compared to single and bi-exponential fitting. Therefore, fixation of the one of the fluorescence lifetimes significantly relaxes the requirements on the minimum photon numbers and speeds up the computational times. This approach is more appropriate for the cells with increased NADPH level, for example, for stem cells undergoing adipogenic differentiation [25]. In general, NADPH contribution in FLIM of most cell and tissue types is not taken into account, because of its much lower concentration compared with NADH and insensitivity to perturbations in energy metabolism [26].

There is also possible to use four-exponential model for fitting NAD(P)H fluorescence decay curves bearing in mind that the free NADH has two decay components [27]. However, as in case with three-exponential fitting, one has to fix at least two lifetime components to be able to analyze the data with the limited photon budget.

We have to mention that phasor analysis has become recently a quite effective tool in the evaluation of the fluorescence lifetime data of NADH [28, 29]. However, the detailed description of it is out of scope of this book chapter.

In contrast to NAD(P)H, flavin cofactor in the form of FAD or FMN, more specifically termed a prosthetic group, is permanently linked to the protein [30]. Fluorescence of FAD decays multiexponentially with two major lifetime components—around 7 ps and 2.7 ns in aqueous solution [31]. The presence of two lifetime components is due to existence of FAD in two conformations—"closed" or stacked, in which the coplanar isoalloxazine and adenine rings interact through π-π interactions, resulting in very efficient fluorescence quenching, and "open" or unstacked, in which the two aromatic ring are separated from each other. The majority of FAD molecules are in a closed conformation, that determines low quantum yield of its fluorescence (0.033) [32]. FMN and riboflavin (vitamin B2), the precursor of flavin cofactors, exhibit monoexponential fluorescence decays with lifetime ~4.7 ns and ~5.1 ns, correspondingly [17].

Among flavins, FAD is the most abundant in tissues, while concentration of FMN is typically 3–5 times lower and riboflavin—40-400 times lower [33, 34], although in cultured cells riboflavin concentration can be comparable with FAD [35]. For flavoproteins, the emission maxima and quantum yield are dependent on the nature of the flavin-binding site [36].

It is common for fluorescence decay of the cellular flavins to be best fit by a bi-exponential model, where the short lifetime component (~0.3 to 0.4 ns, ~75 to 85%) is consistent with the quenched state of FAD and the long lifetime component (~2.5 to 2.8 ns, ~15 to 25%) contains contributions from unquenched FAD, FMN, and riboflavin. Given that the discrimination of different flavins within cells is problematic, biochemical interpretation of the fluorescence lifetime measurements remains a challenge.

Furthermore, in cancer cells and tumors with mitochondrial respiration malfunction fluorescence emission from flavins is often fairly weak and insufficient to collect the required number of photons for FLIM. Therefore, most of the metabolic studies with FLIM are concentrated on NAD(P)H.

3.3 Technical Basis of Time-Correlated Single Photon Counting

Time-correlated single photon counting (TCSPC) is a technique that allows to record photons, measure their time relative to the excitation pulse, and build up a histogram of the photon times [37]. The technique uses the fact that for low level, high-repetition-rate signals the probability to detect one photon per laser period is extremely low. That means that there are many signal periods that do not contain photons, and only few contain just one.

To measure fluorescence decay curves, one excites the sample with a train of short pulses (typically in the ps or fs range). Then emitted fluorescence photons are registered by a detector operating in a single photon counting mode, and their time with respect to the excitation pulse is recorded. With each pulse, more and more photons are recorded and the statistical distribution of the photons within the laser period is built up. The principle of TCSPC is shown in Fig. 3.1.

This method of decay curves measurements is very effective as all the registered photons contribute to the statistical distribution. The efficiency in this case is much higher than for recording techniques that shift a time-gate over the optical waveform [38].

Another advantage of using TCSPC for fluorescence decay measurements compared to other methods is that its time resolution is better than the width of the single-electron response of the detector, as the arrival times of the detector pulses

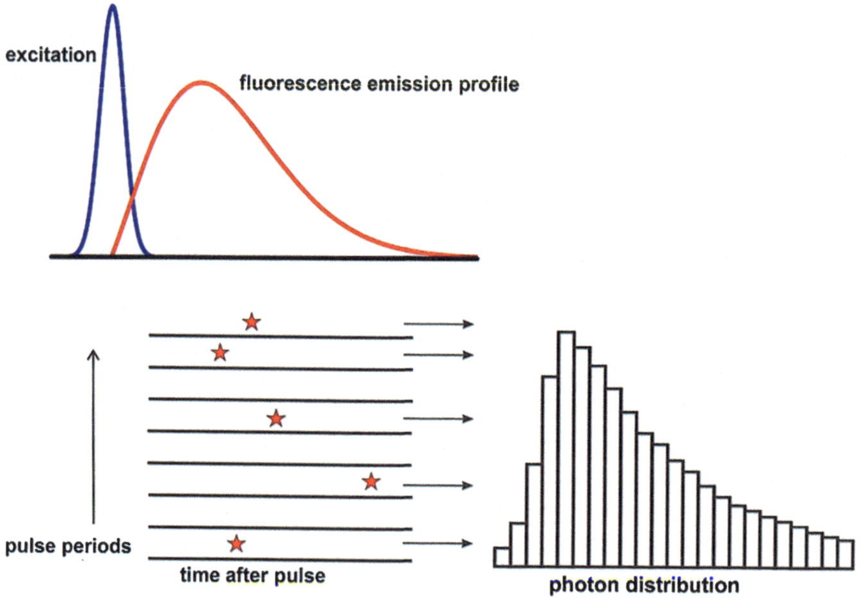

Fig. 3.1 The principle of time-correlated single photon counting technique

can be measured with an accuracy much better than the width of the pulses. That in its turn means that the instrument response function determined by the transient time spread of the pulses at the detector can be much shorter than the single photon response of the detector. Finally, TCSPC technique is not sensitive to the waveform changes during the recording process as it measures an average waveform.

On the other side, if a time-gating approach is used, this may result in a distorted waveform in this case.

The photon distribution can be built up not only with respect to the time after the excitation pulse (one-dimensional TCSPC), but also with respect to other parameters (multi-dimensional TCSPC), like the spatial coordinates, the wavelength, the time from the start of experiment, polarization, etc. Depending on which and how many additional parameters are used, different photon distributions are obtained. The data that will be presented later in this book chapter is based on the TCSPC measurements recorded with one or two variable parameters. In the first case (one parameter-time after the excitation laser pulse) the experiments represent single point, or fluorescence lifetime spectroscopy measurements. In the latter case (two parameters, and that are the time after the excitation pulse in each laser period and the spatial coordinates within an image area), the measurements result in the building up fluorescence lifetime image, or FLIM.

3.4 FLIM-Microscopy

The principle of multi-dimensional TCSPC in laser scanning microscopy (LSM) is shown in Fig. 3.2.

In LSM, the sample is scanned with a focused beam of a pulsed laser (usually femtosecond or picosecond). Unlike one-dimensional TCSPC, the multi-parameter TCSPC involves determination of the laser beam position in addition to counting the number of photons in each period of laser pulses and estimation of the arrival times at the detector. The spatial coordinates are determined by transferring scan synchronization signals from the scanner to the single photon counting module: a "frame" pulse indicates the beginning of a new frame, a "line" pulse—the beginning

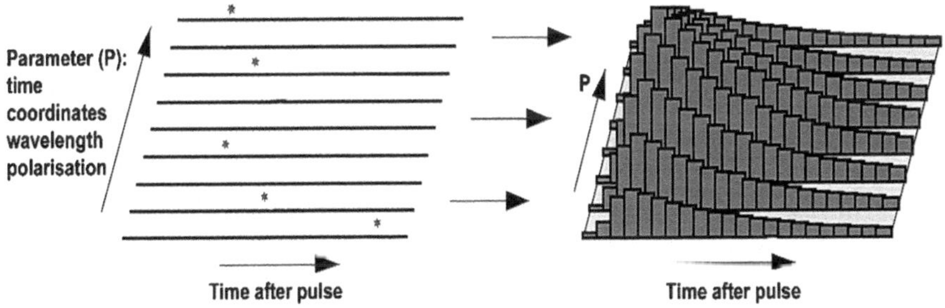

Fig. 3.2 Concept of multi-dimensional TCSPC

of a new line, and a pixel pulse—the transition to the new pixel within the line. The spatial coordinates of the photons are determined by counting these pulses.

Based on the registered parameters (number of photons, photon arrival times, and spatial coordinates), spatial-temporal photon histograms which are in the basis of the time-resolved fluorescence microscopy, or fluorescence lifetime imaging are built up. If other photon parameters (e.g., wavelength) are registered in addition to spatial coordinates, this represents the three-parameter version of FLIM—spectral time-resolved fluorescence microscopy [39, 40].

From the point of view of biological studies, it is essential that the TCSPC-based FLIM allows identification of curves with complex (multi-exponent) profiles of fluorescence decay. It is also important that this method is insensitive to fluctuations in the laser power and fully compatible with fast scanning realized in modern LSM: scanning is continued until the photon statistics sufficient for the generation of reliable results of image analysis is accumulated. It should be mentioned that while photon statistics is not so essential in the standard fluorescence microscopy, the number of photons is a defining factor for accurate data analysis in the time-resolved fluorescence microscopy.

Apparently, the more complex the exponential descriptions of the decay curves are, the higher number of photons must be accumulated for correct estimation of fluorescence lifetimes. That is why the times for data accumulation in the TCSPC-based FLIM are longer than in the standard LSM.

Furthermore, the counting rate in the TCSPC-based systems is limited by the so-called accumulation effects [37]. The essence of these effects lies in the fact that if a very powerful laser radiation hits the sample and the rate of fluorescence photon emission is very high, the probability of the second photon arrival at the detector over one laser period (which means that this photon will be "ignored" by the electronics) increases, which can alter the decay curve profile. To avoid this, the rate of photon emission by the sample should not exceed 10% of the laser pulse repetition rate. This mode is realized in the majority of LSM setups to prevent the photodegradation of biological samples.

Laser scanning microscopes that allow to do FLIM experiments are typically equipped with either ps or fs lasers. While one-photon excitation with ps lasers makes a system cheaper, there are number of advantages of using two-photon excitation with fs lasers. First, multiphoton processes are excited only at the focal spot where the photon flux density is high enough for the nonlinear process to occur, eliminating the need for the confocal detector pinhole. Second, the illuminating laser is typically operates in the near infrared rather than in the visible or ultraviolet spectral regions. Using such long excitation wavelengths results in a deeper penetration due to lower scattering and causes less photodamage to the samples [41].

The first metabolic study with the use of multiphoton TSCPC-based FLIM-microscopy was reported by Bird et al. in 2005 [42]. The authors measured fluorescence lifetimes and ratio of free and protein-bound NADH in live cultured human breast cells upon chemical perturbations.

Multiphoton FLIM-microscopy of NADH was reported to detect the changes in metabolism with neoplasia development *in vivo* [43]. Using hamster cheek pouch model of carcinogenesis, the authors showed that the fluorescence lifetime of protein-bound NADH and its contribution to the fluorescence signal could serve as a marker that allowed differentiation between normal and pre-cancer tissue. Later the ability of FLIM-microscopy to distinguish cancer cells from normal ones on the basis of autofluorescence was demonstrated in different systems—from cultured cells of different origin to tumor samples. For example, this was shown for leukemia cells *in vitro* [44], breast cancer cells [21, 45], squamous carcinoma cells [46], head and neck squamous cell carcinoma [47] and for human lung cancer *ex vivo* [48], experimental glioma tissue [49], and mouse breast tumors [50]. Several of these studies [21, 44, 45, 47, 48] observed a shift towards shorter mean fluorescence lifetime and a higher ratio of free/bound NADH in malignant cells, which correlated with their glycolytic phenotype, while others reported on elongation of the NADH lifetime in cancer [45, 46, 49, 50], suggesting an important role of OXPHOS. Combination of metabolic FLIM-microscopy with phosphorescence lifetime measurements from oxygen-sensitive probes was shown to be promising approach to explore a relation between metabolic phenotypes and oxygen tension [51, 52].

A significant advantage of FLIM-microscopy over other FLIM modalities is the ability to produce a (sub)cellular resolution, which is essential in terms of metabolic heterogeneity at the cellular level.

As revealed by two-photon FLIM-microscopy, NAD(P)H exhibits a heterogeneous fluorescence lifetime (i.e., quantum yield) throughout living cells. The appreciable differences in NAD(P)H lifetime were shown between different cellular compartments; particularly, the relative contribution of the protein-bound species in mitochondria is larger than in cytosol or in nucleus [18, 21, 53]. Segmented cell analysis of FLIM images was proposed as a relevant approach to separate mitochondrial OXPHOS from cytosolic glycolysis in cancer cells [18].

Specifics of cellular metabolism enables FLIM-microscopy to distinguish between different subpopulations of cells that constitute the tumor and visualize the changes accompanying cellular interactions. For example, Heaster et al. demonstrated the feasibility of using the NAD(P)H and FAD fluorescence lifetimes and intensity-based redox ratio to discriminate proliferating, quiescent, and apoptotic cell populations in an acute myeloid leukemia model and generated classification algorithms based on a combination of all measured variables to improve separation between cell groups [54]. Using FLIM we also observed significant metabolic differences between proliferating cells of the outer layers and quiescent cells of the inner layers of multicellular spheroids (Fig. 3.3B) [55]. The common observation is that proliferating cancer cells use more glycolytic metabolism compared to quiescent cells and, consequently, display increased contribution of free NAD(P)H; other parameters can differ as well. Apoptotic cells show increased mean fluorescence lifetimes of NAD(P)H and FAD, increased lifetimes of the long components NAD(P)H and FAD, decreased contribution of free NAD(P)H, and decreased contribution of short lifetime FAD, indicating an elevated level of OXPHOS and binding cofactors to different enzymes [54, 56, 57].

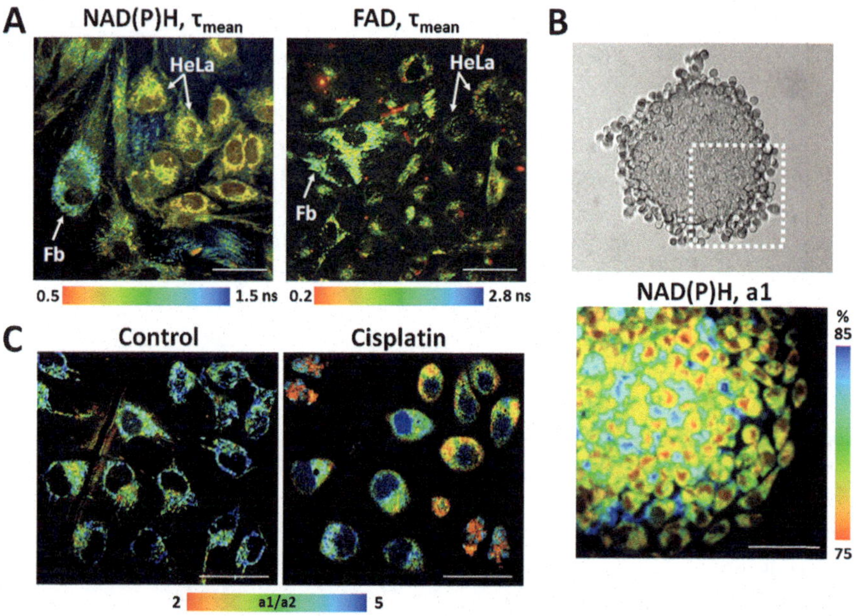

Fig. 3.3 Metabolic heterogeneity at the cellular level visualized by two-photon FLIM-microscopy. (**A**) The NAD(P)H-τ_m and FAD-τ_m FLIM images of co-culture of HeLa cancer cells and human skin fibroblasts. (**B**) The transmitted-light (upper) and NAD(P)H – a1% fluorescence lifetime (bottom) images of HeLa tumor spheroid. FLIM corresponds to the square area marked with a dashed white line. (**C**) The varying responses to chemotherapy with cisplatin in HeLa cells. The NAD(P)H-a1/a2 images are shown for the same field of view before treatment (control, left image) and at 6 h exposure (cisplatin, right image). Scale bar: 50 μm

Trinh et al. identified a significant shift towards longer NADH fluorescence lifetimes, suggesting an increase in the fraction of protein-bound NADH, in the invasive stem-like tumor-initiating cell subpopulation relative to the tumor mass-forming cell subpopulation of malignant gliomas [58].

The metabolic crosstalk between cancer cells and cancer-associated fibroblasts—the main stromal cell type of solid tumors—was examined in several studies with the use of FLIM. Specifically, it was shown that metabolic changes occurred in cancer cells and normal fibroblasts as a result of their interaction in a co-culture model; we observed a shift towards glycolysis in HeLa cancer cells, and from glycolysis to OXPHOS in fibroblasts (Fig. 3.3A) [59]. Walsh et al. found that isolated tumor-associated fibroblasts displayed decreased redox ratio NAD(P)H/FAD and increased mean fluorescence lifetimes of NAD(P)H and FAD compared to 3D organoids generated from primary pancreatic ductal adenocarcinoma [60]. Analysis of the metabolic status of cancer-associated and normal fibroblasts using FLIM revealed increased metabolic activity of fibroblasts derived from patient's colon tumor with a shift to more oxidative metabolism compared to dermal fibroblasts [61].

Cellular-level metabolic heterogeneity was visualized by FLIM-microscopy in tumor models *in vivo* in a few studies. The elegant experiments by Szulczewski et al. demonstrate that at least two different sets of abundant cell types can be distinguished in a live mouse tumor by their endogenous fluorescence: cancer cells with high NAD(P)H fluorescence intensity and stromal cells with high FAD intensity. The latter were found predominantly outside of the tumor and concentrated mainly in the stroma, along collagen fibers, and verified to be tumor-associated macrophages. Notably, macrophages had significantly shorter τ mean of NAD(P)H with a greater fractional component of free NADH, indicating glycolytic-like metabolism [62]. In our previous study on human cervical carcinoma xenografts, we also detected a cellular population with greater contribution of free NAD(P)H in a collagen-rich zones compared to majority of cells comprising the tumor mass [63]. Yet, the interpretation of *in vivo* data remains challenging due to the difficulty of identifying specific cell types within tumor and a variety of factors that collectively determine the metabolic profile of cells.

Multiple studies demonstrate that anti-tumor therapies induce changes in the optical redox ratio and fluorescence lifetimes of NAD(P)H and FAD, and the extent of these changes correlate with responsiveness to the treatment. For example, in the study by Shah et al. head and neck squamous cell carcinoma cell lines were treated with cetuximab (anti-EGFR antibody), BGT226 (PI3K/mTOR inhibitor), or cisplatin (chemotherapy) [47]. Results showed a decreased redox ratio NAD(P)H/FAD and decreased contribution of free NADH-a1 with BGT226 and cisplatin treatment and decreased FAD-a1 with cisplatin treatment, which agreed with decreased proliferation and glycolytic rates. Treatment with cetuximab had no effects on the metabolic imaging variables neither on proliferation nor on glycolysis rate. Similar changes in redox ratio and NAD(P)H lifetime were detected in our study on human cervical cancer cells and xenografts after therapy with cisplatin, which correlated with inhibition of cancer cells growth [64]. Walsh et al. detected decreased redox ratio NAD(P)H/FAD and decreased mean fluorescence lifetime (τ_{mean}) NAD(P)H and FAD in HER2-overexpressing human breast cancer xenografts in mice treated with trastuzumab (herceptin, HER2 inhibitor). This response was confirmed with tumor growth curves and stains for Ki67 and cleaved caspase-3 [45]. In general, targeted therapy promoted more complex metabolic rearrangements, not easily interpreted, due to alterations at the level of signaling pathways and genes involved in regulation of cellular metabolism. In a recent work, we observed increase in the NAD(P)H τ_{mean} m and decrease in the free/bound NAD(P)H ratio in mouse colorectal tumors after chemotherapy with each of three drugs—cisplatin, paclitaxel, and irinotecan [65]. We should note that the treatment with different chemotherapeutic agents caused similar changes in the optical metabolic parameters. These changes serve as an indicator of the switch to a more oxidative/less glycolytic metabolism, which is rational for cells with disrupted cell division and decreased proliferative capacity.

It is important to mention that the changes in cellular metabolism, resolved by FLIM, precede manifestations of cell death and tumor size reduction, which allows to consider autofluorescence parameters as early indicators of drug efficacy.

With FLIM-microscopy longitudinal tracking of cells can be performed in the course of treatment, and subpopulations with different drug sensitivities can be identified on the basis of optical metabolic metrics. In our previous study, we monitored the metabolic activity of the same cervical cancer cells using fluorescence intensity and lifetime measurements of NAD(P)H and FAD in response to paclitaxel or cisplatin treatment (Fig. 3.3C) [55]. Among the optical metrics (the fluorescence intensity-based redox ratio FAD/NAD(P)H, and the fluorescence lifetimes of NAD(P)H and FAD), the fluorescence lifetime of NAD(P)H was the most sensitive to resolve heterogeneous drug response. It was found that metabolic changes developed faster in the more responsive (dying) cells. In the recent studies, the measurements of NAD(P)H, FAD, and tryptophan (Trp) lifetimes and their enzyme-bound fractions in individual cancer cells allowed to map heterogeneity in response to doxorubicin treatment [66, 67]. By measuring the optical redox ratio NAD(P)H/FAD and fluorescence lifetimes of NAD(P)H and FAD, Shah et al. observed increased metabolic heterogeneity in FaDu tumor xenografts of mice treated with cetuximab or cisplatin compared with untreated control [68]. To quantify the level of cellular metabolic heterogeneity and make quantitative comparisons across treatment groups, they developed a heterogeneity index that incorporates the number of subpopulations, evenness between subpopulations, and distance between subpopulations in frequency histograms for each optical metabolic imaging variable.

A series of studies was performed by Melissa Skala' group to validate microscopic metabolic imaging of primary tumor organoid cultures to accurately predict drug response [60, 69, 70]. The possibility to detect early, heterogeneous treatment response, on a label-free basis with cellular resolution makes FLIM-microscopy an attractive platform for testing drugs on patient-derived cancer cells for individualized treatment planning.

3.5 Time-Resolved Fluorescence Spectroscopy

The measurement of autofluorescence signals from the coenzymes for comprehensive tumor analysis *in vivo* is technically challenging. While the signals from the surface of the tumors can be analyzed rather easily, the analysis of the luminescence from the volume of the tumor is hindered by strong absorption and scattering in the UV and blue spectral region. Even in red and near-infrared regions scattering in the tissue still deteriorate the signals. Thus, luminescence from tissues can be retrieved only within 1 mm from a surface irrespective of a wavelength and makes the quantitative analysis of NAD(P)H/FAD fluorescence rather difficult.

Fiber-optic based spectroscopic sensing and imaging have been previously employed for chemical quantification of the tissues, including FLIM and Raman spectroscopy [71, 72]. Fiber-optic probes have also been used for *in vivo* depth-resolved neuron-activity mapping, both with two-photon and one-photon configurations [73, 74]. One of the first applications of time-resolved spectroscopy for skin cancer based on analysis of autofluorescence from NAD(P)H and using a fiber was described in the work of P.A.A. de Beule et al. from P. French group [75]. The authors presented an approach based on analysis of both fluorescence spectrum and fluorescence lifetime of endogenous fluorophores that allowed to discriminate basal cell carcinoma skin cancer from healthy skin tissue in *ex vivo* human skin lesions.

An optical fiber integrated in a thin needle that is inserted into a tumor is a promising approach for the delivery and collection of the luminescence signal from the internal layers of the tissue. Such an attempt has been demonstrated for two-photon probing of xenograft tumors in mice using MCA 207 cells expressing green fluorescent protein (GFP) [76]. Measuring the fluorescence of intrinsic fluorophores poses a few challenges. The quantum yields of NADH and FAD are low, and so are the intensities [77]. A second problem is fluorescence from the optical fibers. The excitation wavelength is in the ultraviolet or blue range of the optical spectrum. Fluorescence from the fibers at these wavelengths can easily be stronger than that from the sample. Two-photon excitation may be employed to overcome the difficulties generated by autofluorescence from fibers, but at the expense of costs and bulkiness of the setups. Alternatively, one may use hollow waveguide (HW) [78] or an air-core photonic crystal fibers (PCF) [79]. In both cases, the excitation beams are guided through the air-core, thus, the background signals originating from the core material are not inherently generated. However, the HW has a disadvantage of a low numerical aperture (NA), which reduce the collection efficiency of the signal. Moreover, the air holes of the HW and PCF should not be directly in contact with the liquids to avoid damage due to the capillary action. Finally, there are no readily available PC fibers in the UV region and, in general, they are very expensive.

The problems associated with the time-resolved fluorescence measurements with a fiber have been addressed in our recent paper [80]. The authors proposed a fiber-optic configuration that allowed to overcome the problem with background autofluorescence from the fiber core and enable measurement of weak fluorescence from NAD(P)H and phosphorescence from oxygen sensors from the deep layers of a tumor *in vivo*. The generalized scheme of the experimental setup is presented in Fig. 3.4.

Using this system, spectroscopic measurements of NAD(P)H fluorescence lifetime were performed in model experiment in solutions and in colorectal tumors in mice *in vivo*. The mean fluorescence lifetime in tumors was considerably shorter, and the relative contribution of free NAD(P)H in tumors was greater than in muscle, indicating a shift to more glycolytic metabolism in tumor tissue as compared to muscle. Simultaneous measurements of phosphorescence lifetime of oxygen-sensitive probe administrated into the tissue showed increased values in tumor reflecting reduced oxygenation, which correlated with its glycolytic state.

In a recent study, we have used fiber-optic-based spectroscopy to investigate metabolic response of the tumors to chemotherapy with irinotecan. *In vivo* fluorescence time-resolved sensing of NAD(P)H showed decreased contribution from the unbound NAD(P)H fraction in the treated tumors in mice compare with untreated control (Fig. 3.5), which is in agreement with our studies using FLIM-microscopy.

Although the examples of time-resolved measurements of intrinsic fluorescence using optical fiber-based systems are rather few, this approach has promises for development of clinical FLIM endoscopy.

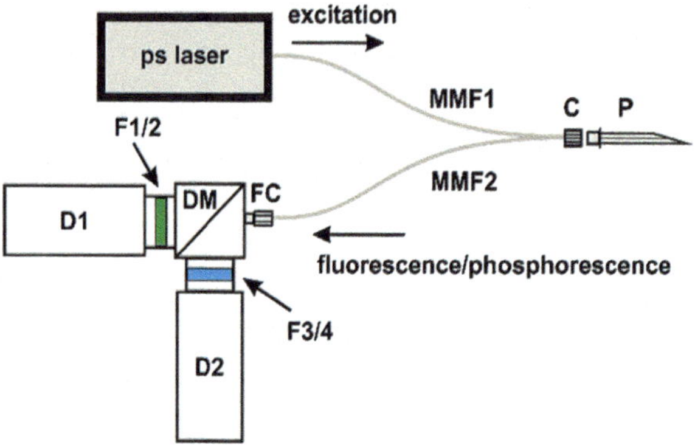

Fig. 3.4 Experimental setup for fluorescence/phosphorescence spectroscopy measurements. MMF1: Multimode fiber (core diameter: 50 μm, cladding diameter: 65 μm); MMF2: Multimode fiber (core diameter: 200 μm, cladding diameter: 220 μm); C: miniature connector; P: exchangeable probe with a multimode fiber (core diameter: 300 μm, cladding diameter: 350 μm) in a needle G26) (see Visualization 1); FC-FC connector; D1 and D2: HPM-100-40 detector. F1/2: emission filters 510LP and 632/90 BP. F3/4: 405LP and 450/60BP. DM: 510 LP dichroic mirror. (Reprinted with permission from [80])

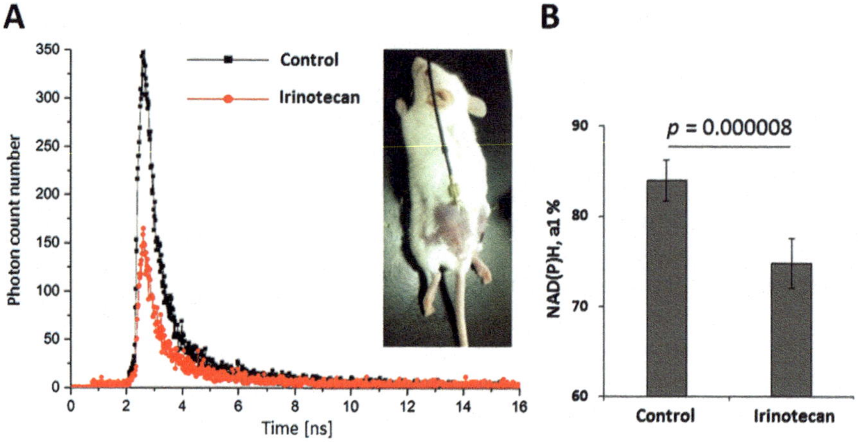

Fig. 3.5 *In vivo* one-photon fluorescence time-resolved spectroscopy of NAD(P)H in mouse colorectal tumors in response to irinotecan. (**A**) Fluorescence decay curves of NAD(P)H in the specific spots in subcutaneous tumors. (**B**) Relative amplitude of free NAD(P)H (a1%) calculated from the *in vivo* decay curves. Mean ± SD, $n = 3$ tumors, 5–7 measurements from each tumor. Measurements were performed on the 14th day of tumor growth

3.6 FLIM-Macroscopy

Extending FLIM from microscopy to macroscopy is an extremely important issue in cancer studies. While it is essential to have a subcellular resolution to follow metabolic processes on a single cell level, it is also critical to understand metabolic status of a living system at the large scale. Intratumoral metabolic heterogeneity at the cellular level is, to a considerable degree, determined by heterogeneous microenvironment within a tumor, namely, the architecture of extracellular matrix, organization of vasculature, infiltration of immune cells and fibroblasts, and interaction with surroundings. As a result, heterogeneity of the metabolic picture on the macroscopic level can be even more pronounced than on the cellular level.

Although there have been a few systems developed for whole-body small animal FLIM, they are insensitive to tissue autofluorescence and have poor spatial resolution [81–83].

Recently, we have used a confocal macro-FLIM system to record fluorescence of NAD(P)H and to investigate heterogeneity of tumors on macroscale. The operation of the macroscanner is described in detail elsewhere [84, 85]. Briefly, the macroscopic imaging is performed by scanning of the objects placed directly in the image plane of a confocal scan head. To combine NAD(P)H imaging with observation of other spectrally distinct fluorophores, there are two excitation and two detection channels in the system.

The optical principle for time-resolved macroscopic scanning system is shown in Fig. 3.6. The image plane of the scan lens is brought in coincidence with the sample surface. As the galvo-mirrors change the beam angle the laser focus scans across the

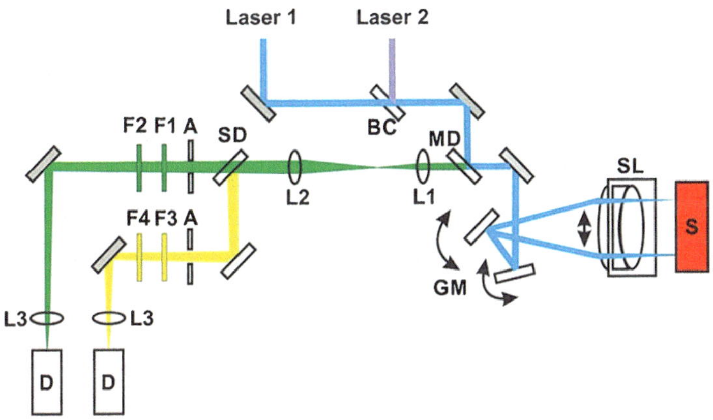

Fig. 3.6 Schematics of the macroscanner for two-channel FLIM. D: HPM-100-40 detector (Becker & Hickl GmbH, Germany), F1, F3: long pass filters; F2, F4: bandpass filters; A: pinhole; L3: detector focusing lenses ($f = 50$ mm); L1&L2 telescope lenses ($f_1 = 7.5$ mm; $f_2 = 45$ mm); *BC* beam combiner mirror, *MD* main dichroic mirror, *GM* galvo-mirrors, *SL* scan lens ($f = 40$ mm); *S* sample. (Reprinted with permission from [85])

sample. Fluorescence signal produced by the sample is collimated by the scan lens, descanned by the galvo-mirrors, and separated from the excitation light by the main dichroic beamsplitter.

Usually, the signal further can be separated into two spectral or polarization channels, and focused into the pinholes. Light passing the pinholes is sent to detectors. The maximum diameter of the image area in the primary image plane of the scanner is about 18 mm. The size of the laser spot in the image plane is about 15 μm. While the numerical aperture of the excitation beam path is determined by the beam diameter (about 1.5 mm) and the focal length of the scan lens (40 mm), the detection path is determined by the scan lens and the aperture of the galvo-mirrors (3 mm). Since the incoming laser beam under-fills the lens aperture, the numerical aperture in excitation is smaller than in the detection.

Overall, collection efficiency of a macroscanner is considerably lower compared to the system consisting of the scanner combined with a microscope. However, in macroscopic imaging one can use much higher laser power, which compensates for low collection efficiency. The power is distributed over a large area so that photobleaching is not a problem. The acquisition time of this system depends on the number of pixels and the desired number of photons for the accurate evaluation of the fluorescence lifetimes.

The performance of metabolic FLIM on a macroscale was demonstrated in mouse tumors *in vivo* and rat glioma samples *ex vivo*.

In the *in vivo* study we visualized NAD(P)H fluorescence lifetime in a whole tumor inoculated subcutaneously in live mice. Since 375 nm wavelength is used for one-photon excitation of NAD(P)H, the skin flap was surgically opened to image tumor. As expected, a solid advanced tumor displayed essential spatial variations of the fluorescence lifetime parameters of NAD(P)H. Comparisons between tumor and muscles showed an increase in the contribution of the free NAD(P)H in tumor tissue, suggesting a shift to glycolytic metabolism, similarly to the above mentioned spectroscopic measurements [84].

A large field of view in the macro-FLIM opens the opportunity to explore metabolism not only in the whole tumor in animal, but also in intraoperational centimeter sized samples of patients' tumors. The metabolic differences between cancer and adjacent normal tissue, if detected by FLIM at the macroscale, can serve for delineation of tumor margins, which is extremely important in glioma surgery. We have made an attempt to evaluate the capability of macroscopic FLIM to delineate glioma margins in rat models on the basis of NAD(P)H fluorescence lifetime parameters recently (Fig. 3.7) [86]. It was found that, in comparison with normal brain tissue, anaplastic astrocytoma had longer values of mean fluorescence lifetime and fluorescence lifetime of a bound fraction of NAD(P)H. Moreover, we have observed decreased contribution from a free NAD(P)H. This shows the potential of macro-FLIM for tumor delineation. The changes in the optical metabolic readouts in the brain tumors are consistent with studies that report on elevated levels of fatty acid oxidation and glutaminolysis in glioma cells [8, 87]. With further development, macro-FLIM can find use in the clinic as a sensitive and precise method for identifying the edges of tumors during surgery.

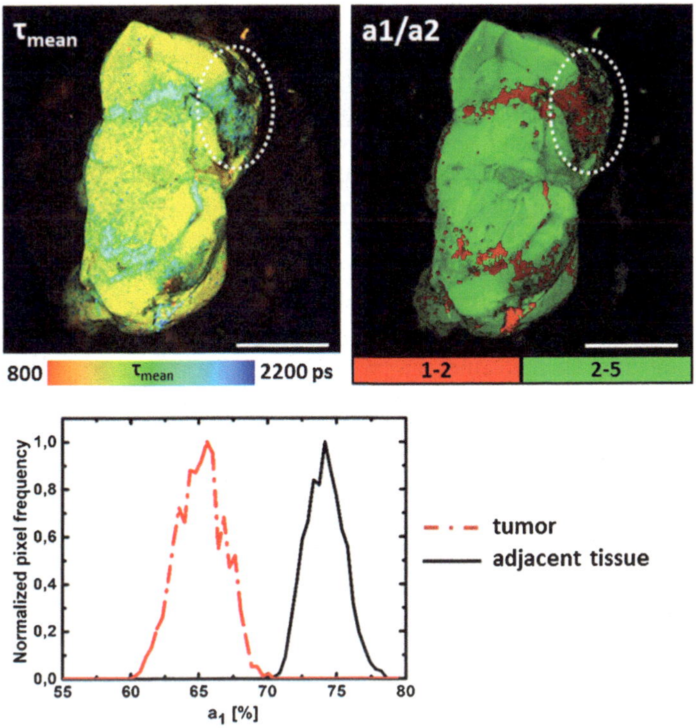

Fig. 3.7 Macro-FLIM of glioma in rat brain. Representative fluorescence lifetime images of NAD(P)H in freshly excised brain with anaplastic astrocytoma. τ_{mean} is an amplitude weighted fluorescence lifetime. a1/a2 is free/bound ratio. In the histogram of the normalized pixel frequency red and black lines represent the distribution of free NAD(P)H in the tumor and in the adjacent tissue, respectively. The tumor is marked with a white dashed line. Collection time: 120 s. Scale bar: 5 mm

3.7 Summary

A significant progress in the field of cancer metabolism led to understanding that the changes in metabolism in cancer cells are not a simple consequence of the uncontrolled proliferation, but are rather a coordinated and complementary program that promote tumor growth. An increased interest in tumor metabolism in recent years has been driven by the idea to exploit the altered and heterogeneous metabolism of cancer cells to develop personalized therapeutic approaches and to find new targets for therapy. Novel imaging technologies that allow detection of intrinsic fluorescence and measurement of its lifetime open the possibility to monitor metabolic state of living cells and tissues non-invasively, in real time. FLIM based on TCSPC is one of such technologies. Metabolic imaging done by fluorescence lifetime and intensity measurements is often associated with Fluorescence Lifetime Imaging Microscopy. Consequently, it is often believed that FLIM of metabolism is related to the microscopic level. This is certainly not correct, because now TCSPC-based fluorescence lifetime measurements can be also realized on the macroscale. Recently developed macro-FLIM allows to interrogate metabolism on the scale of a whole tumor. It gives an opportunity not only to observe the heterogeneity of the tumor on a large scale, which is important from a fundamental point of view, but also to determine the surgical margins of tumors with high precision, and this is essential for clinical applications. Another method to probe tumor metabolism is not related to imaging option, but rather to time-resolved fluorescence single point measurement via a fiber. While this approach does not deliver a map of metabolic state of the tumor, it offers an opportunity to explore its deep layers, since strong scattering and absorption of an excitation light at the surface of the tissue is not an issue any more. Combination of all the modalities may give new insights in the complicated picture of tumor metabolism, not only being limited to experimental models of cancer but also extending to patient's tumors. Moreover, correlation of metabolic measurements with other physiological parameters measured using spectrally distinct exogenous sensors, i.e. multi-parametric recordings, will enhance their value and provide more details of fundamental aspects of cancer biology.

Acknowledgments The authors are thankful to Dr. Varvara Dudenkova, Dr. Nadezhda Ignatova, Dr. Elena Kiseleva, Irina Druzhkova, Liubov Shimolina, Alena Gavrina and Dmitry Reunov from Privolzhsky Research Medical University for their kind assistance in these studies. The studies are supported by the Russian Science Foundation (project # 20-65-46018).

References

1. Warburg, O., Wind, F., Negelein, E.: The metabolism of tumors in the body. J. Gen. Physiol. **8**(6), 519–530 (1927)
2. Bensinger, S.J., Christofk, H.R.: New aspects of the Warburg effect in cancer cell biology. Semin. Cell Dev. Biol. **23**(4), 352–361 (2012). https://doi.org/10.1016/j.semcdb.2012.02.003
3. Coller, H.A.: Is cancer a metabolic disease? Am. J. Pathol. **184**(1), 4–17 (2014). https://doi.org/10.1016/j.ajpath.2013.07.035
4. Solaini, G., Sgarbi, G., Baracca, A.: Oxidative phosphorylation in cancer cells. Biochim. Biophys. Acta Bioenerg. **1807**(6), 534–542 (2011). https://doi.org/10.1016/j.bbabio.2010.09.003
5. Viale, A., Corti, D., Draetta, G.F.: Tumors and mitochondrial respiration: a neglected connection. Cancer Res. **75**(18), 3687 (2015). https://doi.org/10.1158/0008-5472.CAN-15-0491
6. Hui, S., Ghergurovich, J.M., Morscher, R.J., Jang, C., Teng, X., Lu, W., Esparza, L.A., Reya, T., Le, Z., Yanxiang Guo, J., White, E., Rabinowitz, J.D.: Glucose feeds the TCA cycle via circulating lactate. Nature. **551**(7678), 115–118 (2017). https://doi.org/10.1038/nature24057
7. Pavlides, S., Whitaker-Menezes, D., Castello-Cros, R., Flomenberg, N., Witkiewicz, A.K., Frank, P.G., Casimiro, M.C., Wang, C., Fortina, P., Addya, S., Pestell, R.G., Martinez-Outschoorn, U.E., Sotgia, F., Lisanti, M.P.: The reverse Warburg effect: aerobic glycolysis in cancer associated fibroblasts and the tumor stroma. Cell Cycle. **8**(23), 3984–4001 (2009). https://doi.org/10.4161/cc.8.23.10238
8. Cluntun, A.A., Lukey, M.J., Cerione, R.A., Locasale, J.W.: Glutamine metabolism in cancer: understanding the heterogeneity. Trends Cancer. **3**(3), 169–180 (2017). https://doi.org/10.1016/j.trecan.2017.01.005

9. Carracedo, A., Cantley, L.C., Pandolfi, P.P.: Cancer metabolism: fatty acid oxidation in the limelight. Nat. Rev. Cancer. **13**(4), 227–232 (2013). https://doi.org/10.1038/nrc3483
10. Potter, M., Newport, E., Morten, K.J.: The Warburg effect: 80 years on. Biochem. Soc. Trans. **44**(5), 1499–1505 (2016). https://doi.org/10.1042/BST20160094
11. Altman, B.J., Rathmell, J.C.: Metabolic stress in autophagy and cell death pathways. Cold Spring Harb Perspect Biol. **4**(9), 008763 (2012). https://doi.org/10.1101/cshperspect.a008763
12. Green, D.R., Galluzzi, L., Kroemer, G.: Cell biology. Metabolic control of cell death. Science. **345**(6203), 1250256 (2014). https://doi.org/10.1126/science.1250256
13. Zaal, E.A., Berkers, C.R.: The influence of metabolism on drug response in cancer. Front. Oncol. **8**, 500 (2018). https://doi.org/10.3389/fonc.2018.00500
14. Scholz, R., Thurman, R.G., Williamson, J.R., Chance, B., Bücher, T.: Flavin and pyridine nucleotide oxidation-reduction changes in perfused rat liver. I. Anoxia and subcellular localization of fluorescent flavoproteins. J Biol Chem. **244**, 2317–2324 (1969)
15. Chance, B.: Spectrophotometry of intracellular respiratory pigments. Science. **120**(3124), 767–775 (1954)
16. Chance, B., Schoener, B., Oshino, R., Itshak, F., Nakase, Y.: Oxidation-reduction ratio studies of mitochondria in freeze-trapped samples. NADH and flavoprotein fluorescence signals. J Biol Chem. **254**(11), 4764–4771 (1979)
17. Nakabayashi, T.: In: Ghukasyan, V.V., Heikal, A.A. (eds.) Intracellular Autofluorescent Species: Structure, Spectroscopy, and Photophysics. CRC Press, Boca Raton (2014)
18. Wallrabe, H., Svindrych, Z., Alam, S.R., Siller, K.H., Wang, T., Kashatus, D., Hu, S., Periasamy, A.: Segmented cell analyses to measure redox states of autofluorescent NAD(P)H, FAD & Trp in cancer cells by FLIM. Sci Rep. **8**(1), 79 (2018). https://doi.org/10.1038/s41598-017-18634-x
19. Lakowicz, J.R., Szmacinski, H., Nowaczyk, K., Johnson, M.L.: Fluorescence lifetime imaging of free and protein-bound NADH. Proc. Natl. Acad. Sci. U. S. A. **89**(4), 1271–1275 (1992)
20. Blacker, T.S., Marsh, R.J., Duchen, M.R., Bain, A.J.: Activated barrier crossing dynamics in the non-radiative decay of NADH and NADPH. Chem. Phys. **422**, 184–194 (2013). https://doi.org/10.1016/j.chemphys.2013.02.019
21. Yu, Q., Heikal, A.A.: Two-photon autofluorescence dynamics imaging reveals sensitivity of intracellular NADH concentration and conformation to cell physiology at the single-cell level. J. Photochem. Photobiol. B. **95**(1), 46–57 (2009). https://doi.org/10.1016/j.jphotobiol.2008.12.010
22. Sharick, J.T., Favreau, P.F., Gillette, A.A., Sdao, S.M., Merrins, M.J., Skala, M.C.: Protein-bound NAD(P)H lifetime is sensitive to multiple fates of glucose carbon. Sci Rep. **8**(1), 5456 (2018). https://doi.org/10.1038/s41598-018-23691-x
23. Ladokhin, A., Brand, L.: Evidence for an excited-state reaction contributing to NADH fluorescence. J. Fluoresc. **5**(1), 99–106 (1995). https://doi.org/10.1007/BF00718787
24. Blacker, T.S., Mann, Z.F., Gale, J.E., Ziegler, M., Bain, A.J., Szabadkai, G., Duchen, M.R.: Separating NADH and NADPH fluorescence in live cells and tissues using FLIM. Nat. Commun. **5**, 3936 (2014). https://doi.org/10.1038/ncomms4936
25. Meleshina, A.V., Dudenkova, V.V., Shirmanova, M.V., Shcheslavskiy, V.I., Becker, W., Bystrova, A.S., Cherkasova, E.I., Zagaynova, E.V.: Probing metabolic states of differentiating stem cells using two-photon FLIM. Sci. Rep. **6**, 21853 (2016). https://doi.org/10.1038/srep21853
26. Liu, Z., Pouli, D., Alonzo, C.A., Varone, A., Karaliota, S., Quinn, K.P., Münger, K., Karalis, K.P., Georgakoudi, I.: Mapping metabolic changes by noninvasive, multiparametric, high-resolution imaging using endogenous contrast. Sci Adv. **4**(3), 9302 (2018). https://doi.org/10.1126/sciadv.aap9302
27. Yaseen, M.A., Sakadžić, S., Wu, W., Becker, W., Kasischke, K.A., Boas, D.A.: *In vivo* imaging of cerebral energy metabolism with two-photon fluorescence lifetime microscopy of NADH. Biomed. Opt. Express. **4**(2), 307–321 (2013). https://doi.org/10.1364/BOE.4.000307

28. Stringari, C., Cinquin, A., Cinquin, O., Digman, M.A., Donovan, P.J., Gratton, E.: Phasor approach to fluorescence lifetime microscopy distinguishes different metabolic states of germ cells in a live tissue. Proc Natl Acad Sci U S A. **108**(33), 13582 (2011). https://doi.org/10.1073/pnas.1108161108
29. Ma, N., Digman, M.A., Malacrida, L., Gratton, E.: Measurements of absolute concentrations of NADH in cells using the phasor FLIM method. Biomed. Opt. Express. **7**(7), 2441–2452 (2016). https://doi.org/10.1364/BOE.7.002441
30. Starbird, C.A., Maklashina, E., Cecchini, G., Iverson, T.M.: Flavoenzymes: covalent versus noncovalent. In: eLS. Wiley, Chichester (2015)
31. van den Berg, P.A.W., Feenstra, K.A., Mark, A.E., Berendsen, H.J.C., Visser, A.J.W.G.: Dynamic conformations of flavin adenine dinucleotide: simulated molecular dynamics of the flavin cofactor related to the time-resolved fluorescence characteristics. J. Phys. Chem. B. **106**(34), 8858–8869 (2002). https://doi.org/10.1021/jp020356s
32. Islam, S.D.M., Susdorf, T., Penzkofer, A., Hegemann, P.: Fluorescence quenching of flavin adenine dinucleotide in aqueous solution by pH dependent isomerisation and photo-induced electron transfer. Chem. Phys. **295**, 137–149 (2003). https://doi.org/10.1016/j.chemphys.2003.08.013
33. Rivlin, R.S., Hornibrook, R., Osnos, M.: Effects of riboflavin deficiency upon concentrations of riboflavin, flavin mononucleotide, and flavin adenine dinucleotide in Novikoff hepatoma in rats. Cancer Res. **33**(11), 3019–3023 (1973)
34. Pinto, J., Huang, Y.P., Rivlin, R.S.: Inhibition of riboflavin metabolism in rat tissues by chlorpromazine, imipramine, and amitriptyline. J. Clin. Invest. **67**(5), 1500–1506 (1981)
35. Huhner, J., Ingles-Prieto, A., Neususs, C., Lammerhofer, M., Janovjak, H.: Quantification of riboflavin, flavin mononucleotide, and flavin adenine dinucleotide in mammalian model cells by CE with LED-induced fluorescence detection. Electrophoresis. **36**(4), 518–525 (2015). https://doi.org/10.1002/elps.201400451
36. Munro, A.W., Noble, M.A.: In: Chapman, S.K., Reid, G.A. (eds.) Fluorescence Analysis of Flavoproteins, pp. 25–48. Humana Press, Totowa (1999)
37. Becker, W.: In: Castleman, A.W., Toennies, J.P., Zinth, W. (eds.) Advanced Time-Correlated Single Photon Counting Techniques. Springer, Berlin (2005)
38. Philip, J., Carlsson, K.: Theoretical investigation of the signal-to-noise ratio in fluorescence lifetime imaging. J. Opt. Soc. Am. A Opt. Image Sci. Vis. **20**(2), 368–379 (2003). https://doi.org/10.1364/JOSAA.20.000368
39. Becker, W., Bergmann, A., Biskup, C., Zimmer, T., Klöcker, N., Benndorf, K., Becker, A., Gmbh, H.: Multi-wavelength TCSPC lifetime imaging. Proc. SPIE. **4620**, 470679 (2002). https://doi.org/10.1117/12.470679
40. Becker, W., Bergmann, A., Biskup, C.: Multispectral fluorescence lifetime imaging by TCSPC. Microsc. Res. Tech. **70**(5), 403–409 (2007). https://doi.org/10.1002/jemt.20432
41. Cheong, W.F., Prahl, S.A., Welch, A.J.: A review of the optical properties of biological tissues. IEEE J. Quantum Electron. **26**(12), 2166–2185 (1990). https://doi.org/10.1109/3.64354
42. Bird, D.K., Yan, L., Vrotsos, K.M., Eliceiri, K.W., Vaughan, E.M., Keely, P.J., White, J.G., Ramanujam, N.: Metabolic mapping of MCF10A human breast cells via multiphoton fluorescence lifetime imaging of the coenzyme NADH. Cancer Res. **65**(19), 8766–8773 (2005). https://doi.org/10.1158/0008-5472.can-04-3922
43. Skala, M.C., Riching, K.M., Gendron-Fitzpatrick, A., Eickhoff, J., Eliceiri, K.W., White, J.G., Ramanujam, N.: *In vivo* multiphoton microscopy of NADH and FAD redox states, fluorescence lifetimes, and cellular morphology in precancerous epithelia. Proc. Natl. Acad. Sci. U. S. A. **104**(49), 19494–19499 (2007). https://doi.org/10.1073/pnas.0708425104
44. Lee, D.H., Li, X., Ma, N., Digman, M.A., Lee, A.: Rapid and label-free identification of single leukemia cells from blood in a high-density microfluidic trapping array by fluorescence lifetime imaging microscopy. Lab Chip. **18**(9), 1349–1358 (2018). https://doi.org/10.1039/C7LC01301A

45. Walsh, A.J., Cook, R.S., Manning, H.C., Hicks, D.J., Lafontant, A., Arteaga, C.L., Skala, M.C.: Optical metabolic imaging identifies glycolytic levels, subtypes, and early-treatment response in breast cancer. Cancer Res. **73**(20), 6164–6174 (2013). https://doi.org/10.1158/0008-5472.can-13-0527
46. Ruck, A., Hauser, C., Mosch, S., Kalinina, S.: Spectrally resolved fluorescence lifetime imaging to investigate cell metabolism in malignant and nonmalignant oral mucosa cells. J Biomed Opt. **19**(9), 96005 (2014). https://doi.org/10.1117/1.jbo.19.9.096005
47. Shah, A.T., Demory Beckler, M., Walsh, A.J., Jones, W.P., Pohlmann, P.R., Skala, M.C.: Optical metabolic imaging of treatment response in human head and neck squamous cell carcinoma. PLoS One. **9**(3), 90746 (2014). https://doi.org/10.1371/journal.pone.0090746
48. Wang, M., Tang, F., Pan, X., Yao, L., Wang, X., Jing, Y., Ma, J., Wang, G., Mi, L.: Rapid diagnosis and intraoperative margin assessment of human lung cancer with fluorescence lifetime imaging microscopy. BBA Clin. **8**, 7–13 (2017). https://doi.org/10.1016/j.bbacli.2017.04.002
49. Leppert, J., Krajewski, J., Kantelhardt, S.R., Schlaffer, S., Petkus, N., Reusche, E., Huttmann, G., Giese, A.: Multiphoton excitation of autofluorescence for microscopy of glioma tissue. Neurosurgery. **58**(4), 759–767 (2006). https://doi.org/10.1227/01.neu.0000204885.45644.22
50. Conklin, M.W., Provenzano, P.P., Eliceiri, K.W., Sullivan, R., Keely, P.J.: Fluorescence lifetime imaging of endogenous fluorophores in histopathology sections reveals differences between normal and tumor epithelium in carcinoma *in situ* of the breast. Cell Biochem. Biophys. **53**(3), 145–157 (2009). https://doi.org/10.1007/s12013-009-9046-7
51. Kalinina, S., Breymayer, J., Schafer, P., Calzia, E., Shcheslavskiy, V., Becker, W., Ruck, A.: Correlative NAD(P)H-FLIM and oxygen sensing-PLIM for metabolic mapping. J. Biophotonics. **9**(8), 800–811 (2016). https://doi.org/10.1002/jbio.201500297
52. Shcheslavskiy, V.I., Neubauer, A., Bukowiecki, R., Dinter, F., Becker, W.: Combined fluorescence and phosphorescence lifetime imaging. Appl Phys Lett. **108**(9), 091111 (2016). https://doi.org/10.1063/1.4943265
53. Li, D., Zheng, W., Qu, J.Y.: Time-resolved spectroscopic imaging reveals the fundamentals of cellular NADH fluorescence. Opt. Lett. **33**(20), 2365–2367 (2008)
54. Heaster, T.M., Walsh, A.J., Zhao, Y., Hiebert, S.W., Skala, M.C.: Autofluorescence imaging identifies tumor cell-cycle status on a single-cell level. J Biophotonics. **11**(1), e201600276 (2018). https://doi.org/10.1002/jbio.201600276
55. Lukina, M.M., Dudenkova, V.V., Ignatova, N.I., Druzhkova, I.N., Shimolina, L.E., Zagaynova, E.V., Shirmanova, M.V.: Metabolic cofactors NAD(P)H and FAD as potential indicators of cancer cell response to chemotherapy with paclitaxel. Biochim. Biophys. Acta Gen. Subj. **1862**(8), 1693–1700 (2018). https://doi.org/10.1016/j.bbagen.2018.04.021
56. Sergeeva, T.F., Shirmanova, M.V., Zlobovskaya, O.A., Gavrina, A.I., Dudenkova, V.V., Lukina, M.M., Lukyanov, K.A., Zagaynova, E.V.: Relationship between intracellular pH, metabolic co-factors and caspase-3 activation in cancer cells during apoptosis. Biochim. Biophys. Acta, Mol. Cell Res. **1864**(3), 604–611 (2017). https://doi.org/10.1016/j.bbamcr.2016.12.022
57. Shirmanova, M.V., Sergeeva, T.F., Gavrina, A.I., Dudenkova, V.V., Lukyanov, K.A., Zagaynova, E.V.: Multiparametric analysis of cisplatin-induced changes in cancer cells using FLIM. Proc. SPIE. **10498**, 1049807 (2018). https://doi.org/10.1117/12.2293996
58. Trinh, A.L., Chen, H., Chen, Y., Hu, Y., Li, Z., Siegel, E.R., Linskey, M.E., Wang, P.H., Digman, M.A., Zhou, Y.H.: Tracking functional tumor cell subpopulations of malignant glioma by phasor fluorescence lifetime imaging microscopy of NADH. Cancers. **9**(12), 168 (2017). https://doi.org/10.3390/cancers9120168
59. Druzhkova, I.N., Shirmanova, M.V., Lukina, M.M., Dudenkova, V.V., Mishina, N.M., Zagaynova, E.V.: The metabolic interaction of cancer cells and fibroblasts - coupling between NAD(P)H and FAD, intracellular pH and hydrogen peroxide. Cell Cycle. **15**(9), 1257–1266 (2016). https://doi.org/10.1080/15384101.2016.1160974
60. Walsh, A.J., Castellanos, J.A., Nagathihalli, N.S., Merchant, N.B., Skala, M.C.: Optical imaging of drug-induced metabolism changes in murine and human pancreatic cancer organoids reveals heterogeneous drug response. Pancreas. **45**(6), 863–869 (2016). https://doi.org/10.1097/mpa.0000000000000543

61. Druzhkova, I.N., Lukina, M.M., Dudenkova, V., Shimolina, L.E., Zagaynova, E.V., Shirmanova, M.V.: Insight into microenvironment of tumor on the microscopic level with a focus on cancer-associated fibroblasts. Proc. SPIE. **10685**, 106852R (2018). https://doi.org/10.1117/12.2307096
62. Szulczewski, J.M., Inman, D.R., Entenberg, D., Ponik, S.M., Aguirre-Ghiso, J., Castracane, J., Condeelis, J., Eliceiri, K.W., Keely, P.J.: *In vivo* visualization of stromal macrophages via label-free FLIM-based metabolite imaging. Sci. Rep. **6**, 25086 (2016). https://doi.org/10.1038/srep25086
63. Shirmanova, M., Sergeeva, T., Druzhkova, I., Meleshina, A., Lukina, M., Dudenkova, V., Shcheslavskiy, V., Becker, W., Belousov, V., Mishina, N., Zagaynova, E.: In: König, K. (ed.) Metabolic Shifts in Cell Proliferation and Differentiation: Applications in Biology and Medicine. De Gruyter, Berlin (2018)
64. Shirmanova, M.V., Druzhkova, I.N., Lukina, M.M., Dudenkova, V.V., Ignatova, N.I., Snopova, L.B., Shcheslavskiy, V.I., Belousov, V.V., Zagaynova, E.V.: Chemotherapy with cisplatin: insights into intracellular pH and metabolic landscape of cancer cells *in vitro* and *in vivo*. Sci. Rep. **7**(1), 8911 (2017). https://doi.org/10.1038/s41598-017-09426-4
65. Lukina, M.M., Dudenkova, V.V., Shimolina, L.E., Snopova, L.B., Zagaynova, E.V., Shirmanova, M.V.: *In vivo* metabolic and SHG imaging for monitoring of tumor response to chemotherapy. Cytometry A. **95**(1), 47–55 (2019). https://doi.org/10.1002/cyto.a.23607
66. Alam, S.R., Wallrabe, H., Svindrych, Z., Chaudhary, A.K., Christopher, K.G., Chandra, D., Periasamy, A.: Investigation of mitochondrial metabolic response to doxorubicin in prostate cancer cells: an NADH, FAD and tryptophan FLIM assay. Sci. Rep. **7**(1), 10451 (2017). https://doi.org/10.1038/s41598-017-10856-3
67. Cao, R., Wallrabe, H., Siller, K., Rehman Alam, S., Periasamy, A.: Single-cell redox states analyzed by fluorescence lifetime metrics and tryptophan FRET interaction with NAD(P)H. Cytometry A. **95**(1), 110–121 (2019). https://doi.org/10.1002/cyto.a.23711
68. Shah, A.T., Diggins, K.E., Walsh, A.J., Irish, J.M., Skala, M.C.: *In vivo* autofluorescence imaging of tumor heterogeneity in response to treatment. Neoplasia. **17**(12), 862–870 (2015). https://doi.org/10.1016/j.neo.2015.11.006
69. Walsh, A.J., Cook, R.S., Sanders, M.E., Aurisicchio, L., Ciliberto, G., Arteaga, C.L., Skala, M.C.: Quantitative optical imaging of primary tumor organoid metabolism predicts drug response in breast cancer. Cancer Res. **74**(18), 5184–5194 (2014). https://doi.org/10.1158/0008-5472.CAN-14-0663
70. Shah, A.T., Heaster, T.M., Skala, M.C.: Metabolic imaging of head and neck cancer organoids. PLoS One. **12**(1), 0170415 (2017). https://doi.org/10.1371/journal.pone.0170415
71. Cheng, S., Rico-Jimenez, J.J., Jabbour, J., Malik, B., Maitland, K.C., Wright, J., Cheng, Y.-S.L., Jo, J.A.: Flexible endoscope for continuous *in vivo* multispectral fluorescence lifetime imaging. Opt. Lett. **38**(9), 1515–1517 (2013). https://doi.org/10.1364/OL.38.001515
72. Dochow, S., Ma, D., Latka, I., Bocklitz, T., Hartl, B., Bec, J., Fatakdawala, H., Marple, E., Urmey, K., Wachsmann-Hogiu, S., Schmitt, M., Marcu, L., Popp, J.: Combined fiber probe for fluorescence lifetime and Raman spectroscopy. Anal. Bioanal. Chem. **407**(27), 8291–8301 (2015). https://doi.org/10.1007/s00216-015-8800-5
73. Cui, G., Jun, S.B., Jin, X., Pham, M.D., Vogel, S.S., Lovinger, D.M., Costa, R.M.: Concurrent activation of striatal direct and indirect pathways during action initiation. Nature. **494**(7436), 238–242 (2013). https://doi.org/10.1038/nature11846
74. Doronina-Amitonova, L.V., Fedotov, I.V., Ivashkina, O.I., Zots, M.A., Fedotov, A.B., Anokhin, K.V., Zheltikov, A.M.: Implantable fiber-optic interface for parallel multisite long-term optical dynamic brain interrogation in freely moving mice. Sci. Rep. **3**, 3265 (2013). https://doi.org/10.1038/srep03265
75. De Beule, P.A., Dunsby, C., Galletly, N.P., Stamp, G.W., Chu, A.C., Anand, U., Anand, P., Benham, C.D., Naylor, A., French, P.M.: A hyperspectral fluorescence lifetime probe for skin cancer diagnosis. Rev. Sci. Instrum. **78**(12), 123101 (2007). https://doi.org/10.1063/1.2818785

76. Thomas, T.P., Myaing, M.T., Ye, J.Y., Candido, K., Kotlyar, A., Beals, J., Cao, P., Keszler, B., Patri, A.K., Norris, T.B., Baker Jr., J.R.: Detection and analysis of tumor fluorescence using a two-photon optical fiber probe. Biophys. J. **86**(6), 3959–3965 (2004). https://doi.org/10.1529/biophysj.103.034462
77. Lakowicz, J.R.: Principles of Fluorescence Spectroscopy, 3rd edn. Springer, Boston (2006)
78. Sun, Y., Chen, W.L., Lin, S.J., Jee, S.H., Chen, Y.F., Lin, L.C., So, P.T.C., Dong, C.Y.: Investigating mechanisms of collagen thermal denaturation by high resolution second-harmonic generation imaging. Biophys. J. **91**(7), 2620–2625 (2006). https://doi.org/10.1529/biophysj.106.085902
79. Doronina-Amitonova, L.V., Fedotov, I.V., Ivashkina, O.I., Zots, M.A., Fedotov, A.B., Anokhin, K.V., Zheltikov, A.M.: Enhancing the locality of optical interrogation with photonic-crystal fibers. Appl Phys Lett. **101**(2), 021114 (2012). https://doi.org/10.1063/1.4727901
80. Lukina, M., Orlova, A., Shirmanova, M., Shirokov, D., Pavlikov, A., Neubauer, A., Studier, H., Becker, W., Zagaynova, E., Yoshihara, T., Tobita, S., Shcheslavskiy, V.: Interrogation of metabolic and oxygen states of tumors with fiber-based luminescence lifetime spectroscopy. Opt. Lett. **42**(4), 731–734 (2017). https://doi.org/10.1364/OL.42.000731
81. Bloch, S., Lesage, F., McIntosh, L., Gandjbakhche, A., Liang, K., Achilefu, S.: Whole-body fluorescence lifetime imaging of a tumor-targeted near-infrared molecular probe in mice. J. Biomed. Opt. **10**(5), 054003 (2005). https://doi.org/10.1117/1.2070148
82. Rice, W.L., Shcherbakova, D.M., Verkhusha, V.V., Kumar, A.T.: *In vivo* tomographic imaging of deep-seated cancer using fluorescence lifetime contrast. Cancer Res. **75**(7), 1236–1243 (2015). https://doi.org/10.1158/0008-5472.CAN-14-3001
83. Sinsuebphon, N., Rudkouskaya, A., Barroso, M., Intes, X.: Comparison of illumination geometry for lifetime-based measurements in whole-body preclinical imaging. J. Biophotonics. **11**(10), 201800037 (2018). https://doi.org/10.1002/jbio.201800037
84. Shcheslavskiy, V.I., Shirmanova, M.V., Dudenkova, V.V., Lukyanov, K.A., Gavrina, A.I., Shumilova, A.V., Zagaynova, E., Becker, W.: Fluorescence time-resolved macroimaging. Opt. Lett. **43**(13), 3152–3155 (2018). https://doi.org/10.1364/OL.43.003152
85. Zherdeva, V.V., Kazachkina, N.I., Shcheslavskiy, V.I., Savitsky, A.P.: Long-term fluorescence lifetime imaging of a genetically encoded sensor for caspase-3 activity in mouse tumor xenografts. J. Biomed. Opt. **23**(3), 1–11 (2018). https://doi.org/10.1117/1.JBO.23.3.035002
86. Shirmanova, M.V., Maria, L., Kisileva, E.B., Fedoseeva, V.V., Dudenkova, V.V., Zagaynova, E.V., Wolfgang, B., Shcheslavskiy, V.I.: Interrogation of glioma metabolism on macroscale by FLIM. Proc. SPIE. **10882**, 1088209 (2019). https://doi.org/10.1117/12.2511475
87. Strickland, M., Stoll, E.A.: Metabolic reprogramming in glioma. Front. Cell Dev. Biol. **5**, 43 (2017). https://doi.org/10.3389/fcell.2017.00043

Chapter 4
Optical Imaging of Exosomes for Cancer Diagnosis, Monitoring, and Prognosis

Natalia V. Yunusova, Alexey V. Borisov, and Yury V. Kistenev

4.1 Exosomes: Composition and Function

Extracellular vesicles (EVs) participate in cancer progression and metastasis by transferring bioactive molecules between cancer and various cells in the local and distant microenvironments.

Two subtypes of microscopic EVs are considered: microvesicles and exosomes, which vary significantly in morphology, biophysical characteristics (shape, size, density), biogenesis, and functions.

N. V. Yunusova
Tomsk National Research Medical Center of the Russian Academy of Sciences, Tomsk, Russian Federation

Department of Biochemistry and Molecular Biology with a Course of Clinical Laboratory Diagnostics, Siberian State Medical University, Tomsk, Russian Federation
e-mail: bochkarevanv@oncology.tomsk.ru

A. V. Borisov
Department of General and Experimental Physics, National Research Tomsk State University, Tomsk, Russian Federation

Department of Physics with a Course of Higher Mathematics, Siberian State Medical University, Tomsk, Russian Federation

Laboratory of Biophotonics, National Research Tomsk State University, Tomsk, Russian Federation
e-mail: borisov@phys.tsu.ru

Y. V. Kistenev (✉)
Department of Physics with a Course of Higher Mathematics, Siberian State Medical University, Tomsk, Russian Federation

Laboratory of Biophotonics, National Research Tomsk State University, Tomsk, Russian Federation
e-mail: yuk@iao.ru

V. V. Tuchin et al. (eds.), *Multimodal Optical Diagnostics of Cancer*,
https://doi.org/10.1007/978-3-030-44594-2_4

Exosomes are nanosized EVs (with a diameter of 30–100 nm) that are secreted by almost all types of cells. Exosomes are found in biological fluids, including blood, saliva, urine, seminal fluid, amniotic liquid, ascites, bronchoalveolar lavage fluid, synovial fluid, breast milk, and cerebrospinal fluid. The size of exosomes seems to depend on cell origin, being ~50 nm in reticulocytes, ~100 nm in B cell lymphoma cells, or 80–120 nm in adipocytes [1].

The formation of exosomes is preceded by the strengthening of plasma membrane microdomains carrying clathrin. The endosomal sorting complex required for transport turns the membrane invaginations into early endosomes. The multivesicular bodies are formed in early endosomes during their re-invagination. These bodies either merge with lysosomes and degrade or they merge with the cytoplasmic membrane, and exosomes are released into extracellular space [2].

Microvesicles (ectosomes) are EVs with diameter from 100 to 1000 nm. The mechanism of microvesicles formation includes the formation of protrusions of the plasma membrane, followed by their pinching into the extracellular space.

Biogenesis and release of exosomes and microvesicles are presented in Fig. 4.1 [3].

Tetraspanins are a protein superfamily that organizes membrane microdomains by forming clusters and interacting with a large variety of transmembrane and cytosolic signaling proteins [4]. The ability of tetraspanins to interact with molecules and to form supramolecular complexes determines their ability of influence on a wide range of biological and pathological processes, for example, exosome biogen-

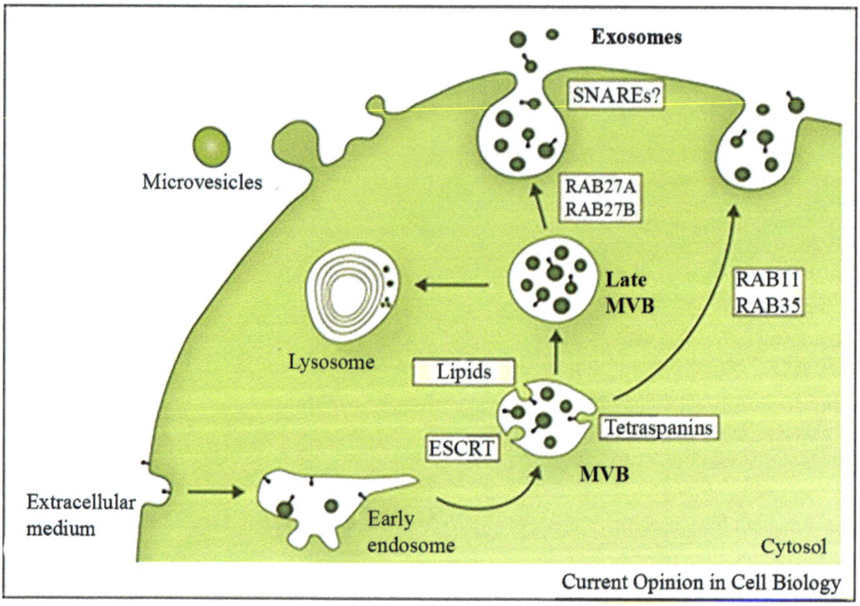

Fig. 4.1 Biogenesis and release of exosomes and microvesicles [3] (reprinted under license of Elsevier, License Number 4545321351777)

esis, exosomal proteins, and miRNA sorting, binding and uptake of exosomes by recipient cells, ability of exosomes to present antigens in the context of a subsequent immune response [5].

Exosomes have a large content of the cytoskeleton and actin-binding proteins (F-and G-actin, cofilin-1, profilin-1, tubulin), GTPases of the Rab superfamily and annexins, which promote membrane fusion.

Exosomes also carry specific proteins, reflecting their origin from a defined cell type. In particular, exosomes derived from cells of the immune system are rich by class II histocompatibility proteins; exosomes derived from melanoma cells contain, for example, a tumor-associated MART antigen; ascitic exosomes in patients with epithelial ovarian cancer carry EpCAM [1, 6].

4.1.1 Exosome Composition

4.1.1.1 Proteins

The exosome proteins include tetraspanins, integrins, heat shock proteins (HSP60, HSP70, HSP90) (Fig. 4.2) [4].

The study of saliva revealed two types of exosomes. The dominant component of type I exosomes of saliva is shown to be the coenzyme, derived from B-lymphocytes and widely represented in all the salivary glands. Carbonyl anhydrase 6 was detected only in type II exosomes [7].

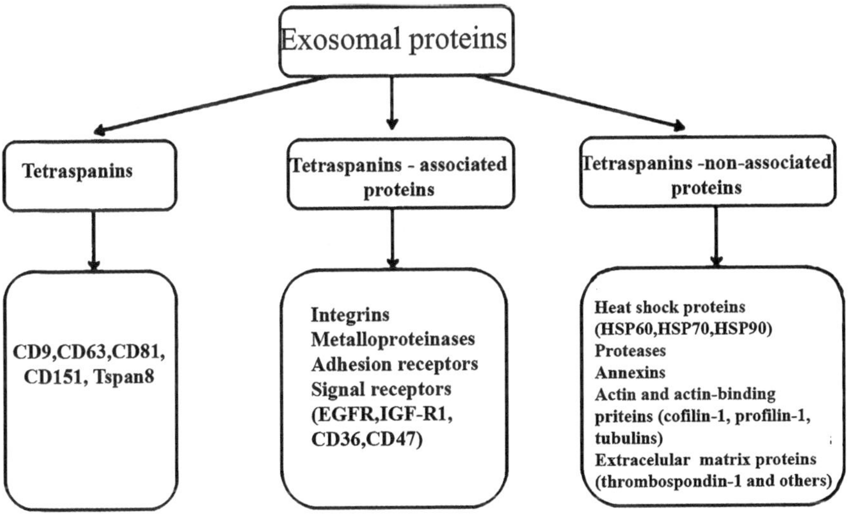

Fig. 4.2 Tetraspanins, tetraspanin-associated and tetraspanin-non-associated proteins in exosomes [4]

Exosomes carry a complex of proteases and their activators (metalloproteinases ADAM10, MMP2, and MMP9), pregnancy-associated protein PAPP-A, proheparanase, extracellular inducer MMP (EMMPRIN, CD147), urokinase-type plasminogen activator (uPA) [4, 5, 8–10].

4.1.1.2 Lipids

Sphingomyelin, phosphatidylcholine, phosphatidylethanolamine, gangliosides, and phosphatidylinositol, as well as lysophosphatidic acid, are the main lipids of EVs. Sphingomyelin and cholesterol are shown to be responsible for the exosomes stability, while the gangliosides prevent the consumption of exosomes by cells of the reticuloendothelial system [11, 12]. Lysophosphatidic acid in the composition of the exosomes induces fusion and rearrangement of intracellular membranes in the biogenesis process of VV [1].

Exosomal lipids are shown to be involved in the process of tumor progression and the formation of chemoresistance [13], as well as in the pathogenesis of some neurodegenerative diseases [14].

4.1.1.3 Nucleic Acids

Exosomes contain both ribo- and deoxyribonucleic acids (RNA and DNA). It has been found that serum and urine exosomes contain a significant number of other types of RNA besides matrix and microRNA, namely tRNA, rRNA, snRNA, piRNA, and scaRNA [15]. The ExoCarta database containing the library of detected lipids, proteins, mRNA, and miRNA in the composition of exosomes includes more than 760 miRNAs and 1600 mRNAs (Fig. 4.3) [16].

The exosomal DNA was shown to contain genetic material of all chromosomes, the exosomes of the blood of oncological patients contain mutant forms of KRAS and p53 tumor DNA [17]. Thus, exosomal DNA of biological fluids of cancer patients can be an additional source of diagnostic material.

4.1.1.4 Techniques for Exosome Isolation

Most of the exosome detection techniques require their initial isolation from the complex biological milieu using separation methods such as differential ultracentrifugation, ultrafiltration, precipitation using water-excluding polymers, immunoaffinity capture, and microfluidics-based techniques [18–20].

Ultracentrifugation is a gold standard of exosome isolation and represents one of the most commonly used techniques. Ultrafiltration using nanomembrane is faster and does not rely on specialized equipment; however, the shear stress might cause deterioration, exosomes might be lost due to trapping in the pores of the filters. Exosome precipitation using polymers such as PEG is more efficient than ultracentrifugation and nanomembrane concentration.

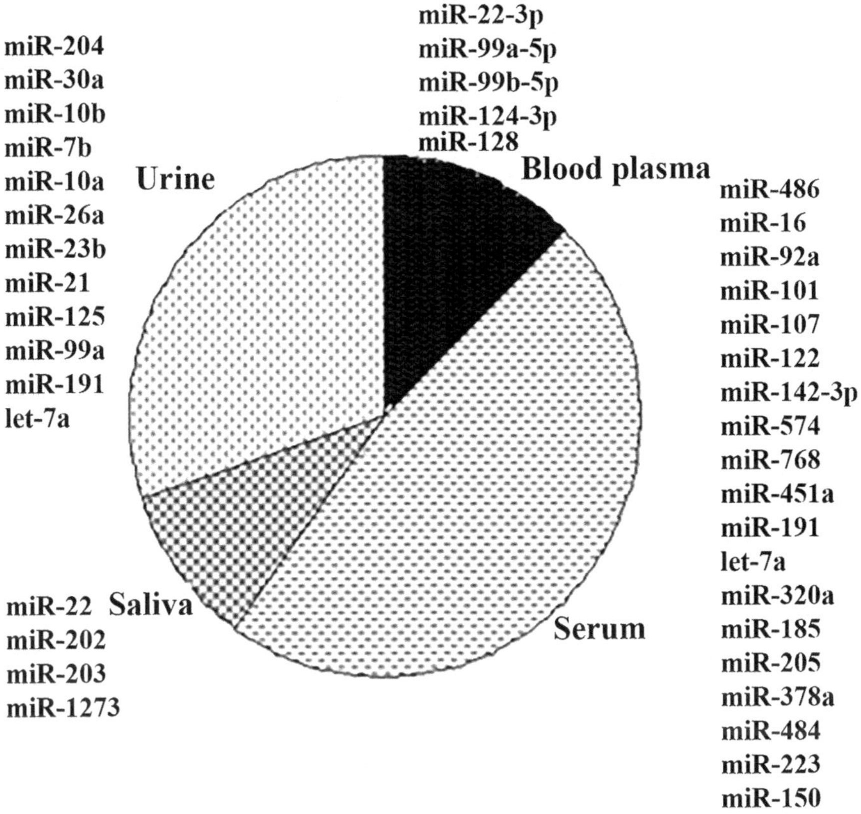

Fig. 4.3 Most common exosomal microRNAs in various biological fluids [16] (reprinted under license of Springer Nature, License Number 4545980782857)

Immunological separation based on selective recognition of proteins on the membrane of exosomes using corresponding antibodies-modified magnetic beads enables specific and fast purification of exosomes; however, it is difficult to apply to large-volume samples. Microfluidic devices in combination with size/immunoaffinity-based techniques as well as acoustic/electrokinetic sorting provide fast separation of exosomes from large number of samples [18].

Simple purification by differential ultracentrifugation is not sufficient to qualify vesicles as exosomes and a combination of quantitative protein composition and morphological and physical criteria must be used to identify exosomes among other EVs.

The use of enzyme inhibitors and/or preservatives in the collection of biological material, centrifugation conditions (number of centrifugations, speed, duration, type of rotor, temperature, etc.), type of filter, and pore size during ultrafiltration affect the efficiency of exosomes isolation and their functional activity [21].

4.1.2 Exosome Functions

Researchers' interest in exosomes increased significantly after the ability of exosomes to incorporate and transfer their contents into recipient cells (hematopoietic, dendritic, mesothelium, and tumor cells, also endotheliocytes and fibroblasts) was discovered. The main mechanisms for the absorption of exosomes include:

- ligand–receptor interactions without membrane fusion (antigen presentation),
- the fusion of the exosome membranes and target cells,
- internalization of exosomes by endocytosis (all fragments of exosomes are translocated into the cell), and
- the action of the components of exosomes on the cell after is their lysis in the extracellular environment (at low pH, for example, in the tumor microenvironment).

The most important physiological functions of exosomes are presented below.

4.1.2.1 Intercellular Communication

Intercellular communication is a necessity for the functioning of a multicellular organism and can be carried out directly by intercellular contacts, and a means of transferring secreted molecules by exocytosis. Mechanisms of intercellular communication carried out through EVs are actively studied (Fig. 4.4). It has been established that exosomes, penetrating recipient cells, can cause cascade changes in the cells on the genomic (due to DNA integration) and epigenomic (due to changes in the expression/content of proteins, miRNA, and so on) levels [20].

4.1.2.2 Immunomodulatory Function

In a normal pregnancy, exosomes derived from syncytiotrophoblast cells participate in the formation of a common immunosuppressive background [22]. Exosomes, derived from tumor cell lines, have been shown to contain molecules capable of modulating the function of natural killer (NK) cells [23, 24]. It was shown that tumor exosomes induce apoptosis of activated CD8+ T cells, promote differentiation of T-helper cells to T-regulatory cells, which in turn is resistant to exosome-induced apoptosis [1]. The large amount of immunoglobulin A on the membrane of exosomes of saliva allows suggesting that pathogens associated with exosomal IgA can be delivered to the tonsils or dendritic cells. In addition, exosomes of type II saliva contain a number of proteases (CD26) that regulate the functions of dendritic cells. Thus, saliva exosomes appear to play an important role in eliminating pathogens of various nature and modulating the function of the local immune system [7].

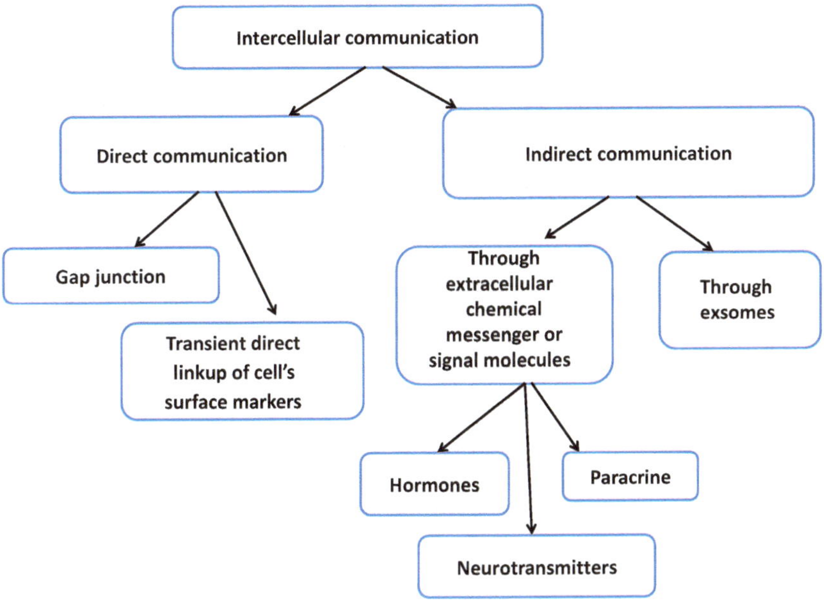

Fig. 4.4 Mechanisms of intercellular communication

4.1.2.3 Induction of Angiogenesis and Stromal Remodeling

The induction of angiogenesis and stromal remodeling by exosomes is mainly associated with exosomal tetraspanins. Currently, there is evidence that tetraspanins in the composition of the exosomes lead to the induction of angiogenesis factors (VEGF, VEGFR1, VEGFR2, uPA, MMP-2, MMP-9) and inhibition of expression of antiangiogenic factors in peritoneal macrophages and mesothelium cells [25].

4.1.2.4 The Effect on Cell Motility

Cell movements affect the formation of many pathological conditions of vascular diseases, chronic inflammatory processes, and degenerative diseases. Hoshino et al. [26] showed that the invasive cell line of squamous cell carcinoma formed multiple invadopodia and they are the main sites for exosome-containing multivesicular endosomes. Inhibition of the formation of invadopodia significantly reduces the secretion of exosomes into the incubation medium. The addition of purified exosomes to the incubation medium or the inhibition of their biogenesis or secretion significantly influences on formation and functioning of the invadopodia of tumor cells: the formation of invadopodia, their stabilization, exocytosis of proteases from invadopodia for the degradation of the extracellular matrix [26]. Thus, a certain synergism is noted in the processes of formation and functioning of invadopodia and EVs secretion in highly invasive tumor cell lines.

4.2 Exosomes in the Diagnosis, Monitoring, and Prognosis of Malignant Tumors

Tumor cells, as well as the tumor microenvironment, are prone to produce EVs, involved in both tumorigenesis and cancer progression. Moreover, exosomes derived from cancer cells transfer physiological information to normal cells initiating their transformation into malignant cells. Cells from the tumor microenvironment generate EVs, which increase angiogenesis of new blood vessel generation necessary to supply oxygen and nutriments facilitating the growth of cancer cells. Markers of oncological diseases (TAX + exosomes, miR-34a down expression in exosomes, miR-15a down expression in exosomes, miR-23a exosomes, and others) were identified and the functional role of extracellular vesicles in metastasis, genetic instability, the oncogenic transformation of fatty stem cells, oncological cells proliferation and migration were demonstrated [1].

The potential of exosomes and other explosives of blood plasma, urine, saliva for the diagnosis of cancer is sufficiently high [27–31]. It is proposed to use exosomal biomarkers as predictive markers during targeted therapy [32, 33], as predictors of the effectiveness of neoadjuvant chemotherapy [34] and as prognostic markers [35].

The role of exosomes in the pathogenesis of malignant tumors is extremely diverse. Fundamentally, it can be noted:

- tumor exosomes constitute a significant part of the exosome population circulating in the blood of patients with malignancies;
- in general, exosomes circulating in the blood of patients have a depressant effect on antitumor immunity;
- tumor exosomes stimulate metastasis;
- tumor exosomes are involved in the process of tumor adaptation to therapy, the development of chemo- and hormone resistance [36, 37];
- the level of circulating exosomes in patients with malignant neoplasms correlated with the effect of antitumor therapy.

Potential ways of using exosomes in clinical oncology is presented in Fig. 4.5.

4.3 Imaging of Exosomes

After purification, the exosomes should be characterized [1]. The most frequently used methods for exosome analysis are presented in Table 4.1 [2]. A more rigorous exosome characterization must take into account molecular content in combination with those methods.

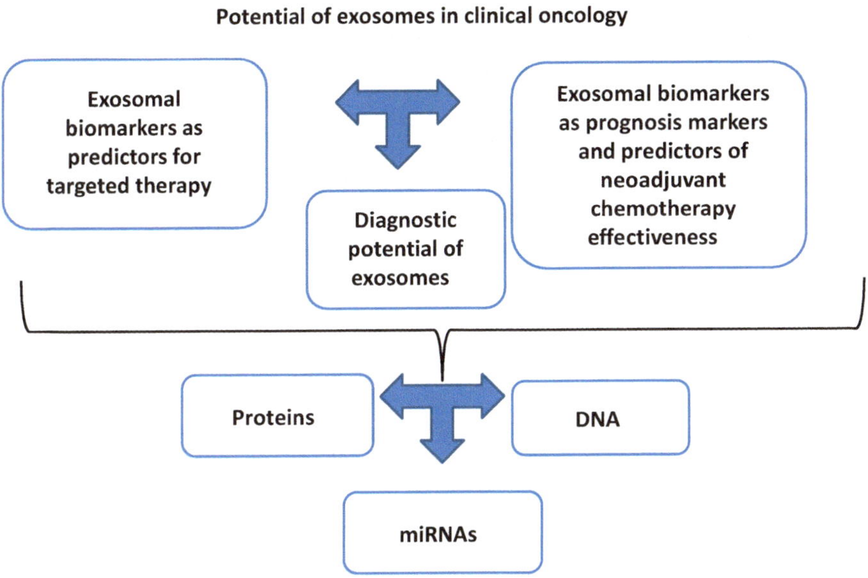

Fig. 4.5 Potential ways of using exosomes in clinical oncology

Table 4.1 The most common methods for exosome analysis [2]

Technique	Effects used	Useful information
Flow cytometry	Scattering of lights and fluorescence	Phenotype, absolute number, and size
Dynamic light scattering (DLS)	Scattering of light as a function of time	Size and distribution
Nanoparticle tracking analysis (NTA)	Scattering of light	Size, size distribution, concentration, phenotype
Transmission electron microscopy (TEM)	Scattering of electron beam	Morphology, size
Atomic force microscopy (AFM)	Interaction forces between the probing tip and surface	Three-dimensional topography, diameter
Nanoparticle tracking analysis	Brownian motion of exosomes in suspension	Hydrodynamics sizes
Infrared spectroscopy	Resonant absorption	Protein composition
Raman scattering or surface-enhanced Raman spectroscopy (SERS)	Raman scattering	Protein composition
Two-photon microscopy	Second-harmonic generation, two photon autofluorescence	Protein composition

4.3.1 Dynamic Light Scattering

Dynamic light scattering (DLS) is based on monochromatic laser radiation scattering on medium inhomogeneity due to Brownian exosome motion. Due to the interference of radiation scattered from various particles, the autocorrelation function of the intensity spectra brings information about exosome size. The advantage of the DLS method is its ability to measure particles ranging from 1 nm to 6 µm. However, DLS allows reliable data only when one type of exosomes is present in the suspension. The method is less accurate in polydispersed suspensions of exosomes. In such cases, the obtained profile of exosome size is strongly influenced by larger exosomes, since they scatter more light. Therefore, when larger vesicles are present in the suspension, even in a low quantity, the detection of smaller events becomes problematic [38].

4.3.2 Transmission Electron Microscopy

Morphology of isolated exosomes can be analyzed by transmission electron microscopy (TEM). TEM imaging is based on electron beam interaction with a sample. The TEM resolution is of 1 nm due to a short wavelength of the electron waves [2]. TEM allows to carry out an ultrastructural study of isolated exosomes, size distribution, and membrane integrity [3].

This method has essential limitations of samples properties, including thickness, contrast, etc. The results of TEM exosome analysis strongly depend on the protocol of sample preparation. In this regard, the Society for the Study of EVs strongly recommends that the articles provide detailed information on the methodological approaches used to isolate EVs/exosomes and also characterize the selected vesicles by electron or atomic force microscopy [39].

Negative staining allows increasing the contrast of exosome TEM images. The colored background is achieved by adding neutral solutions containing heavy metal atoms that are not connected with biological objects [5].

Figure 4.6 presents the results of TEM analysis of exosomes specimens extracted from the blood plasma of patients with malignant neoplasms, as well as from the blood plasma and ascites of ovarian cancer patients [40]. Venous blood (9 mL) from all subjects was collected in K_3EDTA spray-coated vacutainers, immediately mixed using a rotary mixer for vacutainers. Ascites fluid (18–20 mL) was collected in K_3EDTA spray-coated vacutainers during the laparoscopic staging of ovarian cancer patients. Blood and ascites fluid samples from illness patients were obtained before any treatment. Blood was centrifuged at 290 g for 20 min at 4 °C, blood plasma was transferred into a new tube and centrifuged a second time at 1200 g for 20 min at 4 °C. For isolation of exosomes, plasma samples were centrifuged at 17,000 g for 20 min at 4 °C to pellet cell debris. Supernatants were diluted as 1:5 by phosphate-buffered saline (PBS) and were passed through 100-nm pore-size

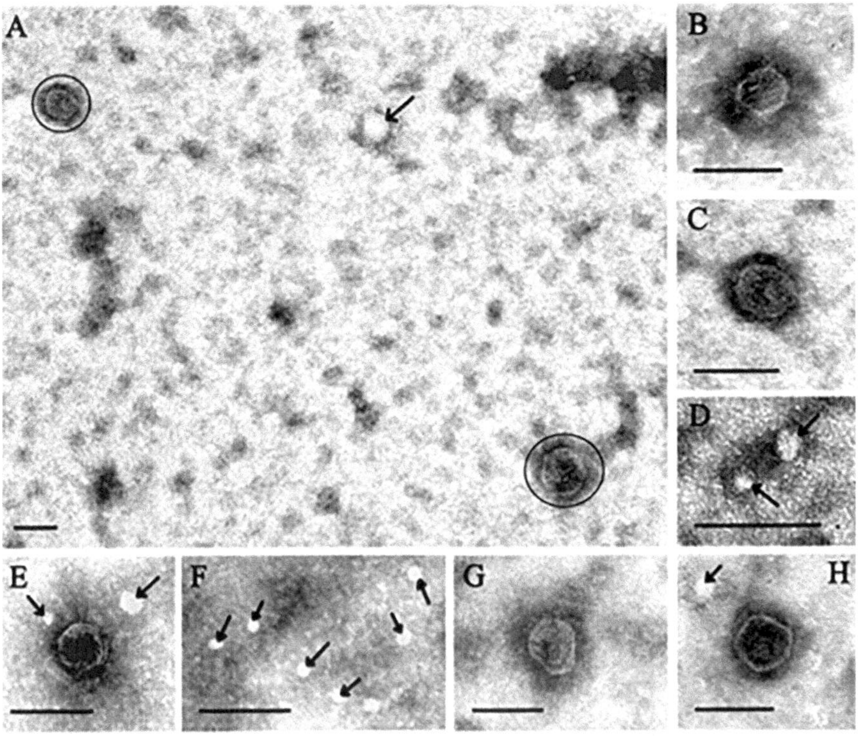

Fig. 4.6 Electron microscopic images of exosomes and "non-vesicle" structures isolated from: (**A**, **C**, **D**) blood plasma of patients with ovarian cancer, (**B**) ascitic fluid of patients with ovarian cancer, (**E**, **F**) blood plasma of healthy donors, (**G**) blood plasma of patients with squamous cell carcinoma of the head and neck, (**H**) blood plasma of patients with colorectal cancer. Arrows show "non-vesicle" structures, ellipses—exosomes. Scale bars correspond to 100 nm. Negative staining by phosphotungstic acid was used [40] (reprinted under license of AIP Publishing, License Number 4546000387063)

filter, then the filtrates were ultracentrifuged at 100,000 g for 90 min at 4 °C, ultracentrifugation and re-suspending steps were repeated three times. The supernatant was removed and the pellet containing plasma exosomes was resuspended in 150 μL of PBS.

Ascites fluids were centrifuged at 900 g for 20 min at 4 °C, the supernatant was transferred into a new tube and centrifuged to pellet cell debris, then supernatants were diluted as 1:3 by PBS and were filtered and ultracentrifuged under the same condition listed above. The supernatant was removed and the pellet containing ascites fluid exosomes was resuspended in 150 μL of PBS.

For negative staining, a drop of exosomes was incubated for 1 min on a copper grid, covered with formvar film, which was stabilized by carbon. Then the grids were exposed for 5–10 s on a drop of 0.5% uranyl acetate or 2% phosphotungstic acid.

The grids, holding adsorbed exosomes, were examined by transmission electron microscope JEM 1400 (Jeol, Japan) supplied with digital camera Veleta (Olympus Corporation, Japan). The measurements were made directly on the camera screen using iTEM (Olympus Corporation, Japan) software.

Obvious structures of a cup-shaped low electron density with a preserved membrane were found in the preparations of exosomes. Their morphology was typical for exosomes of patients with different malignant neoplasms. In the preparations, there were also particles called "non-vesicles," which were very low or low-density lipoproteins [6].

In [2], isolated exosomes from mouse melanoma B16F0 cells were resuspended in cold DPBS containing 2% paraformaldehyde. Exosome samples were mounted on copper grids, fixed by 1% glutaraldehyde in cold DPBS for 5 min to stabilize the immunoreaction, washed in sterile distilled water, contrasted by the uranyl-oxalate solution at pH 7 for 5 min, and embedded by methyl cellulose-UA.

In the study [6], the standard method for cell preparation (fixation, dehydration, embedding, and sectioning) was applied to reduce the exosome drying effect. Using plastic embedding allowed to reduce artifacts caused by denaturation. Glutaraldehyde (GA) was used for the cross-linking (covalent interactions between amino groups) to provide chemical fixation. Osmium tetroxide was used for fixation of lipids and contrast improving.

4.3.3 Nanoparticle Tracking Analysis

A method of nanoparticle tracking analysis (NTA) allows one visualizing and analyzing exosomes in suspension based on the dependence of nanoparticles Brownian motion rate on their size, modal value, and size distribution. The standard NTA equipment includes a laser source, a microscope connected to a CCD or CMOS camera, a hydraulic pump, and a measuring chamber (Fig. 4.7) [38].

However, NTA allows obtaining only evaluative parameters as it operates with hydrodynamics sizes of particles, which can significantly exceed the physical particle sizes. The device also lacks selectivity in structural analysis of particles as it registers all particles and aggregates in the suspension [41].

4.3.4 Atomic Force Microscopy

Atomic force microscopy (AFM) is a very high-resolution type of scanning probe microscopy, which reveals and registrations interactions between a probing tip and a sample surface, when the surface is explored by a sharp tip of small spring-like cantilever. The deviation of the cantilever associated with the interaction of forces when the tip approaches very close to the sample is recorded by the detection sys-

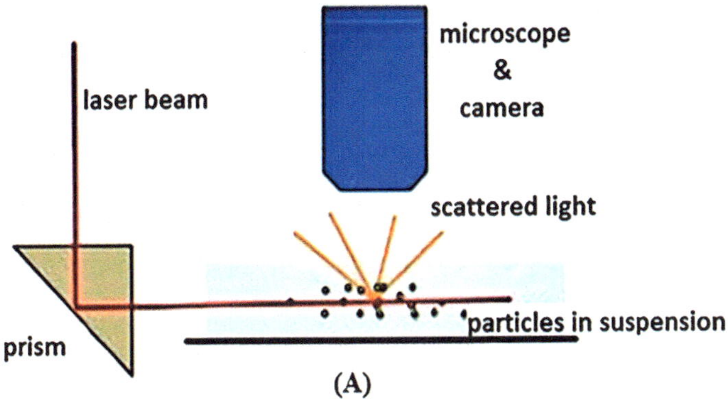

Fig. 4.7 The principle of the nanoparticle tracking analysis (NTA) [38]

tem. Such a system usually consists of a laser and a position-sensitive detector (Fig. 4.8) [38].

To obtain a real three-dimensional image of the surface topography with high resolution, it is necessary to fix all the vesicles on an atomic-flat surface, such as mica. Note that to avoid flattening of the vesicles and changing their shape when interacting with mica, functionalization of the mica surface by molecules is usually used.

EV associated with the surface of specific monoclonal antibodies can be used to collect quantitative information about their specific interaction with the surface of the substrate [24, 25]. Using this approach, one can detect the presence of specific proteins with better resolution than when labeled with an immunogold [26].

However, for a more specific characterization of EV, including the analysis of miRNA, lipids, and proteins, a combination of several methods with the addition of omics technology is necessary.

4.3.5 Infrared Spectroscopy

Infrared (IR) spectroscopy is a non-destructive method that can investigate cells and extracellular vesicles. It acts on the principle of vibrating molecular bonds and the resulting absorption wavelengths, which depend on the involved atoms and strength of intermolecular interactions, determine the chemical profile of a specific material. IR spectroscopy of extracellular vesicles allows one to obtain overall information about the molecular components and their structures [42]. IR spectra of different extracellular vesicles derived from Jurkat cell line, together with the spectrum of the original cell line, are shown in Fig. 4.9A [42]. Almost any biological sample spectra are characterized by amide absorption peaks, one around 1651 cm^{-1} (termed amide I, originating mainly from C=O stretching vibrations of the protein-peptide back-

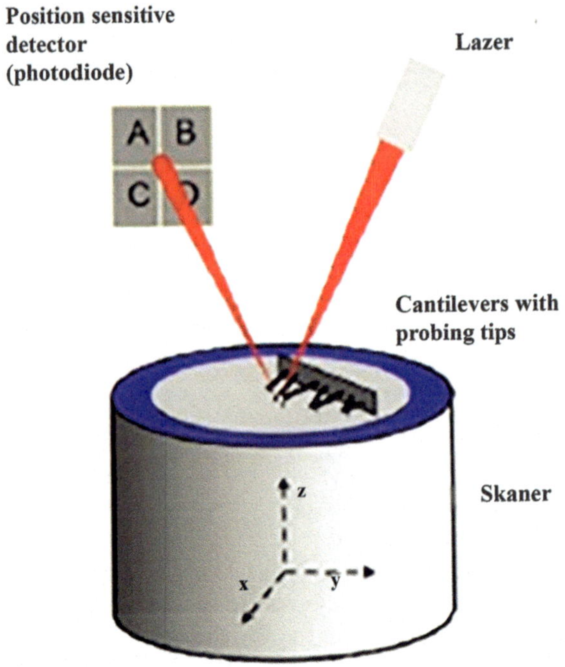

Fig. 4.8 Schematic illustration of atomic force microscopy [38]

bone) and another at 1540 cm^{-1} (termed amide II, arising from N–H bending vibrations of the peptide groups). Amide A termed strong band at 3285 cm^{-1} superposed with the broad, overlapped –OH stretching vibrations treats to the N–H stretching vibrations of the peptide groups of proteins.

The distinct lipid absorption near 1738 cm^{-1} originates from the ester groups of phospholipids, triglycerides, and cholesterol esters, together with the dominant antisymmetric and symmetric stretching vibrations of the lipid acyl CH2 groups corresponding to the bands at 2924 and 2850 cm^{-1}, accordingly. Bands from the spectral region of 1200–950 cm^{-1} generally are featured to the stretching vibrations of the phosphodiester groups of phospholipids and the C–O–C stretching vibrations of phospholipids, triglycerides, and cholesterol esters. This spectral region is completely masked by the broad phosphate vibrations bands of the isotonic PBS buffer [42].

A detailed examination of the spectral region from 1800 to 1350 cm^{-1} for various EV subpopulations (Fig. 4.9B) shows slight changes in the shape and relative intensity of the tensile vibrations of amide I and the lipid-linked C=O ester. The

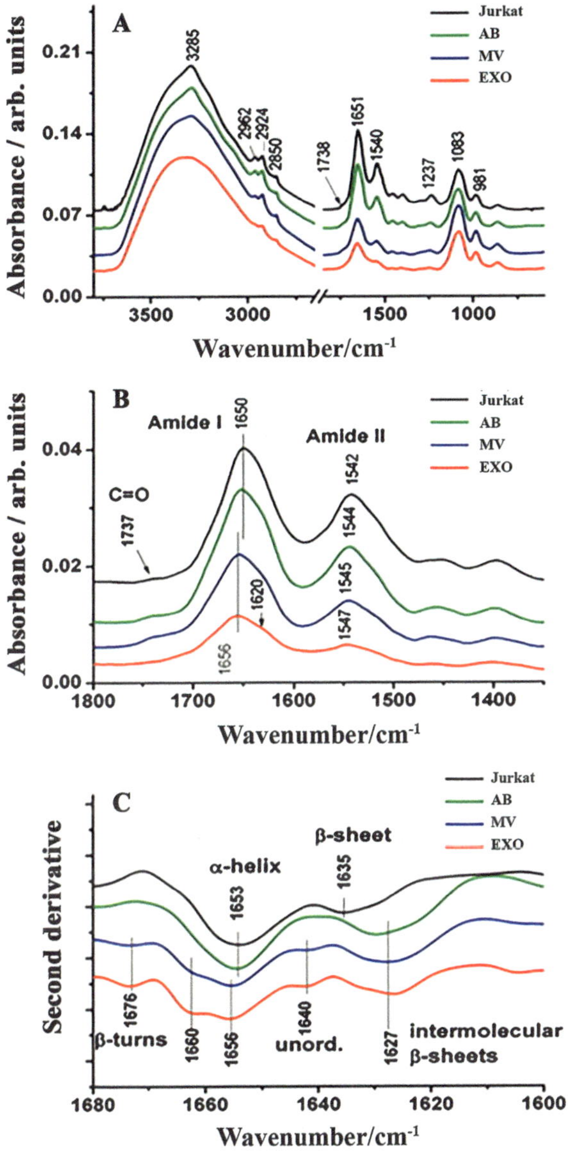

Fig. 4.9 (**A**) Representative ATR-FTIR spectra of EVs isolated by differential centrifugation from Jurkat cell line. (**B**) Representative ATR-FTIR spectra of EVs after PBS buffer subtraction in the 1800–1350 cm^{-1} wavenumber region: CO stretching from lipid esters, amide I and amide II bands of proteins. (**C**) Second derivative IR spectra of the amide I region for protein secondary structure assessing. Jurkat T cell line cell culture (Jurkat), apoptotic bodies (AB), microvesicles (MV), exosomes (EXO) [42] (reprinted under license of Elsevier, License Number 4545310060059)

amide I region (1700–1600 cm^{-1}) is used for the secondary structure determination of the proteins. The broad shell of the amide I can be divided into individual band components, which can be used to characterize the α-helical, β-sheet, random, etc. content of the given proteins [43].

Protein secondary structure was assessed by second derivative IR spectra of amide I band (Fig. 4.9C). Regarding the EVs, exosomes changes in protein composition can be witnessed favoring β-turns (band component at 1676 cm^{-1}) and unordered protein motifs (1640 cm^{-1}) at the expense of β-sheet conformations (1635 cm^{-1}) [44]. The arising band component around 1627 cm^{-1} in MV and EXO spectra, characteristic of non-native intermolecular β-sheets, suggests the presence of aggregated proteins or apolipoproteins [44]. The new band component at 1660 cm^{-1} might be related to triple-helix structure characteristic for some immune complexes [45] or associated with nucleic acid (RNA) content of the EVs [46].

The authors [42] also calculated the "spectroscopic protein-to-lipid ratio" (P/ Lspectr) as a ratio of amide I intensity to total integrated intensity of CH_2/CH_3 stretching vibration from 3040 to 2700 cm^{-1}, which was associated with lipid content. The process of P/Lspectr calculation is illustrated in Fig. 4.10 [42].

Differences between exosomes isolated from the saliva of oral cancer (OC) patients and healthy individuals (HI) were evaluated by the Fourier-transform IR spectroscopy (FTIR) in attenuated total reflection (ATR) mode (FTIR-ATR) [47]. These differences occurred in biomolecules that constitute carbohydrates, nucleic

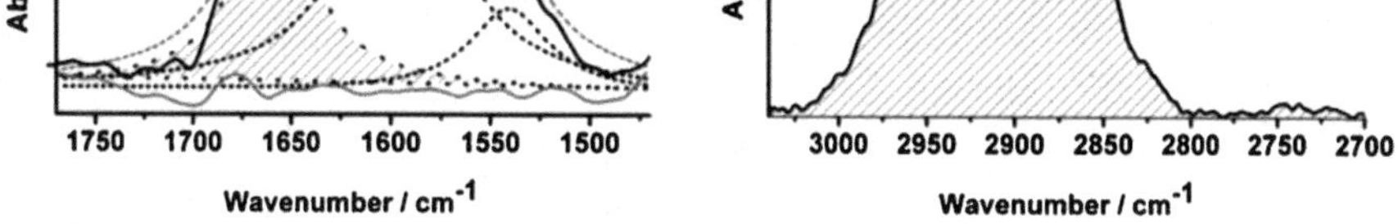

Fig. 4.10 Selected wavenumber regions of a "contaminated" Jurkat cell derived exosome (isolation JK4) used for P/L determination protocol: (**A**) amide I and amide II wavenumber region (1770–1470 cm^{-1}) deconvoluted by curve fitting with Lorentz-function (band denoted by dotted lines), (**B**) CH stretching region (3040–2700 cm^{-1}) acting for lipid components [47] (reprinted under Elsevier license, License Number 4545310060059)

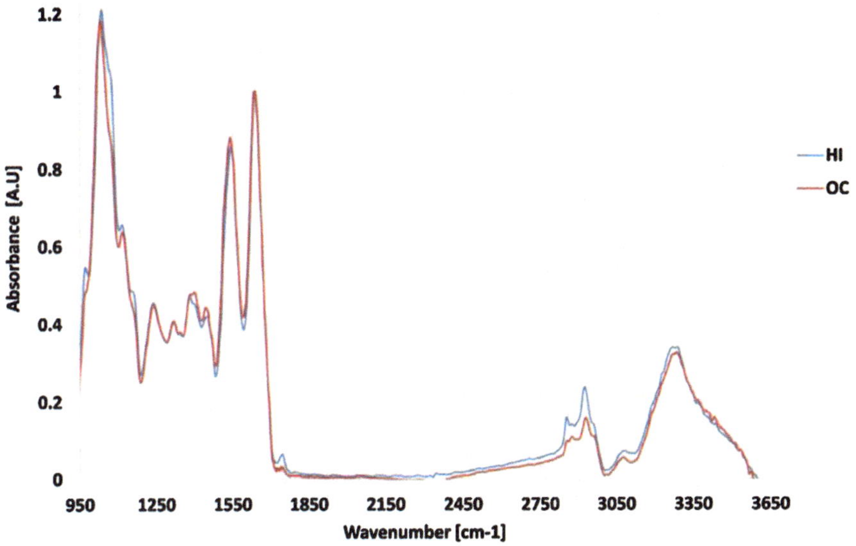

Fig. 4.11 The entire average absorbance spectra in the range of 950–3650 cm^{-1} of the OC and HI exosomes [47] (reprinted under license of Springer Nature, License Number 4545321072023)

acids, proteins, and lipids. The entire absorbance spectra (range 950–3650 cm^{-1}) of the OC and HI exosomes are shown in Fig. 4.11 [47].

Bands at 1033 cm^{-1} are responsible for the vibrational modes of $-CH_2OH$ groups and the C–O stretching vibration coupled with C–O bending of the C–OH groups of carbohydrates (i.e., glucose, fructose, and glycogen) [48]. The peak of HI exosomes in this region was slightly higher. HI and OC exosomes showed a peak of absorbance intensity at 1072 cm^{-1}. It is the region of the symmetric stretching modes of the phosphodiester groups of cellular nucleic acids of DNA and RNA [48, 49]. The spectral peaks at 2924 cm^{-1} and 2854 cm^{-1} of HI exosomes are associated with the absorption bands of asymmetric and symmetric C–H stretching vibrations of CH_2 and CH_3 methylene groups, which are contained in fatty acids within cellular membranes [48, 49]. The band at 1743 cm^{-1} is attributed to C=O stretching vibration in cellular lipids [48–50]. The OC exosome spectrum was of lower intensity in this region. The protein spectra are positioned in the region of 1300–1800 cm^{-1} wavenumbers [48]. Bands in this region represent mainly amide II, likely attributed to the region of transmembrane proteins [51]. In general, the absorbance spectrum of the OC exosomes was of higher intensity compared to the HI exosomes.

4.3.6 Surface-Enhanced Raman Scattering

Raman scattering is a kind of inelastic scattering of optical radiation by molecules. The frequency shifts of scattered light are caused by vibrations of specific chemical groups and can identify the molecular composition of a sample. Thus, Raman scattering holds a unique signature based on the molecular compound of a sample [52]. Hence, Raman spectroscopy can differentiate exosomes as a function of the membrane lipid/protein content along with other various surface modifications.

Raman characterization of optically trapped biological particles in liquids was performed for cellular organelles [53], entire cells [54], single unilamellar lipid vesicles [55], for liposomal membrane composition [56] and single bacterial spores [57].

The main disadvantage of Raman spectroscopy is significantly smaller Raman scattering cross-section compared with resonance absorption. To overcome it, various plasmonic sensing strategies are used, including surface plasmon resonance (SPR) [58, 59], surface-enhanced Raman scattering (SERS) [60–62]. Nanoplasmonic sensing is based on plasmon nanoparticle local refractive index change at contact with a detected object, producing spectral shift [63]. Nanoplasmonic sensing is label-free, does not require complex instrumentation, and can be easily designed to allow multiplex detection using small sample volumes. Since nanoplasmonic sensors essentially operate as optical sensors, the measured parameter is the dry mass of target molecules. As a result, nanoplasmonic sensors are less likely to suffer from solution-based interferences (i.e., effects of bound or coupled solvents) [64].

The size of the exosomes closely corresponds to the distance of SPR perception; therefore, as a rule, the detection of the exosomes by SPR is carried out without labeling [65, 66]. The SPR-based label-free detection platform was developed, which consists of antibody microarray, combined with SPR imaging [67]. Specific to exosome transmembrane proteins antibodies were localized on gold surfaces on the chip. The laser beam passes through the coupling prism at a fixed angle of incidence, and the reflected beam intensity is recorded by a CCD camera. The latter is directly correlated with the amount of captured exosomes.

Various techniques have been introduced to improve the SPR detection performance. Rupert et al. [68] described a dual-wavelength SPR for determining not only the presence but also the size and concentration of extracellular vesicles subpopulations.

Noble metal nanoparticles (Au/Ag) or rough nanostructures can be used to enhance the Raman scattering. This phenomenon is called surface-enhanced Raman scattering, the corresponding technique is called surface-enhanced Raman spectroscopy (SERS) [69–71]. SERS is widely applied for many years; however, only recently, it was employed to analyze exosomes [72, 73].

Two different approaches are typically used for SERS-based characterization of biomolecules and nanosize bioparticles, in particular, exosomes: label-free detection and indirect detection using SERS tags [74].

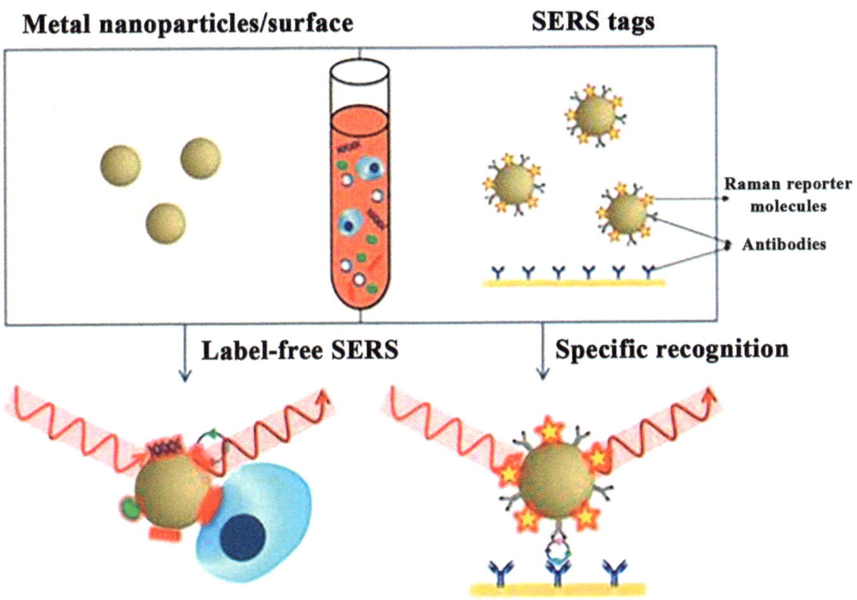

Fig. 4.12 SERS-based liquid biopsy analysis using a label-free SERS approach (left) or SERS tags (right). In label-free SERS, the spectroscopic signal results from analyte adsorption onto the SERS substrate, whereas in SERS tags-based specific recognition assays, the spectroscopic signal results from the reporter molecules on the SERS tags [18]

In label-free SERS detection (Fig. 4.12, left), samples for analysis directly adsorb to the surface of metallic nanostructures and intrinsic fingerprint spectra of the biomolecules are obtained [18, 72, 75, 95]. The main problem is associated with essential RS enhancement that depends strongly on the distance between the molecule and the SERS surface and disappears at distances more than a few nanometers [76]. Also, only Raman modes, which are perpendicular to the surface molecular vibrations, can be increased. Consequently, an overall SERS vibrational spectrum is fragmented, distorted, and often difficult to interpret. Thereby, only biomolecules that are attached to or incorporated into the particle's membrane can be detected and analyzed. Complex spectral analysis is usually a prerequisite to interpret the spectral information for discriminating biomolecules at different status or cells/microorganisms of different species [77].

According to the "indirect" approach [78, 79], SERS nanotags functionalized with biorecognition molecules, which specifically bind to a specific target, are used as quantitative reporters. SERS tags (Fig. 4.12, right) are typically composed of metallic nanoparticles coated with Raman reporter molecules emitting strong and distinct Raman signals. By conjugating specific recognition molecules such as antibodies or aptamers, SERS tags can be used as optical labeling tools for indirect sensing/imaging of the target biomolecules *in vitro* and *in vivo* [80, 81].

The "indirect" SERS strategy is very sensitive and competes directly with fluorescent molecular probes. Compared to traditional external labeling reagents like organic dyes or fluorophores, SERS tags offer advantages such as ultrasensitivity, tremendous multiplexing capacity, high photostability, the need of only a single laser to excite all SERS labels, and minimized interference by autofluorescence from cells/tissues [74, 82]. However, general information about the particle's constituent biomolecules is lost; only the amount of target molecules is assessed via the signal from SERS nanotags.

In the past decade, both label-free SERS detection and SERS tags have been increasingly applied to liquid biopsy analysis, providing qualitative and quantitative information for cancer diagnosis, prognosis, and real-time monitoring of therapy response.

The detection of exosomes using a SERS-based detection platform was mostly conducted in solution. For example, Zong et al. [78] described the facile detection of tumor-derived exosomes using magnetic nanobeads and SERS nanoprobes. The magnetic nanobeads and SERS nanoprobes captured the exosomes by forming a sandwich-type immunocomplex, which was precipitated by a magnet. As a result, high-intensity SERS signals were detected in the precipitates. In the absence of exosomes, there will be no immunocomplex formed, and the SERS signals will be very weak. Stremersch et al. [75] described the method to distinguish exosome-like vesicles derived from different cellular origins, using partial least squares discriminant analysis on the obtained SERS spectra.

The importance of combining SERS fingerprinting with statistical analysis to develop effective cancer diagnosis was pointed out [18]. At the same time, SERS fingerprinting can be conducted either through the identification of specific peaks or through analysis of the entire SERS spectra. Park et al. [72] described exosome pattern-based classification for lung cancer diagnosis. In this case, the detection was performed on a solid support in which exosomes are attached to gold nanoparticle coated glass slides without any biorecognition element. The exosomes from lung cancer cells were distinguished from normal cell exosomes with 95.3% sensitivity and 97.3% specificity.

To characterize exosome SERS signatures, Raman shifts in the range of 719–1800 cm^{-1} were measured for exosomes derived from CD18/HPAF, MiaPaCa, HPDE, and from control (AuNPs) (Fig. 4.13) [83]. Spectra in the 719–1800 cm^{-1} region exhibited peaks corresponding to lipids and proteins, which are the major contributors of exosome surface composition.

High reproducibility and consistency for SERS measurements were observed providing validity to each SERS peak as specific to that sample exosome population and not merely arising from background noise.

This observation highlights the ability of Raman spectroscopy to distinguish exosomes by their lipid composition. The gold nanoparticles can be used to increase the SERS effect [72].

The method of Raman tweezers microspectroscopy (RTM) for extracellular vesicles characterization with particles' size from 50 nm to a few hundreds of nm was described [84]. Figure 4.14 explains the concept of RTM, using the example of

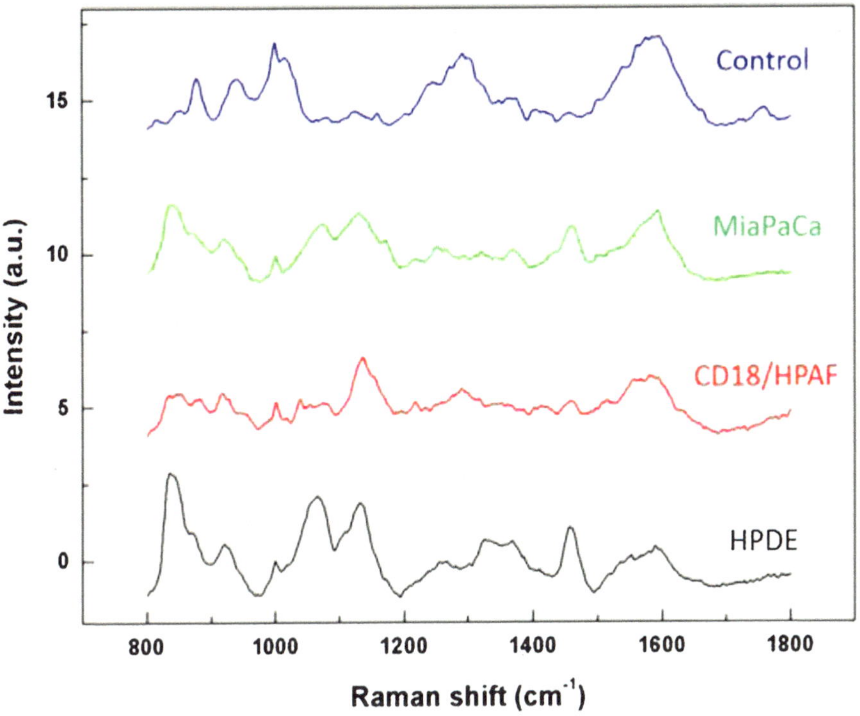

Fig. 4.13 Characteristic SERS of exosomes from the normal and pancreatic cancer cell line. Averaged SERS spectra for the four exosome populations: Gold nanoparticles only were used as a control, and all SERS of exosomes was conducted on sucrose density gradient purified particles [83] (reprinted under license of Elsevier, License Number 4545330658630)

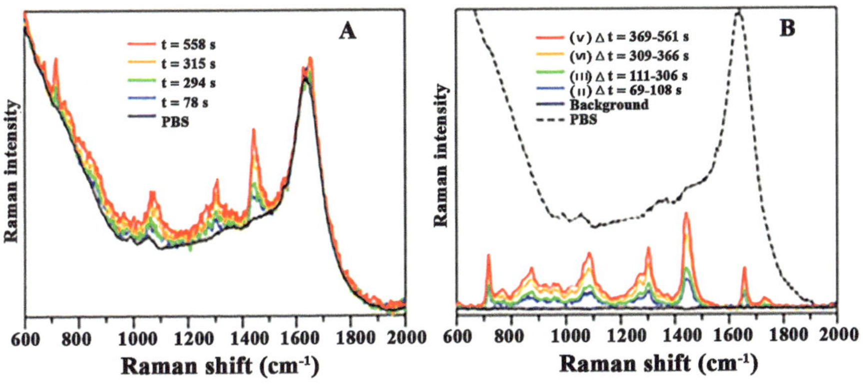

Fig. 4.14 Explanation of the RTM experiment using DOPC liposomes "100 nm." (**A**) Raw Raman spectra recorded at indicated time delays after the start of the experiment, together with an averaged spectrum of PBS; (**B**) averaged Raman spectra of different "particle sets" [84]

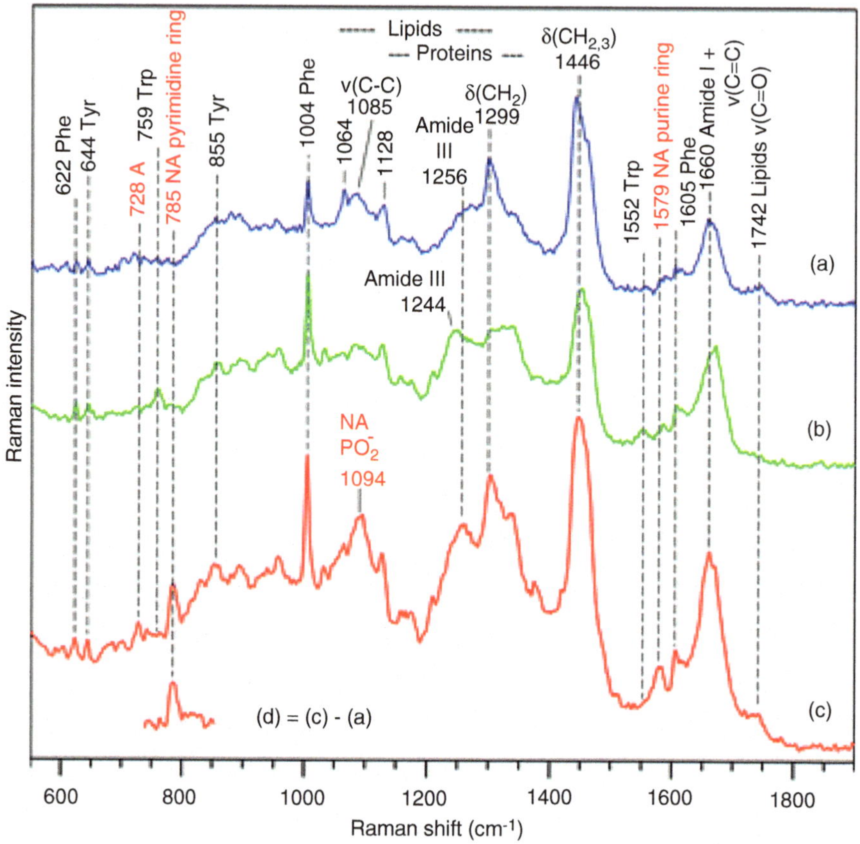

Fig. 4.15 Raman spectra of 3 independent exosome sets (**a–c**) for samples prepared from the urine of donor. The difference spectra (**d**) reveal the presence of nucleic acids in sets (**c**). The water buffer contribution was subtracted [84]

DOPC liposomes "100 nm" with a lipid concentration of 20 μg mL^{-1} in PBS [84]. Liposomes have been chosen instead of exosomes because it is chemically uniform, and their Raman spectra are always the same.

The Raman band assignment for the biomolecular composition of extracellular vesicles is presented in Fig. 4.15, the most-characteristic bands are indicated by vertical dashed lines [84].

Thus, RTM was shown to be useful for identifying various subpopulations that are present in the extracellular vesicles [84]. An approach for analyzing biomolecular components for assessing the major biomolecular contributions of extracellular vesicles (proteins, lipids, nucleic acids, etc.) was reliably obtained with millimolar local concentration for exosomes from human urine.

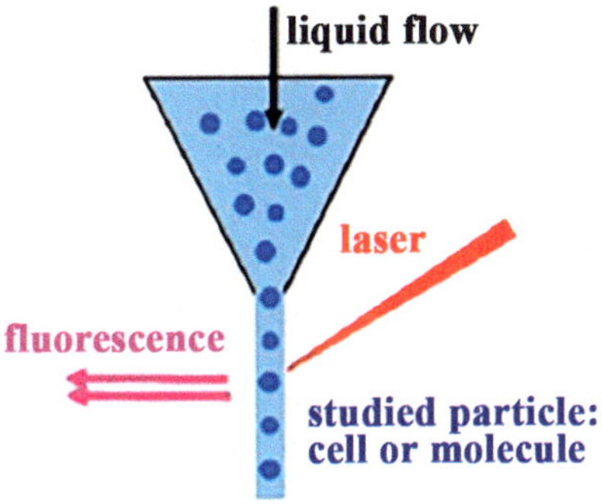

Fig. 4.16 The principle of flow cytometry [38]

4.3.7 Flow Cytometry

If exosomes are labeled with fluorescent dyes that either introduced to their interior or immobilized on their surface, e.g., by monoclonal antibodies, then the fluorescent technique can be used for exosomes imaging. This method is called flow cytometry (Fig. 4.16) [38]. The relative size and granulation of the studied particles, as well as data from fluorescently labeled molecules, can be analyzed using flow cytometry.

Compared with other methods flow cytometry allows to detect rare vesicles, to study a large number of extracellular vesicles, the presence of surface antigens of vesicles that characterize their cellular origin, and to evaluate co-expression of surface markers as well as the detection of individual subpopulations of vesicles.

There is a limitation in exosome size detected, due to limited sensitivity and resolution of flow cytometers. Typical flow cytometers can detect exosomes larger than 500 nm, few of them can detect up to 200 nm exosomes [38]. Newer models allow detecting particles up to 150 nm.

Forward scattering (FSC) signal is frequently used. On the other hand, side-scattered light (15–150°) provides information about smaller exosomes. The last generation of flow cytometers uses multiple angles for FSC detection, which results in better resolution of particles [38].

The technique with latex particles coated with monoclonal antibodies is most often used for exosomes imaging [38]. Visualization of exosomes absorption by endotheliocytes, macrophages, hepatocytes, tumor, and other types of cells by confocal laser microscopy is quite common [85]. Exosomes isolated by various methods are suitable for this study. 3D data reconstruction allowed researchers to demonstrate EVs internalization and cellular localization. EVs internalization has been observed using confocal microscopy after staining with different fluorescent lipid membrane dyes including rhodamine, DiD, Dil, PKH26, PKH67. Membrane-permeable compounds are also used for EVs staining [85].

In addition to the limitation of the size detection, conventional flow cytometers can identify multiple vesicles as a single one at high concentrations of exosomes.

The reason is that every vesicle provides too weak fluorescence to measure the separate event.

Another problem related to exosome flow cytometry analysis is connected with the determination of their size. Due to light scattering, it depends not only on the particle diameter but also on the refractive index, absorption coefficient, and particle shape properties [38].

Colorectal cancer (CRC) remains one of the most common tumors of the gastrointestinal tract and ranks 2–3 in the structure of cancer morbidity in most of the countries of the world, both in men and women [86]. ADAM10 and ADAM17 are transmembrane "molecular scissors" [87]. We have evaluated the level of ADAMs in exosomes from CRC patients with stage T2-4N0-2M0–1 ($n = 60$, 58.6 ± 1.6 years) depending on invasion and metastasis. Control subjects (CSs) had neither colorectal cancer nor other malignant neoplasms, but CSs colorectal polyps were presented.

Blood plasma exosomes were isolated using ultrafiltration with ultracentrifugation. Venous blood (18 mL) from CSs and colorectal cancer patients (CRCPs) was collected in K_3EDTA spray-coated vacutainers. The blood cells were pelleted by centrifugation for 20 min at 1200 g (bucket rotor, Labofuge 400R, Thermo Fisher) and 4 °C. To remove the cell debris, plasma samples were centrifuged at 17,000 g (angular rotor, centrifuge 5415R, Eppendorf) and 4 °C for 20 min. To remove vesicles more than 100 nm, the supernatant was diluted five-fold with PBS (10 mM phosphate buffer, 0.15 M NaCl, pH 7.5) and filtered through 100 nm pore-size filter (Minisart high flow, 16,553-K, Sartorius). For the exosome precipitation, the filtrate was centrifuged at 100,000 g (bucket rotor, Optima XPN 80, Beckman Coulter, USA) and 4 °C for 90 min, the pellet was resuspended in 10 mL PBS and twice centrifuged under the same conditions. The isolated exosomes were resuspended in 200 μl of PBS, were aliquoted, frozen in liquid nitrogen and stored at −80 °C.

The 4 μm-diameter aldehyde/sulfate latex beads (Thermo Fisher Scientific, USA) were incubated with anti-CD9 (ab134375, Abcam) antibodies at room temperature for 14 hours at gentle agitation. The aliquots of exosomes (about 30 μg exosomal protein) were incubated with antibody-coated latex beads in 100 μl of PBS at 4 °C for 14 hours at gentle agitation. The reaction was blocked with 0.2 M glycine for 30 min at 4 °C. The exosomes–antibody–bead complexes were washed twice with washing buffer (2% exosome depleted bovine serum in PBS), were incubated with a blocking immunoglobulin G (BD Biosciences, USA) at room temperature for 10 min with washing. Then there was incubation with FITC-conjugated anti-tetraspanins (CD63, CD81, CD24) antibodies (BD Biosciences, USA) at 4 °C for 50 min. The complexes were washed twice with washing buffer and acquired per sample in a FACS Canto II (BD Biosciences), and data were analyzed using FACS Diva 6.1 Software. The median fluorescence intensity (MFI) of the exosomes was compared with the isotypic control (BD Biosciences, USA).

The isolated exosomes were characterized by flow cytometry. A combination of conjugated and unconjugated antibodies made it possible to identify different subpopulations of exosomes (Fig. 4.17).

The CD9/CD24 subpopulation in plasma exosomes predominated in CPs and all subgroups of CRCPs. There was a difference in MFI CD9/CD24 exosomes between

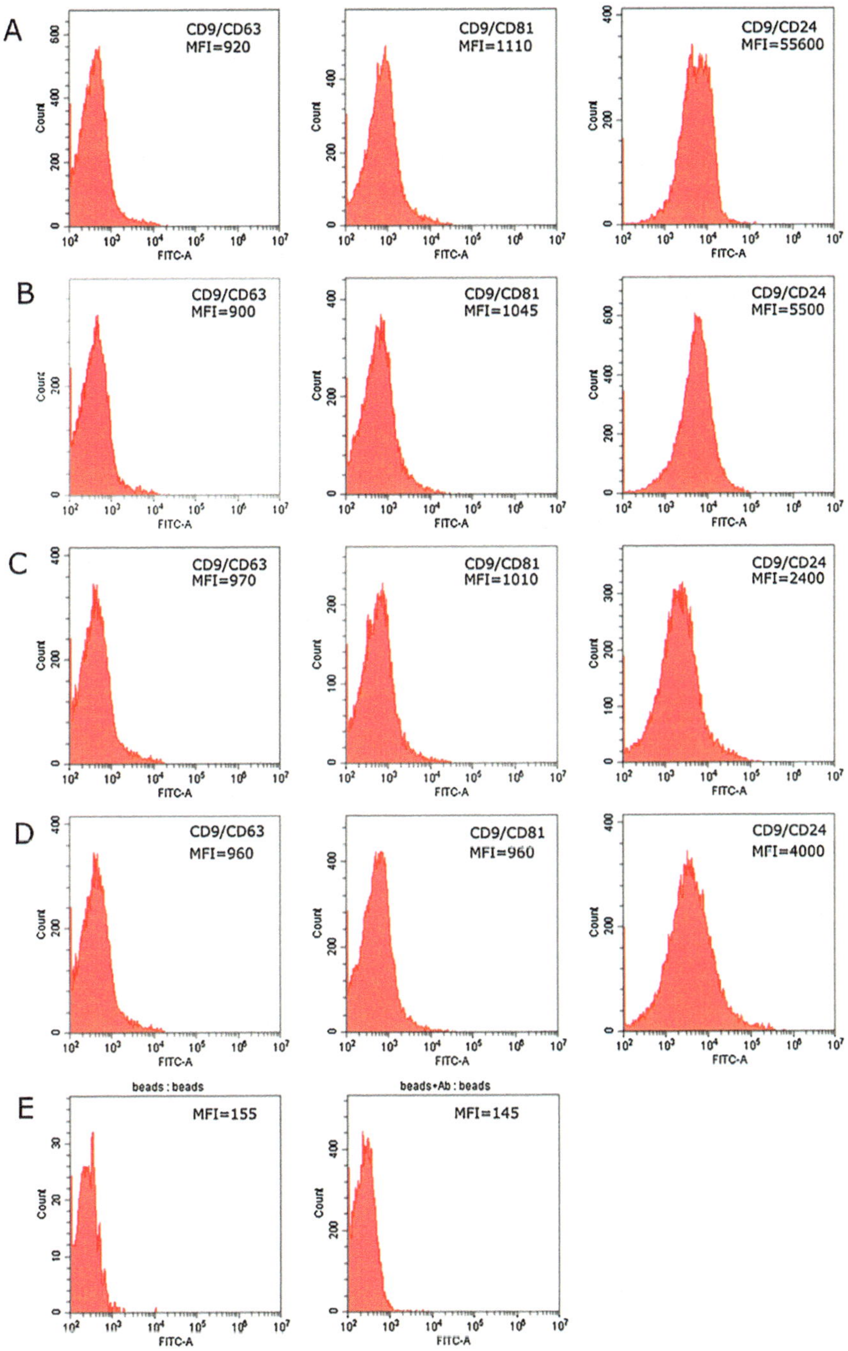

Fig. 4.17 Expression of CD63, CD81, and CD24 on CD9-positive plasma exosomes of CSs (**A**), stage II CRCPs (**B**), stage III CRCPs (**C**), and metastatic CRCPs (**D**). Isotype control and negative control (latex beads labeled anti-CD9 with anti CD81 FITC antibody) (**E**). Mean MFI are shown

stage III and stage II as well as between non-metastatic and metastatic CRCPs. A NanoOrange Protein kit was used to determine the protein concentration. There was no statistically significant difference in the level of exosomal protein between CSs and CRCPs.

Aliquots of exosomes (about 30 μg exosomal protein) were incubated with 3×10^5 latex beads labeled anti-CD9 antibody in 150 μL of PBS at 4 °C overnight at gentle agitation, blocked in 0.2 M glycine for 30 min, then stained with anti-ADAM10 (CD156c)-PE (5 μL on test, BioLegend, USA) and anti-ADAM17/TACE antibody (dilution 1:10, Lifespan Biosciences, USA) for 20 min at room temperature. Then complexes were stained with anti-Rabbit IgG secondary antibody, Alexa Fluor 488 (dilution 1:3000, Thermo Fisher Scientific, USA). Single beads were gated and acquired in a Cytoflex (Beckman Coulter, USA).

We used anti-CD9 coated latex beads for detection ADAM10/ADAM17 subpopulations of exosomes in CSs and CRCPs plasma (Fig. 4.18A) [35]. Distribution of subpopulations of ADAM10/ADAM17 of CSs and of CRCPs was different (Fig. 4.18B–D). The ADAM10-/ADAM17- population predominated in plasma exosomes of CRCPs and the level of ADAM10+ exosomes was significantly higher in exosomes of CSs compared with CRCPs (Fig. 4.18D).

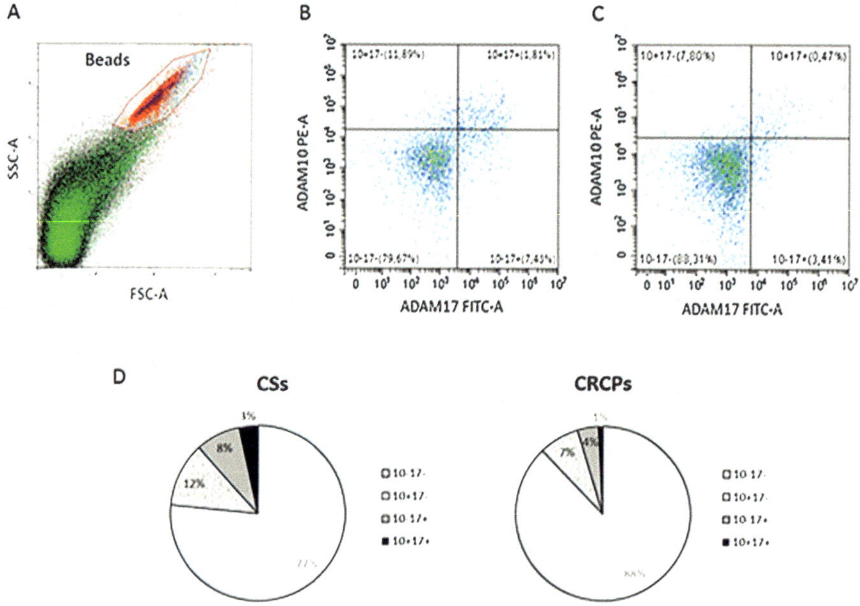

Fig. 4.18 Differences between ADAM10/ADAM17 exosomes are present in the plasma of CSs and CRCPs. (**A**) Forward scatter area (FSC-A) and side scatter area (SSC-A) dot plot representing exosomal samples with aldehyde/sulfate latex beads. (**B**, **C**) Double labeling ADAM10 versus ADAM17 of plasma exosomes of CSs (**B**) and CRCPs (**C**). (**D**) Data show the percentage of ADAM10/ADAM17 subpopulations of plasma exosomes. Flow cytometry analysis of plasma exosomes [35]

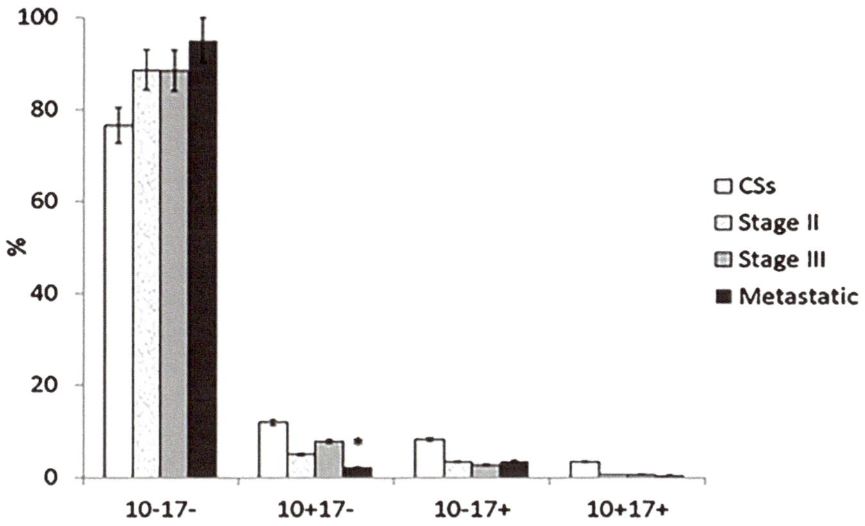

Fig. 4.19 ADAM10/ADAM17 subpopulations (in percentage) of plasma exosomes of CRCPs, depending on the stage of colorectal cancer. ∗- difference was significant ($p < 0.05$) compared to CRCPs with stage III [35]

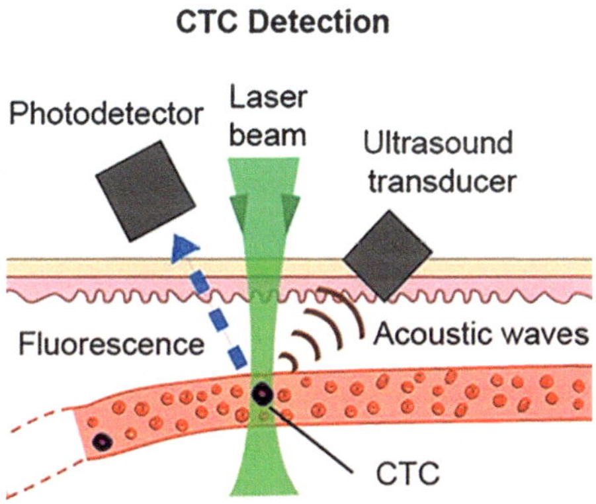

Fig. 4.20 *In vivo* integrated photoacoustic and fluorescence flow cytometry. Schematic for simultaneous detection of circulating cells (e.g., circulating tumor cells including exosomes) with both absorption and fluorescence properties [88] (reprinted under license of Elsevier, License Number 4639160402261)

The statistically significant decrease in the level of ADAM10+/ADAM17- exosomes was revealed in patients with metastatic CRC compared to CRCPs with stage III (Fig. 4.19) [35].

The photoacoustic flow cytometry (PAFC) is a promising approach for high sensitivity exosome detecting [88]. PAFC is based on the irradiation of selected vessels using short laser pulses followed by time-resolved detection of laser-induced acoustic waves with an ultrasound transducer gently held against the skin (Fig. 4.20) [88].

The general principle of photoacoustic flow cytometry is as follows: cells in blood vessels (including exosomes) absorb variated on amplitude light on a specific wavelength, and then acoustic waves are generated due to thermo-acoustic effect. These waves can be registered by an acoustic receiver, which is placed on a patient's body surface. PAFC is a combination of laser and ultrasound techniques, which combines high sensitivity and spectral specificity of optical methods and high spatial resolution and depth penetration of ultrasound methods. PAFC molecular specificity is provided by label-free intrinsic absorption spectroscopic contrast.

The use of a single technique limits the range of detectable cells with different optical properties. Conventional flow cytometry usually combines fluorescent and scattering detection methods for increasing the number of detectable cells. Using a combination of the methods described above, Nolan et al. [89] provided a unique diagnostic platform for the detection and identification of nanoparticles directly in the bloodstream. This platform offers the opportunity to study almost the entire blood volume (up to 3–5 liters), which allows a significant increase in the sample volume and, therefore, the chances of detecting markers of diseases. Thus, it was shown the *in vivo* flow cytometry platform for the detection of tumor-associated particles, including exosomes at early disease stages [89].

4.3.8 Two-Photon Microscopy

Along with other methods of fluorescent imaging of extracellular vesicles, two-photon microscopy is a promising tool for imaging exosomes. Several different non-linear processes can occur when light interacts with matter. Fluorescence excitation by two-photon absorption is extensively used in biomedical imaging. To promote the molecule to an excited state, two photons that arrive "simultaneously" (within ~0.5 fs) at a molecule combine their energies, which then proceeds along the normal fluorescence-emission. Similarly, three or more photons can combine to cause excitation. The efficiency of multiphoton absorption depends on the physical properties of the molecule and the spatial and temporal distribution of the excitation light.

In two-photon imaging, two near-infrared (NIR) (or longer wavelengths) photons are usually used to excite a fluorophore within a visible range [90]. The use of NIR-excitation of light in two-photon imaging allows you to improve the depth of visualization to several hundred micrometers.

During the past few years, two-photon excited photodynamic therapy has been performed on *in vitro* cells, based on various photosensitizers and nanoparticles, under the 700–900 nm fs excitation.

Macklin et al. [91] used multiphoton microscopy with fluorescence lifetime imaging microscopy (FLIM) to track the distribution of PKH67-labeled HiMet-C6 extracellular vesicles in mice after intravenous injection. The authors note that the fluorescence signals of PKH67-labeled extracellular vesicles in phosphate-buffered saline (PBS) *in vitro* could only be detected at 920 nm excitation and emission channel from 450 to 515 nm. The pseudo-color was based on the average fluorescence lifetimes (μm) in individual pixels.

The spatial distribution of the FLIM signal of PKH67-labeled extracellular vesicles in the mouse lung is shown in Fig. 4.21 [91]. The fluorescence signal from collagen second-harmonic generation mainly appears in the 350–450 nm spectral

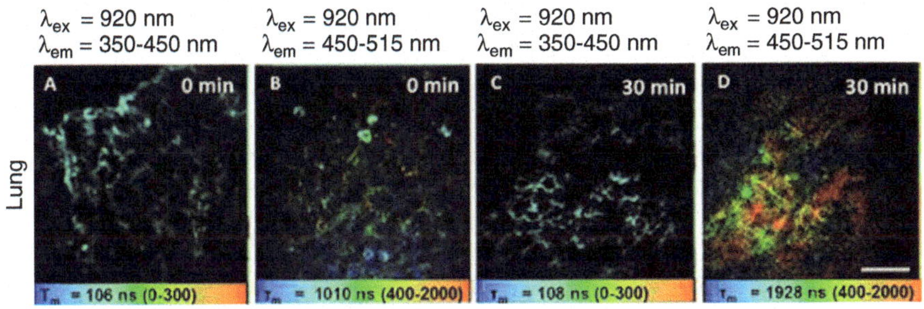

Fig. 4.21 HiMet-C6 EVs migrate to the lung. (**A–D**) Pseudocolored τm fluorescence lifetime image of PKH67-labeled HiMet-C6 EVs in mouse lung before (**A**, **B**) and 30 min after bolus injection (**C**, **D**) [91]

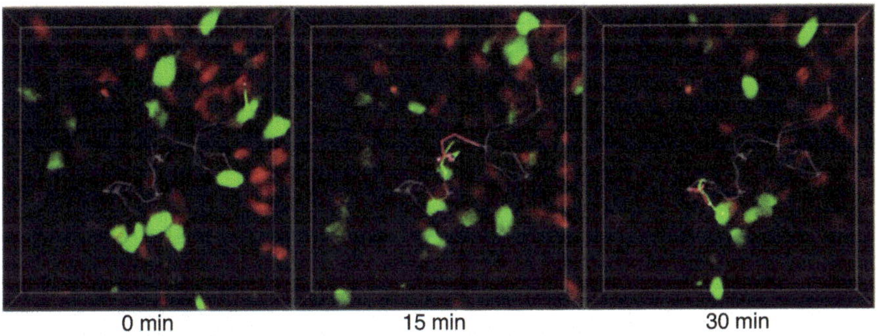

Fig. 4.22 Motility of $CD8^+$ T cells during ovalbumin-specific or non-specific interactions. $CD8^+$ T cells (red) follow targeted $CD4^+$ T cells (green) in their migration. The pathways of a targeted $CD4^+$ T (green/gray line) and a $CD8^+$T cell (red/gray line) remaining interlocked for 30 min are shown [92] (reprinted under license of Springer Nature, License Number 4545331401388)

range. The spectral range of 450–515 nm captured the fluorescence signals of PKH67-labeled extracellular vesicles as well as autofluorescence signals from flavin adenine dinucleotide in cells. The image color changed to orange in the pulmonary capillary (Fig. 4.21B, D) after injection of PKH67-labeled extracellular vesicles and μm increased significantly (from 1089 ± 75 ns to 1663 ± 168 ns, $P < 0.05$) compared to the pre-injection value, demonstrating the localization of the fluorescence signals of PKH67-labeled extracellular vesicles in the lung.

Therefore, it was demonstrated *in vivo* the specific localization of HiMet-C6 extracellular vesicles to lung tissue which shows the lung as the major site for the development of secondary metastases in osteosarcoma [91].

The first visual evidence on targeting $CD4^+$ T cell-based vaccine to cognate $CD8^+$ T cells *in vivo* via exosomal peptide/major histocompatibility complex was provided using two-photon microscopy [92]. The pathways of a targeting $CD4^+$ T cell and an OTI $CD8^+$ T cell remained interconnected during the recording period (Fig. 4.22), indicating that $CD4^+$ specific $CD4^+$ T cells directly interact with the cognate $CD8^+$ T cells *in vivo*.

The FLIM refers to the average time the molecule stays in its excited state before emitting a photon, which is an intrinsic property of a fluorophore. Fluorescence lifetime is sensitive to the local environment, including pH, refractive index, temperature, and insensitive to change in concentration and laser excitation intensity. FRET (fluorescent resonant energy transfer) technology is used as a useful tool of fluorescence microscopy to quantify the dipole–dipole interaction [93]. In this case, one of the molecules is a donor, and the other is an acceptor. This process leads to acceptor fluorescence at the expense of the donor. As the speed of this process depends on the distance between the objects, it allows measuring the distance between the molecules. In most cases, this distance is about 2–5 nm.

Wong et al. [94] carried out monitoring oncogenic receptor signal rewiring using exosomal FRET/FLIM to predict clinical outcome. The receptors of interest were labeled with extracellular Anti-EGFR-IgG-Alexa 546 and Anti-HER3-IgG-Cy5. FRET/FLIM measures the fluorescence lifetime of the donor in the absence and presence of the acceptor and allows for the distinction between average FRET efficiency and FRET subpopulation, independent of the local concentration and stoichiometry of donor and acceptor. The non-interacting donors do not undergo FRET and thus emit fluorescence with the lifetime of the unquenched donor. The donors undergoing FRET expose shortened fluorescence decay.

4.4 Conclusion

Exosomes are microscopic extracellular vesicles (EVs) with a diameter of 30–100 nanometers secreted by all type of cells in various biological fluids, related with many physiological (inflammation, aging, antigen presentation) and pathological processes (pathology of pregnancy, cardiovascular, autoimmune, neurodegenerative diseases, cancer formation, and cancer metastasis). Exosomes provide intercellular communication, immunomodulatory function, induction of angiogenesis and stromal remodeling, cell motility. Cells from the tumor microenvironment generate EVs, which increase angiogenesis of new blood vessel generation necessary to supply oxygen and nutriments facilitating the growth of cancer cells. Thus, exosomes can be considered as predictive cancer markers during targeted therapy, as predictors of the effectiveness of neoadjuvant chemotherapy and as prognostic markers.

After isolation, the exosomes are usually characterized using techniques such as flow cytometry, dynamic light scattering, nanoparticle tracking analysis, transmission electron microscopy, atomic force microscopy, nanoparticle tracking analysis, infrared spectroscopy, Raman scattering or surface-enhanced Raman spectroscopy, two-photon microscopy to evaluate the number of exosomes or their expression levels of disease-related proteins. A more rigorous exosome characterization demands a combination of similar methods with taking into account molecular content.

Compared with other methods flow cytometry allows to detect rare vesicles, to study a large number of extracellular vesicles, the presence of surface antigens of vesicles that characterize their cellular origin, and to evaluate co-expression of surface markers as well as the detection of individual subpopulations of vesicles. The technique with latex particles coated with monoclonal antibodies is most often used for exosomes imaging.

Fourier transform infrared (FTIR) spectroscopy analysis of saliva-derived exosomes isolated from oral cancer healthy patients in attenuated total reflection (ATR) mode FTIR-ATR was carried out.

In the past decade, both label-free SERS detection and SERS tags have been increasingly applied to liquid biopsy analysis, providing qualitative and quantitative information for cancer diagnosis, prognosis, and real-time monitoring of therapy response.

Two-photon microscopy is a promising tool for imaging exosomes. FRET/FLIM fluorescence assay of circulating exosomes extracted from blood serum allows evaluating FRET (fluorescent resonant energy transfer) technology is used as a useful tool for analyzing the dipole–dipole interaction.

Thus, the optical imaging of extracellular vesicles, including exosomes, is an intensively developing scientific field promising for cancer diagnosis, monitoring, and prognosis.

References

1. Malloci, M., Perdomo, L., Veerasamy, M., Andriantsitohaina, R., Simard, G., Martínez, M.C.: Extracellular vesicles: mechanisms in human health and disease. Antioxid. Redox Signal. **30**(6), 813–856 (2019). https://doi.org/10.1089/ars.2017.7265
2. Principe, S., Hui, A.B., Bruce, J., Sinha, A., Liu, F.F., Kislinger, T.: Tumor-derived exosomes and microvesicles in head and neck cancer: implications for tumor biology and biomarker discovery. Proteomics. **13**(10–11), 1608–1623 (2013). https://doi.org/10.1002/pmic.201200533
3. Kowal, J., Tkach, M., Théry, C.: Biogenesis and secretion of exosomes. Curr. Opin. Cell Biol. (2014). https://doi.org/10.1016/j.ceb.2014.05.004
4. Yunusova, N.V., Tugutova, E.A., Tamkovich, S.N., Kondakova, I.V.: The role of exosomal tetraspanins and proteases in tumor progression. Biomed. Khim. **64**(2), 123–133 (2018). https://doi.org/10.18097/PBMC20186402123
5. Andreu, Z., Yáñez-Mó, M.: Tetraspanins in extracellular vesicle formation and function. Front. Immunol. (2014). https://doi.org/10.3389/fimmu.2014.00442. eCollection 2014
6. Rupp, A.K., Rupp, C., Keller, S., Brase, J.C., Ehehalt, R., Fogel, M., Moldenhauer, G., Marmé, F., Sültmann, H., Altevogt, P.: Loss of EpCAM expression in breast cancer derived serum exosomes: role of proteolytic cleavage. Gynecol. Oncol. **122**(2), 437–446 (2011). https://doi.org/10.1016/j.ygyno.2011.04.035
7. Ogawa, Y., Miura, Y., Harazono, A., Kanai-Azuma, M., Akimoto, Y., Kawakami, H., Yamaguchi, T., Toda, T., Endo, T., Tsubuki, M., Yanoshita, R.: Proteomic analysis of two types of exosomes in human whole saliva. Biol. Pharm. Bull. **34**(1), 13–23 (2011)
8. Marimpietri, D., Petretto, A., Raffaghello, L., Pezzolo, A., Gagliani, C., Tacchetti, C., Mauri, P., Melioli, G., Pistoia, V.: Proteome profiling of neuroblastoma-derived exosomes reveal the expression of proteins potentially involved in tumor progression. PLoS One. **8**(9), e75054 (2013). https://doi.org/10.1371/journal.pone.0075054
9. Keller, S., König, A.K., Marmé, F., Runz, S., Wolterink, S., Koensgen, D., Mustea, A., Sehouli, J., Altevogt, P.: Systemic presence and tumor-growth promoting effect of ovarian carcinoma released exosomes. Cancer Lett. **278**(1), 73–81 (2009). https://doi.org/10.1016/j.canlet.2008.12.028

10. Runz, S., Keller, S., Rupp, C., Stoeck, A., Issa, Y., Koensgen, D., Mustea, A., Sehouli, J., Kristiansen, G., Altevogt, P.: Malignant ascites-derived exosomes of ovarian carcinoma patients contain CD24 and EpCAM. Gynecol. Oncol. **107**(3), 563–571 (2007)
11. Yoon, Y.J., Kim, O.Y., Gho, Y.S.: Extracellular vesicles as emerging intercellular communicasomes. BMB Rep. **47**(10), 531–539 (2014)
12. Kim, C.W., Lee, H.M., Lee, T.H., Kang, C., Kleiman, H.: Extracellular membrane vesicles from tumor cells promote angiogenesis via sphingomyelin. Cancer Res. **62**(2), 6312–6317 (2002)
13. Beloribi-Djefaflia, S., Siret, C., Lombardo, D.: Exosomal lipids induce human pancreatic tumoral MiaPaCa-2 cells resistance through the CXCR4-SDF-1α signaling axis. Onco. Targets. Ther. **2**(1), 15–30 (2015)
14. Grey, M., Dunning, C.J., Gaspar, R., Grey, C., Brundin, P., Sparr, E., Linse, S.: Acceleration of α-synuclein aggregation by exosomes. J. Biol. Chem. **290**(5), 2969–2982 (2015). https://doi.org/10.1074/jbc.M114.585703
15. Li, M., Zeringer, E., Barta, T., Schageman, J., Cheng, A., Vlassov, A.V.: Analysis of the RNA content of the exosomes derived from blood serum and urine and its potential as biomarkers. Philos. Trans. R. Soc. Lond. Ser. B Biol. Sci. **369**, 1652 (2013). https://doi.org/10.1098/rstb.2013.0502
16. Tamkovich, S.N., Tutanov, O.S., Laktionov, P.P.: Exosomes: generation, structure, transport, biological activity, and diagnostic application. Biochem. Moscow Suppl. Ser. A. **10**, (2016). https://doi.org/10.1134/S1990747816020112
17. Kahlert, C., Melo, S.A., Protopopov, A., Tang, J., Seth, S., Koch, M., Zhang, J., Weitz, J., Chin, L., Futreal, A., Kalluri, R.: Identification of double- cancer. J. Biol stranded genomic DNA spanning all chromosomes with mutated KRAS and p53 DNA in the serum exosomes of patients with pancreatic. J. Biol. Chem. **289**(7), 3869–3875 (2014). https://doi.org/10.1074/jbc.C113.532267
18. Zhang, Y., Mi, X., Tan, X., Xiang, R.: Recent progress on liquid biopsy analysis using surface-enhanced Raman spectroscopy. Theranostics. **9**(2), 491–525 (2019). https://doi.org/10.7150/thno.29875
19. Lötvall, J., Hill, A.F., Hochberg, F., Buzás, E.I., Di Vizio, D., Gardiner, C., Gho, Y.S., Kurochkin, I.V., Mathivanan, S., Quesenberry, P., Sahoo, S., Tahara, H., Wauben, M.H., Witwer, K.W., Théry, C.: Minimal experimental requirements for definition of extracellular vesicles and their functions: a position statement from the International Society for Extracellular Vesicles. J. Extracell Vesicles. **3**, (2014). https://doi.org/10.3402/jev.v3.26913
20. Konoshenko, M.Y., Lekchnov, E.A., Vlassov, A.V., Laktionov, P.P.: Isolation of extracellular vesicles: general methodologies and latest trends. Biomed. Res. Int. **2018**, 27 (2018). https://doi.org/10.1155/2018/8545347
21. Lässer, C., Alikhani, V.S., Ekström, K., Eldh, M., Paredes, P.T., Bossios, A., Sjöstrand, M., Gabrielsson, S., Lötvall, J., Valadi, H.: Human saliva, plasma and breast milk exosomes contain RNA: uptake by macrophages. J. Transl. Med. **14**(9), 9. https://doi.org/10.1186/1479-5876-9-9
22. Dragovic, R.A., Southcombe, J.H., Tannetta, D.S., Redman, C.W.G., Sergeant, I.L.: Multicolor flow cytometry and nanoparticle tracking analysis of extracellular vesicles in the plasma of normal pregnant and pre-eclamptic women. Biol. Reprod. **89**(6), 151 (2013). https://doi.org/10.1095/biolreprod.113.113266
23. Baginska, J., Viry, E., Paggetti, J., Medves, S., Berchem, G., Moussay, E., Janji, B.: The critical role of the tumor microenvironment in shaping natural killer cell-mediated anti-tumor immunity. Front. Immunol. **4**, 490 (2013). https://doi.org/10.3389/fimmu.2013.00490
24. Whiteside, T.L.: Immune modulation of T-cell and NK (natural killer) cell activities by TEXs (tumour-derived exosomes). Biochem. Soc. Trans. **41**(1), 245–251 (2013). https://doi.org/10.1042/BST20120265
25. Gesierich, S., Berezovskiy, I., Ryschich, E., Zöller, M.: Systemic induction of the angiogenesis switch by the tetraspanin D6.1A/CO-029. Cancer Res. **66**(14), 7083–7094 (2006)

26. Hoshino, D., Kirkbride, K.C., Costello, K., Clark, E.S., Sinha, S., Grega-Larson, N., Tyska, M.J., Weaver, A.M.: Exosome secretion is enhanced by invadopodia and drives invasive behavior. Cell Rep. **5**(5), 1159–1168 (2013). https://doi.org/10.1016/j.celrep.2013.10.050
27. Wang, X., Kaczor-Urbanowicz, K.E., Wong, D.T.: Salivary biomarkers in cancer detection. Med. Oncol. **34**(1), 7 (2017). https://doi.org/10.1007/s12032-016-0863-4
28. Srivastava, A., Moxley, K., Ruskin, R., Dhanasekaran, D.N., Zhao, Y.D., Ramesh, R.: A Non-invasive liquid biopsy screening of urine-derived exosomes for miRNAs as biomarkers in endometrial cancer patients. AAPS J. **20**(5), 82 (2018). https://doi.org/10.1208/s12248-018-0220-y
29. Zhan, Y., Du, L., Wang, L., Jiang, X., Zhang, S., Li, J., Yan, K., Duan, W., Zhao, Y., Wang, L., Wang, Y., Wang, C.: Expression signatures of exosomal long non-coding RNAs in urine serve as novel non-invasive biomarkers for diagnosis and recurrence prediction of bladder cancer. Mol. Cancer. **17**(1), 142 (2018). https://doi.org/10.1186/s12943-018-0893-y
30. Puhka, M., Takatalo, M., Nordberg, M.E., Valkonen, S., Nandania, J., Aatonen, M., Yliperttula, M., Laitinen, S., Velagapudi, V., Mirtti, T., Kallioniemi, O., Rannikko, A., Siljander, P.R., Af Hällström, T.M.: Metabolomic profiling of extracellular vesicles and alternative normalization methods reveal enriched metabolites and strategies to study prostate cancer-related changes. Theranostics. **7**(16), 3824–3841 (2017). https://doi.org/10.7150/thno.19890
31. Rodríguez, M., Bajo-Santos, C., Hessvik, N.P., Lorenz, S., Fromm, B., Berge, V., Sandvig, K., Linē, A., Llorente, A.: Identification of non-invasive miRNAs biomarkers for prostate cancer by deep sequencing analysis of urinary exosomes. Mol. Cancer. **16**(1), 156 (2017). https://doi.org/10.1186/s12943-017-0726-4
32. Yang, Y.N., Zhang, R., Du, J.W., Yuan, H.H., Li, Y.J., Wei, X.L., Du, X.X., Jiang, S.L., Han, Y.: Predictive role of UCA1-containing exosomes in cetuximab-resistant colorectal cancer. Cancer Cell Int. **18**(164), 2018 (2018). https://doi.org/10.1186/s12935-018-0660-6.eCollection
33. Klump, J., Phillipp, U., Follo, M., Eremin, A., Lehmann, H., Nestel, S., von Bubnoff, N., Nazarenko, I.: Extracellular vesicles or free circulating DNA: where to search for BRAF and cKIT mutations? Nanomedicine. **14**(3), 875–882 (2018). https://doi.org/10.1016/j.nano.2017.12.009
34. Stevic, I., Müller, V., Weber, K., Fasching, P.A., Karn, T., Marmé, F., Schem, C., Stickeler, E., Denkert, C., van Mackelenbergh, M., Salat, C., Schneeweiss, A., Pantel, K., Loibl, S., Untch, M., Schwarzenbach, H.: Specific microRNA signatures in exosomes of triple-negative and HER2-positive breast cancer patients undergoing neoadjuvant therapy within the GeparSixto trial. BMC Med. **16**(1), 179 (2018). https://doi.org/10.1186/s12916-018-1163-y
35. Tugutova, E.A., Tamkovich, S.N., Patysheva, M.R., Afanas'ev, S.G., Tsydenova, A.A., Grigor'eva, A.E., Kolegova, E.S., Kondakova, I.V., Yunusova, N.V.: Relation of tetraspanin-associated and tetraspanin-non-associated proteases in exosomes with stage and metabolic syndrome in colorectal cancer patients. APJCP. **20**(3), 809–815 (2019)
36. Stope, M.B., Klinkmann, G., Diesing, K.: Heat shock protein HSP27 secretion by ovarian cancer cells is linked to intracellular expression levels, occurs independently of the endoplasmic reticulum pathway and HSP27's phosphorylation status, and is mediated by exosome liberation. Dis. Markers. **2017**, 12 (2017). https://doi.org/10.1155/2017/1575374
37. Semina, S.E., Scherbakov, A.M., Vnukova, A.A.: Exosome-mediated transfer of cancer cell resistance to antiestrogen drugs. Molecules. **23**(4), 829 (2018). https://doi.org/10.3390/molecules23040829
38. Szatanek, R., Baj-Krzyworzeka, M., Zimoch, J., Lekka, M., Siedlar, M., Baran, J.: The methods of choice for extracellular vesicles (EVs) characterization. Int. J. Mol. Sci. **18**, 1153 (2017). https://doi.org/10.3390/ijms18061153
39. Wu, Y., Deng, W., Klinke II, D.J.: Exosomes: improved methods to characterize their morphology, RNA content, and surface protein biomarkers. Analyst. **140**(19), 6631–6642 (2015). https://doi.org/10.1039/c5an00688k
40. Yunusova, N.V., Tamkovich, S.N., Stakheeva, M.N., et al.: The characterization of exosomes from biological fluids of patients with different types of cancer. AIP Conf. Proc. **1882**, 020080 (2017). https://doi.org/10.1063/1.5001659

41. Grigor'eva, A.E., Dyrkheeva, N.S., Bryzgunova, O.E., Tamkovich, S.N., Chelobanov, B.P., Ryabchikova, E.I.: Contamination of exosome preparations, isolated from biological fluids. Biomed Khim. **63**(1), 91–96 (2017). https://doi.org/10.18097/PBMC20176301911/
42. Mihály, J., Deák, R., Szigyártó, I.C., Bóta, A., Beke-Somfai, T., Varga, Z.: Characterization of extracellular vesicles by IR spectroscopy: fast and simple classification based on amide and C–H stretching vibrations. Biochim. Biophys. Acta Biomembr. **1859**(3), 459–466 (2017). https://doi.org/10.1016/j.bbamem.2016.12.005
43. Goormaghtigh, E., Cabiaux, V., Ruysschaert, J.M.: Determination of soluble and membrane protein structure by Fourier transform infrared spectroscopy. III. Secondary structures. Subcell. Biochem. **23**, 405–450 (1994)
44. Barth, A.: Infrared spectroscopy of proteins. Biochim. Biophys. Acta. **1767**, 1073–1101 (2007)
45. Smith, K., Haris, P.I., Chapman, D., Reid, K.B.M., Perkins, S.J.: β-sheet secondary structure of the trimeric globular domain of C1q of complement and collagen types VIII and X by Fourier-transform infrared spectroscopy and averaged structure predictions. Biochem. J. **301**, 249–256 (1994)
46. Brauns, E.B., Dyer, R.B.: Time-resolved infrared spectroscopy of RNA folding. Biophys. J. **89**, 3523–3530 (2005). https://doi.org/10.1529/biophysj.105.061531
47. Zlotogorski-Hurvitz, A., Dekel, B.Z., Malonek, D., Yahalom, R., Vered, M.: FTIR-based spectrum of salivary exosomes coupled with computational-aided discriminating analysis in the diagnosis of oral cancer. J. Cancer Res. Clin. Oncol. **145**(3), 685–694 (2019). https://doi.org/10.1007/s00432-018-02827-6
48. Simonova, D., Karamancheva, I.: Application of Fourier transform infrared spectroscopy for tumor diagnosis. Biotechnol. Biotechnol. Equip. **27**, 4200–4207 (2013)
49. Gao, Y., Huo, X., Dong, L., et al.: Fourier transform infrared microspectroscopy monitoring of 5-fluorouracil-induced apoptosis in SW620 colon cancer cells. Mol. Med. Rep. **11**, 2585–2589 (2015)
50. Li, Q., Hao, C., Kang, X., et al.: Colorectal cancer and colitis diagnosis using Fourier transform infrared spectroscopy and an improved k-nearest-neighbor classifier. Sensors. **17**, 2739 (2017)
51. Dave, N., Lorenz-Fonfria, V.A., Leblanc, G., Padros, E.: FTIR spectroscopy of secondary-structure reorientation of melibiose permease modulated by substrate binding. Biophys. J. **94**, 3659–3670 (2008)
52. Dustin, W., Shipp, F.S., Notingher, L.: Raman spectroscopy: techniques and application in the life sciences. Adv. Opt. Photon. **9**(2), 315–428 (2017)
53. Ajito, K., Torimitsu, K.: Laser trapping and Raman spectroscopy of single cellular organelles in the nanometer range. Lab Chip. **2**, 11–14 (2002)
54. Xie, C., Dinno, M.A., Li, Y.Q.: Near-infrared Raman spectroscopy of single optically trapped biological cells. Opt. Lett. **27**(4), 249–251 (2002)
55. Cherney, D.P., Conboy, J.C., Harris, J.M.: Optical-trapping Raman microscopy detection of single unilamellar lipid vesicles. Anal. Chem. **75**(23), 6621–6628 (2003)
56. Sanderson, J.M., Ward, A.D.: Analysis of liposomal membrane composition using Raman tweezers. Chem. Commun. **9**, 1120–1121 (2004)
57. Chan, J.W., Esposito, A.P., Talley, C.E., Hollars, C.W., Lane, S.M., Huser, T.: Reagentless identification of single bacterial spores in aqueous solution by confocal laser tweezers Raman spectroscopy. Anal. Chem. **76**(3), 599–603 (2004)
58. Masson, J.-F.: Surface plasmon resonance clinical biosensors for medical diagnostics. ACS Sensors. **2**, 16–30 (2017)
59. Homola, J.: Surface plasmon resonance sensors for detection of chemical and biological species. Chem. Rev. **108**, 462–493 (2008)
60. Smolsky, J., Kaur, S., Hayashi, C., Batra, S., Krasnoslobodtsev, A.: Surface-enhanced Raman scattering-based immunoassay technologies for detection of disease biomarkers. Bios. **7**, 7 (2017)

61. Cialla-May, D., Zheng, X.S., Weber, K., Popp, J.: Recent progress in surface-enhanced Raman spectroscopy for biological and biomedical applications: from cells to clinics. Chem. Soc. Rev. **46**, 3945–3961 (2017)
62. Henry, A.-I., Sharma, B., Cardinal, M.F., Kurouski, D., Van Duyne, R.P.: Surface-enhanced Raman spectroscopy biosensing: *in vivo* diagnostics and multimodal imaging. Anal. Chem. **88**, 6638–6647 (2016)
63. Willets, K.A., Duyne, R.P.V.: Localized surface plasmon resonance spectroscopy and sensing. Annu. Rev. Phys. Chem. **58**, 267–297 (2007)
64. Ferhan, A.R., Jackman, J.A., Park, J.H., Cho, N.-J.: Nanoplasmonic sensors for detecting circulating cancer biomarkers. Adv. Drug Deliv. Rev. **125**, 48–77 (2018). https://doi.org/10.1016/j.addr.2017.12.004
65. Grasso, L., Wyss, R., Weidenauer, L., et al.: Molecular screening of cancer-derived exosomes by surface plasmon resonance spectroscopy. Anal. Bioanal. Chem. **407**, 5425–5432 (2015)
66. Hosseinkhani, B., van den Akker, N., et al.: Direct detection of nano-scale extracellular vesicles derived from inflammation-triggered endothelial cells using surface plasmon resonance. Nanomedicine. **13**, 1663–1167 (2017)
67. Zhu, L., Wang, K., Cui, J., et al.: Label-free quantitative detection of tumor-derived exosomes through surface Plasmon resonance imaging. Anal. Chem. **86**(17), 8857–8864 (2014)
68. Rupert, D.L.M., Shelke, G.V., Emilsson, G., et al.: Dual-wavelength surface plasmon resonance for determining the size and concentration of sub-populations of extracellular vesicles. Anal. Chem. **88**, 9980–9988 (2016)
69. Etchegoin, P.G., Le Ru, E.C.: A perspective on single molecule SERS: current status and future challenges. Phys. Chem. Chem. Phys. **10**(40), 6079–6089 (2008)
70. Li, L., Hutter, T., Steiner, U., Mahajan, S.: Single molecule SERS and detection of biomolecules with a single gold nanoparticle on a mirror junction. Analyst. **138**(16), 4574–4578 (2013)
71. Bonifacio, A., Cervo, S., Sergo, V.: Label-free surface-enhanced Raman spectroscopy of biofluids: fundamental aspects and diagnostic applications. Anal. Bioanal. Chem. **407**, 8265–8277 (2015)
72. Park, J., Hwang, M., Choi, B., Jeong, H., Jung, J.H., Kim, H.K., et al.: Exosome classification by pattern analysis of surface-enhanced Raman spectroscopy data for lung cancer diagnosis. Anal. Chem. **89**(12), 6695–6701 (2017)
73. Lee, C., Carney, R.P., Hazari, S., Smith, Z.J., Knudson, A., Robertson, C.S., et al.: 3D plasmonic nanobowl platform for the study of exosomes in solution. Nanoscale. **7**(20), 9290–9297 (2015)
74. Lane, L.A., Qian, X., Nie, S.: SERS nanoparticles in medicine: from label-free detection to spectroscopic tagging. Chem. Rev. **115**, 10489–10529 (2015)
75. Stremersch, S., Marro, M., Pinchasik, B.-E., Baatsen, P., Hendrix, A., De Smedt, S.C., Loza-Alvarez, P., Skirtach, A.G., Raemdonck, K., Braeckmans, K.: Identification of individual exosome-like vesicles by surface enhanced Raman spectroscopy. Small. **12**, 3292–3301 (2016)
76. Cialla, D., Pollok, S., Steinbrucker, C., Weber, K., Popp, J.: SERS-based detection of biomolecules. Nano. **3**, 383–411 (2014)
77. Zheng, X.S., Jahn, I.J., Weber, K., Cialla-May, D., Popp, J.: Label-free SERS in biological and biomedical applications: recent progress, current challenges and opportunities. Spectrochim. Acta A Mol. Biomol. Spectrosc. **197**, 56–77 (2018)
78. Zong, S., Wang, L., Chen, C., Lu, J., Zhu, D., Zhang, Y., Wang, Z., Cui, Y.: Facile detection of tumor-derived exosomes using magnetic nanobeads and SERS nanoprobes. Anal. Methods. **8**, 5001–5008 (2016)
79. Lee, C., Carney, R., Lam, K., Chan, J.W.: SERS analysis of selectively captured exosomes using an integrin specific peptide ligand. J. Raman Spectrosc. **48**, 1771–1776 (2017)
80. Wang, Y., Yan, B., Chen, L.: SERS tags: novel optical nanoprobes for bioanalysis. Chem. Rev. **113**, 1391–1428 (2013)

81. Zavaleta, C.L., Smith, B.R., Walton, I., et al.: Multiplexed imaging of surface enhanced Raman scattering nanotags in living mice using noninvasive Raman spectroscopy. Proc. Natl. Acad. Sci. U. S. A. **106**, 13511–13516 (2009)
82. Jamieson, L.E., Asiala, S.M., Gracie, K., Faulds, K., Graham, D.: Bioanalytical measurements enabled by surface-enhanced Raman scattering (SERS) probes. Annu Rev Anal Chem (Palo Alto, Calif). **10**, 415–437 (2017)
83. Carmicheal, J., Hayashi, C., Huang, X., et al.: Label-free characterization of exosome via surface enhanced Raman spectroscopy for the early detection of pancreatic cancer. Nanomedicine. **16**, 88–96 (2019). https://doi.org/10.1016/j.nano.2018.11.008
84. Kruglik, S.G., Royo, F., Guigner, J.-M., Palomo, L., Seksek, O., Turpin, P.-Y., Tatischeff, I., Falcón-Pérez, J.M.: Raman tweezers microspectroscopy of: Circa 100 nm extracellular vesicles. Nanoscale. **11**(4), 1661–1679 (2019). https://doi.org/10.1039/c8nr04677h
85. Durak-Kozica, M., Baster, Z., Kubat, K., Stępień, E.: 3D visualization of extracellular vesicle uptake by endothelial cells. Cell. Mol. Biol. Lett. **23**, 57 (2018). https://doi.org/10.1186/s11658-018-0123-z
86. Yunusova, N.V., Tamkovich, S.N., Stakheeva, M.N., et al.: The characterization of exosome from blood plasma of patients with colorectal cancer. AIP Conf. Proc. **1760**, 020070 (2016). https://doi.org/10.1063/1.4960289
87. Matthews, A.L., Noy, P.J., Reyat, J.S., Tomlinson, M.G.: Regulation of a disintegrin and metalloproteinase (ADAM) family sheddases ADAM10 and ADAM17: the emerging role of tetraspanins and rhomboids. Platelets. **28**(4), 333–341 (2017). https://doi.org/10.1080/09537104.2016.1184751
88. Galanzha, E.I., Zharov, V.P.: Photoacoustic flow cytometry. Methods. **57**(3), 280–296 (2012). https://doi.org/10.1016/j.ymeth.2012.06.009
89. Nolan, J., Sarimollaoglu, M., Nedosekin, D.A., Jamshidi-Parsian, A., Galanzha, E.I., Kore, R.A., Griffin, R.J., Zharov, V.P.: *In vivo* flow cytometry of circulating tumor-associated exosomes. Anal. Cell. Pathol. **2016**, 1628057 (2016)
90. Wu, Y., Ali, M.R.K., Chen, K., et al.: Gold nanoparticles in biological optical imaging. Nano Today. **24**, 120–140 (2019). https://doi.org/10.1016/j.nantod.2018.12.006
91. Macklin, R., Wang, H., Loo, D., Martin, S., et al.: Extracellular vesicles secreted by highly metastatic clonal variants of osteosarcoma preferentially localize to the lungs and induce metastatic behaviour in poorly metastatic clones. Oncotarget. **7**, 43570–43587 (2016). https://doi.org/10.18632/oncotarget.9781
92. Wang, L., Xie, Y., Ahmed, K.A., et al.: Exosomal pMHC-I complex targets T cell-based vaccine to directly stimulate CTL responses leading to antitumor immunity in transgenic FVBneuN and HLA-A2/HER2 mice and eradicating trastuzumab-resistant tumor in athymic nude mice. Breast Cancer Res. Treat. **140**(2), 273–284 (2013)
93. Cheng, P.-C.: The contrast formation in optical Microscop. In: Pawley, J.B. (ed.) Handbook of Biological Confocal Microscopy, 3rd edn, pp. 162–206. Springer, New York (2006). https://doi.org/10.1007/978-0-387-45524-2_8
94. Wong, F., Coban, O., Weitsman, G., Ng, T.: Integrating imaging, exosome and protein network rewiring information to track early tumour evolution of resistance mechanisms. Converg. Sci. Phys. Oncol. **3**(1), 013004 (2017)
95. Kim, N., Thomas, M.R., Bergholt, M.S., et al. Surface enhanced Raman scattering artificial nose for high dimensionality fingerprinting. Nature communications. **207**(11), 1–12 (2020). https://doi.org/10.1038/s41467-019-13615-2

Part II
Diffuse Spectroscopy and Fluorescence Analysis for Cancer Detection

Chapter 5
Functional Near-Infrared Spectroscopy in Cancer Diagnostics

Teemu Myllylä and Vesa Korhonen

5.1 Introduction

Functional near-infrared spectroscopy (fNIRS), or diffuse optical spectroscopy (DOS), has been utilized for over three decades as a non-invasive tool for monitoring blood circulation related parameters, and particularly oxygenation changes in the cerebral cortex that are linked to brain function [1]. The principle of fNIRS is based on wavelength-dependent absorption. Each chromophore (a light absorbing molecule) has a distinct absorption spectrum determined by its content and energy level structure. These individual spectra can be used as a profile for identifying individual compounds in a substance [2]. Most applications exploit the spectrum range between 650 nm and 950 nm, where light attenuation in tissue is sufficiently low to enable light to penetrate deeper into tissue. A typical fNIRS measurement setup consists of near-infrared (NIR) receivers/sources and fiber optic acccssories. As light source, these setups employ a LED, laser, laser diode or broadband thermal source, using a fiber optic probe to deliver light on the patient's skin [3, 4].

Different applications require a different fNIRS probe source and detector separation distance. In breast cancer detection, for instance, the distance ranges from 28 to 40 mm, which enables tumor detection at 10–40 mm depth in breast tissue [5, 6]. In brain imaging, the source–detector distance tends to be 30–40 mm to allow sensing of gray matter. In general, a longer separation distance enables deeper photon penetration in tissue, albeit at the cost of reduced signal-to-noise ratio and spatial accuracy [7].

T. Myllylä (✉)
University of Oulu, Oulu, Finland
e-mail: teemu.myllyla@oulu.fi

V. Korhonen
Oulu University Hospital, Oulu, Finland
e-mail: vesa.korhonen@oulu.fi

V. V. Tuchin et al. (eds.), *Multimodal Optical Diagnostics of Cancer*,
https://doi.org/10.1007/978-3-030-44594-2_5

Once fNIRS raw signals have been detected, we may apply the modified Beer–Lambert law to calculate quantitative tissue concentrations of deoxyhemoglobin (HbR), oxyhemoglobin (HbO), water, and lipids. Concentrations of specific substances in a tissue can be optically determined using a relative spectroscopic measurement at two or more wavelengths [8, 9].

5.2 fNIRS Cancer Diagnostics

In differentiating between cancerous and normal tissue with label-free fNIRS techniques, the most important chromophores are HbO, HbR, lipid, and water, or a combination of these [10]. In addition, they can be used to determine the tissue optical index

$$\mathrm{TOI} = \frac{\left[\mathrm{HbR}\right]\mathrm{x}\left[\mathrm{water}\right]}{\left[\mathrm{lipid}\right]}. \tag{5.1}$$

These provide a basis for fNIRS based diagnosis and therapy monitoring of several cancers. TOI is a composite index showing contrasts in metabolically active regions developed to detect breast tumor locations [11, 12]. Spatial variation in TOI enables rapid pinpointing of areas with maximum lesion optical contrast.

5.2.1 Brain Cancer

fNIRS has been used to some degree in brain cancer research. For instance, it may have potential in preoperative localization of tumors [13] and in therapy monitoring. In comparison with fMRI, one of the main benefits of fNIRS is that it allows measuring concentration changes in both HbO and HbR, while blood oxygenation level-dependent contrast functional fMRI (BOLD-fMRI) is only sensitive to changes in HbR concentration in blood. fNIRS activation studies show that, in normal adults, neuronal activation decreases HbR and increases both HbO and HbT [14, 15]. However, this is not always the case with brain tumors, as Fujiwara et al. pointed out in 2004, when they measured evoked cerebral blood oxygenation (CBO) in brain tumors [16] in combination with BOLD-fMRI. They found that, on the lesion side, fNIRS revealed a decrease in HbR in five patients and an increase in seven patients, even though both HbO and HbT typically increased during a hand-grasping task in both groups, indicating the occurrence of a regional cerebral blood flow increase in response to neuronal activation. The non-lesion side demonstrated more typical findings, including a decrease in HbR and an increase in both HbO and HbT in the primary sensorimotor cortex of all patients. The fNIRS device used by the group included four laser diodes directed on the primary sensorimotor cortex with a source–detector distance of 3 or 4 cm. Repeated six times, the activation paradigm consisted of 40 s of rest, alternating with 40 s of self-paced hand grasping. In summary, their results indicated that false-negative activations in the BOLD-fMRI of patients with brain tumors were associated with a rise in HbR during activation measured by fNIRS, even though there was an increase in regional cerebral blood flow. As a result, important cortical activation areas may be overlooked in such patients [16].

Saxena et al. studied the vascular status and pathophysiological changes that occur during tumor vascularization in an orthotopic brain tumor model. They monitored concentrations of HbO, HbR, and water within the tumor region and found a direct correlation between tumor size, intratumoral microvessel density, and tumor oxygenation. The relative decrease in tumor oxygenation with growth indicated that, although blood vessels infiltrate the tumor region and proliferate in it, a hypoxic trend is clearly present [17]. Further, a preclinical study on a murine model of head and neck cancer suggests that monitoring of tumor oxygenation status could be used to predict treatment outcome in solid tumors [18].

More recently, fNIRS has been utilized before neurosurgery to map language areas in patients suffering from tumors such as glioma [19, 20] and to assess oxygen status in patients with glioblastoma [21, 22]. Presurgical evaluation of language functions is of great importance in brain tumor patients in order to spare these areas during surgery and to reduce the risk of postsurgical language deficits [19, 20]. fNIRS offers a valuable and easy tool for identification of essential language functions, especially in children. Sato et al. have also demonstrated that expressive and receptive language functions can be identified separately using fNIRS in presurgical glioma patients [20]. fNIRS revealed increases of HbO and decreases of HbR activity in language areas elicited by a verb generation and a story listening task. These results were completely consistent with those of the WADA test, which is commonly used to look at language and memory on one side of the brain at a time. fNIRS measurements were carried out using a commercial Hitachi ETG-4100 Spectrometer with 695 and 830 nm wavelength lasers at a source–detector distance of 3 cm and a sampling rate of 10 Hz. For each hemisphere, 8 emitters and 8 detectors were placed alternately on a head shell constituting 24 measuring channels. To determine areas associated with language functions, all channels were mounted on both hemispheres and controlled such that the area of measurement covered the inferior frontal regions and superior temporal regions. More recently, Callagher et al. showed that fNIRS permits identification of language networks in children at rest, i.e., without them needing to perform a task [19]. They recorder fNIRS data during rest and during an expressive language task in a single recording session using a frequency domain (FD) fNIRS device (Imagent, ISS Inc) equipped with 8 detectors and 64 sources (32 at 690 nm and 32 at 830 nm). The light sources and detectors were held in place using a comfortable, age-adapted helmet with source–detector distances varying between 2 and 5 cm, thus optimizing the recording of brain signals at various depths in the region of interest, particularly the anterior and posterior language-related regions, Broca's and Wernicke's areas and the Brodmann area. Results obtained in resting state were compared to those of a more conventional task-based fNIRS measurement. This comparison indicated very good correspondence between both approaches for language localization (dice similarity coefficient = 0.81 ± 0.13) and hemispheric language dominance (kappa = 0.86, $p < 0.006$). This makes fNIRS technique an even more valuable and easy to use tool for language mapping in clinical populations, including children and patients with cognitive and behavioral impairments.

In their review characterizing hypoxia in glioblastoma, Corroyer-Dulmont et al. concluded that PET imaging seems the most relevant tool for the purpose [22]. However, they also contended that fNIRS offers a simple and non-invasive, albeit indirect, method of characterizing blood oxygenation levels.

5.2.2 *Breast Cancer*

In detecting breast cancer, fNIRS utilizes HbO, HbR, HbT, water, and lipid concentrations and an increasing array of extrinsic organic compounds available in the NIR optical window [23–26]. These techniques have been applied to non-invasive measurements of subtle physiological differences in healthy breast tissue [27–29], to detecting and localizing tumors [5, 6, 23], and to determining tumor responses to neoadjuvant chemotherapy [5, 30–32]. Enabled by a high tumor tissue signal-to-noise ratio, optical cancer detection has found increasing use as a supplementary diagnostic tool for breast cancer [5, 33].

Cancer regions typically show significantly higher concentrations of HbT and water and significantly lower lipid concentration compared to surrounding healthy breast tissue [29, 33–35]. A preclinical study by Orlova et al. demonstrated that tumor oxygenation and hemoglobin content are the key indicators of the tumor status. These can be measured using fNIRS techniques. Further, they concluded that tumor oxygenation decrease is not caused by the reduction of oxygen delivery to the tumor tissue rate, but, more likely, because of the increase of tissue oxygen consumption and decreased blood outflow rate [36].

Table 5.1 presents the specific parameters and chromophores used in fNIRS human breast cancer studies in 10 selected research articles.

Anderson et al. studied chromophore concentration and hemoglobin saturation contrast in breast cancer using broadband optical mammography [33]. As light source, a broadband optical mammography device utilizes an arc lamp whose emission is first spectrally filtered (400–1000 nm) to reject ultraviolet and infrared light

Table 5.1 Breast cancer studies with background information, including patient group, wavelength, and technique information as well as reported results regarding intrinsic physiological properties

	Subjects	Wavelengths	Technique	Reported change in response (if any)						Ref
				HbO	HbR	HbT	Water	Lipid	Spo2	
1	29 (26) breast cancer patients	Broadband 650–1000 nm	CW	↑	↑	↑	↑	↓	↓	[33]
2	12 malignant breast tumor patients + 1undergoing neoadjuvant chemotherapy	Broadband 650–1000 nm	FD	↑	↑		↑	↓	not reliable	[5]
3	6 large breast carcinomas. 2 get neoadjuvant chemotherapy.	780 nm, 830 nm, 660 nm	AM/FD +Ultra - sound			↓				[30]
4	3 ductal carcinomas + 9 healthy controls	FD: 661 nm,761 nm,785 nm, 808 nm,826 nm,849 nm + CW: 903 nm,912 nm, 948 nm	FD + CW			↑	↑	↓		[37]
5	58 state 2/3 malignant breast tumors	Broadband 650–1000 nm	FD + IM and CW	↑ (50%)	↑ (50%)		↑ (50%)	↓ (20%)		[34]
6	11, (1 fibroadenoma, 1invasive ductal carcinoma and 9 of 11 negative mammographic findings)	750 nm, 800 nm	CW			↑				[38]
7	breast of 49 women	760 nm, 780 nm, 830 nm, 850 nm	TD			↑ malign ↑benign	↑ malign ↑benign		↓malign ↑benign	[35]
8	38 patients, variety of benign and malignant lesions	32 channel, 780 nm & 815 nm	TR, optical images						↓ malign ↑ fibroadenoma	[39]
9	125 subjects, with negative findings and 51 breasts with lesions	RF unit: 685 nm & 830 nm MUX: 685 nm, 810 nm, 830 nm	TOBI (RF + CW)			↑				[29]
10	154 patients, 87 histologically validated carcinomas	670 nm & 785 nm + 843 nm or 883 nm	TD	not significant		↑			Both higher or lower ↑ ↓	[40]

and then focused onto a 3 mm diameter illumination fiber bundle delivering light to the breast [37, 38]. A collection optical fiber (diameter 5 mm) bundle is located on the opposite side of the breast, always collinear with the illumination fiber. The breast is gently compressed between two parallel glass plates, whose distance is recorded for every examination. This stabilizes the position of the breast during the scan without any discomfort to the subject. Having imaged a total of 29 breast cancer patients (3 were excluded from analysis) with the device, Anderson et al. found that, relative to surrounding healthy tissue, the cancer region exhibited significantly higher concentrations of HbT and water and significantly lower lipidconcentration and hemoglobin oxygen saturation. Furthermore, they demonstrated a significant correlation between tumor optical contrast and the grade of breast cancer as quantified by the Nottingham histologic score. This means that optical signatures may represent metabolic and morphological features and indicate the aggressive potential of the tumor. To conclude, their key finding is that breast cancer produces a marked reduction in the oxygen saturation of hemoglobin, which can be used to enhance the information content of optical mammograms.

Tromberg et al. as well used broadband fNIRS to detect tumors in pre-menopausal women and to monitor neoadjuvant chemotherapy in breast cancer treatment [5]. They used a laser breast scanner in conjunction with a handheld probe, in essence, a bedside-capable system combining FD photon migration with steady-state tissue spectroscopy. They measured NIR absorption from 650 nm to 1000 nm and reduced scattering spectra of breast tissue *in vivo*. At a source–detector distance of 2.8 cm, the measuring probe was moved along a linear grid of steps spaced 1 cm apart on both the tumor region (previously identified) and the contralateral normal side. Diagnosed with a malignant breast tumor, the 12 pre-menopausal subjects ranged from 30 to 39 years in age. The results established a statistically significant contrast between normal and tumor regions of tissue for HbR (increased), HbO (increased), water (increased), and lipids (decreased). However, tissue hemoglobin saturation did not prove a reliable parameter for distinguishing between tumor and normal tissue. In addition, results from one patient, undergoing neoadjuvant chemotherapy for locally advanced breast cancer, showed a 50% decrease in TOI within 1 week in response to received chemotherapy. All of these results suggested the potential of diffuse optics to be utilized in the development of new strategies for individualized cancer therapy.

Wang et al. studied *in vivo* quantitative imaging of normal and cancerous breast tissue using broadband diffuse optical tomography (DOT) [39]. Their DOT system combined FD and CW measurements to image normal and malignant breast tissue. Owing to detector features, FD acquisitions were limited to wavelengths of less than 850 nm (661 nm, 761 nm, 785 nm, 808 nm, 826 nm, and 849 nm), whereas light at longer wavelengths, up to 948 nm (903 nm, 912 nm, and 948 nm), was measured in CW mode with a CCD-coupled spectrometer. Having combined the two data sets, Wang et al. processed them into a single spectrally constrained reconstruction. They mapped concentrations of hemoglobin, water, and lipid, as well as scattering parameters in the breast. Nine healthy and asymptomatic subjects were imaged to evaluate their intrasubject and intersubject variability. The recorded data showed physiologically expected trends. Next, three patients with invasive ductal carcinoma were imaged and compared to the control data. The added CW data revealed an increase in water and a decrease in lipid content within the patients' malignancies. Furthermore, these areas showed a 1.5–2 fold contrast increase in hemoglobin and water values. As a conclusion, relative to stand-alone FD data, *in vivo* breast imaging with instrumentation that combines FD and CW NIR data acquisition in a single spectral reconstruction produces more accurate hemoglobin, water, and lipid results.

Cerussi et al. studied *in vivo* absorption, scattering, and physiological properties of 58 state 2/3 malignant breast tumors by broadband fNIRS using a handheld device [28, 34]. The fNIRS equipment combines multi-frequency intensity-modulated and continuous wave (CW) NIR light to quantify tissue absorption and scattering spectra within the spectral range of 650–1000 nm. Values of such intrinsic physiological properties as HbO, HbR, water, lipid, and scatter power provide detailed information on breast physiology. Scanning was performed in eight positions over tumor and contralateral healthy breast tissue for each subject. Cerussi et al. found statistically significant variation in intrinsic physiological properties between malignant and normal tissue for all subjects regardless of patient age or tumor size/type stratification. Compared to normal breast tissue, malignant tumors displayed reduced lipid content and increased water, HbR, and HbO. Furthermore, functional perturbations caused by a tumor were significantly larger than corresponding variations in normal tissue. In addition, the TOI derived from intrinsic physiological properties doubled the contrast difference between malignant tumors and intrinsic tissue properties. In summary, we may conclude that functional perturbations have an effect on intrinsic optical signals characteristic of a malignant transformation.

Furthermore, in the study by Chance et al. it was shown that fNIRS provides a rapid, safe, and patient-compliant method for detecting breast cancer with a very high ROC/AUC score (95%) for a population of breast tumors [6]. All in all, 116 subjects (44 were cancer-verified by biopsy and histopathology) were measured using a handheld NIR breast cancer detector pad with a 3-wavelength (760 nm, 805 nm, and 850 nm) LED and 8 detectors. The source–detector distance was 4 cm and the pad was placed on the subject's breast. HbT and its relative oxygenation were calculated in the breast with cancer and compared to the contralateral breast in a 2D nomogram to enable diagnostic evaluation.

In a pilot study, Gu et al. demonstrated the ability to quantitatively image cysts ranging from 1 to 4 cm in diameter and differentiate them from solid tumors in the breast with DOT [40]. Six cases were studied using compact parallel-detection DOT system. Cysts could be differentiated from solid breast tumors by a lower absorption and scattering coefficient compared with surrounding normal tissue. Solid tumors, on the other hand, manifest higher concurrent absorption and scattering than normal tissue.

5.3 fNIRS Monitoring in Cancer Therapy

Cancer therapy treatments produce extensive changes in the physiological and morphological properties of tissues, including angiogenesis, hypoxia, alterations in cell nuclear size and density, and denaturation [41]. Consequently, great clinical interest has been attached to measuring and tracking these biomarkers and changes. A key challenge involves developing tailored cancer therapy, geared toward considering individual biological responses during therapy.

fNIRS offers an effective technique for monitoring such biomarkers, particularly tumor hypoxia and responses to chemo- and radiotherapy, which involves modulating tumor oxygenation to increase or decrease tumor hypoxia [47]. However, this measurement is often performed by magnetic resonance imaging (MRI), which may not be an ideal solution. Howe et al. used fNIRS to determine absolute hemoglobin and changes in HbR and HbO in subcutaneous tumors in rodents. To better interpret MRI data, they paid particularly attention to factors that alter blood flow and oxygenation. Both carbogen and O_2 breathing produced a significant reduction in HbR and an increase in HbO, but a negligible change in HbT. This contrasts with N_2 breathing in terminal anoxia and intravenous hydralazine regime, which produced a negligible increase in HbR, but large reductions in HbO and HbT. Since HbT is proportional to blood volume, they suggested that large blood volume drops are likely to cause reduced arterial blood pressure. It then follows that MRI techniques that measure the $R2*$ relaxation rate, which varies linearly with total HbR, will underestimate the effects of hypotensive agents with increasing tumor hypoxia [42].

A pilot study by Sunar et al. in 2006 investigated how patients with head and neck tumors responded to chemo-radiation therapy. Blood flow and oxygenation were measured using diffuse reflectance spectroscopy (DCS) and fNIRS. The DCS setup was a four-channel system with a long coherence length laser operating at 785 nm. The shortest and largest separation between source and detector fibers was 2 and 3 cm, respectively. Also comprising four channels, the fNIRS instrument operated at the wavelengths of 690 nm, 785 nm, and 830 nm using four source–detector distances (1.8 cm, 2.2 cm, 2.6 cm, and 3 cm) to quantify oxygenation parameters. Significant changes in relative blood flow, tissue oxygen saturation, and HbT were observed even in the first 2 weeks of treatment. The study protocol consisted of preradiation measurements providing baseline data and weekly follow-up optical measurements, conducted just before each new weekly treatment. Their preliminary results suggest diffuse optics-based therapy monitoring may have clinical promise [21].

To monitor tumor neoadjuvant chemotherapy (NAC) responses in women with locally advanced breast cancer, Schaafsma et al. performed optical mammography, combining fNIRS with standard treatment monitoring by dynamic contrast-enhanced MRI (DCE-MRI). Their fNIRS (or DOS) system consisted of four individual pulsed diode lasers operating at 690 nm, 730 nm, 780 nm, and 830 nm. Light was collected by a mobile detector in a 1 cm-X constellation composed of five optical fibers and detected by a photomultiplier. Each breast scan took approximately 10–20 min. Based on measured absorption and scattering, the group obtained accurate estimates of HbO, HbR, HbT, water, lipids, scattering amplitude (SA) and power (SP). A significant difference between responders and nonresponders was found using fNIRS and these differences continued during treatment. The study concluded that fNIRS allows tumor response assessment and is capable of differentiating between responders and nonresponders after the first chemotherapy cycle and halfway through the treatment. Further, fNIRS proved as effective as DCE-MRI in predicting tumor responses halfway through treatment [43].

Kesler et al. suggest that the default mode network, a network of interacting brain regions that is distinct from other networks in the brain [44], may be a potential biomarker of chemotherapy-related brain injury. Chronic medical conditions and/or their treatments may interact with aging to alter or even accelerate brain senescence. A case in point is adult onset cancer, a disease associated with aging. Emerging evidence suggests that diffuse, subtle brain injury may follow cancer chemotherapy. Currently, the primary model for studying these "chemobrain" effects is breast cancer. Treatment is often followed by a range of widespread changes in brain structure and function as well as impairment of integrated cognitive skills, making it likely that large-scale brain networks, such as the default mode network, are involved. The default mode network is vulnerable to aging and sensitive to toxicity and disease states. Additionally, increased inflammation and oxidative stress are believed to raise toxicity levels in the default mode network during chemotherapy [45]. Biomarkers of default mode network connectivity, measured by fNIRS in combination with other neuroimaging techniques, could help ward off chemotherapy-related cognitive decline.

Kiviniemi et al. monitored primary central nervous system lymphoma patients, who received chemotherapy aided by opening of the blood-brain barrier (BBB). Monitoring BBB opening is of great interest in terms of brain drug delivery. Although proper opening of the BBB is crucial for successful treatment, there was no method for real-time clinical monitoring. To remedy that, Kiviniemi et al. presented a combined method based on direct-current electroencephalography (DC-EEG) and fNIRS. Using a source–detector distance of 3 cm, the setup employed fNIRS at 660 nm and 830 nm with one channel positioned on the forehead of test subjects beneath an electroencephalography (EEG) cap. fNIRS detected a remarkable multiphasic response during carotid artery infusions beginning with mannitol-bolus induced dilution of blood and ending in a prolonged change in the HbO/HbR ratio [46]. A similar effect was recently confirmed using a mouse model, see Fig. 5.1.

The prolonged fall in HbR level is particularly interesting, because it cannot be explained merely by hyperemia-induced dilution. More likely, since HbR is not being produced, it reflects a temporary cessation of oxygen consumption, which may actually be a sign of BBB opening.

An analysis of the results revealed that they were concordant between both carotid artery BBB disruptions. However, because a frontal fNIRS cannot monitor the vertebral artery territory, less prominent, delayed responses were recorded in the vertebral artery after mannitol infusion. Furthermore, the very prolonged cerebrovascular responses, caused by carotid artery infusion, were strikingly different compared to the gradually fading DC shifts, which were measured simultaneously. Consequently, brain hemodynamics measured by fNIRS reflect BBB opening and gradual closing, making the method applicable for monitoring BBB opening during brain chemotherapy.

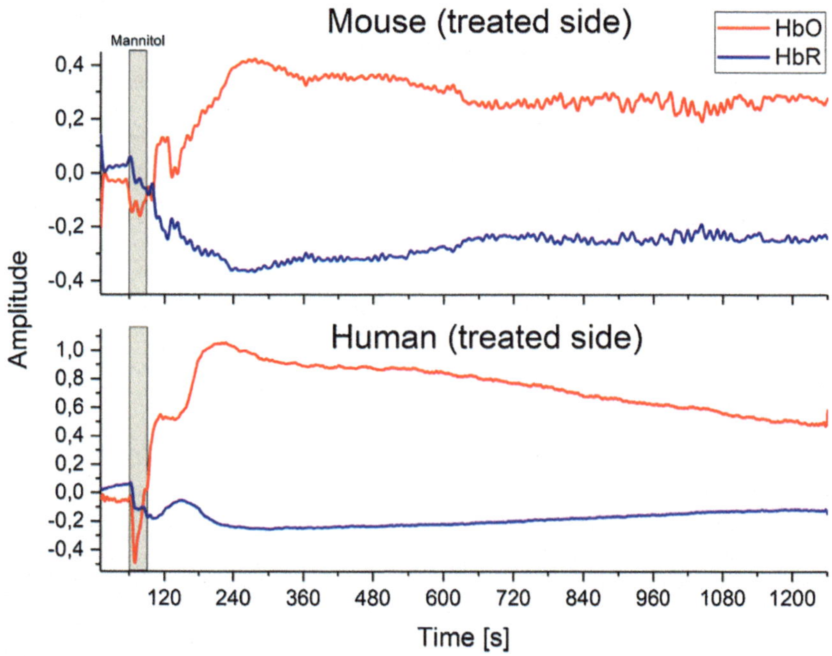

Fig. 5.1 Monitoring mannitol-infused BBBD with fNIRS. HbO and HbR responses to mannitol are similar for both mouse and human

References

1. Boas, D.A., Elwell, C.E., Ferrari, M., Taga, G.: Twenty years of functional near-infrared spectroscopy: introduction for the special issue. Neuroimage. **85**(1), 1–5 (2014)
2. Tuchin, V.V.: Tissue Optics: Light Scattering Methods and Instruments for Medical Diagnosis. SPIE Press, Bellingham (2007)
3. Myllylä, T., Korhonen, V., Kiviniemi, V., Tuchin, V.: Experimental studies with selected light sources for NIRS of brain tissue: quantifying tissue chromophore concentration. In: Optical Techniques in Neurosurgery, Neurophotonics, and Optogenetics II, p. 93051S (2015)
4. Myllylä, T., Korhonen, V., Surażyński, Ł., Zienkiewicz, A., Sorvoja, H., Myllylä, R.: Measurement of cerebral blood flow and metabolism using high power light-emitting diodes. Measurement. **58**, 387–393 (2014)
5. Tromberg, B.J., Cerussi, A., Shah, N., Compton, M., Durkin, A., Hsiang, D., Butler, J., Mehta, R.: Imaging in breast cancer: diffuse optics in breast cancer: detecting tumors in pre-menopausal women and monitoring neoadjuvant chemotherapy. Breast Cancer Res. **7**(6), 279 (2005)
6. Chance, B., Nioka, S., Zhang, J., Conant, E.F., Hwang, E., Briest, S., Orel, S.G., Schnall, M.D., Czerniecki, B.J.: Breast cancer detection based on incremental biochemical and physiological properties of breast cancers: a six-year, two-site study 1. Acad. Radiol. **12**(8), 925–933 (2005)
7. Korhonen, V.O., Myllylä, T.S., Kirillin, M.Y., Popov, A.P., Bykov, A.V., Gorshkov, A.V., Sergeeva, E.A., Kinnunen, M., Kiviniemi, V.: Light propagation in NIR spectroscopy of the human brain. IEEE J. Sel. Top. Quant. Electron. **20**(2), 289–298 (2013)

8. Chung, S., Cerussi, A., Klifa, C., Baek, H., Birgul, O., Gulsen, G., Merritt, S., Hsiang, D., Tromberg, B.: *In vivo* water state measurements in breast cancer using broadband diffuse optical spectroscopy. Phys. Med. Biol. **53**(23), 6713 (2008)
9. Myllylä, T., Harju, M., Korhonen, V., Bykov, A., Kiviniemi, V., Meglinski, I.: Assessment of the dynamics of human glymphatic system by near-infrared spectroscopy. J. Biophotonics. **11**(8), e201700123 (2018)
10. Kondepati, V.R., Heise, H.M., Backhaus, J.: Recent applications of near-infrared spectroscopy in cancer diagnosis and therapy. Anal. Bioanal. Chem. **390**(1), 125 (2008)
11. Yazdi, H.S., O'Sullivan, T.D., Leproux, A., Hill, B., Durkin, A., Telep, S., Lam, J., Yazdi, S.S., Police, A.M., Carroll, R.M.: Mapping breast cancer blood flow index, composition, and metabolism in a human subject using combined diffuse optical spectroscopic imaging and diffuse correlation spectroscopy. J. Biomed. Opt. **22**(4), 045003 (2017)
12. O'Sullivan, T.D., Leproux, A., Chen, J., Bahri, S., Matlock, A., Roblyer, D., McLaren, C.E., Chen, W., Cerussi, A.E., Su, M.: Optical imaging correlates with magnetic resonance imaging breast density and reveals composition changes during neoadjuvant chemotherapy. Breast Cancer Res. **15**(1), R14 (2013)
13. Obrig, H.: NIRS in clinical neurology—a 'promising' tool? Neuroimage. **85**, 535–546 (2014)
14. Kleinschmidt, A., Obrig, H., Requardt, M., Merboldt, K., Dirnagl, U., Villringer, A., Frahm, J.: Simultaneous recording of cerebral blood oxygenation changes during human brain activation by magnetic resonance imaging and near-infrared spectroscopy. J. Cereb. Blood Flow Metab. **16**(5), 817–826 (1996)
15. Villringer, A., Chance, B.: Non-invasive optical spectroscopy and imaging of human brain function. Trends Neurosci. **20**(10), 435–442 (1997)
16. Fujiwara, N., Sakatani, K., Katayama, Y., Murata, Y., Hoshino, T., Fukaya, C., Yamamoto, T.: Evoked-cerebral blood oxygenation changes in false-negative activations in BOLD contrast functional MRI of patients with brain tumors. Neuroimage. **21**(4), 1464–1471 (2004)
17. Saxena, V., Gonzalez-Gomez, I., Laug, W.E.: A noninvasive multimodal technique to monitor brain tumor vascularization. Phys. Med. Biol. **52**(17), 5295 (2007)
18. Vishwanath, K., Klein, D.H., Chang, K., Schroeder, T., Dewhirst, M.W., Ramanujam, N.: Quantitative optical spectroscopy can identify long-term local tumor control in irradiated murine head and neck xenografts. J. Biomed. Opt. **14**(5), 054051 (2009)
19. Gallagher, A., Tremblay, J., Vannasing, P.: Language mapping in children using resting-state functional connectivity: comparison with a task-based approach. J. Biomed. Opt. **21**(12), 125006 (2016)
20. Sato, Y., Uzuka, T., Aoki, H., Natsumeda, M., Oishi, M., Fukuda, M., Fujii, Y.: Near-infrared spectroscopic study and the Wada test for presurgical evaluation of expressive and receptive language functions in glioma patients: with a case report of dissociated language functions. Neurosci. Lett. **510**(2), 104–109 (2012)
21. Sunar, U., Quon, H., Durduran, T., Zhang, J., Du, J., Zhou, C., Yu, G., Choe, R., Kilger, A., Lustig, R.A.: Noninvasive diffuse optical measurement of blood flow and blood oxygenation for monitoring radiation therapy in patients with head and neck tumors: a pilot study. J. Biomed. Opt. **11**(6), 064021 (2006)
22. Corroyer-Dulmont, A., Chakhoyan, A., Collet, S., Durand, L., MacKenzie, E.T., Petit, E., Bernaudin, M., Touzani, O., Valable, S.: Imaging modalities to assess oxygen status in glioblastoma. Front. Med. **2**, 57 (2015)
23. Nioka, S., Chance, B.: NIR spectroscopic detection of breast cancer. Technol. Cancer Res. Treat. **4**(5), 497–512 (2005)
24. Srinivasan, S., Pogue, B.W., Jiang, S., Dehghani, H., Kogel, C., Soho, S., Gibson, J.J., Tosteson, T.D., Poplack, S.P., Paulsen, K.D.: *In vivo* hemoglobin and water concentrations, oxygen saturation, and scattering estimates from near-infrared breast tomography using spectral reconstruction 1. Acad. Radiol. **13**(2), 195–202 (2006)
25. Karellas, A., Vedantham, S.: Breast cancer imaging: a perspective for the next decade. Med. Phys. **35**(11), 4878–4897 (2008)

26. Van de Ven, S., Elias, S., Van den Bosch, M., Luijten, P., Mali, W.T.M.: Optical imaging of the breast. Cancer Imaging. **8**(1), 206 (2008)
27. Cerussi, A.E., Berger, A.J., Bevilacqua, F., Shah, N., Jakubowski, D., Butler, J., Holcombe, R.F., Tromberg, B.J.: Sources of absorption and scattering contrast for near-infrared optical mammography. Acad. Radiol. **8**(3), 211–218 (2001)
28. Carpenter, C.M., Srinivasan, S., Pogue, B.W., Jiang, S., Dehghani, H., Paulsen, K.D.: 3-D MR guided NIRS: optimization of computation and breast interface for *in vivo* imaging. Biomed. Opt. BSuB4 (2008)
29. Fang, Q., Selb, J., Carp, S.A., Boverman, G., Miller, E.L., Brooks, D.H., Moore, R.H., Kopans, D.B., Boas, D.A.: Combined optical and X-ray tomosynthesis breast imaging. Radiology. **258**(1), 89–97 (2011)
30. Zhu, Q., Kurtzman, S.H., Hegde, P., Tannenbaum, S., Kane, M., Huang, M., Chen, N.G., Jagjivan, B., Zarfos, K.: Utilizing optical tomography with ultrasound localization to image heterogeneous hemoglobin distribution in large breast cancers. Neoplasia. **7**(3), 263–270 (2005)
31. Choe, R.: Diffuse optical tomography and spectroscopy of breast cancer and fetal brain. Citeseer. (2005)
32. Jakubowski, D.B., Cerussi, A.E., Bevilacqua, F.P., Shah, N., Hsiang, D., Butler, J.A., Tromberg, B.J.: Monitoring neoadjuvant chemotherapy in breast cancer using quantitative diffuse optical spectroscopy: a case study. J. Biomed. Opt. **9**(1), 230–239 (2004)
33. Anderson, P.G., Kainerstorfer, J.M., Sassaroli, A., Krishnamurthy, N., Homer, M.J., Graham, R.A., Fantini, S.: Broadband optical mammography: chromophore concentration and hemoglobin saturation contrast in breast cancer. PLoS One. **10**(3), e0117322 (2015)
34. Cerussi, A.E., Shah, N.S., Hsiang, D., Durkin, A., Butler, J.A., Tromberg, B.J.: *In vivo* absorption, scattering, and physiologic properties of 58 malignant breast tumors determined by broadband diffuse optical spectroscopy. J. Biomed. Opt. **11**(4), 044005 (2006)
35. Intes, X.: Time-domain optical mammography SoftScan: initial results 1. Acad. Radiol. **12**(8), 934–947 (2005)
36. Orlova, A., Kirillin, M.Y., Volovetsky, A., Shilyagina, N.Y., Sergeeva, E., Golubiatnikov, G.Y., Turchin, I.: Diffuse optical spectroscopy monitoring of oxygen state and hemoglobin concentration during SKBR-3 tumor model growth. Laser Phys. Lett. **14**(1), 015601 (2016)
37. Yu, Y., Liu, N., Sassaroli, A., Fantini, S.: Near-infrared spectral imaging of the female breast for quantitative oximetry in optical mammography. Appl. Optics. **48**(10), D225–D235 (2009)
38. Yu, Y., Sassaroli, A., Chen, D.K., Homer, M.J., Graham, R.A., Fantini, S.: Near-infrared, broad-band spectral imaging of the human breast for quantitative oximetry: applications to healthy and cancerous breasts. J. Innov. Opt. Health Sci. **3**(04), 267–277 (2010)
39. Wang, J., Jiang, S., Li, Z., diFlorio-Alexander, R.M., Barth, R.J., Kaufman, P.A., Pogue, B.W., Paulsen, K.D.: *In vivo* quantitative imaging of normal and cancerous breast tissue using broadband diffuse optical tomography. Med. Phys. **37**(7 Part 1), 3715–3724 (2010)
40. Gu, X., Zhang, Q., Bartlett, M., Schutz, L., Fajardo, L.L., Jiang, H.: Differentiation of cysts from solid tumors in the breast with diffuse optical tomography 1. Acad. Radiol. **11**(1), 53–60 (2004)
41. Yu, B.: Quantitative optical spectroscopy and imaging for cancer diagnosis and treatment monitoring. In: 2016 Progress in Electromagnetic Research Symposium (PIERS), pp. 2322–2322 (2016)
42. Howe, F.A., Connelly, J.P., Robinson, S.P., Springett, R., Griffiths, J.R.: The effects of tumour blood flow and oxygenation modifiers on subcutaneous tumours as determined by NIRS. Adv Exp Med Biol. 75–81 (2005)
43. Schaafsma, B.E., van de Giessen, M., Charehbili, A., Smit, V.T., Kroep, J.R., Lelieveldt, B.P., Liefers, G., Chan, A., Löwik, C.W., Dijkstra, J.: Optical mammography using diffuse optical spectroscopy for monitoring tumor response to neoadjuvant chemotherapy in women with locally advanced breast cancer. Clin. Cancer Res. **21**(3), 577–584 (2015)
44. Buckner, R.L., Andrews-Hanna, J.R., Schacter, D.L.: The brain's default network. Ann. N. Y. Acad. Sci. **1124**(1), 1–38 (2008)

45. Kesler, S.R.: Default mode network as a potential biomarker of chemotherapy-related brain injury. Neurobiol. Aging. **35**, S11–S19 (2014)
46. Kiviniemi, V., Korhonen, V., Kortelainen, J., Rytky, S., Keinänen, T., Tuovinen, T., Isokangas, M., Sonkajärvi, E., Siniluoto, T., Nikkinen, J.: Real-time monitoring of human blood-brain barrier disruption. PLoS One. **12**(3), e0174072 (2017)
47. Myllylä, T., Karthikeyan, P., Honka, U., Korhonen, V., Karhula, S. S., & Nikkinen, J. (2020, April). Cerebral haemodynamic effects in the human brain during radiation therapy for brain cancer. In Tissue Optics and Photonics (Vol. 11363, p. 1136307). International Society for Optics and Photonics

Chapter 6
Breathomics for Lung Cancer Diagnosis

Yury V. Kistenev, Alexey V. Borisov, and Denis A. Vrazhnov

6.1 Volatile Metabolites for Lung Cancer Diagnosis

Lung cancer(LC) is the second most common cancer in both men and women (not counting skin cancer). Lung cancers are about 13% of all new cancers. The American Cancer Society estimated that about 228,150 new lung cancer cases (116,440 in men and 111,710 in women) and 142,670 lung cancer deaths (76,650 in men and 66,020 in women) occurred in the USA in 2019 [1].

Cancer survival depends mainly on stage at diagnosis. It is typically silent in its early stages as a result of which most of the cases are diagnosed in later stages when treatment is ineffective. As an example, for LC patients diagnosed in 2003–2009, the 5-year survival rate was 54% for stage 1 disease (localized), 26% for stage 2 and

Y. V. Kistenev (✉)
Laboratory of Biophotonics, National Research Tomsk State University, Tomsk, Russian Federation

Department of Physics with a Course of Higher Mathematics, Siberian State Medical University, Tomsk, Russian Federation
e-mail: yuk@iao.ru

A. V. Borisov
Laboratory of Biophotonics, National Research Tomsk State University, Tomsk, Russian Federation

Department of General and Experimental Physics, National Research Tomsk State University, Tomsk, Russian Federation
e-mail: borisov@phys.tsu.ru

D. A. Vrazhnov
Laboratory of Biophotonics, National Research Tomsk State University, Tomsk, Russian Federation

Laboratory of Molecular Imaging and Photoacoustics, Institute of Strength Physics and Materials Science of Siberian Branch Russian Academy of Sciences, Tomsk, Russian Federation

V. V. Tuchin et al. (eds.), *Multimodal Optical Diagnostics of Cancer*,
https://doi.org/10.1007/978-3-030-44594-2_6

3 (regional), and 4% for stage 4 (distant). However, only 15% of cases were diagnosed at stage 1, while 22% were diagnosed at stages 2 and 3 and 57% were diagnosed at stage 4 [2].

Control of metabolites in exhaled air produced by biochemical reactions in cells being called "breathomics" provides the ability to predict disease before the appearance of the clinical features [3].

Screening LC diagnosis based on control of volatile metabolites in breath air is very attractive. The potential advantages of this approach are non-invasiveness, ease of use, and minimum cost of diagnosis tests, suitability for long-time monitoring. The exhaled air has less complex composition compared to blood serum or urine, allows to analyze components present in it, without obligatory preliminary preparation of a sample, in contrast to the analysis of blood. Some exhaled air components are closely correlated with their concentrations in blood, which eliminates the need for blood sampling for analysis [4].

Exhaled air contains endogenous or exogenous origin compounds at ppbv–pptv concentration range, in addition to nitrogen, oxygen, carbon dioxide, water vapor, and inert gases. The endogenous compounds include inorganic gases such as NO, CO; volatile organic compounds (VOCs) such as ethane, pentane, acetone, isoprene, acetaldehyde, methanol, ethanol, and other alcohols and alkanes; 2-propanol, sulfur-containing compounds such as dimethyl sulfide, methyl, ethyl, mercaptanes, and carbon disulfide; and nitrogen-containing substances such as ammonia and dimethyl/trimethylamine [5]. Ulanowska et al. [6] measured concentration levels of different VOCs (acetaldehyde, ethanol, acetonitrile, butane, furan, propanal, acetone, carbon disulfide, 2-Propanol, dimethyl sulfide, 1-Propanol, isoprene, pentane, 2-Methylfuran, 3-Buten-2-one, 2-Butanone, ethyl acetate, 2-Methylpentane, 3-Methylpentane, benzene, hexane, 2-Pentanone, pentanal, 2,5-Dimethylfuran, hexanal, toluene, 2-Methylheptane, heptane, p-Xylene, o-Xylene, 4-Methyloctane, ethylbenzene) in exhaled air of healthy volunteers, including smokers.

Typical VOCs in breath air for LC patients are presented in Table 6.1. It should be pointed out that a number of these VOCs have been registered in the breath of smokers and passive smokers, including hydrocarbons, furan, acetonitrile, benzene, 3-methylfuran, 2,5-dimethylfuran, 2-butanone, octane, and decane [7].

Breath air analysis can be used both for understanding specific biochemical processes and as a diagnostic tool. The absolute identification of VOCs is not strictly necessary in clinical activity, and probabilistic discrimination of biomarker profiles can be used for purposes of diagnosis [16]. As an example of "profiling" approach, the analysis of VOCs profile in breath air of LC patients and healthy volunteers was carried out [4]. The use of 15 specific identified VOCs was shown to provide sensitivity at the level of 71% (21 markers provide 80% sensitivity) and 100% specificity in comparison with healthy volunteers. Potential biomarkers included alcohols, aldehydes, ketones, hydrocarbons.

A significant increase in level of 30 VOCs in exhaled air of LC patients (193 people) in relation to healthy volunteers (211 people) was registered by Altorki et al. [17], including isopropyl alcohol, 2,3-hexanedione, camphor, benzophenone, derivatives of tetroxane, benzene, anthracene, benzoic acid, furan, esters, and several others components. The result was presumably due to the activation of

Table 6.1 VOCs concentrations in exhaled air of LC patients, healthy smokers, and healthy volunteers

No	Biomarker	Description	Level of concentrations			References, comments
			Lung cancer	Smokers	Control	
1	1,2,4-trimethyl benzene	An organic compound with the chemical formula $C_6H_3(CH_3)_3$. Classified as an aromatic hydrocarbon	Increasing in comparison with control	Not available	Not available	[8, 9]
2	2,4-dimethyl heptane	Formula: C_9H_{20}	Increasing in comparison with control	Not available	Not available	[8]
3	3-hydroxy-2-butanone	Acetoin, Linear formula $C_8H_{16}O_4$	Median—9.23, ng/L Quartile range—10.96, ng/L	Not available	Median—1.29, ng/L Quartile range—2.01, ng/L	[3, 8][a]
4	Acetone	Acetone (propanone) is the organic compound with the formula $(CH_3)_2CO$	16.75 (11.61–26.12), nmol/L	6.10 (4.18–10.30), nmol/L	15.82 (9.82–22.15), nmol/L	[8, 10][b]
5	Benzene	Benzene is classed as an aromatic hydrocarbon chemical formula C_6H_6	0.11 (0.11–0.11), nmol/L	0.12 (0.11–0.24), nmol/L	0.11 (0.11–0.11), nmol/L	[8, 10][b]
6	1-Butanol	Primary alcohol with a 4-carbon structure and the chemical formula C_4H_9OH	Median—9.67, ng/L Quartile range—12.62, ng/L	Not available	Median—2.18, ng/L Quartile range—2.06, ng/L	[11][a]
7	Butane	Butane is an alkane with the formula C_4H_{10}	0.15 (0.13–0.36), nmol/L	0.13 (0.13–0.25), nmol/L	0.31 (0.14–0.58), nmol/L	[8][b]
8	Cyclohexane	A cycloalkane with the molecular formula C_6H_{12}	0.20 (0.00 ± 1.71) ppb	Not available	0.10 (0.00 ± 0.48) ppb	[12][c]
9	Decane	Decane is an alkane hydrocarbon with the chemical formula $C_{10}H_{22}$	4.25 (0.06 ± 62.9) ppb 568.0 (277.9–1321.6), 10^{-12} M	239.2 (60.0–884.0), 10^{-12} M	2.88 (0.26 ± 18.5) ppb 208.7 (14.3–405.5), 10^{-12} M	[12][c] [13][d]
10	Ethanol	A simple alcohol with the chemical formula C_2H_5OH	14.5 (0.00–160) ppb	Not available	13.5 (0.00–68.7) ppb	[12][c]
11	Heptane	Heptane is the straight-chain alkane with the chemical formula $H_3C(CH_2)_5CH_3$ or C_7H_{16}.	0.17 (0.00–7.36) ppb	Not available	0.15 (0.00–6.26) ppb	[8, 12][c]

(continued)

Table 6.1 (continued)

No	Biomarker	Description	Level of concentrations			References, comments
			Lung cancer	Smokers	Control	
12	Hexanal	An alkyl aldehyde, Chemical formula $C_6H_{12}O$	0.59 (0.38–0.86), nmol/L	0.31 (0.31–0.31), nmol/L	0.63 (0.38–2.32), nmol/L	[8, 10][b]
13	Isoprene	A common organic compound with the formula $CH_2 = C(CH_3) - CH{=}CH_2$	43.4 (2.88–202) ppb	Not available	31.5 (3.76–184) ppb	[12][c]
14	Isopropanol	Isopropyl alcohol	6.47 (3.31–13.40), nmol/L	1.06 (0.18–1.88), nmol/L	1.41 (0.72–2.32), nmol/L	[10][b]
15	Nonane	A linear alkane hydrocarbon with the chemical formula C_9H_{20}	0.10 (0.00–18.0) ppb	Not available	0.07 (0.00–1.35) ppb	[9][c]
16	Octane	A hydrocarbon and an alkane with the chemical formula C_8H_{18},	61.0 (22.4–112.9), 10^{-12} M	33.5 (19.7–57.8), 10^{-12} M	20.2 (4.0–50.8), 10^{-12} M	[8, 14][d]
17	1-propanol	1-propanol is a primary alcohol with the formula $CH_3CH_2CH_2OH$	2599.50 (1279.00–8536.25), nmol/L	286.50 (0.00–1217.13), nmol/L	434.50 (0.00–1295.88), nmol/L	[6][b]
18	Pentane	An organic compound with the formula C_5H_{12}	647.5 (361.3–1112.5), 10^{-12} M	511.4 (241.3–1128.3), 10^{-12} M	268.0 (107.7–462.7), 10^{-12} M	[8, 13][d]
19	Pentamethyl heptane	Alkane with the formula $(CH_3)_2CH\text{-}[CH(CH_3)]_3\text{-}CH(CH_3)_2$	2.5 (1.2–9.7), 10^{-12} M	5.8 (1.2–16.5), 10^{-12} M	0.9 (0.1–2.6), 10^{-12} M	[13][d]
20	Propanal	Chemical formula C_3H_6O	0.34 (0.00–0.37), nmol/L	0.00 (0.00–0.34), nmol/L	0.00 (0.00–0.00), nmol/L	[10]
21	Styrene	An organic compound with the chemical formula $C_6H_5CH{=}CH_2$	17.9 (8.5–37.2), 10^{-12} M	7.2 (2.8–41.6), 10^{-12} M	12.3 (5.3–21.8), 10^{-12} M	[8, 13][d]
22	Toluene	An aromatic hydrocarbon	0.19 (0.00–1.13) ppb	Not available	0.26 (0.00–1.82) ppb	[12][c]
23	Undecane	A liquid alkane hydrocarbon with the chemical formula $CH_3(CH_2)_9CH_3$	(0.1–8.51), ng/mL	Not available	Not available	[15]
24	Xylene	An isomer of dimethylbenzene, the formula $(CH_3)_2C_6H_4$,	0.16 (0.00–5.60) ppb	Not available	0.07 (0.00–1.40) ppb	[12][c]

[a]Lung cancer patients with stage 1–2
[b]Expiratory concentrations in alveolar samples of a single breath
[c]Values are median (range)
[d]Concentrations expressed in shape of median values (25th–75th percentile)

cytochrome P450, which affects the concentrations of these substances in the exhaled air. There was no significant difference between smokers and non-smokers.

A significant increase in the concentration of butanol-1 and 3-hydroxy-2-butanone in the exhaled air of LC patients was found in a comparative analysis of the 68 VOCs in the exhaled air from 43 patients with non-small cell lung cancer (NSCLC) and 41 healthy volunteers [11]. The increase of these components in the exhaled air was more significant in pulmonary adenocarcinoma than in squamous cell lung cancer. As these VOCs are the products of butane oxidation, it indicates an increase in oxidative processes during cancer development.

Gleeson et al. [9] reported that 22 selected VOCs allow distinguishing LC patients and healthy people with a sensitivity of 100% and specificity of 81%. These VOCs included 3-methyloctan, 3-methylnonane, isoprene, cyclohexane, heptanal, hexanal, and derivatives of heptane, decane, benzene. The increase of their concentration can be partly associated with oxidative stress. There was no dependence of the VOCs content on the stage of disease and smoking.

Poli et al. [13] achieved correct classification for 80% of LC patients using polynomial logistic regression analysis of 13 VOCs in breath air (isoprene, 2-methylpentane, pentane, ethylbenzene, xylene, trimethylbenzene, toluene, benzene, heptane, decane, styrene, octane, pentamethylheptan). The 2-methylpentane level was higher for NSCLC patients in comparison with healthy volunteers and patients with chronic obstructive pulmonary disease (COPD). There was a significant decrease in isoprene and decane concentrations after surgical treatment.

Ager et al. [18] stated that 21 VOCs provided a sensitivity of 80% and specificity of 100% of LC patients diagnosis in comparison with healthy volunteers. It was registered a significant decrease in the concentrations of isoprene, acetone, and methanol in the LC patients exhaled, while the concentrations of 2-butanone, benzaldehyde, 2,3-butanone, 2-butanone, 1-propanone, acetophenone, cyclopentene, tetramethylcarbamide, butyl acetate, etc. were significantly increased concerning the healthy volunteers.

According to the paper [19], potential biomarkers of LC are pentane, isoprene, acetone, benzene.

Recently, 112 potential biomarkers of LC in exhaled air have been registered [20]. They included 36 hydrocarbons, 7 alcohols, 8 aldehydes, 2 acids, 12 ketones, 12 aromatic compounds, 2 heterocycles, 2 nitriles, 5 terpenes, 9 esters, 1 sulfide, 2 halogenated compounds, and 15 compounds from other chemical classes. The registered hydrocarbons included 2-methyl-propane, 5-methyl-tridecane; the registered alcohols included 1-octene-3-ol. The registered aldehydes included pentanal, hexanal, octanal, noanal, examples of ketones are 6-methyl-5-hepten-2-OH, the example of aromatic compounds are benzophenone mixture, the example of terpenes are trans-caryophyllene.

A systematical review of various sources, including PubMed, EMBASE, Cochrane databases, of the current knowledge on exhaled VOCs with respect to their potential clinical use for LC and other diseases diagnosis was performed [21]. Seventy-three studies were included, counting in total 3952 patients and 2973 healthy volunteers. Various research groups demonstrated that VOCs profiles could accurately distinguish patients with pulmonary disease from healthy control.

6.2 Experimental Equipment for Profiling of VOCs Breath Air

6.2.1 Gas Chromatography

Gas chromatography is based on a variety of sample components extraction time in the flow of the carrier gas in a column. Thus, each compound exits the column at a specific time (known as the retention time). The column is typically coiled and very thin (0.25 mm internal diameter) allowing even tens of meters in length to be housed within a relatively small temperature-controlled oven. Longer columns (~30 m) are used for metabolic studies.

Gas chromatography with mass spectrometry detection (GC-MS) became a "gold standard" in VOCs analysis [3] due to its low limit of detection (LOD), from 100 ppm to 1 ppb or less [22]. GC-MS is especially effective for analysis of organic compounds in biological gas samples. For example, the LOD for dichloromethane by this method is about 0.1 ppt [23].

The main drawbacks of gas chromatography breathomics are its rather high complexity, high cost of consumables, special skills of the personnel.

6.2.2 Direct Injection Mass Spectrometry

Selected ion flow tube mass spectrometry (SIFT-MS) is based on chemical ionization using molecular ions to transfer charge onto the target compound. The chemical ionization allows reducing fragmentation of the latter in comparison with many other types of ionization. SIFT-MS provides direct analysis without sample preconcentration, is suitable for real-time monitoring, and is slightly influenced by humidity. The LOD of the SIFT-MS Voice200Ultra (Syft Technologies Ltd) is better than 1 pptv.

Proton Transfer Reaction Mass Spectrometry (PTR-MS) is a chemical ionization mass spectrometric technique, which allows measuring trace concentration components in gas mixtures, including human breath. PTR-MS provides fast analysis of exhaled breath without prior sample preparation [24]. To increase accuracy, the duration of the measurement process should be increased, but for breath-to-breath resolution, the time window for measurement should be relatively short. The PTR-QMS 300 instrument (IONICON Analytik GmbH) provides LOD<300 pptv [25]. A disadvantage of PTR-MS is that it is suitable only for molecules with a proton affinity higher than that of water. Additionally, SIFT-MS and PTR-MS do not detect as many compounds such as GC-MS [0].

6.2.3 Ion Mobility Spectrometry

The ion mobility spectrometry (IMS) used to detect substances of very small concentrations [26]. In IMS system, a radioactive source ionizes the molecules in a gas sample. As a result, molecules drift into an electric field inside the so-called drift cell. Each type of molecule has a specific drift velocity in the air and may, therefore, be identified. Gas Chromatograph coupled with Ion Mobility Spectrometer (GC-IMS) by Gesellschaft für Analytische Sensorsysteme mbH provides a typical value of LOD near the low ppbv-range. However, this method has a lack of selectivity in some cases, because, during an analysis of multicomponent samples, different interactions of ion/molecule produce complex, practically non-solvable spectra.

6.2.4 The "Electronic Nose" Equipment

The conception of "electronic nose" devices was evolved in the early 1980 and corresponded to the system of chemical sensors, each of which measures the concentration of a particular substance. The term "electronic noses" was introduced by Gardner and Bartlett in 1994 [27]. These sensor devices are capable of detecting a wide diversity of chemical species and mixtures of compounds present in headspace volatiles of sampled air, including VOCs. The example of the "e-nose" is "Cyranose 320," consisting of 32 polymer chemiresistors [27]. Conducting polymers (CP) have a moderately sensitive detection limit of 0.1–100 ppm for various volatile substances such as organic acids, alcohols, esters, and alkanes depending on the water vapor content [28].

Metal oxide surfaces (MOS) sensors arc based on oxidation by gaseous molecules at a high temperature (250–450 °C), which causes electron transfer from the molecules to the metal oxide structure. This results in a change in electrical conductivity. CP-sensors are low power-consuming instruments with very good sensitivity and reproducibility at room temperatures, but they have high sensitivity to moisture, inactivation by certain strongly polar analytes and relatively low sensor life (compared to MOS sensors) that tend to have a much longer sensor life, but operate at higher temperatures and with greater power demand. CP-sensors and some types of MOS sensors also sometimes exhibit problems of sensor compatibility and uniformity [29].

The breath VOCs are studied to identify the early stage of lung cancer using chemiresistors coated with gold nanoparticles. Metallic nanoparticles are synthesized in a two-phase system which is composed of aqueous and non-aqueous solutions. These solutions are combined with techniques of extraction of ions. The metallic mono-layered nanoparticles are then coated with a hydrophobic layer of alkanethiols. First, the ions from an aqueous solution are transferred to a toluene solution. Then the reduction of gold is done with an aqueous borohydride solution which is later capped with thiols. Lastly, the gold nanoparticles are extracted and purified from thiols. This technique is shown to be highly accurate and is relatively easy to operate [30].

Sensor technology for electronic olfaction that offers the potential to develop miniature sensor chips deploying hundreds of diverse and sensitive sensors based on DNA-decorated semiconducting single-walled carbon nanotubes was presented [31]. Because the DNA oligomers have tightly bound water molecules associated with their phosphate groups and other moieties in the nucleotide backbone, the DNA coated nanotube sensors are not responsive to water vapor, which is often a problem with other sensor technologies of direct breath analysis. The authors have made sensor devices sensitive to some components of human breath (organic acids, trimethylamine).

In common, the disadvantages of chemical sensors are a short life time, low specificity due to reaction not only on a given chemical compound but sensible to nearly all compounds of the same chemical family, such as organic solvents, fatty acids, sulfurous gases, etc. [32].

To overcome these drawbacks, new types of sensors and sensor coatings are being developed, such as chemical sensors that change their color when certain VOCs appear (colorimetric sensors) or change the frequency of quartz resonator, etc. [33]. Quartz microbalance (QMB) sensors contain a quartz crystal with piezoelectric properties. This technique is based on the variation in mechanical oscillations of the crystal and the resonant frequency of the electric circuit following the attachment of VOCs. QMB sensors have very good sensitivity but are not as easily effective for the development of "Breathprints" [29].

6.2.5 The Absorption Spectroscopy

Absorption spectroscopy is based on resonant absorption of optical radiation by a molecule. LOD of absorption laser spectrometers with absorbing Bougher's cuvette depends on the optical thickness of the studied mixture. Multi-pass cuvettes are used for registration of low concentrations of components in breath air.

Cavity ring-down spectroscopy method (CRDS) was proposed by A. O'Keefe and D. A.G. Deacon in 1988 for precision measurements of absorption coefficients with short laser pulses [34]. It was demonstrated that CRDS sensitivity can achieve $\sim 10^{-10}$ cm^{-1} for mirrors with reflectivity R $\geq$ 0.9999. A significant drawback of CRDS is the technical complexity of wavelength tuning because the high reflectivity value may be achieved only in a narrow spectral interval [35].

Laser photoacoustic spectroscopy system (LPAS) is one of the most sensitive approaches of laser absorption spectroscopy to gas analysis, especially with the use of coherent radiation sources and intracavity photoacoustic detection. LPAS has a very low detection limit. For example, LPAS gas analyzer with an intracavity acoustic cell provides the measurement of ethylene down to 6 pptv [36, 37]. Sample preconcentration for LPAS analysis is not needed, because the photoacoustic signal is proportional to the absorbed volume fraction of laser energy, which can be increased by the power of used laser source.

The sensitivity of LPAS is strongly influenced by the photoacoustic cell design as it can operate either in a nonresonant mode or as an acoustic resonator. Nonresonant operation means that the light modulation frequency is below the lowest resonance frequency of the cell. In this case, acoustic wave distribution within the cell is almost spatially independent and resonant amplification of the photoacoustic signal is not used. When the exciting light is modulated at a resonance frequency of the cell, the generated photoacoustic signal is amplified by the quality factor (Q-factor) of the acoustic resonance. Q-factor can be up to several hundred [38]. The most frequently used types of resonant LPAS detectors are based on Helmholtz resonators, one-dimensional cylindrical resonators, and cavity resonators [38–40].

Nonlinear effects in optical parametric oscillator (OPO) are one of the most widespread ways to generate tunable coherent radiation in the wide spectral range. We developed the LaserBreeze gas analyzer based on an LPAS method and OPO with the tuning range of 2.5–10.7 μm [41]. The scheme of the LaserBreeze is shown in Fig. 6.1. Laser source includes two OPOs. The first one is based on fan-out periodically poled lithium niobate structure (PPLN), which provides wavelength tuning in the spectral range 2.5–4.5 μm. The second OPO is based on mercury thiogallate crystals HgGa2S4 (HGS) and has wavelength tuning range 4.45 μm–10.7 μm. Both OPO were pumped by Nd:YLF laser. The switching between two OPO is realized by the motorized translation stage. The linewidth of laser radiation is about 3–4 cm^{-1}. Resolution of wavelength scanning is around 7 nm/s. The photoacoustic detector (PAD) is based on double channel Helmholtz resonator with Q-factor ~40 and fundamental resonance frequency ~1700 Hz. Data from the pyroelectric detector (PD) are used to normalize the PAD signal relatively to laser radiation power. The thermostating at the temperature 40 °C ± 0.2 °C was applied to avoid temperature drift of the OPO parameters and water vapor condensation on the PAD inner walls.

The necessary volume of studied sample is not more than 50 cm^3 and the concentration sensitivity of the LaserBreeze is not worse than 1×10^{-3} ppm. The procedure of sensitivity estimation was described in [42]. PAD was preliminarily cleared by pumping of N_2. The measurements of noise signal value U_N were continued for 3 min. The average value $\langle U_N \rangle$and standard deviation δU_N were calculated. Then, PAD was filled by calibration gas mixture including tested gas with known concentration n and N_2. The concentration of tested gas was chosen to provide useful signal value U_S over U_N in 2–3 times. The measurements procedure was the same as for noise level one. The following equation was used to calculate signal/noise value (S/N).

$$S/N = \frac{U_s}{U_N + \delta U_N},$$

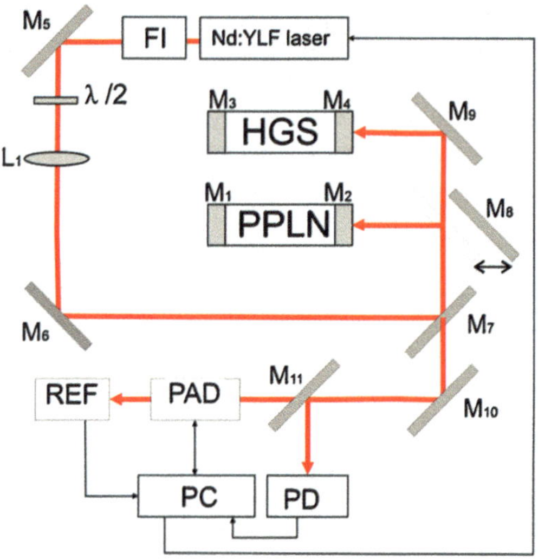

Fig. 6.1 Experimental set-up of LaserBreeze. *FI* Faraday Isolator, M_i mirrors, *PC* Personal Computer, λ/2 is half wave plate, *PD* pyroelectric detector, *PAD* photoacoustic detector, *REF* reference cell, L_1 lens, *HGS* mercury thiogallate crystals HgGa2S4 based OPO, *PPLN* periodically poled lithium niobate based OPO

where $<U_s>$ is the average value of the useful signal. The sensitivity n_o was determined by the formula:

$$n_o = \frac{n}{S/N}.$$

In order to provide wavelength calibration, we use the reference cell (REF) filled by a gas mixture of the compounds having strong absorption lines in known wavelengths within the LaserBreeze tuning range. The absorption spectrum of the REF gas mixture is shown in Fig. 6.2.

The LaserBreeze allows to detect more than 20 molecular biomarkers which have absorption lines in the mentioned spectral range, including acetone (C_3H_6O), acetylene (C_2H_2), ammonia (NH_3), butane (C_4H_{10}), carbon dioxide (CO_2), 13 isotope of carbon dioxide ($13CO_2$), carbon monoxide (CO), ethane (C_2H_6), ethanol (C_2H_5OH), ethyl acetate ($C_4H_8O_2$), ethylene (C_2H_4), formaldehyde (CH_2O), methane (CH_4), methanol (CH_3OH), nitrogen dioxide (NO_2), nitrogen oxide (NO), nitrous oxide (N_2O), pentane (C_5H_{12}), propane (C_3H_8), sulfur dioxide (SO_2). The relative error in determining of VOCs concentrations is not more than 30%.

On the whole, laser photoacoustic spectroscopy looks very attractive for routine clinical breathomics applications due to simplicity and low cost of analysis, no special requirements for sample preparation and ability to operate without special knowledge.

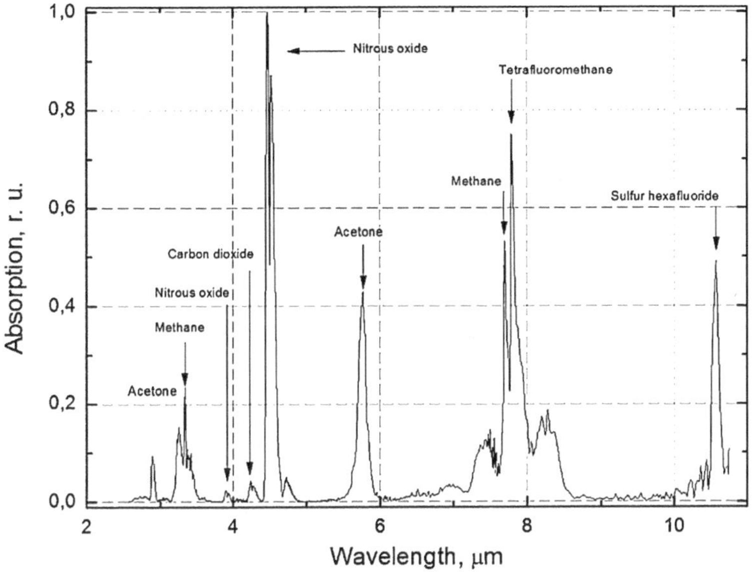

Fig. 6.2 Absorption spectrum of the gas mixture in the reference cell in the spectral range 2500–10,700 nm

6.3 Breath Air Sampling

Equipment for breath air testing varies, but the basic principles keep the same. All systems include a source of test gas, a method for measuring inhaled and exhaled breath air (spirometers, pneumotachometers near the mouthpiece or a bag-inbox), and gas analyzers (single-sample analyzers or continuous high-speed analyzers) [43].

The breath air sampling procedure to identify VOCs is increasingly important. There are two main approaches to sampling breath: direct (real-time) and off-line analysis. Direct analysis of the exhaled air is carried out without prior concentration or storage of samples. Direct methods are of interest for continuous monitoring of certain compounds. Physiological parameters such as the exhaled CO_2 partial pressure, temperature, flow rate, or the exhaled O_2 partial pressure can be used to control the sampling process. Typically, these parameters are used for recognition of respiratory phases and should be monitored with a time resolution of at least 200 ms, regardless of whether direct sampling (real-time) or controlled (alveolar) sampling followed by analysis [44]. Identifying a set of VOCs is time-consuming and cannot be done in real-time using direct methods.

At off-line analysis by chromatography methods, the pre-concentration, as a rule, is used. The review of pre-concentration breath sample techniques was presented [45]. Gordin and Amirav developed the "Snifprobe" [46], which consists of a small-length capillary or porous-layer open tubular (PLOT) column. Breath was drawn through the Snifprobe for 5 s, after which the entire column segment was inserted into the GC injector for thermal desorption. A sorbent material is used to sample from a bag or chamber containing a breath sample. The pre-concentration is realized based on traditional enrichment methods, such as solid-phase extraction followed by thermal desorption of the mixture before analysis. The drawbacks of sorbent use are high sample volumes (typically 100–5000 mL), which are necessary for reliable and sensitive analysis, as well as potential problems associated with high water content. In principle, distributed solid-phase microextraction can overcome the above problems and meets the requirements of controlled and rapid sampling and reliable pre-concentration [44].

Several devices and techniques have been developed to concentrate VOCs onto thermal desorption (TD) tubes and subsequent laboratory analysis. Those optimum parameters for TD were the subject of the investigation [47]. The experiments were conducted to investigate the fraction of breath sampled (whole breath including mouth air versus lower respiratory exhaled breath); breath sample volume (125, 250, 500 and 1000 mL); and breath sample flow rate (400, 200, 100 and 50 mL min^{-1}). The exhaled breath samples were collected using "Respiration Collector for *In Vitro* Analysis" (ReCIVATM) (Owlstone Medical, Cambridge, UK) in combination with a dedicated clean air supply "Clean Air Supply Pump for ReCIVA" (CASPER) (Owlstone Medical, Cambridge, UK).

The four-piece TD tube assembly was inserted into a clean mask for each study participant and then attached to the ReCIVA device ensuring that the TD tube and mask assembly were seated correctly within the device (Fig. 6.3).

This study indicated that the increase in sample volume had improved VOCs detection. However, the influence of the fraction of exhaled breath and the flow rate depends on the target VOCs measured. While the concentration of potential volatile biomarkers for esophago-gastric cancer was not significantly different between the whole and lower airway exhaled breath, the level of other VOCs was varied. Also, the recovery of some VOCs such as phenols and acetone from TD tubes was lower when breath sampling performed at a higher flow rate, but the majority of other VOCs were not affected.

It has been shown that the flow rates pumping exhaled breathing patterns onto TD tubes do not significantly affect the measured concentration of VOCs [47], except acetone and phenol, the level of which decreased at higher flow rates. According to Doran et al. [47] the midrange flow rate (e.g., 200 mL min^{-1}) would be optimum for system performance.

It should be taken into account that in both methods exhaled air includes a portion of dead space air—the air from the nasopharynx, trachea, bronchi, where no gaseous exchange takes place between inhaled air and blood, and alveolar air originating from the lower airways, where this gaseous exchange occurs. Therefore, the concentration of the endogenous compounds that are of interest for diagnostics is relatively high only in alveolar air. The typical value of the volume of exhalation for an adult is approximately 500 mL, of which, 150 mL is the "dead volume" and 350 mL is the air from the alveoli.

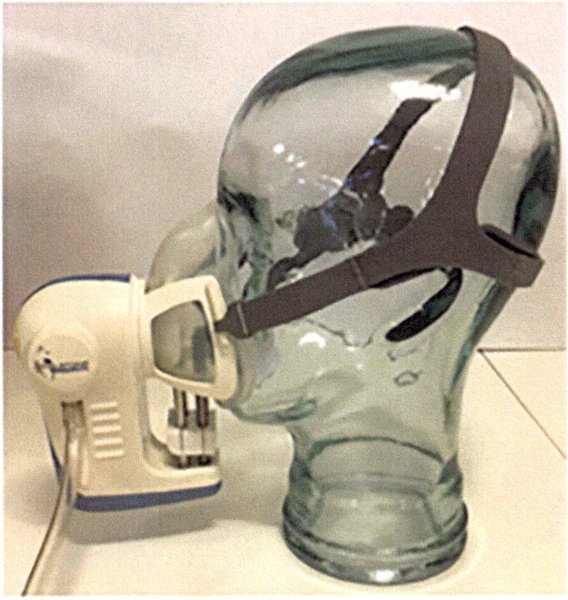

Fig. 6.3 Collection system using ReCIVA device [47]

To select an alveolar part of the breath, it is proposed to use a metal or plastic cylinder opened from the distal end, which allows the exhaled air to escape into the surrounding space [48]. The breath air collector has a cylindrical shape with a diameter of 2.4 cm and a length of 100 cm, wherein distant from the patient a portion of the breath sample in the cylinder is a "dead volume" and the near is the air from alveoli of lungs. There is a channel located in the near to the patient part of the cylinder, and only part of the sample corresponding to the air from the alveoli is taken.

In the cylindrical sampler for breathing tests [49], there are two membrane valves attached to the base 5, one for the inhaled air and the other for the exhaled air (see Fig. 6.4). When inhaling, the first membrane (1) opens, the second one (2) is closed, organizing the incoming flow through the holes (3). When exhaling, the first membrane closes, the second opens, forming a separate flow passing through the hole (4). The elastic properties of the second membrane are selected so that at the end of the exhalation it automatically closes, ending a separate respiratory cycle. Thus, there is no "dead volume" in the device.

The device for sampling exhaled air is shown in Fig. 6.5 [50]. The sampling container (8) includes a vacuum cell (7) hermetically sealed with a rubber stopper (6). The patient exhales air into the sample tube (1), which first inflates the bag (2) attached to the other end of the tube. The bag is designed to collect portions of air from the oral cavity, nasopharynx, trachea, bronchi ("dead space air"). The filling of the bag renders the removal of the uninformative portions of the exhaled air from the sample. A small hole in the bag (3) allows the patient to continue the exhalation process. The decrease in the exhalation rate allows more precise control of the process. After filling the bag, the patient or service personnel shall place the sample container in the discharge duct (5). The needle (4) available in the discharge channel

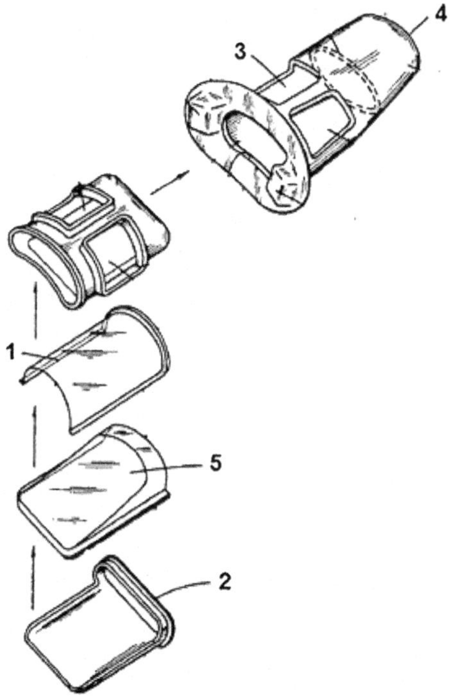

Fig. 6.4 Device for carrying out of respiratory tests [49]

pierces a rubber stopper, and a portion of air through the needle enters the cell due to vacuum. After disconnecting the container, the hole from the needle in the rubber stopper is closed due to the elasticity of the material.

Exhaled air can be sampled in two ways: mixed expiratory sampling and end-tidal sampling. Mixed expiratory sampling entails collecting total breath, including the air contained in the upper airways which experience no gas exchange with the blood. End-tidal sampling involves the collection of only end-tidal air, which contains most of the chemical information on blood composition. End-tidal sampling (collecting breath only at the end of exhalation) has proven successful because samples are less likely to be diluted by mixing with dead space volume (inspired air not taking place in gas exchange) and ambient air [51].

The need for standardization in sampling has been growing with the development in the field of breath research. Sampling devices for analysis of the exhaled air have to meet a number of requirements [52–54].

The following factors are critical for reproducibility of results of NO concentration measuring:

1. Exclusion of nasal NO. Closure of the velopharyngeal aperture during exhalation is one of the ways to minimize nasal NO leakage. This can be achieved by resistance to exhalation. It has been estimated that resistance to exhalation should be at least 5 cm of water column [55]. At the same time, pressures greater than 20 cm of the water column can be uncomfortable for the patient and should be avoided.

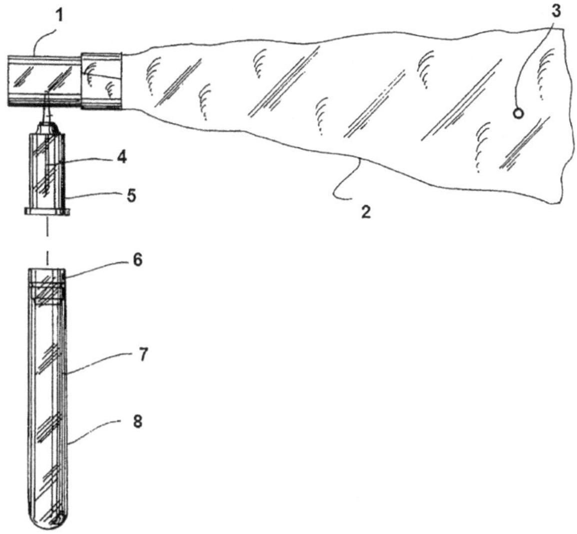

Fig. 6.5 Exhaled air sampling device [50]

2. Standardization of exhalation flow rate. Exhaled NO plateau values vary considerably with exhalation flow rate. Low flow rates (<0.1 L/s) amplify the measured NO concentrations. The flow rate of 0.05 L/s was found to be a reasonable compromise between measurement sensitivity and patient comfort [55].

Performance standards for equipment for single-breath determination of carbon monoxide uptake in the lungs were defined in [42]. The volume measurement accuracy should be the same as that determined by ATS/ERS for spirometry, that is, ±3%, regardless of a gas mixture, the direction of gas flow (e.g., inhaled or exhaled), or pulsatile flow pattern. Gas analyzer accuracy is important in some circumstances, such as measuring CO "back pressure" (the exhaled fraction of CO when no CO has been inhaled).

Only the ratio of the alveolar to inhaled CO and tracer gas is needed for calculation of the diffusing capacity of the lungs for CO (DLCO). Thus, the analyzers must primarily be able to produce an output for measured exhaled CO and tracer gas that is a linear extrapolation between the inhaled (test gas) concentrations and zero (no CO or tracer gas present in the analyzers).

Since the measured DLCO is very sensitive to errors in relative gas concentration, nonlinearity for the analyzers should not exceed 0.5% of full scale, i.e., once the analyzers have been adjusted to zero, with no test gas present and have scaled using test gas concentrations.

If CO_2 and/or H_2O interfere with the gas analyzer performance, this effect can be minimized by their removing from the test gases before they pass through the gas analyzer. The other remedy for CO_2 and/or H_2O analyzer interference is to characterize the effect of these gases on analyzer output aside, and then adjust the output of the analyzers for the presence of the interfering gas species.

Breath air collection for future off-line analysis can be realized by the following methods [56]:

1. Using sampling bags, such as Tedlar® gas sampling bags (PVF), Mylar gas sampling bags, and polyester aluminum (PEA) sampling bags. These sampling bags are cheap and chemically stable and can interface with clinical respiratory equipment. However, these bags may have the risk of leakage or VOC sorption (e.g., Tedlar® bags may be permeable to formaldehyde) and may suffer from ultraviolet degradation.
2. Flow reactor. This method is realized by exhaling air into a glass cylinder, which is linked to an analysis system. After each measurement, the cylinder can be purged with nitrogen for cleaning. As the sample volumes can be examined precisely every time, the reproducibility of this method is ensured. The cylinder is inert and can avoid water condensation. However, this equipment is expensive, and it requires a constant flow of inert gas, such as N_2. It is not suitable for sample storage.
3. Bio-VOC™ breath sampler (Fig. 6.6). This method is realized by exhaling into a one-way valve that is connected to a Teflon® bulb. After breath collection, the internal standard (IS) can be added into the device, and the exhaled VOCs and IS can be extracted by Solid-Phase Microextraction (SPME); the SMPE fiber should be put into the Bio-VOC™ for a certain period of time, and then thermally desorbed in the GC set-up. Bio-VOC™ is cheap, inert, and user-friendly. It can trap the last portion of exhaled air and avoid upper respiratory or oral contamination. But it can only collect 150 mL of end-tidal breath, so breath samples may vary according to patients' lung volume.
4. Phillips et al. [54] reported an example of the breath collection apparatus, which has a long-length tube as the breath reservoir, and a small-length tube affixed at the end as the sorbent trap to capture the VOCs. A flowmeter and a digital timer are also incorporated into the apparatus. This apparatus is portable and user-friendly. It can have separate traps and thus can collect different portions of the exhaled breath. Although it is portable and user-friendly, the size of this apparatus is quite large and the cost might be high.
5. Gastight syringe (GTS). The GTS is a widely used transfer medium for VOCs collection and analysis. The sorptive loss of the highly volatile compounds, such as aldehydes, ketones, esters, alcohols, and aromatic hydrocarbons, is significantly low. But conversely, it is not suitable for the collection of semi-volatile compounds, such as carboxyls and phenols, because there may be a sorptive loss due to contact with the inner surfaces of the GTS, and the sorptive losses will increase with the increase of molecular weight and boiling point of the VOCs.

Exhaled air is saturated with water vapor that often interferes with the measurement of the analyzed volatile components. The water vapor content can be decreased through bonding with chemical absorbers or freezing. The latter way seems to be more reasonable as it does not require any expendable materials. Based on the above, a system for gas mixture dehumidification and preparation of patient exhaled air samples was created [57]. The block diagram of the system air tract is shown in Fig. 6.7, the general view, in Fig. 6.8.

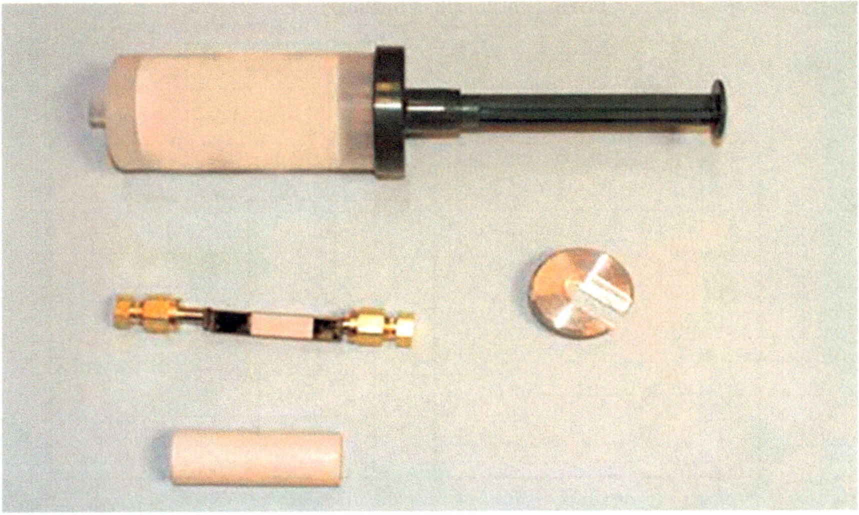

Fig. 6.6 The Bio-VOC™ breath sampler with Supelco solid-phase microextraction fiber holder 57,330 U

The system includes:

- dehumidification block;
- sampling and flow normalization block;
- differential manometer;
- air pump (as an initiator of air consumption).

The dehumidification block uses Peltier elements. The airflow rate through the removable tube significantly influences its final humidity. At a tube diameter of 6 mm, the output water vapor concentration does not exceed 1.5 g/m^3 at an input concentration of 15–17 g/m^3 and an airflow rate of 0.017–0.07 L/s, which corresponds to the recommended exhalation rates. An increase in the airflow rate results in increased humidity.

The block of sampling and normalization of the exhaled airflow is intended for separation of a required portion of air, exhaled by a patient, and providing for uniform air flowing at a rate required for gas analyzer operation. The spirometer calibrating injector, installed vertically, serves as the ballast reservoir. The ballast reservoir controls the uniformity of the airflow, entering the gas analyzer, during its periodic filling through the patient's respiration. Air is expelled from the injector to the line and further to the gas analyzer under the action of the load, pressing the valve. The load choice (0.5 or 1.5 kg) controls the resistance to the patient's exhalation and its frequency required for maintenance of the stationary measurement mode.

The use of two rotameters provides monitoring and measurement of the airflow rate in a range of 0.03–0.8 L/s.

The differential manometer is intended for measurement of the resistance to the patient's exhalation.

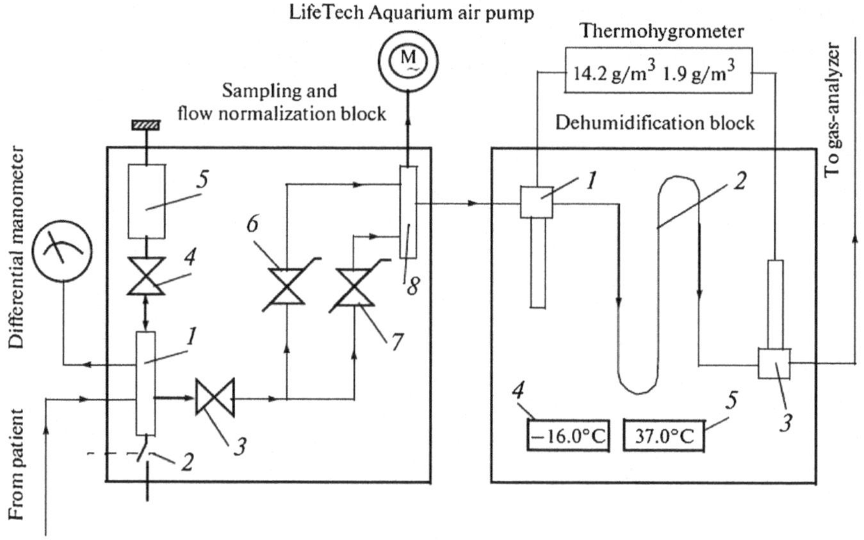

Fig. 6.7 Sampling and flow normalization block: input collector (1); exhaust valve (2); gate of air line (3); gate of ballast reservoir (4); ballast reservoir (5); low pass rotameter (6); high pass rotameter (7); outlet collector (8). Dehumidification block: input humidity sensor (1); cooled tube (2); output humidity sensor (3); thermometers for cooled and heated parts (4 and 5)

Fig. 6.8 General view of the device: air pump (1); differential manometer (2); sampling and flow normalization block (3); dehumidification block (4); thermo-hygrometer (5)

Water vapor condenses on cool surfaces potentially leading to the partial transfer of the volatile components from gas to liquid thereby distorting the measurement result. Due to recent technological advancements, the exhaled breath analysis has moved beyond measuring VOCs in the gas phase only into the measurement of semi-volatiles and dissolved compounds in aerosolized droplets in exhaled breath condensate (EBC) and in exhaled breath vapor (EBV). Aerosolized droplets in EBC can be captured by a variety of methods and analyzed for a wide range of biomarkers, such as metabolic end products, proteins, cytokines, and chemokines, with expanding possibilities. EBV sampling can detect additional compounds not detected in EBC and may provide greater sensitivity as a sampling method, expanding the spectrum of breath sampling [47].

One of the main sources of measured VOCs is exogenous compounds present in the exhaled air. To reduce their influence, one should discriminate impacts from compounds of exogenous origin and compounds of endogenous origin (i.e., produced inside the body by physiological or pathological metabolism).

"Clean air supply" is one of such methods when the inspired air has been passed through a carbon-based scrubber to minimize the impact of environmental contamination on the exhaled breath sample [45]. This will minimize the effect of environmental contamination from volatile compounds with rapid wash-out rates in the body but will not be the case for volatiles with longer retention time and from long term environmental exposure.

In other studies, subjects have been required to remain in the sampling environment for a set time to allow equilibration with the ambient air. Simultaneously, a background sample of the ambient air should be collected. A positive alveolar gradient (the difference between concentration in the breath and concentration in the ambient air) suggests that VOCs was produced *in vivo*, while a negative alveolar gradient indicates the source of the VOCs is external to the body [56]. Such procedures make the process more time-consuming for subjects and require a single sampling location for optimal comparisons of multiple subjects [45].

6.4 Methods of Data Analysis

The development of computer-aided diagnostic (CAD) systems has become a hot topic in recent years. This fact is due to recent advances in machine learning algorithms, namely Deep learning, Big data processing. Development of deep neural networks for image processing and recognition stimulated hardware development, including GPU, FPGA, ASICS devices. Open-source machine learning frameworks (TensorFlow, Keras, etc.) are available for large community usage and come with a lot of tutorials and examples which made them ready to work from a box. Many image data sets (CT, X-Ray, etc.) are available for processing by everybody so the threshold of entry is low now. One should mention that these methods are primarily focused on image processing, while signal processing of spectral data is not so easy to deal with. Data analysis of spectral data, for instance, requires special knowledge about data preprocessing (fitting and interpolation, noise removal) and feature selection along with classification. These pattern-recognition-based techniques provide probabilistic discrimination of biomarker profiles, which forms the basis for assessing diagnostic accuracy [58]. Machine learning allows one to discover functional relationships from examples based on features rather than from manual verification of entire experiments. Compared to conventional approaches, these methods are more efficient in handling multidimensional data analysis such as distinguishing phenotypes that are defined by a high number of features [59].

The key challenge to use machine learning for medical diagnosis is the existence of latent dependences between measured features set and human state variations due to pathological processes. On the other hand, medical diagnosis requires very high sensitivity and specificity of the trained classifier which is difficult to achieve because of lack of representativity of training data sets comparing with annotated image sets used by Google, Amazon for face and object recognition.

Digital presentation of an experimental data is a base for effective data preprocessing to remove artifacts, to select useful parts of an experimental data and to improve greatly the quality of the future analysis.

Typical digital data preprocessing includes denoising, smoothing, feature extraction, classification, and diagnosis.

Let us emphasize that objects under study are represented by respective mathematical models called feature vectors. Further discussion will briefly cover machine learning topic in application to the spectral data. Machine learning methods are of two kinds—supervised and unsupervised learning (also known as clusterization). Supervised learning implies that there exist mapped data with an expert evaluation of the object's belongings to some given class. Such data set allows developing classification rule. By applying this rule to newly obtained data, one can classify it to one of the aforementioned classes. The state-of-the-art methods to generate classification rule are Support Vector Machine [60], Naïve Bayes classifier [61], Artificial Neural Networks [62], Random forests [63], Boosting [64]. In contrast to classification, unsupervised learning assumes you do not have any expert evaluations; instead, you have a priori knowledge about a number of classes or their metric properties in the feature space. Unsupervised learning allows estimating data layout in the feature space that provides the potential possibility of class separability in some metric. These methods, for example, include k-means [65], hierarchical, and c-means clustering [66, 67]. Machine learning biomedical tests showed that Support Vector Machine with different kernels outperforms other methods [68]. Many researches demonstrated high classification precision of artificial neural networks, but there is a significant risk of overtraining, so special validation methods should be used to control training process in this case, for instance, k-fold validation algorithm [69]. As it was shown in [70], quality performance of classification rule can be increased by data preprocessing, allowing to remove outliers which make a negative impact to classification. It should be noted that there is a known problem of "curse of dimensionality" for high dimension data. It is occurring because of data features dimension increasing causes exponential grow of the volume of the training set for successful classification. That is why feature space dimension reduction often carries out before classification. Thereby, a combination of data preprocessing and dimension reduction can boost classifier overall accuracy. Besides the development of classification rule, researches often need to know what components of feature vector play most valuable role for classification. Special methods which address the problem of feature selection are based on random forest like BORUTA package [71] or extended data analysis via principal components analysis aggregated with criteria of significance scores based on explained variance are performed [72].

Let us discuss the preprocessing stage. Research on spectral data highly depends on acquisition device properties and often need aligning (interpolation) along the wavelength axis to make data comparable. Gathered data is high-dimensional and has oscillations with different amplitudes. Data noise includes the following sources: thermal noise in electronic units (modeled as additive white Gaussian noise), signal drift, background extraction noise, interference fridges in optical components, atmospheric pressure changes [73, 74]. A common strategy for noise removal is baseline correction, interference fridge removal, and high-frequency denoising [73]. Interference fridges are noticeable in tunable diode laser absorption spectrometers so these methods are out of our scope, but can be found in papers [75, 76]. Baseline correction is usually performed by data smoothing (e.g., Savitzky–Golay filter [77]) followed by polynomial fitting methods. The most popular high-frequency denoising approaches are rank filtration, wavelet, and Fourier filtering in the spectral domain.

Rank filters use a series of neighboring points, which define sliding window size. If the size of the window is odd, the filter is known as the median. The median filter is robust and has edge-preserving property. The main drawback of the rank filters is a difficulty of estimation of proper window size.

Another state-of-the-art method for data preprocessing is based on signal decomposition. These methods are very effective against noise with high frequencies and often include Fourier or wavelet filtration. Fourier transform-based filters are fast and effective, but they also remove meaningful information at high frequencies. Wavelet transform filters allow keeping specific waves in a signal, which look similar to mother wavelet. The main advantages of wavelet transformations are as follows [78].

Unlike the Fourier transform, wavelets can be well localized in time and frequency. Wavelets help to identify and describe some hidden signal characteristics, in particular, its symmetry. If it is necessary to analyze different information in the signal, wavelets allow one to consider the specified scale conversion levels (filtration). There is a wide range of mother wavelets with various degree of smoothness. The disadvantage of the wavelet signal processing is a relative computational complexity and the difficulty of the correct choice of the mother wavelet.

Another promising group of filtration technique is based on the solution of differential equations. For example, original, noisy image is considered as boundary conditions of 2D diffusion equation. By solving it numerically one can obtain a smoothed image, similar to the result of the application of Gauss filter to initial image. The methods of diffusion filtration (MDF) of 1-D and 2-D signals have been actively developing during the past two decades and currently offer a set of effective algorithms sufficient to extract the content-relevant information from the initial array of noisy and distorted data, i.e., allow within certain limits to manage the data processing depending on the required conditions. MDFs are of considerable interest in the analysis of medical images [79, 80]. The disadvantage of MDF is that it does not allow to both clear input signal from the noise and to preserve the content-relevant information, since Gaussian filtering simultaneously smooths out not only the noise but also informative part of the fast varied signal.

As noted above, the dimensionality reduction is one of the key steps in data preprocessing. In general, all dimensionality reduction methods can be classified into continuous linear, nonlinear, and discrete ones.

The Principal Component Analysis (PCA) [81] is a well-known and relatively simple linear algorithm of dimension reduction. This reduction is provided by a selection of a due number of principal components. The number of necessary principal components is defined by the level of explained variance of initial data like it is shown in Fig. 6.9.

Computation of principal components can be realized either by calculation of eigenvectors and eigenvalues of covariation matrix of input data or by singular value decomposition of the data. PCA does not always provide effective feature space dimension reduction. Straight lines and planes are not always providing a good approximation of the spatial distribution of data in a feature space. For instance, data can be aligned along some axis, though this line position might be very complex in feature space. To work with such curvilinear principle components, method of principal manifolds and another extension of PCA to the nonlinear case have been developed.

A number of nonlinear methods of dimension reduction have been developed, for example, geometric methods of features selection, the method of nonlinear dimensional reduction, the method of local isomaps [82, 83]. The latter allows one to reconstruct low-dimensional nonlinear structures in multidimensional data sets, but it is possible to lose significant information when the size of the neighborhood in the data set is larger than the distance between the elements of the structure.

The nonlinear kernel PCA is based on linear operations under data preliminary transformed by a nonlinear kernel. The Maximum Variance Unfolding (MVU) is based on the convex optimization of the objective function and is useful for multidimensional data analysis [84, 85].

The method of diffuse mappings is based on using the family of embeddings of the data set into a Euclidean space (possibly of minimal dimension), whose coordinates can be calculated with the help of eigenvectors and eigenvalues of the diffusion operator [86]. The Euclidean distance between points in an immersed manifold is interpreted as the "diffusion distance" between the probability distributions concentrated at these points. By combining the local similarities on different scales, the diffusion maps provide the global description of the data set. It should be noted that in comparison with other methods, the diffuse mapping algorithm is noise-proof.

Nonlinear methods are able to operate with complex varieties of the real data which are nonlinear in the sense that they do not form the linear space but can be regarded as a geometric manifold. In particular, this is the case of real data with strong nonlinear variety. It should be noted that nonlinear methods demonstrate efficiency on artificial data sets, but on the real data, the dimension reduction is less convincing since the application of this or another method depends on the nature of the analyzed data.

Discrete methods include the so-called filters, i.e., algorithms based on the selection of a subset of the original set of characteristics (Pearson's criterion, mutual information based on the Shannon information criterion and the Kullback–Leibler divergence). Among the discrete methods, there are the filters, the methods of "wrappers" (the classifier is considered as a black box with the input of the generated feature sets and the result of classification is evaluated), and "built-in" methods

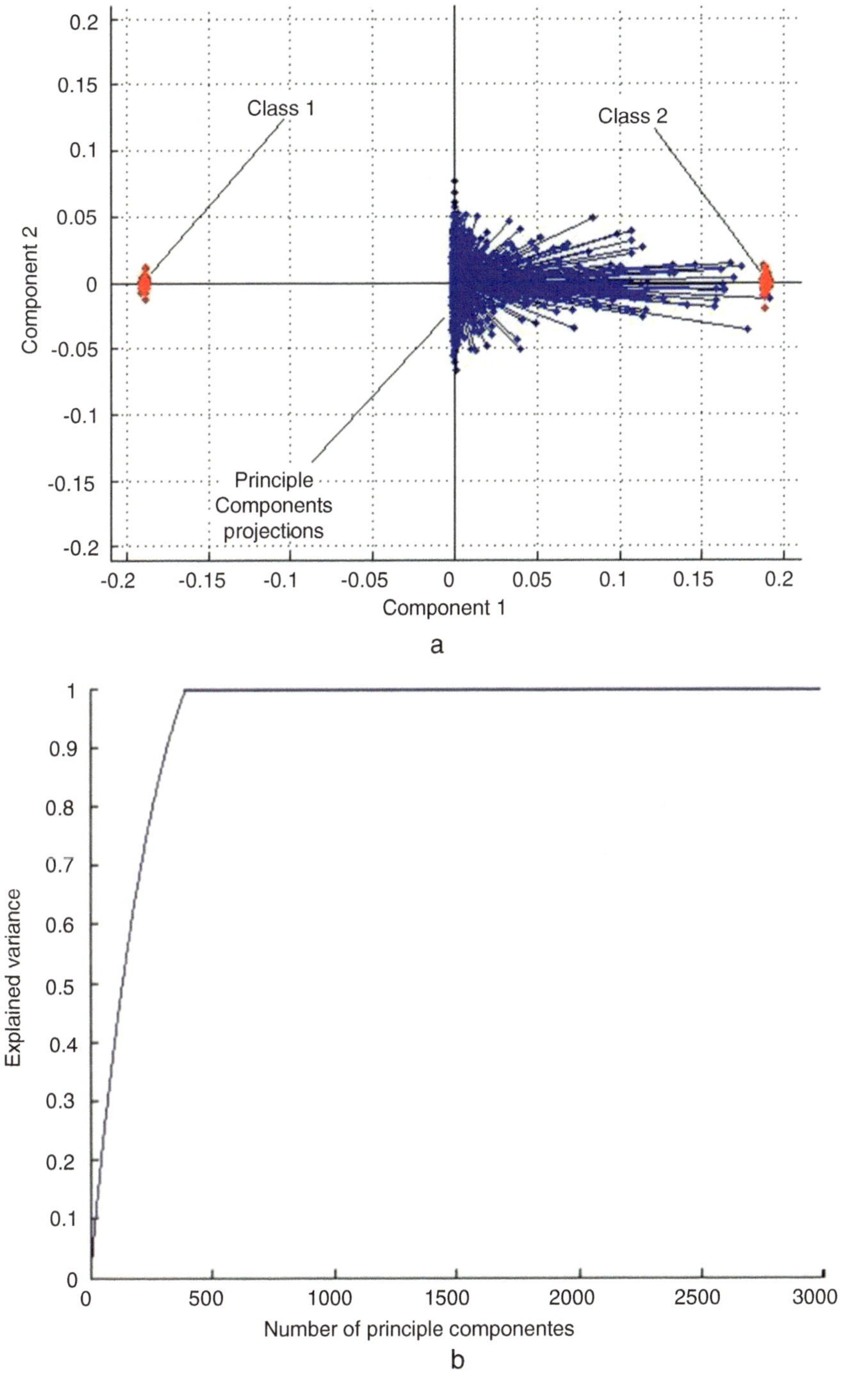

Fig. 6.9 PCA application example: projections of data on the first two principle components (**a**); the dependence of explained variance on the number of principal components (**b**)

that optimize the methods of "wrappers" to reduce the number of repeated classifications.

As aforementioned, the last step for medical diagnosis is the construction of a classification rule. Here, we focus on classifiers, which show the best accuracy for spectral data analysis.

Support vector machine (SVM) belongs to supervised learning methods, based on two object groups separation by a hyperplane in a feature space. Supervised learning is based on using the stage of algorithm training on a part of the initial data. The main objective of SVM is a selection of the optimal position of hyperplane which allows separating classes with maximal precision. To do it, separating hyperplane should be chosen in such way that distance between nearest points located by each side of these plane will be maximal. This distance is called gap and points—support vectors. In other words, separating hyperplane should provide a maximum gap for better confident class separation. When this task cannot be solved linearly, the "kernel trick" can be used. This implies the use of SVM with special functions similar to well-known kernels from theory if integral equations. Kernels allow projecting original feature space into another one with a higher dimension. The goal of this operation to make class linearly separable in a new higher dimension feature space.

Random forest method is an ensemble classifier (like AdaBoost and other boosting methods) proposed by Leo Breiman in 1999 [87]. Breiman defined Random forests as a classifier, which consist of an aggregation of decision trees $\{h(x, \theta_k), k = 1, \ldots\}$, where $\{\theta_k\}$ are independent identically distributed random vectors and each decision tree classifier votes for a most popular class for input vector x [88]. This algorithm is characterized by its simplicity of implementation and good generalization properties. Also, Random forests allow to obtain not only the object's mark of belongings to a given class, but even measure the confidence of classification.

Artificial neural networks (ANNs) reborn in 2010 in the form of convolutional neural networks with significant results in image recognition using a huge number of images in teaching set (field of deep learning and big data). Benefits of ANNs are connected with their property to adapt to input data without specification needs and ability to approximate any function with a given precision. ANNs transform data nonlinearly that gives us a very flexible tool for complex natural data modeling. Besides, ANNs can estimate a posterior probability and can be used for the generation of statistical classification rules for medical diagnostics. ANNs may consist of an arbitrary number of neurons, grouped into single or multiple hidden layers. The choice of ANN's configuration depends on a specific task and may vary from tens to thousands in spectra classification tasks [89]. Originally, most popular ANNs were based on Kohonen's self-organizing maps and backpropagation [90].

The main drawback of ANNs is non-availability of guarantee positive training result because there is no explicit way of choosing an algorithm's configuration parameters to get a proper result. But, it is compensated by the effectiveness of big data classification.

Methodological problems of data analysis and different approaches to their solution were subjects of several discussions [91]. A summary of these discussions is presented in Table 6.2, where: N is a sample size, S is the specificity, P is the sensitivity, p is the feature vector dimension.

To train the classifier, a part of the data should be used (training set). When classifiers are trained, there is a danger that the classifier will be too well adjusted for the training set, which will lead to the impossibility of correctly classifying new (unseen) data. This problem is called "overtraining" or "overfitting" of the classifier. Deciding on the quality of the resulting classifier on the basis of a test on the training set may lead to the fact that retraining (if it exists) may not be detected. A more adequate approach is to use a test suite-a set of data classified by class, but not used in the training process.

In spite of examples of successful application of machine learning for medical diagnosis, there are risks associated with applying these methods as a "black box" to perform diagnosis. A flexible learning system in high-dimensional feature space can behave unexpectedly and this can be difficult to detect [92]. Thus, an instrumental or computer-stage reduction of a feature space under controlled conditions to understand driven factors for pathological stage data variations is very important.

6.5 Results and Discussion

The results of exhaled breath samples (EBS) analysis using GC-MS and LPAS with extra-wide tuning range in combination with machine learning for lung cancer diagnosis are presented.

The study has been involved lung cancer (LC) patients (n = 18), patients with chronic obstructive pulmonary disease (COPD) (n = 22), patients with pneumonia (n = 21), and a control group of healthy nonsmoking volunteers (n = 39). Diagnoses have been established by standard clinical methods. Patients with severe comorbidities or an unconfirmed clinical diagnosis were excluded from the study.

COPD patients were men of the average age 67.8 ± 9.7 years; 10 from 12 are smokers with average smoking of 42 ± 13 years. LC patients were men of average age 61.5 ± 4.8 years; 8 of 9 patients are smokers with average smoking of 44.9 ± 8.2 years. The average age of healthy volunteers in a control group was 21.5 ± 1.6 years. Smokers and individuals with a disease of the bronchopulmonary, cardiovascular, digestive, endocrine, reproductive, and urinary system were excluded from the control group.

Research Protocol was approved by the Ethics Committee of the Siberian State Medical University (Tomsk, Russia).

All measurements were carried out at room temperature (variations were 20–25 °C) and humidity (50–60%). The sampling was produced before eating or 2 h thereafter. Just before sampling, participants rinsed the mouth with pure water without any special cleaning of the oral cavity. The EBS were collected in disposable plastic containers (syringe) with a volume of 150 mL and analyzed using the LaserBreeze gas analyzer. Additionally, EBS were collected into the Bio-VOC breath sampler with Supelco solid-phase microextraction fiber holder 57,330 U

Table 6.2 Statistical methods for data analysis [91]

Method	Applicability	S/P	*N*/*p*
Statistical hypothesis testing, t-test	Test of morbidity as a null hypothesis; if multiple testing is performed Mann–Whitney U-test method is applied	S	*N* depends on testing conditions $p = 1$
Rank tests, Spearman's rank correlation, Pearson's chi-squared test	Small samples ($N < 30$) can determine rank data scale	–	$p = 2$
Principal components analysis (PCA)	Multivariate method of feature selection, which allows reducing feature vector dimension by removing non-informative components	–	Usually, $N/p > 10$, but can be applied when $N < p$
Linear discriminant analysis (LDA)	Decision-making method based on a finding of linear combinations of feature vectors components most suitable for object separation in feature space. The method is close to PCA	S(P)	Usually $N > p$, but the method can be extended for cases when $N < p$
Decision trees	An intuitively simple prognostic model of disease based on binary feature separation. Simplicity is achieved by lowering the precision of prediction. Algorithms, as a rule, include inner cross-validations or training/testing data set for decision-making. Method's main drawback is a high dispersion: Relatively small modifications of input data lead to great output changes in prediction.	P	$p > N$
Classifiers of feature space (SVM, RVM, etc.)	Applied for decision-making about types of diseases. These methods are based on splitting the set of feature vectors into classes. Data preprocessing can significantly improve prediction precision. The method is computationally complex, yet it produces better results than decision trees and rank methods	P	$p > N$
Canonical correlational analysis (CCA)	Applied to solve one of the most complex statistical problems—Determination of relations between two or more set of feature vectors, which describe an object and independent detectors. This method allows increasing precision and reliability of the results, obtained by other means. Application in combination with PCA can significantly improve the precision and quality of disease prediction results. By computing principle components for each set of variables and correlations between obtained components, one can reveal complex relations between these sets	P	$p > N$

(Fig. 6.10). A participant did some calm breaths through a sterile plastic tube into the sample container. The "dead volume" was exhaled outside the container.

The EBS from the Bio-VOC breath sampler were analyzed by gas chromatograph Finnigan Trace GC with MS detector Finnigan Trace DSQ. The extraction time was 30 min.

Qualitative determination of VOCs in EBS was carried out under the following conditions: ionization method was electron impact; Supel-QTM column PLOT (manufacturer Thermo Scientific) of 30 m long and 0.32 mm inner diameter was used; evaporator temperature −200 °C; interface temperature −200 °C; temperature of thermostat −40 °C during 1 min, then temperature was increased to 250 °C with speed 10 °C/min; carrier gas was helium; range of masses scan was 50–650 a.e.m.

To validate the suitability of the Bio-VOC breath sampler and plastic containers for the sampling of the EBS, we had filled both containers by nitrogen of 99% purity and analyzed the content by GC-MS technique. No peaks have been observed on chromatograms, which indicates that used samplers do not contribute any additional substances.

GC-MS identification of EBS substances was carried out according to their retention time by comparison of measured chromatograms with the spectra from the NIST MS Search 2.0 library. Standard gas samples were used for calibration. Examples of typical chromatograms for EBS for a lung cancer patient, COPD patient, and a healthy volunteer are presented in Figs. 6.10, 6.11, and 6.12, respectively.

The square of chromatographic peak was considered as a substance concentration. Mean values of VOCs concentrations in EBS for groups under study are presented in Fig. 6.13. Concentrations of Benzene are expressed as 10^{-2}.

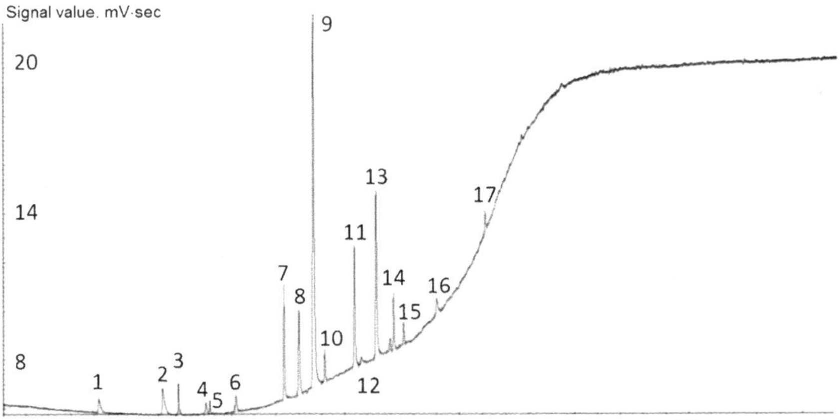

Fig. 6.10 The chromatogram of exhaled air of a lung cancer patient: 1-methanol; 2-ethanol; 3-acetonitrile; 4-acetone; 5-methylene chloride; 6-pentane; 7-ethyl acetate; 8-hexane; 9-benzene; 10-chloropropylene oxide; 11-N-ethylformamide; 12-octane; 13-toluene; 14-butyl acetate; 15-chlorobenzene; 16-o-xylene; 17-decane

According to the results, ethanol, chloropropylene oxide, and N-ethylformamide are not detected in EBS of pneumonia patients, chloropropylene oxide is not detected in EBS of COPD patients; chloropropylene oxide, O-xylene were detected only in EBS of LC patients. Methylene chloride, pentane, acetonitrile, toluene are inherent for all groups but with various concentrations, therefore, they can be used only in a pattern-recognition approach.

The SVM with Gaussian radial basis function (RBF) kernel [60] was used to construct a predictive model for the binary classification using EBS analysis by GC-MS. The results of the model validation are presented in Table 6.3 [93].

The LPAS analysis of EBS was carried out by the LaserBreeze gas analyzer. The machine learning pipeline was realized in three steps. First, the informative feature extraction from absorption spectra of EBS have been produced using PCA processing. Second, the binary SVM "One-vs-One" classification [94] has been carried out [95]. The "bootstrap aggregation" method was used to improve the quality of the model. According to this approach, a random separation of initial data on teaching and testing sets have been repeated. Then the results of testing of a set of the predictive models were averaged. The model validation results are shown in Table 6.4.

Finally, "One-vs-One" classifiers have been used to construct the differential diagnosis by enumeration of these classifiers for the feature vector of an object under study. The diagnostic decision was based on the output which was selected two and more times. If this condition was not met, the diagnosis was considered as not set (see Table 6.5).

Lung cancer diagnosis accuracy was 95.7% that confirms the high quality of the proposed predictive model.

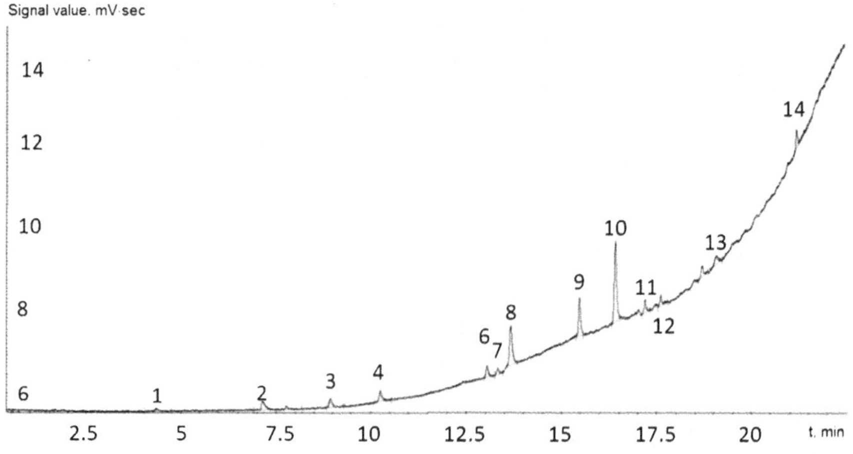

Fig. 6.11 The chromatogram of exhaled air of a COPDpatient: 1-methanol; 2-ethanol; 3-acetonitrile; 4-acetone; 5-pentane; 6-hexane; 7-benzene; 8-N-ethylformamide; 9- toluene; 10-butyl acetate; 11-chlorobenzene; 12-o-xylene; 13-decane

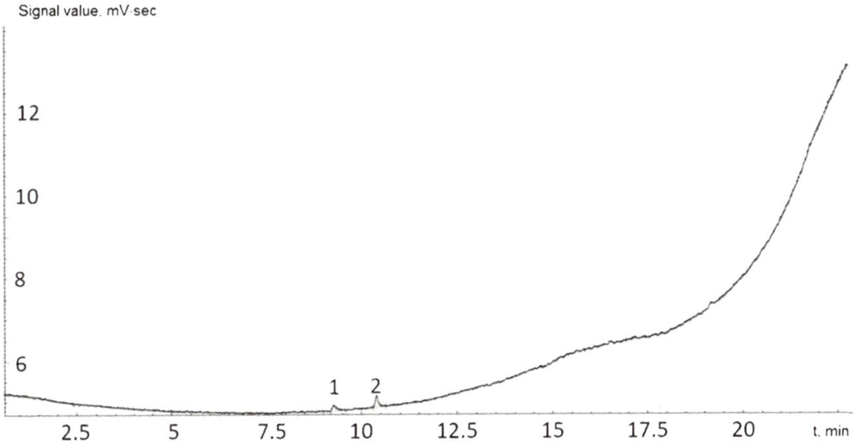

Fig. 6.12 The chromatogram of exhaled air of a healthy volunteer: 1-methylene chloride; 2-pentane

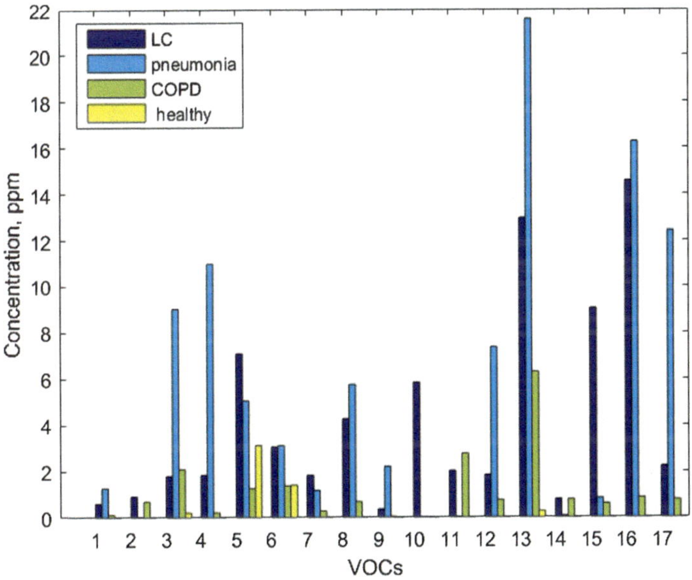

Fig. 6.13 VOCs concentrations in EBS: 1-Methanol, 2-Ethanol, 3-Acetonitrile, 4-Acetone, 5-Methylene chloride, 6-Pentane, 7-Ethyl acetate, 8-Hexane, 9-Benzene∗, 10-Chloropropylene oxide, 11- N-ethylformamide, 12-Octane, 13-Toluene, 14-Butyl acetate, 15-Chlorobenzene, 16-O-xylene, 17-Decane

Table 6.3 The results of validation of the predictive model for binary classification of the groups under study [93]

Groups	Kernel parameter value	Sensitivity		Specificity	
		Mean	Dispersion	Mean	Dispersion
LC—COPD	3.5803	0.9620	0.0094	0.9900	0.0024
LC—Pneumonia	18.805	0.9143	0.0339	0.9463	0.0156
LC—Healthy volunteers	0.6560	0.9400	0.0116	0.9999	0.0001

Table 6.4 The results of RBF kernel "One-vs-One" SVM models validation [95, 96]

Pairwise classification	Kernel parameters	Sensitivity		Specificity	
		Mean value	Dispersion	Mean value	Dispersion
COPD-Pneumonia	1.2041	0.95	0.0016	0.95	0.0012
Pneumonia- healthy volunteers	0.5641	0.96	0.0009	0.92	0.0019
COPD-healthy volunteers	1.2414	0.86	0.0022	0.83	0.0020
LC- Pneumonia	0.7152	0.96	0.0014	0.93	0.0012
LC- COPD	1.2216	0.98	0.0003	0.94	0.0007
LC-healthy volunteers	0.2698	0.96	0.0011	0.90	0.0013

Table 6.5 Results of differential diagnosis of the representatives from testing set [95]

Group	Diagnosis					
	Set right		Set not right		Did not set	
	Mean	Dispersion	Mean	Dispersion	Mean	Dispersion
Lung cancer	0.9565	0.0013	0.0341	0.0011	0.0094	0.0013
COPD	0.8112	0.0091	0.0981	0.0082	0.0907	0.0047
Pneumonia	0.8412	0.0048	0.0991	0.0032	0.0597	0.0025
Healthy volunteers	0.8946	0.0038	0.0901	0.0024	0.0153	0.0018

6.6 Conclusion

EBS content analysis is a promising tool screening diagnosis. Taking into account low specificity of typical molecular biomarkers in EBS, VOCs profile analysis using pattern-recognition-based methods should be applied. We used IR LPAS and GC-MS methods to provide spectral analysis of EBS. According to the typical machine learning pipeline, the analysis of measured spectra was based on reduction of the dimension of the feature space using principal component analysis; after that, the dichotomous classification was carried out using a Support Vector Machine. The estimated average sensitivity of exhaled breath sample analysis by the LPAS in dichotomous classification was not worse than 90%, the analogous results of analysis by GC-MS were 68%. The accuracy of multiclass classification of patients with several pulmonary diseases using a set of binary SVM "One-vs-One" classifiers based on EBS IR absorption spectra data is high enough for use in routine practices, especially for screening tests.

The future steps in bringing this technology to clinics should include the design of cost-effective and informative measurement devices, for example, specialized medical LPAS equipment, standardization of the sampling, accumulation spectral information about EBS of patients with confirmed a diagnosis and finding and discovery of most specific profiles of biomarkers, development of effective methods of data analysis and classification.

References

1. Siegel, R.L., Miller, K.D., Jemal, A.: Cancer statistics. CA A Cancer J Clin. **69**, 7–34 (2019). https://doi.org/10.3322/caac.21551
2. Torre, L.A., Siegel, R.L., Jemal, A.: Lung cancer and personalized medicine. In: Ahmad, A., Gadgeel, S. (eds.) Advances in Experimental Medicine and Biology, Lung Cancer Statistics, vol. 893, pp. 1–19. Springer, Cham (2016). https://doi.org/10.1007/978-3-319-24223-1_1
3. Beale, D.J., Jones, O.A.H., Karpe, A.V., Dayalan, S., Yuan Oh, D., Kouremenos, K.A., Ahmed, W., Palombo, E.A.: A review of analytical techniques and their application in disease diagnosis in breathomics and salivaomics research. Int. J. Mol. Sci. **18**(1), 24 (2017). https://doi.org/10.3390/ijms18010024
4. Cao, W., Duan, Y.: Breath analysis: potential for clinical diagnosis and exposure assessment. Clin. Chem. **52**(5), 800–811 (2006). https://doi.org/10.1373/clinchem.2005.063545
5. Smith, D., Amann, A.: Breath Analysis for Clinical Diagnosis and Therapeutic Monitoring. World Scientific, Singapore (2005). https://doi.org/10.1142/9789812701954
6. Ulanowska, A., Kowalkowski, T., Trawinska, E., Buszewski, B.: The application of statistical methods using VOCs to identify patients with lung cancer. J. Breath Res. **5**, 046008 (2011)
7. Buszewski, B., Ulanowska, A., Ligor, T., Denderz, N., Amann, A.: Analysis of exhaled breath from smokers, passive smokers and non-smokers by solid-phase microextraction gas chromatography/mass spectrometry. Biomed. Chromatogr. **23**(5), 551–556 (2009)
8. Zhou, J., Huang, Z.-A., Kumar, U., Chen, D.D.Y.: Review of recent developments in determining volatile organic compounds in exhaled breath as biomarkers for lung cancer diagnosis. Anal. Chim. Acta. **996**, 1–9 (2017). https://doi.org/10.1016/j.aca.2017.09.021
9. Gleeson, K., Phillips, M., et al.: Volatile organic compounds in breath as markers of lung cancer: a cross-sectional study. Lancet. **353**(9168), 1930–1933 (1999)
10. Kischkel, S., Miekisch, W., Sawacki, A., Straker, E.M., Trefz, P., Amann, A., Schubert, J.K.: Breath biomarkers for lung cancer detection and assessment of smoking related effects — confounding variables, influence of normalization and statistical algorithms. Clin. Chim. Acta. **411**, 1637–1644 (2010)
11. Song, G., Qin, T., et al.: Quantitative breath analysis of volatile organic compounds of lung cancer patients. Lung Cancer. **67**(2), 227–231 (2010)
12. Oguma, T., Nagaoka, T., Kurahashi, M., Kobayashi, N., Yamamori, S., Tsuji, C., et al.: Clinical contributions of exhaled volatile organic compounds in the diagnosis of lung cancer. PLoS One. **12**(4), e0174802 (2017). https://doi.org/10.1371/journal.pone.0174802
13. Poli, D., Carbognani, P., Corradi, M., Goldoni, M., Acampa, O., Balbi, B., Bianchi, L., Rusca, M., Mutti, A.: Exhaled volatile organic compounds in patients with non-small cell lung cancer: cross sectional and nested short-term follow-up study. Respir. Res. **6**(1), 71 (2005). https://doi.org/10.1186/1465-9921-6-71

14. Chen, X., Cao, M., Li, Y., Hu, W., Wang, P., Ying, K., Pan, H.: A study of an electronic nose for detection of lung cancer based on a virtual SAW gas sensors array and imaging recognition method. Meas. Sci. Technol. **16**(8), 1535–1546 (2005)
15. Yu, H., Xu, L., Wang, P.: Solid phase microextraction for analysis of alkanes and aromatic hydrocarbons in human breath. J. Chromatogr. B. **826**, 69–74 (2005)
16. van der Schee, M.P., et al.: Breathomics in lung disease. Chest. **147**(1), 224–231 (2015)
17. Altorki, N., Phillips, M., et al.: Detection of lung cancer using weighted digital analysis of breath biomarkers. Clin. Chim. Acta. **393**(2), 76–84 (2008)
18. Ager, C., Bajtarevic, A., et al.: Noninvasive detection of lung cancer by analysis of exhaled breath. BMC Cancer. **9**, 348 (2009)
19. Laskowski, D., Deffenderfer, O., Machado, R.F., et al.: Detection of lung cancer by sensor array analyses of exhaled breath. Am. J. Respir. Crit. Care Med. **171**, 1286–1291 (2005)
20. Mochalski, P., Ruzsanyi, V., Broza, Y.Y., et al.: Assessment of the exhalation kinetics of volatile cancer biomarkers based on their physicochemical properties. J. Breath Res. **8**(1), 016003 (2014)
21. van de Kant, K.D.G., van der Sande, L.J.T.M., Jöbsis, Q., van Schayck, O.C.P., Dompeling, E.: Clinical use of exhaled volatile organic compounds in pulmonary diseases: a systematic review. Respir. Res. **13**(117), (2012). https://doi.org/10.1186/1465-9921-13-117
22. Koczulla, R., Dragonieri, S., Schot, R., Bals, R., Gauw, S.A., Vogelmeier, C., Rabe, K.F., Sterk, P.J., Hiemstra, P.S.: Comparison of exhaled breath condensate pH using two commercially available devices in healthy controls, asthma and COPD patients. Respir. Res. **10**, 78 (2009)
23. Obersteiner, F., Bönisch, H.A.: Engel an automated gas chromatography time-of-flight mass spectrometry instrument for the quantitative analysis of halocarbons in air. Atmos. Meas. Tech. **9**, 179–194 (2016)
24. Schwarz, K., Filipiak, W., Amann, A.: Determining concentration patterns of volatile compounds in exhaled breath by PTR-MS. J. Breath Res. **3**, 027002 (2009). https://doi.org/10.1088/1752-7155/3/2/027002
25. Lindinger, W., Hansel, A., Jordan, A.: On-line monitoring of volatile organic compounds at pptv levels by means of proton-transfer-reaction mass spectrometry (PTR-MS) medical applications, food control and environmental research. Int. J. Mass Spectrom. Ion Process. **173**(3), 191–241 (1998)
26. Cumeras, R., Figueras, E., Davis, C.E., Baumbach, J.I., Gràcia, I.: Review on ion mobility spectrometry. Part 1: current instrumentation. Analyst. **140**(5), 1376–1390 (2015). https://doi.org/10.1039/c4an01100g
27. Wilson, A.D., Baietto, M.: Advances in electronic-nose technologies developed for biomedical applications. Sensors. **11**, 1105–1176 (2011). https://doi.org/10.3390/s110101105
28. Bikov, A., et al.: Established methodological issues in electronic nose research: how far are we from using these instruments in clinical settings of breath analysis? J. Breath Res. **9**, 034001 (2015)
29. Alphus, M., Wilson, D.: Advances in electronic-nose technologies for the detection of volatile biomarker metabolites in the human breath. Metabolites. **5**, 140–163 (2015). https://doi.org/10.3390/metabo5010140
30. Fernandes, S., Venkatesh, B., Sudarshan, G.: Early detection of lung cancer using nano-nose - a review. Open Biomed. Eng. J. **9**, 228–233 (2015)
31. Gelperin, A., Johnson, A.T.C.: Nanotube-based sensor arrays for clinical breath analysis. J. Breath Res. **2**, 037015 (2008). https://doi.org/10.1088/1752-7155/2/3/037015
32. Kuske, M., Romain, A.-C., Nicolas, J.: Microbial volatile organic compounds as indicators of fungi. Can an electronic nose detect fungi in indoor environments? Build. Environ. **40**(6), 824–831 (2005)
33. Greiter, M.B., Keck, L., Siegmund, T., Hoeschen, C., Oeh, U., Paretzke, H.G.: Differences in s. Diabetes Technol. Ther. **12**(6), 455–463 (2010)
34. O'Keefe, A., Deacon, D.A.G.: Cavity ring-down optical spectrometer for absorption measurements using pulsed laser sources. Rev. Sci. Instrum. **59**, 2544 (1988). https://doi.org/10.1063/1.1139895

35. Chow, K.K., Short, M., Zeng, H.: A comparison of spectroscopic techniques for human breath analysis. Biomed. Spectrosc. Imaging. **1**, 339–353 (2012)
36. de Gouw, J.A., Te, S., Hekkert, L., Mellqvist, J., Warneke, C., Atlas, E.L., Fehsenfeld, F.C., Fried, A., Frost, G.J., Harren, F.J.M., et al.: Airborne measurements of ethene from industrial sources using laser photo-acoustic spectroscopy. Environ. Sci. Technol. **43**(7), 2437–2442 (2009)
37. Bijnen, F.G.C., Reuss, J., Harren, F.J.M.: Geometrical optimization of a longitudinal resonant photoacoustic cell for sensitive and fast trace gas detection. Rev. Sci. Instrum. **67**, 2914 (1996)
38. Miklós, A., Hess, P., Bozóki, Z.: Application of acoustic resonators in photoacoustic trace gas analysis. Rev. Sci. Instrum. **72**(4), 1937–1955 (2001)
39. Zéninari, V., et al.: Photoacoustic detection of methane in large concentrations with a Helmholtz sensor: simulation and experimentation. Int. J. Thermophys. **37**(1), 1–11 (2016)
40. Zéninari, V., et al.: Helmholtz resonant photoacoustic cell for spectroscopy of weakly absorbing gases and gas analysis. Atmos. Oceanic Opt. **12**(10), 928–940 (1999)
41. Karapuzikov, et al.: LaserBreeze gas analyzer for noninvasive diagnostics of air exhaled by patients. Phys. Wave Phenom. **22**(3), 189–196 (2014)
42. Lee, C.-M., et al.: High-sensitivity laser photoacoustic leak detector. Opt. Eng. **46**(6), 065002 (2007)
43. MacIntyre, N., Crapo, R.O., Viegi, G., et al.: Standardisation of the single-breath determination of carbon monoxide uptake in the lung. Eur. Respir. J. **26**, 720–735 (2005)
44. de Lacy Costello, B., Miekisch, W., Schubert, J., Amann, A., et al.: The human volatilome: volatile organic compounds (VOCs) in exhaled breath, skin emanations, urine, feces and saliva. J. Breath Res. **8**(3), 034001 (2014)
45. Martin, A.N., Farquar, G.R., Jones, A.D., Frank, M.: Human breath analysis: methods for sample collection and reduction of localized background effects. Anal. Bioanal. Chem. **396**, 739–750 (2010). https://doi.org/10.1007/s00216-009-3217-7
46. A. Gordin, A. Amirav, SnifProbe: new method and device for vapor and gas sampling, J. Chromatogr. A V. 903, 1–2, 2000, p. 155-172, https://doi.org/10.1016/S0021-9673(00)00877-3.
47. Doran, S.L.F., Romano, A., Hanna, G.B.: Optimisation of sampling parameters for standardised exhaled breath sampling. J. Breath Res. **12**, 016007 (2018). https://doi.org/10.1088/1752-7163/aa8a46
48. M. Phillips, US Patent 6,726,637 B2: Breath collection apparatus, 6 Dec 2001
49. McKenna, US Patent 5255687: Zero dead space respiratory exercise valve, 21 Jan 1992
50. S. D. Hamilton, US Patent 5467776: Air sampling device and method for sampling exhaled air, 28 Jul 1993
51. Francesco, F.D., Loccioni, C., Fioravanti, M., Russo, A., Pioggia, G., et al.: Implementation of fowler's method for end-tidal air sampling. J. Breath Res. **2**, 037009 (2008)
52. Kistenev, Y.V., Karapuzikov, A.A.: Methods of spectral analysis of exhaled air suitable for routine diagnostics of diseases of the respiratory system. Adv. Biomat. Dev. Med. **2**, 79–87 (2015)
53. Paschke, K.M., Mashir, A., Dweik, R.A.: Clinical applications of breath testing. Med. Rep. **2**, 56–62 (2010)
54. Phillips, m.: method for the collection and assay of volatile organic compounds in breath. Anal. Biochem. **247**, 272–278 (1997)
55. Silkoff, P.E.: Recommendations for standardized procedures for the online and offline measurement of exhaled lower respiratory nitric oxide and nasal nitric oxide in adults and children. Am. J. Respir. Crit. Care Med. **160**(6), 2104–2117 (1999)
56. Sun, X., Shao, K., Wang, T.: Detection of volatile organic compounds (VOCs) from exhaled breath as noninvasive methods for cancer diagnosis. Anal. Bioanal. Chem. **40**, 2759–2780 (2016). https://doi.org/10.1007/s00216-015-9200-6
57. Kistenev, Y.V., Kuryak, A.N., Makogon, M.M., Ponomarev, Y.N.: The system for dehumidifi cation of samples in laser gas analysis. Atmos. Oceanic Opt. **25**(1), 92–95 (2012). https://doi.org/10.1134/S102485601201

58. Kistenev, Y.V., Borisov, A.V., Kuzmin, D.A., Penkova, O.V., Kostyukova, N.Y., Karapuzikov, A.A.: Exhaled air analysis using wideband wave number tuning range infrared laser photoacoustic spectroscopy. J. Biomed. Opt. **22**(1), 017002 (2017). https://doi.org/10.1117/1.JBO.22.1.017002
59. Casiraghi, E., Huber, V., Frasca, M., Cossa, M., Tozzi, M., et al.: A novel computational method for automatic segmentation, quantification and comparative analysis of immunohistochemically labeled tissue sections. BMC Bioinform. **19**(10), 357, 75–357, 91 (2018). https://doi.org/10.1186/s12859-018-2302-3
60. Vapnik, V., Cortes, C.: Support Vector Networks. Kluwer Academic Publishers, Boston (1995)
61. Murphy, K.P., et al.: Naive Bayes Classifiers, vol. 18. University of British Columbia, Vancouver (2006)
62. Khan, J., et al.: Classification and diagnostic prediction of cancers using gene expression profiling and artificial neural networks. Nat. Med. **7**(6), 673 (2001)
63. Pal, M.: Random forest classifier for remote sensing classification. Int. J. Remote Sens. **26**(1), 217–222 (2005)
64. Scholz, M., Klinkenberg, R.: Boosting classifiers for drifting concepts. Intell Data Anal. **11**(1), 3–28 (2007)
65. Hartigan, J.A., Wong, M.A.: Algorithm AS 136: A k-means clustering algorithm. J Royal Statis Soc Ser. C (Appl. Statis.). **28**(1), 100–108 (1979)
66. Johnson, S.C.: Hierarchical clustering schemes. Psychometrika. **32**(3), 241–254 (1967)
67. Bezdek, J.C., Ehrlich, R., Full, W.: FCM: The fuzzy c-means clustering algorithm. Comput. Geosci. **10**(2-3), 19–203 (1984)
68. Borisov, A.V., et al.: Development of Classification Rules for a Screening Diagnostics of Lung Cancer Patients Based on the Spectral Analysis of Metabolic Profiles in the Exhaled Air. In: Proceedings of the Scientific-Practical Conference" Research and Development-2016", pp. 573–580. Springer, Cham (2018)
69. Amari, S., et al.: Asymptotic statistical theory of overtraining and cross-validation. IEEE Trans. Neural Netw. **8**(5), 985–996 (1997)
70. Kistenev, Y.V., et al.: Wavelet based de-noising of breath air absorption spectra profiles for improved classification by principal component analysis. AIP Conf. Proc. **1688**(1), 030010 (2015)
71. Kursa, M.B., Rudnicki, W.R.: Feature selection with the Boruta package. J. Stat. Softw. **36**(11), 1–13 (2010)
72. Kistenev, Y.V., et al.: Applications of principal component analysis to breath air absorption spectra profiles classification. In: International Conference on Atomic and Molecular Pulsed Lasers XII, vol. 9810, p. 98101Y (2015)
73. Zhang, T., et al.: Mathematical methods and algorithms for improving near-infrared tunable diode-laser absorption spectroscopy. Sensors. **18**(12), 4295 (2018)
74. Feher, M., Martin, P.A.: Tunable diode laser monitoring of atmospheric trace gas constituents. Spectrochim. Acta A Mol. Biomol. Spectrosc. **51**(10), 1579–1599 (1995)
75. Yang, R., et al.: A dual-tone modulation method to reduce the background fluctuation in tunable diode laser absorption spectroscopy. Optik-Int. J. Light Electron Optics. **142**, 608–614 (2017)
76. Kagawa, N., Osami, W., Koga, R.: Suppression of the etalon fringe in absorption spectrometry with an infrared tunable diode laser. Opt. Eng. **36**(9), 2586–2593 (1997)
77. Vivó-Truyols, G., Schoenmakers, P.J.: Automatic selection of optimal Savitzky–Golay smoothing. Anal. Chem. **78**(13), 4598–4608 (2006)
78. Addison, P.S.: The Illustrated Wavelet Transform Handbook: Introductory Theory and Applications in Science, Engineering, Medicine and Finance, p. 368. CRC Press, Boca Raton (2002)
79. Zhang, D., Yan, P., Suzuki, K., Shen, D., Wang, F.: Machine Learning in Medical Imaging: First International Workshop Machine Learning in Medical Imaging, p. 262. Springer, Berlin (2013)

80. J.C. Caicedo, J.G. Moreno, E.A. Niño, F.A. González, Medical image retrieval using latent semantic kernels, MIR'10, March 29–31, Philadelphia, Pennsylvania, USA, (2010)
81. Pearson, K.: On lines and planes of closest fit to systems of points in space. Philos. Mag. **2**(11), 559–572 (1901). https://doi.org/10.1080/14786440109462720
82. Verleysen, M., Lee, J.A.: Nonlinear Dimensionality Reduction. Springer, New York (2007)
83. Saxena, A., Gupta, A., Mukerjee, A.: Non-linear dimensionality reduction by locally linear isomaps. Lect. Notes Comput. Sci. **3316**, 1038–1043 (2004)
84. Sha, F., Saul, L.K., Weinberger, K.Q.: Learning a kernel matrix for nonlinear dimensionality reduction. In: Proceeding ICML '04 Proceedings of the twenty-first international conference on Machine learning, vol. 106, (2004)
85. McDonald, R.P.: Factor Analysis and Related Methods. Psychology Press, New York (1985)
86. Lafon, S., Coifman, R.R.: Diffusion maps. Appl. Comput. Harmon. Anal. **21**(1), 5–30 (2006)
87. Breiman, L.: Random Forests, vol. 94720, p. 567. Technical Report Statistics Department University of California, Berkeley (1999)
88. Breiman, L.: Random forests. Mach. Learn. **45**(1), 5–32 (2001)
89. Gasteiger, J., Zupan, J.: Neural Networks for Chemists: An Introduction, 1st edn. VCH, Weinheim (1993)
90. Kohonen, T.: Self-Organization and Associative Memory, 8th edn. Springer, Berlin (1989)
91. de Bruijne, M.: Machine learning approaches in medical image analysis: from detection to diagnosis. Med. Image Anal. **33**, 1–4 (2016). https://doi.org/10.1016/j.media.2016.06.032
92. Chen, S., Liu, H., Zeng, X., Qian, S., Yu, J., Guo, W.: Image Classification Based on Convolutional Denoising Sparse Autoencoder. Math. Probl. Eng. **2017**, 16 (2017). https://doi.org/10.1155/2017/5218247
93. V.I. Skomoroschenko, O.V. Penkova, Yu. V. Kistenev, A.V. Borisov, Identification of the most specific volatile metabolites by gas chromatography in samples of exhaled air from patients with lung cancer and healthy volunteers. Bull. Tomsk State Univ. Chem. 7, 45–54 (2017).
94. Aly, M.: Survey on Multiclass Classification Methods. Technical report. California Institute of Technology, Pasadena (2005)
95. Kistenev, Y.V., Borisov, A.V., Nikolaev, V.V., Vrazhnov, D.A., Kuzmin, D.A.: Laser photoacoustic spectroscopy applications in breathomics. J. Biomed. Photon. Eng. **5**(1), 010303 (2019)
96. Borisov, A.V., Kistenev, Y.V., Kuzmin, D.A., Nikolaev, V.V., Shapovalov, A.V., Vrazhnov, D.A.: Development of Classification Rules for a Screening Diagnostics of Lung Cancer Patients Based on the Spectral Analysis of Metabolic Profiles in the Exhaled Air. In: Anisimov, K.V., et al. (eds.) Proceedings of the Scientific-Practical Conference "Research and Development - 2016", pp. 573–580 (2017). https://doi.org/10.1007/978-3-319-62870-7_60

Chapter 7
Diagnostics of Pigmented Skin Tumors Based on Light-Induced Autofluorescence and Diffuse Reflectance Spectroscopy

Ekaterina G. Borisova and Petranka Troyanova

7.1 Light-Induced Autofluorescence Spectroscopy (LIAFS) of Skin

Light-induced autofluorescence spectroscopy (LIAFS) is a sensitive, fast, and non-invasive tool for diagnosis of cancerous lesions. It could be applied for *in situ* detection of tumors during primary clinical observations or as add-on measurement modality for evaluation of therapeutic treatment of neoplastic pathologies, as well as in monitoring of lesions' recurrence in control clinical follow-ups. LIAFS of cutaneous tissues is based on the detection of emitted light from the endogenous fluorescent compounds, called fluorophores [1–4].

LIAFS is a very attractive tool for early cancer diagnosis because of its high sensitivity, easy-to-use measurement methodology, no need for contrast agents to be applied to the tissue examined, real-time measurement capabilities, and non-invasive detection per tumor [5, 6]. It allows differentiation based on differences in biochemical content and metabolic state of the pathology. However, when the lesion is highly pigmented, the resulting fluorescence signal is too weak to be used for diagnosis. In such cases, exogenous fluorescence markers [7] may be applied or diffuse reflection techniques could be used for obtaining significant signal from highly absorbing lesion areas [8].

Despite all the excellent features of the autofluorescence spectroscopy technique, to date, no reliable and versatile system for fluorescence detection of skin cancer has emerged in the medical market.

E. G. Borisova (✉)
Institute of Electronics, Bulgarian Academy of Sciences, Sofia, Bulgaria

Saratov State University, Saratov, Russian Federation
e-mail: borisova@ie.bas.bg

P. Troyanova
University Hospital "Tsaritsa Yoanna – ISUL", Sofia, Bulgaria

V. V. Tuchin et al. (eds.), *Multimodal Optical Diagnostics of Cancer*,
https://doi.org/10.1007/978-3-030-44594-2_7

Problems with the development of such a diagnostic skin cancer fluorescence detection system are related to the wide variety of benign and malignant skin pathologies. For example, basal cell carcinoma (BCC) lesions have more than 15 subtypes, squamous cell carcinoma (SCC) lesions about 10 different subtypes, all of them have a variety of benign, and dysplastic forms with different features, including fluorescence properties, at different stages of lesion growth [6].

The major advantage is the fact that one could use LIAFS to evaluate the stage of the lesion development, but the limitation is that the evaluation result should be compared with a wide variety of other options, such as lesion type, growth stage, and even the patient's general conditions, such as the influence of drugs and nutrition, ages, skin phototype, anatomic place, etc. [9, 10].

A wide variety of studies are being done to determine the optical properties and sources of the autofluorescence signal coming from human skin tissues that can be used to diagnose skin cancer [5, 11–13]. Endogenous fluorophores with diagnostic value are amino acids, structural proteins, coenzymes, lipids, some vitamins, and endogenous porphyrins. Aromatic amino acids, phenylalanine, tryptophan, and tyrosine, fluoresce when ultraviolet (UV) excitation is applied (in the spectral range 260–300 nm). Structural proteins, collagen and elastin and their cross-links, are responsible for most of the skin's autofluorescence endogenous signal when excitation applied is in the UV region at 320–400 nm. Using the same excitation region, autofluorescence signals of coenzymes such as nicotinamide adenine dinucleotide (NADH) and flavins can be obtained, which are indicators of metabolic changes that can be observed in the skin pathological tissue. Two major approaches could be used in steady-state regime to obtain fluorescence emission from biological tissue (1) application of one single excitation wavelength in the short-wavelength region (UV-blue), where most of the diagnostically important fluorophores have absorption bands, or (2) application of different excitation wavelengths in a scanning regime. In the second case the so-called excitation–emission matrices (EEMs) are developed. EEM measurements are based on the scanning with a specific step of the excitation and detection of a set of emission spectra corresponding to each applied excitation wavelength. Obtained spectral maps reveal specific fluorescence intensity "islands," which correspond to particular fluorophore with specific excitation and emission maximum, allowing addressing of the fluorophore content by type and quantity in the investigated tissue. These EEMs allow obtaining complete autofluorescent response from the investigated tissue and if a broad enough excitation range is applied the whole set of endogenous fluorophores existing in the sample investigated could be found and addressed. Significant drawback of the EEMs mode of fluorescent detection is the time-consuming measurements regime and specific requirements and more complicated equipment that allow exciting and detecting the fluorescence signal in such broad spectral range. Both steady-state modalities have their place in the clinical practice. If preliminary information about searched fluorophores with diagnostic value for a given type of pathology exists, then single wavelength excitation mode could give high diagnostic accuracy for discrimination of this type of lesion from other cutaneous formations. Table 7.1 presents the major diagnostically important skin fluorophores and their excitation and emission

Table 7.1 Diagnostically important endogenous cutaneous fluorophores used in cancer detection studies

Compound	Excitation max (nm)	Emission max (nm)
Amino acids		
Phenylalanine	260 nm	280 nm
Tyrosine	275 nm	300 nm
Tryptophan	280 nm	350 nm
Structural proteins		
Collagen	320–350 nm	400–440 nm
Elastin	290–325 nm	340, 400 nm
Collagen cross-links	380–420 nm	440–500 nm
Elastin cross-links	320–360, 400 nm	480–520 nm
Keratin	450–470 nm	500–530 nm
Enzymes and coenzymes		
NADH	290, 350–370 nm	440–460 nm
NADPH	340 nm	460 nm
FAD, flavins	450 nm	500–540 nm
Vitamins		
Vitamin A	327 nm	510 nm
Vitamin D	390 nm	480 nm
Vitamin K	335 nm	480 nm
Lipids		
Phospholipids	435 nm	540, 560 nm
Lipofuscin	340–390 nm	430–460, 540 nm
Ceroid	340–395 nm	430–460, 540 nm
Porphyrins		
Porphyrins	400–450, 630 nm	635–690, 704 nm

maxima, whose fluorescence signals are valuable for the cutaneous tumors detection and monitoring of tumor treatment, according to the existing literature [5, 6, 12–15].

The autofluorescence technique is applied to diagnose and differentiate non-malignant and systematic skin pathologies such as vitiligo and psoriasis [16], as well-localized lesions, including benign, dysplastic, and malignant tumors [5, 6, 13, 17].

Different research groups have reported a variety of results on the applicability of autofluorescence to skin tumor detection and differentiation. Some of these reports are extremely promising presenting high diagnostic sensitivity and specificity, but others in opposite reported moderate values of the statistical diagnostic parameters for lesion discrimination. In steady-state regime of fluorescence detection the main differences discussed are due to alterations in the emission intensity for some of the endogenous fluorophores, which could be used as diagnostic indicators of tissue malignization. However, these differences are strongly dependent on the type of tumor compared with normal skin, patients' phototype, stage of lesions' growth, used excitation light, giving contradictive results in different measurement sets. For example, Sterenborg et al. [18] in 1994 examined the possibility of autofluorescence spectroscopy to detect non-melanoma skin cancer (NMSC) using excitation at 375 nm. Investigators did not observe significant differences in the fluorescence spectra shapes or significant differences in fluorescence intensity values for tumor and normal skin and reported the method as suboptimal for tumor detection.

On the contrary, Brancaleon and co-authors [19] found significant differences, observing a higher intensity of fluorescence in non-melanoma tumors vs. normal skin, using UV excitation for tryptophan residues, which may be a result of thickening of the epidermis in the lesion area. In contrast, many researchers reported in later works that the intensity of fluorescence maxima associated with collagen and collagen cross-links are lower by intensity in the tumors in comparison with normal skin, due to erosion and degradation of extracellular matrix.

Panjepour et al. [20] found a correlation between the diagnostic accuracy of cancer detection and the patient's skin phototype using 410 nm excitation to detect non-melanoma tumors *in vivo*. With increasing of the cutaneous pigmentation, the accuracy of diagnosis for the detection and differentiation of normal skin from tumors drops from 93% for phototype I to 78% for phototype III patients with non-melanoma malignancies. Na et al. [13] observed a low fluorescence signal in basal cell carcinoma tumors, compared to normal skin fluorescence, which was also observed by the group of Zeng et al. [21] and our own observations [6, 22, 23]. However, the reasons for such fluorescence decrease vary from the extracellular matrix structural alterations, to accumulation of absorbing compounds, such as melanin and oxidized keratin in the lesions' area. Objectively, the alterations in the observed fluorescence could be due to a combination of several factors influencing morphology and biochemical content of the tissue during malignization process [24–28].

Therefore, when detecting autofluorescence *in vivo*, researchers should consider the effect of skin absorbers, which distort the spectral shape of the autofluorescence signal coming from the tissue due to its re-absorption. The most typical skin absorbers that affect fluorescence in the UV-visible spectral region are melanin, bilirubin, oxy- and deoxy-hemoglobin. Melanin is a typical skin pigment with unique absorption properties, since its absorption decreases exponentially from the UV to the near-infrared spectral region, with no specific absorption bands. The hemoglobin in its two forms, oxidized and deoxidized, has absorption maxima in the region of about 400–420 nm, and two maxima for the oxidized form at 545 and 573 nm and one broad maximum at 550–580 nm for the deoxidized form, respectively. Bilirubin, a product of red blood cell breakdown, is a pigment commonly seen in the skin when some liver pathology have place, such as bilirubinemia in newborns or liver infections. This pigment gives the yellow color to human skin and its absorption maximum is in the blue spectral region—about 460 nm [29]. Nevertheless that this pigment in small concentrations could be found in the skin, it is not typically discussed in the literature, as an indicator of cancerous alterations of the skin, and its diagnostic value for malignancy detection is suboptimal.

The absorption of hemoglobin is expressed in the fluorescence spectra detected *in vivo* as additional minima of the fluorescence signal corresponding to the absorption maxima at 420 and 540–580 nm, respectively, for both oxidized and non-oxidized forms. The melanin absorption leads to a general decrease of the autofluorescence signal, more pronounced in the short-wavelength region [6, 16, 17, 24].

These two pigments, melanin and hemoglobin, are the most important ones that influence absorption properties of the cancerous lesions investigated, when pigmented skin neoplasia must be diagnosed using fluorescence or diffuse reflectance spectroscopic techniques. When skin pathologies examined are highly pigmented, fluorescence signals coming from the tissues could be too weak to be detected or the signal-to-noise ratio is so low that it does not allow proper analysis of the spectra obtained. In this case, it is better to apply diffuse reflectance spectroscopic technique [30].

7.2 Diffuse Reflectance Spectroscopy (DRS) of Skin

Diffuse reflectance spectroscopy (DRS) is mainly responsible for the morphological information that can be obtained from tissues and for the content and distribution of tissue light-absorbing compounds. The backscattering light intensity and the spectral distribution of the detected signal can give information about the size and distribution of the scatterers (cells, nuclei, etc.) inside of the tissue. Since the detected diffuse reflectance is a superposition of backward diffuse scattering and absorption of tissue pigments in the volume irradiated, the resultant spectrum obtained also reveals information about size and distribution of the scatterers, as well of absorbing compounds in the biological tissues, such as hemoglobin and melanin in the skin and its pathologies [2, 25].

To take full advantage of reflectance spectroscopic technique, the spectral characteristics must be correlated with the morphology and biochemical composition of the tissue examined. Diffuse reflectance and backscattering spectroscopy have been applied to evaluate skin color and erythema doses [31] to diagnose skin cancer [32, 33]. The results confirm that tissue reflectance spectroscopy provides valuable information on the condition of the tissues.

Anisotropy and phase measurements are also used to quantify the absorption and scattering factors induced by biological pigments and structures, respectively [34, 35]. Spatially resolved diffuse reflectance measurements are used to determine the optical scattering and absorption coefficients of biological tissues [36, 37], which fully describe the optical properties of the tissue in question and could be used as discrimination parameters for evaluation of tissue type and lesion appearance. Examination of the reflectance spectra obtained from microscopic tissue volumes can provide information on the fine structure, size, and shape of nuclei and other cellular organelles. This evaluation is a useful tool for *in vivo* diagnosis of precancerous changes—enlarged nuclei are primary indicators of cancer, dysplasia, and cell regeneration in most human tissues [38, 39].

At the macroscopic level, reflectance spectroscopy is used to detect skin tissue stretching and roughness [40], which is an important problem in plastic surgery. Changes in the diffusive reflectivity of the skin due to mechanical deformation can be used to assess tension in the wound and tissue expanders and assist surgeons in treating wounds by minimizing scar tissue.

Measurement of the skin color by optical reflectance spectroscopy is another important part of the study of skin tissue properties. Medical doctors have used skin color as an indicator of many pathological conditions for centuries. Such measurements are used for initial diagnosis, therapeutic follow-up instrument, as well for drug and cosmetics application effects evaluation and monitoring [41–43].

Diffuse reflectance spectroscopy can provide information not only on tissue morphological information. In many cases, it is used to obtain indirectly the pigment content in the biological tissues examined. The absorption of various skin pigments, bilirubin [44], blood [45], melanin [46], can provide valuable information about pathological conditions, such as high skin bilirubin levels in newborns that could cause permanent brain damage [5]. Reflectance spectra correlate to the skin color and minimum erythema dose [41] (redness [45] or pigmentation [46] measurements), melanin and hemoglobin indices can be calculated [42, 43] based on such measurements. The absorption of these pigments causes significant distortions in the spectral shape of the diffusely reflected signals coming from the skin and such distortions could have a high diagnostic value. DRS in skin cancer studies is mainly applicable to melanin-pigmented skin pathologies detection, including malignant melanoma (MM) lesions, as well as in a combination simultaneously with fluorescence spectroscopy that allows increased diagnostic accuracy overall for all pathologies examined.

DRS is one of the optical techniques with a promising future for its application in skin lesion research. Reflectance spectroscopy has been used to study pigmented skin lesions [46–48], which makes it possible to distinguish between melanoma and benign pigmented areas of the skin (e.g., moles, freckles, etc.). Many people have pigmented skin spots that can be confused with melanoma; therefore, differential diagnosis is very important. For the experienced dermatologist, the distinction between melanoma and other lesions is straightforward, but more difficult for melanomas in the early stages of their growth.

Diffuse reflection signals can be applied to the absolute determination of the absorption and scattering coefficients of biological tissue and lead to reliable results over a wide range of wavelengths. Reflectance spectroscopy of biological tissues has many advantages related to the relative simplicity of the technique used, the ability to quantify pigments and other chromophores contained in the tissue of interest, and the ability to apply a non-invasive real-time diagnostic method to determine the type of tissue pathology.

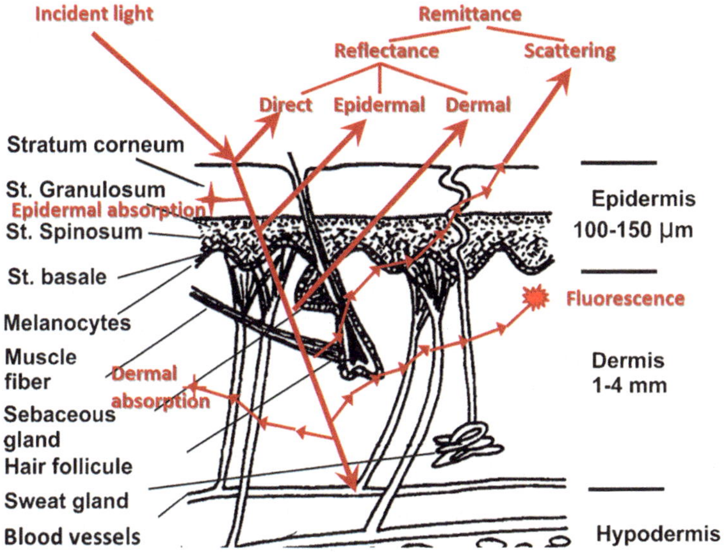

Fig. 7.1 Schematic cross section of human skin—description of light interaction with skin layers. Reflection remittance is the total radiation returned from the skin tissue. (Reprinted with permission from Ref. [49])

The reflection spectra obtained from the tissue examined have a specular and diffuse component, see Fig.7.1. Direct reflection from the surface of the skin is called specular (or regular) reflection and is associated only with the differences of the refractive index between the air and the epidermis layer. In practice, this also strongly depends on the exact surface conditions and illumination due to surface optical irregularitics and roughness [49]. Another component consists of a specular reflectance of deeper skin layers, and a backward diffusely scattering light from the tissue volume. Tissue scattering is due to discontinuities in the refractive index at microscopic level from the cell membrane and organelles, as well as due to collagen, elastin, and keratin fibrils in the extracellular matrix. The resulting reflectance spectrum is adjusted with respect to the incident light spectrum, taking into account the influence of different skin absorbers. As above mentioned the tissue absorption in the visible and infrared spectral regions is due to pigments, such as hemoglobin, bilirubin, and melanin. In infrared spectral region, the water absorption is predominant and determines the tissue optical properties. Nucleic acids, amino acids, urocanic acid, proteins, melanin absorb light in the ultraviolet spectral region [48] and, respectively, determine the optical properties in short-wavelength optical region.

When researcher measures *in vivo* a diffuse reflectance spectrum from a patient's skin using an optical fiber probe, it actually measures the remittance part of the spectra. The influence of the geometry of the optical probe of the skin can be enormous [50] and should be taken into account in the analysis of spectral results.

7.3 Optical Biopsy of Skin Tumors—Principles and Applications

Optical spectra provide biochemical and morphological information on the tissue under study based on its absorption, elastic and non-elastic (Raman) scattering properties [1, 2, 17]. Fluorescence, absorption, and diffuse scattering spectroscopy have been widely used as probes to obtain information on physical, chemical, or physiological processes in tissues, including neoplastic alterations. Optical diffuse tomography and optical coherence tomography are used as effective imaging methods for the detection of tissue structures. It is proposed that these methods be used by the medical community to extend the capabilities of standard diagnostic modalities already introduced in clinical practice, such as X-ray, magnetic resonance imaging, and ultrasound imaging.

Relatively new term used in medical practice to describe spectral techniques used for the early diagnosis of various tissue pathologies *in vivo* is the so-called optical biopsy. The painless, immediate diagnosis based on spectroscopic evaluation of tissue state will soon become a reality. Such form of optical diagnosis is preferable to the removal of several cubic millimeters of tissue needed for histological examination of the traditional biopsies—followed by few days delay required for tissue samples preparation, microscopy slides development, and histological examination. Obviously using scanners, for early diagnosis of the lesions (abnormal tissues), before they are visible to the eye, a total optical examination of the area can be performed instead of random, hit, and missed sample selections. In general, the predictive accuracy of optical biopsy is also better than the prediction based on standard biopsy alone [3, 4, 51]. On top of that, the optical biopsy equipment only requires a learning curve from a few practice trials, compared to the years of training required for some more conventional techniques. LIAFS and DRS are two of the most promising spectroscopic techniques proposed to be introduced into medical diagnostic practice as the main clinical tools for "optical biopsy." The diagnostically important features that could be detected using these two techniques are related to the absorption and emission properties of endogenous chromophores (absorbers and fluorophores). Figure 7.2 presents the spectral regions for the main tissue absorbers and fluorescent compounds, which are reported as diagnostically—relevant, when optical spectroscopy tools are applied for tissue evaluation.

LIAFS can be used to quantify differences between normal and abnormal tissues *in vivo*, providing an appropriate method for detecting pathological lesions in real time. DRS also allows the pathological areas to be distinguished from the normal tissues. These two techniques provide complementary information to each other. LIAFS has revealed very high sensitivity for the diagnosis and differentiation of low-pigmented pathologies. LIAFS is suboptimal for diagnosing and differentiation of highly pigmented lesions, such as dysplastic nevi, or pigmented malignant melanoma due to the low level of fluorescence detected in such lesions. DRS is an indispensable tool for highly pigmented lesions but reveals moderate sensitivity and specificity for non-pigmented or low-pigmented skin neoplasia. Therefore, many

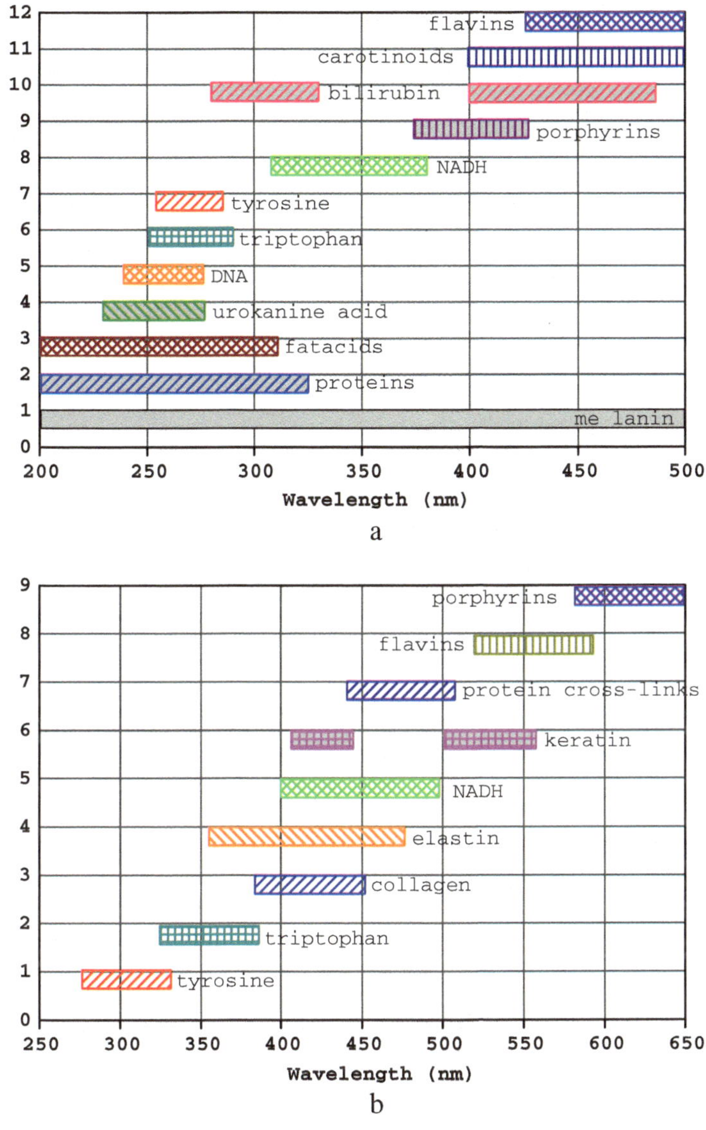

Fig. 7.2 (**a**) Absorption spectral ranges for the most typical endogenous tissue absorbers; (**b**) fluorescent emission spectral ranges for the most typical endogenous tissue fluorophores

research groups using only one of these techniques may report controversial comments about their applicability to be used as a clinical tool for early detection and differentiation of skin lesions. Their results are highly dependent on the group of lesions studied and the technique used. When attempting to monitor and diagnose pigmented pathologies and distinguish malignant melanoma lesions from benign and dysplastic nevi, for example, using reflectance spectroscopy, such groups report excellent results (greater than 90% sensitivity and specificity) [47, 52]. At the same time, if pigmented basal cell carcinoma is detected only by a fluorescence spectroscopic approach, the results of the diagnostic analysis will be very moderate and will not reveal the diagnostic capacity of this detection technique, which results in its poor efficiency and poor clinical relevance [18].

Therefore, in recent years there has been an increasing interest in the combined usage of light-induced autofluorescence and diffuse reflectance spectroscopy to distinguish tumor from normal surrounding tissue. Autofluorescence and diffuse reflectance spectroscopy have been applied in several systems for *in vivo* and *in vitro* studies. A combination of these types of measurements has been used for better understanding of the optical properties of normal and abnormal tissues and for increasing the sensitivity and specificity of lesion diagnosis from the point of view of their clinical feasibility and applicability as add-on or primary diagnostic tool in onco-dermatology practice [5, 53, 54].

The obtained sensitivity and specificity values may also strongly depend on the spectral characteristics taken into account in differentiation algorithms applied by different researchers. The reflection measurements used to determine different pigmented skin lesions use specific spectral characteristics obtained from normal skin, benign, and malignant lesions, namely the mean value of intensity on a specific wavelength (s), the slope of the spectrum in one or more spectral ranges, integral value of the reflected signal for a specific wavelength region, etc. Additional discrimination algorithms based on principal component analysis (PCA), support vector machines (SVM) analysis, neural networks (NN), etc. are also applied to differentiate normal from abnormal tissues evaluating spectral data from both addresses. Depending on the comparison between different special characteristics used by different scientists, the sensitivity (SE) and specificity (SP) of the reflectance spectroscopy technique can vary over a wide range. Thus, the SE:SP values reported in various works are 76:87 [48], 80:46 [55], 83.6:90.8 [47], 89:88 [56], 90.3:77.4 [52], and 91:84 [47]. In our own studies, we observed this peculiarity and investigated the influence on the diagnostic values of sensitivity and specificity depending on the spectral technique or their combination applied to diagnose a given set of lesions [57]. Best results are obtained when the combination of laser-induced autofluorescence and diffuse reflectance spectroscopy is commonly applied to detect malignant melanoma lesions. The achieved diagnostic accuracy is 90%, the sensitivity and specificity values are 100% and 93.3%, respectively, which is an excellent result compared to the diagnostic values achieved by other non-invasive clinical techniques (about 60%–70% for diagnostic accuracy) [58].

Our own studies dedicated to skin cancer detection and differentiation of cutaneous tumors by type and severity are a part of clinical trials for the introduction of a spectral diagnostic system for the detection of skin cancer in the general practice of dermatological departments of Bulgarian hospitals. Experimental spectroscopic studies of *ex vivo* skin samples, obtained after surgical excisions, with a variety of malignancies, including pigmented basal cell carcinoma, compound, dermal, dysplastic nevi, and pigmented malignant melanoma lesions are presented and spectral sources of the signals detected are discussed. Some of our recent results and observations from the clinical practice in application of LIAFS and DRS techniques *in vivo* on patients used as an "optical biopsy" tool for skin cancer primary diagnostics in clinical environment are also given below.

7.4 Experimental Applications of LIAFS and DRS for Skin Cancer Diagnosis

7.4.1 Skin Cancer Samples

The presented results are a part of the laboratorial experiments and pre-clinical trial for the introduction of a spectroscopic diagnostic system for optical biopsy of skin cancer in the general clinical practice of the dermatological unit of the University Hospital "Tsaritsa Yoanna—ISUL."

Skin tumor samples are two large groups: *ex vivo* tissue samples consisted of tumor and healthy skin areas, obtained during surgical excisions of the lesions in the surgical department of the hospital and *in vivo* patients' lesions spectroscopic measurements using LIAFS and DRS techniques for detection of spectral features with diagnostic value, useful for the development of detection and discrimination algorithms for validation of cutaneous pigmented neoplasia.

In both groups are included pigmented skin lesions as follows: benign ones (compound nevi (CN)), dysplastic ones (dysplastic nevi (DN)), and malignant ones (pigmented basal cell carcinoma (BCC), pigmented malignant melanoma (MM)). From clinical point of view the most important discrimination is by pairs pigmented BCC-MM and DN-MM. In the case of differentiation of BCC from MM, different treatment for carcinoma and melanoma lesions is a crucial point required precise and proper diagnosis. For the second case, when DN has to be differentiated from MM, as the dysplastic form (DN) is a precursor of severe malignant lesion (MM), than early changes related to malignization process started in this cutaneous formation.

Compound nevi are the most common type of pigmented skin lesions, benign melanocytic proliferations, which increase in number and size with ages, usually with oval or circular form and well-defined borders in the dermal layer of the skin [58–60].

Dysplastic nevi are described usually as a continuum between common benign nevi and malignant melanoma, as they have intermediate morphological and biological properties. These melanocytic nevi are characterized by architectural disorder and cytologic atypia, similar to dysplastic lesions in other organs and should be considered important because of their association with an increased risk for malignization and transformation to melanoma lesions [61].

Basal cell carcinoma is the most common malignant tumor of skin, comprising about 70% of cutaneous cancers. Pigmented basal cell carcinoma is a clinical and histological variant of basal cell carcinoma that exhibits increased pigmentation and

Table 7.2 Lesions investigated *ex vivo* and *in vivo* by tumor type using LIAFS and DRS

Lesion type	Compound nevi	Dysplastic nevi	Pigmented BCC	Malignant melanoma
Ex vivo spectral measurements	3	6	6	8
In vivo spectral measurements	126	41	18	62

it is a relatively rare variant of BCC lesion with about 6–8% of BCC cases. The increased pigmentation make this lesion similar to MM pathology and nevertheless that histology examination could relatively easily differentiate it from MM lesion, the possibility that observed pigmented lesion could be MM make the diagnostic procedure more complicated. According existing clinical practice MM lesions are verified by histology as a gold standard technique just before surgical excision procedure, while BCC could be treated by local chemo- or radio-therapy, with good success rate, without need of radical surgical intervention.

Pigmented MM lesions are one of the most severe types of tumors in humans and comprising about 10% of cutaneous cancers. MM develops from the pigment-containing cells placed on epidermal–dermal junction, known as melanocytes. Fast growth, expressed high metastatic activity after just a few months after the appearance of the suspicious pigmented spot, high mortality rate make them lesions with a very bad prognosis for the patients. Even standard biopsy procedure is dangerous, if a doubt about MM lesion has place, as it could lead to spreading of tumor cells through blood vessels. Therefore, the differentiation of MMs from all other types of skin malignancies or its dysplastic and benign forms is a primary goal of all onco-dermatology techniques developed for non-invasive detection of cutaneous lesion type [58–61].

For *ex vivo* studies, the procedure of obtaining the investigated samples included their excision during surgery for removal of cutaneous lesions. After the surgical removal lesions were divided into two parts: for histological and for spectral analysis. For spectral measurements, the biological samples were transported in isothermal conditions and safe-keeping solution from the hospital to the spectral laboratory, where their fluorescence and diffuse reflectance properties were investigated.

For *in vivo* studies on site in the dermatology clinic prior to each spectroscopic measurement, clinical observation and dermatoscopic evaluation of the lesion of interest are made by an experienced dermatologist. After initial medical and spectroscopic measurements, cytological and/or histological samples were obtained from each lesion. In comparison, diagnosis results are used as the "gold standard" in comparison to all the data obtained (Table 7.2). All subjects were Caucasian volunteers—patients of the University Hospital "Tsaritsa Yoanna—ISUL" with skin phototypes I, II, and III according to Fitzpatrick's classification. All patients received and signed written informed consent and the research was approved by the Ethics Committee of University Hospital "Tsaritsa Yoanna-ISUL," Sofia (#286/24.07.2012).

7.4.2 Ex Vivo LIAFS and DRS Skin Cancer Measurements

Spectrofluorimeter Fluorolog 3 (HORIBA Jobin Yvon, France) was used for the fluorescence measurements of the surgically removed tissue samples. The excitation light source is a xenon lamp with 300 W output optical power, performance range of 200–650 nm. The detector is a photomultiplier tube with performance range of 220–850 nm for fluorescence detection. Since our tissue samples *ex vivo* varied by shape and dimensions, their fluorescence properties were investigated with additional fiber-optical module F-3000, which allows investigation outside of the sample chamber of the spectrofluorimeter. Measurements of the fluorescence signals obtained in EEM regime were performed with excitation in 280–440 nm spectral range and emission observed between 300 nm and 700 nm. DRS measurements were performed with a broad light source (380–900 nm) halogen lamp (HL-2000, OceanOptics Inc., Dunedin, USA) used for irradiation of tissue samples and fiber probe 6 + 1 geometry for detection of diffuse reflectance signals from the tissue areas investigated. Reflectance measurements were performed with the same equipment, as used for *in vivo* skin cancer studies. After performing both LIAFS and DRS spectroscopic measurements for healthy, benign (if any), and cancerous areas of the skin samples, they were stored in formalin solution.

7.4.3 In Vivo LIAFS and DRS Skin Cancer Measurements

Initially, the lesions were classified visually by an experienced dermatologist and dermatoscopically using the ABCD evaluation criteria. The second stage is the detection of a lesion and surrounding normal autofluorescence of the skin, using different excitation wavelengths, namely 365, 385, and 405 nm, obtained with several narrow-band light diodes used as excitation sources. An optical fiber probe is used to deliver light from LEDs (in fluorescent mode) or from halogen lamp (in diffuse reflectance mode) and to collect emitted signals from the skin surface. The fiber probe consists of 7 fibers in circular geometry. The central fiber is used to detect the fluorescence and reflectance signals and is coupled to a microspectrometric system USB4000 (OceanOptics Inc., Dunedin, USA), and the surrounding six fibers are used to transmit excitation light from the LEDs for fluorescence and in DRS test mode, the broadband halogen lamp is used as a light source replacing LEDs. In this way, we obtain skin diffuse reflectance spectra for the region of 380–900 nm.

Multiple spectra are measured from each suspicious area and for the normal skin surface and averaged to reduce the impact of lesion/tissue inhomogeneity. These averaged spectra from the healthy skin are used as an indicator of spectral alterations in the pathological areas as well for evaluation of the effects associated with inter- and intra-patients' spectral variations.

7.5 Light-Induced Autofluorescence Spectroscopy of Pigmented Skin Tumors

EEM data of normal skin, benign nevi, dysplastic nevi, and malignant melanoma lesions in 2-D spectral map format are presented in Fig.7.3. Two specific "islands" are observed I—in the excitation range of 280–320 nm and emission range of 300–380 nm, and second island consisting from two overlapped areas: IIA—340–360 nm excitation and 400–440 nm emission ranges and IIB—360–420 nm excitation and 440–520 nm emission.

The changes in the EEM autofluorescence maps of skin samples are a consequence of malignant transformations of cells, leading to differences in the fluorescence properties of the skin. Short-wavelength EEM "island" I corresponds to the amino acids tryptophan and tyrosine fluorescence emission and it is observably higher in malignant tissue versus normal skin. In the case of tumor lesion appearance, the amino acids fluorescent emission rapidly raised, due to increased growth of neoplasia itself, required more structural units for its development and due to demolition of some part of the structural proteins in the extracellular matrix of the cancerous lesion, due to uncontrolled growth of the pathology. This is correlated with the alterations foreseen in the emission peaks of collagen and NADH in shorter

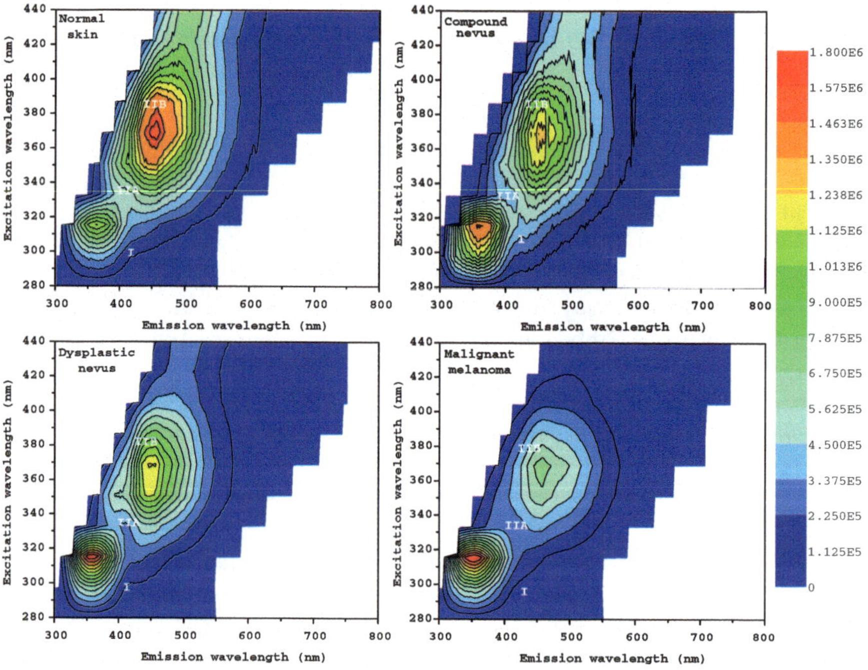

Fig. 7.3 Excitation–emission matrices of normal skin, benign nevus, dysplastic nevus, and malignant melanoma

wavelength region of the "island," and protein cross-links in longer wavelength part of the tissue spectra detected, which revealed significant decrease by intensity (II EEM "island").

Weak signal observed from co-enzyme NADH (region IIA) during *ex vivo* tissue samples study could not be used as diagnostically significant indicator for the tissue alterations due to rapid decrease of this metabolic indicator emission after surgical excision procedure. It could be done correctly only during *in vivo* tissue investigations, or if the period between the surgical excision and spectral measurement itself is known and used for corrections of the NADH fluorescence intensity dynamic decrease.

The structural protein collagen, one of the most informative fluorophores and a main component of extracellular matrix of the skin, plays a crucial role in the degradation and penetration of basal membrane. During transformation from normal to malignant tissue, the enzyme metalloproteinase collagenase is responsible for degradation of collagen and changes in the collagen cross-links, which lead to a general decrease of the fluorescence signal from this protein and its cross-links. Cross-links emission corresponds to the IIB area of the EEM "islands" observed in normal, benign, and malignant skin.

Pigmented BCC lesions revealed very similar to MM lesions by spectral shape fluorescence signals from coenzymes and collagen cross-links spectral area (see Fig.7.4). Significant difference is observed in the ratio between emission of amino acids and proteins, as the fluorescent signal detected from BCC lesions in the region

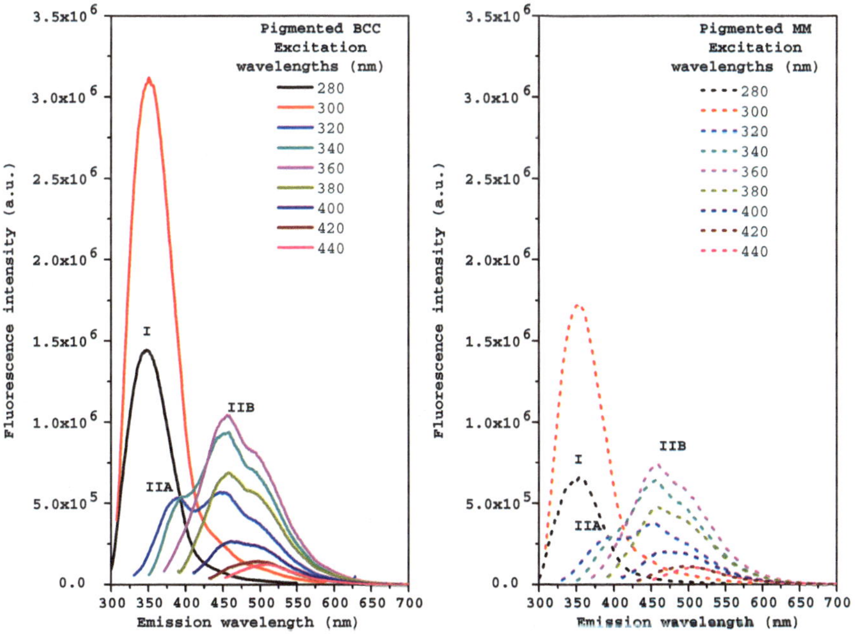

Fig. 7.4 Fluorescence EEM data in spectral curves format for pigmented BCC and MM lesions, averaged by pathology type for the skin samples investigated *ex vivo*

of 300–340 nm, corresponding to the tryptophan and tyrosine is higher in comparison with the same type of signal from melanoma lesions.

Pigmentation of BCC lesions is addressed to melanin as well in melanoma lesions, based on melanocytes malignization. The spectral shape of the fluorescent signal is relatively close for both pathologies and emission maxima observed can be addressed to the same types of endogenous fluorophores—amino acids tyrosine and tryptophan, collagen and its cross-links, NADH and flavins.

That similarity is an indicator for common sources of endogenous emission signal (same type of fluorophores) in both pathologies. However, the concentration of the cutaneous pigment melanin is rather different in both types of lesions, which is well pronounced in the emission spectra obtained from pigmented BCC and MM lesions. In case of BCC the general fluorescence intensity is higher than from MM for whole spectral range, especially pronounced in shorted wavelength range, which correspond to the exponential decay of the absorption of melanin from ultraviolet to infrared spectral region. At the same time, the autofluorescence spectra of both malignancies show a decrease in the concentration of collagen in malignant tissue compared to the normal skin, which is well corresponding with the literature data.

EEM data could allow finding all endogenous fluorescent compounds and could be used for fundamental experimental studies of skin alteration during lesion development. However, some limitations must be taken into account for *in vivo* measurements. The ultraviolet exposure of the skin up to ~360 nm is not optimal, due to the harmful effect of such short-wavelength light irradiation, including increased probability for erythema, DNA and RNA damages, mutagenic and carcinogenic effects of UV radiation. Nevertheless of the diagnostic value of the fluorescent emission of amino acids, which could be observed in 300–340 nm region after excitation in the 280–320 nm region during *in vivo* investigations this spectral range is not used for LIAFS measurements of skin *in vivo*. The excitation wavelength applied to patients *in vivo* is chosen to be higher than 360 nm, to reduce the possible harmful effects on the skin due to spectroscopic observations. DRS technique could be also applied in UV region, but the same safety reasons limit the light spectral range that could be used for skin cancer diagnosis.

Long measurement time and sophisticated equipment required for EEMs development are also suboptimal for application in clinical environment. For example, for the development of detailed EEM (Fig. 7.5a) with wide excitation and emission ranges and fine step of 10 nm for different applied excitation wavelengths required about 20 minutes for measurements, which is not a convenient time period for single point measurement for elder or disabled patients. Requirements to compare normal and abnormal tissue spectra double the measurement time as well. Therefore, the number of excitation wavelengths has to be reduced to the few ones, which correspond to specific maxima of excitation of diagnostically important fluorophores, namely collagen, elastin, keratin, NADH, collagen cross-links, flavins, and endogenous porphyrins. This reduction leads to a usage of 365, 385, and 405 nm excitation wavelengths for LIAFS measurements *in vivo,* see Fig. 7.5b. Light sources could be lasers or light-emitting diodes, as well filtered broad spectrum sources, such as high-pressure xenon lamps. LEDs are the most convenient from the

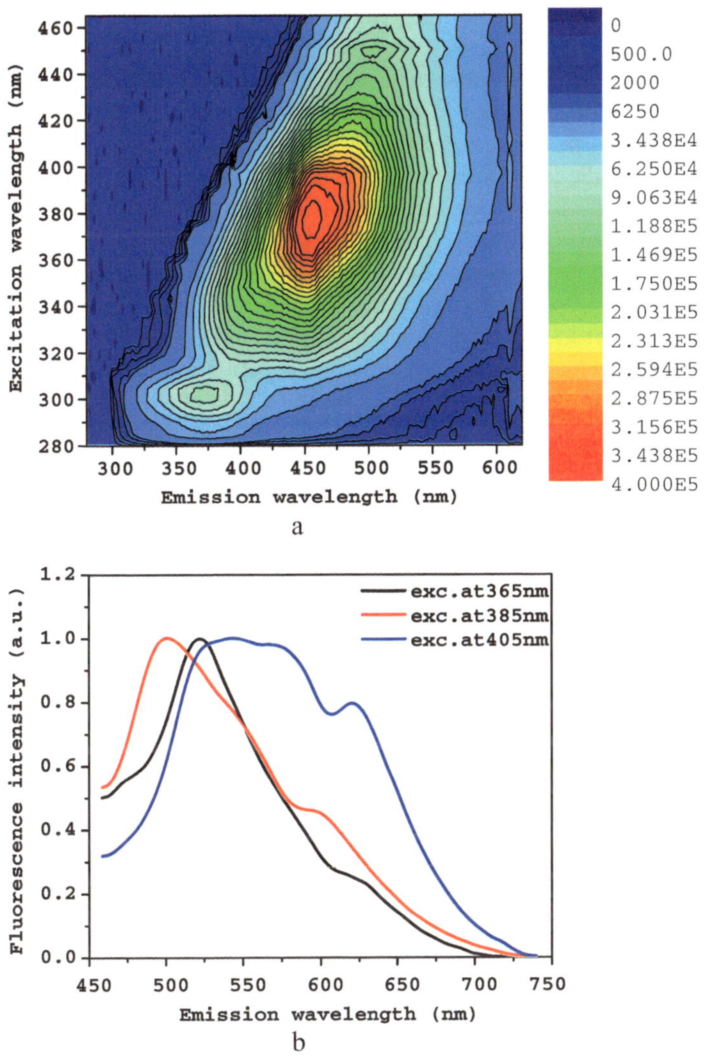

Fig. 7.5 (**a**) Excitation–emission matrix of normal skin *in vivo* for the excitation applied from 280 to 460 nm with a step of 10 nm and detection range from 290 to 650 nm; (b) normalized with respect to maximum fluorescence spectra of normal human skin from the same volunteer at 365, 385, and 405 nm LEDs used as excitation sources for LIAFS measurements *in vivo*

list of light sources mentioned, electrically and light-safety sources of narrow-band light in broad spectral range from deep UV to mid- and far-infrared region and therefore were also chosen in the current *in vivo* studies for pigmented skin tumors evaluation using LIAFS technique.

Each autofluorescence spectrum recorded *in vivo* is a superposition of the fluorescence spectra of endogenous fluorophores existing in the tissue [16, 17, 51] distorted by the re-absorption of the photons by tissue pigments, mainly blood and melanin. The normal skin fluorescence spectral shape does not usually represent significant differences from patient to patient with the same skin phototype. Changes in intensity are more pronounced due to the different phototypes of the skin and anatomical areas, as in both cases different levels of pigmentation of melanin can be observed. The slight differences in the spectral form are found only in the case of fluorescence spectra of the skin of the palms relative to other anatomical sites, in which the lack of melanin leads to a deeper penetration of the excitation and, accordingly, of the emission detected. In this case, the effect of the re-absorption of hemoglobin on the fluorescence from a deeper dermal layer is mostly expressed. This effect has been considered in detail in our previous work [6]. The autofluorescence spectra of normal human skin using different excitation sources that emit different wavelengths also reveal changes in the spectral shape due to different excitation efficiencies of tissue fluorophore type that can be excited in a given spectral range. This effect results in a set of autofluorescence spectra typical of normal skin tissues that have been discussed in the literature and observed in our own studies [62]. In comparison of fluorescence spectra obtained from normal skin using different excitation wavelengths is clearly observed appearance of new emission maxima and changes in fluorescence intensities, depending on absorption for a given excitation wavelength and exact fluorophore compound. Figure 7.4 presents comparison of EEM fluorescent data and fluorescence spectra normalized with respect to maximum, received on three discrete excitation wavelengths applied on normal skin *in vivo*. Normalization in Fig. 7.4b is applied to better detect changes in the spectral shape.

When the excitation wavelength increases, new fluorophores are involved in the formation of the fluorescence spectrum. This is associated with a deeper penetration of excitation with the wavelength [1, 4], with higher absorption at a given wavelength for some additional fluorophores and with differences in the influence of pigments, especially hemoglobin—the minima observed at 545 and 575 nm.

However, the light-induced autofluorescence spectra of normal skin for a given excitation wavelength from different patients with the same phototype and from the same anatomical region are very similar [62] and show very good reproducibility in spectral form and intensity characteristics.

Similar trends are observed when pathological changes occur, as spectral forms and intensity trends are similarly affected for a given type of pathology. For example, basal cell carcinoma lesions always show fluorescence intensity lower than the surrounding normal skin tissue, whereas squamous cell carcinoma usually has fluorescence brighter than the surrounding skin. This observation can be used to distinguish between these two types of non-melanoma malignancies [21, 25, 53].

Malignant melanoma lesions do not have significant spectral shape changes associated with the appearance of new fluorophores in their cells, but they always have extremely low fluorescence and can be distinguished from dysplastic nevi with a relatively good diagnostic accuracy about 70%, which could be improved if reflectance spectral features of the lesions are also evaluated and then can reach 90% [57].

The appearance of endogenous porphyrins is usually clearly expressed in the advanced stage of basal cell carcinoma lesions growth. There is a well-pronounced signal of the porphyrin fluorescence that is characteristic for protoporphyrin IX and other endogenous porphyrins, with peaks in the region of 635 nm and 700–710 nm [22]. This specificity of the autofluorescence signal for the early and advanced stages of tumors is useful for treatment planning for patients with multiple lesions, which required to be treated sequentially due to significant health issues and reduced patient immune response in general. Endogenous porphyrins' signals were not observed in our investigations in the melanin-pigmented compound nevi, dysplastic nevi, and malignant melanoma lesions.

7.6 Diffuse Reflectance Spectroscopy of Pigmented Skin Tumors

Figure 7.6a shows diffuse reflectance spectra of healthy skin from different skin areas of one volunteer. The medial part of the forearm is an area where the skin is not exposed to the sunlight and, respectively, less tanned than other skin areas, so the level of melanin produced by the skin's response to sunlight is low. The lateral part is the most pigmented part of the human forearm due to the nearly permanent exposure to a sunlight (increased melanin level) and therefore has a lower reflectance intensity than the medial part where the melanin level is much lower. The skin of the palm is the part of the hand without melanin pigmentation, but with well-pronounced pigmentation coming from the hemoglobin in dermal part of the skin tissue. In fact, the reflectance spectra have specific minima at 400–420 nm, 543 nm, and 575 nm, related to hemoglobin absorption, which are clearly observed in the palm reflectance spectrum. These minima are also present in the forearm reflection spectra, but they are overlapped by the absorption of melanin from the epidermal layer. The decrease in the reflected signal from the lateral forearm throughout the spectral region compared to the medial part of the forearm is caused by the absorption of melanin.

Figure 7.6a illustrates the strong influence of the location of the measurement points on the measured reflectance spectra from normal skin.

The spectra obtained from normal skin in identical anatomical locations of different patients have similar spectral shape characteristics, but differ in the reflection intensity at different wavelengths, depending on the specific phototype of the patient's skin. However, the spectra obtained from normal skin regions near the examined pigmented lesion vary significantly from patient to patient due to different spectral properties of the skin at different anatomical sites (differences in the epidermal layers in different anatomic areas, different level of the melanin pigmentation, respectively, absorption of the epidermal layer, or different concentrations of oxy- and deoxy-hemoglobin in the dermal blood vessels). Figure 7.6b shows the averaged spectra of normal skin near the lesion examined in seven different patients.

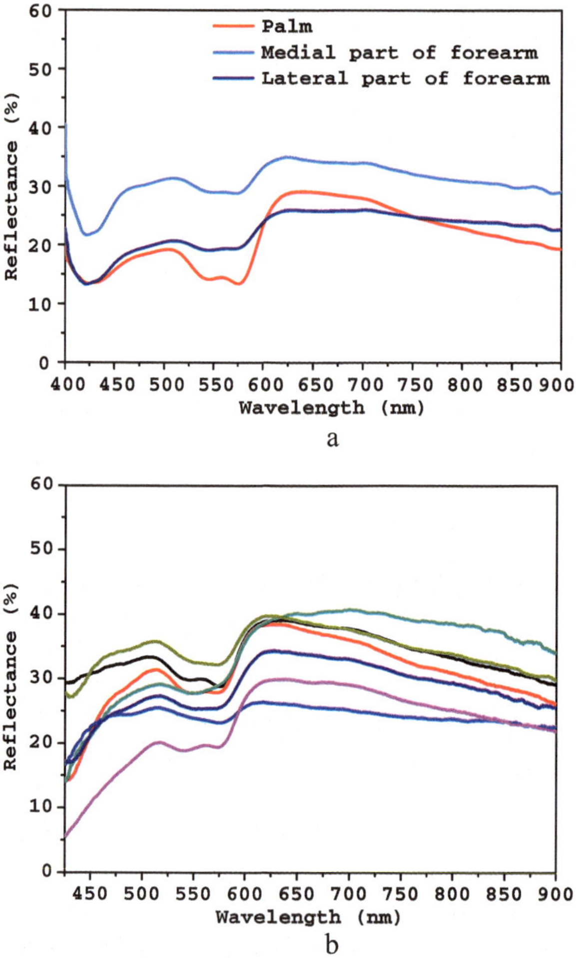

Fig. 7.6 (**a**) Normal skin reflectance spectra from the medial and lateral parts of the forearm and the palm for one healthy volunteer, phototype I; (**b**) randomly chosen reflectance spectra of normal skin for seven different patients, phototypes I–III

These individual differences are related to skin phototype, patient age, and specific area of measurement on the skin surface and may affect the skin spectra of the lesion, especially in the case of not very pigmented moles and spots. Thus, when developing a diagnostic algorithm, the normal skin characteristics that are specific to each patient and each position of the lesion under study should be included.

The mean values of the averaged reflectance spectra of melanin-pigmented nevi and melanoma, as well as comparison between pigmented BCC and MM reflectance spectra are presented in Fig. 7.7. The benign compound nevus reflectance spectrum shows a significant decrease in the entire spectral region, best expressed in the blue region where melanin has stronger absorption than in the red region.

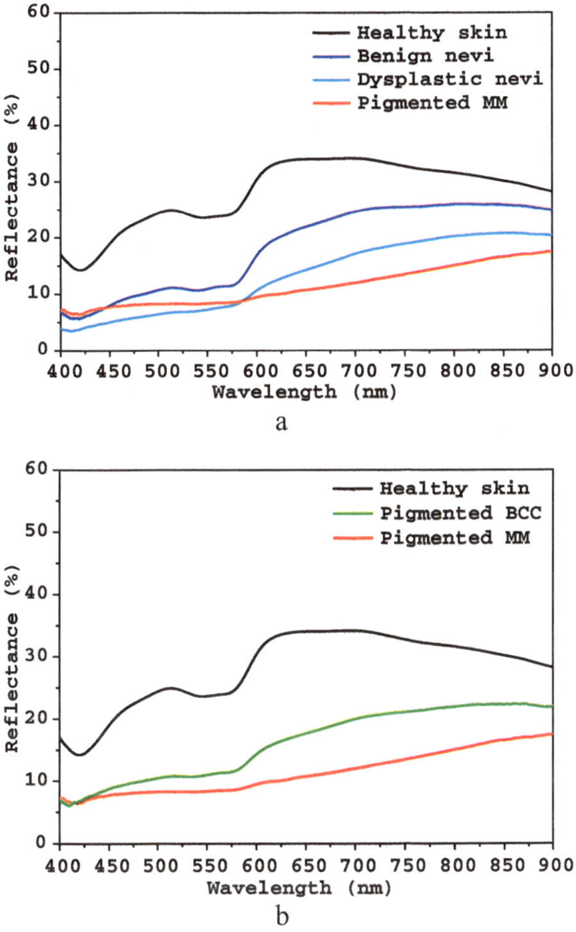

Fig. 7.7 (**a**) Averaged reflectance spectra obtained from normal skin compared to benign compound nevus, dysplastic nevus, and malignant melanoma; (**b**) Averaged reflectance spectra of normal skin, pigmented basal cell carcinoma, and malignant melanoma

Similar results are observed in the case of dysplastic nevus, but the reflectance signal intensity is lower. The spectrum of malignant melanoma has the lowest total reflectance from all types of lesions. Benign and dysplastic melanin-pigmented nevi revealed diffuse reflectance spectra that are significantly different from those of melanoma, which allow discriminating easily with high diagnostic accuracy the nevi from melanoma lesions. Pigmented BCC lesions also have diffuse reflectance spectra shapes significantly different from pigmented melanoma lesions, which allow using the reflectance for discrimination between pigmented basal cell carcinoma and cutaneous melanoma. However, the reflectance of pigmented BCC is close to dysplastic nevi by intensity and shape, which lead to some decrease of overall diagnostic accuracy, when pigmented BCC is diagnosed using DRS technique solely.

One characteristic that the reflection spectra of all pigmented lesions show is a gradual decrease in the hemoglobin absorption peaks at 420, 540, and 575 nm as they move from normal skin to benign and dysplastic nevus to malignant melanoma, in which case lowest values are observed. Also characteristic for melanin-pigmented lesions is the change in sign slope in the spectral region 650–900 nm from that of normal skin reflection spectra. These spectral characteristics allowed to distinguish all types of pigmented skin lesions from normally pigmented surrounding skin, as well as to distinguish benign and dysplastic nevi from malignant melanoma. However, the differences are not significant enough to allow a clear simultaneous differentiation between the three pigment lesions examined. Diagnostic parameters achieved when DRS solely is used for MM lesions discrimination from other pigmented cutaneous lesions are sensitivity—92.4% and specificity—77.6% [63].

We demonstrated a significant improvement of the diagnostic statistical values of sensitivity, specificity, and diagnostic accuracy, based on common usage of LIAFS and DRS data for pigmented cutaneous lesions diagnosis [57]. Combination of LIAFS and DRS used for the development of discrimination algorithm between normal skin, benign, and dysplastic nevi and malignant melanoma lesions allow achieving sensitivity (SE) of 100%, specificity (SP) of 93.3%, and diagnostic accuracy up to 90% for MM lesion proper determination in the frames of a procedure of primary clinical observation using "optical biopsy" spectroscopic tool for skin investigation. In comparison, diagnosis based on ABCD criteria during dermatoscopic observations allow to achieve SE: SP values reported in various works vary from 68% to 92,9% for sensitivity (SE) and from 32,2% to 99% for specificity (SP) for MM lesions diagnosis. These values are in a strong dependence on the experience of the dermatologist applying the dermatoscopic ABCD-technique for lesion evaluation [58–60].

7.7 Conclusions

Melanoma incidence and mortality rates are increasing in many countries. There is much evidence in clinical practice that standard biopsy may be the cause of cancer cell proliferation and is not recommended when suspicion about possible MM lesion occurs during initial diagnosis of new patient. In this context, the development of non-invasive, rapid, and reliable methods such as optical biopsy based on autofluorescence and diffuse-reflectance measurements of the tissue properties is of considerable importance. The clinical applications of optical biopsy *in vivo* are based on non-invasive extractive information on the tissue absorption, fluorescence, and scattering optical properties by highly sensitive measurements of fluorescence and/or diffuse reflected light. These spectral properties are related to the morphology or biochemical content of the cutaneous tissue investigated. Therefore, spectroscopic techniques based on detecting chemical compounds, such as fluorescence, or on detecting of morphological structures, such as diffuse-reflectance, could be extremely sensitive and appropriate tools for development of add-on technique for skin cancer diagnosis. Autofluorescence and diffuse reflectance spectroscopy of human skin, combined as a common tool for optical biopsy, are very prominent for the real-time determination of existing pathological conditions.

Most important for any diagnostic procedure applied in onco-dermatology is its ability to differentiate malignant from non-malignant cutaneous lesions. 100% of lesions could be predicted if such procedure is absolutely accurate. However, every diagnostic test is imperfect in its own way – one procedure could miss many disease cases but make a just a few false-positive diagnoses, another one could miss just a few lesions, but the number of false positive results could be excessively higher. Using autofluorescence detection of benign and malignant pigmented skin lesions, we obtain very good diagnostic parameters for distinguishing non-melanoma lesions *in vivo* from other simulating benign and malignant pathologies. Using diffuse reflectance spectroscopy, we obtain an excellent tool for differentiating melanin-pigmented pathologies, but it is a tool with moderate sensitivity for the detection of non-melanoma lesions. When these two detection techniques are applied in the common examination of given lesion and multispectral algorithms for diagnosis and differentiation are applied, one could rapidly increase the diagnostic accuracy of the combined method, receiving "optical biopsy" tool.

Based on the observed spectral changes, the results of the current pre-clinical trial, we could assume that the LIAFS and DRS measurements can provide useful information on given lesions that could be transformed into diagnostic algorithms for clinical use. The clinical trial is currently underway and with the expansion of the spectral database for major skin benign, dysplastic, and malignant pathologies, we expect to receive an objective tool to detect and evaluate the skin neoplasia, based on its spectral properties.

Acknowledgments This work is supported by the National Science Fund of Bulgarian Ministry of Education and Science under grant #DMU-03-46/2011 and grant #KP06-N28/11/2018.

References

1. Svanberg, S.: Environmental and medical applications of photonic interactions. Phys. Scr. **T110**, 39–50 (2004)
2. Wang, L., Wu, H.: Biomedical Optics: Principles and Imaging. Wiley, Hoboken (2007)
3. Sinichkin, Y., Utz, S., Mavliutov, A., Pilipenko, H.: *In vivo* fluorescence spectroscopy of the human skin: experiments and models. J. Biomed. Opt. **3**, 201–208 (1998)
4. Kollias, N., Zonios, G., Stamatas, G.: Fluorescence spectroscopy of skin. Vib. Spectrosc. **28**, 17–24 (2002)
5. Bigio, J., Mourant, J.R.: Ultraviolet and visible spectroscopies for tissue diagnostics: fluorescence spectroscopy and elastic-scattering spectroscopy. Phys. Med. Biol. **42**, 803–814 (1997)

6. Borisova, E., Avramov, L., Pavlova, P., Pavlova, E., Trchromeoyanova, P.: Qualitative optical evaluation of malignancies related to cutaneous phototype. Proc. SPIE - Dyn. Fluctuations Biomed. Photon. VII. **7563**, 7563–750X (2010)
7. Borisova, E., Bliznakova, I., Mantareva, V., Angelov, I., Avramov, L., Petkova, E.: In: Cao, M.Y. (ed.) Current Management of Malignant Melanoma, p. 141. InTech, New York (2011)
8. Pavlova, P., Borisova, E., Avramov, L., Petkova, E., Troyanova, P.: In: Murph, M. (ed.) Melanoma in the Clinic - Diagnosis, Management and Complications of Malignancy, p. 87. InTech, New York (2011)
9. Deev, A.I., Kozhukhova, E., Tyurin-Kuzmin, A., Vladimirov, Y.: Age dependence on skin autofluorescence. Bull. Exp. Biol. Med. **127**, 317–319 (1999)
10. Na, R., Stender, I., Ma, L., Wulf, H.: Autofluorescence spectrum of skin: component bands and body site variations. Skin Res. Tech. **6**, 112–117 (2000)
11. Andersson, R., Parish, J.: The optics of human skin. J. Invest. Dermatol. **77**, 13–19 (1981)
12. Liu, Q.: Role of optical spectroscopy using endogenous contrasts in clinical cancer diagnosis. World J. Clin. Oncol. **10**, 50–63 (2011)
13. Na, R., Stender, I., Wulf, H.: Can autofluorescence demarcate basal cell carcinoma from normal skin? A comparison with protoporphyrin IX fluorescence. Acta Derm. Venerol. **81**, 246–249 (2001)
14. Pena, A., Strupler, M., Boulesteix, T., Godeau, G., Schanne-Klein, M.-C.: Spectroscopic analysis of keratin endogenous signal for skin multiphoton microscopy. Opt. Express. **13**, 6268–6274 (2005)
15. Ramanujam, N.: Fluorescence spectroscopy of neoplastic and non-neoplastic tissues. Neoplasia. **2**, 89–117 (2000)
16. Borisova, E., Avramov, L.: Laser system for optical biopsy and *in vivo* study of the human skin. Proc. SPIE. **4397**, 405–409 (2000)
17. Bachmann, L., Zezell, D., da Costa Ribeiro, A., Gomes, L., Ito, A.: Fluorescence spectroscopy of biological tissues—a review. Appl. Spectr. Rev. **41**, 575–590 (2006)
18. Sterenborg, H., Motamedi, M., Wagner, R., Duvic, M., Thomsen, S., Jacques, S.: *In vivo* fluorescence spectroscopy and imaging of human skin tumours. Lasers Med. Sci. **9**, 191–201 (1994)
19. Brancaleon, L., Durkin, A., Tu, J., Menaker, G., Fallon, J., Kollias, N.: *In vivo* fluorescence spectroscopy of nonmelanoma skin cancer. Photochem. Photobiol. **73**, 178–183 (2001)
20. Panjepour, M., Julius, C., Phan, M., Vo-Dinh, T., Overholt, S.: Laser-induced fluorescence spectroscopy for *in vivo* diagnosis of non-melanoma skin cancers. Lasers Surg. Med. **31**, 367–373 (2002)
21. Zeng, H., Mclean, D., MacAulay, C., Palcic, B., Lui, H.: Autofluorescence of basal cell carcinoma. Proc. SPIE. **3245**, 314–317 (1998)
22. Borisova, E., Carstea, E., Cristescu, L., Pavlova, E., Hadjiolov, N., Troyanova, P., Avramov, L.: Light-induced fluorescence spectroscopy and optical coherence tomography of basal cell carcinoma. J. Innov. Opt. Health Sci. **2**, 261–268 (2009)
23. Borisova, E., Troyanova, P., Avramov, L.: Optical biopsy of non-melanin pigmented cutaneous benign and malignant lesions. Proc. SPIE. **6257**, (2006). 62570U-1-8
24. Borisova, E., Nikolova, E., Troyanova, P., Avramov, L.: Autofluorescence and diffuse reflectance spectroscopy of pigment disorders in human skin. JOAM. **10**, 717–722 (2008)
25. Borisova, E., Troyanova, P., Avramov, L.: Optical biopsy of non-melanin pigmented cutaneous benign and malignant lesions. Proc. SPIE. **6257**, (2006). 0U-1
26. Dramicanin, T., Lenhardt, L., Zekovic, I., Dramicanin, M.: Biophysical characterization of human breast tissues by photoluminescence excitation-emission spectroscopy. J. Res. Phys. **36**, 53–62 (2012)
27. E. Drakaki, E. Kaselouris, M. Makropoulou, A. Serafetinides, A. Tsenga, A. Stratigos, A.Katsambas, Ch. Antoniou, Laser-induced fluorescence spectroscopy for *ex vivo* diagnosis and classification of basal cell skin carcinoma. Skin Pharmacol. Physiol. 22(3), 158–165 (2009)
28. Han, S.H., Song, T.K.: *In vivo* fluorescence spectroscopic monitoring of radiotherapy in cancer treatment. Int. J. Cancer Ther. Oncol. **3**(1), 03013 (2015)

29. Lamola, A., Russo, M.: Fluorescence excitation Spectrum of bilirubin in blood: a model for the action spectrum for phototherapy of neonatal jaundice. Photochem. Photobiol. **90**(2), 294–296 (2014)
30. Arroyo-Camarena, S., Domınguez-Cherit, J., Lammoglia-Ordiales, L., Fabila-Bustos, D., Escobar-Pio, A., Stolik, S., Valor-Reed, A., de la Rosa-Vazquez, J.: Spectroscopic and imaging characteristics of pigmented non-melanoma skin Cancer and melanoma in patients with skin phototypes III and IV. Oncol. Ther. **4**, 315–331 (2016)
31. Rigal, J., Abella, M., Giron, F., Caisey, L., Lefebvre, M.: Development and validation of new skin chart. Skin Res. Tech. **13**, 101–109 (2007)
32. Wallace, V., Crawford, D., Mortimer, P., Ott, R., Bamber, J.: Spectrophotometric assessment of pigmented skin lesions: methods and feature selection for evaluation of diagnostic performance. Phys. Med. Biol. **45**, 735–741 (2000)
33. Marchesini, R., Bono, A., Baroli, C., Lualdi, M., Tomatis, S., Cascinelli, N.: Optical imaging and automated melanoma detection: questions and answers. Melanoma Res. **12**, 279–287 (2002)
34. Chance, B., Cope, M., Gratton, E., Ramanujam, N., Tromberg, B.: Phase measurement of light absorption and scatter in human tissue. Rev. Sci. Instrum. **69**, 3457–3481 (1998)
35. Marquez, G., Wang, L., Lin, S., Schwartz, J., Thomsen, S.: Anisotropy in the absorption and scattering spectra of chicken breast tissue. Appl. Opt. **37**, 798–804 (1998)
36. Kienle, A., Lilge, L., Patterson, M., Hibst, R., Steiner, R., Wilson, B.: Spatially resolved absolute diffuse reflectance measurements for noninvasive determination of the optical scattering and absorption coefficients of biological tissue. Appl. Opt. **35**, 2304–2314 (1996)
37. Marquez, G., Wang, L.: White light oblique incidence reflectometer for measuring absorption and reduced scattering spectra of tissue-like turbid media. Opt. Express. **1**, 454–460 (1997)
38. Perelman, L., Backman, V., Wallace, M., Zonios, G., Manoharan, R., Nustrat, A., Shields, S., Seiler, M., Lima, C., Hamano, T., Itzkan, I., Van Dam, J., Crawford, M., Feld, M.: Observation of periodic fine structure in reflectance from biological tissue: a new technique for measuring nuclear size distribution. Phys. Rev. Let. **80**, 627–630 (1998)
39. Yang, C., Perelman, L., Wax, A., Dasari, R., Feld, M.: Feasibility of field-based light scattering spectroscopy. J. Biomed. Opt. **5**, 138–143 (2000)
40. Federici, J., Guzelsu, N., Lim, H., Januzzi, G., Findley, T., Chaudhry, H., Ritter, A.: Noninvasive light-reflection technique for measuring soft-tissue stretch. Appl. Opt. **38**, 6653–6660 (1999)
41. Park, B.S., Youn, J.I.: Topographic measurement of skin color by narrow-band reflectance spectrophotometer and minimal erythema dose (MED) in Koreans. Skin Res. Tech. **4**, 14–17 (1998)
42. Angelopoulou, E.: Understanding the color of human skin. Proc. SPIE. **4299**, 243–252 (2001)
43. Clarys, P., Alewaeters, K., Lambrecht, R., Barel, A.: Skin color measurements: comparison between three instruments: the Chromameter®, the DermaSpectrometer®, and the Mexameter®. Skin Res. Tech. **6**, 230–238 (2010)
44. Jacques, S., Saidi, I., Ladner, A., Oelberg, D.: Developing an optical fiber reflectance spectrometer to monitor bilirubinemia in neonates, in "laser-tissue interactions". Proc. SPIE. **2975**, 115–124 (1997)
45. Lock-Andersen, J., Gniadecka, M., Olivarius, F., Dahlstrom, K., Wulf, H.: Skin temperature of UV-induced erythema correlated to laser Doppler flowmetry and skin reflectance measured redness. Skin Res. Tech. **4**, 41–48 (1998)
46. Lock-Andersen, J., Olivarius, F., Hedersdal, M., Poulsen, T., Wulf, H.: Minimal erythema dose in UV-shielded and UV-exposed skin predicted by skin reflectance measured pigmentation. Skin Res. Tech. **5**, 88–95 (1999)
47. Wallace, V., Crawford, D., Mortimer, P., Ott, R., Bamber, J.: Spectrophotometric assessment of pigmented skin lesions: methods and feature selection for evaluation of diagnostic performance. Phys. Med. Biol. **45**, 735–751 (2000)

48. Wallace, V., Bamber, J., Crawford, D., Ott, R., Mortimer, P.: Classification of reflectance spectra from pigmented skin lesions, a comparison of multivariate discriminant analysis and artificial neural networks. Phys. Med. Biol. **45**, 2859–2871 (2000)
49. Borisova, E., Pavlova, P., Pavlova, E., Troyanova, P., Avramov, L.: Optical biopsy of human skin – a tool for cutaneous tumours' diagnosis. Int. J. Bioautomation. **16**(1), 53–72 (2012)
50. Borisova, E., Troyanova, P., Avramov, L.: Influence of measurement geometry on the human skin reflectance spectra detection. Proc. SPIE. **6734**, 6734–6716 (2007)
51. Tuchin, V.: Handbook of Biomedical Diagnostics. SPIE Press, Bellingham (2002)
52. Marchesini, R., Cascinelli, N., Brambilla, M., Clemente, C., Mascheroni, L., Pignoli, E., Testori, A., Venturoli, D.: *In vivo* spectrophotometric evaluation of neoplastic and non-neoplastic skin pigmented lesions II: discriminant analysis between nevus and melanoma. Photochem. Photobiol. **55**, 515–522 (1992)
53. Zeng, H., MacAulay, C., Palcic, B., McLean, D.: A computerized autofluorescence and diffuse reflectance spectroanalyser system for *in vivo* skin studies. Phys. Med. Biol. **38**, 231–240 (1993)
54. Palmer, G., Marshek, C., Vrotsos, K., Ramanujam, N.: Optical methods for fluorescent and diffuse reflectance measurements of tissue biopsy simples. Las. Surg. Med. **30**, 191–200 (2002)
55. Farina, B., Bartoli, C., Bono, A., Colombo, A., Lualdi, M., Tragni, G., Marchesini, R.: Multispectral imaging approach in the diagnosis of cutaneous melanoma: potentiality and limits. Phys. Med. Biol. **45**, 1243–1254 (2009)
56. Tomatis, S., Bartoli, C., Bono, A., Cascinelli, N., Clemente, C., Marchesini, R.: Spectrophotometric imaging of cutaneous pigmented lesions: discriminant analysis, optical properties and histological characteristics. J. Photochem. Photobiol. B: Biol. **42**, 32–39 (1998)
57. Borisova, E., Troyanova, P., Pavlova, P., Avramov, L.: Diagnostics of pigmented skin tumors based on laser-induced autofluorescence and diffuse reflectance spectroscopy. Quantum Electron. **38**(6), 597–605 (2008)
58. Witkowski, A., Ludzik, J., Arginelli, F., Bassoli, S., Benati, E., Casari, A.: Improving diagnostic sensitivity of combined dermoscopy and reflectance confocal microscopy imaging through double reader concordance evaluation in telemedicine settings: a retrospective study of 1000 equivocal cases. PLoS One. **12**(11), e0187748 (2017)
59. Yoo, J., Rigel, D.: Comparing diagnostic sensitivity and specificity for pigmented lesions in clinical dermatologists versus a multispectral digital dermoscopy system. JAAD. **64**(Suppl I), AB122 (2011)
60. Piccolo, D., Ferrari, A., Peris, K., Daidone, R., Ruggeri, B., Climenti, S.: Dermoscopic diagnosis by a trained clinician vs. a clinician with minimal dermoscopy training vs. computer-aided diagnosis of 341 pigmented skin lesions: a comparative study. British J. Dermatol. **147**, 481–486 (2002)
61. Goldstein, A., Tucker, M.: Dysplastic nevi and melanoma. Cancer Epidemiol. Biomark. Prev. **22**(4), 528–532 (2013)
62. Bliznakova, I., Borisova, E., Troyanova, P., Momchilov, N., Avramov, L.: Autofluorescence spectroscopy for noninvasive skin phototypes differentiation. Proc. SPIE. **6604**, 6604–6629 (2007)
63. Borisova, E., Trojanova, P., Avramov, L.: Reflectance measurements of skin lesions – non-invasive method for diagnostic evaluation of pigmented neoplasia. Proc. SPIE. **5862**, 20A1–20A11 (2005)

Part III
Raman Spectroscopy for Cancer Diagnostics

Chapter 8
Raman Spectroscopy and Advanced Statistics for Cancer Diagnostics

Nicole M. Ralbovsky and Igor K. Lednev

8.1 Introduction

Whether it be indirectly or directly, cancer has a vast impact on the majority of society. In 2018 alone, there was approximately 18.1 million newly diagnosed cases of cancer and an estimated 9.6 million deaths due to cancer, worldwide [1]. Cancer is the second-leading cause of death in the USA, where national expenses for care totaled $147.3 billion in 2017 [2].

Diagnosing the various forms of cancer often requires a myriad of methods. The main diagnostic approaches involve lab tests for blood, urine, and other body fluid samples; imaging tests, such as a CT scan, PET scan, X-ray, ultrasound, or MRI scan; and biopsy with either a needle, endoscope, or via surgery [2]. Detailed information regarding diagnosing cancers can be found elsewhere [3]—however, it is important to note that many cases of cancer can go undiagnosed or are misdiagnosed, and in some cases healthy individuals are even falsely diagnosed. Further, the time of diagnosis plays a crucial role in the survival rate of the afflicted. Early detection and diagnosis of cancer typically improves an individual's prognosis and increases the chances for successful treatment by allowing for care to be administered at the earliest opportunity. While early diagnostics and screening methods do exist, not all results are definitive or accurate. Even more, certain exams are invasive, expensive, and not accessible to all who require them. Financial burdens, as well as geographic and sociocultural barriers, prevent large groups of people from

---

This chapter is adapted from Spectrochimica Acta Part A: Molecular and Biomolecular Spectroscopy, 219, N.M. Ralbovsky, I.K. Lednev, Raman spectroscopy and chemometrics: A potential universal method for diagnosing cancer, 463–487, Copyright (2019), with permission from Elsevier.

N. M. Ralbovsky · I. K. Lednev (✉)
Department of Chemistry, University at Albany, State University of New York, Albany, NY, USA
e-mail: iednev@albany.edu

V. V. Tuchin et al. (eds.), *Multimodal Optical Diagnostics of Cancer*,
https://doi.org/10.1007/978-3-030-44594-2_8

seeking proper diagnostic opportunities. Another pressing issue, which is of further interest here, exists—there is no singular universal method that can accurately diagnose all forms of cancer early on. As such, there is a crucial unmet need for developing the first universal method for the non-invasive, inexpensive, and accurate diagnosis of all cancers which can be made accessible to all individuals who require testing. This chapter will discuss how the combination of Raman spectroscopy (RS) and advanced statistical analysis (or, chemometrics) has emerged with a strong potential to solve this imposing issue.

RS is advantageous over other techniques used for disease diagnostics due to its ability to produce a spectral "fingerprint" which specifically represents the total biochemical composition of a sample. As quoted by Mann and Vickers, RS "is unusually, if not uniquely, suited to be the process control star of the next century" [4]. This is because "the intrinsic selectivity of RS allows for accurate identification of organic, inorganic and biological species, an advantage that is lacking in many other analytical techniques, such as ultraviolet absorbance and fluorescence spectroscopies" [5]. RS provides considerable detail regarding the biochemical composition of a sample, and is thus able to detect changes that occur in biological samples during the onset and progression of a disease. RS has the ability to be non-invasive and has the potential for *in vivo* use, which makes it a much more appealing technique for diagnosing diseases over other methods, such as biopsies. It is much less expensive than imaging tests, and it is objective, making it a better choice than some diagnostic methods which require subjective analysis of the results. RS goes beyond simply ruling out other possible diagnoses, as it has the potential to definitively determine both the presence and the stage of disease progression. What's more, RS is a fast, easy-to-use, and reliable technique that can be easily incorporated into clinical settings, making it an exceptionally valuable diagnostic tool.

While RS has a high level of chemical specificity, the changes that occur between spectra of different classes of samples can oftentimes be minute and difficult to visually observe. Thus, advanced statistical analysis, or chemometrics, is utilized to better understand the information found within the obtained data. Chemometrics is essentially "the art of extracting chemically relevant information from data produced in chemical experiments." [6] Chemometrics is suitable to use with RS because the spectral data exists in the form of a data matrix consisting of wavenumbers (cm^{-1}) and corresponding intensities for each spectrum. Different algorithms can be applied to the spectral data matrix for the purpose of building statistical models. These models identify the most useful chemical data and separate it from less informative data and insignificant noise, all while learning how to recognize patterns and similarities within the data matrix. In this way, the models are able to learn to recognize similarities and differences between either labeled or unlabeled data, which it can then use to return either a qualitative or quantitative response.

Many chemometric algorithms exist to answer a wide array of questions a researcher may have. Notably, the two main categories of statistical models that can be built include supervised and unsupervised techniques. Unsupervised algorithms do not utilize sample labels or user-defined information when the model is being constructed [7]. Examples of unsupervised modeling include principal component analysis (PCA) and hierarchical clustering analysis (HCA). These models are exploratory types of analysis and are not used for classification but rather can be used to display similarities or differences between groups of data. On the other hand, supervised models take into account user-defined labels, known as classes, for all samples in order to build prediction models. Supervised models can be further split into regression models or classification techniques. A regression model is often used for calibration purposes and will give a quantitative answer, while a classification model will give a qualitative response, such as the classification of a spectrum. A common regression model is multiple linear regression (MLR), while various classification techniques include partial least squares discriminant analysis (PLS-DA), support vector machine discriminant analysis (SVM-DA), and artificial neural networks (ANN).

Because of the significant ability to pick out important information and recognize patterns and similarities within sets of data, chemometric techniques are ideal for analyzing Raman spectral data. Specifically, these advanced statistical methods are used in this chapter for the purpose of understanding the spatial distribution of biochemical components within a sample, identifying potential biomarkers, differentiating healthy biological samples from diseased ones, and for determining the stage of a disease, all for the purpose of diagnosing cancer.

This chapter presents research published between 2014 and 2018. Specifically, articles that focus on the application of RS combined with chemometrics for diagnosing cancer are considered. Modifications of RS will be considered, such as incorporating the use of fiber-optic techniques, which have not been well reviewed in the past. Studies using all forms of biological materials (cells, tissue, bodily fluids) will be included, and studies will not be limited by the Raman spectral range examined (i.e., high wavenumber region versus the fingerprint region). Research that does not utilize any advanced statistical techniques will not be considered. It is anticipated that in order for the RS methodology to be implemented clinically, an automatic data analysis procedure will be required to interpret the Raman spectral data and to make a diagnosis. Statistical analysis can be performed using software which can lead to automatic and definitive diagnoses in real-time, making diagnosing cancer much more accurate, rapid, and inexpensive. Furthermore, the goal of this chapter is to show that spontaneous RS is sufficient enough for cancer diagnostics, and that more complicated or expensive technology is not required to achieve highly accurate diagnoses. In this chapter, we aim to support and buttress the claim that RS in combination with chemometrics has a strong potential to be implemented as a novel universal method for diagnosing all forms of cancer in the near future.

8.2 Discussion

8.2.1 Spontaneous Raman Spectroscopy

Spontaneous Raman spectroscopy utilizes a monochromatic laser beam to radiate the sample being studied. Inelastically scattered light which interacts with molecular vibrations of the sample will be detected by the instrument. The outcome is a very specific spectral fingerprint of the sample. Spontaneous RS is uniquely suitable for characterizing microheterogeneous environments; specifically, the collection of multiple spectra from a single sample will allow for the detection and spatial distribution of biochemical components within a sample to be determined [8]. RS can be used to collect and process spectral information obtained from multiple positions on a sample with the purpose of providing statistically significant characterization of a sample's heterogeneity and multicomponent composition. By collecting multiple spectra, biomolecules present at high local concentrations can be detected, allowing for identification of potential biomarkers, including those present at low average concentrations [9]. A great advantage of RS resides in its ability to probe the entire biochemical composition of a sample, thus producing a spectroscopic signature for different disease states which are based on the simultaneous integration of multiple biomarkers. This capability significantly improves the sensitivity and selectivity of the diagnostic technique. The following studies have capitalized on the advantages of spontaneous Raman spectroscopy for the purpose of diagnosing various cancers through analysis of either tissue, cells, or bodily fluids.

8.2.1.1 Tissue

Tissue is frequently analyzed in disease diagnostic studies due to its ability to indicate the presence of cancer in the body. As such, tissue is frequently biopsied and thus readily available for *in vitro* Raman spectroscopic analysis and especially for the purpose of diagnosing cancer.

In a study conducted by Kalkanis et al., 95 regions from 40 tissue samples were analyzed to distinguish normal brain tissue from glioblastoma multiforme (GBM) and necrosis using discriminant function analysis (DFA), achieving 99.6% and 97.8% accuracy in the training and validation datasets, respectively (Fig. 8.1) [10]. On the other hand, an average 87.6% accuracy for diagnosing a tissue sample as

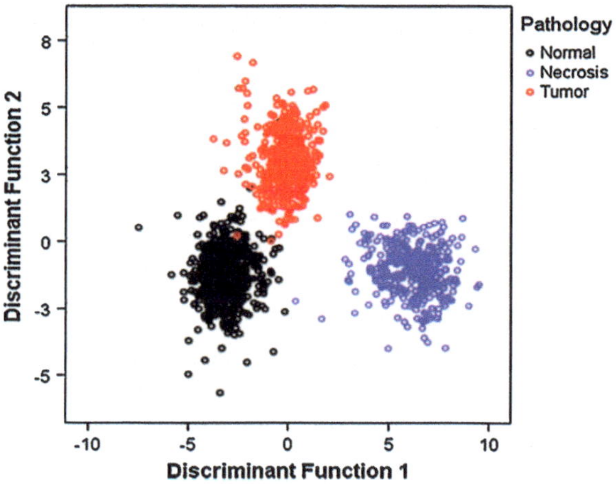

Fig. 8.1 Plot of discriminant function analysis scores for training data (Reprinted with permission from [10])

originating from a healthy donor or from an individual with a brain tumor was achieved through analysis of only 20 tissue samples by a learning vector quantization neural network (LVQNN) technique [11]. The apparent role of number of samples and chemometric technique used for obtaining successful results is already well demonstrated.

Several attempts for using RS to diagnose breast cancer have been made in the last 4 years. The carbonate intercalation signature in type II microcalcifications in tissue, a common diagnostic feature of breast cancer, was used to demonstrate the differences between benign and malignant breast lesions. Raman decision algorithms were developed to distinguish between benign and malignant lesions with type II microcalcifications. The differences in carbonate intercalation could differentiate benign and malignant lesions; specifically, empirical decision algorithms based on carbonate and cytoplasmic protein content achieved 77–83% accuracy for discrimination [12]. Raman spectroscopic analysis of 39 breast tissue samples was employed to understand the differences between normal, atypical ductal hyperplasia, ductal carcinoma *in situ*, and invasive ductal carcinoma lesions of the breast. A support vector machine (SVM) diagnostic model was built using the radial basis function (RBF) with leave-one-out cross-validation (LOO-CV) and achieved an overall accuracy of 74.39% for identifying a sample as belonging to one of the four classes [13]. Fallahzadeh et al. aimed to diagnose breast cancer by using ant colony optimization (ACO) to find the most useful Raman features for discrimination. With five spectral features selected by ACO, the algorithm could correctly classify the 11 tissue samples as normal, benign, or cancerous with 87.7% accuracy [14]. Based on the results of these small studies, RS analysis of tissue samples is not suggested as the most optimal method for diagnosing breast cancer. Results are greatly improved when cells are instead analyzed, as is later discussed.

The mortality rate of cervical cancer can be reduced if the disease is detected in the premalignant stage. As such, Rashid et al. utilized Raman spectral mapping to elucidate biochemical changes associated with premalignant stages of the cancer. When analyzed by K-means cluster analysis (KCA), cervical biopsies classified as negative for intraepithelial lesion and malignancy were divided into three different layers—stroma, basal/para-basal, and superficial—based on differences in collagen, DNA bases, and glycogen spectral features. For low-grade and high-grade squamous intraepithelial lesion (SIL) samples, KCA clustered regions of the basal layer together with the superficial layer. When morphological changes were not apparent, PCA could identify biochemical changes associated with the cancer, creating a useful method for detecting premalignant changes in cervical tissue [15]. Raman spectral mapping was further used to understand the differences between neoplasia and malignancy of cervical tissues. Gradual biochemical changes associated with cancer progression were identified using PCA and KCA, including changes in glycogen, collagen, lipids, protein, carotene, and the nucleus to cytoplasm ratio [16]. While both of the previous studies demonstrated the usefulness of RS to detect biochemical differences between different samples, Daniel et al. went on to improve the usefulness of RS by obtaining quantifiable results through analysis of spectral data by PCA in combination with ANN. The method could classify tissue as normal, premalignant, or malignant with an overall accuracy of 99%. Following this, well differentiated, moderately differentiated, and poorly differentiated squamous cell carcinoma (SCC) samples were investigated using PCA combined with linear discriminant analysis (LDA), achieving 94% diagnostic accuracy [17].

Tissue obtained from healthy donors and from colorectal cancer patients was investigated by several groups. In one proof-of-concept experiment, different excitation wavelengths were used to study the disease. Near-infrared (NIR) Fourier transform-Raman (1064 nm), NIR-visible-Raman (785 nm) and visible-Raman (532 nm) excitation wavelengths were used to collect spectra from 14 samples. Each of the three sets of spectra was analyzed using PCA, and partial spectral differences in each dataset were observed between the normal and diseased samples. Interestingly, when the datasets were combined, the clearest separation between the two classes was seen [18]. While the previous study was useful for establishing biochemical differences between groups, Li et al. obtained quantitative results using ACO-SVM. ACO identified five diagnostically important Raman bands, which were then used to build the SVM diagnostic model. Results showed 93.2% accuracy for identifying colorectal cancer in 44 patients [19]. Two different chemometric systems were used in an additional study to evaluate which could best diagnose colorectal cancer based on tissue analysis of 81 patients. PCA-LDA and PLS-DA models were built and validated using leave-one-patient-out cross-validation (LOPO-CV). PLS-DA performed better, achieving a diagnostic accuracy of 84.3%, which was an improvement over the 79.2% accuracy achieved by PCA-LDA [20].

Tissue samples from patients with early-stage (stage 0 or I) esophageal cancer were examined *ex vivo*; Raman bands that showed a statistically significant difference in band intensity, determined using a *t*-test, were analyzed using LDA. The stage of tissue was correctly predicted with 81.0% sensitivity and 94.0% specificity [21]. However, it should be noted that the algorithm was not tested with comparison to healthy tissue samples.

Several studies were aimed at identifying gastric cancer. In the first study, which used a significant number of samples as well as a robust validation method, Jin et al. analyzed 105 tissue samples from cancerous and pre-cancerous lesions and normal gastric mucosa (NGM). Raman spectra showed differences between the samples related to protein, nucleic acid, and lipid content. Using PCA-LDA with LOO-CV, an average sensitivity of 88.9% and specificity of 94.6% were achieved for discriminating the three classes [22]. Yao et al. achieved 91.7% accuracy for distinguishing normal gastric tissue from cancerous tissue using Fisher discriminate analysis (FDA); however, only 12 samples were analyzed [23]. Two studies were conducted by Hsu and co-workers with a similar goal. In the first, PCA could differentiate the four main histological types of gastric adenocarcinoma (AC), including papillary adenocarcinoma (PAC), tubular adenocarcinoma (TAC), mucinous adenocarcinoma (MAC), and signet ring cell adenocarcinoma (SRC) by analyzing 79 tissue samples. PCA distinguished all gastric AC types from NGM in a binary system. SRC and MAC were able to be differentiated from TAC and PAC; however, TAC and PAC showed no significant differences between each other. Furthermore, LDA scatter plots successfully differentiated all gastric AC types from NGM [24]. In the second study, the results improved to show gastrointestinal stromal tumors (GISTs) could be differentiated from AC and benign lesions from 119 patients. PCA-LDA was employed with CV, achieving an average sensitivity of 99.67%, specificity of 95.45%, and accuracy of 98.32% for distinguishing GISTs, gastric AC, and NGM; this information can be used to help clinicians determine an appropriate treatment path [25]. Based on the range of experiments performed, RS is clearly capable of detecting gastric cancer within tissue samples.

To diagnose nasopharyngeal carcinoma (NPC), tissue samples were collected from 15 individuals with the cancer and from 15 healthy donors. Investigators generated four models using PCA-LDA to discriminate spectra collected from NPC tissue and healthy tissue at various depths of the sample. Each model achieved greater than 95% sensitivity and specificity, with the exception of the deepest level of tissue [26]. Another group also utilized PCA-LDA with LOO-CV to distinguish normal tissue from NPC tissue, achieving a sensitivity of 81% and specificity of 87%. When the method was coupled with PLS, the sensitivity and specificity increased to 85% and 88%, respectively, showing how the chemometric technique selected plays a role in performance success [27]. Mian et al. constructed tissue engineered models of normal, dysplastic, and head and neck SCC using corresponding cell lines. PCA was used to analyze the Raman spectral data collected from the tissues to determine the maximum variance between the groups. LDA was then used to test the discriminatory capacity of the data and classify the tissue samples as normal, dysplastic, or cancerous. Predictions showed an average specificity of 70% and sensitivity of 100% in a binary model of normal versus cancerous tissue. For differentiating dysplastic versus cancerous tissue, 90% sensitivity and 98% sensitivity were achieved. To further validate the study, predictions were made for 12 blinded samples, obtaining 75% specificity for predicting normal tissue, 90% sensitivity for dysplastic tissue, and 98% sensitivity for cancer tissue [28].

Pence et al. utilized two different excitation wavelengths (785 and 1064 nm) to study a total of 15 healthy, AC, and hepatocellular carcinoma (HCC) liver tissue samples for the purpose of diagnosing liver cancer. Spectral data collected using the 1064 nm excitation wavelength was classified using sparse multinomial logistic regression (SMLR); data collected using the 785 nm laser suffered from an intense and highly variable fluorescence background that dominated the Raman spectra and was thus not analyzed further. Two different models were generated using the data collected with 1064 nm excitation. The binary model (healthy versus all tumor tissue) showed 100% sensitivity and 89% specificity and the tertiary model (AC versus healthy versus HCC tissue) achieved an average accuracy of 75.67% [29]. Notably, greater success for diagnosing liver cancer was achieved using Raman hyperspectral imaging, as is later discussed.

Wang et al. aimed to understand mutations of the epidermal growth factor receptor (EGFR), its relation to lung AC, and its potential to be used in future diagnostic studies through RS analysis of 156 lung AC tissues. Tissue samples of carriers without the mutation showed increased levels of amino acids and DNA, whereas samples from donors with a specific mutation group, L858R, exhibited increased

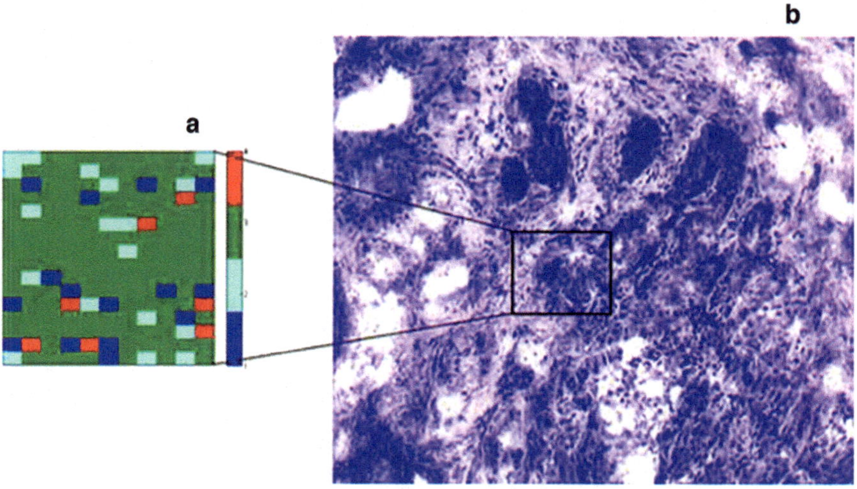

Fig. 8.2 Reconstructed Raman mapping (**a**) and their respective hematoxylin and eosin-stained image (**b**) of blinded sample; the box indicates the site of the Raman image (Reprinted with permission from [32])

arginine levels. To predict to which class a sample belonged, PCA-SVM with LOO-CV was used. The L858R and E19del mutation groups were differentiated from wild-type EGFR tissue with 87.8% accuracy; considering the sample-size and the level of accuracy, this study provides significant evidence for a novel lung cancer screening method based on RS analysis of EGFR mutations [30].

Oral tumor and healthy tissue (10 samples each) were qualitatively analyzed using KCA and PCA; KCA was used to generate Raman maps which correlated to the sample's histopathology. In healthy tissue sections, stratification of epithelial layers was observed. Each of the three layers detected within the normal epithelium tissue was successfully distinguished from the tumor section using PCA. In the unhealthy tissue samples, inflammatory regions of tumor cells and tumor-stroma regions were detected; while not providing quantifiable results, this study shows how Raman mapping can provide novel insight for understanding pathological states [31]. Continuing in the same manner, Raman maps of normal and cancerous oral tissue were obtained by Daniel et al. The maps showed an increase in glycogen, lipid, and protein content within the healthy tissue and an increase in nucleic acid content in the cancerous tissue. Similarly, PCA and KCA were again used to demonstrate the distribution of biochemical components within the samples. Dissimilarly, to improve the usefulness of the results, a discrimination line was computed, resulting in 98.9% accuracy for discerning the two groups. Raman spectral data was collected from a blind sample, which was then subjected to histological evaluation. A Raman image was generated, and the sample was determined to be dysplastic, which was confirmed by H&E staining (Fig. 8.2) [32]. In a third study, Raman spectral data was collected from 24 tissue samples of 14 donors with oral SCC and analyzed using multivariate curve resolution (MCR). The spectral maps of the tissue were automatically and objectively compared through spectral matching of the MCR decomposed Raman spectra and the Raman spectrum of keratin, a biomarker of oral SCC. The oral SCC tissues were correctly identified with 77–92% sensitivity and 100% specificity, with the difference in sensitivity level depending on how positivity was defined [33].

RS was used to understand the pathological changes occurring in ovarian tissue for the purpose of distinguishing adenoma and early AC from benign tumors. Using PLS-DA and LOO-CV, the discrimination model provided an accuracy of 85.2% for diagnosing ovarian cancer [34].

Raman spectra from 25 malignant and benign pheochromocytoma and paraganglioma (PPGL) tissue samples were identified using PCA-LDA with a sensitivity of 80.0% and specificity of 100.0%. PPGLs are tumors that arise from adrenal or extra-adrenal chromaffin tissues. Notably, the obtained results were higher than those obtained using the pheochromocytoma of the adrenal gland scaled score, which is a current method for distinguishing between benign and malignant PPGLs [35].

In a thorough study performed by Liu et al., discriminate analysis (DA) with LOO-CV was applied to spectral data collected from tissue of 63 different patients to determine if RS could distinguish malignant and benign renal tumors using biopsy specimens. Results showed success not only in separating tumor and normal tissue samples (82.53% accuracy) but also in distinguishing malignant and benign tumors (91.79% sensitivity and 71.15% specificity) and low-grade and high-grade tumors (86.98% accuracy). Oncocytoma and angiomyolipoma, two different forms of benign tumors, were successfully differentiated from clear cell renal carcinoma with 100% and 89.25% accuracy, respectively, and subtypes of cell carcinoma were distinguished from each other with an accuracy of 93.48%. Notably, Raman spectroscopic analysis further resulted in successful diagnoses for 7 of 11 cases whose diagnoses were missed during biopsy, illustrating an improvement of the RS methodology over current diagnostic methods [36].

A selective-sampling method was used to collect Raman spectra of tissue samples from individuals with basal cell carcinoma (BCC) and healthy volunteers. A multinomial logistic regression classifier indicated 100% sensitivity and 92.9% specificity for correct classification of an independent set of skin tissue samples [37]. In a large study by Zhao et al., wavenumber selection based analysis was used to diagnose skin cancer. Multivariate techniques PCA-general discriminant analysis (GDA) and PLS with LOO-CV were employed; both were capable of classifying 645 lesions (including skin cancer, pre-cancer, and benign skin lesions) from 573 patients with skin cancer [38]. Interestingly, skin cancer is not well studied using biological fluids or cell samples, suggesting the greatest success for diagnosing skin cancer is through tissue analysis.

Raman spectral data collected from a total of 30 normal thyroid, goiter, and thyroid cancer tissue samples were analyzed by PCA and LDA in combination with CV and binary logistic regression (BLR). The results of LDA with CV showed normal versus cancerous tissues reached a discriminant value of 78.3%; goiter versus cancerous tissue reached a discriminant value of 75%; and normal versus goiter tissue reached a discriminant value of 68% when the spectral region was limited to

1200–1400 cm^{-1}. The results of the BLR model showed the same three groups each achieved greater than 80% concordance [39]. Senol et al. diagnosed papillary thyroid carcinoma (PTC) using an orthogonal PLS algorithm which discriminated 23 tumor and healthy tissue samples with 100% sensitivity and 81.8% specificity for the calibration dataset; the root mean square error of CV was about 47.8%, which is considered low [40]. Using 28 samples (18 for the calibration dataset and 10 for the test dataset) Palermo et al. was better able to differentiate healthy parathyroid tissue and parathyroid adenoma using PLS-DA, achieving 100% accuracy. Further, chief cell adenoma and oxyphil cell adenoma were distinguished from each other with 100% of oxyphil and 99.8% of the chief cell adenoma samples correctly predicted during external validation [41]. When observed together, these studies demonstrate the interesting effect that different chemometric techniques can have on developing successful prediction algorithms.

The Raman spectral data of biopsies from 27 women suspected to have vulval lichen sclerosus (LS), a condition associated with an increased risk of developing vulval carcinoma, were analyzed using PCA-LDA with LOPO-CV. LS tissue was separated from tissue of other inflammatory vulval conditions with 91.0% sensitivity and 80.0% specificity [42]. It is important to note that a comparison to healthy vulval tissue was not taken into consideration.

As these studies clearly show, tissue samples have an immense potential to diagnose various forms of cancer when studied by spontaneous RS and chemometrics. However, the collection of tissue samples can be invasive and uncomfortable for the afflicted patient. On the other hand, RS analysis of tissue samples can be used to confirm typical histopathological diagnosis, which can oftentimes be hindered by subjective and experience-based analysis, making RS advantageous for incorporating into diagnostic procedures. Further *in vivo* analysis is required, and has been conducted, to better understand the capacity of RS for diagnosing cancer, as is later discussed.

8.2.1.2 Cells

Cytology has been widely used for diagnosing cancer—this is most likely due to the fact that cytology specimens are usually easier to obtain while causing less discomfort, cost less money, and are less likely to result in complications when compared to biopsied tissue samples. In this regard, several recent studies have successfully applied spontaneous RS analysis of cells in combination with chemometrics for diagnosing cancers.

Kerr et al. conducted four different experiments to evaluate the potential of RS to diagnose bladder cancer. In each experiment, various standard clinical procedures were used in order to prepare the cell samples for analysis. Spectral data from each experiment was analyzed through PCA-LDA with LOO-CV, with each experiment achieving greater than 88% sensitivity and specificity. Importantly, it was determined that none of the standard procedures that was tested significantly impacted the methodology's ability to diagnose bladder cancer, setting the foundation for RS analysis of cells to be used under a wide variety of clinical settings for diagnostic purposes [43].

Four brain cancer GBM cell lines were obtained from four different patients who each had grade IV astrocytoma. Raman spectra of single cells from each cell line were investigated using multivariate analysis. While this study did not focus on discriminating healthy and diseased states, the results do show that cell lines were similar among all four afflicted patients, thus confirming the reliability of RS analysis of cells for cancer diagnostics and staging for future studies [44].

RS was popularly used to examine cells for the purpose of detecting and diagnosing breast cancer. The results are generally an improvement over those achieved when tissue was analyzed and provide additional information regarding the effect of drugs, which is generally difficult or impossible to do through analysis of other biological samples. Marro et al. utilized RS to study cells undergoing an epithelial-to-mesenchymal transition, a process indicative of breast cancer metastasis. MCR was used to determine how the transition affected the lipid profiles of the cells; specifically, the transition resulted in increased levels of tryptophan and maintenance of a low fatty acid content as compared to highly metastatic cells. PLS-DA successfully discriminated cells within various stages of the transition process, achieving 94% sensitivity and 100% specificity, providing the ability to identify breast cancer in the earliest stages of malignancy [45]. Bi et al. studied the overexpression of human epidermal growth factor receptor 2 (HER2), which is associated with increased chances of developing breast cancer. Three different cell lines were studied—BT474 (HER2 positive breast cancer cell), MCF-10A (HER2 negative healthy control cell), and HER2+ MCF-10A (HER2 positive healthy control cell). The data was analyzed using lasso and elastic-net generalized linear models with CV, which achieved an average 99.8% sensitivity and 99.6% specificity for separating the three cell lines. Following this, Raman spectra of 104 Lapatinib-treated and 104 Lapatinib-resistant breast cancer cells were collected. Lapatinib, a tyrosine kinase inhibitor, is a common drug used to treat breast cancer patients. Significant differences between the spectral signatures of the two cell lines were observed, revealing vital biochemical information which could potentially identify cells resistant to important cancer-fighting drugs as well as demonstrating a novel method for studying the response of cancer cells to therapeutic interventions [46]. In a different study, the effect of pentoxifylline, a drug used to treat muscle pain, on human breast cancer cells was examined. Spectral changes suggested a linear relationship between alterations in DNA, protein, and lipid content with drug dosage. Further, PC-LDA with LOO-CV could separate the control group from cells treated with different levels of pentoxifylline, providing an opportunity to monitor changes occurring within cell lines as a result of medications [47]. Talari and co-workers published two studies on analyzing breast cancer cell lines. In one, a combination of PCA and LDA differentiated two different breast cancer cell lines and one normal breast cell line with 100% sensitivity and 91% specificity [48]. In a later study, normal proliferating, hypoxic, and necrotic regions of a T-47D human breast cancer spheroid model were analyzed by RS to identify chemical changes that occur as the regions progress to necrosis. PCA showed lipid, amide I and III, and nucleic acid content differ significantly between the three regions, providing information for understanding the progression of cells to necrosis [49]. Winnard Jr. and researchers studied organ-specific isogenic metastatic breast cancer cell lines. PLS-DA with LOO-CV was used to classify the different cell lines with 96.8% accuracy; SVM was also used and provided similar results [50].

Ramos et al. evaluated the potential of RS to screen for cervical cancer using cell samples. Both the cervical intraepithelial neoplasia (CIN) and the SIL terminology systems for classifying cervical cancer cells were tested in the process. Biochemical fingerprints of normal and abnormal cell samples were used for discrimination by PCA. Subsequently, PCA-LDA models with LOO-CV were built for classification using either CIN or SIL terminology. The model built using SIL terminology, which characterizes lesions into low-grade and high-grade categories, achieved an average 93.45% sensitivity and 97.55% specificity. The model using CIN terminology gave better results—CIN divides classification of cells into three grades where CIN1 corresponds to mild dysplasia, CIN2 to moderate dysplasia, and CIN3 to both severe dysplasia and carcinoma *in situ*. The CIN model reached an average sensitivity of 96.3% and specificity of 98.27% [51]. Notably, the terminology system used plays a small but identifiable role in the classification efficiency of the chemometric technique employed in this study.

Hundreds of live colorectal cancer cells, derived from primary and secondary tumor cells of the same patient, were studied by Gala de Pablo et al. Using PCA-LDA with CV to analyze the RS data, an accuracy of 98.7% was achieved for classifying cells as either SW480 or SW620 (Fig. 8.3). These results were better than those achieved with SVM and discriminant trees, illustrating the usefulness of certain chemometric techniques over others. PCA-LDA was also used to classify HL60, HT29, HCT116, SW620, and SW480 cells with 92.4% accuracy [52].

Efforts to diagnose non-Hodgkin lymphoma was assessed through examination of normal B-cells and non-Hodgkin lymphoma B-cells using asymmetric least squares (ALS) baseline correction and PCA. K-nearest neighbor (KNN) was used

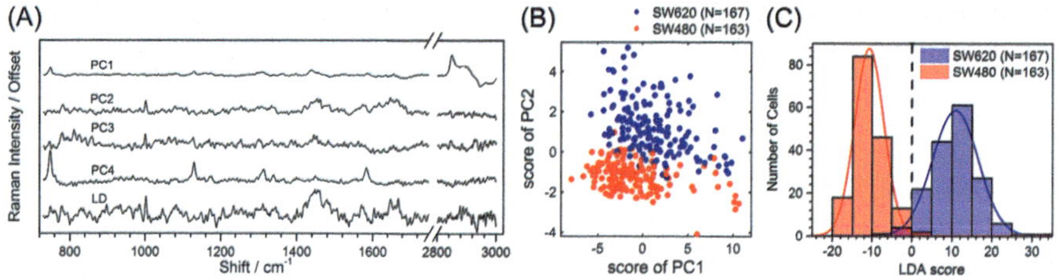

Fig. 8.3 PCA/LDA results. (**A**) Shape of the PCs 1 to 4 and of the LD (**B**) 2D plot of the scores for the first two PCs. (**C**) Histogram of the individual cell scores when projecting the cell data onto the LD from (**A**) with a vertical dashed line at the point of best separation. LD = linear discriminant; LDA = linear discriminant analysis; PC = principal component; PCA = principal component analysis (Reprinted with permission from [52])

to confirm the discriminatory powers of PCA, resulting in 100% accuracy, thus providing a potentially novel method for diagnosing the cancer [53]. It was further found that the Raman spectral data from peripheral blood mononuclear cells (PBMC) could be used to discriminate a significantly pure population of T-cell lymphocytes from other PBMC myeloid cells. Several classifiers, including PCA-LDA, SVM, and Random Forests (RF), were used for discrimination. SVM built using RBF performed the most optimally, achieving 98% sensitivity and 92% specificity. This study demonstrates fundamental differences between myeloid cells and lymphocytes which can be used to identify different PBMC subtypes for diagnostic functions, as well as the importance of testing different chemometric techniques for the purpose of optimizing diagnostic capabilities [54].

Carvalho and researchers collected Raman spectra of the nucleoli, nuclei and cytoplasm of oral epithelial cancer and pre-cancerous cell lines, as well as from normal oral epithelial primary cell cultures. PCA exhibited significant differences between the cell lines, and contributions from nucleic acids and proteins of nucleolar and nuclear sites and from lipids of the cytoplasmic area were primarily responsible for discrimination. This study shows the ability of RS analysis of cells to uncover incredibly useful information regarding cellular components which cannot be achieved through tissue or biological fluid analysis, and that can contribute significantly toward diagnosing cancer [55].

Cisplatin-resistant and cisplatin-sensitive ovarian carcinoma cells were discriminated using PCA-LDA with LOO-CV. Cisplatin, an anti-cancer chemotherapy drug, is often used to treat ovarian cancer. Using the classifier, a diagnostic accuracy of 82% was obtained [56]. These results are similar to those obtained from the previously described analysis of tissue samples; however, the analysis of cells provides the added benefit of monitoring the effect of anti-cancer drugs.

Corsetti et al. analyzed the Raman fingerprints of normal and metastatic hormone-resistant prostate cancer cells by PCA-LDA with CV, which reliably distinguished the two with 95% sensitivity and 88% specificity [57]. Alternatively, Olmos et al. aimed to understand the effect of the pesticide Aldrin on human prostate cancer cells. A portion of prostate cancer cells were exposed to Aldrin, which has been shown to increase the risk of developing prostate cancer in men exposed to it. To assess the differences between the normal and treated prostate cell populations, PLS-DA with CV identified biomarkers associated with pollutant stress, and the best classifier built achieved 91.3% specificity and 80.0% sensitivity for distinguishing the two cell classes [58]. Most recently, deep-ultraviolet Raman spectroscopy (excitation wavelength of 198 nm) also showed that normal human primary prostate epithelial cells and grade IV adenocarcinoma PC-3 prostate cancer cells could be successfully differentiated [59]. The results obtained here are comparable to those obtained using both high-throughput methods and fiber-optic probes, which are further discussed, suggesting the powerful diagnostic capability of spontaneous RS analysis of cells.

The combination of RS and chemometrics has the ability to detect chemical signatures of cells in order to quickly and accurately diagnose various types of cancer. Cells provide unique information regarding the mixtures of metabolites present at a single point in the lifetime of the cell and can be used to probe cellular components which cannot be accessed in other biological samples. Furthermore, cells can be manipulated and exposed to different drugs as well as carcinogens in order to better understand the pathology of cancer as well as the effect of drugs on cancer, providing advantageous and unique information which cannot be easily accomplished using other biological samples. While the reported results are promising, it should be noted that in several of the previously reviewed studies, there is a slight problem of the number of samples analyzed—that is, Raman spectra were collected from a significant number of cells, but not a significant number of donors. This, however, is a straightforward criticism to address in future work; as such, because of the other advantages which outweigh this small issue, the potential for RS analysis of cells to diagnose cancer should not be disregarded.

8.2.1.3 Body Fluids

In an effort to create a more simple and less-invasive sample collection procedure, many studies have focused on studying various body fluids, including blood, urine, and saliva, for cancer diagnostics. These body fluids provide biochemical information which can be used not only for identifying cancer but also for determining the stage of the cancer. Body fluid analysis tends to be less costly and is a much more appealing option for reoccurring testing due to the ease of non-invasive collection; thus, many researchers have used RS and advanced statistical techniques to analyze body fluids for cancer diagnostic purposes.

Blood serum of 35 subjects with meningioma was investigated by RS and compared to blood serum collected from 35 control subjects. Through PCA and PC-LDA followed by LOO-CV, healthy and meningioma subjects were correctly classified with efficiency levels of 92% and 80%, respectively. Similar results were also obtained for identifying different grades of meningioma [60].

Blood serum and urine were both studied in an attempt to diagnose cervical cancer. In one study, González-Solís et al. utilized PCA to distinguish serum samples from 19 cervical cancer patients, 3 pre-cancer individuals, and 20 healthy controls. Differences in Raman spectra indicated a high amount of carotenoids and intense protein contribution in the control serum and higher concentrations of glutathione and tryptophan in the disease serum (Fig. 8.4). Using a LOPO-CV technique, 100% sensitivity and specificity were achieved [61]. Pappu et al. investigated 27 urine samples collected from healthy volunteers and patients with cancer. Using an LDA diagnostic algorithm with CV, 100% accuracy was achieved for discrimination [62]. Interestingly, these studies suggest that regardless of the body fluid type analyzed, markedly successful results are obtained for diagnosing cervical cancer.

The potential to diagnose colon cancer using blood serum was examined in a large study with 75 healthy volunteers, 65 colon cancer patients, and 60 post-operation colon cancer patients. Differences in Raman spectra were assigned to changes due to nucleic acids, amino acids, and chromophores. PCA and KNN

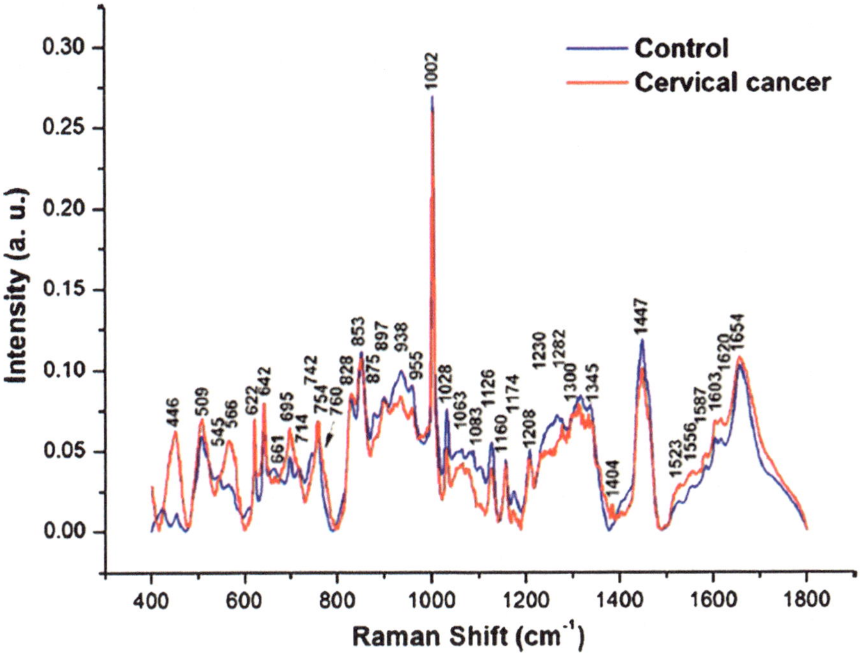

Fig. 8.4 Mean Raman spectra of the control and cervical cancer serum samples (Reprinted with permission from [61])

analyses were used to discriminate between the three classes, resulting in 91.0% accuracy [63].

Khan et al. used RS analysis of blood serum to diagnose NPC. PCA was used to highlight spectral differences and SVM with RBF and CV classified the serum as belonging to either the pathological class or the healthy class with 93% accuracy. Although a smaller dataset was used, these results are generally an improvement over those achieved through RS analysis of tissue [64].

Happillon et al. diagnosed chronic lymphocytic leukemia based on Raman spectral analysis of blood smears collected from 27 healthy volunteers and 49 individuals with the disease. Two SVM models were built with CV—the first could discriminate between the two main subpopulations of leukemia (lymphocytes and polymorphonuclears) with sensitivity and specificity levels both over 98.5%. The second SVM model discriminated neoplastic and healthy lymphocyte spectra with an average sensitivity of 88% and specificity of 91% [65]. These results are relatively comparable to those obtained through the analysis of cells, indicating this method should be considered further due to easier sample collection.

Interestingly, blood serum was used in two different studies to successfully diagnose lung cancer, suggesting the advantage of blood serum over other body fluids for this purpose. Li et al. tested several different modeling techniques to distinguish blood serum of 29 healthy donors and of 68 donors with lung cancer. Uncorrelated linear discriminant analysis (ULDA) and LDA in combination with multiple scatter correction (MSC) pretreatment could each make the distinction with 100% sensitivity and specificity each. Interestingly, MSC combined with PLS-DA was unsuccessful in achieving the goal, further demonstrating the significance of chemometric technique selection [66]. Wang et al. analyzed 91 blood serum samples from healthy individuals and individuals with varying stages of non-small cell lung cancer (stages I–IV), and found LDA with CV could distinguish the five different groups with an overall accuracy of 92% [67].

Urine samples were obtained from patients with oral cancer and from healthy donors; the corresponding Raman spectra were analyzed using PCA-LDA with LOO-CV. The model achieved 98.6% sensitivity and 87.1% specificity, with an overall accuracy of 93.7% for identifying the cancer patients [68]. Pachaiappan et al. utilized both blood plasma and saliva to diagnose oral cancer. In one study, the blood plasma of 30 healthy individuals, 27 patients with oral sub mucous fibrosis, and 34 with oral SCC was analyzed by PCA-LDA. The algorithms could separate the normal group from the premalignant group with 96.3% sensitivity and 80.0% specificity and the normal group from the malignant group with 91.2% sensitivity and 96.7% specificity [69]. Saliva of 83 individuals from the same aforementioned groups was also subjected to analysis via PCA-LDA with LOO-CV. The algorithms separated normal from premalignant samples with 96.4% sensitivity and 70.2% specificity and normal from malignant samples with 93.8% sensitivity and 95.7% specificity [70]. These studies show that regardless of body fluid analyzed, high levels of performance can be achieved for diagnosing oral cancer.

Body fluid analysis is advantageous over analysis conducted using other biological materials for many significant reasons. Collection of body fluids is considerably less-invasive, and even non-invasive in certain cases; it is inexpensive and the process is quick, which allows for rapid results as well as repeat analyses as necessary, and can be conducted during routine exams. Biological fluids provide a great amount of biochemical information regarding the composition of the sample and have a great potential to diagnose all forms of cancer when analyzed by RS.

8.2.1.4 Spontaneous Raman Spectroscopy with Expanded Raman Spectral Range

The majority of the aforementioned studies using spontaneous RS in combination with chemometrics have focused on analyzing the "fingerprint" (FP) region of the Raman spectral data range. The FP region usually refers to the section of Raman spectral bands existing between 400 and 1800 cm^{-1}. It has been discovered that a wider Raman spectral range, which includes the high wavenumber (HWN) region, provides additional information that can be used for many analytical purposes, including disease diagnostics and biomarker detection. The HWN region of spectral data usually refers to the spectral range between 2800 and 3600 cm^{-1} which is found to contain important contributions from water, various C–H bond vibrational modes of lipids and proteins, as well as other N–H and O–H bond vibrations of biomolecules. Notably, the HWN region does not usually suffer as much from autofluorescence signal as the FP region. Several studies have exhibited the usefulness of this region for diagnosing cancers.

Several studies have used the HWN region to successfully investigate oral cancer. Barroso et al. aimed to differentiate healthy tissue from oral SCC tumor tissue within 14 patients. Various bands attributed to water were used to quantify the water content in each sample. Specifically, the bands located between 3350 and 3550 cm^{-1}, for O–H-stretching vibrations, and 2910 and 2965 cm^{-1}, for C–H-stretching, were used. It was found that the water content values determined for the oral SCC samples were significantly higher than the healthy tissue values (Fig. 8.5). A receiver operating characteristic (ROC) curve determined that, using a water content cutoff value of 69%, tumor tissue could be identified with 99% sensitivity and 92% specificity [71]. In a different study, Pachaiappan et al. performed PCA-LDA with LOPO-CV of the HWN region (here, 2500–3500 cm^{-1}) of Raman spectra from blood plasma of 64 individuals. Results showed that oral malignancy could be identified with 92.2% accuracy for the training dataset and 84.4% accuracy for the CV dataset. Analysis of the HWN region allowed researchers to discover additional lipid and water spectral contributions useful for distinguishing the two classes [72]. Further, the HWN region of spectral data for 197 urine samples collected from healthy subjects, oral premalignant, and malignant patients was analyzed using PCA-LDA with LOO-CV. Three different models were built—normal and oral premalignant subjects were classified with 94.9% accuracy, normal and oral malignant groups with 92.1% accuracy, and all three groups with 91.2% accuracy for CV [73]. In a fourth study, Carvalho et al. showed that the HWN region of Raman spectra could differentiate the nucleolus, nucleus, and cytoplasmic areas of oral epithelial cancer, dysplastic, and normal epithelial primary cell lines. The combination of PCA and feature discriminate analysis showed that the cell type could be identified with 99.9% sensitivity and 97.4% specificity using the cytoplasm, 100% sensitivity and 99.1% specificity using the nucleus, and 100% sensitivity and 95.4% specificity using the nucleoli [74]. These studies clearly show the usefulness of the HWN region of Raman spectral data for diagnosing oral cancer; interestingly, these results are generally either comparable or an improvement over those obtained through analysis of other biological samples using only the FP region.

Melanoma and benign melanocytic lesions suspected of melanoma were investigated by Santos et al. Raman bands in the range of 2840–2930 cm^{-1} displayed significant spectral differences between the two groups; PCA-LDA with LOPO-CV of this region could distinguish samples which were considered difficult to distinguish by trained dermatologists. A ROC curved was used to set an optimal discrimination threshold; results showed that melanoma and benign melanocytic lesions often misdiagnosed as being melanoma could be discriminated based on the information found in the C–H-stretching region of HWN Raman data, thus suggesting the potential of the method for improving clinical diagnosis of skin malignancies [75].

While the HWN region provides novel useful and unique information, in several other recent studies, analysis of both the FP and HWN regions was considered for

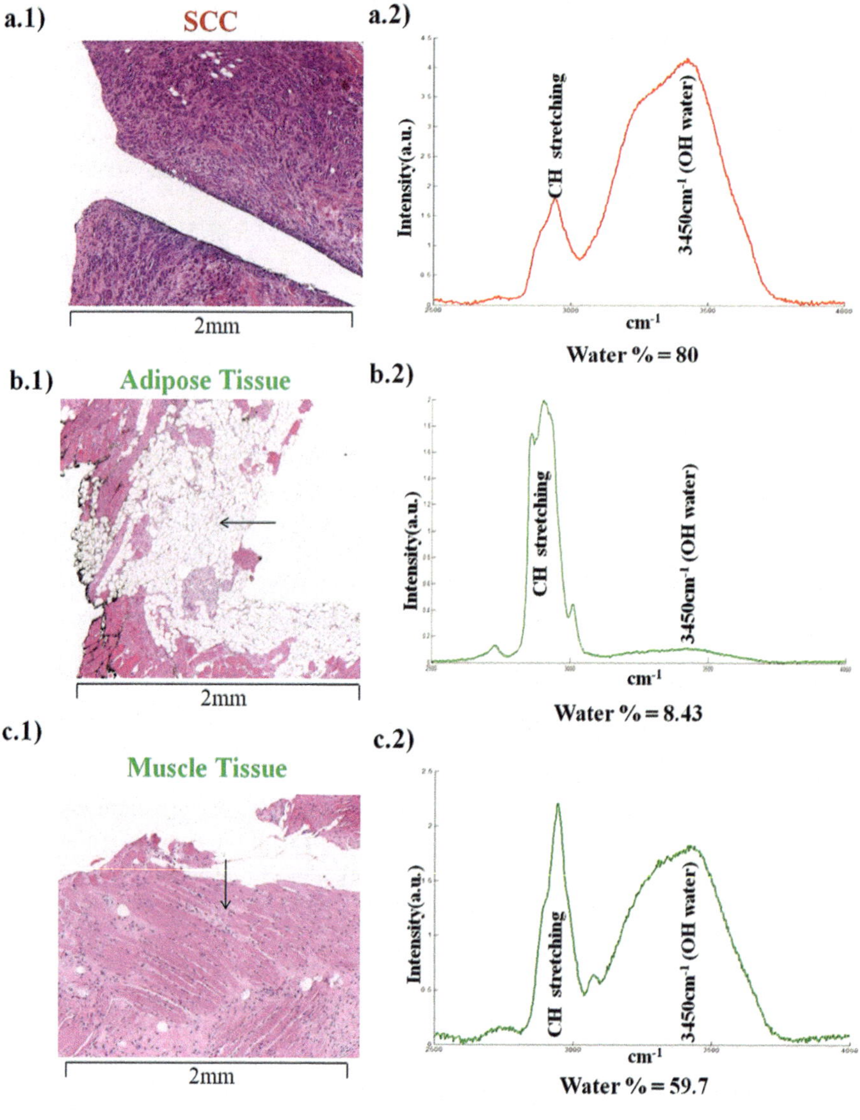

Fig. 8.5 Examples of HWNR spectra measured in (**a.1**) H&E stained thin tissue section of SCC, (**a.2**) typical Raman spectrum of SCC, (**b.1**) H&E stained thin tissue section showing adipose tissue (arrow), (**b.2**) Raman spectrum of adipose tissue, (**c.1**) H&E stained thin tissue section showing muscle tissue (arrow) and (**c.2**) representative Raman spectrum of muscle (Reprinted with permission from [71])

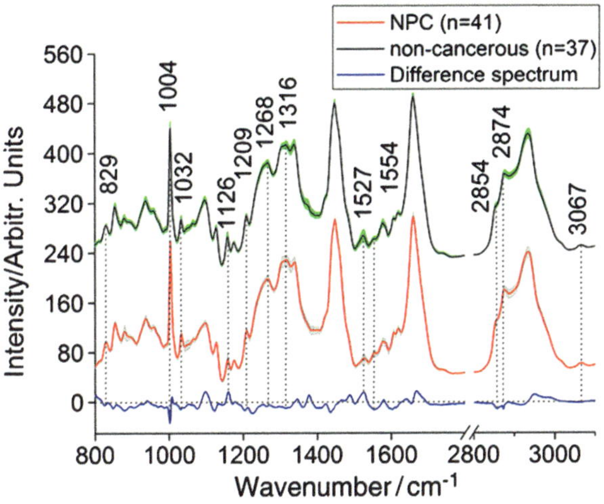

Fig. 8.6 Comparison of the mean intensities of FP/HW Raman spectra from NPC tissue (red line, $n = 41$) versus that of non-cancerous tissue (black line, $n = 37$) in the FP and HW spectral regions. For better visualization, the mean Raman spectra of nasopharyngeal non-cancerous tissue are shifted vertically. The shaded areas represent the respective standard deviations of the means. ((non-cancerous)-cancerous)∗2 was used to represent the corresponding mean difference spectrum (blue line), which is also shown at the bottom. The Raman spectral ranging from 1800 cm^{-1} to 2800 cm^{-1} was not shown by using the broken interval (—//—) to indicate which does not contain tissue biochemical information (Reprinted with permission from [77])

diagnostic purposes. The best results are most consistently seen when the two spectral regions are considered together.

The diagnostic potential of RS for gastric cancer was evaluated through a comparison of the FP and HWN regions. Raman spectra were collected from normal mucosa and gastric cancer tissue areas. Diagnostic algorithms were generated using PLS-DA with LOPO-CV, which yielded 94.59% sensitivity and 86.48% specificity for the FP region and 81.08% sensitivity and 71.05% specificity for the HWN region. Although both regions provide useful information, here, the FP region was better able to diagnose gastric cancer [76].

Huang et al. obtained FP and HWN region spectral data of nasopharyngeal tissue. Non-cancerous tissue was differentiated from cancerous tissue using only FP spectral data (800–1800 cm^{-1}), only HWN spectral data (2700–3100 cm^{-1}), and an integrated FP/HWN dataset (Fig. 8.6). The results, using PCA-LDA with LOPO-CV, showed the samples could be distinguished with 87.8% sensitivity and 86.5% specificity for the FP region, 85.4% sensitivity and 91.9% specificity for the HWN region, and 95.1% sensitivity and 89.2% specificity for the integrated dataset, thus demonstrating the potential of both FP and HWN regions to diagnosis NPC [77]. Sun et al. performed a study with a similar goal, with spectral data collected from biopsy tissue smear samples of 74 patients in the regions of 800–1800 cm^{-1} and 2800–3100 cm^{-1}. Using PCA-LDA of the combined FP/HWN regions dataset, 87.2% sensitivity and 85.7% specificity were achieved for classifying a sample as belonging to the NPC group or to the non-cancerous group [78]. Clearly, the HWN region provides unique additional information useful for diagnosing cancer based on RS and chemometrics.

8.2.1.5 Raman Hyperspectral Imaging

Spontaneous RS in combination with chemometrics has shown great potential for generating a diagnosis based on the analysis of biological specimens; however, it is important to consider the advantages of Raman hyperspectral imaging (HSI). Raman HSI utilizes an imaging camera to collect additional information regarding the sample being analyzed; as such, the result is the combination of spectral information with spatial information. Raman spectral information is collected from each pixel of an image. The spectral signature from each pixel, or small volume of the sample, depends on the biochemical components present in that small volume—these components can vary within the sample itself and between different samples (i.e., tissue from a healthy donor or tissue from a donor with cancer). In Raman HSI, a three-dimensional (x, y, λ) hyperspectral data cube is formed where the x and y components are spatial dimensions and the λ component is the spectral signature. Together, this information can generate an image which provides information regarding the distribution of biochemical components within the sample. In fact, the hyperspectral images are useful for depicting the relative concentrations of various biomarkers in a biological sample, potentially indicating which areas of the sample, if any, are affected by the disease in question. Furthermore, Raman hyperspectral images of tissue in particular can be compared to the tissue after it has been stained using hematoxylin and eosin (H&E) staining. Thus, Raman HSI is useful for medical diagnostics as it can confirm the presence of a disease, distinguish between normal and diseased samples, and distinguish between disease stages, all through objective analysis. The following studies employ Raman HSI for the purpose of understanding the distribution of biochemical components within samples in order to identify cancer. Additionally, some studies go a step further and utilize advanced statistical analysis to build algorithms for quantitative diagnosis of various cancers.

In a study by Kast et al., the concentrations of Raman spectral bands corresponding primarily to lipid and protein content (1004, 1300:1344, and 1660 cm^{-1}) were imaged across forty brain tissue sections diagnosed as normal, GBM, necrosis, or infiltrating GBM. The goal was to understand the boundaries that exist between gray matter, white matter, and diseased tissue in an attempt to develop a novel method for rapid and non-destructive imaging of brain tissue for cancer diagnosis. The resulting Raman imaging maps corresponded with adjacent H&E-stained sections and could therefore successfully discriminate between the various regions of brain tissue [79].

In the last few years, several manuscripts were published concerning applying Raman HSI for breast cancer analysis. These studies were able to pinpoint valuable differences in biochemistry between diseased and healthy samples, which can be more easily detected with the advantages of HSI. In one study, tumor regions of breast cancer tissue were discriminated from healthy tissue based on altered concentrations of nucleic acids, collagen, and fat as determined by Raman HSI and KCA. Furthermore, LDA could diagnose ductal carcinoma in breast tissue samples with 95.6% sensitivity and 96.2% specificity. Fresh samples were then subjected to Raman imaging using a selective-sampling strategy in order to decrease data acquisition time based on auto fluorescence imaging (AFI); results were in agreement with the diagnosis made by conventional histopathology [80]. In a different study, live non-malignant, mildly malignant, and malignant breast cancer cells as well as breast cancer tissue were analyzed. Results from Raman HSI suggested that lipid droplets in the various cell lines differ not only in concentration but also in biochemical composition, suggesting their potential role in breast cancer pathology. Differences were observed in the lipid composition within breast epithelial cells as well as in breast tissue. Further, PCA displayed identifiable differences in the Raman signatures of the cells, suggesting a method for predicting the state of the oncogenic pathway [81]. Brozek-Pluska et al. showed that RS and Raman HSI could detect relative amounts of acetylated and methylated lysine, which have been previously designated as biomarkers for breast cancer. The stretching vibration of the acetyl group observed near 2938–2942 cm^{-1} and of the methyl group around 2970 cm^{-1} allowed these molecular changes occurring in human breast tissue cancer cells to be monitored. Further, PLS-DA with CV provided 85.3% sensitivity and 91.3% specificity for detecting cancer [82].

Vanna et al. successfully distinguished the four subtypes of acute myeloid leukemia (AML), which include myeloblasts, promyelocytes, abnormal promyelocytes, and erythroblasts. Bone marrow samples of seven patients, each affected with one of the four AML subtypes, were collected. For each cell isolated from the bone marrow aspirate, 4096 spectra were collected in order to generate Raman images which could accurately demonstrate morphological features. When the Raman images were analyzed by HCA, automatic discrimination and localization of the nucleus, cytoplasm, myeloperoxidase-containing granules, and hemoglobin was achieved. The images provide additional biochemical information than what could be obtained using only spontaneous RS. Following this, the average Raman fingerprint of each cell was analyzed by PCA-LDA with LOO-CV. Myeloblasts, promyelocytes (both abnormal and normal), and erythroblasts were differentiated with 100% accuracy. Normal and abnormal promyelocytes were correctly classified with 95% accuracy, and all four subtypes could be classified with 98% accuracy [83].

Interestingly, when Raman HSI was used to study liver cancer specimens, the results were a great improvement and provided useful biochemical information as compared to those results obtained through spontaneous RS tissue analysis. Two liver cancer cell lines, HepG2—including HepG2 cells in different cellular growth phases—and SK-Hep1, were analyzed by Tolstik et al. The collected spectral data was used to generate color-coded images which were analyzed by HCA and PCA; this provided significant information regarding the biochemical composition of the samples. Spectral differences were mainly attributed to higher expression of unsaturated fatty acids in the HCC cells as well as during the proliferation phase of cellular growth. Through SVM analysis with CV, previously unknown cells were classified as belonging to one of the two cell lines with 93% accuracy. Predictions of the unknown proliferation phase for HepG2 cells showed 100% sensitivity and 98% specificity. Raman HSI uniquely provides information regarding cell type and proliferation behavior, which are essential tools in identifying features of malignant tumors [84]. In a second study by the same group, Raman imaging of liver tissue was used to identify molecular information beneficial for diagnosing liver cancer. The most notable difference between HCC and fibrosis regions of tissue was found to be due to fatty acids, especially palmitic acid. A RF model with CV classified malignant and non-malignant tissue regions with 86% accuracy [85]. More recently, Ryabchykov et al. discriminated three different cell lines (HepG2, nondifferentiated hepatic stem cell line HepaRG, and differentiated hepatocyte-like HepaRG) using Raman HSI. KCA was used to visualize clusters of different cell components within the cells. Following this, a three-class LDA with LOO-CV model was constructed to achieve cell line classification, reaching 96% accuracy [86].

Raman HSI shows great capabilities for detecting oral cancer, in both paraffin-free and paraffin-embedded tissue. Oral SCC and healthy tissue samples were analyzed to assess the potential of RS to perform discrimination tasks at the histological level. 127 Raman images were generated from 25 unstained thin tissue sections; the images were comparable to corresponding histological evaluation obtained through H&E staining. After imaging, the spectra were labeled as cancerous or as a surrounding healthy tissue structure (squamous epithelium, connective tissue, adipose tissue, muscle, gland, or nerve) (Fig. 8.7). LDA models were built to analyze the labeled spectra for classification purposes. A total of six binary LDA models were built to distinguish oral SCC spectra from each of the surrounding healthy tissue structures, achieving an overall average accuracy of 93.17% [87]. In a unique study, Meksiarun et al. aimed to first understand if multivariate methods could extract the paraffin component of paraffin-embedded oral cancer tissue spectra. Typically, oral SCC tissue will be removed from a patient, fixed with formalin, and embedded in paraffin to prevent degradation. However, the Raman spectral features of paraffin overlap with main Raman spectral tissue bands, including the amide I and III bands. Three methods were tested for their ability to remove the paraffin spectral features while maintaining the integrity of the rest of the Raman spectrum, including PLS, independent component (IC) analysis, and IC-PLS. All methods were successful, however, PLS and IC-PLS were the most successful at removing the paraffin spectral component while still maintaining spectral integrity of the cancer tissue. The paraffin-removed spectra obtained via IC-PLS were analyzed by PCA to construct Raman images. Main Raman markers for discriminating healthy and malignant tissue were found to be collagen, phosphate, and DNA. The produced Raman images showed similarity to H&E stained tissue, thus demonstrating the ability of Raman HSI to diagnose oral cancer in paraffin-embedded tissue [88].

Human prostatic cells were collected and analyzed using Raman HSI. An emphasis was placed on the C–H vibration region (2800–3100 cm^{-1}) of the spectra due to its ability to pinpoint the main differences between normal and tumor cell lines. PCA was used for image processing and identified protein and lipid fractions which were important for differentiation. A self-modeling curve resolution (SMCR)

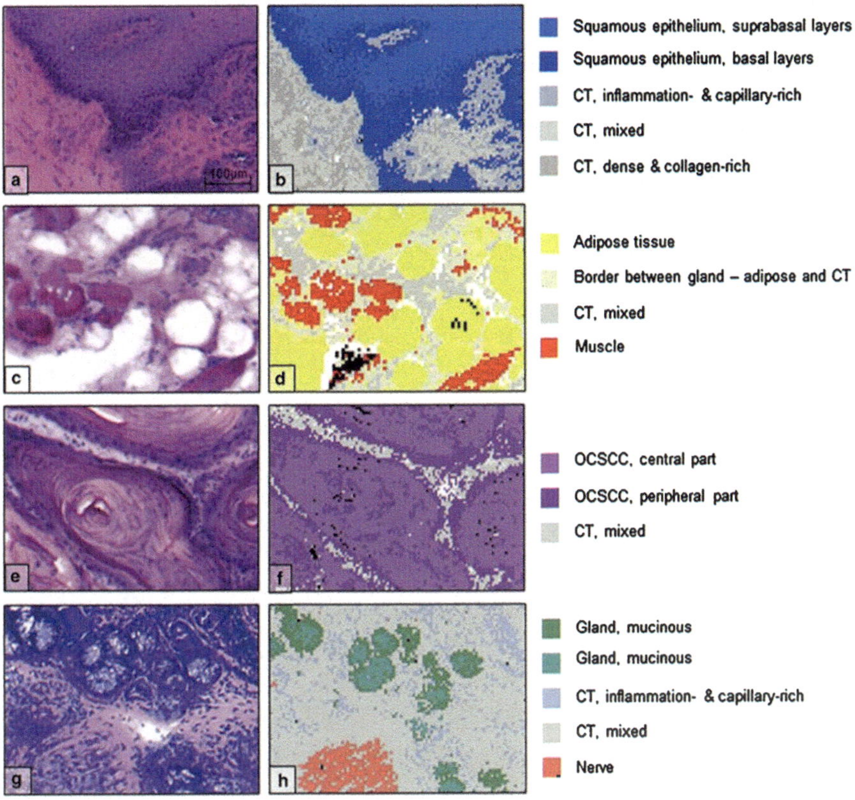

Fig. 8.7 H&E stained sections and corresponding pseudo-color Raman images. H&E-stained tissue sections (**a**, **c**, **e**, and **g**) and corresponding pseudo-color images (**b**, **d**, **f**, and **h**). The K-means cluster averages were annotated as one of the following tissue structures: OCSCC (central part, peripheral part, or n.o.s.), squamous epithelium (superficial layers, suprabasal layers, or basal layers), CT (dense and collagen-rich, mixed, or inflammation- and capillary-rich), gland (mucinous or serous), muscle, adipose tissue, or nerve. CT = connective tissue; n.o.s. = not otherwise specified; OCSCC = oral cavity squamous cell carcinoma (Reprinted with permission from [87])

algorithm was also employed and revealed tumor cells experience a 97% increase of the lipid fraction with respect to the control cells. Analysis by least squares curve fitting gave reproducible results for identifying differences at the molecular level between normal and tumor cells [89].

Raman imaging was performed on healthy and neoplastic thyroid tissue to improve the diagnosis of PTC (Fig. 8.8). Biochemical features of PTC were characterized by the significant presence of carotenoids in comparison to healthy tissue. LDA with LOO-CV was applied to estimate tissue classification. Healthy and PTC thyroid tissue were discriminated with 100% accuracy and classical and follicular variants of PTC were discriminated with 95% accuracy [90]. The performance of this study is generally an improvement over analysis of tissue by spontaneous RS alone.

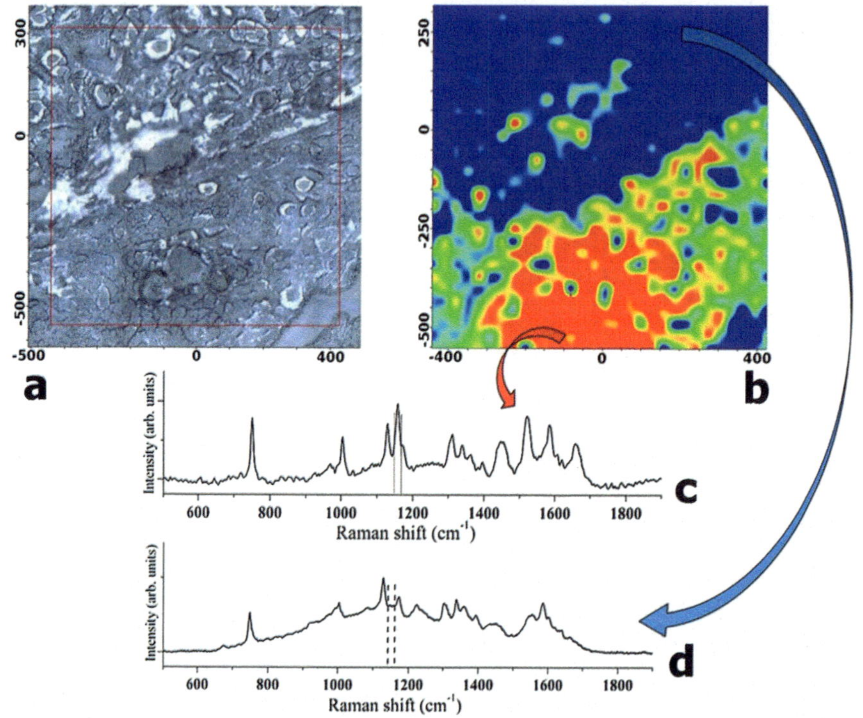

Fig. 8.8 Typical example of Raman chemigram map (1156 cm^{-1} band reference) of a mixed zone of thyroid tissue (blue-healthy; red-yellow-green–PTC): (**a**) dark field optical image, (**b**) Raman map, (**c**) average reference Raman spectrum corresponding to healthy tissue, (**d**) average reference Raman spectrum corresponding to PTC tissue. The red square on the right side (**a**) corresponds to the investigated tissue area shown on the left (**b**). The scale bars are expressed in μm (reprinted with permission from [90])

Raman HSI capitalizes on the advantages of spontaneous Raman spectroscopy while providing additional tools that can be used in diagnosing cancer. The images produced are comparable to those produced by H&E staining and the information provided is useful for understanding presence of a disease as well as the stage of a disease. Importantly, Raman HSI is able to identify incredibly valuable biochemical differences between healthy and diseased samples, further enabling the identification of biochemical changes that occur during pathogenesis as well as potential novel biomarkers that have not yet been considered. What's more, Raman HSI opens to the door for *in vivo* applications where the images can indicate tumor location which can be useful for surgical procedures.

8.2.2 Spontaneous Raman Spectroscopy Combined with Other Analytical Techniques

In an attempt to increase the amount of useful information obtained for cancer diagnostics, some research efforts have focused on combining RS with other analytical techniques. Ideally, these additional methods will provide complimentary information to that obtained by Raman spectroscopic analysis and will increase the confidence and statistical significance of the methodology for diagnosing cancer.

RS was used to study 12 healthy and 30 tumor bladder tissue samples. Using HCA and differences in peak ratios, the tissue type could be classified with 96.7% sensitivity and 66.7% specificity. Major differences between the two classes included higher tryptophan, cholesterol, and lipid content levels in healthy tissue, and increased levels of nucleic acids, collagen, and carotenoids in bladder tumor tissue. High-performance liquid chromatography (HPLC), an analytical technique useful for separating, identifying, and quantifying individual components within a mixture, was employed to analyze carotenoids extracted from the two tissue types. While the Raman spectra reflect contribution due to carotenoids, HPLC was able to further narrow down this contribution to a specific biomarker; it was found that β-carotene was the major carotenoid present in tumor tissue, marking the first time this biomarker has been identified for bladder cancer [91].

RS and Raman HSI were combined with atomic force microscopy (AFM) to discriminate brain tumor from normal brain tissue samples. AFM, a type of scanning probe microscopy, was used to obtain nanomechanical properties to form images of healthy and cancerous brain tissue, while RS was used to glean information regarding the biochemical composition of the tissues. High-grade medulloblastoma (grade IV) and non-tumor samples from tissue of the central nervous system were compared. After analyzing the Raman spectra and images, it was determined that proteins within medulloblastoma tumors exist in the β-sheet conformation at enhanced levels and in the α-helix conformation at decreased levels as compared to proteins within normal tissue. Upon comparison of Raman peak ratios, it was discovered that in normal brain tissue, the relative amount of lipids compared to proteins is considerably higher. Mechanical indentation by AFM discovered that medulloblastoma tissue mechanical properties are strongly heterogeneous. Lastly, RS data was analyzed using PLS-DA with CV, indicating 96.3% sensitivity and 92% specificity for separating the two tissue types. Through combination of Raman HSI and AFM, the biochemical and nanomechanical signatures obtained have the potential to identify biomarkers associated with the development of brain cancer [92]. Although these results are comparable to studies which use only spontaneous RS, the added information that is obtained improves the usefulness of the methodology for diagnosing brain cancer.

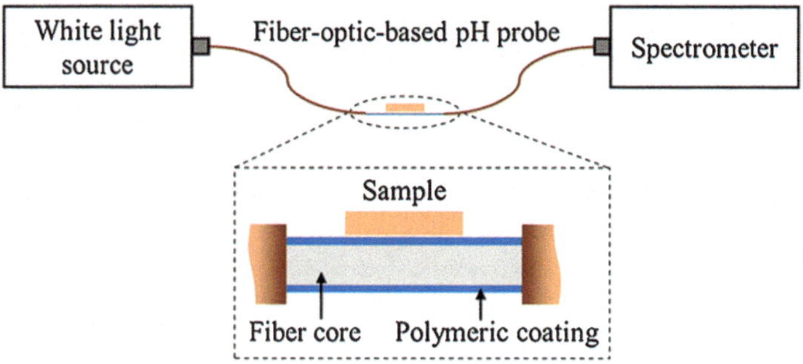

Fig. 8.9 Experimental setup using the fiber-optic-based pH probe for measuring the transmission spectra varying according to the pH level of the normal and cancerous breast tissue samples. The intensity of transmission spectra varies due to the absorbance change of the polymeric coating layer, which consists of neutral red/poly(acrylic acid) bilayers (Reprinted with permission from [93])

A novel approach combining spontaneous RS and optical pH sensing was used to differentiate healthy and cancerous breast tissue. To better prepare the Raman spectroscopic method for *in vivo* cancer detection, pH sensing can be first used to detect areas of tissue with lower pH levels, which is associated with cancer, thus ideally increasing the accuracy of the method as opposed to just using RS (Fig. 8.9). Fiber-optic-based Raman and pH probes were used to evaluate tissue samples; the pH sensing is based upon the pH level's dependence on the optical transmission spectrum. Raman spectra were collected first, followed immediately by collection of transmission spectra using the optical pH probe. The Raman spectra were combined with the transmission spectra from the same sample. PC-LDA with LOO-CV was employed for classification, first using only Raman spectra and then using the combined pH-Raman spectra. When Raman spectra were analyzed alone, the algorithm achieved 100% sensitivity and 91.5% specificity. When the algorithm analyzed the combined pH-Raman spectra, 100% sensitivity and 98% specificity were achieved, indicating the added advantage of pH sensing for diagnosing breast cancer using RS [93].

Both Raman and infrared (IR) spectroscopies were used in combination in several studies for the purpose of identifying various types of cancer. IR spectroscopy, another vibrational spectroscopic technique, is known to provide complimentary information to that obtained by RS. Specifically, IR spectroscopy analyzes the interaction of IR light with a molecule, generating an IR spectrum of energy that is absorbed or transmitted by the molecule as a function of either frequency or wavelength of light. The spectral information can then be used to identify and study the sample. The vibrational signatures of 164 invasive ductal carcinoma and invasive lobular carcinoma breast tissue samples were analyzed by both Raman and IR spectroscopies for the purpose of discriminating non-cancerous and cancerous tissue. Here, KCA followed by PCA and PLS-DA with CV were used to analyze the Raman spectral data. Raman imaging identified differences in spectral regions corresponding to vibrations of carotenoids, fatty acids, and proteins between normal and cancerous tissue, while IR spectra depict differences in proteins and phospholipids. Results of statistical analysis showed 84.7% sensitivity and 71.9% specificity for determining if breast tissue displayed either normal biochemistry or cancer pathology [94]. Owens et al. aimed to determine whether attenuated total reflection Fourier-transform infrared (ATR-FTIR) spectroscopy or RS could better characterize the biomolecular signatures of blood plasma or serum collected from patients with ovarian cancer as compared to healthy controls. FTIR is used to simultaneously collect data over a wide spectral range; the ATR attachment allows for surface properties of a sample to be measured rather than bulk properties, thus decreasing the potential for strong attenuation of the IR signal in samples that are highly absorbent. Here, 60 blood samples were analyzed using ATR-FTIR spectroscopy, while only 8 samples were studied using RS. All spectra were subjected to PCA-LDA, which showed statistically significant differences between healthy and cancerous samples using both spectroscopic methods. A SVM classifier successfully differentiated Raman spectral data of blood plasma with 74% accuracy; notably, the IR spectral data of blood plasma was successfully classified with 93.3% accuracy. It was further found that blood plasma was better suited for diagnostic discrimination than blood serum. Although ATR-FTIR spectroscopy is shown here to better diagnose ovarian cancer, one should consider the different sample sizes used in each part of the experiment [95]. In another study, Raman and ATR-FTIR spectroscopies were used to determine if either could identify the primary site of a metastatic tumor. Metastases were obtained from primary lung and colorectum AC as well as from metastatic melanoma. PCA-LDA determined points of dissimilarity between spectra; PCA in combination with a linear discriminate classifier (LDC) calculated classification accuracy. In a three-class algorithm built using Raman spectral data, 69% accuracy for predicting colorectal AC, 69% for lung AC, and 72% for melanoma were achieved. Using ATR-FTIR spectral data, 60% accuracy for predicting colorectal AC, 59% for lung AC, and 47% for melanoma were achieved. Interestingly, combination of the two AC groups improves results to 85% accuracy for predicting AC and 75.4% for melanoma using the Raman data and to 96% accuracy for AC and 72% for melanoma using the ATR-FTIR data [96]. In general, IR spectroscopy performs similarly to, if not better than, RS in these studies. It should be noted that differences in sample sizes may play a role, and that spontaneous RS has already been shown in other studies to successfully diagnose these same cancers.

In a unique study by Tatarkoič et al., blood plasma samples from 55 individuals were investigated using a combination of electronic circular dichroism (ECD), Raman optical activity (ROA), and conventional Raman and FTIR spectroscopies for the purpose of diagnosing colon cancer. ECD is a useful technique for analyzing stereochemistry; an ECD spectrum is the difference between absorption of left and right circularly polarized lights due to electronic transitions in the UV or visible regions of the spectrum [97]. Similarly, ROA measures the difference in intensity of Raman scattered left and right circularly polarized light which arises because of molecular chirality [98]. These techniques help to provide more specific information regarding the biochemical composition of a sample in order to better increase the ability to identify cancer with chemometrics. The results of LDA showed that, for each of the individual methods, limited discrimination between control group subjects and patients with colon cancer was achieved. However, when spectra from all methods were combined and again evaluated using LDA with LOO-CV, sensitivity and specificity reached 93% and 81%, respectively, with an overall accuracy of 87% for discriminating the two classes of blood plasma samples [99]. Despite the combination of so many techniques, the performance of the model is not necessarily a significant improvement over those built using spontaneous RS data of various biological samples.

Lin et al. also used the combination of several different techniques, this time for diagnosing NPC. Here, a 4-modality endoscopy system comprised of white light imaging (WLI), AFI, diffuse reflectance spectroscopy (DRS), and RS was used for *in vivo* NPC detection. WLI can locate suspicious lesions, but has low diagnostic sensitivity and relies on subjective analysis. AFI, which has a higher diagnostic sensitivity, has the ability to monitor biochemical changes that occur in tissue based on the fluorescence profile of internal fluorophores which are associated with cancer progression. DRS can improve AFI by providing morphological and functional quantitative information regarding the tissue samples. RS and DRS data were collected from patients with NPC and from healthy subjects under the assistance of AFI and WLI. When the combined DRS/RS dataset was applied to PCA-LDA, the algorithm achieved 98.6% sensitivity and 95.1% specificity for separating the two groups of tissue samples, showing the usefulness of combining multiple methods to improve results [100].

The combination of RS and AFI was used in multiple studies to diagnose skin cancer. Zakharov et al. used fluorescence analysis first to quickly scan large areas of tissue samples for abnormality detection; when malignancy was suspected, Raman spectral analysis of the tissue was performed. Quadratic discriminant analysis (QDA) of the data provided a diagnosis of malignant melanoma with 89% sensitivity and 87% specificity [101]. In a proceeding study, Raman and auto fluorescence (AF) spectroscopies were used to identify skin neoplasms as melanoma, BCC, or benign tumors. Here, the Raman and AF signals were combined and analyzed via PLS-DA with LOO-CV. Results showed 98.3% accuracy for separating malignant and benign tumors [102]. Similarly, Bratchenko et al. differentiated skin melanoma and BCC tissue samples through the combination of Raman and AF spectra (Fig. 8.10). When considered separately, neither set of spectra was able to exceed 79% accuracy; however, PCA-DA analysis of a combined spectral dataset with six selected spectral features provided 97.3% accuracy for malignant skin detection [103]. Interestingly, these studies each show how AF can be used to increase the reliability of the RS method for diagnosing skin cancer.

While RS is oftentimes suitable for identifying cancer by itself, analysis can oftentimes be improved when additional methodologies are combined, as is displayed by the previously mentioned studies. However, it is important to note that the combination of multiple techniques increases the level of complexity of the methodology as well as potentially increasing the time, cost, and effort required to

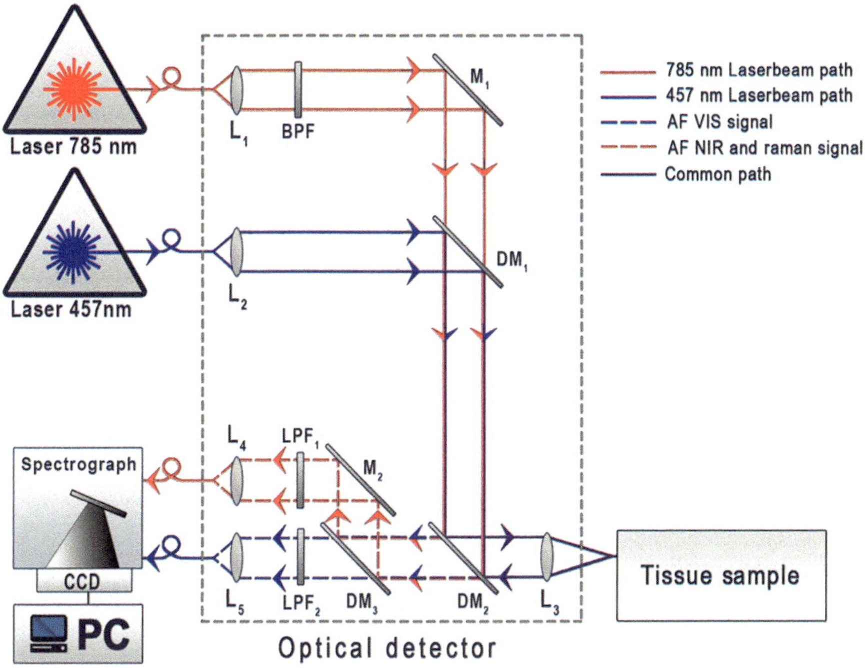

Fig. 8.10 Experimental setup: L_1, L_2, L_4, and L_5: matching lenses, L_3: focusing lens, BPF: band-pass filter, M_1 and M_2: mirrors, DM_1, DM_2, and DM_3: dichroic mirrors, and LPF_1 and LPF_2: long-pass filters (Reprinted with permission from [103])

achieve a diagnosis, indicating the importance of considering the costs and the benefits that accompany a more complex methodology system.

8.2.3 Modifications of Spontaneous Raman Spectroscopy

While conventional spontaneous RS has shown it is capable of diagnosing different cancers, some studies have advocated for the implementation of modifications of the technique. Variations of spontaneous RS have been proposed as effective methods to enhance diagnostic efforts. Those methods considered here are simple modifications of conventional RS; inclusion of techniques such as surface-enhanced and tip-enhanced RS are beyond the scope of this review chapter.

Fullwood et al. employed immersion Raman spectroscopy (IRS) to investigate brain cancer. Because IRS utilizes a specific immersion lens, the lens can have direct contact with a specific liquid; in this study, the liquid used was deionized water which covered the tissue sample being studied. Immersion of the sample in liquid protects the tissue from potential photo-damage and increases the spectral quality by reducing contribution of stray light to the spectral background. Both spontaneous RS and IRS data were collected from 48 tissue samples. It was determined that a lower background contribution was observed in the IRS data as compared to the RS data. A PC-LDA diagnostic algorithm was therefore built using the IRS data which could successfully discriminate between normal, GBM, and metastatic brain tissue spectra. Following this, researchers effectively distinguished different primary sites of cancerous tissue and investigated the biochemical differences between primary and metastatic cancer using samples from the same patient [104].

The majority of previously mentioned manuscripts have used dispersive RS; in a paper published in 2017, Fourier transformation (FT)-NIR RS was used to diagnose oral epithelial dysplasia. FT-NIR RS excites samples using a laser, such as the Nd:YAG used in this study, at a wavelength of 1064 nm; excitation in the IR region of light helps eliminate fluorescence but provides a weaker Raman signal. To adjust for this, an interferometer is used to convert the Raman signal to an interferogram which allows the entire Raman spectrum to be collected simultaneously by the detector, improving the signal-to-noise ratio. The FT algorithm then converts the interferogram to a conventional Raman spectrum. In this study, the goal was to differentiate normal oral mucosa, oral SCC, and dysplastic tissue samples. After spectra were collected, a SVM classifier was built and results were verified using PCA-LDA. Through SVM, accuracies for distinguishing mild, moderate, and severe dysplasia from oral SCC were 100%, 44.44%, and 71.15%, respectively. PCA-LDA analysis did not allow for successful discrimination of the stages, either, suggesting the need for improvements to the classification system. However, PCA-LDA could still identify biochemical discrepancies between normal, oral SCC, and dysplastic tissue samples [105]. Interestingly, these results are not necessarily an improvement of those performed using spontaneous RS analysis of biological specimens.

Coherent anti-Stokes Raman scattering (CARS) imaging was used in an attempt to diagnose both bladder and lung cancer. Similar to spontaneous RS, CARS is sensitive to molecular vibrational modes. Dissimilarly, three laser beams will each emit photons of particular frequencies to produce a coherent optical signal, at the anti-Stokes frequency, with the goal of producing a much stronger signal as compared to normal RS. Weng et al. used CARS to collect cellular-level images of normal and cancerous lung tissue samples. A deep convolutional neural network (DCNN) learning algorithm automatically differentiated normal, small cell carcinoma, AC, and SCC lung images with 89.2% accuracy [106]. Yosef et al. collected both CARS and second harmonic generation (SHG) images. The CARS imaging of urine sediments was used to preselect urothelial cancer cells. Next, Raman HSI of the cells was performed (Fig. 8.11). Through HCA, it was found that the cancer cells displayed a decrease in glycogen and an increase in fatty acid levels as compared to healthy controls. A RF classifier was built which could identify cancerous urothelial cancer cells based on the analysis of full cells or cytoplasm with 100% accuracy and based on nuclei with 90% accuracy after LOPO-CV [107]. The results of using CARS for diagnosing cancer are inconsistent, seeming to depend on the type of sample analyzed and the cancer being targeted.

Shifted-excitation Raman difference spectroscopy (SERDS) was employed as a label-free and non-invasive method for diagnosing oral SCC. During SERDS

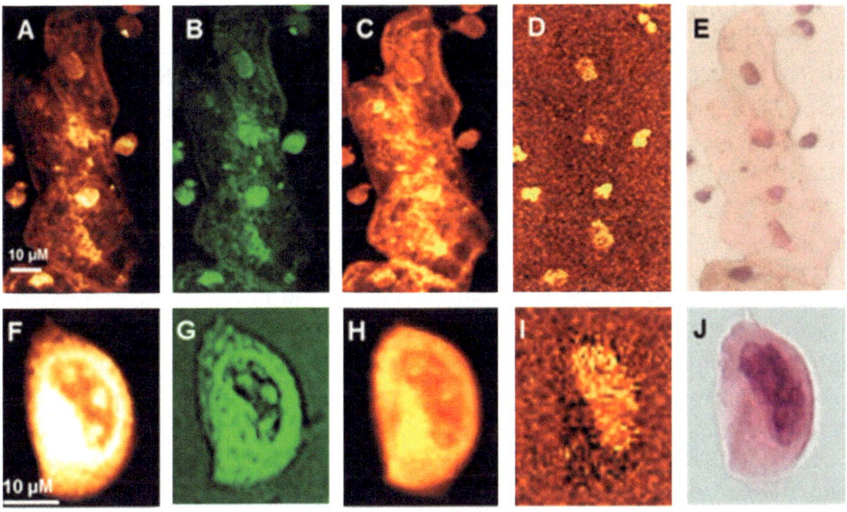

Fig. 8.11 Different imaging techniques applied to normal (**A–E**) and high-grade cancerous (**F–J**) urothelial cells in urine sediments: (**A**, **F**) SHG images, (**B**, **G**) CARS images, (**C**, **D**, **H**, **I**) integrated Raman intensity of cells in the (**C**, **H**) 2800–3050 cm^{-1} and (**D**, **I**) in 785–805 cm^{-1} regions, and (**E**, **J**) H&E-stained images (Reprinted with permission from [107])

measurements, spectra are first collected when the wavelength is set to a particular number (here, 783 nm). Then, the excitation wavelength undergoes a small shift to a second number (here, 785 nm), and a second spectral dataset is acquired. The dataset collected at each wavelength is averaged and the mean spectrum using the first excitation wavelength is subtracted from the mean spectrum collected using the second, ideally removing any contribution from fluorescence emission. In this manner, 72 SERDS spectra were collected, one from each of three different physiological tissue points and three different pathological lesions from 12 different patients. The SERDS spectra of malignant and benign tissues were discriminated using PCA-LDA, which achieved 86.1% sensitivity and 94.4% specificity for diagnosing oral SCC [108]. Although these results indicate success, it should be noted that many other research groups were able to accomplish similarly effective outcomes with much more simple RS technology.

Polarized Raman spectroscopy (PRS) was used by Daniel et al. for observing the biomolecular structural changes that occur in cervical cancer tissue samples. PRS is observed as the result of polarized light interacting with vibrating molecules, where the polarization is either parallel or perpendicular to the excitation laser's intrinsic polarization. Here, PRS provided information regarding the differences in orientation of biomolecules such as tyrosine, collagen, and DNA between normal and malignant cervical tissue samples. Depolarization ratios were analyzed by LDA with CV, yielding sensitivity, specificity and accuracy levels of 96%, 97.2%, and 96.7%, respectively. This is an improvement over the 92% sensitivity, 72.2% specificity, and 80.3% accuracy achieved using only spontaneous RS [109].

Stimulated Raman scattering (SRS) microscopy was used to reveal the infiltration of brain tumors in fresh surgical specimens collected from 22 neurosurgical patients. The basic mechanism of SRS is similar to spontaneous RS; however, SRS can enhance the signal of specific vibrational transitions due to the introduction of a second photon, a Stokes photon at a particular frequency, which can stimulate a specific molecular transition. By maintaining the pump laser beam at a constant frequency and scanning the Stokes laser beam, the spectral fingerprint of the sample can be improved. Here, results of SRS were in near-perfect agreement with results of H&E light microscopy. The data was interpreted using quasi-likelihood generalized additive models. Based on cellularity, axonal density, and protein/lipid ratios observed in SRS images, the classifier could successfully detect tumor infiltration with a sensitivity of 97.5% and specificity of 98.5%. The classifier was also able to distinguish between various categories of tumor infiltration including normal to minimal hypercellularity, infiltrating glioma, or dense glioma with high levels of accuracy [110]. Stimulated Raman histology (SRH) was used in a complimentary study for the intra-operative diagnosis of pediatric-type brain tumors. Based on RF analysis, 25 pediatric-type surgical specimens were correctly classified as normal versus lesional tissue and low-grade versus high-grade tumors all with 100% accuracy [111]. These results are generally better than those obtained through analysis of various biological samples by spontaneous RS for brain cancer diagnosis.

High-throughput (HT) RS was used for rapid screening of blood plasma samples collected from prostate cancer patients and healthy volunteers. In general, HT screening methods have the ability to automatically control and conduct millions of tests with a specific goal, saving time and effort for the user. Medipally et al. developed a HT-RS method which was optimized through testing a series of different instrumental and sample preparation parameters (Fig. 8.12). Once adjusted, the method was able to automatically record multiple Raman spectra from each of the well throughputs in a 94-well plate. To test the method, Raman spectra were obtained for blood plasma collected from 10 healthy volunteers and 10 prostate cancer patients using both 785 and 532 nm excitation. The best results were seen using the 785 nm excitation, with PCA-LDA yielding 96.5% sensitivity and 95% specificity after CV, demonstrating the ability of HT screening methods to be successfully incorporated with RS methodology [112].

A unique study performed by Stables et al. classified brain tumor spectra using spontaneous RS in combination with sound and listening tests. Metastatic brain cancer, glioblastoma, and non-cancer tissue samples were analyzed using RS. Three different chemometric techniques (SVM, KNN, and LDA) with CV were evaluated for their potential to identify brain cancer within the tissue samples using a feature extraction approach. Compared to using PCA for spectral dimensionality reduction, the feature extraction approach increased classification accuracy of the KNN classifier by 25% to 91.02% and of the SVM classifier by 26.25% to 97.01%. For LDA, the classification accuracy decreased from 96.54% to 95.38%. The results suggest feature extraction to be a more effective approach as opposed to dimensionality reduction for classification efficiency. Sonification was then used on the reduced Raman dataset of extracted features. Frequency modulation synthesis was used to

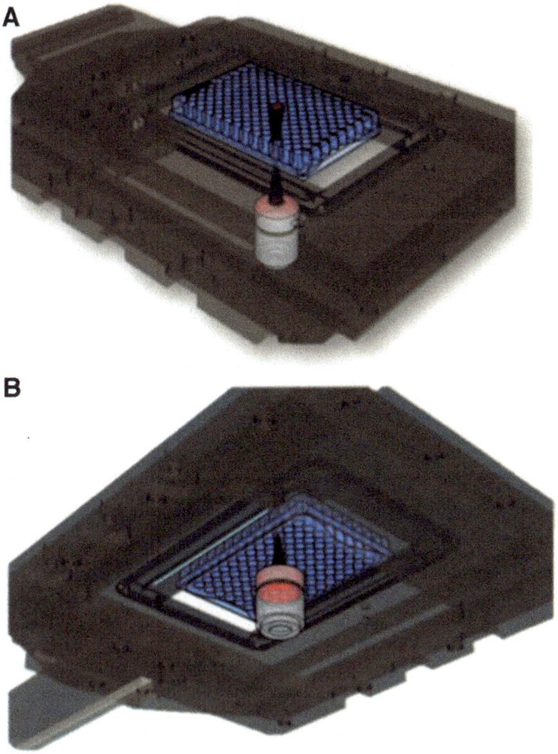

Fig. 8.12 Schematic representation of HT-Raman spectroscopy method. (**A**) Top view, (**B**) bottom view (these schematics are developed using Google Sketch up software) (Reproduced from [112] with permission from The Royal Society of Chemistry)

generate audio clips for each tissue sample based on the subset of extracted features, thus giving each one its own sound timbre, with similar tissue types having similar timbres. Listening tests were implemented with 25 participants, and based on the sound timbres, a mean classification accuracy of 71.1% was achieved, presenting a novel tool which can be used in addition to RS for clinicians to generate a diagnosis during endoscopic procedures [113].

Interestingly, not all modifications of spontaneous RS were completely effective in improving diagnostic accuracy. It should be observed that, on the other hand, there were some studies which did provide improvements. However, with the addition of more sophisticated methodology comes an increase in difficulty for bringing the technology into clinical settings. The more complicated the method, the less likely it is to be introduced as a new technology for universal cancer detection. While these aforementioned studies provide unique variations of RS, spontaneous RS alone has still shown great success in diagnosing cancers, suggesting the previously summarized alterations of the methodology may not ultimately be necessary for bringing the method to clinical settings.

8.2.4 Fiber-Optic Studies

To reach the ultimate goal of *in vivo* diagnoses, probes have been increasingly incorporated into RS studies. Special instrument setups have been created in which a fiber-optic probe can analyze tissue *in vivo* and collect Raman spectral data. Fiber-optic probes have the advantage of being less bulky and less expensive than typical Raman spectrometers; probes can be used intra-operatively, preventing the need for additional biopsy or *ex vivo* studies. They provide a shorter analysis time than typical histopathological examinations of biopsied tissue do, while still objectively capturing vital biochemical compositional changes that occur during disease progression. Probes can also allow for the identification of tumors, signaling where a surgeon should make excisions. When Raman spectral data collected through probes is analyzed using advanced statistical methods, research scientists are able to greatly reduce false positive biopsy results and increase the ease of and success of diagnosing cancers. Recent research which has incorporated probes into Raman spectroscopic systems, through either *in vivo* or *ex vivo* studies, for the purpose of diagnosing cancer are reviewed here. A schematic of a general fiber-optic probe setup is seen in Fig. 8.13.

Chen et al. used a low-resolution fiber-optic Raman sensing system to evaluate its diagnostic potential for *ex vivo* identification of different bladder pathologies. Spectra of 32 normal bladder tissue and low- and high-grade tumor bladder tissues were analyzed using a PCA fed ANN with CV. An overall accuracy of 93.1% was obtained for predicting to which class a sample belonged, introducing the possibility for further experiments to be successfully conducted *in vivo* [114].

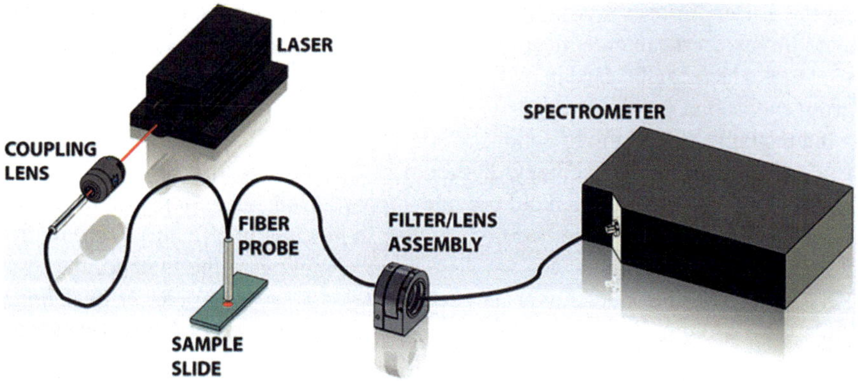

Fig. 8.13 Schematic of the experimental setup showing the 785 nm laser directed into the Raman probe via the 10× objective lens. The probe illuminates the tissue sample and collects the scattered light. The elastically scattered signal is removed via a long pass filter in the filter/lens assembly before the light is transmitted into the Maya Pro 2000 NIR spectrometer for dispersion and storage (Reprinted with permission from [153])

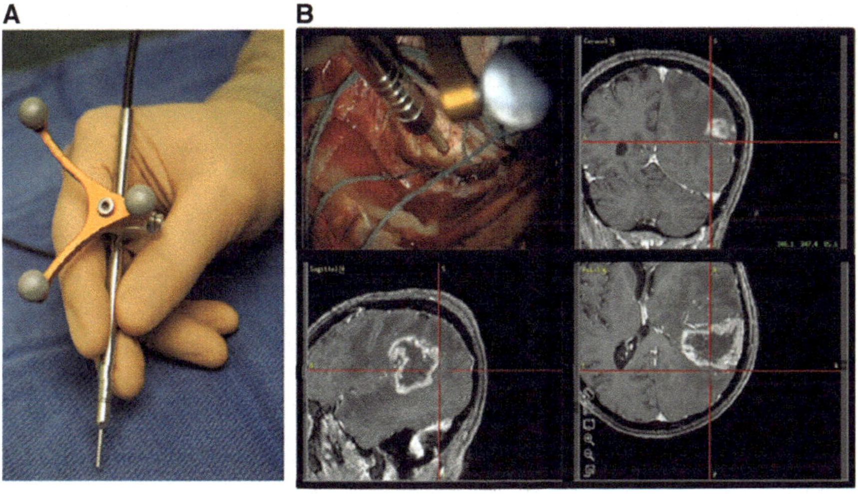

Fig. 8.14 Raman spectroscopy system for intra-operative detection. (**A**) Photograph of the hand-held contact probe, with the attached neuronavigation tracking unit. (**B**) Illustration of the probe being used intra-operatively, with the neuronavigation system showing the location of the tip of the probe (cross hairs) on the preoperative magnetic resonance images (Reprinted with permission from [115])

Three different studies show the strength of Raman probe systems for diagnosing brain cancer *in vivo*. A handheld RS probe system was used to collect *in vivo* spectral data from normal, cancerous, and necrotic brain tissue of ten patients (Fig. 8.14). Using PCA and a boosted trees (BT) classification algorithm with LOO-CV, an accuracy of 87% for distinguishing necrosis from tumor and normal brain tissue was achieved [115]. Jermyn et al. used a handheld contact RS probe to differentiate normal brain, dense cancer, and normal brain invaded by cancer cells with 93% sensitivity and 91% specificity, using the BT machine learning method with CV. The RS probe system was also able to detect previously unidentifiable invasive brain cancer cells in patients with grade II through IV gliomas, showing the usefulness of fiber-optic probes for detecting cancerous cells which are oftentimes missed during normal surgery [116]. In a following study, RS data was collected intra-operatively from 17 patients with grades II through IV gliomas. Both BT and ANN were used for classifying the spectra. ANN performed better than BT when algorithms for distinguishing cancer from normal brain were built including light artifacts but performed the same when they were built excluding light artifacts due to operating room sources. Specifically, when light artifacts were excluded from the spectra, ANN achieved 92% classification accuracy, an improvement over 90% accuracy achieved when light artifacts were not excluded [117].

Li and co-researchers performed several studies using a miniature Raman spectrometer equipped with a fiber-optic probe for the purpose of diagnosing breast cancer. In the earliest study, 16 breast tissue samples were analyzed and an adaptive weight k-local hyperplane (AKWH) algorithm was used for differentiation. Three different data processing schemes were generated based on varied splitting of the Raman spectral dataset; on average, the AWKH algorithm gave a 95.8% accuracy for classifying breast tissue as either cancerous or healthy [118]. The same samples were then analyzed using an adaptive net analyte signal AWKH pattern recognition method. Again, three different data processing mechanisms were generated based on different splitting of the Raman spectral dataset; the average accuracy of classification was 94.83% [119]. In their last study, new normal and malignant breast tissue samples were obtained, with the cancerous tissue existing at various stages of the disease. An adaptive local hyperplane K-nearest neighbor method was used for binary classification, achieving 93.2% accuracy [120]. While these successful results are generally comparable to those previously reviewed, they indicate the vital diagnostic potential of RS to be used intra-operatively.

A fiber-optic Raman system was used to obtain 68 spectra from benign and low- and high-grade SIL of 25 cervical tissue specimens. Multiclass PLS-DA with LOPO-CV showed an average sensitivity of 86.6% and specificity of 93.6% for classification [121]. Shaikh et al. performed two studies to explore *in vivo* classification of normal and cervix tumor tissue Raman spectra. First, 314 Raman spectra were collected from 63 subjects; the data was subjected to PC-LDA with LOO-CV, and classification efficiency reached 96.7% and 100% for the normal and cancerous conditions, respectively [122]. In the second study, PC-LDA was used to distinguish between normal and cancerous tissue as well as tissue collected from the vagina of both healthy controls and cancer patients, in an attempt to design an internal control. PC-LDA could classify normal and tumor spectra with 97% efficiency. When a PC-LDA algorithm was built to discriminate between all controls (normal cervix, and vagina of tumor and normal subjects) high misclassification levels were seen, suggesting similarities in biochemical composition among the control samples. Results of classification between tumor tissue and all controls support the idea of using the vagina as an internal control in cervical cancer diagnostics [123].

Wood et al. evaluated biopsy samples collected during colonoscopy using probe-based RS. The *in vitro* study examined 356 colon biopsies, including from normal colon mucosa, hyperplastic polyps (HP), adenomatous polyps, AC, and ulcerative colitis specimens. PC-LDA with LOO-CV was used to make two-group and three-group classification systems. For the binary models, accuracies ranged between 72.1% and 95.9% with ten-second acquisition times and between 61.5% and 95.1% with one-second acquisition times. For the tertiary model, normal tissue, adenomas, and AC tissue were identified with an overall accuracy of 74.1% for the ten-second acquisition time and 63.5% for the one-second acquisition time [124]. Raman fiber-optic measurements of colon biopsy samples, which were categorized as AC, tubular adenomas (TA), HP, and normal tissue, were analyzed from 151 patients. A SVM classifier was trained and validated using a LOPO-CV approach. For classifying AC versus normal tissue, 75% accuracy was achieved. To improve results, three different methods for outlier identification were applied: One Class Classification with SVM, Local Outlier Factor, and Refinement of Training Data (RoTD). The best improvement was seen with RoTD, which increased the accuracy of AC versus normal tissue classification to 81%. To classify high-risk (AC and TA) and low-risk

(HP and normal tissue) lesions, the SVM model without outlier identification reached 71% accuracy; with RoTD, accuracy increased to 77% [125]. Although other studies performed using typical spontaneous RS were more successful, it is important to note the beginning successes of an *in vivo* approach toward diagnosing colorectal cancer using RS and chemometrics.

A custom-built fiber-optic endoscopic Raman probe was used to analyze 673 *ex vivo* esophageal tissue samples from patients with Barrett's esophagus (BE). BE is known to increase the risk of developing esophageal cancer. The tissue was evaluated with PCA-fed LDA with LOPO-CV, which discriminated BE-associated high-grade dysplasia (HGD) and AC from low-grade dysplasia, nondysplastic BE, and normal squamous esophagus with 86% sensitivity and 88% specificity. AC was differentiated from normal squamous esophagus with 94% sensitivity and 91% specificity. Finally, BE and gastric mucosa were differentiated with 96% sensitivity and 92% specificity [126]. A beveled fiber-optic confocal Raman probe was evaluated for *in vivo* diagnosis of BE using epithelial tissue from 373 patients, obtained at endoscopy. Trichotomous probabilistic PLS-DA was used to discriminate columnar-lined epithelium, nondysplastic BE, and HGD BE. For *in vivo* detection of HGD BE, 87.0% sensitivity and 84.7% specificity were attained [127]. In another study, a Raman endoscopic probe measured 673 *ex vivo* benign and esophageal cancer specimens from 62 patients. The results of using a semi-supervised LDA technique, where some of the data is labeled and some is left unlabeled, was compared to standard (supervised) LDA results. Identification of intestinal metaplasia versus dysplasia improved from sensitivity and specificity levels of 73% and 78% with standard PCA-LDA to 78% and 84% for the semi-supervised method. Similarly, performance for differentiating intestinal metaplasia and low-grade dysplasia increased from 44% and 66% using standard PCA-LDA to 63% and 72% sensitivity and specificity levels, respectively, with semi-supervised LDA [128]. In a different study, Maeda et al. performed *ex vivo* experiments using a portable Raman system equipped with a micro-Raman probe. Spectra collected of normal and early-stage (stage 0) cancerous regions within six esophageal samples were analyzed by PC-LDA, which predicted the tissue type with 80% accuracy [129]. Interestingly, these studies all used either PCA-LDA or PLS-DA, showing that the number of samples and the method of sample probing can have a significant impact on the results of a study.

A fiber-optic depth-resolved NIR Raman endoscopic technique was integrated with diagnostic algorithms for *in vivo* epithelial diagnosis of gastric cancer with the assistance of wide-field imaging techniques. Generated diagnostic models using probabilistic PLS-DA with LOPO-CV identified gastric dysplasia with 81.3% sensitivity and 88.3% specificity [130]. Wang et al. compared the performance of two different endoscope-based fiber-optic Raman probe methods. Beveled and volume Raman probes were used for real-time *in vivo* detection of gastric dysplasia. The beveled probe consists of a central flat fiber used for laser light delivery, surrounded by 18 beveled collection fibers positioned in a ring formation; the volume probe also consists of a central flat fiber for excitation but is surrounded by 18 flat collection fibers positioned in a ring formation. A total of 1050 Raman spectra of normal and dysplastic sites were collected from 66 gastric patients using the beveled Raman probe, while 1913 Raman spectra were collected from 98 gastric patients using the volume Raman probe. PLS-DA with LOPO-CV yielded diagnostic accuracies of 93.0% and 88.4% for the beveled and the volume fiber-optic probes, respectively, suggesting the beveled probe is better suited for further studies [131].

A miniature fiber-optic probe was used to investigate NPC in patients. Raman spectra were collected from nasopharynx tissue of patients with newly diagnosed NPC, post-irradiated nasopharynx (received radiotherapy greater than 6 months ago), or normal nasopharynx. A posterior probability model using PLS distinguished normal nasopharynx and NPC with 91% sensitivity and 95% specificity; the same method could distinguish post-irradiated nasopharynx versus NPC tissue with 77% sensitivity and 96% specificity [132]. A Raman spectrometer with a beam-steered fiber-optic probe was used to detect normal parotid gland and parotid gland tumors, including pleomorphic adenoma, Warthin's tumor, and mucoepidermoid carcinoma, for the purpose of identifying head and neck cancer. SVM with CV was used to distinguish each parotid gland tumor type against normal parotid glands, achieving an average accuracy of 99.43%. Three additional binary models were then built to distinguish the three tumor types from each other, achieving an average accuracy of 97.23% [133]. Here, it is observed that fiber-optic probes are successful for detection of head and neck cancers through both *in vivo* and *ex vivo* studies.

Lung cancer was studied using an endoscopic RS system. Spectra were collected *in vivo* from 280 tissue sites (including 72 HGD/malignant lesions and 208 normal/benign lesions) of 80 patients. Using stepwise multiple regression PLS with LOO-CV, HGD and malignant lesions were detected with 90% sensitivity and 65% specificity [134].

Oral cancer has been widely studied using Raman fiber-optic systems. In one study, Raman spectra were collected from the oral cavity of 18 human subjects *in vivo*, and premalignant/malignant lesions were correctly distinguished from normal and benign tissue with 100% sensitivity and 77% specificity using PCA-LDA

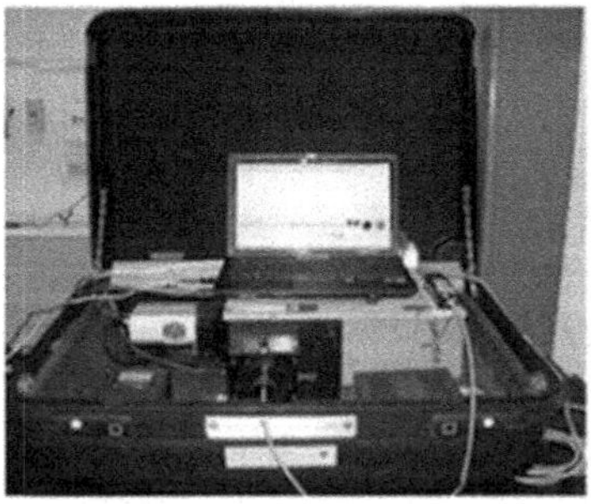

(a)

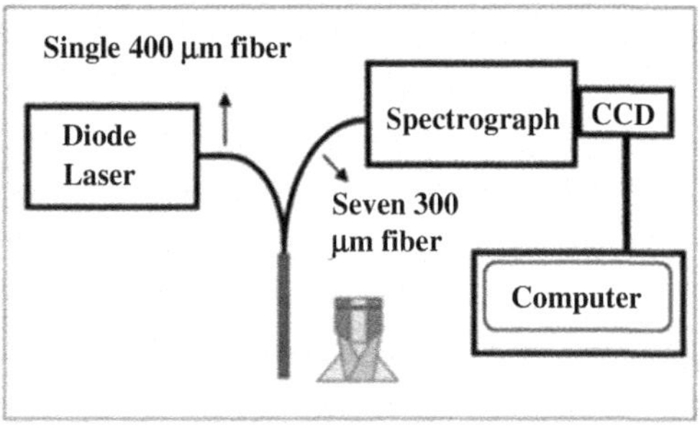

(b)

Fig. 8.15 (**a**) A photograph and (**b**) a schematic of the portable clinical Raman spectroscopy system for *in vivo* Raman measurements (Reprinted with permission from [136]) with LOO-CV [135]. Krishna et al. aimed to diagnose malignant and potentially malignant lesions of the oral cavity from 28 healthy volunteers and 171 patients. *In vivo* Raman spectra were collected from normal oral mucosa sites as well as histopathological sites including oral SCC, oral sub-mucous fibrosis, and leukoplakia using a portable clinical RS system (Fig. 8.15). A maximum representation and discrimination feature coupled with SMLR provided an average accuracy of 85.25% for classifying all four sites based on LOPO-CV; a binary model resulted in a sensitivity and specificity each of 94% for discriminating normal tissue spectra from all abnormal tissue spectra [136]. Research led by Sahu resulted in four studies published on diagnosing oral malignancies. In the earliest study, *in vivo* Raman spectra were collected from sera of buccal mucosa, tongue cancer, and healthy subjects using a fiber-optic Raman microprobe. Through PC-LDA with LOO-CV, binary models showed normal and cancer serum groups could be differentiated with about 70% classification efficiency and buccal mucosa and tongue cancer groups with about 68% efficiency [137]. Raman spectra were collected from oral exfoliated cells from healthy volunteers, healthy tobacco users, and subjects with oral cancer (from both tumor and healthy sites). PCA and PC-LDA showed distinct differences between the two healthy groups and the two cancer groups. Furthermore, PC-LDA with LOO-CV showed about 67% efficiency for predicting to which class the samples belonged, based on spectra-wise classification. Using a patient-wise approach, about 69% classification efficiency was achieved [138]. Following this, serum was collected from 22 oral cancer patients before and after surgery and analyzed using a Raman microprobe. PC-LDA followed by LOO-CV was again employed for discrimination, providing 78% classification efficiency for distinguishing recurrence and non-recurrence groups after surgery. The same method could distinguish recurrence and non-recurrence groups before surgery with only about 53% efficiency [139]. In the last study by Sahu et al., oral exfoliated samples were analyzed from healthy volunteers, healthy volunteers with tobacco habits, and patients with oral premalignant conditions (OPC) by the Raman microprobe. In the three-group model, OPC spectra were classified with 77% and 70% sensitivity for PC-LDA with spectra-wise and patient-wise CV methods, respectively. The sensitivity improved to 86% (spectra-wise) and 83% (patient-wise) using a binary model [140]. Interestingly, each of the studies performed by Sahu et al. show the ability of fiber-optic probes to be used for *ex vivo* studies performed on samples other than tissue. Yasser et al. analyzed the Raman spectra from parental oral cancer cell lines and from two different developed radio-resistant sublines using a fiber-optic microprobe system. Spectral differences were observed between the three different cell lines, and PCA showed distinct clustering, depicting the ability of RS to predict radioresistance in cells, which can be used for improved prognosis of oral cancer [141]. Notably, the best results for diagnosing oral cancer were achieved using the probe systems *in vivo* rather than *ex vivo*.

Spectra of fresh and non-processed post-prostatectomy specimens were collected using a macroscopic handheld RS probe. The areas of the tissue were labeled with tissue type (extra-prostatic or prostatic), malignant or benign, cancer grade (grade groups I–V), and tissue glandular level. Neural networks were used to classify the spectra in binary models. Prostate and extra-prostatic tissue were distinguished with 82% sensitivity and 83% specificity, whereas benign and malignant tissue were correctly classified with 87% sensitivity and 86% specificity. Benign spectra were differentiated from each of the five cancer grade groups in multiple binary models, achieving an average sensitivity of 81.8% and specificity of 85.2% [142]. Silveira Jr. et al. collected 160 spectra from 16 benign tissue and 16 prostate cancer tissue samples. A discrimination model was built using Euclidean distance based on the relative concentrations of phosphatidylcholine and water in the tissue samples. The two kinds of tissues were discriminated with 74% accuracy [143]. While these *ex vivo* studies are less successful than others previously reviewed, the results are still accomplished enough to indicate the potential for the method to be used intra-operatively.

A Raman instrument equipped with a fiber-optic probe was used to collect spectra *in vivo* from 137 lesions in 76 skin cancer patients; biopsies of the lesions were classified as malignant melanoma (MM), non-melanoma pigmented lesion (PL), BCC, actinic keratosis (AK), and SCC. The collected data was analyzed by PCA, and LOO-logistic regression classifiers were built, the results of which were compared to the histopathology of the lesions. The sensitivity and specificity for binary classification of MM versus PL were 100% and 100%; of SCC and BCC versus AK was 95% and 71%; and of AK, SCC, and BCC versus normal tissue was 90% and 85%, respectively [144]. Zakharov et al. investigated the potential for diagnosing malignant tumors in both skin and lung tissue. 40 *ex vivo* lung tissue samples and 50 *in vivo* skin tumor samples were investigated through a combination of LDA, QDA, and SVM. It was discovered that MM could be diagnosed with 88.9% sensitivity and 87.8% specificity, lung AC with 100% sensitivity and 81.5% specificity, and lung SCC with 90.9% sensitivity and 77.8% specificity [145]. In a different study, lesions suspected of being MM, BCC, or SCC were subjected to *in vivo* Raman spectral analysis through a fiber-coupled probe. Non-melanoma skin cancers were discriminated from normal skin through PLS-DA with accuracies of 73% (BCC) and 85% (SCC). MM and pigmented nevi (moles) were discriminated with 91% accuracy [146]. A dispersive spectrometer connected to a Raman probe collected data from non-melanoma (BCC and SCC), pre-cancerous (AK), and benign lesions and from normal tissue. Using PCA-DA and PLS-DA algorithms, non-melanoma and pre-cancerous lesions were differentiated from benign and normal tissue with accuracies of 82.8% and 91.9%, respectively [147]. Zhao et al. used PC-GDA and PLS, built with selected wavenumber windows, to classify 645 cases of pre-cancerous, benign, and skin cancer lesions. Malignant and benign skin lesions could be identified *in vivo* with high levels of diagnostic accuracy [148]. The success for skin cancer detection *in vivo* through Raman probe systems is clear; these results are comparable with those from other studies, and indicate that the method should be strongly considered for real-time diagnosis of skin cancer in clinical settings.

8.2.4.1 Fiber-Optic Studies with Expanded Raman Spectral Range

In several additional papers, the HWN region of spectral data collected using Raman probe systems was considered in addition to the FP region for real-time diagnoses of cancers. Specifically, a fiber-optic Raman endoscope was used to collect *in vivo* Raman spectra in the FP (800–1800 cm^{-1}) and HWN (2800–3600 cm^{-1}) regions from colorectal tissue. Raman measurements were made at five different anatomical locations of normal colorectal tissue and PLS-DA with LOPO-CV was used to identify the different tissue sites. An average sensitivity of 29.27% and specificity of 83.51% were achieved, indicating low levels of inter-anatomical molecular variability between normal colorectal tissue areas. For discriminating between normal tissue areas and tissue affected by colorectal cancer, PLS-DA with LOPO-CV of the FP/HWN dataset was again performed, attaining a diagnostic accuracy of 88.8% [149]. In a follow-up study, researchers again simultaneously acquired *in vivo* FP and HWN region Raman spectra from colorectal tissue. Adenoma and HP were differentiated with 90.9% sensitivity and 83.3% specificity using PLS-DA with LOPO-CV, which is superior to results achieved using only FP or only HWN region spectral data [150]. Both studies show the advantages of collecting FP/HWN spectral data via Raman probes for intra-operative diagnoses of colorectal cancer. *In vivo* diagnosis of esophageal SCC at the time of clinical endoscopy was investigated by Wang et al. FP and HWN region Raman spectra were collected from 48 patients using the developed fiber-optic RS technique. Through PLS-DA with LOPO-CV, a sensitivity of 92.7% and specificity of 93.6% for esophageal SCC identification were achieved. Again, these results were found to be superior to those obtained using only FP or HWN region spectral data [151]. Wang et al. applied their rapid fiber-optic RS technique for diagnosing gastric pre-cancer during endoscopic examination. FP/HWN region Raman spectra from normal, dysplasia, and AC tissue sites were collected. PLS-DA with LOPO-CV reached an average sensitivity of 88.67% and specificity of 92.53% for detecting each of the three groups. A binary model could detect gastric dysplasia with 90.9% sensitivity and 95.9% specificity [152]. Further, the discrimination of diseased tissue and adjacent healthy tissue from patients who have head and neck cancer was accomplished using a wide Raman spectral range of 100–4300 cm^{-1}. PCA showed effective separation between healthy controls and malignant tissue samples, which included SCC and tonsil SCC; the separation was better observed through analysis of the full spectrum then it was of only the FP region [153].

Lin et al. utilized fiber-optic RS to develop a method for *in vivo* diagnosis of NPC at the time of endoscopy using the FP and HWN regions. Spectral data was collected from 204 different tissue sites of 95 subjects; PCA-LDA with LOO-CV provided a diagnostic accuracy of 93.1% [154]. In another study, Lin et al. acquired FP and HWN region spectral data of 101 healthy and diseased tissue sites from 60 patients with laryngeal cancer undergoing endoscopic examination. Here, PLS-DA with LOPO-CV could discriminate the two classes with an accuracy of 91.1% [155]. In both of those studies, and in general, the results were improvements over those achieved using just FP or HWN region spectral data.

The combination of Raman spectroscopic analysis with optical probing systems provides the first crucial step toward bringing the methodology to the clinical setting. By allowing for collection of spectral data *in vivo*, the need for additional sample collection which can be time-consuming, expensive, and invasive is eliminated. The spectral data can still be analyzed using a wide variety of chemometric techniques, as was shown, in order to develop an automatic diagnostic system which can be incorporated into clinical settings for quick and accurate diagnoses. Although not every study was an improvement upon those performed without using probes, each of the aforementioned projects serves to indicate that the technology is capable of use within clinical settings. This is an exciting first step toward introducing RS as a universal method for cancer detection that can be used *in vivo* and provide accurate results in real-time.

8.3 Critical Evaluation

There has been a vast amount of research published on utilizing Raman spectroscopy and advanced statistical analysis for the purpose of diagnosing cancer. Obviously, the methodology has great potential. Regardless of the biological sample analyzed, the exact variation of RS used, or the statistical technique applied—it is impossible to disagree that based on the incredible amount of research conducted and published, RS and advanced statistical analysis have a great potential for creating the first universal method for cancer detection.

While the potential of the method is obvious, it is important to note that some results published in the aforementioned studies may overstress their significance, and as such there is a need to address the risk associated with overestimating the capabilities of the methodology based on the reported results. A small number of studies do not report quantitative results, and focus more on the qualitative success of the methodology; while these findings are still important for indicating the ability of RS to detect biochemical differences between different sample types, it is necessary to remember that quantitative results are necessary for supporting efforts to bring the methodology into clinical settings. Several studies suffered from too small of a dataset to be considered significant. Some research achieved sensitivity, specificity, or accuracy levels which are not necessarily an improvement of those achieved using current methods for diagnosing cancer. The balance between number of samples used and performance results is something that needs to be kept in mind when evaluating the significance of different experiments with comparison to each other. Ideally, the most reliable studies are those which use a statistically significant number of samples and achieve impressive performance levels. Further, the methods of validation for several experiments are considered "internal" validation—this means that the model was tested with the same spectral data that was also used to build it. This can lead to the potential for the model to "over-fit" itself to the data it sees, preventing it from being able to accurately predict spectral data from new unknown samples.

On the other hand, many more studies can be considered reliable—those experiments that utilized external validation are more trustworthy than those that did not. External validation utilizes an independent spectral dataset, which the algorithm has not yet seen, in order to test the performance of the model. In this regard, there are much lower chances of the model becoming over-fit, and the results are abundantly more dependable. Furthermore, while a small number of samples is useful to accomplish a proof-of-concept study, many more published manuscripts used a statistically significant number of samples which also obtained impressive levels of accuracy. Most importantly, the number of strong research studies heavily outweigh those that are less ideal.

Based on the extensive amount of recent evidence which suggests the potential for RS to diagnose cancer, not to mention the plethora of significant research published greater than 4 years ago, one important question remains: why has this methodology not yet been introduced into clinical settings as a solitary method for diagnosing cancer? Although the answer is complicated and multifaceted, the truth remains that research has already shown RS is being used intra-operatively. In fact, many of the previously reviewed experiments performed using fiber-optic probes were conducted *in vivo*. In 2013, Kallaway et al. summarized the clinical usage of Raman spectroscopy for diagnosing colorectal, esophagus, breast, and bladder diseases and cancers [156]. Pence et al. detail the necessary considerations required for clinical implementation of RS as well as review large (considered as greater than 50 samples) *in vivo* applications of the method [157]. More recently, Santos et al. greatly covered the translation of RS into clinical settings for detecting cancer, reviewing many studies which successfully detected cancer with high levels of accuracy under conditions that resemble the intended clinical environment [158].

Although research repeatedly advocates for the potential of the method, as Santos importantly points out, there needs to be an increase in communication as well as trust between spectroscopists and clinicians in order to bring this methodology closer to the goal of true clinical applications. Understanding the methods behind the technology as well as understanding the needs of clinicians are important areas that are still being developed. However, the stepping stones have been laid, and there remains a clear path forward for introducing Raman-based technology to clinical oncology departments for cancer detection purposes.

Lastly, there is an essential need to consolidate the classification techniques used in order to unite the automatic analysis and disease detection portion of the methodology. A multiclass discrimination technique will need to be developed and fine-tuned which considers all cancers, or at least considers many groups of similar cancers. Based on typical expectations of the medical field, a major clinical trial needs to be conducted with a wide cohort of participants ranging in age, gender, race, ethnicity, location of residence, prescribed medications, and comorbidities. While research has repeatedly shown success on a smaller scale, a successful large-scale clinical trial is needed in order to convince all involved parties of the method's capabilities. Although there is work to still be done, RS has shown every indication that it has the potential to be used in clinical settings in the future for universal cancer detection.

8.4 Conclusion

Cancer affects millions of individuals every year and is a leading cause of death worldwide. Because of its association with mortality, it is vitally important to diagnose cancer as early on in its progression as possible. Early diagnoses provide the best chance for the afflicted individual to seek effective treatment options. While many methods exist to diagnose cancers individually, there is a strong unmet need to accurately, definitively, and with minimal invasiveness diagnose all cancers using one universal method. This chapter proposes Raman spectroscopy as a potential solution for this task. In countless different ways, the combination of Raman spectroscopy with chemometric analysis has proven its usefulness for diagnosing cancer. Innumerous studies have been published on over twenty different forms of cancer in the last few years alone. Regardless of the sample studied or which variation of the technique is used, if a probe is used to collect data or a second analytical technique is used to provide complimentary information, the obvious capability of RS for diagnosing cancers cannot be denied. The abundance of information that is obtained through Raman spectral data provides not only helpful material for classification purposes, but also delivers insight into the biochemical composition of samples, revealing both useful information as well as potential biomarkers indicative of different cancers. The plethora of herein reported studies have repeatedly shown evidence that RS is sensitive, specific, and overall a reliable technique for differentiating healthy samples from diseased samples. The next step for this methodology is to unite researchers to work toward conducting a large-scale clinical trial where a combined algorithm can be developed which can diagnose all forms of cancer. The results of such a trial would need to be confirmed using already established methods for cancer diagnosis; however, if such a trial is successful, it is possible that Raman spectroscopy with advanced statistical analysis may become the first singular universal method for diagnosing cancer.

References

1. Bray, F., et al.: Global cancer statistics 2018: GLOBOCAN estimates of incidence and mortality worldwide for 36 cancers in 185 countries. CA Cancer J. Clin. **68**(6), 394–424 (2018)
2. Siegel, R.L., Miller, K.D., Jemal, A.: Cancer statistics, 2019. CA Cancer J. Clin. **69**(1), 7–34 (2019)
3. Smith, R.A., et al.: Cancer screening in the United States, 2018: a review of current American Cancer Society guidelines and current issues in cancer screening. CA Cancer J. Clin. **68**(4), 297–316 (2018)
4. Mann, C.K., Vickers, T.J.: In: Lewis, I.R., Edwards, H.G.M. (eds.) Handbook of Raman Spectroscopy: From the Research Laboratory to the Process Line, pp. 251–274. Marcel Dekker, Inc., New York (2001)
5. Williams, T.L., Collette, T.W.: In: Lewis, I.R., Edwards, H.G.M. (eds.) Handbook of Raman Spectroscopy: From the Research Laboratory to the Process Line, pp. 683–731. Marcel Dekker, Inc., New York (2001)

6. Wold, S.: Chemometrics; what do we mean with it, and what do we want from it? Chemom. Intel. Lab. Syst. **30**(1), 109–115 (1995)
7. Geladi, P.: Chemometrics in spectroscopy. Part 1. Classical chemometrics. Spectrochim. Acta B: Atomic Spectrosc. **58**(5), 767–782 (2003)
8. Sikirzhytski, V., Sikirzhytskaya, A., Lednev, I.K.: Multidimensional Raman spectroscopic signatures as a tool for forensic identification of body fluid traces: a review. Appl. Spectrosc. **65**(11), 1223–1232 (2011)
9. Muro, C.K., et al.: Forensic body fluid identification and differentiation by Raman spectroscopy. Forensic Chem. **1**, 31–38 (2016)
10. Kalkanis, S.N., et al.: Raman spectroscopy to distinguish grey matter, necrosis, and glioblastoma multiforme in frozen tissue sections. J. Neurooncol. **116**(3), 477–485 (2014)
11. Liu, T., et al.: Evaluation of Raman spectra of human brain tumor tissue using the learning vector quantization neural network. Laser Phys. **26**(5), 055606 (2016)
12. Sathyavathi, R., et al.: Raman spectroscopic sensing of carbonate intercalation in breast microcalcifications at stereotactic biopsy. Sci. Rep. **5**, 9907 (2015)
13. Han, B., et al.: Differences and relationships between normal and atypical ductal hyperplasia, ductal carcinoma *in situ*, and invasive ductal carcinoma tissues in the breast based on Raman spectroscopy. Appl. Spectrosc. **71**(2), 300–307 (2017)
14. Fallahzadeh, O., Dehghani-Bidgoli, Z., Assarian, M.: Raman spectral feature selection using ant colony optimization for breast cancer diagnosis. Lasers Med. Sci. **33**(8), 1799–1806 (2018)
15. Rashid, N., et al.: Raman microspectroscopy for the early detection of pre-malignant changes in cervical tissue. Exp. Mol. Pathol. **97**(3), 554–564 (2014)
16. Daniel, A., et al.: Biochemical assessment of human uterine cervix by micro-Raman mapping. Photodiagnosis Photodyn. Ther. **17**, 65–74 (2017)
17. Daniel, A., Aruna, P., Ganesan, S.: Near-infrared Raman spectroscopy for estimating biochemical changes associated with different pathological conditions of cervix. Spectrochim. Acta A Mol. Biomol. Spectrosc. **190**, 409–416 (2018)
18. Synytsya, A., et al.: Raman spectroscopy at different excitation wavelengths (1064, 785 and 532 nm) as a tool for diagnosis of colon cancer. J. Raman Spectrosc. **45**(10), 903–911 (2014)
19. Li, S., et al.: Identification and characterization of colorectal cancer using Raman spectroscopy and feature selection techniques. Opt. Express. **22**(21), 25895–25908 (2014)
20. Liu, W., et al.: Raman spectroscopy in colorectal cancer diagnostics: comparison of PCA-LDA and PLS-DA models. J. Spectrosc. **2016**, 1603609 (2016)
21. Ishigaki, M., et al.: Diagnosis of early-stage esophageal cancer by Raman spectroscopy and chemometric techniques. Analyst. **141**(3), 1027–1033 (2016)
22. Jin, S., Mao, H.: Near-infrared Raman spectroscopy for diagnosis of gastric cancer. J. South. Med. Univ. **34**(3), 391–395 (2014)
23. Yao, C., et al.: A preliminary study of protein alterations accompanied with malignant transformation of gastric mucosa using Raman spectroscopy. Chongqing Med. J. **29**, 3875–3878 (2014)
24. Hsu, C.-W., et al.: Novel method for differentiating histological types of gastric adenocarcinoma by using confocal Raman microspectroscopy. PLoS One. **11**(7), e0159829 (2016)
25. Hsu, C.-W., et al.: Differentiating gastrointestinal stromal tumors from gastric adenocarcinomas and normal mucosae using confocal Raman microspectroscopy. J. Biomed. Opt. **21**(7), 075006 (2016)
26. Li, J.H., Li, W.T., Zhang, G.H.: Detection of nasopharyngeal carcinoma using deep NIR Raman spectroscopy. Laser Phys. **24**(12), 125601 (2014)
27. Li, Y., et al.: Rapid detection of nasopharyngeal cancer using Raman spectroscopy and multivariate statistical analysis. Mol. Clin. Oncol. **3**(2), 375–380 (2015)
28. Mian, S.A., et al.: Raman spectroscopy can discriminate between normal, dysplastic and cancerous oral mucosa: a tissue-engineering approach. J. Tissue Eng. Regen. Med. **11**(11), 3253–3262 (2017)

29. Pence, I.J., et al.: Discrimination of liver malignancies with 1064 nm dispersive Raman spectroscopy. Biomed. Opt. Express. **6**(8), 2724–2737 (2015)
30. Wang, L., et al.: Evaluation of Raman spectroscopy for diagnosing EGFR mutation status in lung adenocarcinoma. Analyst. **139**(2), 455–463 (2014)
31. Behl, I., et al.: Raman microspectroscopic study of oral buccal mucosa. In: SPIE BiOS. International Society for Optics and Photonics, San Francisco (2014)
32. Daniel, A., et al.: Raman mapping of oral tissues for cancer diagnosis. J. Raman Spectrosc. **45**(7), 541–549 (2014)
33. Chen, P.-H., et al.: Automatic and objective oral cancer diagnosis by Raman spectroscopic detection of keratin with multivariate curve resolution analysis. Sci. Rep. **6**, 20097 (2016)
34. Lu, S.-y., et al.: Raman spectroscopy in ovarian cancer diagnostics. Spectrosc. Spectr. Anal. **37**(6), 1784–1788 (2017)
35. Du, Y., et al.: A novel diagnostic method of Raman spectroscopy for malignant pheochromocytoma/paraganglioma. RSC Adv. **6**(103), 101178–101184 (2016)
36. Liu, Y., et al.: Renal mass biopsy using Raman spectroscopy identifies malignant and benign renal tumors: potential for pre-operative diagnosis. Oncotarget. **8**(22), 36012–36019 (2017)
37. Kong, K., et al.: Increasing the speed of tumour diagnosis during surgery with selective scanning Raman microscopy. J. Mol. Struct. **1073**, 58–65 (2014)
38. Zhao, J., et al.: Wavenumber selection based analysis in Raman spectroscopy improves skin cancer diagnostic specificity. Analyst. **141**(3), 1034–1043 (2016)
39. Medeiros Neto, L.P., et al.: Micro-Raman spectroscopic study of thyroid tissues. Photodiagnosis Photodyn. Ther. **17**, 164–172 (2017)
40. Senol, O., et al.: Application of photonics in diagnosis of papillary thyroid carcinoma tissues through Raman spectroscopy-assisted with chemometrics. Anal. Lett. **51**(1–2), 229–235 (2018)
41. Palermo, A., et al.: Raman spectroscopy applied to parathyroid tissues: a new diagnostic tool to discriminate normal tissue from adenoma. Anal. Chem. **90**(1), 847–854 (2018)
42. Frost, J., et al.: Raman spectroscopy and multivariate analysis for the non invasive diagnosis of clinically inconclusive vulval lichen sclerosus. Analyst. **142**(8), 1200–1206 (2017)
43. Kerr, L.T., et al.: Methodologies for bladder cancer detection with Raman based urine cytology. Anal. Methods. **8**(25), 4991–5000 (2016)
44. Banerjee, H.N., et al.: Deciphering the finger prints of brain cancer glioblastoma multiforme from four different patients by using near infrared Raman spectroscopy. J. Cancer Sci. Ther. **7**(2), 44–47 (2015)
45. Marro, M., et al.: Molecular monitoring of epithelial-to-mesenchymal transition in breast cancer cells by means of Raman spectroscopy. Biochim. Biophys. Acta (BBA)-Mol. Cell Res. **1843**(9), 1785–1795 (2014)
46. Bi, X., et al.: Evaluating HER2 amplification status and acquired drug resistance in breast cancer cells using Raman spectroscopy. J. Biomed. Opt. **19**(2), 025001 (2014)
47. Goel, P.N., et al.: Investigating the effects of Pentoxifylline on human breast cancer cells using Raman spectroscopy. J. Innov. Opt. Health Sci. **8**(02), 1550004 (2015)
48. Talari, A.C.S., et al.: Raman spectroscopic analysis differentiates between breast cancer cell lines. J. Raman Spectrosc. **46**(5), 421–427 (2015)
49. Talari, A.C.S., et al.: Analyzing normal proliferating, hypoxic and necrotic regions of T-47D human breast cancer spheroids using Raman spectroscopy. Appl. Spectrosc. Rev. **52**(10), 909–924 (2017)
50. Winnard Jr., P.T., et al.: Organ-specific isogenic metastatic breast cancer cell lines exhibit distinct Raman spectral signatures and metabolomes. Oncotarget. **8**(12), 20266–20287 (2017)
51. Ramos, I., et al.: Raman spectroscopy for cytopathology of exfoliated cervical cells. Faraday Discuss. **187**, 187–198 (2016)
52. Gala de Pablo, J., et al.: Biochemical fingerprint of colorectal cancer cell lines using label-free live single-cell Raman spectroscopy. J. Raman Spectrosc. **49**(8), 1323–1332 (2018)

53. Shiramizu, B., et al.: Unique Raman spectroscopic fingerprints of B-cell non-Hodgkin lymphoma: implications for diagnosis, prognosis and new therapies. J. Biol. Med. Sci. **2**(1), 105 (2018)
54. Maguire, A., et al.: Competitive evaluation of data mining algorithms for use in classification of leukocyte subtypes with Raman microspectroscopy. Analyst. **140**(7), 2473–2481 (2015)
55. Carvalho, L.F.C.S., et al.: Raman micro-spectroscopy for rapid screening of oral squamous cell carcinoma. Exp. Mol. Pathol. **98**(3), 502–509 (2015)
56. Moradi, H., et al.: Raman micro-spectroscopy applied to treatment resistant and sensitive human ovarian cancer cells. J. Biophotonics. **10**(10), 1327–1334 (2017)
57. Corsetti, S., et al.: Raman spectroscopy for accurately characterizing biomolecular changes in androgen-independent prostate cancer cells. J. Biophotonics. **11**(3), e201700166 (2018)
58. Olmos, V., et al.: Preprocessing tools applied to improve the assessment of Aldrin effects on prostate cancer cells using Raman spectroscopy. Appl. Spectrosc. **72**(3), 489–500 (2018)
59. Ralbovsky, N.M., et al.: Deep-Ultraviolet Raman spectroscopy for cancer diagnostics: a feasibility study with cell lines and tissues. Cancer Stud. Mol. Med. - Open J. **5**(1), 1–10 (2019)
60. Mehta, K., et al.: An early investigative serum Raman spectroscopy study of meningioma. Analyst. **143**(8), 1916–1923 (2018)
61. González-Solís, J.L., et al.: Cervical cancer detection based on serum sample Raman spectroscopy. Lasers Med. Sci. **29**(3), 979–985 (2014)
62. Pappu, R., et al.: Raman spectroscopic characterization of urine of normal and cervical cancer subjects. In: SPIE BiOS. International Society for Optics and Photonics, San Francisco (2017)
63. Li, X., et al.: Raman spectroscopy combined with principal component analysis and k nearest neighbour analysis for non-invasive detection of colon cancer. Laser Phys. **26**(3), 035702 (2016)
64. Khan, S., et al.: Optical screening of nasopharyngeal cancer using Raman spectroscopy and support vector machine. Optik-Int. J. Light Electron Opt. **157**, 565–570 (2018)
65. Happillon, T., et al.: Diagnosis approach of chronic lymphocytic leukemia on unstained blood smears using Raman microspectroscopy and supervised classification. Analyst. **140**(13), 4465–4472 (2015)
66. Li, Y., et al.: Pattern recognition methods combined with Raman spectra applied to distinguish serums from lung cancer patients and healthy people. J. Biosci. Med. **5**(9), 95–105 (2017)
67. Wang, H., et al.: Screening and staging for non-small cell lung cancer by serum laser Raman spectroscopy. Spectrochim. Acta A Mol. Biomol. Spectrosc. **201**, 34–38 (2018)
68. Brindha, E., et al.: Raman spectroscopic characterization of urine of normal and oral cancer subjects. J. Raman Spectrosc. **46**(1), 84–93 (2015)
69. Pachaiappan, R., et al.: Near infrared Raman spectroscopic characterization of blood plasma of normal, oral premalignant and malignant conditions—a pilot study. J. Raman Spectrosc. **46**(9), 735–743 (2015)
70. Pachaiappan, R., et al.: Near-infrared Raman spectroscopic characterization of salivary metabolites in the discrimination of normal from oral premalignant and malignant conditions. J. Raman Spectrosc. **47**(7), 763–772 (2016)
71. Barroso, E.M., et al.: Discrimination between oral cancer and healthy tissue based on water content determined by Raman spectroscopy. Anal. Chem. **87**(4), 2419–2426 (2015)
72. Pachaiappan, R., et al.: High wavenumber Raman spectroscopic characterization of normal and oral cancer using blood plasma. In: SPIE BiOS. International Society for Optics and Photonics, San Francisco (2017)
73. Brindha, E., et al.: High wavenumber Raman spectroscopy in the characterization of urinary metabolites of normal subjects, oral premalignant and malignant patients. Spectrochim. Acta A Mol. Biomol. Spectrosc. **171**, 52–59 (2017)
74. Carvalho, L.F.C.S., et al.: Raman spectroscopic analysis of oral cells in the high wavenumber region. Exp. Mol. Pathol. **103**(3), 255–262 (2017)

75. Santos, I.P., et al.: Raman spectroscopic characterization of melanoma and benign melanocytic lesions suspected of melanoma using high-wavenumber Raman spectroscopy. Anal. Chem. **88**(15), 7683–7688 (2016)
76. Zhou, X., et al.: Evaluation of the diagnostic potential of *ex vivo* Raman spectroscopy in gastric cancers: fingerprint versus high wavenumber. J. Biomed. Opt. **21**(10), 105002 (2016)
77. Huang, W., et al.: Study of both fingerprint and high wavenumber Raman spectroscopy of pathological nasopharyngeal tissues. J. Raman Spectrosc. **46**(6), 537–544 (2015)
78. Sun, L., et al.: Preliminary study of differentiating smears from cancerous and non-cancerous nasopharyngeal tissue using confocal Raman spectroscopy. J. Cancer Res. Clin. Oncol. **142**(4), 823–831 (2016)
79. Kast, R., et al.: Identification of regions of normal grey matter and white matter from pathologic glioblastoma and necrosis in frozen sections using Raman imaging. J. Neurooncol. **125**(2), 287–295 (2015)
80. Kong, K., et al.: Towards intra-operative diagnosis of tumours during breast conserving surgery by selective-sampling Raman micro-spectroscopy. Phys. Med. Biol. **59**(20), 6141–6152 (2014)
81. Abramczyk, H., et al.: The role of lipid droplets and adipocytes in cancer. Raman imaging of cell cultures: MCF10A, MCF7, and MDA-MB-231 compared to adipocytes in cancerous human breast tissue. Analyst. **140**(7), 2224–2235 (2015)
82. Brozek-Pluska, B., Kopeć, M., Abramczyk, H.: Development of a new diagnostic Raman method for monitoring epigenetic modifications in the cancer cells of human breast tissue. Anal. Methods. **8**(48), 8542–8553 (2016)
83. Vanna, R., et al.: Label-free imaging and identification of typical cells of acute myeloid leukaemia and myelodysplastic syndrome by Raman microspectroscopy. Analyst. **140**(4), 1054–1064 (2015)
84. Tolstik, T., et al.: Discrimination and classification of liver cancer cells and proliferation states by Raman spectroscopic imaging. Analyst. **139**(22), 6036–6043 (2014)
85. Tolstik, T., et al.: Classification and prediction of HCC tissues by Raman imaging with identification of fatty acids as potential lipid biomarkers. J. Cancer Res. Clin. Oncol. **141**(3), 407–418 (2015)
86. Ryabchykov, O., et al.: Raman spectroscopic investigation of the human liver stem cell line HepaRG. J. Raman Spectrosc. **49**(6), 935–942 (2018)
87. Cals, F.L.J., et al.: Investigation of the potential of Raman spectroscopy for oral cancer detection in surgical margins. Lab. Invest. **95**, 1186–1196 (2015)
88. Meksiarun, P., et al.: Comparison of multivariate analysis methods for extracting the paraffin component from the paraffin-embedded cancer tissue spectra for Raman imaging. Sci. Rep. **7**, 44890 (2017)
89. Musto, P., et al.: Hyperspectral Raman imaging of human prostatic cells: an attempt to differentiate normal and malignant cell lines by univariate and multivariate data analysis. Spectrochim. Acta A Mol. Biomol. Spectrosc. **173**, 476–488 (2017)
90. Rau, J.V., et al.: Raman spectroscopy imaging improves the diagnosis of papillary thyroid carcinoma. Sci. Rep. **6**, 35117 (2016)
91. Zha, W.L., et al.: HPLC assisted Raman spectroscopic studies on bladder cancer. Laser Phys. Lett. **12**(4), 045701 (2015)
92. Abramczyk, H., Imiela, A.: The biochemical, nanomechanical and chemometric signatures of brain cancer. Spectrochim. Acta A Mol. Biomol. Spectrosc. **188**, 8–19 (2018)
93. Kim, S., et al.: Dual-modal cancer detection based on optical pH sensing and Raman spectroscopy. J. Biomed. Opt. **22**(10), 105002 (2017)
94. Surmacki, J., et al.: The lipid-reactive oxygen species phenotype of breast cancer. Raman spectroscopy and mapping, PCA and PLSDA for invasive ductal carcinoma and invasive lobular carcinoma. Molecular tumorigenic mechanisms beyond Warburg effect. Analyst. **140**(7), 2121–2133 (2015)

95. Owens, G.L., et al.: Vibrational biospectroscopy coupled with multivariate analysis extracts potentially diagnostic features in blood plasma/serum of ovarian cancer patients. J. Biophotonics. **7**(3–4), 200–209 (2014)
96. Bury, D., et al.: Phenotyping metastatic brain tumors applying spectrochemical analyses: segregation of different cancer types. Anal. Lett. **52**(4), 575–587 (2018)
97. Polavarapu, P.L.: Determination of molecular stereochemistry using optical rotatory dispersion, vibrational circular dichroism and vibrational Raman optical activity. In: Busch, K.W., Busch, M.A. (eds.) Chiral Analysis, pp. 461–504. Elsevier (2006)
98. Ostovar Pour, S., et al.: Raman optical activity. In: Polavarapu, P.L. (ed.) Chiral Analysis (Second Edition): Advances in Spectroscopy, Chromatography and Emerging Methods, pp. 249–291. Elsevier (2018)
99. Tatarkovič, M., et al.: The potential of chiroptical and vibrational spectroscopy of blood plasma for the discrimination between colon cancer patients and the control group. Analyst. **140**, 2287–2293 (2015)
100. Lin, D., et al.: Autofluorescence and white light imaging-guided endoscopic Raman and diffuse reflectance spectroscopy for *in vivo* nasopharyngeal cancer detection. J. Biophotonics. **11**(4), e201700251 (2018)
101. Zakharov, V.P., et al.: Combined Raman spectroscopy and autofluorescence imaging method for *in vivo* skin tumor diagnosis. In: SPIE Optical Engineering + Applications. International Society for Optics and Photonics, San Diego (2014)
102. Khristoforova, Y.A., et al.: *In vivo* diagnostics of malignant and benign tumors with low-cost Raman spectrometer. In: Frontiers in Optics 2017. Optical Society of America, Washington, D.C. (2017)
103. Bratchenko, I.A., et al.: Combined Raman and autofluorescence *ex vivo* diagnostics of skin cancer in near-infrared and visible regions. J. Biomed. Opt. **22**(2), 027005 (2017)
104. Fullwood, L.M., et al.: Investigating the use of Raman and immersion Raman spectroscopy for spectral histopathology of metastatic brain cancer and primary sites of origin. Anal. Methods. **6**(12), 3948–3961 (2014)
105. Li, B., et al.: Evaluating oral epithelial dysplasia classification system by near-infrared Raman spectroscopy. Oncotarget. **8**(44), 76257–76265 (2017)
106. Weng, S., et al.: Combining deep learning and coherent anti-Stokes Raman scattering imaging for automated differential diagnosis of lung cancer. J. Biomed. Opt. **22**(10), 106017 (2017)
107. Yosef, H.K., et al.: Noninvasive diagnosis of high-grade urothelial carcinoma in urine by Raman spectral imaging. Anal. Chem. **89**(12), 6893–6899 (2017)
108. Knipfer, C., et al.: Raman difference spectroscopy: a non-invasive method for identification of oral squamous cell carcinoma. Biomed. Opt. Express. **5**(9), 3252–3265 (2014)
109. Daniel, A., et al.: Polarized Raman spectroscopy unravels the biomolecular structural changes in cervical cancer. Spectrochim. Acta A Mol. Biomol. Spectrosc. **152**, 58–63 (2016)
110. Ji, M., et al.: Detection of human brain tumor infiltration with quantitative stimulated Raman scattering microscopy. Sci. Transl. Med. **7**(309), 309ra163 (2015)
111. Hollon, T.C., et al.: Rapid intraoperative diagnosis of pediatric brain tumors using stimulated Raman histology. Cancer Res. **78**(1), 278–289 (2018)
112. Medipally, D.K.R., et al.: Development of a high throughput (HT) Raman spectroscopy method for rapid screening of liquid blood plasma from prostate cancer patients. Analyst. **142**(8), 1216–1226 (2017)
113. Stables, R., et al.: Feature driven classification of Raman spectra for real-time spectral brain tumour diagnosis using sound. Analyst. **142**(1), 98–109 (2017)
114. Chen, H., et al.: Identification and characterization of bladder cancer by low-resolution fiber-optic Raman spectroscopy. J. Biophotonics. **11**(9), e201800016 (2018)
115. Desroches, J., et al.: Characterization of a Raman spectroscopy probe system for intraoperative brain tissue classification. Biomed. Opt. Express. **6**(7), 2380–2397 (2015)
116. Jermyn, M., et al.: Intraoperative brain cancer detection with Raman spectroscopy in humans. Sci. Transl. Med. **7**(274), 274ra19 (2015)

117. Jermyn, M., et al.: Neural networks improve brain cancer detection with Raman spectroscopy in the presence of operating room light artifacts. J. Biomed. Opt. **21**(9), 094002 (2016)
118. Li, Q., Gao, Q., Zhang, G.: Classification for breast cancer diagnosis with Raman spectroscopy. Biomed. Opt. Express. **5**(7), 2435–2445 (2014)
119. Li, Q., Hao, C., Xu, Z.: Diagnosis of breast cancer tissues using 785 nm miniature Raman spectrometer and pattern regression. Sensors. **17**(3), 627 (2017)
120. Li, Q.-B., et al.: Discrimination of breast cancer from normal tissue with Raman spectroscopy and chemometrics. J. Appl. Spectrosc. **82**(3), 450–455 (2015)
121. Duraipandian, S., et al.: Near-infrared Raman spectroscopy for assessing biochemical changes of cervical tissue associated with precarcinogenic transformation. Analyst. **139**(21), 5379–5386 (2014)
122. Shaikh, R.S., et al.: *In vivo* Raman spectroscopy of cervix cancers. In: SPIE BiOS. International Society for Optics and Photonics, San Francisco (2014)
123. Shaikh, R.S., et al.: *In vivo* Raman spectroscopy of human uterine cervix: exploring the utility of vagina as an internal control. J. Biomed. Opt. **19**(8), 087001 (2014)
124. Wood, J.J., et al.: Evaluation of a confocal Raman probe for pathological diagnosis during colonoscopy. Colorectal Dis. **16**(9), 732–738 (2014)
125. Petersen, D., et al.: Raman fiber-optical method for colon cancer detection: cross-validation and outlier identification approach. Spectrochim. Acta A Mol. Biomol. Spectrosc. **181**, 270–275 (2017)
126. Almond, L.M., et al.: Endoscopic Raman spectroscopy enables objective diagnosis of dysplasia in Barrett's esophagus. Gastrointest. Endosc. **79**(1), 37–45 (2014)
127. Bergholt, M.S., et al.: Fiberoptic confocal Raman spectroscopy for real-time *in vivo* diagnosis of dysplasia in Barrett's esophagus. Gastroenterology. **146**(1), 27–32 (2014)
128. Lloyd, G.R., et al.: Utilising non-consensus pathology measurements to improve the diagnosis of oesophageal cancer using a Raman spectroscopic probe. Analyst. **139**(2), 381–388 (2014)
129. Maeda, Y., et al.: Measurement of the human esophageal cancer in an early stage with Raman spectroscopy. In: SPIE BiOS. International Society for Optics and Photonics, San Francisco (2014)
130. Bergholt, M.S., et al.: Real-time depth-resolved fiber optic Raman endoscopy for *in vivo* diagnosis of gastric precancer. In: SPIE BiOS. International Society for Optics and Photonics, San Francisco (2014)
131. Wang, J., et al.: Comparative study of the endoscope-based bevelled and volume fiber-optic Raman probes for optical diagnosis of gastric dysplasia *in vivo* at endoscopy. Anal. Bioanal. Chem. **407**(27), 8303–8310 (2015)
132. Ming, L.C., et al.: Real time near-infrared Raman spectroscopy for the diagnosis of nasopharyngeal cancer. Oncotarget. **8**(30), 49443–49450 (2017)
133. Yan, B., et al.: An intraoperative diagnosis of parotid gland tumors using Raman spectroscopy and support vector machine. Laser Phys. **24**(11), 115601 (2014)
134. McGregor, H.C., et al.: Real-time endoscopic Raman spectroscopy for *in vivo* early lung cancer detection. J. Biophotonics. **10**(1), 98–110 (2017)
135. Guze, K., et al.: Pilot study: Raman spectroscopy in differentiating premalignant and malignant oral lesions from normal mucosa and benign lesions in humans. Head Neck. **37**(4), 511–517 (2015)
136. Krishna, H., et al.: *In vivo* Raman spectroscopy for detection of oral neoplasia: a pilot clinical study. J. Biophotonics. **7**(9), 690–702 (2014)
137. Sahu, A., et al.: Classification of oral cancers using Raman spectroscopy of serum. In: SPIE BiOS. International Society for Optics and Photonics, San Francisco (2014)
138. Sahu, A., et al.: Raman spectroscopy and cytopathology of oral exfoliated cells for oral cancer diagnosis. Anal. Methods. **7**(18), 7548–7559 (2015)
139. Sahu, A., et al.: Recurrence prediction in oral cancers: a serum Raman spectroscopy study. Analyst. **140**(7), 2294–2301 (2015)

140. Sahu, A., et al.: Raman exfoliative cytology for oral precancer diagnosis. J. Biomed. Opt. **22**(11), 115003 (2017)
141. Yasser, M., et al.: Raman spectroscopic study of radioresistant oral cancer sublines established by fractionated ionizing radiation. PLoS One. **9**(5), e97777 (2014)
142. Aubertin, K., et al.: Mesoscopic characterization of prostate cancer using Raman spectroscopy: potential for diagnostics and therapeutics. BJU Int. **122**(2), 326–336 (2018)
143. Silveira Jr., L., et al.: Discrimination of prostate carcinoma from benign prostate tissue fragments *in vitro* by estimating the gross biochemical alterations through Raman spectroscopy. Lasers Med. Sci. **29**(4), 1469–1477 (2014)
144. Lim, L., et al.: Clinical study of noninvasive *in vivo* melanoma and nonmelanoma skin cancers using multimodal spectral diagnosis. J. Biomed. Opt. **19**(11), 117003 (2014)
145. Zakharov, V.P., et al.: Two-step Raman spectroscopy method for tumor diagnosis. In: SPIE Photonics Europe. International Society for Optics and Photonics, Brussels (2014)
146. Schleusener, J., et al.: *In vivo* study for the discrimination of cancerous and normal skin using fibre probe-based Raman spectroscopy. Exp. Dermatol. **24**(10), 767–772 (2015)
147. Silveira, F.L., et al.: Discrimination of non-melanoma skin lesions from non-tumor human skin tissues *in vivo* using Raman spectroscopy and multivariate statistics. Lasers Surg. Med. **47**(1), 6–16 (2015)
148. Zhao, J., et al.: Real-time Raman spectroscopy for automatic *in vivo* skin cancer detection: an independent validation. Anal. Bioanal. Chem. **407**(27), 8373–8379 (2015)
149. Bergholt, M.S., et al.: Characterizing variability of *in vivo* Raman spectroscopic properties of different anatomical sites of normal colorectal tissue towards cancer diagnosis at colonoscopy. Anal. Chem. **87**(2), 960–966 (2015)
150. Bergholt, M.S., et al.: Simultaneous fingerprint and high-wavenumber fiber-optic Raman spectroscopy enhances real-time *in vivo* diagnosis of adenomatous polyps during colonoscopy. J. Biophotonics. **9**(4), 333–342 (2016)
151. Wang, J., et al.: Simultaneous fingerprint and high-wavenumber fiber-optic Raman spectroscopy improves *in vivo* diagnosis of esophageal squamous cell carcinoma at endoscopy. Sci. Rep. **5**, 12957 (2015)
152. Wang, J., et al.: Fiber-optic Raman spectroscopy for *in vivo* diagnosis of gastric dysplasia. Faraday Discuss. **187**, 377–392 (2016)
153. Holler, S., et al.: Raman spectroscopy of head and neck cancer: separation of malignant and healthy tissue using signatures outside the “fingerprint” region. Biosensors. **7**(2), 20 (2017)
154. Lin, K., et al.: Real-time *in vivo* diagnosis of nasopharyngeal carcinoma using rapid fiber-optic Raman spectroscopy. Theranostics. **7**(14), 3517–3526 (2017)
155. Lin, K., et al.: Real-time *in vivo* diagnosis of laryngeal carcinoma with rapid fiber-optic Raman spectroscopy. Biomed. Opt. Express. **7**(9), 3705–3715 (2016)
156. Kallaway, C., et al.: Advances in the clinical application of Raman spectroscopy for cancer diagnostics. Photodiagnosis Photodyn. Ther. **10**(3), 207–219 (2013)
157. Pence, I., Mahadevan-Jansen, A.: Clinical instrumentation and applications of Raman spectroscopy. Chem. Soc. Rev. **45**(7), 1958–1979 (2016)
158. Santos, I.P., et al.: Raman spectroscopy for cancer detection and cancer surgery guidance: translation to the clinics. Analyst. **142**(17), 3025–3047 (2017)

Chapter 9
Combination of Spontaneous and Coherent Raman Scattering Approaches with Other Spectroscopic Modalities for Molecular Multi-contrast Cancer Diagnosis

Christoph Krafft and Jürgen Popp

9.1 Introduction

Raman spectroscopy is a widespread method in analytical chemistry. The principle is based on inelastic scattering of monochromatic light. Its advantages include that it probes inherent molecular vibrations without sample preparation, without extrinsic markers, in aqueous environment, in a non-destructive way, and at a diffraction-limited resolution in the submicrometer range. The spectrum of molecular vibrations provides a specific fingerprint of the analyte. Due to these advantages, Raman spectroscopic approaches draw attention as clinical tool for cancer diagnosis and other applications in histopathology and cytopathology. Cancer, a synonym for a malignant tumor, is defined as a group of diseases involving abnormal cell growth with the potential to invade or spread to other parts of the body. These contrast with benign tumors that lack the ability to invade neighboring tissue or metastasize. Many types of benign tumors have the potential to become cancerous (malignant) through a process known as tumor progression. For this reason and other possible negative health effects, some benign tumors are removed by surgery. In general, main scope of spectroscopic modalities in the context of cancer is to distinguish normal from abnormal tissue and determine the malignancy of abnormal tissue. The implementation as imaging technique enables visualization of tumor borders which is particularly interesting during surgery to guide resection of cancerous mass of cells.

C. Krafft (✉)
Leibniz Institute of Photonic Technologies, Jena, Germany
e-mail: christoph.krafft@leibniz-ipht.de

J. Popp
Leibniz Institute of Photonic Technologies, Jena, Germany

Institute of Physical Chemistry and Abbe Center of Photonics, Jena, Germany

V. V. Tuchin et al. (eds.), *Multimodal Optical Diagnostics of Cancer*,
https://doi.org/10.1007/978-3-030-44594-2_9

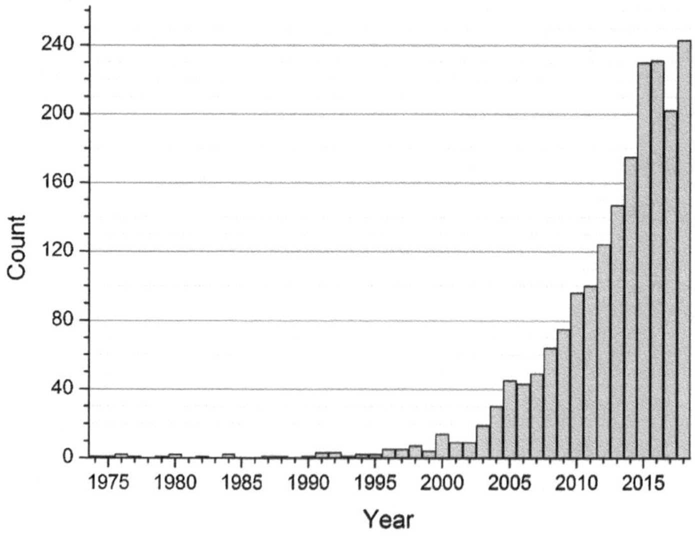

Fig. 9.1 Number of papers per year in the Medline database searching for "Raman spectroscopy AND cancer" from 1974 to 2018

The keywords "Raman spectroscopy and cancer" return more than 2000 hits in the Medline database. Although the earliest paper was published in the year 1974, the number of papers per year remained below 10 until 1999 and reached more than 240 in 2018 (see Fig. 9.1).

The relatively small number of cancer-related Raman papers before the year 2000 demonstrates that applications of Raman scattering approaches in biomedicine were limited due to three main challenges for which solutions have been developed since 2000. First, Raman scattering cross sections of most biomolecules such as proteins and nucleic acids are weak. Examples with higher Raman cross sections are lipids, calcified material such as hydroxyapatite in bone and teeth, and molecules with chromophores such as carotenes or heme groups. Consequently, Raman spectroscopy of most tissues and cells has relatively low sensitivity, and applications suffered from weak band intensities, long exposure times, and low throughput for screening cells or extended tissue regions. Progress in lasers, optical filters, microscopes, diffraction gratings, and detectors significantly improved the sensitivity of Raman instruments to a level that enabled groundbreaking research in cancer diagnosis. Signal enhancement effects can be utilized to further increase the sensitivity. The most frequently applied signal enhancement techniques are surface enhanced Raman scattering (SERS) using plasmonic effects of metal nanoparticles, and coherent Raman scattering (CRS) using picosecond, high intensity laser pulses for excitation. The latter multiphoton technique enables microscopic imaging with megapixel resolution at real-time video frame rates as the integration time for single pixels is in the range of microseconds. The second challenge is that the excitation light can be partly absorbed by biological specimens followed by fluorescence emission. As fluorescence cross sections are several orders of magnitude larger than Raman scattering cross sections, the emission of even trace amounts of fluorophores can obscure the weaker Raman bands. Another risk is that absorption of intense excitation light heats up the sample with possible carbonization. These effects can be minimized by Raman excitation with near infrared (NIR) laser wavelengths between 700 and 1100 nm and detectors with enhanced collection efficiencies in NIR range. The third challenge is that tissues and cells constitute a complex assembly of thousands of different molecules and spectral variations between normal and pathological tissues or cells are often weak and distributed over a broad spectral range. Therefore, careful sample preparation, instrument calibration, data with sufficiently high signal to noise ratio, and multivariate chemometric data processing are required to separate confounding factors, and obtain statistically meaningful and robust results.

Details of the instrumentation are not given here. They can be found in other chapters and recent reviews [1]. Furthermore, Raman spectroscopy and advanced chemometrics for differentiating cancer, intraoperative brain cancer detection, and Raman imaging of the skin are covered in other chapters of this book. Instead, each modality is briefly introduced and typical examples in the context of cancer are described. Finally, complementary optical methods were combined with Raman scattering as multi-contrast modalities.

9.2 Spontaneous Raman Spectroscopy

Before Raman spectroscopic applications to cancer cells and tissues are described, principles of the instrumentation are briefly introduced. Popular excitation sources are continuous-wave lasers with emission between 488 and 1064 nm. Shorter wavelengths λ offer better spatial resolution and higher signal intensities as scattering scales with λ^{-4}. Laser excitation below 600 nm is only appropriate for non-living, fixed cells. For live cell Raman studies, laser excitation below 600 nm is phototoxic because cytochrome proteins absorb in this range, give rise to resonance enhanced Raman bands for a few seconds, and irreversibly degrade afterwards. Laser excitation above 600 nm is not toxic for living cells, even at long exposure times. Laser excitation above 600 nm is also preferred for Raman tissue studies because absorption is minimum, light penetration is maximum, and autofluorescence overlap is usually low. However, the low quantum efficiency of silicon-based CCD detectors above 1000 nm is a disadvantage of the frequently used excitation wavelength 785 nm, in particular for the detection of the high wavenumber range above 2700 cm^{-1}. Raman spectrometers can be coupled with microscopes. The use of high numerical aperture objective lenses enables diffraction-limited spatial resolution and high signal collection efficiency. Raman images can be collected by moving the laser focus over the sample with scanning mirrors or by moving the sample with a motorized stage. Due to the weak Raman cross sections of biological samples, imaging speed is relatively slow because acquisition of thousands of spectra at an exposure time in the range of 1 s per spectrum takes hours. This fact motivates to combine spontaneous Raman spectroscopy with faster imaging modalities which is the content of this chapter. Raman spectrometers can also be coupled to fiber optic probes for clinical applications during open surgery or minimal-invasive endoscopy. A key requirement for Raman probes is to integrate a bandpass filter onto the excitation fiber and a longpass filter onto the detection fibers which is not trivial for probes with diameters below 2 mm.

9.2.1 Raman Spectroscopy of Cancer Cells

Raman spectra of single cells provide a specific fingerprint that can be used to determine cell types (e.g., normal versus cancer), cell state (e.g., normal versus apoptotic), or stage of the cell cycle. Typical eukaryotic cells have diameter between 10 and 20 μm with a 5 μm cell nucleus. Although Raman images can resolve subcellular details that might contribute to the differentiation of cancer versus non-cancer cells, single average spectra are often sufficient for cell classification. An average spectrum can be calculated as a mean from an image, acquired with an expanded laser focus corresponding to the cell size or by moving the laser focus over the cell during signal collection. The latter option was called integrated acquisition mode and was found to give superior classification results for T lymphocyte Jurkat cells, and pancreatic cell lines Capan1 and MiaPaca2 [2].

One promising diagnostic application of Raman spectroscopy is the identification of cancer cells that circulate in body fluids (CTCs, circulating tumor cells) such as blood and urine. Because these cells are extremely rare (10–100 in 1 mL blood containing ca. 5×10^6 leukocytes) model systems were studied that used leukocytes from control patients, and leukemia cells (OCI-AML3) and breast tumor cells (MCF-7 and BT-20) from cultures. Average spectra from Raman images of dried cells and spectra from cells in aqueous buffer that were trapped by the laser tweezers effect are shown in Fig. 9.2.

A comprehensive table with assignment of labeled bands was published [3]. Briefly, main bands were assigned to the peptide backbone of proteins (1658, 1252, 960, 934 cm^{-1}), $CH_{2/3}$ deformation of aliphatic amino acids of proteins (1451, 1341, 1319 cm^{-1}), aromatic amino acids of proteins (1209, 1032, 1004, 854, 830, 758, 644 and 623 cm^{-1}), and nucleic acids (1100 and 785 cm^{-1}). Most band positions for dried and hydrated cells agree within ±2 cm^{-1}. Variations are due to structural changes of biomolecules after dehydration in dried cells, due to different cell types—e.g., evident in the band ratios 1341/1319, due to spectral contributions of water near 1640 cm^{-1}, and due to different baselines, in particular between1200 and 700 cm^{-1}. Support vector machines classified dried cells with an accuracy of 99.7% [3]. This classification model was subsequently applied to identify single cells from an independent mixture of cells based on their Raman spectra. Correct classification was confirmed by fluorescence staining of the cells after the Raman measurement. In a follow-up study, support vector machines were applied to develop a supervised

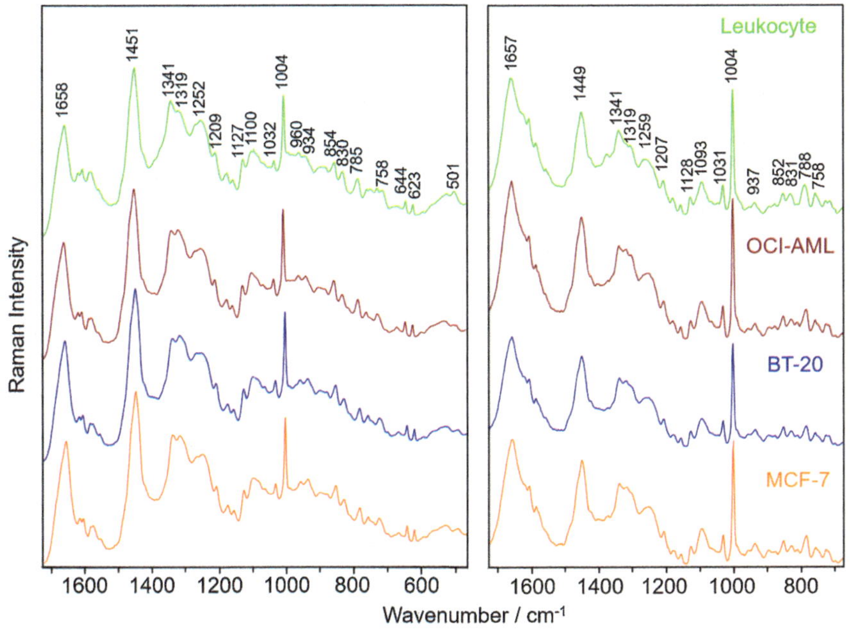

Fig. 9.2 Raman spectra of dried cells (left) and cells in aqueous buffer (right): leukocytes from donors (green), leukemia cells OCI-AML (red), breast cancer cells BT-20 (blue) and MCF-7 (orange) from cultures. Most band positions agree within ±2 cm^{-1}. Variations are due to structural changes of biomolecules after dehydration in dried cells and due to different cell types—e.g., evident in the band ratios 1341/1319. See text for more details

classification model with spectra of 1210 trapped cells originating from three different donors and three independent cultivation batches [4]. Distinguishing tumor cells from healthy cells was achieved with a sensitivity of >99.7% and a specificity of >99.5%. In addition, the correct cell types were predicted with an accuracy of approximately 92%.

Microfluidic chips were developed that combined cell trapping by two 1070 nm fiber lasers with Raman spectroscopy. A first version of the chip was made from glass wafers for Raman spectroscopy with 514 nm excitation, and an overall classification accuracy of 95% was achieved [5]. A second version was made from quartz wafers for Raman spectroscopy with 785 nm excitation because glass shows higher background upon NIR excitation. Classification by linear discriminant analysis gave sensitivity and specificity of 96 and 99%, respectively, determined by iterated tenfold cross validation [6]. However, throughput was low in these approaches using 10 s exposure time per cell and control of the microfluidic flow including sorting of cancer cells was difficult.

Recently, a high throughput screening Raman platform was introduced that collected more than 100,000 spectra to determine the fractions of lymphocytes, monocytes and neutrophils in a small white blood cell count, and more than 1000 cells to determine pancreas cells in leukocytes at mixing ratios 1:1, 1:10, and 1:100. The principle was that cells from suspensions were pipetted on a substrate and sediment. Instead of screening the full slide, cells were located by processing light microscopy images, and single Raman spectra were automatically collected from each cell at 0.5 s exposure time [7].

Raman imaging is able to resolve subcellular details. Cluster analysis can segment Raman images of cells into nucleus, lipid droplets, and cytoplasm. Two papers reported that two liver cancer cell lines Hep-G2 and SK-Hep1 [8] and two isogenic cancer cell lines derived from the MDA-MB-435 breast cancer cell line [9] were classified more accurately if spectra of cytoplasm were used as input instead of average spectra or spectra of the nucleus. It was recently shown that even morphological normal appearing cells could be used to detect high-grade squamous intraepithelial (HSIL) cells by Raman spectroscopy [10]. The results demonstrated that Raman spectra encompass the biochemical signatures of morphologically normal appearing cells to discriminate between negative and HSIL cytology.

9.2.2 Raman Spectroscopy of Cancer Tissues

Raman spectroscopy to detect, diagnose, and delineate cancer tissues is a promising complementary tool in histopathology [11]. Whereas tissue types can be identified from single Raman spectra, Raman imaging is important for three reasons. First, Raman imaging and subsequent comparison with the golden standard, usually a hematoxylin and eosin (H&E) stained tissue section, is the method of choice to develop and validate Raman-based classification models. After validation the model can be applied to classify single spectra. Second, Raman images allow accurate determination of margins between normal and non-normal tissue, e.g., inflammation, dysplasia, tumor, or necrosis. Third, Raman images at high pixel resolution resolve morphological details without staining which give complementary information to the molecular information of the spectra. To avoid long acquisition times to image extended regions of interest, the step or pixel size can be increased to 10 μm, 100 μm, or even more for a coarse overview. However, such an undersampled Raman image harbors the risk to miss small biochemical and morphological details.

Figure 9.3 compares Raman images of dried brain tumor tissue sections and non-dried brain tumor tissue sections in aqueous buffer. The hyperspectral unmixing algorithm vertex component analysis decomposed Raman images at 2 μm step size into the most dissimilar spectral signatures and their abundances. Three components were identified in Raman images of dried tissue sections and were assigned to proteins with DNA, proteins with lipids and cholesterol that correlated with cell nuclei, brain parenchyma, and microcrystals, respectively [12]. Even seven components were identified in Raman images of brain tumors and were assigned to proteins with DNA, proteins, proteins with carotenoids, cholesterol, cholesterol ester, phospholipids, and buffer [13]. The spectral signatures of proteins with DNA component are presented in Fig. 9.3 and the abundance plots resolve the cell nuclei that are not visible in unstained tissue sections. The contrast is higher in non-dried tissue section

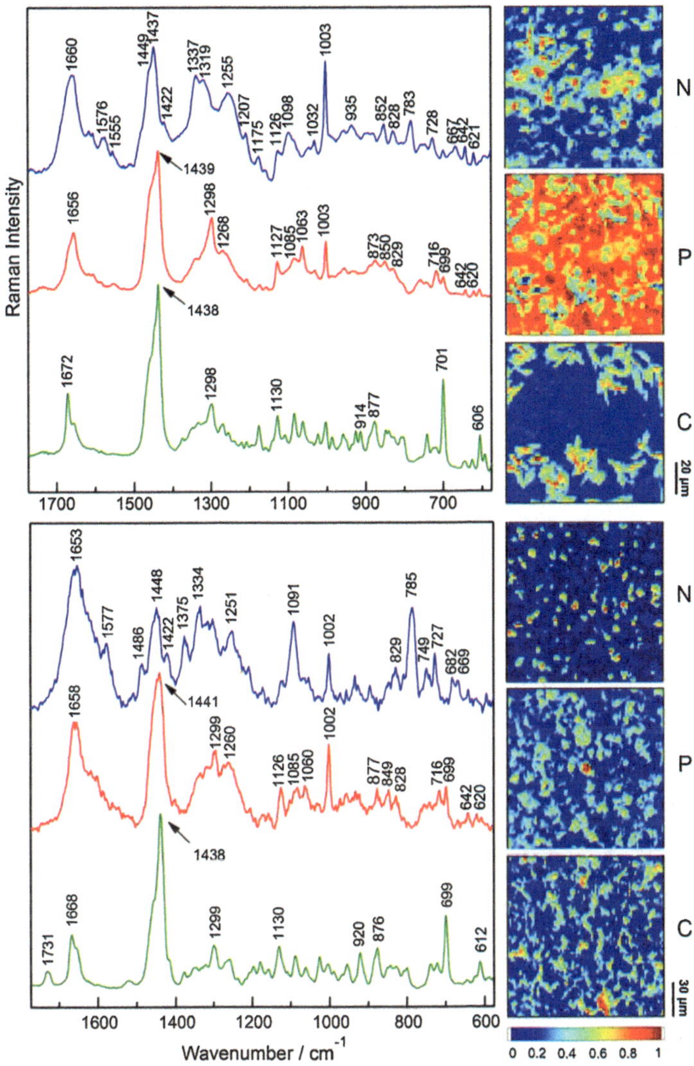

Fig. 9.3 Raman spectra and images of brain tumor tissues section after drying (top) and in aqueous buffer (bottom). Hyperspectral unmixing algorithms decomposed the data sets into three spectral signatures and their abundances: cell nuclei (blue traces, N), proteins and lipids (red traces, P), and cholesterol/cholesterol ester (green traces, C). Distribution of biomolecules, image contrast and spectral features differ between components and tissue preparation. See text for more details

because nucleic acid bands at 1577, 1486, 1422, 1375, 1334, 1091, 829, 785, 749, 727, 682, and 669 cm^{-1} are more intense in hydrated cell nuclei. Upon dehydration Raman intensities of DNA bands decrease and the conformation denatures which causes band shifts from 1091 to 1098 cm^{-1} and 682 to 667 cm^{-1}. The drying process also denatures proteins and induces crystallization of lipids, in particular cholesterol and cholesterol ester as the most hydrophobic ones. Whereas the protein bands were already found for single cells in Fig. 9.2, typical lipid bands are due to CH_2 vibrations (1439 and 1298 cm^{-1}), C–C vibrations (1126 and 1060 cm^{-1}), phosphate group of phospholipids (1085 cm^{-1}) and choline group (877 and 716 cm^{-1}). Cholesterol is characterized by numerous bands with the most intense one at 699 cm^{-1}, and a C=O band at 1731 cm^{-1} points to the ester group in cholesterol ester with further characteristic spectral features. Different protein, lipid, and cholesterol distributions are evident in the abundance plots. In particular, cholesterol formed clusters of microcrystals in dried tissue sections that were absent in hydrated tissue. The decomposed spectral signatures that represent DNA, proteins, cholesterol, and lipids were used as input to determine the tissue composition in primary brain tumors by a least-squares fitting algorithm [14]. It was found that the protein content was independent of the tumor grade, whereas the lipid content decreased and the DNA content increases with increasing tumor grade. Simultaneous morphological analyses revealed that the number of cell nuclei and the cell nuclei to cytoplasm ratios also increase with increasing malignancy. The fact that both chemical parameters obtained from analysis of Raman spectra and morphological parameters obtained from the label-free contrast in Raman images correlate with the tumor grade enables the application of coherent Raman scattering for rapid tissue screening. It will be presented in Sects. 9.4 and 9.5 that CARS and SRS give superior imaging speed compared to spontaneous Raman spectroscopy, but inferior spectral information because only few wavenumbers are probed. FTIR imaging is another vibrational spectroscopic method that was applied to grade brain tissue sections [15] and the results are consistent with the Raman studies mentioned above.

The identification of the primary tumor for brain metastasis by FTIR imaging and Raman imaging is an example how important the full spectral signature is. Brain metastases of primary tumors outside the brain, e.g., from lung cancer, breast cancer, colon cancer, prostate cancer, malignant melanoma and renal cell cancer, are more frequent that primary brain tumors. The diagnosis is straightforward if the primary tumor is known. However, in significant number of cases the primary tumor is not known because it is too small to cause symptoms. Even after extensive screening by standard tomographic procedures, the primary tumor remains unknown in 10% of cases. IR and Raman spectra provide a molecular fingerprint to determine the primary tumor. Whereas a single classifier was used in early IR studies [16, 17] hierarchical classifiers were developed for another set of Raman and IR imaging data [18, 19]. The first classification level identified the tumor spectra, whereas the second level assigned the tumor spectra to the correct primary tumor. It was found that the spectral markers were weak and distributed over a wide spectral range which required the acquisition and analysis of full spectra. Glycogen in metastases of renal cell carcinoma (RCC) is a prominent example. As its content is extremely low in normal brain tissue and other brain metastases, the detection of glycogen which is produced by renal cells can be considered as sensitive and specific marker of RCC. In a similar way, molecular markers exist also for other primary tumors of brain metastases that are less intense, but can still be measured by Raman and IR spectroscopy.

9.3 Surface Enhanced Raman Scattering

Among other factors such as concentration, polarizability, and excitation wavelength, the Raman intensities depend on the electric field strength that analyte molecules experience upon laser excitation. Noble metal nanoparticles and nanostructured surfaces show plasmonic properties when irradiated with wavelengths near plasmonic resonance, and consequently the electric field close to the surface is enhanced. If the analyte molecule is adsorbed or covalently bound to the nanoparticle surface, Raman signal intensities can be amplified by several orders of magnitude. This effect is called surface enhanced Raman scattering (SERS). The effect is roughly divided into (1) a direct or label-free procedure and (2) an indirect detection procedure. In the first procedure, nanoparticles interact with the sample, and the SERS spectra are dominated by cell and tissue components such as proteins, metabolites, and nucleic acids. However, the interpretation of these SERS spectra might be complicated since a high number of various molecules are competing for the free binding sites on nanostructures and contribute to the SERS response. Another complication is that the SERS spectra deviate from Raman spectra making band assignments difficult. The indirect detection employs SERS-active labels or tags, i.e., rational designed metal nanoparticles coated with reporter molecules showing a specific SERS spectrum and, if requested, specific recognition units such as antigens or aptamers for targeting biomolecules. This SERS principle corresponds to fluorescence labeling with advantages of multiplexing capacity due to the much narrower bandwidth and more bands, and higher stability without bleaching. SERS procedures can also be distinguished by the substrates or the noble metals. Many protocols have been described to prepare nanoparticles or nanostructured surfaces made of gold or silver. SERS-active silver nanoparticles show maximum plasmonic resonance near 400 nm, whereas the maximum plasmonic resonance of SERS-active gold nanoparticles is shifted towards 500 nm. After induction of nanoparticles aggregation the maxima are broader and extended to the NIR range with the advantage of deeper light penetration in cells and tissues. A few examples are given here. The reader is referred to a recent review for more details [20].

9.3.1 Labeled, Non-labeled, and Functionalized SERS-Active Nanoparticles for Detection of Cancer Cells

A fast and simple protocol to prepare silver nanoparticles for the direct procedure was published by Leopold and Lendl [21]. It is based on reduction of silver nitrate with hydroxylamine hydrochloride at alkaline pH and at ambient temperature. Their aggregation can easily be induced by adding potassium chloride. After lysing cells by sonication reproducible SERS spectra were collected due to good interaction between biomolecules of cells and these nanoparticles. The SERS spectra of lysed cells are dominated by protein contributions as shown in Fig. 9.4 [22]. Bands at

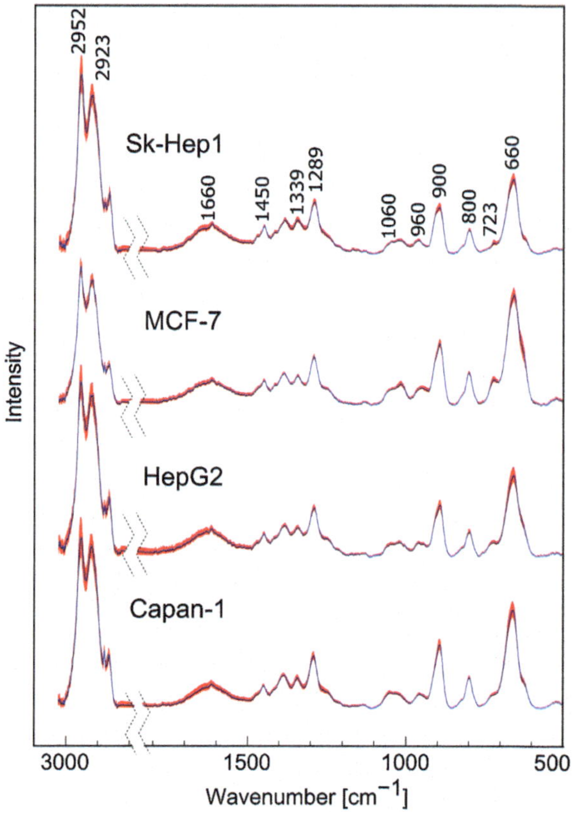

Fig. 9.4 Preprocessed mean SERS spectra and standard deviations of the different cell lysates mixed with silver nanoparticles and KCl. Labeled bands are assigned to whole cell contents including nuclei acids, proteins, and carbohydrates. The low standard deviation values (represented by red shadow) emphasize the high reproducibility of technique (adapted from reference [22])

2952 and 2923 are assigned to valence vibrations of CH_3 groups. The fingerprint range in SERS spectra from 1800 to 500 cm^{-1} strongly deviates from the fingerprint range in spontaneous Raman spectra in Figs. 9.2 and 9.3. Not only the spectra, but—even more importantly—also the variations between different cells were enhanced which enable to distinguish liver cancer cells Sk-Hep1 and HepG2, breast cancer cells MCF-7, and pancreas cancer cells Capan-1. Furthermore, three leukemia cell lines were classified in a lab-on-a-chip approach [23] and the fraction of leukemia cells were determined in control monocytes [24]. As spontaneous, passive uptake of silver nanoparticles by cells was low for intracellular SERS spectroscopy, a method based on electroporation was presented for fast delivery of silver nanoparticles into living cells [25]. An ultrasound-mediated method was developed as another approach to accelerate the transfer of silver nanoparticles into living cells [26]. Rapid acquisition of reproducible SERS spectra was achieved from living human nasopharyngeal carcinoma cell lines (C666 and CNE1) and normal nasopharyngeal cell line (NP69). Tentative assignment of the Raman bands in the measured SERS spectra showed cancer cell specific biomolecular differences, including significantly lower DNA concentrations and higher protein concentrations in cancerous nasopharyngeal cells as compared to those of normal cells. Combined with PCA–LDA multivariate analysis, ultrasound-mediated cell SERS spectroscopy differentiated the cancerous cells from the normal nasopharyngeal cells with high diagnostic accuracy (98.7%).

Silver nanoparticles have toxic effects for intracellular spectroscopic investigations, whereas gold is inert to oxidation and is biocompatible due to less toxicity for cells. An early paper reported spontaneous uptake of gold nanoparticles by cells [27]. However, the collected SERS spectra showed a high variability at each pixel in a SERS image. To target and image specific cancer markers in live cells by SERS, gold/silver core-shell nanoparticles were conjugated with monoclonal antibodies [28]. Live HEK293 cells expressing PLCγ1 were used as the optical imaging target in this report.

The protocol to prepare functionalized gold nanoparticle labels for the indirect procedure started with the controlled aggregation of 60 nm gold nanoparticles followed by adsorption of a monolayer of a reporter molecule (dye or aromatic substance) and a protective shell to prevent further aggregation [29]. The SERS spectrum using these labels is simplified because only bands of the reporter dye are detected, and no bands of cell molecules. Gold cores with hydroxyapatite (HA) shells were introduced as another SERS tag [30]. HA acted both as reporter, whose Raman bands were enhanced by the SERS effect, and biocompatible coating. SERS images demonstrated that this nanocomposite bound to the outer cell membrane.

Further functionalization of nanoparticles included binding of an antibody for specific recognition of cell antigens such as EpCAM. The specific binding to MCF7 breast cancer cells with EpCAM was demonstrated by Raman imaging and dark field microscopy, whereas no binding was observed to leukocytes and human foreskin fibroblast without EpCAM [29]. Injection of MCF7 cells into a microfluidic chip after mixing with these functionalized gold nanoparticles enabled to detect them in the presence of leukocytes without bound nanoparticles and polystyrene beads with intense Raman signals [31]. The acquisition time was set to 25 ms and cancer cells could be enumerated in continuous flow without trapping due to ca. 1000fold SERS enhancement. The delay due to readout time of the CCD detector limited the throughput to 17 spectra per second. As a proof of concept of multiplex capability, three biocompatible SERS nanoparticles/nanotags actively targeted three intrinsic cancer biomarkers, EGFR, CD44, and TGFβRII, in a breast cancer model [32]. Silver-gold nanorods were functionalized with four Raman-active molecules and four antibodies specific to breast cancer markers, i.e., EpCAM, IGF-1 receptor β, CD44, and keratin 18 [33]. Silica-encapsulated SERS tags were modified with three different labels as well as three different antibodies to distinguish human breast cancer cell lines based on phenotypic markers expressed on the cell surface [34]. In addition, SERS tags were successfully applied for the simultaneous detection of three surface proteins on malignant B cells [35].

9.3.2 SERS for Detection of Cancer Tissues

As described in the previous section, Raman spectroscopy amplified by SERS-active nanoparticles is a molecular imaging modality with ultra-high sensitivity and the unique ability to multiplex readouts from different molecular targets using a single wavelength of excitation. This approach holds exciting prospects for a range of applications in medicine, including identification and characterization of malignancy during endoscopy and intraoperative image guidance of surgical resection. The approach using antibodies as specific recognition units was called immuno-SERS microscopy which allows rapid tissue imaging with single nanoparticle sensitivity [36, 37]. Here, small glass-coated clusters were applied as the core of the SERS labels showing excellent SERS signals with only 0.03 s of acquisition. The authors used prostate tissue samples from healthy donors and SERS labels modified with antibodies directed against p63 and PSA. The selective abundance of p63 was shown for basal cells of the epithelium as well as of the PSA protein in the epithelium of the prostate gland.

The topical application and quantification of a multiplexed cocktail of receptor-targeted SERS nanoparticles enabled rapid quantitative molecular phenotyping of the surface of freshly excised tissues to determine the presence of disease [38]. In order to mitigate the ambiguity due to nonspecific sources of contrast such as off-target binding or uneven delivery, a ratiometric method was employed to quantify the specific vs. nonspecific binding of the multiplexed nanoparticles. Validation experiments with human tumor cell lines, fresh human tumor xenografts in mice shown in Fig. 9.5, and fresh human breast specimens demonstrated that this imaging approach of excised tissues agreed with flow cytometry and immunohistochemistry. A potential application is to image excised tissue during tumor resection procedures for identification of residual tumor at the margins and guidance of complete tumor removal in breast-conserving surgeries in less than 15 min.

NIR-SERS nanoprobes for *in vivo* targeting consisted of plasmonic Au/Ag hollow shell assemblies on the surface of silica nanospheres and simple aromatic Raman labels [39]. Plasmonic extinction of NIR-SERS nanoprobes caused enhanced SERS signals in the NIR optical window (700–900 nm), where endogenous tissue absorption coefficients are more than two orders of magnitude lower than those for ultraviolet and visible light. The signals from NIR-SERS dots were detectable from 8 mm deep in animal tissues. Three kinds of NIR-SERS nanoprobes, which were injected into live animal tissues, were detected without spectral overlap, demonstrating their potential for *in vivo* multiplex detection of specific target molecules.

A dedicated small animal Raman imaging instrument was presented that enabled rapid, high-spatial resolution, spectroscopic imaging over a wide field of view (>6 cm^2), with simplified animal handling [40]. Imaging of SERS nanoparticles in small animals demonstrated that this small animal Raman imaging system could detect multiplexed SERS signals in both superficial and deep tissue locations at least an order of magnitude faster than existing systems without compromising sensitivity.

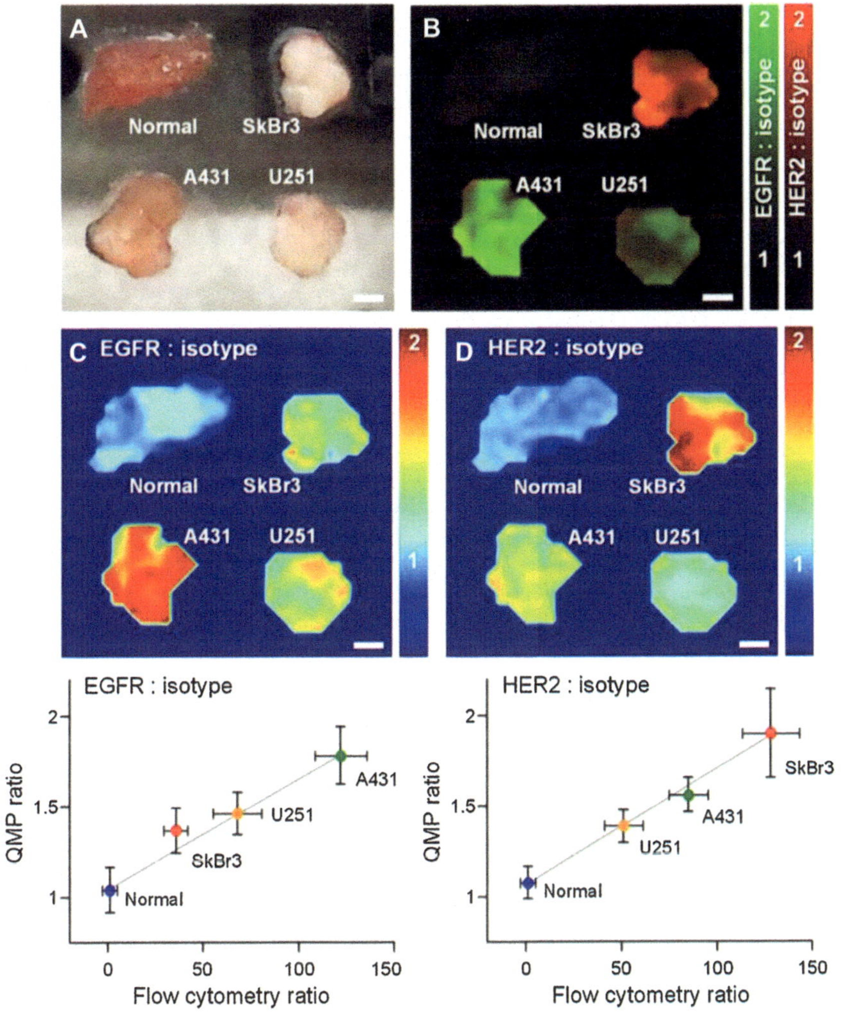

Fig. 9.5 Quantitative molecular phenotyping (QMP) imaging, with 0.5 mm spatial resolution, of tumor xenograft specimens stained with a three-flavor NP mixture (EGFR-NPs, HER2-NPs, and isotype-NPs). (**A**) Photograph of resected tumor xenografts and normal tissue. (**B**) A multiplexed QMP image generated by overlaying the ratiometric images of EGFR-NPs/isotype-NPs (plotted with a green colormap) and HER2-NPs/isotype-NPs (plotted with a red colormap). Images showing the concentration ratio of (**C**) EGFR-NPs/isotype-NPs and (**D**) HER2-NPs/isotype-NPs. The bottom plots show the correlation between the QMP ratio of a particular tissue specimen (in **C**, **D**) and the corresponding fluorescence ratio (targeted NP vs. isotype NP) from flow-cytometry experiments with the cell lines used to generate the various tumor xenografts. $R > 0.98$. Scale bars represent 2 mm (reprinted under a Creative Commons Attribution 4.0 International License from reference [38])

A noncontact fiber optic Raman probe was developed to provide real-time, multiplexed functional information during routine endoscopy [41]. This device was designed for insertion through a clinical endoscope and has the potential to detect and quantify the presence of a multiplexed panel of tumor-targeting functionalized SERS nanoparticles. The Raman instrument was characterized with SERS particles on excised human tissue samples. Detecting 326-fM concentrations of SERS nanoparticles and unmixing 10 variations of co-localized SERS nanoparticles were shown. Another feature was a wide range of working distances from 1 to 10 mm which was necessary to accommodate for imperfect centering during endoscopy and the non-uniform surface topology of human tissue. Using this endoscope as a key part of a multiplexed detection approach could allow endoscopists to distinguish between normal and precancerous tissues rapidly and to identify flat lesions that are otherwise missed.

9.4 Coherent Anti-Stokes Raman Scattering

In coherent anti-Stokes Raman scattering (CARS), two pulses usually denoted as pump (*p*) and Stokes (*S*) laser with the frequencies ω_p and ω_S interact with the sample, and coherently drive molecular vibrations in case the energy difference between pump and Stokes laser matches a Raman transition. Another pump photon ω_p can be subsequently scattered inelastically off this ensemble of coherently driven molecular vibrations generating an anti-Stokes signal ω_{aS}. The implementation of CARS as microscopic contrast modality is achieved by aligning pump and Stokes beam collinearly and focusing these two collinear beams with a high numerical aperture microscope objective lens onto one spot in the sample [42]. The phase matching condition is fulfilled for such tight focusing and the anti-Stokes signal itself is generated only over a very short interaction length of the incident lasers with the sample giving inherent confocality. By coupling the pulses in a laser scanning microscope, the directional and coherent anti-Stokes signal enables a very efficient detection using a photomultiplier tube at microsecond dwell time and pushing the imaging of selective molecular vibrations towards video repetition rates. Although CARS relies on high intensity excitation laser pulses, destructive effects on biological materials are usually not observed because (1) the duration of laser pulses are short (pico- to femtosecond) at MHz repetition rates, (2) photons are just scattered and not absorbed, and (3) NIR lasers are used outside of major electronic absorption. Additionally, CARS avoids interference with disturbing autofluorescence because the signal emerges at the anti-Stokes frequency ω_{aS} that is at shorter wavelengths than the excitation lasers. Whereas laboratory CARS instruments use complex laser systems that consist of a high power pump laser, a titanium-sapphire laser and optic parametric oscillators for frequency tuning, compact fiber lasers were developed for excitation that can be installed on a mobile cart for use in a clinical setting [43]. As this system also included other multiphoton modalities it will be described in more detail in Sect. 9.6.3.

While narrowband, single-band CRS increases the imaging speed, the spectral information content that is collected during the measurement is reduced compared to spontaneous Raman scattering because only the distribution of one specific Raman resonance is highlighted. In order to extract more meaningful information about the chemical composition of complex specimens, spectrally broadband sources for the simultaneous detection of the CRS spectrum called multiplex CRS or rapidly tunable narrow band sources for subsequent acquisition of images at different spectral positions called hyperspectral CRS can be used. Detecting multiplex CRS spectra severely slows down the imaging process since read out times of sensitive CCD detectors in conventional spectroscopy are rather long. More details about laser concepts for multiplex CRS can be found elsewhere [44]. A review summarized progress of biological and clinical applications of CARS and stimulated Raman scattering (SRS, see Sect. 9.5) covering applications such as lipid droplet research, single cell analysis, tissue imaging and multiphoton histopathology of atherosclerosis, myelin sheaths, skin, hair, pharmaceutics, and cancer and surgical margin detection [45]. The history of CRS microscopy, the pros and cons of different modalities, and the current challenges and possible future directions of the field of clinical diagnostics have recently been reviewed [46].

9.4.1 CARS Microscopy of Cancer Cells

The label-free investigation of intracellular lipid droplet genesis, distribution, and motility has been for a long time at the core of CARS microscopy. Intracellular lipid accumulations occur in form of esterified fatty acids in lipid droplets. Because fatty acids have long hydrocarbon chains and are densely packed in cellular lipid droplets the high local concentration is an ideal target for CARS microscopy.

Unsupervised multivariate analysis of CARS data sets was used to visualize subcellular organelles [47]. In addition, a supervised learning algorithm based on the "random forest" ensemble learning method as a classifier was trained with CARS spectra using immunofluorescence images as a reference. The supervised classifier was then used to automatically identify lipid droplets, nucleus, nucleoli, and endoplasmic reticulum in data sets that are not used for training. These four subcellular components were simultaneously and label-free monitored instead of using several fluorescent labels which opens new avenues for investigation of normal and cancer cells.

CARS microscopy was employed to examine the lipid content of 100 CTCs isolated from the peripheral blood of eight metastatic prostate cancer patients, and to evaluate lipid uptake and mobilization kinetics of a metastatic human prostate cancer cell line [48]. On average, CARS signal intensity of prostate CTCs was sevenfold higher than that of leukocytes (see Fig. 9.6). When incubated with human plasma, C4-2 metastatic human prostate cancer cells exhibited rapid lipid uptake kinetics and slow lipid mobilization kinetics. It was concluded that intracellular

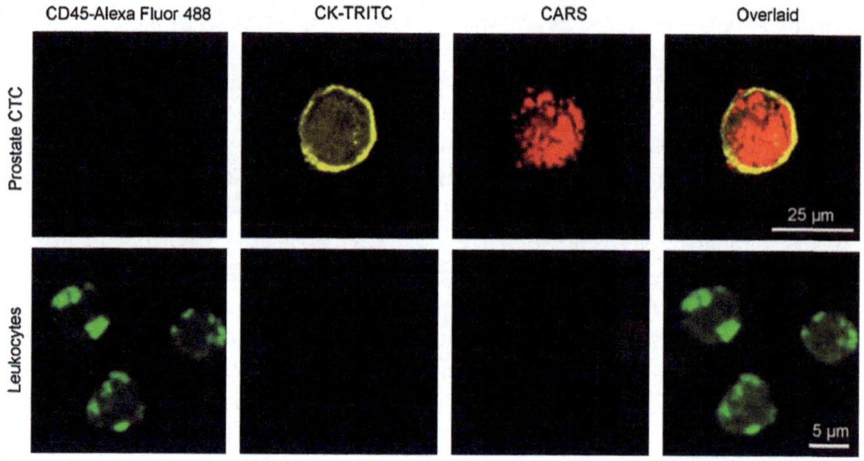

Fig. 9.6 Detection of lipid-rich prostate circulating tumor cells (CTCs) with CARS microscopy. Upper row: Prostate CTC exhibited strong CARS signal arising from intracellular lipid accumulation. Lower row: Leukocytes exhibited weak CARS signal arising mainly from cellular membrane (reprinted under a Creative Commons Attribution 4.0 International License from reference [48])

lipids could serve as a biomarker for prostate CTCs that could be sensitively detected with CARS microscopy in a label-free manner.

Multiplex CARS via supercontinuum excitation was used to study ovarian cancer cells in response to treatment with lysophosphatidic acid (LPA) [49]. Analysis of multiplex CARS images distinguished between molecular components, such as lipids and proteins. The CARS analysis shows a distinct decrease in protein and increase in lipid composition on the surface of LPA-treated cells.

9.4.2 CARS Microscopy of Cancer Tissues

A broad interest exists in pathological diagnostics to develop image analysis tools. To date, the contrast in digital pathology is based on H&E-stained tissue sections and digital slide readers. CARS microscopy is a complementary modality to provide molecular contrast in unstained tissue sections. CARS imaging within the CH stretching region enables detection of single cells and cell nuclei which allows for histopathological grading of tissue. Relevant information such as nucleus-to-cytoplasm ratio, cell density, nucleus size and shape was extracted from CARS images by innovative image processing procedures [50]. In this contribution CARS image contrast was interpreted by direct comparison with Raman imaging. An approach was presented to automatically extract relevant texture features to analyze and predict healthy and tumor regions in CARS images [51]. The approach was transferred to CARS images from basal cell carcinoma skin samples and differentiated cancer tissue from normal tissue using a perceptron algorithm for supervised learning of binary classifiers.

To address the lipid content of different brain tumors (glioblastoma, metastases of melanoma and breast cancer) which were induced in an orthotopic mouse model, cryosections were investigated by CARS imaging tuned to probe CH molecular vibrations [52]. All tumors were characterized by a lower CARS signal intensity than the normal parenchyma. On this basis, tumor borders and infiltrations could be identified with cellular resolution. Quantitative analysis revealed that the tumor-related reduction of CARS signal intensity was more pronounced in glioblastoma than in metastases. Raman spectroscopy enabled relating the CARS intensity variation to the decline of total lipid content in tumors.

Development and application of a knowledge-based CARS microscopy system were reported for label-free imaging, pattern recognition, and classification of cells and tissue structures for differentiating lung cancer from non-neoplastic lung tissues and identifying lung cancer subtypes [53]. The knowledge-based CARS system using a machine learning approach separated normal, non-neoplastic, and subtypes of lung cancer tissues based on extracted quantitative features describing fibrils and cell morphology. Lung cancer was distinguished from normal and non-neoplastic lung tissue with 91% sensitivity and 92% specificity, and small cell carcinomas from non-small cell carcinomas with 100% sensitivity and specificity. CARS imaging was combined with deep learning for automated differential diagnosis of lung cancer [54]. Conventional means of analyzing CARS images requires extensive image processing, feature engineering, and human intervention. This study demonstrated how a deep learning algorithm was applied to automatically differentiate normal and cancerous lung tissue CARS images. The features learned by pre-trained deep neural networks were leveraged and the model was re-trained using CARS images as the input. 89.2% accuracy was achieved in classifying normal, small cell carcinoma, adenocarcinoma, and squamous cell carcinoma lung images.

9.5 Stimulated Raman Scattering

Besides CARS, coherent Raman signal enhancement can also be achieved by stimulated Raman scattering (SRS). In SRS the vibrational transition is driven coherently by two high intensity excitation pulses with a frequency difference that matches a vibrational transition of interest. This results in a signal gain on the Stokes beam and a signal loss on the pump beam. The signal is detected through the modulation of one of the excitation beams and lock-in detection [55, 56]. Either a signal gain on the probe beam is measured or a signal loss on the pump beam. Since the laser power is in the order of several 10 mW, custom large area photodiodes are used for detection instead of PMTs. SRS microscopy overcomes some problems of CARS microscopy. On the one hand, the SRS signal is linearly dependent on excitation intensity and the concentration, whereas the CARS signal scales quadratic with the pump laser intensity and is proportional to the square of the concentration. On the other hand, there is no non-resonant background, which commonly obscures the Raman-resonant CARS signals from many vibrations of the Raman fingerprint region and limits the contrast. Spectroscopically, the SRS spectrum is identical to that of spontaneous Raman, whereas inference between resonant and non-resonant background results in spectral distortions such that the CARS spectrum is different from a Raman spectrum.

9.5.1 SRS Microscopy of Cells

Zhang et al. showed that it is possible to investigate the low-wavenumber region with SRS microscopy, exploiting a much larger number of Raman-active vibrations [57]. Single polytene chromosomes were imaged in salivary glands of drosophila melanogaster and two mammalian cells, i.e., HEK-293, a human embryonic kidney cell line, and MCF-7, a breast cancer carcinoma cell line, during their cell cycles. By plotting the intensities of the 785 cm^{-1} and the 1090 cm^{-1} vibrations, the distribution of polytene DNA in the nucleus was shown, also proteins at the 1655 cm^{-1} amide I vibration, and lipids at 2845 cm^{-1} CH_2 stretch vibration.

SRS allows the targeted observation of specific molecules such as metabolites, lipids, and amino acids in combination with specific tags. These Raman tags offer higher multiplexing ability, smaller size, and less photobleaching than fluorescent labels that harbors the risk to change the functionality of the target molecules or affect the cells. This approach is called bioorthogonal imaging [58]. One of the first application was to track alkyne-bearing drugs in mouse tissues and visualize de novo synthesis of DNA, RNA, proteins, phospholipids, and triglycerides through metabolic incorporation of alkyne-tagged small precursors [59]. A class of polyyne-based materials was engineered for optical supermultiplexing. 20 distinct Raman frequencies between 2000 and 2300 cm^{-1}, called carbon rainbow (Carbow), were achieved through rational engineering of conjugation length, bond-selective isotope doping, and end-capping substitution of polyynes [60]. With further probe functionalization, ten-color organelle imaging was demonstrated in individual living cells with high specificity, sensitivity, and photostability combining five organelle targeted Carbow-based Raman probes and five fluorescent reporters. This approach did not require any unmixing or color compensation in image processing.

Another panel of Raman tags utilized stable isotope labeling and the addition of a single molecular group with a large Raman scattering cross section such as nitrile or alkyne groups [61]. Taking advantage of electronic preresonance SRS (epr-SRS) enabled to get strong SRS signals with very low background, 1000 higher sensitivity than conventional SRS approaching sensitivities of confocal fluorescence. In epr-SRS excitation, the energy of the pump laser is chosen to be close to but slightly lower than that of molecular absorption energy. The resulting epr-SRS microscopy has demonstrated detection sensitivity down to nano-molar concentration with retained narrowband vibrational contrast for imaging and essentially no electronic background and little photobleaching. A palette of triple-bond-conjugated NIR dyes was created that each displays a single peak in the cell-silent Raman spectral window between 2000 and 2300 cm^{-1}. When combined with available fluorescent probes, this palette provides 24 resolvable colors, called Manhattan Raman scattering (MARS), with the potential for further expansion. These dyes were easily separable using epr-SRS microscopy, and they were used for sixteen-color imaging of dye-labeled cells. Eight-color imaging was demonstrated in hippocampal neurons as well as live cell imaging in mammalian and bacterial cells.

9.5.2 SRS Microscopy of Tissues

Microscopic inspection of H&E-stained tissue sections has been the gold standard for histopathological diagnosis of a wide range of diseases. SRS microscopy was developed as an alternative, multicolored way to visualize tissue in real-time. Multicolor images originating from CH_2 stretching vibrations at 2845 cm^{-1} and CH_3 vibrations at 2940 cm^{-1} of lipids and proteins, as well as vibrationally off-resonant two-photon absorption of hemoglobin, were obtained with subcellular resolution from fresh tissue [62]. Figure 9.7 displays SRS images of a live cell and fresh *ex vivo* brain tissue. A pseudo-color scheme could be chosen to mimic the appearance of an H&E-stained micrograph. Such stain-free histopathological images from mouse brain tissue regions, various mouse organs and mouse models of glioma and metastatic breast cancer showed resolutions similar to those obtained by conventional techniques, but do not require tissue fixation, sectioning or staining of the tissue analyzed. High-speed molecular spectral imaging of tissue by 30 frames per second SRS with frame by frame wavelength tunability was described [63]. Various SRS imaging modalities were presented such as two-dimensional spectral imaging of rat liver, two-color three-dimensional imaging of a vessel in rat liver, spectral imaging of several sections of intestinal villi in mouse, and *in vivo* spectral imaging of mouse ear skin. Squamous cell carcinoma (SCC) and healthy skin tissues were studied *ex vivo* using SRS microscopy at 2945 cm^{-1} [64]. The SRS contrast was compared with the contrast obtained in reflectance confocal microscopy and standard histology. The morphological features of SCC tumor seen in the SRS images correlated well with the diagnostic features identified by histological examination. Additionally, SRS exhibited enhanced cellular contrast in comparison to that seen in confocal microscopy.

The *in vivo* identification of surgical margins has been a long-lasting promise from the Raman community. SRS microscopy was shown to rapidly assess tumor margins and cell infiltration of human glioblastoma during a brain surgery [65]. The results obtained with SRS microscopy correlated well with the gold standard of H&E-stained tissue sections for the detection of glioma infiltration. SRS was performed on the CH_2 stretching vibration at 2835 cm^{-1}, representing the lipid fraction, and the CH_3 stretching vibration at 2930 cm^{-1}, representing the protein fraction of the sample. The comparison of the bright field images and SRS images of the infiltrated regions exemplary shows the strength of the label-free approach. A follow-up

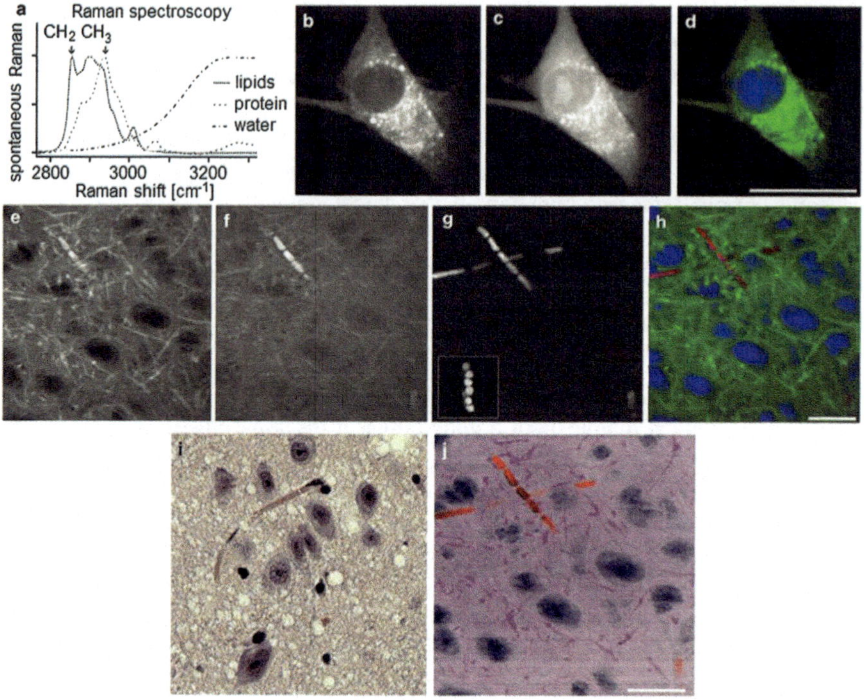

Fig. 9.7 Stain-free histologic imaging with multicolor coherent Raman imaging. (**a**) Vibrational spectra of the major constituents of tissue: lipids, protein, and water. Arrows indicate Raman shifts at which imaging is performed. (**b–d**) stimulated Raman scattering (SRS) images of a live C2C12 mammalian cell acquired at the CH_2-stretching vibration at 2845 cm^{-1} (**b**) and CH_3-stretching vibration at 2940 cm^{-1} (**c**). Multicolor image (**d**) generated from images (**b, c**) with the green channel (CH_2 image) showing the cell-body and the blue channel (thresholded CH_3-CH_2 difference image) highlighting the nuclear morphology including a bright nucleolus. (**e–h**) SRS images of fresh *ex vivo* brain tissue acquired at CH_2-stretching vibration at 2845 cm^{-1} (**e**), CH_3-stretching vibration at 2940 cm^{-1} (**f**), and vibrationally off-resonant showing two-color two-photon absorption of hemoglobin at a sum frequency of 23,700 cm^{-1} (**g**). Multicolor image (**h**) generated from images (**e–g**) with the green channel (CH_2 image) highlighting cytoplasm and myelin sheaths, blue channel (thresholded CH_3-CH_2 difference image) showing the nuclear morphology, and the red channel (hemoglobin image) highlighting the red blood cells. (**i**) Hematoxylin & eosin (H&E)-stained micrograph from the same region in the brain. (**j**) Same multicolor image as (**h**) with a different pseudo-color scheme, chosen to mimic the appearance of an H&E-stained micrograph, illustrates the similar image content and appearance of stain-free images and H&E-stained sections. Scale bar, 25 μm (reprinted with permission from reference [62])

publication detected human brain tumor infiltration with quantitative SRS microscopy [66]. SRS revealed quantifiable alterations in tissue cellularity, axonal density, and protein/lipid ratio, and a classifier was trained based on these features in SRS images to detect tumor infiltration in 22 neurosurgical patients with 97.5% sensitivity and 98.5% specificity.

Tumor metabolism supports the abnormal survival and growth of malignant cells by providing energy, biomolecular precursors, and reducing equivalents.

Bioorthogonal chemical imaging, introduced in Sect. 9.5.1, was applied to visualize newly synthesized biomolecules in tumors after *in vivo* delivery of a glucose analogue labeled with a small alkyne moiety into tumor-bearing mice [67]. Cancer cells with differing metabolic activities could be distinguished. Heterogeneous uptake patterns were observed with clear cell–cell variations in tumor xenograft tissues, neuronal culture, and mouse brain tissues.

9.6 Multimodal Approaches with Spontaneous and Coherent Raman Scattering

A recent review about multimodal spectroscopic imaging of tissue covered besides spontaneous Raman and coherent Raman scattering (CARS, SRS) also optical coherence tomography (OCT), fluorescence lifetime imaging (FLIM), photoacoustic imaging (PAI), confocal/reflectance laser scanning microscopy (CLSM/RLSM) and the multiphoton microscopies two-photon excited fluorescence (TPEF) and second harmonic generation (SHG) [68]. Here, we focus on Raman-based spectroscopies to detect cancer cells and tissues. Each modality offers advantages and disadvantages with respect to speed, spatial resolution, and penetration depth. Furthermore, they probe different properties of cells and tissues. Combination of multiple modalities is an effort to improve the performance of optical spectroscopy methods for cancer detection in a complementary manner.

9.6.1 Multimodal Approaches Involving Spontaneous Raman Spectroscopy

Multimodal instruments combine the biochemical specificity of Raman spectroscopy with the rapid screening capabilities of other label-free high resolution optical imaging like autofluorescence, FLIM, OCT, or multiphoton microscopy. Combinations are beneficial for spontaneous Raman spectroscopy to guide point measurements or small area scans of suspicious lesions in overview images.

9.6.1.1 Raman and Fluorescence

An integrated optical technique was demonstrated for basal cell carcinoma of skin based on autofluorescence imaging and Raman scattering. Automated segmentation of autofluorescence images was used to select and prioritize sparse sampling points for Raman spectroscopy. Raster scanning Raman microspectroscopy of tissue sections is slow with typically 10,000 spectra per mm^2 which is equivalent to 5 h per mm^2. The selective sampling approach reduced the number of Raman spectra to 20

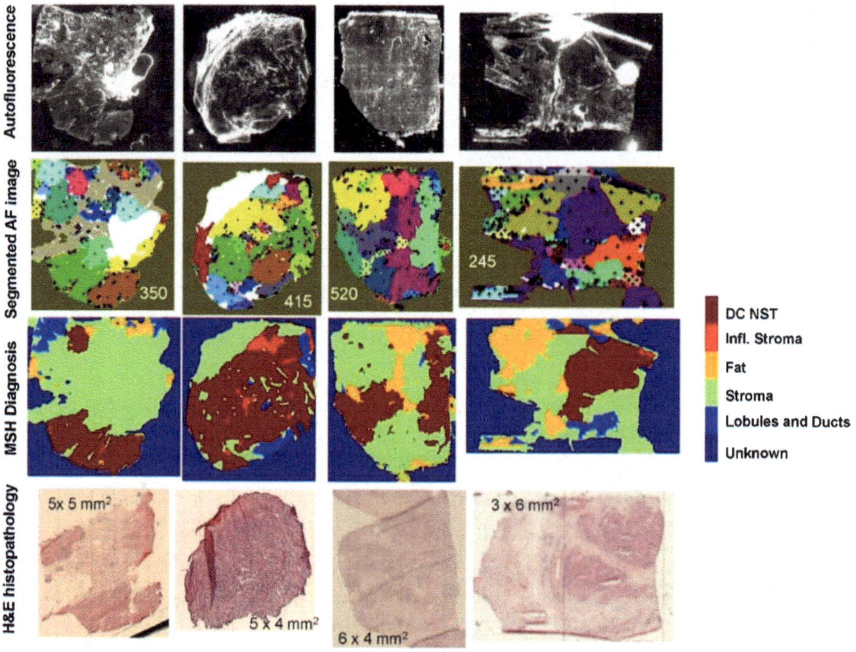

Fig. 9.8 Typical examples of multimodal spectral histopathology diagnosis of ductal carcinoma in breast tissue samples. The number of Raman spectra measured for every sample is included in the segmented autofluorescence (AF) images. Histopathology images for adjacent sections are included for comparison (reprinted under a Creative Commons Attribution 4.0 International License from reference [70])

spectra per mm^2 which is equivalent to an acquisition time of 15 min for 25 mm^2. This automated sampling strategy yielded classification models with 100% sensitivity and 92% specificity, and allowed objective diagnosis faster than frozen section histopathology and faster than IR or Raman imaging alone [69]. The approach was transferred to the diagnosis of mammary ductal carcinoma [70] and basal cell carcinoma in skin tissue samples [71]. Typical examples of multimodal spectral histopathology diagnosis of ductal carcinoma in breast tissue samples are shown in Fig. 9.8. The number of Raman spectra measured for every sample was included in the segmented autofluorescence images and ranged from 245 to 520.

A fiber optic Raman endoscopy system combined Raman spectroscopy at 785 nm excitation under multimodal guidance with white light reflectance imaging, narrow band imaging, and autofluorescence imaging [72]. The multimodal approach was evaluated for real-time *in vivo* diagnosis of cancer in the esophagus during clinical endoscopic examinations. A similar system combined fluorescence, reflectance, and Raman spectroscopy to detect features of vulnerable atherosclerotic plaques and breast cancer *in vivo* and *ex vivo* [73]. Another group designed and characterized an instrument combining Raman, fluorescence, and reflectance spectroscopic modalities for skin cancer applications [74]. Preliminary clinical data indicated the system's ability to measure physiological quantities such as relative collagen and nicotinamide adenine dinucleotide concentration, oxygen saturation, blood volume fraction, and mean vessel diameter. A study determined whether adding Raman spectroscopy to white light bronchoscopy and autofluorescence bronchoscopy has the potential to improve the specificity to detect preneoplastic lesions of the bronchial tree [75]. For this purpose, a Raman system was developed to collect real-time, *in vivo* lung spectra with a fiber optic catheter passed down the instrument channel of a bronchoscope.

Fluorescence lifetime was developed as a complementary technique which detects the fluorescence temporal decay. In addition to the respective excitation and emission spectra, fluorophores can be characterized by their lifetimes. Fluorescence lifetime measurements offer several advantages compared to fluorescence that include invariance to intensity variations of the incoming light, no dependence on the transmission efficiency of the detection system, independence of the local concentration of the fluorophore, insensitivity to moderate levels of photobleaching. The most popular variant uses pulsed laser excitation with MHz repetition rates and time correlated single photon counting [76]. Another variant uses blue lasers, dielectric filters, delay lines, and micro-channel plate detector for sequential registration of four fluorescence lifetimes [77]. As the acquisition speed is fast, fluorescence lifetime images (FLIm) can be registered with scanning mirrors in a microscope or upon moving a fiber optic probe with a motorized arm. A dual modal fiber optic probe was designed for both FLIM and Raman spectroscopy, and its *in vivo* endoscopic application was demonstrated for rat brain tissue under a cranial window [78]. In a follow-up paper, dual modal images were simultaneously acquired from two human coronary specimens [79]. Raman images identified the components triglycerides and cholesterol, carotenoids, and calcium salts. Three channels in lifetime images were assigned to collagen/elastin, elastin/lipids, and extracellular lipids. The results were correlated with the tissue sections along the entire length of the specimens that were stained with H&E, CD68, CD45, and elastic trichrome.

9.6.1.2 Raman and OCT

OCT produces images of comparable large areas across the sample in the range of several millimeters with real-time speed and micrometer spatial resolution. The limiting factor of OCT is its low biochemical specificity as the images map the differences in reflectivity of tissue morphology. Raman spectroscopy can provide complementary specificity. An instrument was developed to combine Raman spectroscopy at 785 nm excitation and 1310 nm OCT system backbone for biochemical and morphological characterization of breast cancer [80] and skin cancer [81]. Advances were made to further integrate the two modalities by using a common detector [82]. This was demonstrated with a different OCT source centered at 855 nm for dissected calvaria of a mouse *ex vivo* and on human skin *in vivo*. A portable Raman-OCT was introduced for clinical use with an amendable probe to investigate skin cancers [83]. The image acquisition time for OCT was improved from one frame per second to eight frames per second allowing for real-time imaging. These instruments enabled sequential acquisition of co-registered OCT and Raman data sets.

Ashok et al. reported an optical approach using Raman spectroscopy and OCT in tandem to discriminate between colonic adenocarcinoma and normal colon [84]. This study collected exclusively *ex vivo* data without spatial co-registration. The chemical information derived from Raman spectroscopy was combined with texture parameters extracted from OCT images. The classifier performed superior in terms of sensitivity and specificity (increase to 94%) using information from both modalities in comparison to using information from only one.

Another OCT-Raman instrument improved the depth sensitivity of Raman spectroscopy combining a time-domain OCT system with a confocal Raman setup [85]. The co-alignment of both modalities is obtained via the sample arm of the OCT system. The acquisition rate for OCT with a 1310 nm excitation source was specified as 125 A-scans per frame at 8 frames/s. Raman spectra at 785 nm excitation in tissue were collected at 10 s integration time. Raman spectra of resected goat mucosal tissue could be obtained from a depth of 900 μm corresponding well with the penetration depth of the OCT subsystem.

9.6.1.3 Raman and Multiphoton Microscopy

Increasing interest in the role of lipids in cancer cell proliferation and resistance to drug therapies motivated the need to develop better tools for cellular lipid analysis. Quantification of lipids in cells is typically done by destructive chromatography protocols that do not provide spatial information on lipid distribution and prevent dynamic live cell studies. Raman-based methods allow the analysis of lipid content in live cells. Raman spectroscopy on single cellular lipid droplets and least-squares fitting of pure fatty acid spectra was used to determine the composition of individual lipid droplets in cells after treatment with different ratios of oleic and palmitic acid [86]. The results of the Raman spectroscopy-based single lipid droplet analysis were validated with results obtained by gas chromatography analysis of millions of cells. The approach was found to accurately predict the relative amount of a specific fatty acid in the lipid droplet. This approach was expanded to investigate the lipid composition in single cellular peroxisomes. Cellular peroxisomes labeled with the green fluorescent protein were localized based on TPEF imaging, and successive Raman spectroscopy of peroxisomes was performed. In some cases, peroxisomes produced a detectable CARS signal, and the peroxisomal Raman spectra exhibited an oleic acid-like signature. Combining micro-Raman spectroscopy with CARS imaging, the process of hormone-mediated lipogenesis was studied [87]. Lipid profiles for breast (T47D, MDA-MB-231) and prostate (LNCaP, PC3) cancer cells were generated upon exposure to medroxyprogesterone acetate (MPA) and synthetic androgen R1881. Hormone-treated cancer cells T47D and LNCaP showed an increased number and size of intracellular lipid droplets and higher degree of saturation than untreated cells. MDA-MB-231 and PC3 cancer cells showed no significant changes upon treatment. Principal component analysis with linear discriminant analysis of the Raman spectra was able to differentiate between cancer cells that were treated with MPA, R1881, and untreated. In summary, both studies used Raman spectroscopy to analyze lipids on the molecular level (e.g., level of unsaturation), whereas multiphoton microscopic images provided morphological information about number, size, and location of lipid droplets.

Towards noninvasive diagnosis of urothelial carcinoma, CARS imaging of urine sediments was used for fast preselection of urothelial cells, where high-grade urothelial cancer cells are characterized by a large nucleus-to-cytoplasm ratio [88]. Then, Raman spectral images were collected from urothelial cells. A supervised classifier was implemented to automatically differentiate normal and cancerous urothelial cells with 100% accuracy. Raman marker bands directly showed decreased content of glycogen and increased content of fatty acids in cancer cells as compared to controls.

9.6.2 *Multimodal SERS Labels*

In addition to the multiplexing capabilities of SERS tags achieved by using various Raman reporter molecules as shown in Sect. 9.3, multimodal approaches, i.e., employing two or more readout modalities gained much attention by researchers for cell and tissue imaging.

To combine the advantages of fluorescence and SERS, dual modal nanoprobes were created by coating gold nanoparticles with a Raman reporter as well as with a fluorescence dye separated by a silica spacer to prevent fluorescence quenching [89]. Fluorescence microscopy is a well-known imaging technique that shows specific protein distributions within cells. However, most currently available fluorescent organic dyes have relatively weak emission intensities and are rapidly photo-bleached. SERS detection is a powerful technique that allows ultrasensitive chemical or biochemical analysis through unlimited multiplexing and single molecule sensitivity. Investigations to target and image the local distribution of specific cancer markers (CD24 and CD44) on living breast cancer cells illustrated the potential of antibody-modified dual tags. A critical point is that the report used both fluorescence and SERS to localize the tags in cells, which gave the same information. In order to provide supplementary information or functionality, the research should point toward the innovative combination of SERS tags with modalities that are not achieved via SERS.

Such a complementary functionality was implemented by employing magnetic SERS labels. Here, Fe_3O_4 cores were modified with SERS-active gold nanoparticles, and the immobilized antibodies on the surface of those hybrid structures allowed specific binding to cancer cells [90]. Thus, the magnetic SERS tags could be used to capture cells and to separate them from a complex matrix at a very low abundance (ca. 10 cells per ml). This assay had the advantages of (1) high sensitivity because the SERS technique itself enabled the detection of very low concentration analytes, (2) high specificity because the enhanced SERS signal arose in sandwich configuration, which was formed in specific antigen–antibody interaction events, and (3) high efficiency because the magnetic SERS tags concentrated the captured cells and separated them from the complex detection system without repetitive washing steps. Alternatively, circulating tumor cells (CTCs) could be detected by using two different labels, magnetic beads to separate or enrich the cells, and SERS tags to detect CTCs with high specificity and sensitivity [91]. The specificity was achieved by aptamers as recognition elements, and rare target cancer cells could efficiently be captured from buffer and whole blood sample with capture efficiency of 73% and 55%, respectively. As a novel strategy SERS nanoprobes were fabricated without expensive and potentially toxic organic dyes [92]. Multifunctional conducting polymer (CP) materials served as both biocompatible surface coatings and NIR-active reporters on the surface of gold nanorods (GNRs). GNR-CPs showed extraordinary photothermal transduction efficiency, indicating the potential for cancer therapy as theranostic agents. A combined photoacoustic/Raman approach was proposed using GNRs as a passively targeted molecular imaging agent [93]. Maximum photoacoustic signal was observed within 3 h after injection for imaging ovarian cancer cell lines in living mice and increased signal persisted for at least two days post-administration. The same molecular imaging agent could be used to clearly visualize the margin between tumor and normal tissue and tumor debulking via SERS imaging. Plasmonic nanoprobes with five modalities, i.e., SERS, magnetic resonance imaging (MRI), computed tomography (CT), two-photon luminescence (TPL), and photothermal therapy (PTT) were developed [94]. The gold nanostars were modified with the Raman reporter for the SERS readout and with Gd^{3+} as MRI contrast agent. In first experiments using BT549 cancer cells for their investigations, the authors demonstrated the potential as platform for pre-operative tumor scanning (MRI and CT), intraoperative tumor detection (SERS and TPL), and PTT as post-operative treatment.

A dual modal fluorescence-Raman endomicroscopic system (FRES) was developed which used fluorescence and SERS nanoprobes (F-SERS-dots) [95]. Real-time, *in vivo*, and multiple target detection of a specific cancer was successful, based on the fast imaging capability of fluorescence signals and the multiplex capability of simultaneously detected SERS signals using an optical fiber bundle for intraoperative endoscopic system. Human epidermal growth factor receptor 2 (HER2) and epidermal growth factor receptor (EGFR) on the breast cancer xenografts in a mouse orthotopic model were detected in a multiplexed way as shown in Fig. 9.9. A follow-up paper targeted EGFR and vascular endothelial growth factor with antibody-conjugated fluorescence and SERS nanoprobes for cancer diagnosis in an orthotopically induced colorectal cancer xenograft model [96].

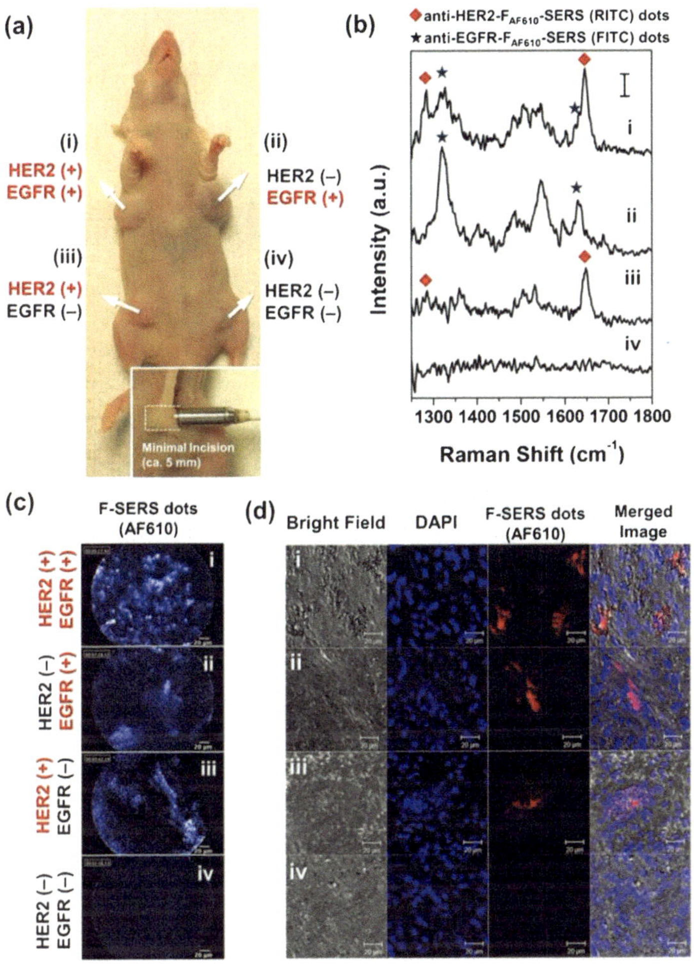

Fig. 9.9 Each tumor site (i) was not treated with any antibody, (ii) pre-treated with the anti-HER2 antibody, (iii) pre-treated with the anti-EGFR antibody, or (iv) pre-treated with both antibodies for blocking specific binding. Then, they were treated with the F-SERS dots (anti-HER2-FAF610-SERSRITC dots and anti-EGFR-FAF610-SERSFITC dots) for the FRES imaging. (**a**) A photograph of the tumor-bearing mouse with the receptor expression status as shown besides the image. The real-time fluorescence images and Raman spectra were simultaneously obtained with an optical fiber bundle probe of the FRES (lower box). (**b**) Fluorescence images were obtained by the FRES with real-time (12 frames/s). The bright area in fluorescence images corresponds to the targeted probes. (**c**) The Raman spectra were obtained by the FRES at a laser power of 2.7 mW and acquisition time of 1 s. The observed Raman bands in the Raman spectra correspond to the RITC (filled diamond) and FITC (filled star) from the F-SERS dots. (**d**) The confocal fluorescence laser scanning (CLSM) images of the tumor sites. The nuclei of the tumor cells stained with 4′,6-diamidino-2-phenylindole (DAPI) dye were shown as blue spots, and the targeted F-SERS dots containing Alexa Fluor 610 were shown as red spots (reprinted under a Creative Commons Attribution 4.0 International License from reference [95])

9.6.3 Multimodal Multiphoton Imaging Using CRS, TPEF and SHG

CRS relies on high intensity excitation picosecond or femtosecond pulses for an efficient signal generation. Other multiphoton effects such as second harmonic generation (SHG) and two-photon excited fluorescence (TPEF) are generated at the same time giving complementary information without the need for external markers and with the possibility of optical sectioning. All modalities can be detected simultaneously using proper filtering and additional detectors for each modality allowing for a straightforward integration of either CARS-TPEF-SHG or SRS-TPEF-SHG. However, the lock-in detection scheme of SRS does not readily allow integration with other nonlinear imaging modalities. Multimodal, nonlinear microscopy in biomedical sciences and the synergy of multiple contrast mechanisms were reviewed [97].

A compact microscope setup for multimodal imaging including CARS-TPEF-SHG was designed for clinical imaging with an enlarged field of view (1.2×1.2 mm^2 for 20× objective lens), the highest reported NIR transmission of 60% along the excitation path, and an alignment free, portable multicolor fiber laser [43]. To minimize the system's complexity, the laser providing picosecond pulses optimized for CARS was operated at a fixed spectral position tuned to the aliphatic CH_2-stretching vibrations at 2850 cm^{-1}. The microscope was optimized for a simultaneous efficient generation of all three signals yielding high quality, high resolution images of morphological and biochemical features of the tissue. *Ex vivo* imaging of clinically relevant entities such as head and neck SCC samples or aorta with atherosclerotic plaques demonstrated the potential of such an easy to handle multimodal setup for clinical applications.

The underlying principle of TPEF is the excitation of an electronic molecular transition by the simultaneous absorption of two photons having twice the excitation wavelength. Due to the nonlinearity, an effective excitation of fluorophores is observed only in the proximity of the nominal focus. Thus, the intrinsic confocality simplifies the experimental setup and reduces out-of-focus bleaching and further allows imaging tissue sections in higher depth and with better axial resolution compared to conventional one-photon fluorescence microscopy. Main tissue autofluorophores are nicotinamide adenine dinucleotide hydrogenase (NADH) and flavin adenine dinucleotide (FAD) that monitor the metabolic activity of the sample. Furthermore, melanin and structural proteins such as elastin, keratin, and collagen that provide information about the extracellular matrix and connective tissue contribute to the autofluorescence.

SHG microscopy is a coherent second-order nonlinear scattering technique. Two photons of a wavelength typically lying in the NIR region simultaneously interact with the tissue, and a photon with exactly half the excitation wavelength is produced. Thus, the process is also referred to as frequency doubling. Molecular groups with a large hyperpolarizability and bulk non-centrosymmetric structures in the focal volume meet the requirements for strong SHG signal generation. As collagen proteins yield intense SHG signals, collagen-rich tissues such as cornea, tendons, arteries, skin, collagen organization or its alteration in the microenvironment of tumors, during fibrosis, and in muscle, bones, cartilages, myosins, and tubulins were investigated by SHG.

A multimodal approach combining CARS, TPEF, and SHG was used to evaluate skin samples of 32 individuals [98]. The multiphoton images enabled to distinguish the main tissue layers epidermis, dermis, and subcutis and to identify tissue structures such as hair follicles, sebaceous and sweat glands, and blood vessels. This approach was also applied to detect and discriminate non-melanoma skin cancer (NMSC), more precisely basal cell carcinoma (BCC) and squamous cell carcinoma (SCC), which together account for 97% of all NMSC [99]. Multimodal multiphoton microscopy was used to study head and neck squamous cell carcinomas [100]. Image analysis using CARS, SHG, and TPEF modalities predicted the diagnosis of tissue sections with an overall accuracy of 90% for a 4-class model (cancer, epithelial tissue, other, background). With a time frame below 20 min this approach can complement conventional staining protocols to assist frozen section analysis during surgical interventions. Colon tissue sections of inflammatory bowel disease (IBD) were investigated using the combination CARS, TPEF, and SHG [101]. Various geometry and intensity related features were extracted from the multimodal images. An optimized feature set was utilized to predict histological index levels. CARS and TPEF microscopy data from 55 lesions of the central nervous system were presented [102]. The generated images demonstrated cytological and architectural features that are required for pathological tumor grading. Furthermore, the data provided information on the molecular content. The imaging modalities CARS, TPEF, and SHG yielded information that could be translated into computational, pseudo-H&E-stained images by multivariate statistics [103]. The performance was demonstrated for images of mouse colon sections. Figure 9.10 compares multimodal acquired and pseudo-H&E generated images with H&E-stained images as examples.

To extend the applicability of multimodal multiphoton imaging for *in vivo* tissue screening, a compact fiber optic probe was developed for simultaneous recording of CARS, TPEF, and SHG [104]. The probe was based on a gradient index lens design and 10,000 coherent light guiding elements preserving the spatial relationship between the entrance and the output of the fiber. The scanning procedure was shifted from the distal to the proximal end of the probe without moving parts or driving current at the probe head. The generated signals were collected in the backward (epi) direction and guided with a multimode fiber to a detection setup. A rigid, compact multimodal endoscope was described and its use was demonstrated with samples of neurosurgical relevance [105]. The endoscope has a diameter of 2.2 mm and a length of 187 mm, and offers a spatial resolution of 750 nm over a field of view of 250 μm making clinical applications during surgical interventions possible.

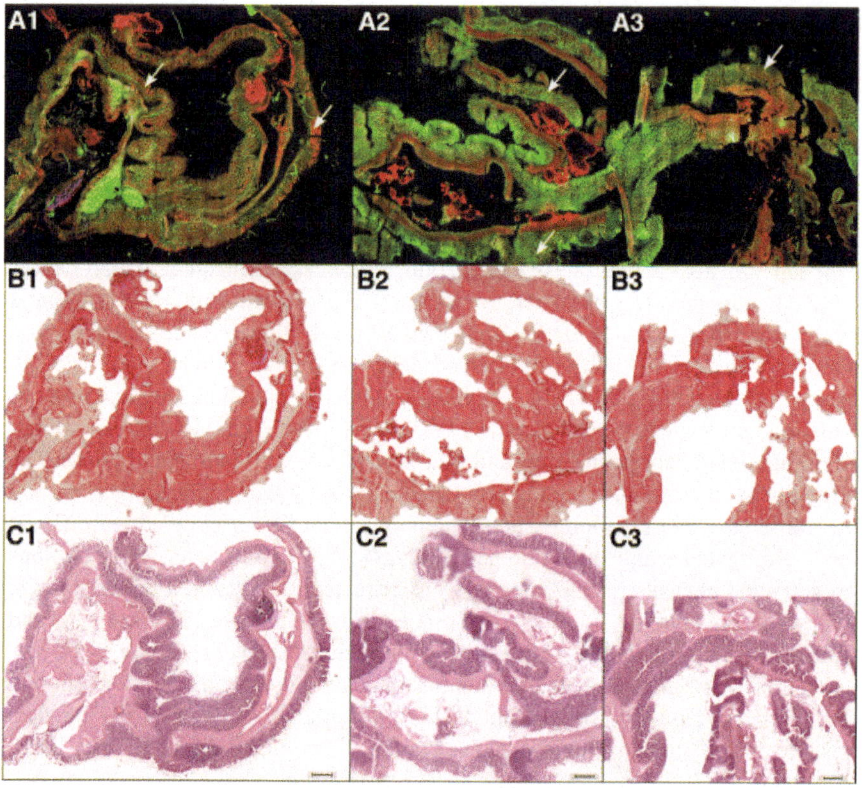

Fig. 9.10 Overview of acquired and generated images of mouse colon sections (3 out of 22 images in total): In row A, multimodal images are displayed. In rows B and C the computationally derived pseudo-HE-stained images based on the multimodal images and the HE-stained image are displayed, respectively. The pseudo-HE images of row B are generated non-invasively allowing for a subsequent analysis by other modalities or stains. The scale bar represents 500 µm (reprinted under a Creative Commons Attribution 4.0 International License from reference [103])

9.7 Outlook

This chapter described the wide field of Raman-based approaches for stand-alone and multimodal optical diagnostics of cancer ranging from single cells, tissues of various organs to small animals and *in vivo* studies of patients. Special attention has to be paid to the development of easy-to-use and robust instrumentation. Moreover, algorithms that provide automatically information about the presence of absence of disease marker will greatly enhance applicability of spectroscopy and imaging tools to the end-user. For a future translation to a clinical environment it is key to design and perform large and representative studies to show the benefits of these Raman-based concepts. Furthermore, such large, ideally multi-center, studies can take advantage of the recent advent of artificial intelligence and machine learning strategies that already demonstrated their power in image analysis and data processing.

This constitutes a great opportunity for a network of researchers and end-users to accomplish these goals.

References

1. Krafft, C., Schie, I., Meyer, T., Schmitt, M., Popp, J.: Chem. Soc. Rev. **45**, 1819 (2016)
2. Schie, I.W., Kiselev, R., Krafft, C., Popp, J.: Analyst. **141**, 6387 (2016)
3. Neugebauer, U., Clement, J.H., Bocklitz, T., Krafft, C., Popp, J.: J. Biophotonics. **3**, 579 (2010)
4. Neugebauer, U., Bocklitz, T., Clement, J.H., Krafft, C., Popp, J.: Analyst. **135**, 3178 (2010)
5. Dochow, S., Krafft, C., Neugebauer, U., Bocklitz, T., Henkel, T., Mayer, G., Albert, J., Popp, J.: Lab Chip. **11**, 1484 (2011)
6. Dochow, S., Beleites, C., Henkel, T., Mayer, G., Albert, J., Clement, J., Krafft, C., Popp, J.: Anal. Bioanal. Chem. **405**, 2743 (2013)
7. Schie, I.W., Rüger, J., Mondol, A.S., Ramoji, A., Neugebauer, U., Krafft, C., Popp, J.: Anal. Chem. **90**, 2023 (2018)
8. Tolstik, T., Marquardt, C., Matthaeus, C., Bergner, N., Bielecki, C., Krafft, C., Stallmach, A., Popp, J.: Analyst. **139**, 6036 (2014)
9. Hedegaard, M., Krafft, C., Ditzel, H.J., Johansen, L.E., Hassing, S., Popp, J.: Anal. Chem. **82**, 2797 (2010)
10. Duraipandian, S., Traynor, D., Kearney, P., Martin, C., O'Leary, J.J., Lyng, F.M.: Sci. Rep. 8, 15048 (2018)
11. Diem, M., Mazur, A., Lenau, K., Schubert, J., Bird, B., Miljkovic, M., Krafft, C., Popp, J.: J. Biophotonics. **6**, 855 (2013)
12. Krafft, C., Belay, B., Bergner, N., Romeike, B.F., Reichart, R., Kalff, R., Popp, J.: Analyst. **137**, 5533 (2012)
13. Bergner, N., Krafft, C., Geiger, K.D., Kirsch, M., Schackert, G., Popp, J.: Anal. Bioanal. Chem. **403**, 719 (2012)
14. Bergner, N., Medyukhina, A., Geiger, K.D., Kirsch, M., Schackert, G., Krafft, C., Popp, J.: Anal. Bioanal. Chem. **405**, 8719 (2013)
15. Sobottka, S.B., Geiger, K.D., Salzer, R., Schackert, G., Krafft, C.: Anal. Bioanal. Chem. **393**, 187 (2009)
16. Krafft, C., Shapoval, L., Sobottka, S.B., Geiger, K.D., Schackert, G., Salzer, R.: Biochim. Biophys. Acta. **1758**, 883 (2006)
17. Krafft, C., Shapoval, L., Sobottka, S.B., Schackert, G., Salzer, R.: Technol. Cancer Res. Treat. **5**, 291 (2006)
18. Bergner, N., Bocklitz, T., Romeike, B.F.M., Reichart, R., Kalff, R., Krafft, C., Popp, J.: Chemom. Intel. Lab. Syst. **117**, 224 (2012)
19. Bergner, N., Romeike, B.F.M., Reichart, R., Kalff, R., Krafft, C., Popp, J.: Analyst. **138**, 3983 (2013)
20. Cialla-May, D., Zheng, X.-S., Weber, K., Popp, J.: Chem. Soc. Rev. **46**, 3945 (2017)
21. Leopold, N., Lendl, B.: J. Phys. Chem. B. **107**, 5723 (2003)
22. Hassoun, M., Schie, I.W., Tolstik, T., Stanca, S.E., Krafft, C., Popp, J.: Beilstein J. Nanotechnol. **8**, 1183 (2017)
23. Hassoun, M., Rüger, J., Kirchberger-Tolstik, T., Schie, I.W., Henkel, T., Weber, K., Cialla-May, D., Krafft, C., Popp, J.: Anal. Bioanal. Chem. 1 (2017)
24. Hassoun, M., Köse, N., Kiselev, R., Tolstik, T., Schie, I., Krafft, C., Popp, J.: Anal. Methods. **10**, 2785 (2018)
25. Lin, J., Chen, R., Feng, S., Li, Y., Huang, Z., Xie, S., Yu, Y., Cheng, M., Zeng, H.: Biosens. Bioelectron. **25**, 388 (2009)

26. Feng, S., et al.: Nanotechnology. **26**, 065101 (2015)
27. Kneipp, K., et al.: Appl. Spectrosc. **56**, 150 (2002)
28. Lee, S., Kim, S., Choo, J., Shin, S.Y., Lee, Y.H., Choi, H.Y., Ha, S., Kang, K., Oh, C.H.: Anal. Chem. **79**, 916 (2007)
29. Freitag, I., Matthäus, C., Csaki, A., Clement, J., Cialla-May, D., Weber, K., Krafft, C., Popp, J.: J. Biomed. Opt. **20**, 055002 (2015)
30. Stanca, S.E., et al.: Nanomed.: Nanotechnol. Biol. Med. **11**, 1831 (2015)
31. Freitag, I., Beleites, C., Dochow, S., Clement, J., Krafft, C., Popp, J.: Analyst. **141**, 5986 (2016)
32. Dinish, U.S., Balasundaram, G., Chang, Y.T., Olivo, M.: Sci. Rep. **4**, 4075 (2014)
33. Nima, Z.A., et al.: Sci. Rep. **4**, 4752 (2014)
34. Lee, S., Chon, H., Lee, J., Ko, J., Chung, B.H., Lim, D.W., Choo, J.: Biosens. Bioelectron. **51**, 238 (2014)
35. MacLaughlin, C.M., Mullaithilaga, N., Yang, G., Ip, S.Y., Wang, C., Walker, G.C.: Langmuir. **29**, 1908 (2013)
36. Salehi, M., Steinigeweg, D., Stroebel, P., Marx, A., Packeisen, J., Schluecker, S.: J. Biophotonics. **6**, 785 (2013)
37. Salehi, M., Schneider, L., Stroebel, P., Marx, A., Packeisen, J., Schluecker, S.: Nanoscale. **6**, 2361 (2014)
38. Wang, Y., et al.: Sci. Rep. **6**, 21242 (2016)
39. Kang, H., et al.: Adv. Funct. Mater. **23**, 3719 (2013)
40. Bohndiek, S.E., Wagadarikar, A., Zavaleta, C.L., Van de Sompel, D., Garai, E., Jokerst, J.V., Yazdanfar, S., Gambhir, S.S.: Proc. Natl. Acad. Sci. U. S. A. **110**, 12408 (2013)
41. Zavaleta, C.L., et al.: Proc. Natl. Acad. Sci. U. S. A. **110**, E2288 (2013)
42. Volkmer, A.: J. Phys. D: Appl. Phys. **38**, R59 (2005)
43. Meyer, T., et al.: Analyst. **138**, 4048 (2013)
44. Gottschall, T., Meyer, T., Schmitt, M., Popp, J., Limpert, J., Tünnermann, A.: TrAC Trends Anal. Chem. **102**, 103 (2018)
45. Schie, I.W., Krafft, C., Popp, J.: Analyst. **140**, 3897 (2015)
46. Zhang, C., Cheng, J.-X.: APL Photonics. **3**, 090901 (2018)
47. El-Mashtoly, S.F., et al.: Biophys. J. **106**, 1910 (2014)
48. Mitra, R., Chao, O., Urasaki, Y., Goodman, O.B., Le, T.T.: BMC Cancer. **12**, 540 (2012)
49. Bailey, K.A., Klymenko, Y., Feist, P.E., Hummon, A.B., Stack, M.S., Schultz, Z.D.: Sci. Rep. **7**, 15295 (2017)
50. Meyer, T., Bergner, N., Medyukhina, A., Dietzek, B., Krafft, C., Romeike, B.F., Reichart, R., Kalff, R., Popp, J.: J. Biophotonics. **5**, 729 (2012)
51. Legesse, F.B., Medyukhina, A., Heuke, S., Popp, J.: Comput. Med. Imaging Graph. **43**, 36 (2015)
52. Uckermann, O., Galli, R., Tamosaityte, S., Leipnitz, E., Geiger, K.D., Schackert, G., Koch, E., Steiner, G., Kirsch, M.: PLoS One. **9**, e107115 (2014)
53. Gao, L.S., et al.: J. Biomed. Opt. **16**, 096004 (2011)
54. Weng, S., Xu, X., Li, J., Wong, S.T.: J. Biomed. Opt. **22**, 106017 (2017)
55. Freudiger, C.W., Min, W., Saar, B.G., Lu, S., Holtom, G.R., He, C., Tsai, J.C., Kang, J.X., Xie, X.S.: Science. **322**, 1857 (2008)
56. Nandakumar, P., Kovalev, A., Volkmer, A.: New J. Phys. **11**, (2009)
57. Zhang, X., Roeffaers, M.B., Basu, S., Daniele, J.R., Fu, D., Freudiger, C.W., Holtom, G.R., Xie, X.S.: ChemPhysChem. **13**, 1054 (2012)
58. Wei, L., Hu, F., Chen, Z., Shen, Y., Zhang, L., Min, W.: Acc. Chem. Res. **49**, 1494 (2016)
59. Wei, L., Hu, F., Shen, Y., Chen, Z., Yu, Y., Lin, C.-C., Wang, M.C., Min, W.: Nat. Methods. **11**, 410 (2014)
60. Hu, F., Zeng, C., Long, R., Miao, Y., Wei, L., Xu, Q., Min, W.: Nat. Methods. **15**, 194 (2018)
61. Wei, L., et al.: Nature. **544**, 465 (2017)
62. Freudiger, C.W., et al.: Lab. Invest. **92**, 1492 (2012)

63. Ozeki, Y., Umemura, W., Otsuka, Y., Satoh, S., Hashimoto, H., Sumimura, K., Nishizawa, N., Fukui, K., Itoh, K.: Nat. Photonics. **6**, 845 (2012)
64. Mittal, R., Balu, M., Krasieva, T., Potma, E.O., Elkeeb, L., Zachary, C.B., Wilder-Smith, P.: Lasers Surg. Med. **45**, 496 (2013)
65. Ji, M.B., et al.: Sci. Transl. Med. **5**, 201ra119 (2013)
66. Ji, M.B., et al.: Sci. Transl. Med. **7**, 309ra163 (2015)
67. Hu, F., Chen, Z., Zhang, L., Shen, Y., Wei, L., Min, W.: Angew. Chem. Int. Ed. **54**, 9821 (2015)
68. Vogler, N., Heuke, S., Bocklitz, T.W., Schmitt, M., Popp, J.: Annu. Rev. Anal. Chem. **8**, 359 (2015)
69. Kong, K., Rowlands, C.J., Varma, S., Perkins, W., Leach, I.H., Koloydenko, A.A., Williams, H.C., Notingher, I.: Proc. Natl. Acad. Sci. U. S. A. **110**, 15189 (2013)
70. Kong, K., Zaabar, F., Rakha, E., Ellis, I., Koloydenko, A., Notingher, I.: Phys. Med. Biol. **59**, 6141 (2014)
71. Takamori, S., Kong, K., Varma, S., Leach, I., Williams, H.C., Notingher, I.: Biomed. Opt. Express. **6**, 98 (2015)
72. Bergholt, M.S., Zheng, W., Lin, K., Ho, K.Y., Teh, M., Yeoh, K.G., So, J.B.Y., Huang, Z.: Technol. Cancer Res. Treat. **10**, 103 (2011)
73. Šćepanović, O. R., Volynskaya, Z., Kong, C-R., Galindo, L. H., Dasari, R. R., Feld M. S. **80**(4), 043103 (2009)
74. Sharma, M., Marple, E., Reichenberg, J., Tunnell, J.W.: Rev. Sci. Instrum. **85**, 083101 (2014)
75. Short, M.A., Lam, S., McWilliams, A.M., Ionescu, D.N., Zeng, H.: J. Thorac. Oncol. **6**, 1206 (2011)
76. Becker, W.: The bh TCSPC Handbook, 8th edn. Becker & Hickl GmbH, Berlin (2019)
77. Marcu, L.: Ann. Biomed. Eng. **40**, 304 (2012)
78. Dochow, S., et al.: Anal. Bioanal. Chem. **407**, 8291 (2015)
79. Dochow, S., et al.: J. Biophotonics. **9**, 958 (2016)
80. Patil, C.A., Bosschaart, N., Keller, M.D., van Leeuwen, T.G., Mahadevan-Jansen, A.: Opt. Lett. **33**, 1135 (2008)
81. Patil, C.A., Kirshnamoorthi, H., Ellis, D.L., van Leeuwen, T.G., Mahadevan-Jansen, A.: Lasers Surg. Med. **43**, 143 (2011)
82. Patil, C., Kalkman, J., Faber, D.J., Nyman, J.S., van Leeuwen, T.G., Mahadevan-Jansen, A.: J. Biomed. Opt. **16**, 011007 (2011)
83. Patil, C.A., Arrasmith, C.L., Mackanos, M.A., Dickensheets, D.L., Mahadevan-Jansen, A.: Biomed. Opt. Express. **3**, 488 (2012)
84. Ashok, P.C., Praveen, B.B., Bellini, N., Riches, A., Dholakia, K., Herrington, C.S.: Biomed. Opt. Express. **4**, 2179 (2013)
85. Khan, K.M., Krishna, H., Majumder, S.K., Rao, K.D., Gupta, P.K.: J. Biophotonics. **7**, 77 (2014)
86. Schie, I.W., Nolte, L., Pedersen, T.L., Smith, Z., Wu, J., Yahiatene, I., Newman, J.W., Huser, T.: Analyst. **138**, 6662 (2013)
87. Potcoava, M.C., Futia, G.L., Aughenbaugh, J., Schlaepfer, I.R., Gibson, E.A.: J. Biomed. Opt. **19**, 111605 (2014)
88. Yosef, H.K., et al.: Anal. Chem. **89**, 6893 (2017)
89. Lee, S., Chon, H., Yoon, S.-Y., Lee, E.K., Chang, S.-I., Lim, D.W., Choo, J.: Nanoscale. **4**, 124 (2012)
90. Qiu, Y., Deng, D., Deng, Q., Wu, P., Zhang, H., Cai, C.: J. Mater. Chem. B. **3**, 4487 (2015)
91. Sun, C., Zhang, R., Gao, M., Zhang, X.: Anal. Bioanal. Chem. **407**, 8883 (2015)
92. Liu, Z., Ye, B., Jin, M., Chen, H., Zhong, H., Wang, X., Guo, Z.: Nanoscale. **7**, 6754 (2015)
93. Jokerst, J.V., Cole, A.J., Van de Sompel, D., Gambhir, S.S.: ACS Nano. **6**, 10366 (2012)
94. Liu, Y., Chang, Z., Yuan, H., Fales, A.M., Vo-Dinh, T.: Nanoscale. **5**, 12126 (2013)
95. Jeong, S., et al.: Sci. Rep. **5**, 9455 (2015)
96. Kim, Y.-i., et al.: Sci. Rep. **7**, 1035 (2017)
97. Meyer, T., Schmitt, M., Dietzek, B., Popp, J.: J. Biophotonics. **6**, 887 (2013)

98. Heuke, S., Vogler, N., Meyer, T., Akimov, D., Kluschke, F., Rowert-Huber, H.J., Lademann, J., Dietzek, B., Popp, J.: Br. J. Dermatol. **169**, 794 (2013)
99. Heuke, S., Vogler, N., Meyer, T., Akimov, D., Kluschke, F., Röwert-Huber, H.-J., Lademann, J., Dietzek, B., Popp, J.: Healthcare, p. 64. Multidisciplinary Digital Publishing Institute (2013)
100. Heuke, S., et al.: Head Neck. **38**, 1545 (2016)
101. Chernavskaia, O., et al.: Sci. Rep. **6**, 29239 (2016)
102. Romeike, B.F., Meyer, T., Reichart, R., Kalff, R., Petersen, I., Dietzek, B., Popp, J.: Clin. Neurol. Neurosurg. **131**, 42 (2015)
103. Bocklitz, T.W., et al.: BMC Cancer. **16**, 534 (2016)
104. Lukic, A., Dochow, S., Bae, H., Matz, G., Latka, I., Messerschmidt, B., Schmitt, M., Popp, J.: Optica. **4**, 496 (2017)
105. Zirak, P., et al.: APL Photonics. **3**, 092409 (2018)

Chapter 10
Raman Spectroscopy Techniques for Skin Cancer Detection and Diagnosis

Ivan A. Bratchenko, Dmitry N. Artemyev, Yulia A. Khristoforova, Lyudmila A. Bratchenko, Oleg O. Myakinin, Alexander A. Moryatov, Andrey E. Orlov, Sergey V. Kozlov, and Valery P. Zakharov

10.1 Introduction

Skin cancer is one of the most common malignancies. Untimely detection of skin pathology may provoke cancer growth and metastases spread in internal organs. Today more than half of skin cancer cases are found at advanced stages, which is a direct threat to the patient's life [1]. Among other tumors, malignant melanoma of the skin occupies a special position. Due to aggressive behavior, melanomas are responsible for more than 85% of total mortality caused by malignant skin tumors. Melanoma of the skin is unevenly distributed, the highest incidence rates are typical for Australia and New Zealand (up to 40 cases per 100 thousand people), the USA (up to 21.1 cases per 100 thousand), some European countries [2]. In 2013 in the USA melanoma was diagnosed in 76600 Americans, and 35% of the patients were younger than 45 years old. In Russia, more than half a million new patients with malignant neoplasms are registered annually, and 14% of these cases are malignant skin neoplasms. However, melanoma of the skin in Russia is less common, about 8500 new cases every year (3.97 new cases per 100 thousand people), but there is a high level of annual increase in the incidence (from 4.55 to 6.1%) in the last decade. In 2017, Samara region held the first place among Russian regions with the highest

I. A. Bratchenko (✉) · D. N. Artemyev · Y. A. Khristoforova · L. A. Bratchenko
O. O. Myakinin · V. P. Zakharov
Department of Laser and Biotechnical Systems, Samara National Research University, Samara, Russian Federation
e-mail: bratchenko@ssau.ru

A. A. Moryatov · S. V. Kozlov
Department of Oncology, Samara State Medical University, Samara, Russian Federation

Samara Regional Clinical Oncology Dispensary, Samara, Russian Federation

A. E. Orlov
Samara Regional Clinical Oncology Dispensary, Samara, Russian Federation

V. V. Tuchin et al. (eds.), *Multimodal Optical Diagnostics of Cancer*,
https://doi.org/10.1007/978-3-030-44594-2_10

skin cancer incidence (18.6% among all cancer localizations) and high incidence rates of malignant tumors (446.6 cases per 100 thousand people) [3].

In the case of malignant melanoma, the survival time of patients directly depends on the stage of the disease. In excision of melanoma at Stage I, the 5-year survival rate of patients is 93–97%, while patients with Stage III melanomas survive the first 5-year period in only 40–78% of cases [4]. One of the main reasons of this unpromising statistics is the low detection rate of skin cancers in preventive examinations. The effectiveness of the skin cancer diagnosis by the general practitioners, as a rule, does not exceed 40–50% [5]. Moreover, a large percentage of false skin cancer diagnoses leads to excessive psychological stress for patients and overcrowding of specialized oncological clinics with patients who have only age skin changes, papillomas, warts, and many other non-cancerous skin diseases [6].

Skin tumors and particularly melanomas are unique tumors, which may be characterized by a variety of clinical features; it is especially difficult to differentiate melanoma from other skin neoplasms in the initial stages of the disease. Potentially dangerous pigment or pigmented skin neoplasms may be identified during visual examination in more than 90% of the population [7]. Long-term observation or inadequate treatment resulting in the rapid growth and dissemination of a neoplasm with a predictable outcome may be a consequence of the failure in melanoma diagnosis. On the contrary, overdiagnosis of melanoma may lead to an unnecessarily high volume of surgical intervention and serious cosmetic defects. Despite the significant advances in systemic drug therapy for skin melanoma, the results of 5-year survival rate depend most of all on timely and effective diagnosis and an adequate amount of surgical treatment [8]. Therefore, the development of methods for early and clarifying diagnosis of skin melanoma remains relevant.

Traditionally, examination of skin lesions is based on visual inspection that can be supplemented with dermatoscopy analysis [9], which improves the quality of diagnostics. At a dermoscopic examination, the physician evaluates the alteration of the shape and pigmentation of the tumor, which are difficult to detect with the naked eye [10]. For example, cross-polarized dermatoscope [11] or immersion dermatoscope [12] acquires pathologies images which allow to involve additional criteria in suspicious skin tissue inspection in comparison with standard ABCD [13] or 7-point checklist [14] rules. Dermatoscopic analysis may include monitoring of more specifically related tumor features like atypical pigmentation and vascular networks as well as the so-called blue-whitish veil, which represents a gray-white pattern with well perceptible blue tone [15]. However, the application of the complex criteria is quite complicated for non-trained person. More precise techniques and complex criteria require more trained and qualified practitioners. Thus, the result of the dermatoscopy analysis directly depends on the qualification of the medical specialist conducting visual diagnostics. A detailed analysis of neoplasm morphological structures and deviations from the normal skin tissues may be performed only by an experienced oncologist with many years of practice [16]. Thus, determining the presence or absence of malignancy is an important task for an oncologist, because the correct preliminary diagnosis is the basis of the further effective tumor treatment.

One possible way to increase the accuracy of early diagnosis is the development of screening analysis that can help general practitioners to determine the presence of malignant skin pathology. There are several skin cancer diagnosis methods and techniques currently used in clinical practice [17]:

Photodynamic Diagnosis In order to detect the presence of tumor cells, a photosensitive marker may be introduced into the particular area of the skin to form a protoporphyrin IX as a result of chemical reactions. Protoporphyrin IX accumulated in the tumor cells, and later after neoplasm illumination the tumor cells emit fluorescence. Thus, the photodynamic diagnosis of skin lesions is based on the fluorescent properties of an exogenous and endogenous compound in response to illumination. However, injection of exogenous fluorophores in patient's body is an obligatory part of diagnostic procedure, which makes photodynamic method unacceptable for mass screening [18, 19].

Sonography Sonography is based on acoustic waves reflection from density inhomogeneities. Usually, this method is used to assess skin lesion depth and margins before biopsy as well as to classify adjacent lymph nodes. It has the ability to differentiate benign from malignant skin cancer. The accurate measurement of skin cancer lesion thickness is one of the main advantages of sonography as it helps to identify margins for excision and correctly plan the operation, while the main disadvantage is the low specificity of sonographic analysis [20].

Electrical Bioimpedance The local electrical impedance of skin is dominated by the high resistance of the stratum corneum and highly depends on stratum corneum hydration or structure, especially at low frequencies. Stratum corneum must be bypassed to measure the living epidermis and dermis. Therefore, the direct contacts of microneedles are used for stratum corneum penetration and measurement of the electrical activity underlying the skin. This technique demonstrates the ability of melanoma detection; however, overall accuracy in tumor identification remains relatively low, making questionable the application of electrical bioimpedance in clinical practice [21].

Thermography Infrared imaging can provide a wealth of information on the processes responsible for heat generation and thermoregulation of the skin, in particular the deviation from normal conditions, often caused by disease. For example, increased metabolic activity of melanoma can be detected by dynamic thermography, which includes preliminary lesion cooling for enhanced temperature differences between the lesion and the surrounding healthy tissue [22]. Dynamic thermography is critical for motion correction, spatial resolution, and temperature sensitivity, especially when discerning small lesions. Therefore, the thermography effectiveness is not enough for discrimination of tumors on early stages and mostly used for tumor staging.

Optical Spectroscopy During multiple scattering in the skin light photons experience many interactions including absorption, elastic and nonlinear scattering, reemission, fluorescence, phosphorescence, and others. Therefore, reflected light flux contains a wealth of information about the biochemical and morphological composition of the tissue, including information about the presence and stage of the disease [23]. However, the extraction of such biochemical information from spectra alterations is a very complicated task. The connection between tissue scattering and emission properties and tissue microstructure greatly depends on the way of their excitation, collection, and processing. It is well known that the combination of multiple techniques may enhance the diagnostic quality and accuracy [17]. In this regard optical methods open a promising way for the development of noninvasive low cost, compact, and portable multimodal device for high sensitive skin cancer diagnosis.

In recent decades, a number of optical methods have been used for cancer detection and imaging, including confocal microscopy [24], optical coherence tomography (OCT) [25], multiphoton tomography (MPT), etc. For example, MPT in combination with fluorescence lifetime imaging (FLIM) may provide label free imaging with subcellular resolution and thereby significantly increase overall accuracy of tumor diagnostics [26]. MPT/FLIM can be performed using fast detection electronics and complicated equipment, which makes it hard to explore on difficult body sites or irregular surfaces [27]. OCT images can be used for analysis of morphological, fractal, and texture features of oncological pathologies [28], but the possibility of tumor diagnosis with these features is still under consideration [29] as malignant and especially premalignant tumors do not have distinctive patterns [30].

It should be noted the screening system must be sensitive to the changes in the skin biochemical composition and able to determine the nature of the neoplasms. One of such approaches is optical biopsy [31] using Raman spectroscopy (RS) and/or autofluorescence (AF) tissue response [32, 33]. RS is based on inelastic scattering of optical radiation when interacting with molecular bonds of tissue components. The recorded Raman signal contains characteristic bands responsible for the presence of certain chemical bonds in the tested object, which makes this spectrum a unique "fingerprint" of each neoplasm [34]. Thus, RS application makes it possible to detect changes in tissue chemical composition as compared to healthy skin. It is important to note that in addition to the Raman shift, the scattered light contains an AF signal obtained after interaction with the tissues. The spectral properties of AF signal depend on the presence of endogenous fluorophores in the tissues [35]. In general, AF tumor studies show lower accuracy in comparison with RS [36], but AF is characterized by much higher intensity, which makes it possible to register signals within a short period and rapidly scan the tumor area. Therefore, there is a need in clinical trials to optimize RS-AF efficiency and validity for cancer diagnostics [37]. In first sections of this chapter, we present the results of RS implementation for cancer detection, and in this case AF always considered to be a noise signal masking the Raman signal. However, the combination of RS and AF is also discussed in the last section as application of joint multimodal techniques for achieving higher diagnostic accuracy [30].

The application of RS for skin cancer study is presented in numerous trials [38–40]. Optical biopsy based on RS and AF application demonstrates a high level of accuracy (exceeding 90%) in the diagnosis of skin cancer [40–43]. Many research teams [39, 44, 45] have been trying to increase the efficiency of diagnostics by using more sensitive (high spectral resolution, deeply cooled detectors, etc.) and, consequently, more expensive equipment. Indeed, the RS and AF methods are based on the detection of spectral differences of tumor and normal tissue, and an increase in spectral resolution should seemingly lead to an increase in the efficiency of cancer type recognition. However, analysis of morphologically different tumors reveals a significant biochemical commonality among them, which may be explained by the relative homogeneity of the enzymatic set and enzyme activity of tumors, as well as the relative similarity of the composition, content, and biochemical activity of specific chemical components [46]. Thus, there is an utmost ("physiological") accuracy of cancer type recognition associated with limitations due to physiological similarity of the tissues chemical composition. The physiological factors become more decisive in the recognition procedure than the spectral resolution of the device. Currently, the high potential of RS provides the possibility of tumor types differentiation with the efficiency based on the registered signal quality [47]. The effectiveness of instrumental diagnostics would depend on detected Raman signal quality, and obviously the signal quality would be higher in high-sensitivity spectrometers than in simplified systems. However, the application of costly stationary systems is not always possible or convenient. Therefore, the effectiveness of Raman signal registration by different techniques is still under consideration [30].

This chapter presents results of skin cancer optical biopsy with variety of RS techniques and instrumentation that have been used in clinical and preclinical trials in recent decades as well as demonstrates application of portable Raman instrumentation in clinical practice.

10.2 Conventional (Spontaneous) Raman Scattering

With the illumination of tissue by monochromatic laser radiation most of the light is scattered on the same frequency. In opposite, inelastic Raman scattering leads to a frequency shift proportional to the energy difference between the vibrational states of the specific molecular bond [48]. An excited molecule can transmit from the virtual state to one of the vibrational states. Such transitions are called Stokes or anti-Stokes if, respectively, the scattered light has a lower or a higher frequency than the incident light. Typically, the molecules are initially in the ground vibrational state, so the Stokes component is usually much stronger and therefore more widely used in the practical applications. Thus, Raman spectrum reveals vibrational and rotational modes of the molecules. The peak positions in the Raman spectrum characterize the molecular composition of the tissue, and the amplitude of each peak depends on the relative concentration of each chemical bond in the molecule.

In spontaneous Raman spectroscopy, as a rule, the Stokes signal is detected under continuous illumination of a laser with a high spectral stability. As mentioned above, the signal intensity increases for short wavelengths, but the intrinsic fluorescence of the tissue also increases, while the depth of light penetration into the tissue decreases due to the absorption of chromophores, in particular hemoglobin and melanin, so that the signal-to-background ratio decreases. In addition, the absorption coefficient $\mu_a(\lambda)$ in the visible range strongly depends on the wavelength and tissue type, which prevents the comparison of the intensities of Raman different peaks. The choice of wavelength usually involves a compromise between the intensity of the Raman signal and tissue AF and depends on the specific tissue and the purpose of the study. The optimal optical windows for RS of skin are between 700 and 900 nm and near 1064 nm [49].

Due to the simplicity of spontaneous Raman signal detection it was widely used in different clinical applications. For example, spontaneous RS was used for evaluation of cosmetology effect of creams on the skin [50], analysis of skin dermatitis [51], numerous studies of skin cancers [52–54], and many other applications [49].

Silveria et al. [55] demonstrated the possibility to classify normal skin from non-melanoma malignant and premalignant lesions by using portable Raman system (830 nm excitation wavelength, CCD camera cooled down to −75 °C temperature, 20 s acquisition time). The results showed the important spectral differences between non-tumor lesions and precancerous tissue for lipids (between 1250 cm^{-1} and 1300 cm^{-1} and at 1450 cm^{-1}) and for proteins (between 870 cm^{-1} and 940 cm^{-1}, 1240 cm^{-1} and 1271 cm^{-1}, and at 1000 cm^{-1} and 1450 cm^{-1}). Figure 10.1 demonstrates mean Raman spectra of tested skin tissues and difference spectra (between tumors and normal skin).

The authors investigated different statistical models (principal component analysis (PCA) or projection on latent structures (PLS) with linear or quadratic discriminant analysis) and achieved 77.8% accuracy in classification of 53 normal skin, 28 basal cell carcinomas (BCC), 7 squamous cell carcinomas (SCC), and 11 actinic keratosis. The highest accuracy of 91.9% (sensitivity 89.1% and specificity 94.3%) was achieved for discrimination of nonmelanoma and premalignant tumor (BCC, SCC, and AK) from normal tissue and benign tumors. However, pointed values of skin cancer classification accuracy may significantly vary from one study to another, so more examples of spontaneous RS application in skin cancer analysis have been presented in Sect. 10.8 dedicated to RS clinical applications.

10.3 Confocal Raman Spectroscopy

RS can be implemented in a confocal configuration to provide optical depth separation by spatially filtering of the collected Raman signal with a pinhole or optical fiber to block the signal out of the focus. This approach improves the transverse and

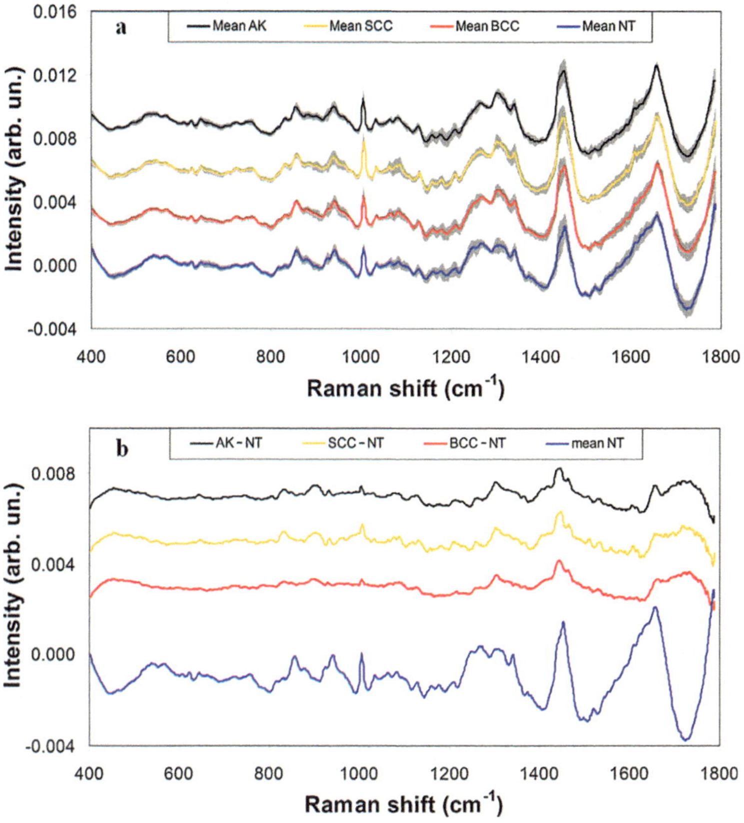

Fig. 10.1 (**a**) Mean Raman spectra of normal and benign skin (non-tumor, NT), basal cell carcinoma (BCC), squamous cell carcinoma (SCC), and precancerous (AK) tissues. The gray shadows represent the standard deviation for each spectral group; (**b**) difference spectra of basal cell carcinoma (BCC), squamous cell carcinoma (SCC), and precancerous (AK) tissues related to non-tumor (NT) (reprinted with permission from [55])

axial resolution up to 2 μm [56]. However, the accumulation time of a signal at each point is usually more than 5 s and increases with the focal depth in the tissue. A raster-scanning confocal RS produces two-dimensional images with a field of view typically in the range of 0.01–1 mm^2. Confocal Raman probes have so far been used mainly for *ex vivo* and *in vitro* studies [57, 58].

Confocal RS became popular in studying skin layers properties [59]. Zhang et al. [60] investigated stratum corneum hydration and collagen content. The results of study [60] and other similar studies may be widely used in wound healing monitoring and also in the determination of skin aging, including clinical and forensic applications [59, 61]. Thus, recent confocal RS techniques allow for the determination of skin layers chemical composition along with measurement of layers thickness [61].

As for skin cancer detection, confocal RS was implemented both for *ex vivo* and *in vivo* studies [62]. At the same time confocal RS implementation frequently requires rather long accumulation time of the signal that makes difficult *in vivo* measurements. Lieber et al. [63] investigated nonmelanoma skin cancer from 19 patients using a handheld Raman microspectroscopy system (only 30 s accumulation time). They classified 21 nonmelanoma cancer (9 BCC, 4 SCC, 8 inflamed scar tissues) from 21 normal skin sections with 100% sensitivity and 91% specificity (resulting in 95% accuracy). The authors assumed that spectral differences between abnormal and normal skin may possibly be due to lipid content that affected the 807–814, 1069–1073, 1321–1325 cm^{-1} Raman peaks and change of the tryptophan that produces a 1542–1556 cm^{-1} Raman peak.

10.4 Coherent and Resonance Raman Spectroscopy

Coherent RS uses two light beams (pump wave and Stokes wave) with frequencies ω_p and ω_s, respectively, such that their difference matches to the frequency of the vibrational mode $\Omega = \omega_p - \omega_s$ for the molecular bond of the interest. The coherent Raman signal for different molecules increases as compared with spontaneous RS, as a rule, up to 10^5 times [64]. Enhanced coherent Raman signal may be especially important for fast image acquisition. The technologies of coherent Raman spectroscopy include stimulated Raman spectroscopy (SRS) and coherent anti-Stokes Raman spectroscopy (CARS) [65]. Both SRS and CARS can be used in strong fluorescent media, which is usually a significant limiting factor for Raman imaging in tissues.

Such approaches enable acquisition of high-resolution images in microscopy techniques. The increase of resolution in microscopic study results in increase of the sensitivity of medical and biological studies [66]. For skin studies, this allows for obtaining *in vivo* histological images of tissues at different depths. Demonstration of such technique was performed by Drutis et al. [67]. The authors investigated native skin of the pig and demonstrated that SRS microscopy provides a unique view of the chemical organization of the skin, rapidly and noninvasively assessing the layers of intact tissue. The pump beam was tuned to 816.0, 810.6, or 781.3 nm for skin imaging, corresponding to Raman shifts of 2850, 2950, or 3340 cm^{-1}, respectively. The main contributors from skin at these vibrational frequencies can be assigned to vibrational modes from lipids, proteins, and water. Figure 10.2 shows SRS microscopic images of skin tissues obtained with the proposed technique. The transition from the stratum corneum into the viable epidermis is discernible in images collected 18 μm below the surface (Fig. 10.2d–f). At this depth, the distinct boundaries between the clusters are visible, such as between the arrows in Fig. 10.2e. The average canyon width in these images is 21.2 ± 4.4 μm. The signal intensity in the protein and lipid images is highest along the canyon walls, as evidenced by the bright lines outlining the canyons (Fig. 10.2d and e). Fine lines are visible between the canyons, as can be seen the arrows in the lipid (2850 cm^{-1}) image (Fig. 10.2e).

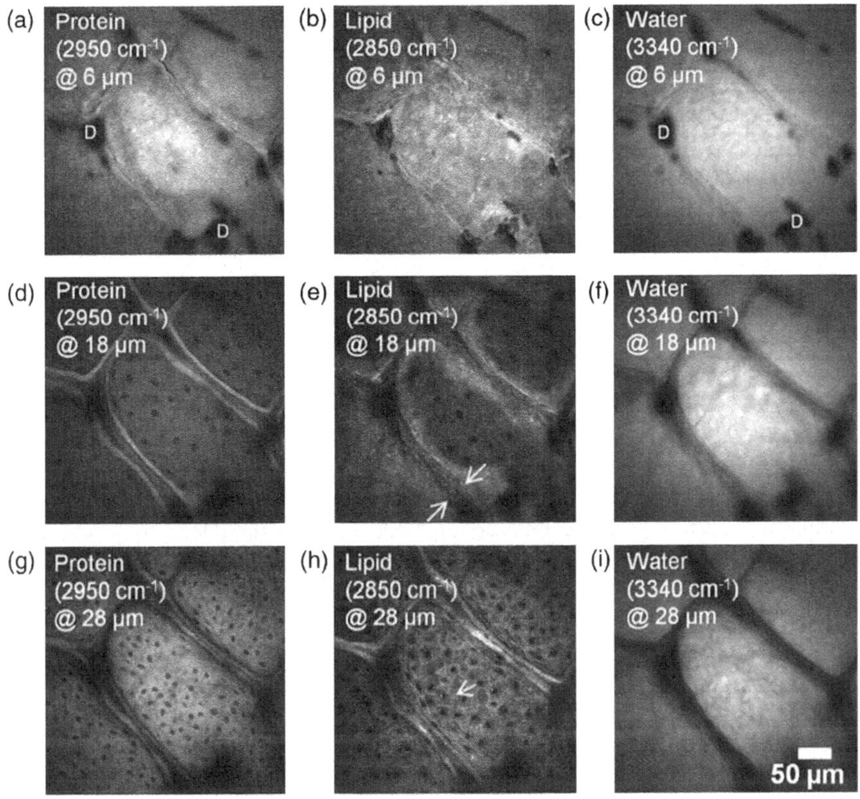

Fig. 10.2 Changes in chemical distribution are visible by collecting depth stacks at multiple wavelengths. SRS images were acquired at 60× magnification from pigskin. The rows correspond to focal planes 6 µm (**a**–**c**), 18 µm (**d**–**f**), and 28 µm (**g**–**i**) below the skin surface. The columns from left to right represent protein, lipid, and water content, respectively. The dark spots labeled "D" in (**a**) and (**c**) are due to optical shadows from air pockets between the skin surface and coverslip. The arrows in (**e**) designate one canyon in the image. Fine lines can also be seen within the canyon. The arrow in (**h**) points to one of the visible cell membranes in the cluster. Scale bar = 50 µm (reprinted with permission from [67])

These are the edges of stratum corneum corneocytes following the contours of the canyon. The water signal is significantly weaker in the canyons than in the cluster regions. In the cluster regions, nuclei are visible as dark ovals in the protein and lipid images (Fig. 10.2d and e), and as bright ovals in the water image (Fig. 10.2f). Around this depth, the distinctive polygonal shape of the cells is revealed as the lipid-rich cell wall membranes appear in the lipid (2850 cm^{-1}) image (Fig. 10.2e) [67]. However, the acquisition time was 30–60 s for only a single wavelength that complicates *in vivo* application of such technique.

CARS allows for imaging of skin tissues. Wang et al. [68] performed *ex vivo* and *in vivo* investigation of mice skin and human melanoma sections with CARS to estimate distribution of pheomelanin in cells and tissues. The authors validated

CARS imaging technique and demonstrated (Fig. 10.3) possibility to track tumor-specific chromophores that may be used in cancer detection and classification. A distinct feature here is a possibility for real-time image acquisition in comparison to previously described coherent RS technique. In this way, modern coherent RS techniques open a new dimension of skin tissue analysis and make possible performing of *in vivo* real-time optical biopsy based on optical histological examination of living tissue. The limitation of such approaches is the necessity of very sensitive laboratory setups, which may be utilized only in lab research. Thus, the use of SRS and CARS is complicated in mass screening applications particularly in skin tumors examinations.

Resonance RS is a type of SRS technique in which only one laser source is used for tissue investigation, and the incident photon energy is equal or close in energy to a molecular electronic transition of specific bond of investigated molecule. The frequency coincidence (or resonance) can lead to greatly enhanced intensity of the RS, and thus, help in revealing of molecules that may hardly be detected using a conventional RS [69]. In skin research most efforts of resonance RS implementations were associated with the analysis of carotenoids content. Carotenoids in the skin play a protective role, mainly photoprotection, protection against erythema, and sunlight damage. Carotenoids are known to quench singlet oxygen and other free radical species, which are generated in the skin by exposure to ultraviolet radiation and can cause skin damage. Several recent studies have examined the potential protective effects of carotenoids against premature photoaging of the skin, marked by signs such as wrinkling, pigmentation, dryness, and inelasticity [70]. Carotenoids absorb the light in the blue-green spectral range and therefore provide resonance absorption. Ar+ lasers operating at 488 nm and 514.5 nm were recently used for measurements of carotenoids [71]. Despite the vast variety of applications for carotenoids monitoring in skin aging and photoaging, application of resonance RS in skin cancer is still limited. The main reason for such limitation is a complex composition of tumors and high AF in the visible spectral range which masks weak Raman peaks.

However, the application of resonance RS still is in interest, and recently Liu et al. [72] tested the application of resonance Raman spectroscopy for *ex vivo* rapid skin cancer diagnosis. The authors measured resonance Raman spectra (30 averages of 1 s accumulations) for 43 BCC and 12 normal tissues with excitation of 532 nm laser. It was achieved the sensitivity, specificity, and accuracy of 93%, 100%, and 94.5%, respectively. The authors suggested that Raman spectra of BCC tissues change significantly in comparison to the spectra of normal skin tissues due to the

Fig. 10.3 (continued) obvious sign of melanin. (**g**) Bright-field trans-illumination image acquired from the microscope eyepiece from an unstained slide of the melanoma area, showing slightly pigmented granular structures (red circle). (**h**) CARS image of the same tumor area compared to (**g**), with the same settings as for (**e**). Saturated bright pheomelanin signals were found (red circle) corresponding to the minimally pigmented region shown in (**g**). (**i, j, k**) Respectively, H&E, trans-illumination, and CARS images of the tumor area of slides from a second amelanotic melanoma patient. Strong pheomelanin signals were again observed (red circles) (reprinted with permission from [68])

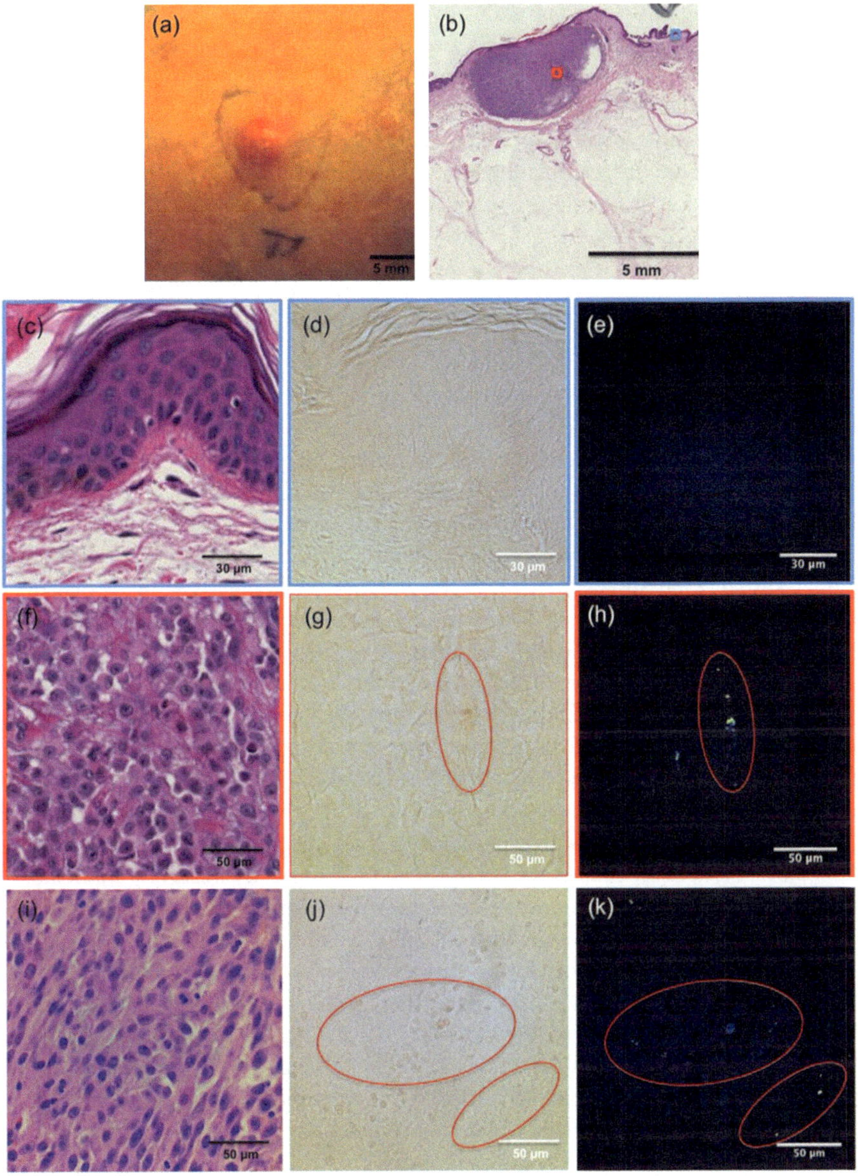

Fig. 10.3 Imaging of human amelanotic melanoma: (**a**) Clinical photograph of one amelanotic melanoma lesion. (**b**) H&E stain of the patient slide (10× magnification). (**c**) Perilesional skin showing normal architecture of both epidermis and dermis. (**d**) Bright-field trans-illumination image acquired from the microscope eyepiece from the perilesional area. (**e**) CARS image of the same perilesional area compared to (**d**) (image acquired with pump beam wavelength at 841 nm ($\omega_p - \omega_s = 2275\ \text{cm}^{-1}$) was subtracted from the image acquired with pump beam wavelength at 855 nm ($\omega_p - \omega_s = 2081\ \text{cm}^{-1}$) to minimize the nonresonant background from structures other than pheomelanin). (**f**) View of the amelanotic melanoma area showing high density of cells with no

changes in the relative concentrations of tryptophan, carotenoids, lipids, and proteins. Figure 10.4 demonstrates the spectra of normal skin and BCC at different depths and reveals significant changes in tumor. It was found correlation between the depth dependence of resonance Raman spectra and the status change of BCC tissue at a molecular level. Only *ex vivo* tests were performed with the proposed technique and future large clinical studies are required to prove the applicability of resonant RS in skin cancer detection and classification.

10.5 Surface-Enhanced Raman Spectroscopy

Low signal intensity is undoubtedly the main limitation of RS. An alternative to coherent signal amplification is the use of the effects of electromagnetic amplification that occur near the metal surface, including metal nanoparticles [73]. This approach is called surface-enhanced Raman spectroscopy (SERS). Under the incident electromagnetic field, metallic nanoparticles generate a localized surface plasmon resonance, which enhances the excitation at the pump frequency and the Raman signal at the Stokes frequency. The gain decreases rapidly with the distance, so that the molecules must be within tens of nanometers of the surface [74]. The limitations of SERS for clinical applications are that implementation depends on the sensitivity of the biomarkers of diseases and the availability of appropriate target fragments, as well as potential toxicity and the need for clinical approval of each contrast agent. It should be noted that since the utilization of SERS technology allows for individual narrow Raman bands registration, SERS application might be an alternative to fluorescence labeling of tissues.

In vivo skin SERS-analysis requires biologically inert nanostructures. Nanoparticles of noble metals, such as gold and silver, satisfy these requirements. A number of studies showed a possibility of nanoparticles of different type synthesis for the application in clinical practice [75, 76]. Kang et al. [77] reported approach to develop NIR-sensitive SERS nanoprobes consisting of the plasmonic Au/Ag hollow-shells assembled with silica nanospheres and simple aromatic compounds for *in vivo* multiplex detection. However, for the moment *in vivo* studies of human tissues by SERS techniques are not approved. Therefore, the assessment of SERS utilization for *in vivo* skin studied is carried out on animal models.

Stuart et al. [78] showed the first *in vivo* application of SERS for transcutaneous glucose measurement. Further, the description of the biocompatible and nontoxic nanoparticles for *in vivo* tumor targeting and detection in live animals based on pegylated gold nanoparticles was presented by Qian et al. [79]. The authors found that single-chain variable fragment antibodies-conjugated gold nanoparticles are able to target tumors *in vivo* in animal models. These particles were 4200 times brighter than near-infrared-emitting quantum dots and allowed spectroscopic detection of even a small tumor (0.03 cm^3).

Pinzaru et al. [80] performed an examination of *ex vivo* mice skin with induced melanoma. The authors proposed conventional RS and SERS molecular character-

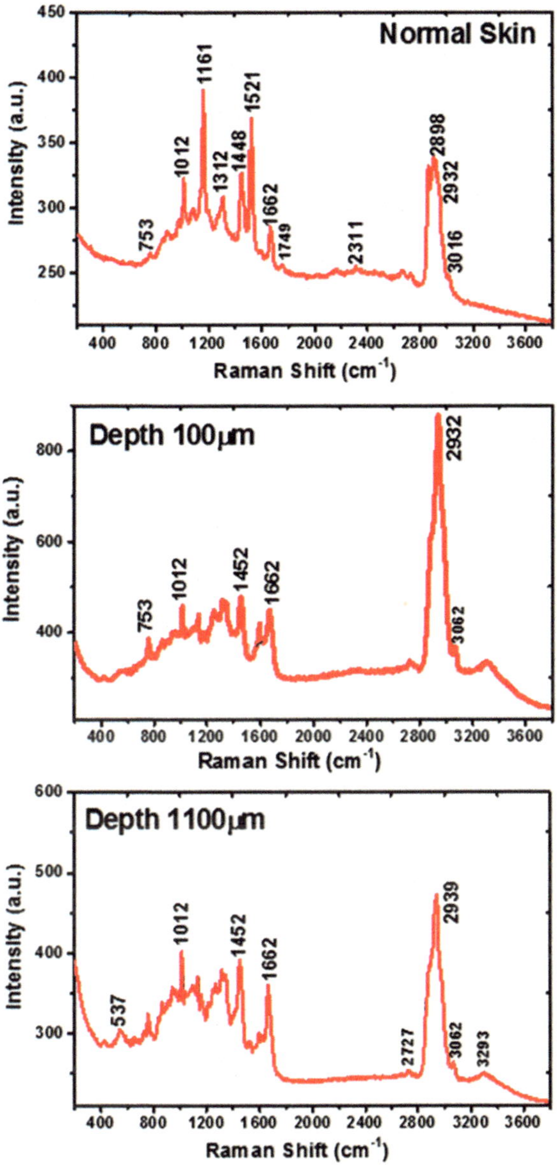

Fig. 10.4 A set of typical resonance Raman raw spectra collected from a horizontal section of normal human skin sample, and a vertically sliced BCC skin sample measured at different depths. (Top) the spectrum was from dermis layer of normal skin showing nine feature peaks; (middle) the spectrum was from the vertically sliced BCC sample at a depth of 100 μm. There are eight characteristic peaks including increased peaks at 753 cm^{-1} and 1589 cm^{-1}, but intense carotenoids peaks at 1161 cm^{-1} and 1521 cm^{-1} disappeared compared to the normal tissue (top); (bottom) the spectrum was from BCC sample at a depth of 1100 μm, substantially similar to the depth of 100 μm, with six Raman peaks, but carotenoids peaks at 1161 cm^{-1} and 1521 cm^{-1} are present and obviously weaker than normal tissue sample (top). Those peaks of 753 cm^{-1} and 1589 cm^{-1} greatly decreased in comparison with the depth of 100 μm. BCC: basal cell carcinoma (reprinted with permission from [72])

ization of melanoma. Raman and SERS spectra of tissues from normal, early stage melanoma and advanced melanoma were acquired using two dispersive spectrometers, a DeltaNu Advantage Raman spectrometer equipped with He-Ne laser operating at 4 mW, 632.8 nm, and a Bruker Senterra dispersive Raman spectrometer using the 785 nm excitation diode laser. The same scientific group presented results of *in vivo* and *ex vivo* reproducible SERS signal collection from mouse skin and utilized it for the differentiation of skin pathologies [81]. In this study, the 785 nm laser line at 100 mW was used for tissue sample analysis. The laser was focused on the mouse skin using the 20× (NA = 0.4) or the 50× (NA = 0.75) objectives, small dosages of 10 mL Ag nanoparticles were subcutaneously injected in mice.

The acquired data were successfully used for confirming the differentiation between the three types of skin pathologies using principal component analysis (PCA) and K-means clustering. The authors used PCA in order to highlight the differences and similarities between the three groups of SERS spectra (healthy, cancerous, and betulin treated). PCA revealed the most informative bands for tissues discrimination: Amide II (1557 cm^{-1}), collagen (1313, 1284 cm^{-1}), DNA (745 cm^{-1}), tyrosine (641 cm^{-1}), phenylalanine (1589 cm^{-1}), Amide I (1539 cm^{-1}), and glycogen (482 cm^{-1}). The presented results show that high-quality SERS spectra could be acquired both for *in vivo* and *ex vivo* examinations of skin tissues. Thus, animal model SERS studies may become the basis of future studies of human skin.

10.6 Spatially Offset Raman Spectroscopy

Spatially offset Raman spectroscopy (SORS) is similar to conventions RS, the difference is that the Raman signal is collected from the deeper layers of the tissue by spatially shifting of the collecting and excitation fibers. The registered photons will then be repeatedly elastically scattered and will pass some distance from the source of illumination. Collecting the Raman signal at various offsets effectively measures the different layers of the examined tissue. SORS typically uses a probe with an illuminating fiber surrounded by the set of collecting fibers with 1–5 mm offset [82, 83]; an offset of 16 mm was used to study bone Raman signal [84]. Since SORS collects a signal from the depth, the signal-to-noise ratio decreases, which leads to an increase in the accumulation time of 30 s [83].

Application of SORS in cancer detection and screening was demonstrated for breast tumor margin evaluation [83], bone cancer detection [85], urology, and other applications [86]. Keller et al. [83] demonstrated the application of SORS in breast tissue discrimination with the sample characterization up to depths of 2 mm, 95% sensitivity and 100% specificity were achieved by the authors in the detection of breast tumors. However, the application of SORS technique in cancer detection now is limited only to cancer margins detection during cancer removal surgery. In addition, a number of questions about its efficacy still are not studied enough, SORS of soft tissues holds considerable promise for biomedical applications [86].

One of the promising future implementations in skin cancer research is the so-called SESORS (combination of SERS and SORS). Yuen et al. [87] demonstrated in a rat model possibility of *in vivo* transcutaneous measurements of glucose that proves that such technique may be a powerful new approach to the challenging problem of *in vivo* metabolite and drug sensing. However, future translation of such techniques into clinics needs further trials to prove its efficiency in medical diagnosis.

10.7 Comparison of Raman Spectroscopy Techniques

Analysis of the applicability of different RS approaches for clinical studies and skin cancer detection (see Table 10.1) demonstrates that each RS technique has its advantages and disadvantages.

Signal registration requires a long collection time for confocal RS and SORS. Thus, their application is limited in mass screening of skin cancer. SERS applications show a high sensitivity and specificity in the detection of specific molecules or cancer biomarkers, which may significantly increase accuracy of cancer detection. However, the implementation of SERS requires specialized substrates or nanoparticles, so it requires additional time-consuming sample preparation and laboratory setups for tissue analysis. Also, despite recent progress in SERS techniques there is still a great challenge in implementation of *in vivo* applications, due to unclear metal nanoparticles toxicity. Resonance RS may be very useful in the analysis of specific bonds or molecules, but in skin cancer analysis this approach is limited due to unpredictable and complex molecular composition of tumors. CARS and SRS have high spatial resolution, low AF background (or even completely absence of AF), and low signal accumulation time, but these techniques are difficult to implement from a technical point of view. Moreover, the application of SRS and CARS requires expensive optical equipment, which is much more complicated than an equipment used for spontaneous Raman signal registration. Joint study of skin tissues by a number of proposed techniques also demonstrates very promising results, but as in case of coherent RS methods, such approaches are high-cost and require wide clinical trials to prove their applicability to skin cancer screening. Thus, the most accessible of the above Raman methods for mass skin cancer studies is the usual spontaneous RS due to the compromise of simplicity and signal accumulation time in comparison to other RS techniques.

10.8 *In Vivo* Clinical Applications

One of the largest clinical studies of skin cancer detection with RS was performed by Zeng Research Group [88, 89]. For example, Lui et al. [89] studied 518 benign and malignant skin lesions of 9 different types, including malignant melanoma

Table 10.1 Comparison of RS techniques applied for skin cancer detection

Raman technique	Light source and visualization	Acquisition time	Signal and background	Notice
Conventional RS	Continuous monochromatic laser light, Raman probe sometimes coupled with an objective Wavelength: 500–1100 nm Mean laser power: 150 mW	>200 ms	Signal: $\omega = \omega_{p-\delta}$ Background: Native tissue fluorescence (autofluorescence—AF)	– High AF background – Low axial resolution – Low cost – Relatively simple – Spot, linear, or matrix visualization
Confocal/ micro RS		>5 s		– High AF background – High axial resolution – Analysis of deep tissue layers (<500 μm) – High collection time
SORS		>5 s		– High AF background – Low axial resolution – Analysis of deep tissue layers (<1–2 mm) – High collection time
SERS	Typically the same as for spontaneous RS, but a source wavelength is dependent on nanoparticles (nanolayers) size and material	>1 s	Signal: $\omega = \omega_{p-\delta}$ Background: AF	– Low AF background – Necessity of nanoparticle – Heterogeneous signal intensity – Necessity of hotspot search – Low collection time
Resonance RS	Typically the same as for spontaneous RS, but a source wavelength is close in energy to a molecular electronic transition of specific bond	>0.3 s	Signal: $\omega = \omega_{p-\delta}$ Background: AF	– High AF background – High intensity of specific bonds – Low collection time

(continued)

Table 10.1 (continued)

Raman technique	Light source and visualization	Acquisition time	Signal and background	Notice
CARS	Two impulse lasers or one tunable impulse laser Wavelength: $\omega_1 = \omega_s$ (690–900 nm) $\omega_2 = \omega_p$ (1150–2300 nm) Mean laser power: 150 mW Impulse: 5 ps Repetition rate: 1–100 MHz	>0.16 μs for one pixel (in microscopic image)	Signal: $\omega = 2\omega_p - \omega_s$ Background: Nonresonant	– Low nonresonant background – High axial resolution – High-cost – Hard to implement (light delivery, miniaturization) – Low collection time
SRS		>40 ms/frame (in microscopic image)	Signal: $\omega = \omega_p - \omega_s$ Background: Absence of AF	– Absence of AF – High axial resolution – High-cost – Hard to implement (light delivery, miniaturization) – Low collection time

Notes: ω detected signal frequency, ω_p pump frequency, ω_s Stokes wave frequency, δ Raman shift

(MM), BCC, SCC, actinic keratoses, atypical nevi, melanocytic nevi, blue nevi, and seborrheic keratosis. It has been shown that Raman spectra of the skin are data-rich and complex, but there are no distinctive Raman peaks or bands that could be uniquely assigned to specific skin cancers. The strongest Raman peak is located around 1445 cm^{-1}with other major Raman bands centered at 855, 936, 1002, 1271, 1302, 1655, and 1745 cm^{-1}. The relatively conservative statistical techniques can be used to extract the diagnostic information embedded within Raman signals and it must be taken into account that the diagnostically useful information may be contained within certain spectral regions. Specifically, spectral features between 1055 cm^{-1} and 1800 cm^{-1} offered best discrimination between melanomas and non-melanoma pigmented lesions, whereas for distinguishing skin cancers from benign lesions overall, the full spectrum from 500 to 1800 cm^{-1} region is preferred. For a 90% sensitivity, specificity rates varied from 63% to 88%. In another study Zhao et al. [90] performed large clinical trial resulted in 88.7% sensitivity and 93.6% specificity for discrimination of 645 cancer and benign lesions. The authors note that all the wavebands contributed to the final diagnosis to some extent, particularly the waveband 1200–1400 cm^{-1} provides more diagnostic capability than other wavebands, but none of them alone could provide diagnostic performance as high as the full wavelength band. The authors used in both studies Raman system with

785 nm laser excitation and CCD camera cooled below −120 °C at 1 s integration time.

Schleusener et al. [91] used Raman system with 785-nm laser excitation and CCD detector cooled to −70 °C (10 s acquisition time) to study skin tumors from 104 patients. Implementation of conventional RS allowed to achieve 74% sensitivity and 66% specificity for discriminating BCC (n = 39) from SCC (n = 29), 80% sensitivity and 79% specificity in decimating of BCC (n = 39) from melanoma (n = 36), 83% sensitivity and 86% specificity in discriminating of SCC (n = 29) from melanoma (n = 36). Moreover, discriminating of melanoma and pigmented nevi resulted in a 91% accuracy.

Raman spectroscopy data analysis provides the possibility of differentiating various tumor types with efficiency based on registered signal quality [92]. With different characteristics of spectroscopic equipment, the effectiveness of instrumental diagnostics would depend on the signal-to-noise ratio (SNR). Obviously, the better SNR would be achieved with high-sensitivity spectrometers than with simplified portable spectral systems. However, the application of costly stationary systems is not always possible or convenient. Moreover, as one may see from the presented above studies, the resulting tumor detection sensitivity is restricted by the physiological similarity of the tissues chemical composition of different tumor types. Therefore, there is some sort of saturation in accuracy dependence from SNR and it is only possible to minimize the acquiring time without a significant increase in accuracy of cancer detection by implementation of high-sensitivity spectrometers in experimental instrumentation.

Previously our group conducted *ex vivo* skin cancer study using RS and AF methods and highly sensitive laboratory spectroscopy system (with camera cooling down to −65 °C). It was achieved 97.3% accuracy of the malignant melanoma diagnosis [93]. For realization of mass human skin screening with the possibility of effective cancer detection it is necessary to use portable cost-effective spectral apparatus. Thus, it was performed clinical trials with complex *in vivo* diagnostic unit based on portable spectrometer with detector cooling down to −15 °C. It has been studied 32 melanoma, 33 pigmented nevus (dysplastic nevus, blue nevus, compound nevus), 18 BCC, 19 benign lesions. The multivariate projection on latent structures with discriminant analysis (PLS-DA) method was used to analyze the collected spectral data. Several classification models were built to distinguish different pathology types. Based on the clinical significance we estimate differentiating models of the skin tumors: (I) malignant tumors (MLG) versus benign tumors (BN), (II) melanoma versus pigmented BN, (III) melanoma versus BCC, (IV) melanoma versus other skin lesions. The variable importance in projection (VIP) scores was estimated. The VIP scores show the importance of the variables for the prediction ability of the model. The higher VIP score value, the more important is the corresponding variable in the built PLS model. In our work, the VIP scores allow finding out informative bands of the neoplasm spectra in regression specification to classify different tumor types. The proposed models help to estimate the contribution of RS to the final accuracy of skin tumors diagnostics. The effectiveness of "optical diagnosis" was compared with the accuracy of the trained oncologist preliminary diagnosis. The registered spectra of neoplasms were preprocessed with baseline removal, smoothing by the Savitzky–Golay method, the standard normal variation method, and centering. The details of the utilized RS setup, Raman spectra processing, and classification models creation may be found elsewhere [30]. This research adhered to the tenets set forth in the Declaration of Helsinki. The protocols of *in vivo* skin tissue diagnostics were approved by the ethical committee of Samara State Medical University. Informed consent of each subject was obtained. Every tumor study was accompanied by histological analysis to make a final diagnosis.

(I) Malignant Versus Benign Tumors

Raw spectra of the neoplasms are a combination of the Raman and AF signals. It was found that AF basically influences detection accuracy in low-frequency spectral range 300–1000 cm^{-1} and the analysis of typical Raman peaks shows that the differences in the 1200, 1249, 1281, 1315, 1392, 1440, 1529, and 1644 cm^{-1} bands contribute significantly to classify the MLG and BN spectra. The calculated PLS-predictors based on the extracted differences for each Raman spectrum are plotted on the box plot diagram (Fig. 10.5). The 100% differentiating accuracy was achieved, and ROC AUC equals 1.

High accuracy of MLG and BN pathologies separation is explained by pathological processes in malignant tumors that lead to the specific changes in the structure and composition of various organic compounds. The development of the malignant skin disease increases the content of metabolic products in the pathological areas of the skin, changing the ratio of protein and lipid concentrations. Proteins predominantly contribute to the appearance of bands in the spectral range 1240–1270, 1340, 1440–1460, and 1665 cm^{-1}, the spectral features arising from the contribution of lipids, mainly of triolein, are observed in the 1271–1301, 1440,

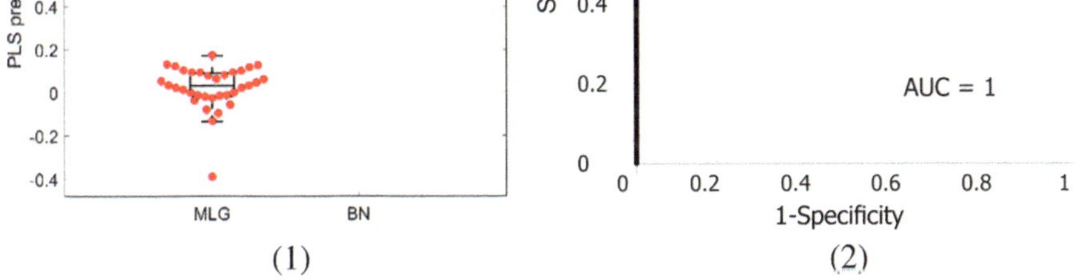

Fig. 10.5 (1) Box-plots and bee swarm plots of PLS-predictors for discriminating malignant and benign lesions and (2) corresponding ROC curve is derived from spreading PLS-predictors

1650–1660 cm^{-1} bands [94]. One of the significant differences between malignant and benign formations is the process of metabolism and destruction of collagen. Cells of malignant tumors form fast-growing, low-differentiated structures, and the development of such structures is accompanied by the increased activity of collagenase [95]. Collagenase destroys the molecular bonds of collagen fibers, and changes in Raman spectra of skin tissue may be observed in 1248, 1454, and 1665 cm^{-1} bands associated with peaks of collagen [96]. Note that the influence of biochemical features of tissue malignization on alteration of the Raman signal also makes it possible to obtain 100% accuracy of differentiation without taking into account the AF spectra. At the same time, as VIP analysis showed, the key diagnostic feature is the intensity difference around the 1644 cm^{-1} band. Contribution to the 1640–1660 cm^{-1} band is made by C=O vibration in Amid I of structural proteins, C=C alkyl stretching of lipids and nucleic acids [95]. The intensity of the Amide I band according to previous studies [33, 97] is lower for the spectra of cancerous tumors compared to the spectra of benign formations and healthy skin.

(II) Melanoma Versus Pigmented Benign Tumors

Melanoma arises from melanocytes, pigmented cells that make it similar to the pigmented benign tumors especially pigmented nevus. However, many of the pigmented lesions are excised without further histological confirmation to be malignant. To avoid missing any potentially dangerous tumors, oncologists may identify a pigmented benign tumor as a melanoma one and refer the patient for additional examinations in a specialized oncological institution for diagnosis confirmation. This leads to a surgical excision of the tumor with a wide (up to 3–5 cm) removal of healthy skin. Also, specialized oncological institutions became overcrowded because of numerous incorrect diagnosis; this leads to the improper financial and time costs, because many other patients at the same time really wait for the treatment. Recent survey showed that the ratio of the total number of excised melanocytic lesions to the number of histopathologically confirmed melanoma is varied from 20–40 in general dermatology practices to 4–18 in specialized skin cancer clinics [98]. Therefore, the meaningful task in the melanoma diagnostics is the improving the early detection of melanoma without any omit of potential suspected tumor while reducing unnecessary biopsies of the benign pigmented lesions. The growing trend to use noninvasive biopsy can be responsible for improving the accurate clinical diagnostics of melanoma.

Spectral bands with important differences between spectra of malignant melanoma and pigmented nevus were revealed by VIP analysis. Diagnostic VIP bands for melanoma versus pigmented nevus correspond to Raman bands of Amide III (1246 cm^{-1}), Amide I (1645 cm^{-1}), and collagens (1745, 1816 cm^{-1} bands). In different studies it is reported about collagen destruction and reduced intensity of the Amide I band because of increased melanin content [40]. The lower intensity in the melanoma spectra at the Amide I in comparison to nevus spectra can be associated with the major differences at the 1645 cm^{-1} in the obtained PLS model. Spectral changes in Amide III region that related to the proteins structure are also effective

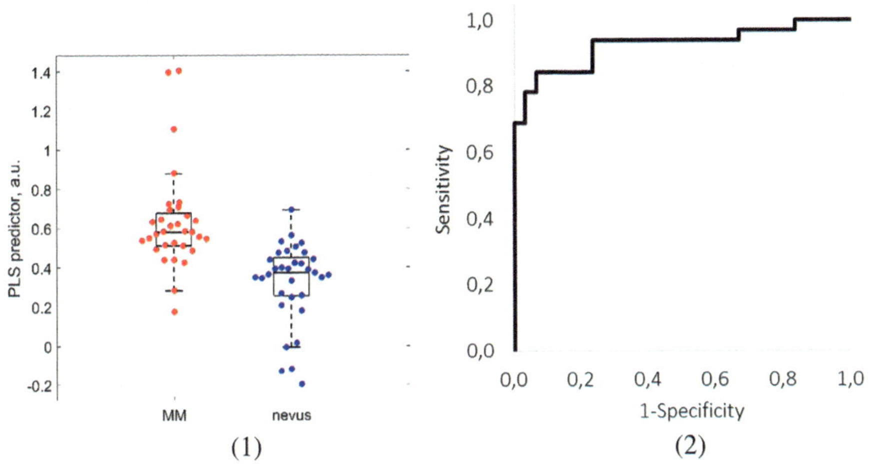

Fig. 10.6 (1) Box-plots and bee swarm plots of PLS-predictors for discriminating melanoma and pigmented nevus and (2) corresponding ROC curve is derived from spreading PLS-predictors

diagnostic parameters for this classification. The Amide I and Amide III are widely used for different tumors classification.

Based on the spectral differences, PLS-predictors were calculated for each spectrum. Figure 10.6 shows the box plot diagram of the PLS-predictors on the basis of the extracted differences of the model. Spectral differences between melanoma and pigmented nevus allow to achieve 84% sensitivity and 82% specificity. ROC AUC was found to be 0.92.

In this differentiating model 5 out of 32 melanoma were incorrectly classified with three of them amelanotic melanoma and one weak pigmented melanoma. The amelanotic melanoma is a type of melanoma without pigmentation. However, other four amelanotic melanoma were correctly classified as melanoma. According to the previous studies [99], the melanin has Raman peak near the 1380 and 1580 cm^{-1} bands and is not observed in our classification model. For this reason, it is not clear whether pigmentation is an important factor to differentiate melanoma from pigmented lesions. In contrast to our results, Philipsen et al. [100] investigated the influence of skin pigmentation on spectral Raman intensity. The authors studied 165 normal skin samples with different pigmentation (Fitzpatrick skin type I–VI), 25 pigmented and unpigmented BCC, 41 pigmented nevi, and 15 pigmented malignant melanoma. The important finding of this study is that the Raman diagnoses are not relying on pigmentation. They claimed that increased skin pigmentation results in a higher spectral background caused by AF, which could be removed by background correction. The authors showed 93.3% sensitivity and 96.4% specificity for melanoma diagnostics, 88% sensitivity and 85.5% specificity for BCC, 87.8% sensitivity and 84.2% specificity for pigmented nevus.

(III) Melanoma Versus Basal Cell Carcinoma

In the malignant versus benign tumors classification, we combined melanoma and BCC spectra as malignant and achieved 100% differentiation accuracy. In melanoma versus BCC classification, we achieved only 75% accuracy (90% sensitivity–67% specificity). ROC AUC of this model equals 0.77 (Fig. 10.7). Therefore, we can assume that RS is more sensitive to the similarity features of the pathological growth despite the differences in the structure and principle of melanoma and BCC tumor cells development.

The important differences between the spectra of the melanoma and BCC were found in 1444, 1535, 1655 cm^{-1} bands. Gniadecka et al. [101] showed that the spectra of melanoma and BCC have similar tendency to increase the spectral intensity in the 1300–1330 cm^{-1} band of lipids. This feature is in good agreement with our results, because VIP score analysis for differentiation of melanoma and BCC did not point this spectral band as important for the model. On the other hand, BCC has much more motility and invasiveness of tumor cells due to the higher level of actin in comparison with melanoma [102, 103]. The bands responsible for the actin content are 1210, 1342, 1452, and 1613 cm^{-1}. In the melanoma vs BCC model the VIP analysis identified the 1207–1217 and 1433–1444 cm^{-1} regions, representing differences in actin concentration. In the melanoma spectra there is a decrease in the intensity in the Amide I region of 1644–1660 cm^{-1}.

(IV) Melanoma Versus Other Skin Lesions

To differentiate melanomas from other skin neoplasms, the VIP analysis extracts 1261, 1330, 1369, 1450, 1510, 1570, 899, 1625, 1730, 1781 cm^{-1} peaks as informative bands. From the distribution of the PLS-predictors, 91% sensitivity and 85%

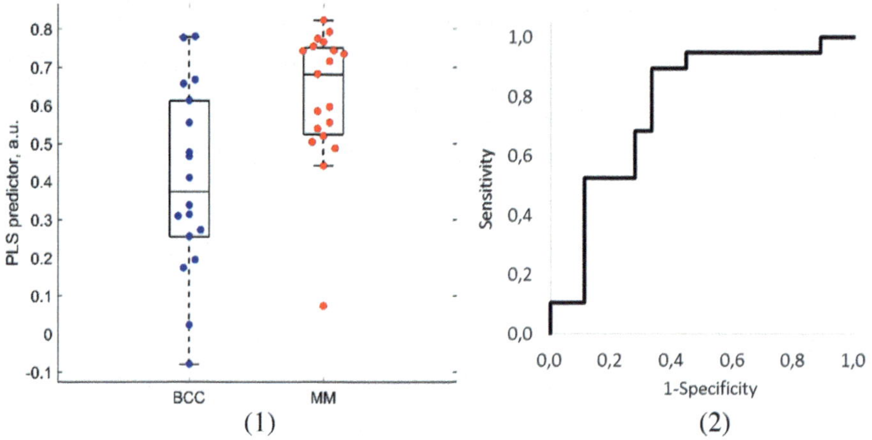

Fig. 10.7 (1) Box-plots and bee swarm plots of PLS-predictors for discriminating melanoma and BCC and (2) corresponding ROC curve is derived from spreading PLS-predictors

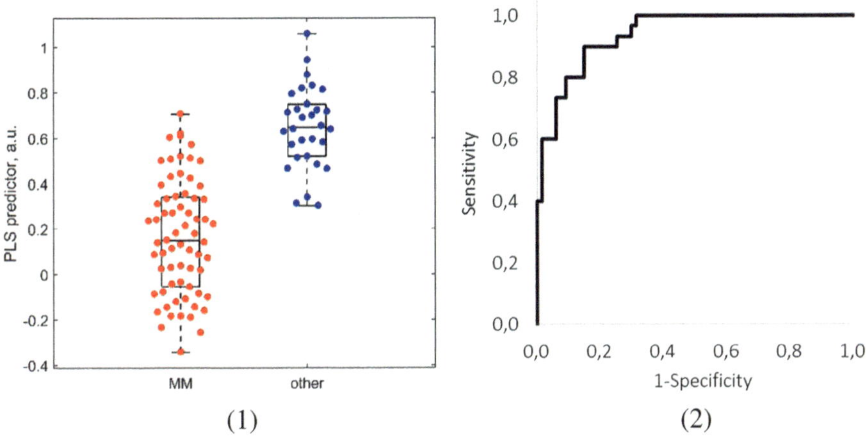

Fig. 10.8 (1) Box-plots and bee swarm plots of PLS-predictors for discriminating melanoma and other skin lesions and (2) corresponding ROC curve is derived from spreading PLS-predictors

specificity were achieved, ROC AUC equaled 0.94 (Fig. 10.8). 29 out of 32 melanoma and 60 out of 71 nonmelanoma were correctly identified. Moreover, 10 out of 11 incorrectly identified nonmelanoma lesions were pigmented nevus.

Oncologist's Diagnostics Accuracy To compare the accuracy of RS study with the accuracy of clinical diagnosis, we assessed the diagnostic accuracy of the skin neoplasm from the primary inspection by an oncologist. 1 out of 19 melanoma was defined preliminary as a benign tumor and another 1 melanoma was defined as a malignant tumor without type specification. Five out of 19 benign tumors were identified as melanoma and another 7 benign tumors—as BCC. The oncologist found suspected melanoma for 4 studied BCC.

In other cohorts 2 out of 32 melanoma were identified as pigmented nevus, 2—as BCC. In addition, 8 out of 33 nevus were incorrectly identified by oncologist as malignant tumor, namely 5 of them as malignant tumors and 3 as malignant tumor without type specification. Accuracy, sensitivity, and specificity of the preliminary diagnostics were calculated and presented in Table 10.2.

The results shown in Table 10.2 demonstrate that the proposed method based on the application of conventional RS may be an effective and rapid approach for skin pathologies classification because of the advantages associated with Raman features and possibility to register optical signal by a single portable device. Simultaneous analysis of changes in endogenous chromophores by RS is a powerful tool for tissue malignancy detection.

Table 10.2 Classification accuracy of PLS-DA models and diagnosis by oncologist

Classification model	Number of samples	Accuracy (sensitivity–specificity)	
		PLS-DA model	Oncologist
Melanoma versus BCC	19 versus 18	78% (90%–67%)	85% (90%–81%)
Melanoma versus nevus	32 versus 33	83% (81%–85%)	80% (87%–73%)
Melanoma versus other neoplasms (BCC+benign)	32 versus 71	87% (91%–85%)	79% (87%–75%)
Malignant (melanoma+BCC) versus benign tumors	37 versus 19	100% (100%–100%)	63% (90%–37%)

10.9 Multimodal Raman Optical Techniques

The accuracy of skin tumors diagnosis by spontaneous RS may be improved by a number of techniques, including application of other Raman modalities (or other excitation wavelengths) or even implementation of other optical approaches. For instance, Feng et al. [96] utilized a Raman system (830 nm excitation wavelength, CCD camera cooled to −30 °C) to study different skin neoplasm types from 199 skin lesions. PCA analysis of RS data demonstrated significant differences between cancer, precancer, benign and normal lesions. Nonmelanoma skin cancers versus precancer lesions model was realized with 95% sensitivity and 21% specificity. The better result was achieved for classification of 12 malignant melanoma versus 17 dysplastic nevus with 95% sensitivity and 100% specificity. This result is very meaningful because high accuracy of melanocytic neoplasms differentiation is challenging due to the similar appearance of melanoma [104]. Cicchi et al. [32] reported 56% sensitivity and 89% specificity for discriminating 10 malignant melanoma and 10 melanocytic nevi in *ex vivo* studies using 785 nm laser for RS excitation. According to the achieved results, the most informative spectroscopic signatures for this discriminating model are in the 1260 cm^{-1} and 1620 cm^{-1} regions. In this study, the combination of visible fluorescence and NIR Raman spectroscopy allowed for improving diagnostic capabilities for discriminating melanoma and nonmelanoma tumors with 89% sensitivity and 100% specificity.

The combination of several optical methods is a good solution to improve diagnostic performance of the cancer detection and classification. Generally, the multimodal system unites different optical methods that demonstrated ability to jointly analyze biochemical, structural, and physiological tissue characteristics.

Lim et al. [40] suggested multimodal diagnosis of melanoma and nonmelanoma skin cancer using reflectance, AF, and RS. The classification of melanoma versus pigmented lesions resulted in 100% accuracy based on the alone RS with important diagnostic parameters in Amide I, 1300–1340 cm^{-1} lipids, Amide III, CH_2 around 1450 cm^{-1}, and spectral bands around 800–1000 cm^{-1}. In classification of nonmelanoma skin cancer versus actinic keratosis the combined approach improved 72% sensitivity and 64% specificity using only RS to 95% sensitivity and 71% specificity for multimodal approach. The main spectral feature of nonmelanoma skin cancer is a lower reflectance intensity because of breakdown of collagen from the progression of malignancy.

In our previous studies, Zakharov et al. [105] combined RS, backscattering spectroscopy (BS), and optical coherence tomography (OCT) methods to detect BCC and melanoma. The combination of RS with either of OCT or backscattering spectroscopy improves the sensitivity and specificity of BCC detection by 20% and 10% compared with only RS implementation. Conversely, the specificity of melanoma diagnosis by combined approach increases by 5% compared with RS, while sensitivity does not change. The results demonstrated that OCT is a highly informative tool for BCC diagnosis, while RS is more accurate for melanoma detection. Also, BS can be used for rapid scanning of tumor boundaries because of processing rate.

In the study by Bratchenko et al. [93] we demonstrated application of RS with AF in visible range (457 nm and 532 nm lasers were used to stimulate AF response). Combination of AF analysis with RS study allows for increasing the sensitivity and specificity of melanoma detection from 80% to 97% due to involvement in joint criteria analysis of the changes in concentration of porphyrins, flavins, and melanin along with RS criteria. High level of malignant skin tumors classification accuracy is indicative of high clinical potential of the proposed method. This method may be used with equal success as for precise tumor type determination as for mass screening surveys. For example, fast analysis of large tissue areas may be performed on the first step only with NIR and visible AF, as the AF signal is much more intense than an RS signal and may be collected during a short time. The accuracy of only AF skin tissue analysis (both in visible and NIR spectral regions) is about 87% and it may be increased by RS analysis of the suspicious areas on the next step. Such joint analysis will be characterized by 100% sensitivity with 96% specificity.

In our latest studies [30] we tried to utilize joint RS and AF NIR analysis for the classification of benign and malignant skin tumors. Typical spectra of BCC and BN formation with VIP for malignant versus benign tumors model are shown in Fig. 10.9. According to VIP analysis, the most informative differences between the malignant ($n = 37$) and benign ($n = 19$) neoplasm spectra are observed in 533, 713, 877, 946, 1048, 1281, 1439, 1562, and 1644 cm^{-1} bands. This model considers both Raman and AF features of the neoplasms. Spectral differences between the malignant and benign tumors allow achieving 100% accuracy, and ROC AUC equals 1.

Santos et al. [104] presented a classification model to distinguish melanoma from benign melanocytic lesions. A high-wavenumber Raman system with 976 nm excitation wavelength was used to register Raman spectra from 24 melanomas and 47 benign melanocytic lesions. The authors suggested increase in lipid content in melanomas in comparison with the other lesions and found the AUC of the ROC curve to be 0.77 for such classification.

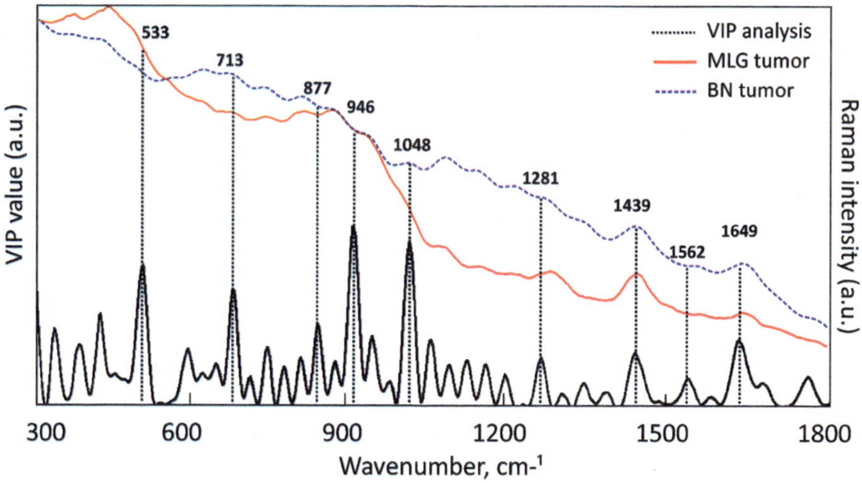

Fig. 10.9 VIP scores for malignant vs benign tumors classification

Gniadecka et al. [101] presented the other work studying high-wavenumber region. The authors investigated melanoma detection using NIR Fourier transform RS with neural network classifier. Neural network analysis allowed for differentiation of 22 melanoma from 41 pigmented nevi, 48 BCC, 23 seborrheic keratosis, 89 normal skin with 85% sensitivity and 99% specificity. The authors suggested that melanoma could be differentiated from pigmented nevi, BCC, seborrheic keratoses, and normal skin due to the decrease in the intensity of the Amide I protein band around 1660 cm^{-1}. Later Philipsen et al. [100] showed application of NIR Fourier transform RS with a 1064 nm excitation source and liquid nitrogen-cooled Ge detector at 8 min acquisition time of spectra. The examination was performed from different skin sites of 55 healthy persons with different skin pigmentation and 72 patients with lesions. The important finding of that work is that RS diagnosis does not depend on tissue pigmentation. Melanoma was classified from normal skin with 93.3% sensitivity and 96.4% specificity, BCC with 88% sensitivity and 85.5% specificity.

Kong et al. [106] reported about scanning Raman microscopy for BBC diagnostics during surgery. This method allowed obtaining diagnostics maps of BCC excised during Mohs micrographic surgery. Moreover, Raman diagnostics maps were in excellent agreement with the gold-standard histopathology, but obtained faster than standard frozen section histopathology. The authors achieved 100% sensitivity and 92.9% specificity for the diagnosis of 22 BCC samples.

Heuke et al. [107] described the study of nonmelanoma skin (BCC and SCC) cancer using the multimodal imaging approach including CARS, second harmonic generation (SHG), and two photon excited fluorescence (TPEF). The author reported that morphological features of cancer site can be identified using CARS. Figure 10.10

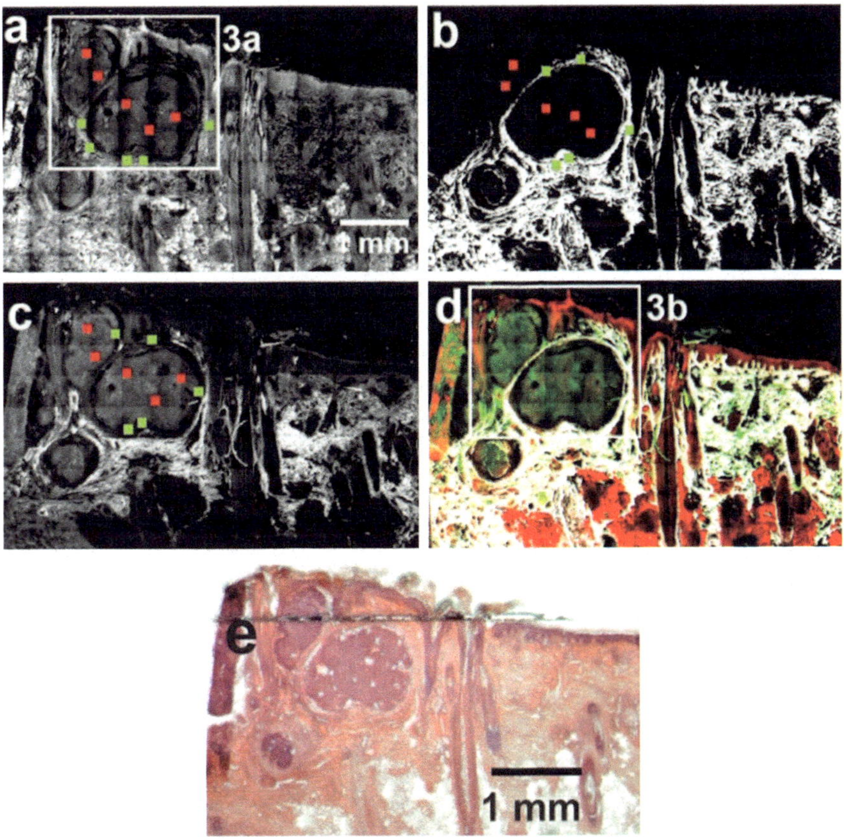

Fig. 10.10 (**a**) CARS; (**b**) SHG; (**c**) TPEF (grayscale); and (**d**) multimodal image (CARS: red; TPEF: green; SHG: gray) as well as (**e**) H&E stained picture of the BBC section. The red (tumorous) and green (non-tumorous) squares locate the position of the 100 × 100 pixel sized areas selected for statistical analysis. (**a**) CARS overview image of a BCC. The BCC tissue appears darker in CARS. In general BCC tissue shows no SHG signal; (**c**) corresponding TPEF image. BCC tissue appears slightly brighter than the adjacent dermal tissue, but darker than elastin fibers; (**d**) Multimodal large overview image, i.e., the combination of the images 2a–c (reprinted with permission from [107])

demonstrates multimodal analysis of histopathological BCC section. The authors state that both BCC and SCC show a lack of SHG and increased TPEF signal for cancerous tissue. Moreover, the distinct appearance of BCC and SCC in CARS is reported. While the BCC regions display a relatively weak CARS signal, the SCC nests show a strong CARS signal compared to the surrounding tissue. Also it is expected to boost *in vivo* studies of this promising technological approach.

10.10 Conclusions and Future Perspectives

For the moment, only conventional RS is widely used for skin cancer biopsy in clinical trials. To develop RS methods for cancer diagnostics, various scientific groups use highly sensitive equipment with a high degree of detector cooling [55, 94, 97, 108–110]. As we can see from the considered studies, basically, the cooling range from −120 °C to −60 °C has been used. In the study by Khristoforova et al. [30] and in the latest clinical trials of our research group described in Sect. 10.8, we obtained a comparable result of skin lesions differentiation using multimodal approach with combined RS and AF multivariate analysis and portable setup with a detector cooled only down to −15 °C. This peculiarity considerably distinguishes the proposed equipment in terms of cost and capability of operation in clinical studies. The use of the proposed portable cost-effective spectrometer is ideally suitable for implementing the idea of skin cancer screening. Compactness and easy transportation can make such systems the most attractive and convenient for mass screening surveys. Thus, the use of a portable Raman spectrometer (to acquire conventional Raman signal) along with the PCA or PLS is a promising approach that may become an indispensable tool to the medical specialists when examining skin lesions for malignancy.

However, optical biopsy of skin cancer may be enhanced by implementation of numerous Raman techniques in addition to conventional RS. These techniques include both methods of tissue chemical composition analysis and imaging techniques. Such techniques as SERS or resonance RS may provide additional information about tissue chemical composition in comparison to conventional RS, and thus make diagnosis more accurate. At the same time, the application of SERS and resonance RS requires more complicated experimental setup and is characterized by complex process of spectra registration. Imaging techniques provide an opportunity to study tumor morphology (SORS) and even provide an opportunity to perform optical histology of studied tissues (CARS and SRS). Optical histology opens a new chance for *in vivo* examination of tumor tissues, as it helps to find the exact tumor type using an analogue of a medical gold standard (histopathological study of tissue). CARS and SRS setups are stationary facilities that restrict wide implementation of such approaches in mass screening applications. However, CARS and SRS may be effectively applied in specialized clinics for exact tumor margins detection during surgery operations. A combination of RS and other optical techniques is very promising in clinical applications due to the increase of diagnostic accuracy. Such multimodal units may include combinations of RS, AF, backscattering spectroscopy, optical coherence tomography, different fluorescent techniques, multiphoton spectroscopy and many others. Limitations of multimodal approaches are high-cost and time-consuming procedures of tissue examination. Nevertheless, recent progress in optical equipment brings closer the opportunity to use CARS, SRS, SERS, and different multimodal approaches in clinical practice.

Acknowledgments The work was supported by the Russian Federation President grant # MK-1888.2019.2 for state support of young scientists (I.A. Bratchenko) and by the Russian Foundation of Basic Research grant # 19-52-18001 Bolg_a (V.P. Zakharov).

References

1. American Cancer Society: Cancer Facts & Figures 2018. American Cancer Society, Atlanta (2018). https://www.cancer.org/research/cancer-facts-statistics/all-cancer-facts-figures/cancer-facts-figures-2018.html. Accessed 22 Sept 1998
2. Bray, F., Ferlay, J., Soerjomataram, I., et al.: Global Cancer Statistics 2018: GLOBOCAN estimates of incidence and mortality worldwide for 36 cancers in 185 countries. CA Cancer J. Clin. **68**(6), 394–424 (2018). https://doi.org/10.3322/caac.21492
3. Kaprin, A.D., et al. (eds.): Malignant Neoplasms in Russia in 2017 (Incidence and Mortality). P.A. Herzen Moscow Research Institute of Oncology, Moscow (2018)
4. Balch, C.M., Gershenwald, J.E., Soong, S.-J., Thompson, J.F., Atkins, M.B., Byrd, D.R., Buzaid, A.C., Cochran, A.J., Coit, D.G., Ding, S., Eggermont, A.M., Flaherty, K.T., Gimotty, P.A., Kirkwood, J.M., McMasters, K.M., Mihm Jr., M.C., Morton, D.L., Ross, M.I., Sober, A.J., Sondak, V.K.: Final version of 2009 AJCC melanoma staging and classification. J. Clin. Oncol. **27**(36), 6199–6206 (2009). https://doi.org/10.1200/JCO.2009.23.4799
5. Eikje, N.S., Aizawa, K., Ozaki, Y.: Vibrational spectroscopy for molecular characterisation and diagnosis of benign, premalignant and malignant skin tumours. Biotechnol. Annu. Rev. **11**, 191–225 (2005). https://doi.org/10.1016/S1387-2656(05)11006-0
6. Youl, P.H., Baade, P.D., Janda, M., Mar, C.B.D., Whiteman, D.C., Aitken, J.F.: Diagnosing skin cancer in primary care: how do mainstream general practitioners compare with primary care skin cancer clinic doctors? Med. J. Aust. **187**(4), 215 (2007). https://doi.org/10.5694/j.1326-5377.2007.tb01202.x
7. Ermakov, A.V.: Oncosurgery. **5**(3), 52–58 (2013)
8. White, R.R., Stanley, W.E., Johnson, J.L., et al.: Long-term survival in 2,505 patients with melanoma with regional lymph node metastasis. Ann. Surg. **235**(6), 879–887 (2002)
9. Carli, P., de Giorgi, V., Chiarugi, A., Nardini, P., Weinstock, M.A., Crocetti, E., Stante, M., Giannotti, B.: Addition of dermoscopy to conventional naked-eye examination in melanoma screening: a randomized study. J. Am. Acad. Dermatol. **50**(5), 683–689 (2004). https://doi.org/10.1016/j.jaad.2003.09.009
10. Vestergaard, M.E., Macaskill, P., Holt, P.E., Menzies, S.W.: Dermoscopy compared with naked eye examination for the diagnosis of primary melanoma: a meta-analysis of studies performed in a clinical setting. Br. J. Dermatol. **159**(3), 669–676 (2008). https://doi.org/10.1111/j.1365-2133.2008.08713.x
11. Alfano, R.R., Budansky, Y., Luo, J.: U.S. Patent 6587711, 2003.
12. Benvenuto-Andrade, C., Dusza, S.W., Agero, A.L.C.: Differences between polarized light dermoscopy and immersion contact dermoscopy for the evaluation of skin lesions. Arch. Dermatol. **143**(3), 329–338 (2007). https://doi.org/10.1001/archderm.143.3.329
13. Abbasi, N.R., Shaw, H.M., Rigel, D.S., Friedman, R.J., McCarthy, W.H., Osman, I., Polsky, D.: Early diagnosis of cutaneous melanoma: revisiting the ABCD criteria. JAMA. **292**(22), 2771–2776 (2004). https://doi.org/10.1001/jama.292.22.2771
14. Walter, F.M., Prevost, A.T., Vasconcelos, J., Hall, P.N., Burrows, N.P., Morris, H.C., Emery, J.D.: Using the 7-point checklist as a diagnostic aid for pigmented skin lesions in general practice: a diagnostic validation study. Br. J. Gen. Pract. **63**(610), e345–e353 (2013). https://doi.org/10.3399/bjgp13X667213
15. Argenziano, G., Fabbrocini, G., Carli, P., De Giorgi, V., Sammarco, E., Delfino, M.: Epiluminescence microscopy for the diagnosis of doubtful melanocytic skin lesions: compar-

ison of the ABCD rule of dermatoscopy and a new 7-point checklist based on pattern analysis. Arch. Dermatol. **134**(12), 1563–1570 (1998). https://doi.org/10.1001/archderm.134.12.1563
16. Chen, S.C., Bravata, D.M., Weil, E., Olkin, I.: A comparison of dermatologists' and primary care physician accuracy in diagnosing melanoma: a systemic review. Arch. Dermatol. **137**(12), 1627 (2001). https://doi.org/10.1001/archderm.137.12.1627
17. Narayanamurthy, V., Padmapriya, P., Noorasafrin, A., Pooja, B., Hema, K., Khan, A.Y.F., Nithyakalyanic, K., Samsurib, F.: Skin cancer detection using non-invasive techniques. RSC Adv. **8**, 28095–28130 (2018). https://doi.org/10.1039/C8RA04164D
18. Papakonstantinou, E., et al. (eds.): Photodynamic Therapy and Skin Cancer, Dermatologic Surgery and Procedures. Pierre Vereecken, IntechOpen (2017). https://doi.org/10.5772/intechopen.70309
19. Fritsch, C., Lang, K., Neuse, W., Ruzicka, T., Lehmann, P.: Photodynamic diagnosis and therapy in dermatology. Skin Pharmacol. Appl. Skin Physiol. **11**(6), 358–373 (1998). https://doi.org/10.1159/000029858
20. Dinnes, J., Bamber, J., Chuchu, N., Bayliss, S.E., Takwoingi, Y., Davenport, C., Godfrey, K., O'Sullivan, C., Matin, R.N., Deeks, J.J., Williams, H.C.: High-frequency ultrasound for diagnosing skin cancer in adults. Cochrane Database Syst. Rev. **12**, CD013188 (2018). https://doi.org/10.1002/14651858.CD013188
21. Moqadam, S.M., Grewal, P.K., Haeri, Z., Ingledew, P.A., Kohli, K., Golnaraghi, F.: Cancer detection based on electrical impedance spectroscopy: a clinical study. J. Electr. Bioimpedance. **9**(1), 17–23 (2018). https://doi.org/10.2478/joeb-2018-0004
22. Herman, C., Cetingul, M.P.: Quantitative visualization and detection of skin cancer using dynamic thermal imaging. J. Vis. Exp. **51**, e2679 (2011). https://doi.org/10.3791/2679
23. Richards-Kortum, R., Sevick-Muraca, E.: Quantitative optical spectroscopy for tissue diagnosis. Annu. Rev. Phys. Chem. **47**, 555–606 (1996). https://doi.org/10.1146/annurev.physchem.47.1.555
24. Ulricha, M., Lange-Asschenfeldt, S., Gonzalez, S.: *In vivo* reflectance confocal microscopy for early diagnosis of nonmelanoma skin cancer. Actas Dermosifiliogr. **103**(9), 784–789 (2012). https://doi.org/10.1016/j.ad.2011.10.017
25. Mogensen, M., Thrane, L., Jørgensen, T.M., Andersen, P.E., Jemec, G.B.E.: OCT imaging of skin cancer and other dermatological diseases. J. Biophotonics. **2**, 442 (2009). https://doi.org/10.1002/jbio.200910020
26. Seidenari, S., Arginelli, F., Dunsby, C., French, P.M.W., König, K., Magnoni, C., Talbot, C., Ponti, G.: Multiphoton laser tomography and fluorescence lifetime imaging of melanoma: morphologic features and quantitative data for sensitive and specific non-invasive diagnostics. PLoS One. **8**(7), e70682 (2013). https://doi.org/10.1371/journal.pone.0070682
27. Seidenarietal, S., Arginelli, F., Bassoli, S., Cautela, J., French, P.M.W., Guanti, M., Guardoli, D., König, K., Talbot, C., Dunsby, C.: Multiphoton laser microscopy and fluorescence lifetime imaging for the evaluation of the skin. Dermatol. Res. Pract. **2012**, 1 (2012). https://doi.org/10.1155/2012/810749
28. Gao, W., Zakharov, V.P., Myakinin, O.O., Bratchenko, I.A., Artemyev, D.N., Kornilin, D.V.: Medical images classification for skin cancer using quantitative image features with optical coherence tomography. J. Innov. Opt. Health Sci. **9**(2), 1650003 (2016). https://doi.org/10.1142/S1793545816500036
29. Raupov, D.S., Myakinin, O.O., Bratchenko, I.A., Zakharov, V.P., Khramov, A.G.: Skin cancer texture analysis of OCT images based on Haralick, fractal dimension, Markov random field features, and the complex directional field features. Proc. SPIE. **10024**, 100244I (2016). https://doi.org/10.1117/12.2246446
30. Khristoforova, Y.A., Bratchenko, I.A., Myakinin, O.O., Artemyev, D.N., Moryatov, A.A., Orlov, A.E., Kozlov, S.V., Zakharov, V.P.: Portable spectroscopic system for *in vivo* skin neoplasms diagnostics by Raman and autofluorescence analysis. J. Biophotonics. **12**(4), e201800400 (2019). https://doi.org/10.1002/jbio.201800400

31. Alfano, R.R., Pu, Y.: In: Jelinkova, H. (ed.) Laser for Medical Applications. Woodhead Publishing Limited, Cambridge (2013)
32. Cicchi, R., Cosci, A., Rossari, S., Kapsokalyvas, D., Baria, E., Maio, V., Massi, D., Giorgi, V.D., Pimpinelli, N., Pavone, F.S.: Combined fluorescence-Raman spectroscopic setup for the diagnosis of melanocytic lesions. J. Biophotonics. **7**(1–2), 86 (2013). https://doi.org/10.1002/jbio.201200230
33. Calin, M.A., Parasca, S.V., Savastru, R., Calin, M.R., Dontu, S.: Optical techniques for the noninvasive diagnosis of skin cancer. J. Cancer Res. Clin. Oncol. **139**, 1083 (2013). https://doi.org/10.1007/s00432-013-1423-3
34. Ferraro, J.R., Nakamoto, K.: Introductory Raman Spectroscopy, p. 135. Academic Press, Amsterdam (1994). https://doi.org/10.1002/jrs.1407
35. Vo-Dinh, T., Cullum, B.: Fluorescence spectroscopy for biomedical diagnostics. In: Biomedical Photonics Handbook. CRC Press, Boca Raton (2003). https://doi.org/10.1201/9780203008997
36. Wachsmann-Hogiu, S., Weeks, T., Huser, T.: Chemical analysis *in vivo* and *in vitro* by Raman spectroscopy—from single cells to humans. Curr. Opin. Biotechnol. **20**(1), 63–73 (2009). https://doi.org/10.1016/j.copbio.2009.02.006
37. Liu, W., et al.: Dual laser-induced fluorescence: progress and perspective for *in vivo* cancer diagnosis. Chin. Sci. Bull. **58**(17), 2003–2016 (2013). https://doi.org/10.1007/s11434-013-5826-y
38. Bodanese, B., Silveira, F.L., Zangaro, R.A., Pacheco, M.T.T., Pasqualucci, C.A., Silveira, L.: Discrimination of basal cell carcinoma and melanoma from normal skin biopsies *in vitro* through Raman spectroscopy and principal component analysis. Photomed. Laser Surg. **30**(7), 381–387 (2012). https://doi.org/10.1089/pho.2011.3191
39. Schleusener, J., Gluszczynska, P., Reble, C., Gersonde, I., Helfmann, J., Fluhr, J.W., Lademann, J., Röwert-Huber, J., Patzelt, A., Meinke, M.C.: *In vivo* study for the discrimination of cancerous and normal skin using fibre probe-based Raman spectroscopy. Exp. Dermatol. **24**(10), 767–772 (2015). https://doi.org/10.1111/exd.12768
40. Lim, L., Nichols, B., Migden, M.R., Rajaram, N., Reichenberg, J.S., Markey, M.K., Ross, M.I., Tunnell, J.W.: Clinical study of noninvasive *in vivo* melanoma and nonmelanoma skin cancers using multimodal spectral diagnosis. J. Biomed. Opt. **19**(11), 117003 (2014). https://doi.org/10.1117/1.JBO.19.11.117003
41. Lieber, C.A., Majumder, S.K., Ellis, D.L., Billheimer, D.D., Mahadevan-Jansen, A.: *In vivo* nonmelanoma skin cancer diagnosis using Raman microspectroscopy. Lasers Surg. Med. **40**(7), 461–467 (2008). https://doi.org/10.1002/lsm.20653
42. Schut, T.C.B., Caspers, P.J., Puppels, G.J., Nijssen, A., Heule, F., Neumann, M.H.A., Hayes, D.P.: Discriminating basal cell carcinoma from its surrounding tissue by Raman spectroscopy. J. Invest. Dermatol. **119**(1), 64–69 (2002). https://doi.org/10.1046/j.1523-1747.2002.01807.x
43. Wang, S., Zhao, J., Lui, H., He, Q., Zeng, H.: *In vivo* near-infrared autofluorescence imaging of pigmented skin lesions: methods, technical improvements and preliminary clinical results. Skin Res. Technol. **19**(1), 20 (2012). https://doi.org/10.1111/j.1600-0846.2012.00632.x
44. Zheng, J., Pang, S., Labuza, T.P., He, L.: Evaluation of surface-enhanced Raman scattering detection using a handheld and a bench-top Raman spectrometer: a comparative study. Talanta. **129**, 79–85 (2014). https://doi.org/10.1016/j.talanta.2014.05.015
45. Sahu, A., Deshmukh, A., Hole, A.R., Chaturvedi, P., Krishna, C.M.: *In vivo* subsite classification and diagnosis of oral cancers using Raman spectroscopy. J. Innov. Opt. Health Sci. **9**(5), 1650017 (2016). https://doi.org/10.1142/s1793545816500176
46. Bolognia, J., Schaffer, J., Cerroni, L.: Section 18: Neoplasms of the Skin. Dermatology: 2-Volume Set, 4th edn. Elsevier (2017)
47. Jermyn, M., Desroches, J., Aubertin, K., St-Arnaud, K., Madore, W.-J., Montigny, E.D., Guiot, M.C., Trudel, D., Wilson, B.C., Petrecca, K., Leblond, F.: A review of Raman spectroscopy advances with an emphasis on clinical translation challenges in oncology. Phys. Med. Biol. **61**(23), R370 (2016). https://doi.org/10.1088/0031-9155/61/23/r370

48. Suhalim, J.L., Boik, J.C., Tromberg, B.J., Potma, E.O.: The need for speed. J. Biophotonics. **5**, 387–395 (2012). https://doi.org/10.1002/jbio.201200002
49. Pence, I., Mahadevan-Jansen, A.: Clinical instrumentation and applications of Raman spectroscopy. Chem. Soc. Rev. **45**(7), 1958–1979 (2016). https://doi.org/10.1039/c5cs00581g
50. Daniel, A., Savarimuthu, W.P., Gowripal, Y., et al.: *In-vivo* study using Raman Spectroscopy to estimate the effect of fairness creams on skin. J. Dermat. Cosmetol. **2**(5), 57–59 (2018). https://doi.org/10.15406/jdc.2018.02.00085
51. Gonzalez, F.J., Alda, J., Moreno-Cruz, B., Martinez-Escaname, M., Ramirez-Elias, M.G., Torres-Alvarez, B., Moncada, B.: Use of Raman spectroscopy for the early detection of filaggrin-related atopic dermatitis. Skin Res. Technol. **17**(1), 45–50 (2011). https://doi.org/10.1111/j.1600-0846.2010.00461.x
52. Zhang, J., Fan, Y., Song, Y., Xu, J.: Accuracy of Raman spectroscopy for differentiating skin cancer from normal tissue. Medicine (Baltimore). **97**(34), e12022 (2018). https://doi.org/10.1097/MD.0000000000012022
53. Kourkoumelis, N., et al.: Advances in the *in vivo* Raman spectroscopy of malignant skin tumors using portable instrumentation. Int. J. Mol. Sci. **16**, 14554–14570 (2015). https://doi.org/10.3390/ijms160714554
54. Feng, X., et al.: Biophysical basis of skin cancer margin assessment using Raman spectroscopy. Biomed. Opt. Express. **10**(1), 104–118 (2019). https://doi.org/10.1364/BOE.10.000104
55. Silveira, F.L., et al.: Discrimination of non-melanoma skin lesions from non-tumor human skin tissues *in vivo* using Raman spectroscopy and multivariate statistics. Lasers Surg. Med. **47**, 6–16 (2015). https://doi.org/10.1002/lsm.22318
56. Shafer-Peltier, K.E., Haka, A.S., Fitzmaurice, M., Crowe, J., Myles, J., Dasari, R.R., Feld, M.S.: Raman microspectroscopic model of human breast tissue: implications for breast cancer diagnosis *in vivo*. J. Raman Spectrosc. **33**, 552–563 (2002). https://doi.org/10.1002/jrs.877
57. Wang, J., Bergholt, M.S., Zheng, W., Huang, Z.: Development of a beveled fiber-optic confocal Raman probe for enhancing *in vivo* epithelial tissue Raman measurements at endoscopy. Opt. Lett. **38**, 2321–233O (2013). https://doi.org/10.1364/OL.38.002321
58. Wood, J.J., Kendall, C., Hutchings, J., Lloyd, G.R., Stone, N., Shepherd, N., Day, J., Cook, T.A.: Evaluation of a confocal Raman probe for pathological diagnosis during colonoscopy. Colorectal Dis. **16**, 732–738 (2014). https://doi.org/10.1111/codi.12664
59. Pezzotti, G., et al.: Raman spectroscopy of human skin: looking for a quantitative algorithm to reliably estimate human age. J. Biomed. Opt. **20**(6), 065008 (2015). https://doi.org/10.1117/1.JBO.20.6.065008
60. Zhang, G., Papillon, A., Ruvolo, E., Barda, P., Kollias, N.: *In vivo* comparative documentation of skin hydration by confocal Raman microscopy, Skin Sensor, Skicon and NovaMeter. Proc. SPIE. 7548 (2010). https://doi.org/10.1117/12.848215
61. Sieg, A.: Chapter 19. Raman spectroscopy. In: Berardesca, E., Maibach, H.I., Wilhelm, K.-P. (eds.) Non Invasive Diagnostic Techniques in Clinical Dermatology, p. 593. Springer-Verlag, Berlin (2014)
62. Förster, M., Bolzinger, M.A., Montagnac, G., Briançon, S.: Confocal Raman microspectroscopy of the skin. Eur. J. Dermatol. **21**(6), 851–863 (2011). https://doi.org/10.1684/ejd.2011.1494
63. Lieber, C.A., Majumder, S.K., Ellis, D.L., Billheimer, D.D., Mahadevan-Jansen, A.: *In vivo* nonmelanoma skin cancer diagnosis using Raman microspectroscopy. Lasers Surg. Med. **40**, 461–467 (2008). https://doi.org/10.1002/lsm.20653
64. Kong, K., Kendall, C., Stone, N., Notingher, I.: Raman spectroscopy for medical diagnostics—from *in-vitro* biofluid assays to *in vivo* cancer detection. Adv. Drug Deliv. Rev. **89**, 121–134 (2015). https://doi.org/10.1016/j.addr.2015.03.009
65. Alfonso-Garcia, A., Mittal, R., Lee, E.S., Potma, E.O.: Biological imaging with coherent Raman scattering microscopy: a tutorial. J. Biomed. Opt. **19**, 71407 (2014). https://doi.org/10.1117/1.JBO.19.7.071407

66. Freudiger, C.W., et al.: Label-free biomedical imaging with high sensitivity by stimulated Raman scattering microscopy. Science. **322**(5909), 1857–1861 (2008). https://doi.org/10.1126/science.1165758
67. Drutis, D.M., Hancewicz, T.M., Pashkovski, E., Feng, L., Mihalov, D., Holtom, G., Ananthapadmanabhan, K.P., Xie, X.S., Misra, M.: Three-dimensional chemical imaging of skin using stimulated Raman scattering microscopy. J. Biomed. Opt. **19**(11), 111604 (2014). https://doi.org/10.1117/1.JBO.19.11.111604
68. Wang, H., Osseiran, S., Igras, V., Nichols, A.J., Roider, E.M., Pruessner, J., Tsao, H., Fisher, D.E., Evans, C.L.: *In vivo* coherent Raman imaging of the melanomagenesis-associated pigment pheomelanin. Sci. Rep. **6**, 37986 (2016). https://doi.org/10.1038/srep37986
69. Alfano, R.R., Tata, D., Cordero, J., Tomashefsky, P., Longo, F.W., Alfano, M.A.: Laser induced fluorescence spectroscopy from native cancerous and normal tissues. IEEE J. Quant. Electron. **20**(12), 1507–1511 (1984). https://doi.org/10.1109/JQE.1984.1072322
70. Mayne, S.T., et al.: Resonance Raman spectroscopic evaluation of skin carotenoids as a biomarker of carotenoid status for human studies. Arch. Biochem. Biophys. **539**(2), 163–170 (2013). https://doi.org/10.1016/j.abb.2013.06.007
71. Darvin, M.E., Gersonde, I., Albrecht, H., Meinke, M., Sterry, W., Lademann, J.: Non-invasive *in vivo* detection of the carotenoid antioxidant substance lycopene in the human skin using the resonance Raman spectroscopy. Laser Phys. Lett. **3**(9), 460–463 (2006). https://doi.org/10.1002/lapl.200610032
72. Liu, C., et al.: A pilot study for distinguishing basal cell carcinoma from normal human skin tissues using visible resonance Raman spectroscopy. J. Cancer Metastasis Treat. **5**, 4 (2019). https://doi.org/10.20517/2394-4722.2018.55
73. Schlücker, S.: Surface-enhanced Raman spectroscopy: concepts and chemical applications. Angew. Chem. Int. Ed. Engl. **53**(19), 4756–4795 (2014). https://doi.org/10.1002/anie.201205748
74. Shanthil, M., Thomas, R., Swathi, R.S., George Thomas, K.: Ag@SiO2 core–shell nanostructures: distance-dependent plasmon coupling and SERS investigation. J. Phys. Chem. Lett. **3**(11), 1459–1464 (2012). https://doi.org/10.1021/jz3004014
75. Xie, J., Zhang, Q., Lee, J.Y., Wang, D.I.C.: The synthesis of SERS-active gold nanoflower tags for *in vivo* applications. ACS Nano. **2**(12), 2473–2480 (2008). https://doi.org/10.1021/nn800442q
76. Yuan, H., Khoury, C.G., Hwang, H., Wilson, C.M., Grant, G.A., Vo-Dinh, T.: Gold nanostars: surfactant-free synthesis, 3D modelling, and two-photon photoluminescence imaging. Nanotechnology. **23**(7), 075102 (2012). https://doi.org/10.1088/0957-4484/23/7/075102
77. Kang, H., Jeong, S., Park, Y., Yim, J., Jun, B.H., Kyeong, S., Lee, Y.S.: Near-infrared SERS nanoprobes with plasmonic Au/Ag hollow-shell assemblies for *in vivo* multiplex detection. Adv. Funct. Mater. **23**(30), 3719–3727 (2013). https://doi.org/10.1002/adfm.201203726
78. Stuart, D.A., Yuen, J.M., Shah, N., Lyandres, O., Yonzon, C.R., Glucksberg, M.R., Van Duyne, R.P.: *In vivo* glucose measurement by surface-enhanced Raman spectroscopy. Anal. Chem. **78**(20), 7211–7215 (2006). https://doi.org/10.1021/ac061238u
79. Qian, X., Peng, X.H., Ansari, D.O., Yin-Goen, Q., Chen, G.Z., Shin, D.M., Nie, S.: *In vivo* tumor targeting and spectroscopic detection with surface-enhanced Raman nanoparticle tags. Nat. Biotechnol. **26**(1), 83–90 (2007). https://doi.org/10.1038/nbt1377
80. Cinta Pinzaru, S., Falamas, A., Dehelean, C.A.: Molecular conformation changes along the malignancy revealed by optical nanosensors. J. Cell. Mol. Med. **17**(2), 277–286 (2013). https://doi.org/10.1111/jcmm.12006
81. Falamas, A., Dehelean, C.A., Cinta Pinzaru, S.: Monitoring of betulin nanoemulsion treatment and molecular changes in mouse skin cancer using surface enhanced Raman spectroscopy. Vib. Spectrosc. **95**, 44–50 (2018). https://doi.org/10.1016/j.vibspec.2018.01.004
82. Macleod, N.A., Goodship, A., Parker, A.W., Matousek, P.: Prediction of sublayer depth in turbid media using spatially offset Raman spectroscopy. Anal. Chem. **80**, 8146–8152 (2008). https://doi.org/10.1021/ac801219a

83. Keller, M.D., Vargis, E., Vargis, E., de Matos Granja, N., Mahadevan-Jansen, A.: Development of a spatially offset Raman spectroscopy probe for breast tumor surgical margin evaluation. J. Biomed. Opt. **16**(7), 077006 (2011). https://doi.org/10.1117/1.3600708
84. Schulmerich, M.V., Cole, J.H., Dooley, K.A., Morris, M.D., Kreider, J.M., Goldstein, S.A., Srinivasan, S., Pogue, B.W.: Noninvasive Raman tomographic imaging of canine bone tissue. J. Biomed. Opt. **13**(2), 0205062008. https://doi.org/10.1117/1.2904940
85. Sowoidnich, K., Churchwell, J.H., Buckley, K., Goodship, A.E., Parker, A.W., Matousek, P.: Spatially offset Raman spectroscopy for photon migration studies in bones with different mineralization levels. Analyst. **142**(17), 3219–3226 (2017). https://doi.org/10.1039/c7an00408g
86. Matousek, P., Stone, N.: Development of deep subsurface Raman spectroscopy for medical diagnosis and disease monitoring. Chem. Soc. Rev. **45**, 1794–1802 (2016). https://doi.org/10.1039/C5CS00466G
87. Yuen, J.M., et al.: Transcutaneous glucose sensing by surface-enhanced spatially offset Raman spectroscopy in a rat model. Anal. Chem. **82**(20), 8382–8385 (2010). https://doi.org/10.1021/ac101951j
88. Wang, W., Zhao, J., Short, M., Zeng, H.: Real-time *in vivo* cancer diagnosis using Raman spectroscopy. J. Biophotonics. **8**(7), 527–545 (2015). https://doi.org/10.1002/jbio.201400026
89. Lui, H., Zhao, J., McLean, D., Zeng, H.: Real-time Raman spectroscopy for *in vivo* skin cancer diagnosis. Cancer Res. **72**(10), 2491–2500 (2012). https://doi.org/10.1158/0008-5472.can-11-4061
90. Zhao, J., Zeng, H., Kalia, S., Lui, H.: Wavenumber selection based analysis in Raman spectroscopy improves skin cancer diagnostic specificity. Analyst. **141**(3), 1034–1043 (2016). https://doi.org/10.1039/c5an02073e
91. Schleusener, J., Gluszczynska, P., Reble, C., Gersonde, I., Helfmann, J., Fluhr, J.W., Meinke, M.C.: *In vivo* study for the discrimination of cancerous and normal skin using fiber probe-based Raman spectroscopy. Exp. Dermatol. **24**(10), 767–772 (2015). https://doi.org/10.1111/exd.12768
92. Jermyn, M., Desroches, J., Aubertin, K., St-Arnaud, K., Madore, W.-J., Montigny, E.D., Guiot, M.-C., Trudel, D., Wilson, B.C., Petrecca, K., Leblond, F.: A review of Raman spectroscopy advances with an emphasis on clinical translation challenges in oncology. Phys. Med. Biol. **61**(23), R370–R400 (2016). https://doi.org/10.1088/0031-9155/61/23/r370
93. Bratchenko, I.A., Artemyev, D.N., Myakinin, O.O., Khristoforova, Y.A., Moryatov, A.A., Kozlov, S.V., Zakharov, V.P.: Combined Raman and autofluorescence *ex vivo* diagnostics of skin cancer in near-infrared and visible regions. J. Biomed. Opt. **22**, 027005 (2017). https://doi.org/10.1117/1.JBO.22.2.027005
94. Silveira, L., Silveira, F.L., Bodanese, B., Zangaro, R.A., Pacheco, M.T.T.: Discriminating model for diagnosis of basal cell carcinoma and melanoma *in vitro* based on the Raman spectra of selected biochemicals. J. Biomed. Opt. **17**(7), 077003 (2012). https://doi.org/10.1117/1.jbo.17.7.077003
95. Mahadevan-Jansen, A., Richards-Kortum, R.: Raman spectroscopy for the detection of cancers and precancers. J. Biomed. Opt. **1**(1), 31 (1996). https://doi.org/10.1117/12.227815
96. Feng, X., Moy, A.J., Nguyen, H.T.M., Zhang, Y., Zhang, J., Fox, M.C., Sebastian, K.R., Reichenberg, J.S., Markey, M.K., Tunnell, J.W.: Raman biophysical markers in skin cancer diagnosis. J. Biomed. Opt. **23**, 1 (2018). https://doi.org/10.1117/1.JBO.23.5.057002
97. Tfayli, A., Piot, O., Derancourt, S., Cadiot, G., Diebold, M.D., Bernard, P., Manfait, M.: *In vivo* analysis of tissue by Raman microprobe: examination of human skin lesions and esophagus Barrett's mucosa on an animal model. Proc. SPIE. **6093**, 609312 (2006). https://doi.org/10.1117/12.645874
98. Argenziano, G., Cerroni, L., Zalaudek, I., Staibano, S., Hofmann-Wellenhof, R., Arpaia, N., Kittler, H.: Accuracy in melanoma detection: a 10-year multicenter survey. J. Am. Acad. Dermatol. **67**(1), 54–59 (2012). https://doi.org/10.1016/j.jaad.2011.07.019

99. Huang, Z., Lui, H., Chen, X.K., Alajlan, A., McLean, D.I., Zeng, H.: Raman spectroscopy of *in vivo* cutaneous melanin. J. Biomed. Opt. **9**(6), 1198 (2004). https://doi.org/10.1117/1.1805553
100. Philipsen, P.A., Knudsen, L., Gniadecka, M., Ravnbak, M.H., Wulf, H.C.: Diagnosis of malignant melanoma and basal cell carcinoma by *in vivo* NIR-FT Raman spectroscopy is independent of skin pigmentation. Photochem. Photobiol. Sci. **12**(5), 770 (2013). https://doi.org/10.1039/c3pp25344a
101. Gniadecka, M., Philipsen, P.A., Wessel, S., Gniadecki, R., Wulf, H.C., Sigurdsson, S., Hansen, L.K.: Melanoma diagnosis by Raman spectroscopy and neural networks: structure alterations in proteins and lipids in intact cancer tissue. J. Invest. Dermatol. **122**(2), 443–449 (2004). https://doi.org/10.1046/j.0022-202x.2004.22208.x
102. Uzquiano, M.C., Prieto, V.G., Nash, J.W., Ivan, D.S., Gong, Y., Lazar, A.J.F., Diwan, A.H.: Metastatic basal cell carcinoma exhibits reduced actin expression. Mod. Pathol. **21**(5), 540–543 (2008). https://doi.org/10.1038/modpathol.3801051
103. Bodanese, B., Silveira, F.L., Zangaro, R.A., Pacheco, M.T.T., Pasqualucci, C.A., Silveira, L.: Discrimination of basal cell carcinoma and melanoma from normal skin biopsies *in vitro* through Raman spectroscopy and principal component analysis. Photomed. Laser Surg. **30**(7), 381–387 (2012). https://doi.org/10.1089/pho.2011.3191
104. Santos, I.P., Caspers, P.J., Bakker Schut, T.C., van Doorn, R., Noordhoek Hegt, V., Koljenovic, S., Puppels, G.J.: Raman spectroscopic characterization of melanoma and benign melanocytic lesions suspected of melanoma using high-wavenumber Raman spectroscopy. Anal. Chem. **88**(15), 7683–7688 (2016). https://doi.org/10.1021/acs.analchem.6b01592
105. Zakharov, V.P., Bratchenko, I.A., Artemyev, D.N., Myakinin, O.O., Kornilin, D.V., Kozlov, S.V., Moryatov, A.A.: Comparative analysis of combined spectral and optical tomography methods for detection of skin and lung cancers. J. Biomed. Opt. **20**(2), 025003 (2015). https://doi.org/10.1117/1.jbo.20.2.025003
106. Kong, K., Rowlands, C.J., Varma, S., Perkins, W., Leach, I.H., Koloydenko, A.A., Notingher, I.: Increasing the speed of tumour diagnosis during surgery with selective scanning Raman microscopy. J. Mol. Struct. **1073**, 58–65 (2014). https://doi.org/10.1016/j.molstruc.2014.03.065
107. Heuke, S., Vogler, N., Meyer, T., Akimov, D., Kluschke, F., Röwert-Huber, H.J., Popp, J.: Detection and discrimination of non-melanoma skin cancer by multimodal imaging. Healthcare. **1**(1), 64–83 (2013). https://doi.org/10.3390/healthcare1010064
108. Schleusener, J., Gluszczynska, P., Reble, C., Gersonde, I., Helfmann, J., Fluhr, J.W., Lademann, J., Röwert-Huber, J., Patzelt, A., Meinke, M.C.: *In vivo* study for the discrimination of cancerous and normal skin using fibre probe-based Raman spectroscopy. Exp. Dermatol. **24**, 767 (2015). https://doi.org/10.1111/exd.12768
109. Zhao, J., Lui, H., Kalia, S., Zeng, H.: Real-time Raman spectroscopy for automatic *in vivo* skin cancer detection: an independent validation. Anal. Bioanal. Chem. **407**(27), 8373 (2015). https://doi.org/10.1007/s00216-015-8914-9
110. Pence, I.J., Patil, C.A., Lieber, C.A., Mahadevan-Jansen, A.: Discrimination of liver malignancies with 1064 nm dispersive Raman spectroscopy. Biomed. Opt. Express. **6**(8), 2724 (2015). https://doi.org/10.1364/boe.6.002724

Part IV
Multimodal Cancer Imaging

Chapter 11
Multimodal Optical Diagnostic in Minimally Invasive Surgery

Elena Potapova, Viktor Dremin, Evgeny Zherebtsov, Andrian Mamoshin, and Andrey Dunaev

11.1 *In Vivo* Multimodal Optical Diagnostic in Abdominal Surgery

Over the past decades, new supplementary techniques for video endosurgical, transcutaneous, minimally invasive, and endovascular interventions have been successfully applied in clinical surgery. In 1987 E.A. Wickham coined the term "minimally invasive procedure" [1], which in many aspects had cornerstoned the dynamical development of that research area [2]. Minimally invasive surgery techniques (MIS) offer ultimate benefits significantly reducing tissue trauma due to minimization and even elimination of surgical cuts, as well as shortening patient's recovery time and lowering the overall treatment costs. One of the key technologies for MIS is the use of remote handling instruments with channels for optical visualization of the operation field. Such revolutionary techniques as natural orifice transluminal endoscopic surgery (NOTES) and single-incision laparoscopic surgery (SILS) can be implemented [2, 3] with the introduction of the remote handling instruments.

The common approach of feedback for a surgeon during the minimally invasive intervention is the use of video image of the operation field [4]. In most cases the peculiarities of the image rotation and data transmission hamper visual perception and adequate judging of the tissue condition [5, 6]. SILS is carried out through a

E. Potapova · A. Dunaev (✉)
Orel State University, Orel, Russian Federation
e-mail: dunaev@bmecenter.ru

V. Dremin · E. Zherebtsov
Orel State University, Orel, Russian Federation

University of Oulu, Oulu, Finland

A. Mamoshin
Orel State University, Orel, Russian Federation

Orel Regional Clinical Hospital, Orel, Russian Federation

V. V. Tuchin et al. (eds.), *Multimodal Optical Diagnostics of Cancer*,
https://doi.org/10.1007/978-3-030-44594-2_11

small incision, allowing surgical instruments to penetrate into the abdominal cavity with synchronous imaging of surgical field. NOTES is an emerging technology within MIS which gives the surgeon feasibility to access the peritoneal cavity via a hollow viscus for diagnostic and therapeutic procedures. Standard laparoscopic and endoscopic instruments provide visualization of the operation field by the white light illumination with the collection of the reflected light through the lens system and capture of the color video frames in real time. White light imaging allows the surgeon to distinguish anatomical structures and evaluate its most significant pathological changes. Modern MIS techniques feature the use of new visualization technologies to establish real-time feedback to help the surgeon with procedure optimization [7]. The application of advanced optical methods in MIS can be a truly game-changing technology for diagnostics and treatment of many types of pathologies while sometimes not too difficult for immediate implementation becoming mainstream research trend.

Detection of subtle lesions at the beginning stage for intraoperative diagnostics is usually a task of great difficulty. The detection of flat adenoma in the colon, as well as early stages of esophagus cancers, requires increased contrast and resolution for the imaging system. Recent advancements in molecular biology and development of new light sources, compact spectral imaging systems have opened new possibilities for effective instant recognition of malignant tissue during endoscopy procedure without conventional biopsy or cytology analysis. The optical biopsy methods implemented during an endoscopic procedure allow better contrasting and resolution of the image and help to obtain biologically and biochemically related information about the mucosa [8, 9]. Leading core technologies in cancer detection are fluorescence endoscopy, optical coherence tomography, confocal microendoscopy, and Raman spectroscopy. Each method has its own advantages and disadvantages but being combined together they can supplement each other. Such multimodal approach benefits better sensitivity and specificity rather than each method being applied separately.

In recent years a significant number of papers have been dedicated to the studies with the multimodal diagnostics during MIS, in particular the clinical diagnosis of abdominal cancer. Nevertheless, the tasks of microvascular network mapping as well as acquiring information about the molecular characteristics of the tumor are still challenging.

In this chapter, we would consider only those MIS instrumental systems that have passed at least the preclinical stage of testing in animal tumor models or have already been introduced into the clinical practice of abdominal surgery. The use of multimodal device for united analysis of white light, bioluminescence, fluorescence imaging, narrow-band light reflection, and Raman scattering allows one not only to obtain macroscopic images of the region of interest but also to visualize cellular and subcellular tissue features with microscopic resolution [10]. In such approach one could initially quickly inspect a large area of tissue for detection of abnormal areas with high diagnostic sensitivity, and then more scrupulously monitor these suspicious areas using high-resolution imaging technologies to eliminate interfering factors, improving the overall specificity of cancer detection [11].

One of the first optical multimodal devices combined methods of fluorescence, reflective, and light scattering spectroscopy with standard endoscopic instrument for evaluation of dysplasia in patients with Barrett's esophagus [12]. In this work, it was demonstrated how the endoscopic examination could use in clinical conditions the combination of fluorescence and reflective spectroscopy for obtaining additional information about the biochemical and morphological state of the tissue and the corresponding changes that occur during the progression of dysplasia. Fluorescence spectroscopy is based on the registration of laser-induced fluorescence from natural fluorophores such as FADH, NAD, tryptophan, collagen, and porphyrins and serves as a sensitive indicator of the biochemical composition of the tissue. Light reflection and scattering spectroscopy provide morphological information about tissue architecture and nuclei of epithelial cells. The scheme of a multimodal endoscopic installation for recording the spectral characteristics of esophagus tissues through the instrument channel of the endoscope is shown in Fig. 11.1.

The described multimodal approach made it possible to achieve a high level of sensitivity and specificity in the classification of nondysplastic, low-grade and high-grade dysplastic tissue with Barrett's esophagus. This is confirmed by clinical studies, which showed that simultaneous analysis of the results obtained by fluorescence, reflective, and light scattering spectroscopy can be used to differentiate high-grade dysplasia from low-grade and nondysplastic sites with high levels of sensitivity and

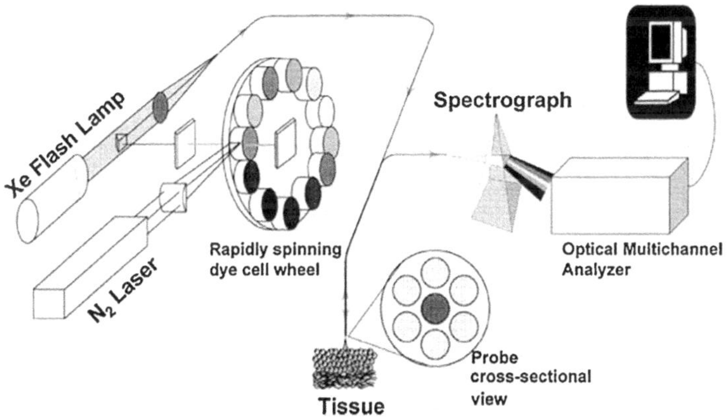

Fig. 11.1 Experimental scheme. During endoscopy, the probe was inserted into the accessory channel of the endoscope and brought into gentle contact with the tissue, thus providing a fixed delivery-collection geometry. The reflected and fluorescence light was collected by the probe and coupled to a spectrograph and detector. The average of 3 sets of spectra from each site was used for analysis. Immediately after the spectral acquisition, the probe was removed, and a slight demarcation remained on the tissue for 30–60 s as a result of the probe contact. This endoscopically apparent marker was used as a guide for taking a biopsy specimen from the same site at which spectra were acquired. The biopsy specimen was interpreted and classified by an experienced gastrointestinal pathologist. If a dysplastic lesion was suspected, the specimen was reviewed and the diagnosis was confirmed by a second gastrointestinal pathologist, following the standard of care (reprinted with permission from [12])

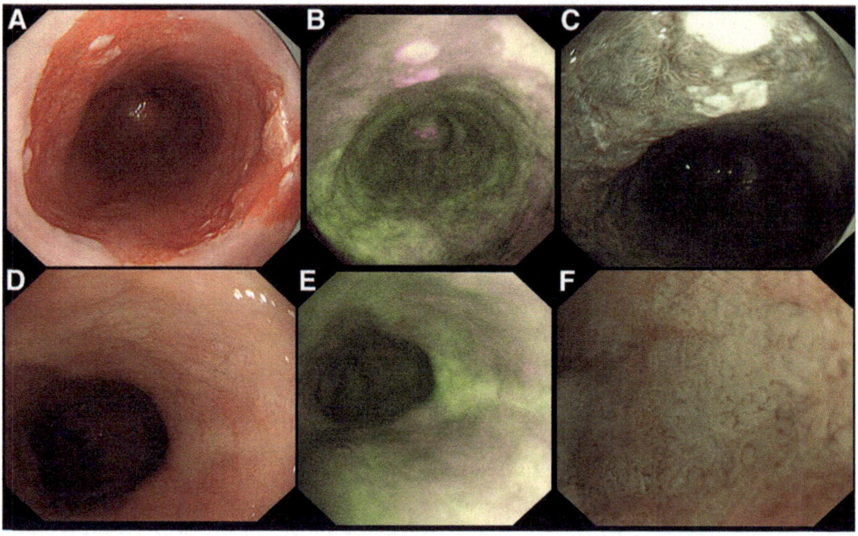

Fig. 11.2 (**A–C**) A lesion containing high-grade dysplasia and early carcinoma is depicted at the 11 o'clock position. (**A**) A more prominent lesion located at the 5 o'clock position is shown. Both lesions were detected with HRE and AFI. NBI showed irregular mucosal and vascular patterns suspicious for dysplasia. (**D** and **E**) Two lesions with high-grade dysplasia/carcinoma located at the 2 and the 5 o'clock positions are shown. These lesions were missed during HRE but were subsequently detected with AFI (**E**). NBI (**F**) showed irregular mucosal and vascular patterns and abnormal blood vessels suspicious for dysplasia (reprinted with permission from [14])

specificity (100%). Additionally, dysplastic and nondysplastic epithelia can be distinguished with very high sensitivity and specificity (93% and 100%, respectively).

Bergman with co-researchers [13, 14] demonstrated one of the first high-resolution multimodal endoscopic imaging systems integrated into a single device for intraoperative diagnostics of abdominal organs, which allows obtaining autofluorescence imaging and narrow-band imaging to detect early neoplasia in patients with Barrett's esophagus. These studies showed that the unification of high-resolution endoscopy (HRE), magnifying endoscopy, autofluorescence imaging (AFI), and narrow-band imaging (NBI) in a single device can significantly increase the sensitivity of the device and reduce false positives. High-resolution endoscopy provides clear imaging of small details that can facilitate the detection of early tumor lesions. Moreover, autofluorescence imaging may be used for determining high-grade intraepithelial neoplasia as areas with reduced autofluorescence have an increased tendency to neoplasm [15]. Autofluorescence imaging is used in a browse mode to draw attention to potentially tumor areas, which are subsequently examined more closely by narrow-band imaging. Narrow-band imaging is a term applied to a specific commercial imaging technique (Olympus America, Center Valley, Pennsylvania) that uses two specific backlight wavelengths. Blue illumination at 414 nm allows for obtaining images of surface capillaries, and green illumination at 540 nm allows for visualization of vessels at a greater depth [16]. Figure 11.2 shows

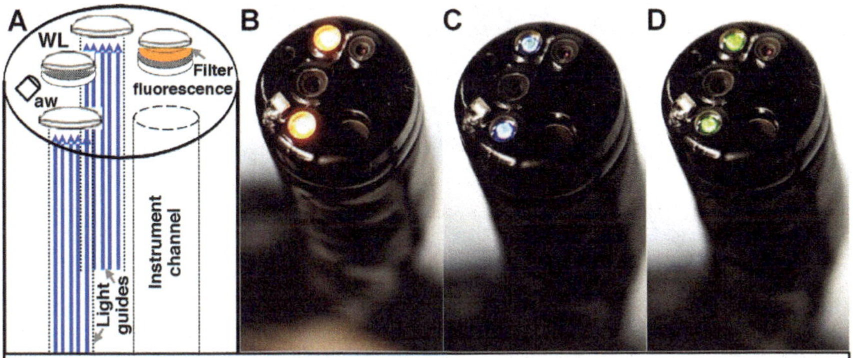

E	White light	Fluorescence	Reflectance
Illumination (nm)	400–700	450–490	540–560
Collection (nm)	400–700	520–600	540–560
Imaging speed (frames/sec)	20	10	10
Divergence angle (deg)	140	140	140
Depth-of-field (mm)	7–100	5–100	5–100

Fig. 11.3 Multimodal video colonoscope: (**A**) Schematic of prototype instrument. Separate standard definition detectors are used to collect either (**B**) white light (WL) or (**C**) fluorescence images. A filter located in between the fluorescence objective and detector passes light from FITC. (**D**) Reflectance images collected with the same detector are co-registered with fluorescence. (**E**) Summary of optical parameters. White light and fluorescence/reflectance images are collected at 20 and 10 frames/s, respectively. aw—air/water nozzle (reprinted with permission from [19])

typical example of application of the described trimodal approach for diagnosis of intraepithelial neoplasia in Barrett's esophagus. It is clearly visible that autofluorescence imaging helps to visualize lesions which were missed during high-resolution white light endoscopy. At the same time, the narrow-band imaging highlights areas of irregular, disturbed mucosal pattern and abundant small abnormal blood vessels.

Later, the effectiveness of the use of the trimodal endoscopy was also shown for the diagnosis and contrast imaging of other tumors, in particular, early neoplasia of the stomach [17], as well as for the detection and differentiation of colon polyps [18].

The multimodal approach may also enhance the sensitivity of endoscopy neoplastic lesions mapping of colorectal cancer (CRC) for subsequent accurate removal. Joshi et al. [19] have developed a multimodal video colonoscope for targeted wide-field detection of non-polypoid colorectal neoplasia. The multimodal video colonoscope (CF-Y0012, Olympus) contains two separate channels for collecting white light and fluorescence images, as shown in Fig. 11.3.

Fluorescence microscopy provides microscopic images using fluorescent dyes at the subcellular level. Diffuse reflection spectroscopy examines the scattering of light from the surface of the mucous membrane and is able to differentiate between normal and tumor tissues due to increased scattering of light from cells that are less organized and have larger nuclei. To obtain fluorescence images, fluorescein

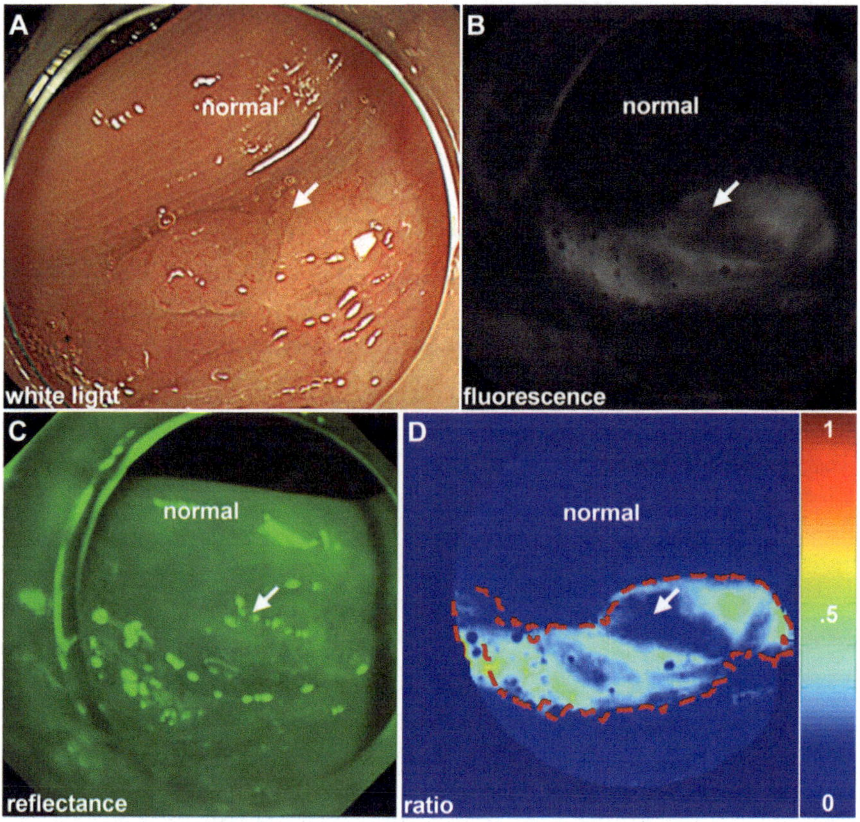

Fig. 11.4 (**A**) A cap is placed on the distal end of the multimodal video colonoscope to flatten mucosal folds for improved visualization. On white light, a non-polypoid lesion (arrow) is seen in the proximal colon that could easily be missed. (**B**) After topical peptide administration, the lesion is clearly visualized with fluorescence. Normal colon shows minimal background. (**C**, **D**) The registered reflectance image was used to produce a ratio-image, shown in pseudo-color, that corrects for distance and geometry. Intensities can be quantified to outline the extent of the lesion (dashed red line) and guide resection (reprinted with permission from [19])

isothiocyanate is injected locally into the mucosa of the proximal colon in concentration through a nebulization catheter passed through the instrument channel.

Processing co-registered images of fluorescence and reflection allows to determine the "red flag" region (dashed red line), which may be shown on the screen as an overlay on the white light image (Fig. 11.4) to assist the physician in performing the biopsy.

Improving the accuracy of endoscopic diagnosis of CRC can be achieved using Raman spectroscopy along with fluorescence imaging [20]. For these purposes, Jeong et al. [21] have developed a dual modal fluorescence-Raman endomicroscopic system (see Fig. 11.5), which used fluorescence and surface-enhanced Raman scattering nanoprobes (F-SERS dots). Fluorescence-Raman endoscopy

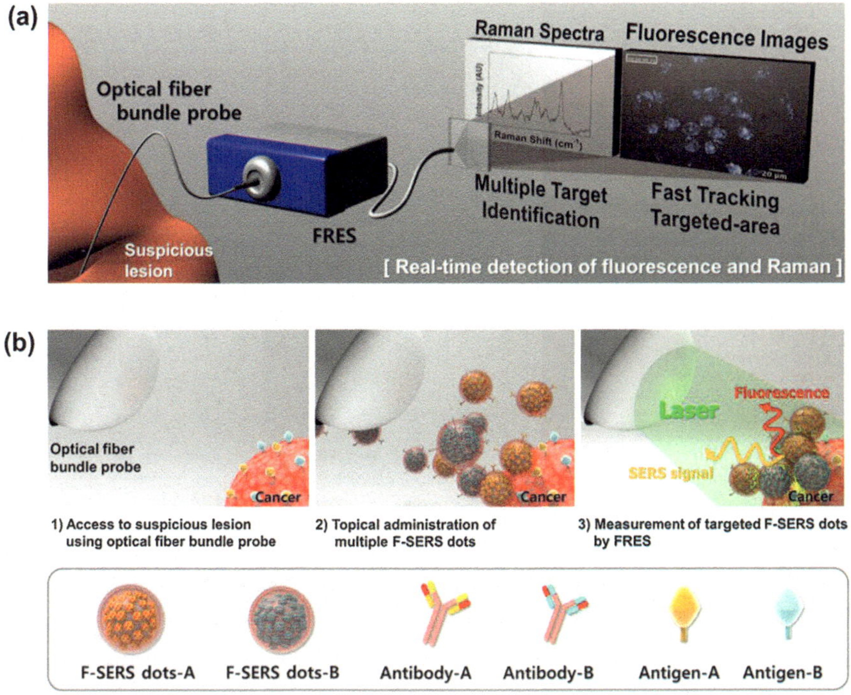

Fig. 11.5 Schematic illustration of real-time multiplexed imaging using the fluorescence-Raman endoscopic system (FRES). (**a**) The mode of dual modal detection with fluorescence and Raman scattering. The real-time fluorescence imaging tracks the locations of the probe-targeted areas, and a concurrent SERS spectral analysis identifies the species of targets. (**b**) Illustration of the *in vivo* multiplexed molecular imaging procedure: First, access to a suspicious lesion via optical images; second, the spray-and-mix multiple F-SERS dots for topical administration; third, a multiplexed measurement of the targeted F-SERS dots with FRES (reprinted with permission from [21])

showed high sensitivity in an orthotopic CRC xenograft model using antibody-conjugated F-SERS dots (Fig. 11.6). Moreover, fluorescence recording was useful for quickly tracking lesions in a large field of view, and Raman signals were used for multiple target identification (both tumor cells and tumor microenvironment) and quantification.

Miller et al. [22] have developed a flexible multispectral scanning system for fiber-optic endoscopy. The system has demonstrated better characteristics in the detection of CRC at early stages (Fig. 11.7).

The suggested methodology integrates the narrow bandwidth visualization, chromoendoscopy, and autofluorescence visualization. The use of chromoendoscopy with nonspecific exogenous dyes allows one to improve the contrast for the visualization of the microanatomical malformations in mucosa. The developed fiber-optic system for endoscopy employs helical scanning by red, green, and blue wavelength (440 nm, 532 nm, 635 nm) and collection of the backscattered radiation by a central ring of fibers. The rotation of the scanning tool allows the system to

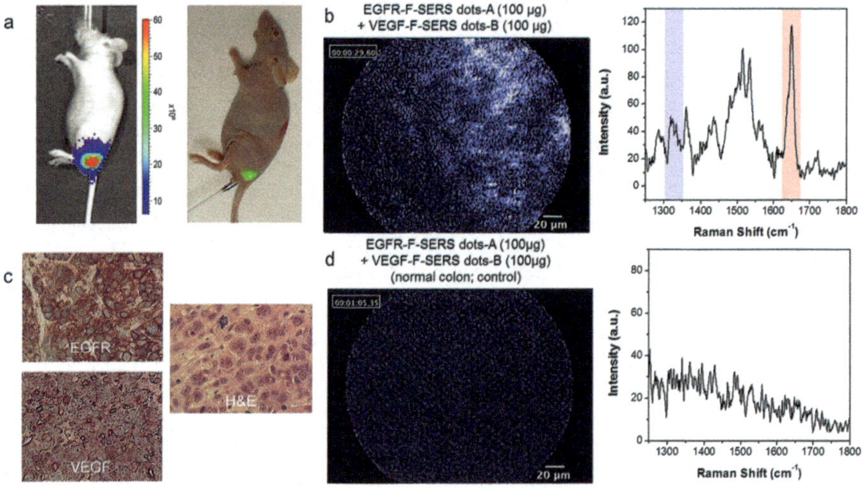

Fig. 11.6 Evaluation of FRES in a real-time endoscopic system. (**a**) Orthotopic CRC xenograft modeling. One week after injecting 1 × 107 HT29-effluc cells, bioluminescence imaging (left) showed moderate-to-high activity in the colorectal area of mice (lateral view; $n = 20$). The FRES imaging method used in a real-time endoscopic system is shown (right). (**b**) Mice with CRC (two weeks after injecting 1 × 107 of HT29-effluc cells) were investigated in a real-time endoscopic study of FRES. After spraying 100 μg each of EGFR-F-SERS-A and VEGF-F-SERS-B dots, fluorescence signals were found as well as two corresponding Raman signals [RITC (-A) and FITC (-B)]. (**c**) Tumors were excised, fixed, and sectioned. EGFR and VEGF positivity was identified by immunohistochemistry (IHC), and tumor cell infiltration was observed by hematoxylin and eosin (H&E) staining. (**d**) After spraying 100 μg each of EGFR-F-SERS-A and VEGF-F-SERS-B dots on normal colons (control), no fluorescence or Raman signals were detected (reprinted with permission from [20])

register 2D maps of the visualized tissue surface. An important part of the technique is the use of highly specific dye proteins binding to the sites of CRC dysplasia. Together with the described scanning fiber-optic endoscope, the dyeing method has shown highly promising results in early detection of tumor with prospects of clinical applications.

For the early CRC cancer screening also a combined use of the endoscopic OCT and fluorescence visualization has been reported in a series of publications [23–27].

The combination of the OCT and fluorescence imaging channels simultaneously provides information on fine morphological structure of the tissues as well as on its molecular characteristics. Mapping of the microvascular structure of the tumor *in vivo* is a quite challenging task. Suggested by Yan Li et al. [27] system with endoscopic OCT and near-infrared (NIR) fluorescence imaging for mapping of the microvasculature is an example of a diagnostic system (Fig. 11.8), which provides a combination of imaging with high frame rate, good spatial resolution and uniformity of the scanning system while using the FDA approved dye.

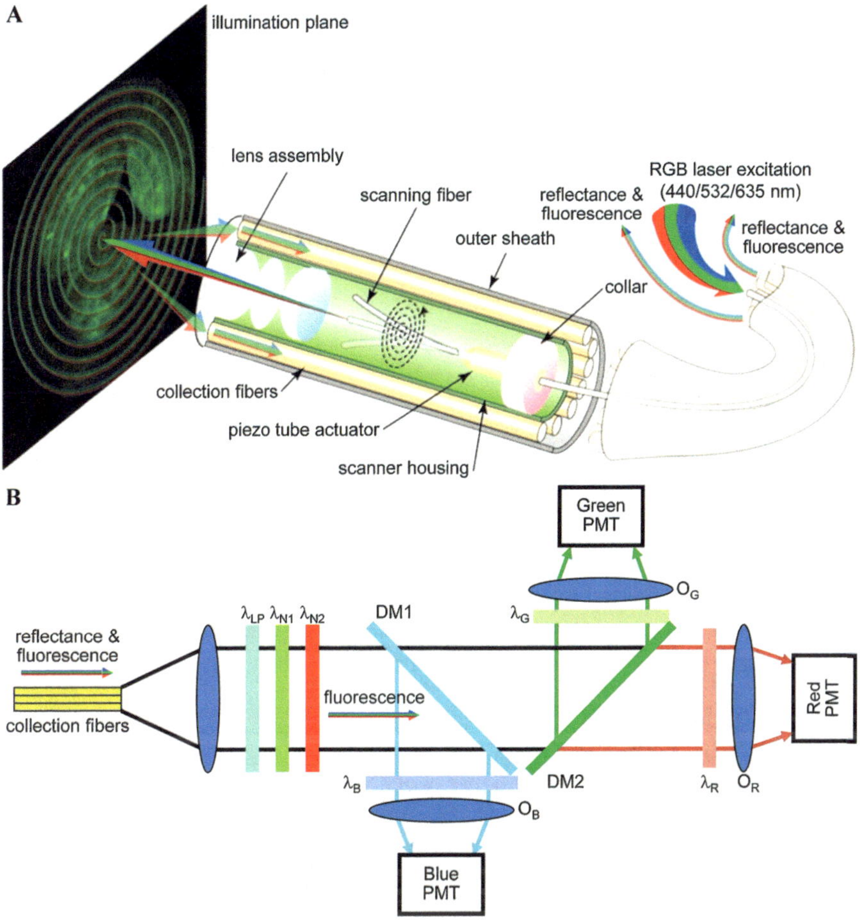

Fig. 11.7 Multispectral scanning fiber endoscope. (**A**) Optical design. RGB laser excitation (440, 532, and 635 nm) is delivered into a single-mode optical fiber that is scanned in a spiral pattern by a piezo tube actuator and focused on to the tissue (illumination plane) by a lens assembly. Fluorescence is collected by a ring of 12 collection fibers mounted around the periphery of the scanner housing, protected by an outer sheath. (**B**) Fluorescence detection. Reflectance from RGB laser excitation is removed using a combination of longpass ($\lambda_{LP} = 450$ nm) and notch ($\lambda_{N1} = 532$ nm and $\lambda_{N2} = 632.8$ nm) filters. Fluorescence is deflected into individual RGB channels using dichroic mirrors DM1 ($\lambda_C = 460$ nm) and DM2 ($\lambda_C = 550$ nm) and an additive dichroic filter set (λ_R, λ_G, and λ_B) prior to detection with PMTs (reprinted with permission from [22])

An example of the combined OCT and NIR-fluorescence images obtained using the described system as well as reference histology slices of the same tissue sample is shown in Fig. 11.9.

In OCT image a mucosa bulging can be clearly identified as well as dark areas of necrosis absorbing bulk of the probing optical radiation. The fluorescence images of the corresponding areas of the wall swelling on the contrary demonstrate increased

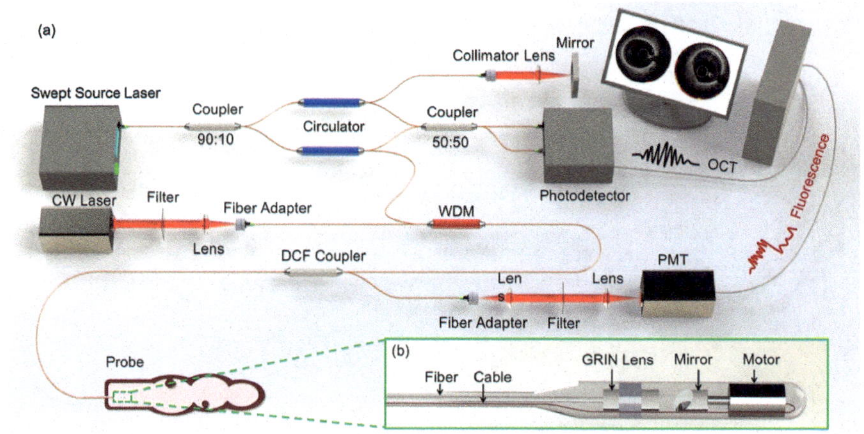

Fig. 11.8 (**a**) Overall design of endoscopic multimodality OCT and NIR-fluorescence system. (**b**) Multimodality imaging probe. WDM: wavelength division multiplexer. PMT: photomultiplier tube. DCF coupler: double clad fiber coupler. OCT: optical coherence tomography. CW: continuous wavelength. GRIN: gradient index (reprinted with permission from [27])

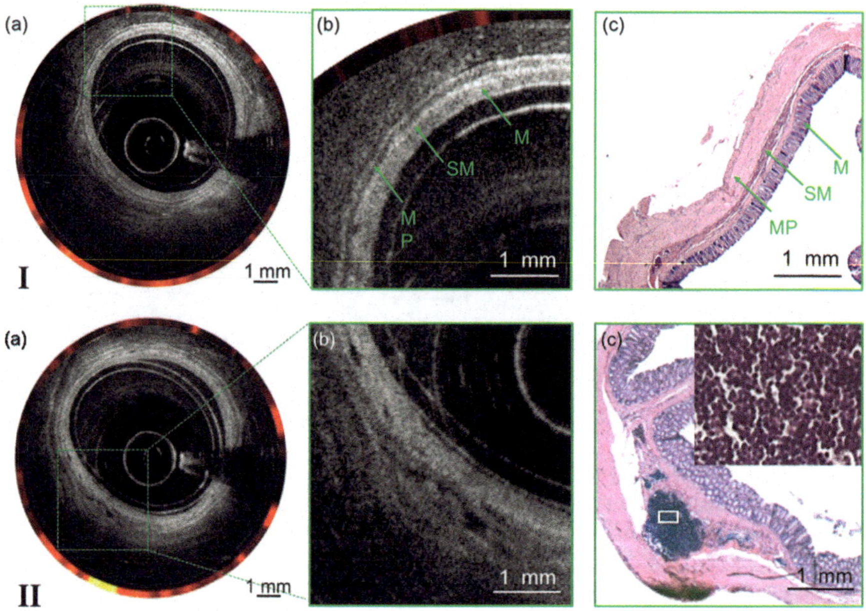

Fig. 11.9 Normal rectum (I) and adenocarcinoma (II): (**a**) The combined OCT and NIR-fluorescence image. (**b**) Enlarged view of the dashed box in (**a**). (**c**) Histology. M: mucosa; SM: submucosa; MP: muscularis propria (reprinted with permission from [27])

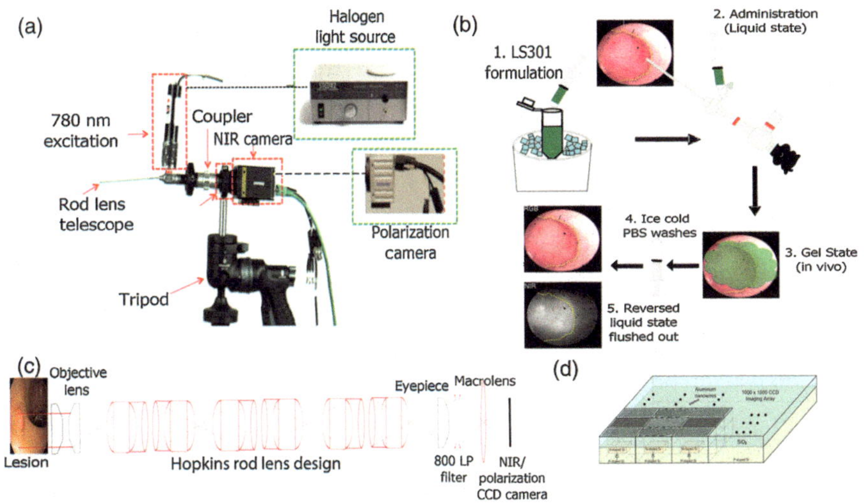

Fig. 11.10 Trimodal color-fluorescence-polarization endoscopy. (**a**) Endoscope setup used to perform white light, fluorescence, and polarization endoscopy. A telescope coupler with a focus ring was used to focus the image onto the image formation plane located at the back of the Karl Storz telescope lens. For white light endoscopy, Xenon Nova 175 was used as a light source and a charge-coupled device (CCD) camera (FluorVivo) with an RGB Bayer filter was used. In NIR-fluorescence mode, a 100-mW 780-nm laser excitation source (inset I) was coupled to the light source channel on the endoscope. An 800-nm long-pass emission filter was placed behind the telescope, as shown in the figure. A CCD camera with NIR capabilities (FluorVivo) was utilized to capture images in NIR-fluorescence mode. In polarization mode, the Xenon broadband light source was used along with a custom-made polarization camera, as shown in inset II. The entire setup was fixed on a tripod and an x-y translation stage was utilized for finer adjustments to avoid causing damage to the walls of the colon. (**b**) Steps involved in the dye administration method utilizing Pluronic F127 gel formulation. (**c**) Schematic of the light path associated with the endoscopy setup presented in (**a**). (**d**) Polarization sensor consisting of an imaging array arranged in blocks of four (two by two) superpixels. Each superpixel consists of four pixels that are composed of nanowire polarization filters with the transmission axis oriented at 0°, 45°, 90°, and 135°. These nanowires are 70-nm wide, 140-nm tall, and spaced 140 nm from center to center (reprinted with permission from [28])

intensity of the emission. The results point out that the multimodal approach is able to simultaneously visualize multilayer structure and microvascular network of colon tissue which is important for the identification and differentiation of healthy tissue, hyperplastic polypuses, and adenocarcinomas.

An interesting trimodal method for detection of flat lesions in colitis-associated cancer has been implemented by a color-fluorescence-polarization endoscopy system [28]. The introduced polarization channel allowed the setup to highlight almost non-transparent to the fluorescence excitation radiation areas and significantly increase the overall sensitivity of the delivering diagnostics. The schematic diagram of the system utilizing the RGB, fluorescent and polarization channels is shown in Fig. 11.10.

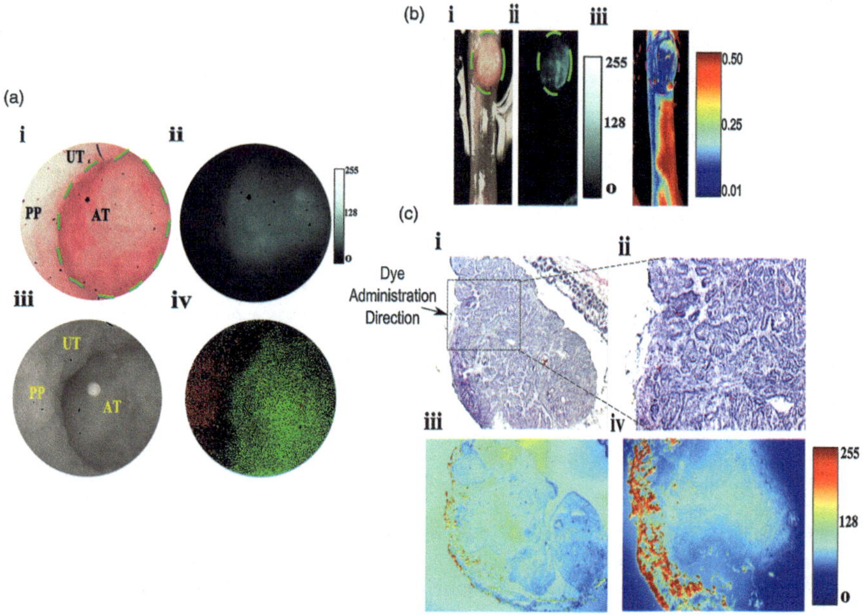

Fig. 11.11 *In vivo* and *in situ* fluorescence and polarization endoscopy of adenomatous tumor-surrounding uninvolved boundary. (**a**) *In vivo* fluorescence and polarization endoscopy. (i) RGB color image of the tumor identified in the distal colon of an AOM-DSS treated mouse. (ii) Fluorescence image clearly identifies boundaries of the tumor. (iii) Grayscale image consisting of degree of linear polarization (DoLP) and angle of polarization information in the polarization mode at 30 fps. Bright spot in image is from the light source. (iv) Thresholded DoLP mask image identifies inflamed regions (reported as DoLP value above 10%; red color) and cancerous regions (DoLP value under 5%; green color). (**b**) *Ex vivo* (i) RGB, (ii) NIR fluorescence, and (iii) polarization images of tumor identified in (**a**). Dotted yellow lines in (i) and (ii) indicate the boundary of the tumor. Thresholded DoLP mask image identifies cancerous regions (DoLP value under 10%; green color). (**c**) Histological validation of adenomatous tumor identified in (**b**), along with corresponding NIR-fluorescence intensity. Dotted rectangle identifies the region under investigation in (ii and iv). Arrow indicates the direction of molecular probe administration. UT: uninvolved tissue, PP: Peyer's patch, AT: adenomatous tumor (reprinted with permission from [28])

An example of the visualization of the adenomatous tumor-surrounding uninvolved boundary using the described system is presented in Fig. 11.11. The introduction of the polarization imaging in the multimodal system provided a much higher contrast of depolarization both in adenomatous tumors and flat lesions, reflecting the significantly impaired structural integrity of tissues under study. The implemented approach made it possible to isolate real-time areas of suspicious colon tissue using molecular fluorescence endoscopy and to scrutinize their condition using polarization imaging, providing high accuracy and specificity of the proposed method of trimodal diagnosis.

The combination of several optical diagnostic methods in a single tool makes it possible to visualize and differentiate normal and pathological tissues in the gastrointestinal tract supplementing the information obtained from a routine examination under white light. The effectiveness of multimodal endoscopes is confirmed by clinical studies demonstrating the significant increase in the sensitivity, accuracy, and specificity of the tumor diagnosis. It seems interesting to use a multimodal approach in laparoscopy for *in vivo* diagnosis of tumors. Currently there only a few investigations in this field. Although there are known publications on fluorescence endoscopy [29], speckle contrast laparoscopy [30, 31], laparoscopic optical coherence tomography imaging [32], applications of hyperspectral imaging during laparoscopic interventions [33, 34], laparoscopic narrow band imaging [35, 36]. Almost new area for multimodal optical diagnostics is the diagnostic intervention through percutaneous drainage channels [37].

In general, the use of combined different optical imaging methods in MIS instruments can improve the diagnosis of tissue lesions, the detection of tumor boundaries, and upgrade the overall success of the surgical intervention.

11.2 Multimodal Fiber-Optic Tool for Interventional Surgery

Interventional radiological procedures, developed in the late 1970s, represent a revolutionary advance in the field of minimally invasive diagnostics and treatment methods with the use of images [38]. This is due to minimal traumatic interventions, low surgical and postoperative pain, short period of hospitalization. Among the advantages of interventional radiological procedures, it should be noted that percutaneous access to the area of interest is carried out using thin atraumatic needles under the control of imaging techniques to determine the optimal safe route of the instrument and to protect large vessels and hollow organs from accidental puncture. The drainage channels are formed by the sequential moving of conical plastic dilators through a metal conductor, accompanied by minimal tissue damage and displacement of vital structures. In the long-term presence of drainage catheters, the drainage channel represents as an isolated connective tissue fistula that allows manipulations, including repeated diagnostic ones, and does not require surgical correction [39, 40]. Image-guided minimally invasive interventions are most commonly used in the histopathological diagnosis and therapy of tumors.

Percutaneous transhepatic biliary drainage is a therapeutic manipulation performed under the control of imaging techniques and involving bile duct cannulation followed by internal/external catheter drainage of bile contents. Percutaneous access to the bile ducts is performed using thin atraumatic needles followed by biliary tree contrast with percutaneous transhepatic cholangiography. After the formation of the drainage channel, a drainage catheter is installed using a sequential moving of conical plastic dilators through a metal conductor into the biliary tract. Percutaneous transhepatic biliary drainage is indicated for decompression of the biliary tract, most often in malignant obstruction, and is usually used for cases where retrograde endoscopic intervention is not effective or there is a high intrahepatic obstruction [41].

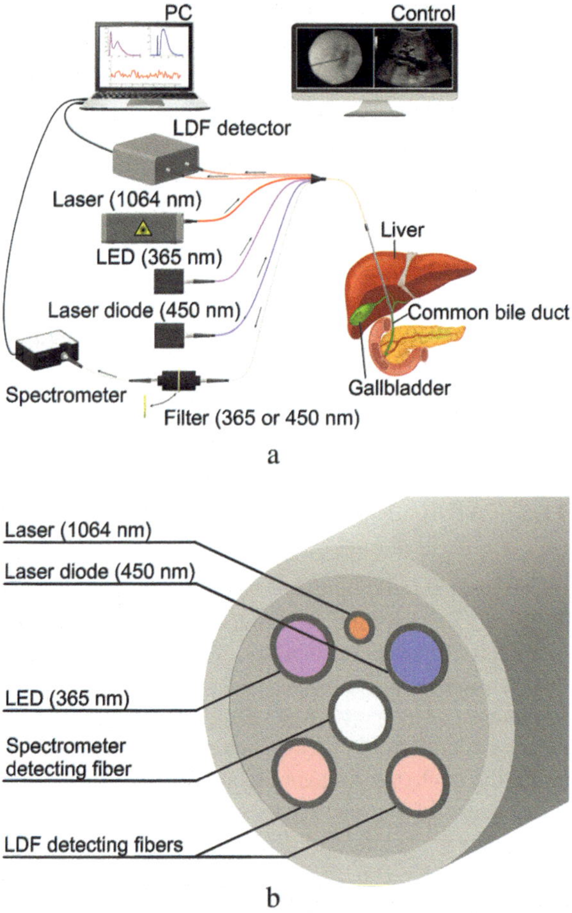

Fig. 11.12 Schematic presentation of installation for experimental studies (**a**) and location of fibers in the optical probe (**b**) [37]

Percutaneous drainage channels provide perfect access for further endoscopic or instrumental examination. A fiber-optic system with an optical probe which can be used for diagnostics during minimally invasive interventions has been developed in STC of Biomedical Photonics (Orel, Russia) [37]. The main units of a device were designed in cooperation with SPE "LAZMA" Ltd. (Moscow, Russia).

The multimodal approach was implemented by combining fluorescence spectroscopy (FS) and laser Doppler flowmetry (LDF) methods in a single fiber-optic probe to record the parameters of endogenous fluorescence and blood perfusion during minimally invasive interventions. The fiber-optic system (Fig. 11.12a) provides measurements of all optical parameters from approximately the same diagnostic volume of the tissue (1–3 mm^3).

A distinctive feature of the system is laparoscopic optical probe (length 30 cm, diameter 3 mm), designed for accessing organs under study through standard instruments for minimally invasive manipulations. The probe contains 6 optical fibers (Fig. 11.12b). A single-mode laser module with a radiation wavelength of 1064 nm was used in LDF channel, LED (365 nm) and laser diode (450 nm) were used for fluorescence excitation. The excitation wavelengths were agreed with the fluorescent properties of the main endogenous fluorophores, such as NADH, FAD, and stromal collagen. Namely, reduced NADH has a maximum of emission at approximately 490 nm under excitation with UV radiation (365 nm). Oxidized FAD fluoresces with peak of emission at approximately 520–540 nm under excitation radiation with wavelength of 450 nm [42].

The probing fiber of LDF channel has diameter of 6 μm, 2 receiving fibers have a diameter of 400 μm. The source-detector distance for LDF channel is 1.5 mm. The diameters of the probing and receiving fibers of FS channel are 400 μm. The numerical aperture of the fibers is 0.22. For safety reasons, as well as to minimize photobleaching effect, the radiation power for a wavelength of 365 nm at the fiber probe output does not exceed 1.5 mW. The output power for an excitation wavelength of 450 nm does not exceed 3.5 mW. The distance between the radiation source and receiver in FS channel is 1 mm. A spectrometer of 350–820 nm detection range is used to record the fluorescence spectra.

The experimental study involved 20 patients aged 67 ± 2 years diagnosed with obstructive jaundice caused by malignant tumors of hepatopancreatobiliary organs. All percutaneous minimally invasive interventions were performed in a specially equipped X-ray operating room. Percutaneous access to the area of interest was carried out under the control of imaging techniques using thin atraumatic puncture needles. The manipulation technique was standard. The position of the distal part of the needle in the area of interest was controlled by the position of the ultrasound mark of the needle tip and the nature of the discharge coming under pressure from the needle cannula or with active aspiration, as well as by the introduction of a contrast agent under fluoroscopic control. Dilatation of the puncture canal to the required diameter was performed using conical plastic dilators. Optical probe for the target parameters registration was set through the formed percutaneous canal of minimum diameter 8F to the area of interest. A drainage catheter was installed on the conductor after the procedure.

In the research process LDF and FS data was received at the area, which was blocked by the malignant tumor, and another area located higher than the first one. In the analysis of normalized spectra obtained at a wavelength of 365 nm, the results were divided into two groups. Half of them referred into the first type (Fig. 11.13a), the remaining spectra were considered as the second type (Fig. 11.13b). The first type spectra demonstrated higher fluorescence intensity in the non-compressed areas. By comparison, the results of the second group showed more intense fluorescence in the blocked area. LDF signals (Fig. 11.13c) were registered in the same areas during 1 min.

During *in vivo* studies, it was not possible to definitely determine whether the exophytic and intracranial growth of tumor had occurred or not. In a FS study, usually higher fluorescence intensities excited by UV radiation were observed in the intact area, rather than in the tumor [44–48].

Thus, we hypothesized that the spectra of the first type demonstrate tumor growth into the walls of the common bile duct. On the other hand, an increase in the

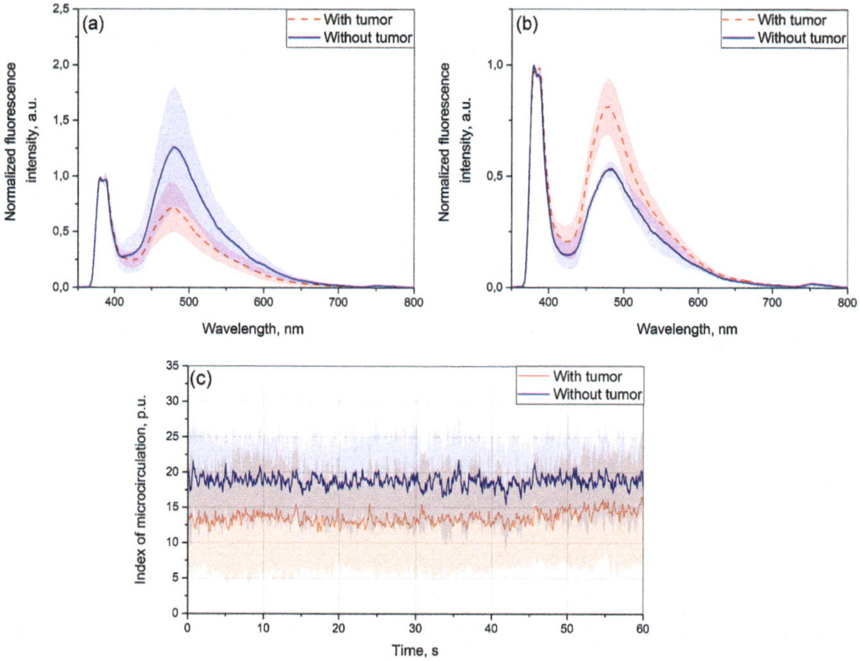

Fig. 11.13 The normalized averaged spectra at excitation wavelength 365 nm of first (**a**) and second (**b**) types, the averaged LDF signals (**c**) of the areas of interest (printed with permission from [43])

fluorescence intensity in the blocked area for the second group may indicate a decrease in blood supply caused by tumor obstruction outside the common bile duct. In addition, it is possible that the increased presence of collagen in the intercellular matrix of the muscle layer of the common bile duct wall or the presence of blood and bile in the studied area could also affect the recorded fluorescence spectra. The obtained LDF data showed a decrease in blood microcirculation in the area of tumor foci compared to the intact area. This can also show the reduced blood supply of the blocked area caused by tumor obstruction.

The high individual variability of the results is due to the influence of a number of factors such as presence of blood and other substances, the state of the pathological process and the disadvantages of existing methods of monitoring the location of the instrument for MIS.

In general, the obtained data showed promising results of a multimodal approach with the joint application of FS and LDF in MIS in oncology of the common bile duct.

11.3 Multimodal Optical Fine-Needle Biopsy

Modern cancer diagnosis requires histological and cytological analysis of tumors. Material sampling for such analysis is often performed using fine-needle aspiration biopsy (FNAB). FNAB is well established for cancer diagnosis without open surgery [49] and it is carried out not only to confirm the initial diagnosis, but also to assess the progression and prognosis of the further course of the disease, as well as to correct the therapy [50].

For accurate insertion of the instrument into tumor tissue, FNAB is performed using imaging tools such as ultrasound, CT, and MRI. However, during the procedure, due to the physiological movement of organs, manipulations preceding FNAB, involuntary movements of the patient, there may be a displacement of tissues from the area defined for the material intake. Also, the size of the affected area may be small or poorly determined by imaging techniques. False negative results of histopathological and cytological studies associated with the sampling of material from an incorrectly defined area of interest as well as the heterogeneity of cancer contribute uncertainty to the diagnostic result and become important limitations of this method. At the same time, the absence of tumor cells in the obtained tissue samples does not guarantee the absence of cancer. Even experienced surgeons using imaging techniques allow up to 15–25% of non-diagnostic biopsy samples from uninformative areas [51–53]. This often leads to a second biopsy, which poses an increased risk to the patient.

Optical technologies can improve targeting to the right areas by analysis of the molecular and morphological structure of biological tissue prior to sampling, provide real-time information on tissue status, thereby decreasing the number of undiagnosed samples, enhancing the sensitivity of FNAB, and reducing the need for repeated and open biopsies for diagnosis. The biopsy taken at the time of surgery operation demands immediate analysis. Otherwise, it necessitates the repeated surgical intervention which usually leads to a worsening of the prognosis of the disease. Development of efficient technology for optical biopsy may offer a solution to the problem. The ability to recognize the suspect tissue in a short time without interruption of the operation is able to increase the effectiveness of the endoscopic surgery of the cancerous tumors.

Problems with the accurate detection and characterization of pancreatic tumors are associated with the relative inaccessibility of the organ, as well as the nonspecific nature of the symptoms of diseases. Lloyd et al. [54] conducted an *in vivo* experimental study to assess the feasibility of multimodal optical spectroscopy in the clinical detection of pancreatic adenocarcinoma. A reflectance and fluorescence lifetime spectrometer (RFLS) and fiber-optic probe were employed to collect optical data from human pancreatic tissues *in vivo* (Fig. 11.14).

The results of the study showed that this technology is a promising method for improving the diagnosis of pancreatic cancer *in vivo* and can improve the effectiveness of the diagnostic procedure for selecting tissue sites for further histological and cytological analysis in suspected pancreatic cancer.

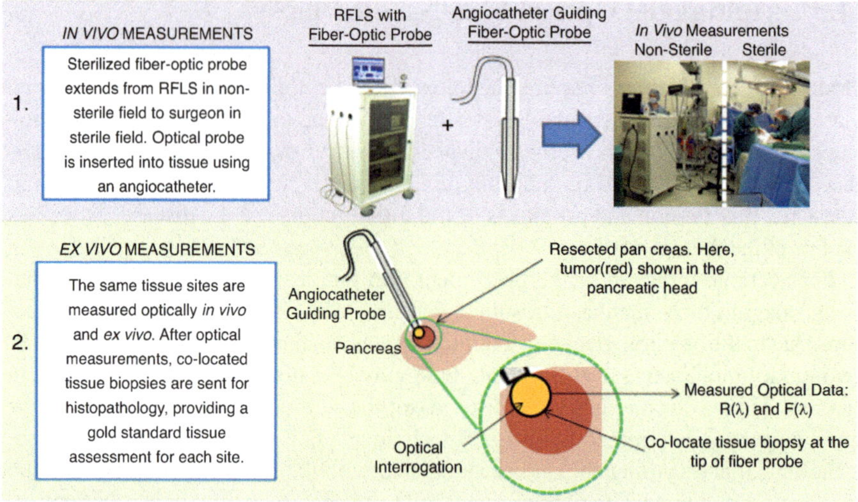

Fig. 11.14 Pancreatic tissue optical measurement protocol, designed to mimic fine-needle aspiration procedures by introducing the optical probe via a hollow angiocatheter. (1) *In vivo*: at this stage of surgery, tissue was still perfused with some blood and was near body temperature. The sterilized fiber-optic probe (6 m length) extended from the RFLS in the non-sterile field to the patient in the sterile field. At each selected site, the surgeon inserted a 14 gauge angiocatheter (B Braun Medical) ~1 cm into the tissue, removed the stylet from the angiocatheter, and inserted the fiber-optic probe. Data acquisition for each site was <45 s, with each modality acquired in <1 s. (2) *In vivo* and *ex vivo* measurements were acquired at the same tissue site by marking the site prior to resection. (2, right) Each optical measurement was estimated to interrogate ~1 mm^3 of tissue. Repeatability was tested by collecting two successive optical measurements at each site, with up to two tissue sites measured per patient. Tissue sites were biopsied for histopathologic analysis (reprinted with permission from [54])

Transbronchial lung biopsies are performed by pulmonologists to diagnose focal and diffuse lung diseases. Harris et al. [55] proposed to use the capabilities of bimodal optical spectroscopic system combining diffuse reflection spectroscopy and diffuse fluorescence spectroscopy to enhance differentiation between malignant and benign tissues. The first studies were conducted *ex vivo*; however, the design of the probe with a diameter of 1 mm suggests that in the future these studies can be carried out *in vivo*. It was confirmed that multimodal optical measurements can provide detailed information about vascular and biochemical differences in lung tissue and reduce the percentage of undiagnosed samples.

Scolaro et al. [56] reported on the development of a molecular imaging needle that includes a dual-modality diagnostic system using optical coherence tomography and fluorescence imaging of labeled antibodies. To evaluate the effectiveness of visualization with the molecular needle there were obtained images of a 15-μm section of human liver fluorescently labeled at the GCTM-5 (Fig. 11.15).

The results showed that the dual-modality system is able to visualize tissues with a sensitivity comparable to desktop fluorescence microscopes and to obtain images

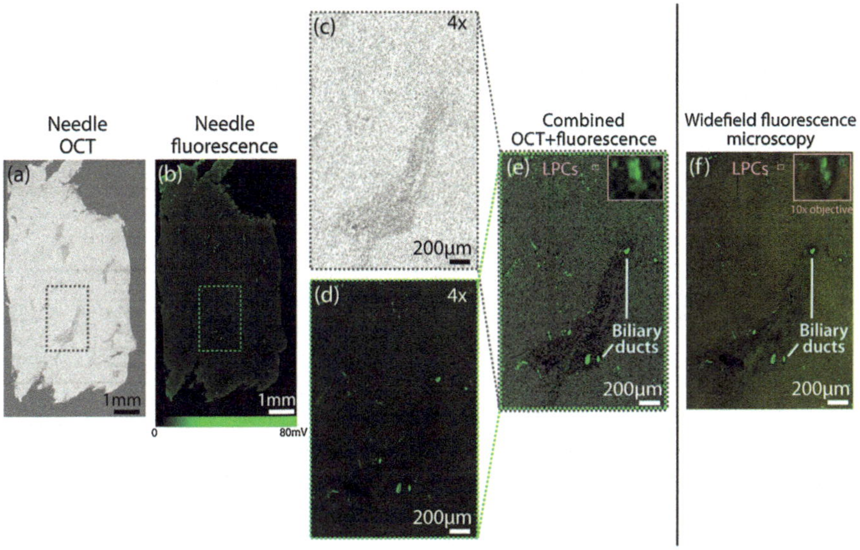

Fig. 11.15 Side-viewing needle images obtained for a section of normal human liver, showing labeled biliary ducts and liver progenitor cells. (**a**) *En face* OCT needle image obtained after averaging 4 co-located OCT A-scans at each radial position. (**b**) Corresponding fluorescence needle image. (**c**) Magnified OCT image of selected region indicated by dashed box in **a**. (**d**) Corresponding magnified fluorescence needle image of selected region indicated by dashed box in **b**. (**e**) OCT + fluorescence magnified image of selected region. Inset: Magnified view of region indicated by small pink box. (**f**) Corresponding standard wide-field fluorescence microscopy image for selected region. Inset: High magnification (10× objective) image of region indicated by pink box confirming liver progenitor cells (reprinted with permission from [56])

of deep tissue at the same time. In the future, such a system can also be used to improve the efficiency of biopsies of various organs.

The stereotactic brain tumor biopsy is an effective procedure for assessing intracranial lesions. However, the ambiguity in the images or brain shift can make sampling less accurate and sometimes non-diagnostic. The use of the intraoperative examination of pathological focus can improve the diagnostic result of this procedure. Haj-Hosseini et al. [57] developed a multimodal fiber-optic probe in order to provide optimal guidance for locating the most likely malignant formations. The proposed multimodal spectroscopy system has two channels implemented in a single probe: fluorescence spectroscopy for the registration of 5-ALA-induced fluorescence protoporphyrin IX (PpIX) and laser Doppler flowmetry for studying the microvascular blood flow (Fig. 11.16).

This technology has been clinically tested. *In vivo* studies have shown the possibility of real-time fluorescence detection during stereotactic biopsy procedures simultaneously with microvascular blood flow recording. This approach can help to identify the optimal positions for biopsy sampling, determining the malignancy of the formation in real time. Also, LDF channel potentially can act as a “vessel

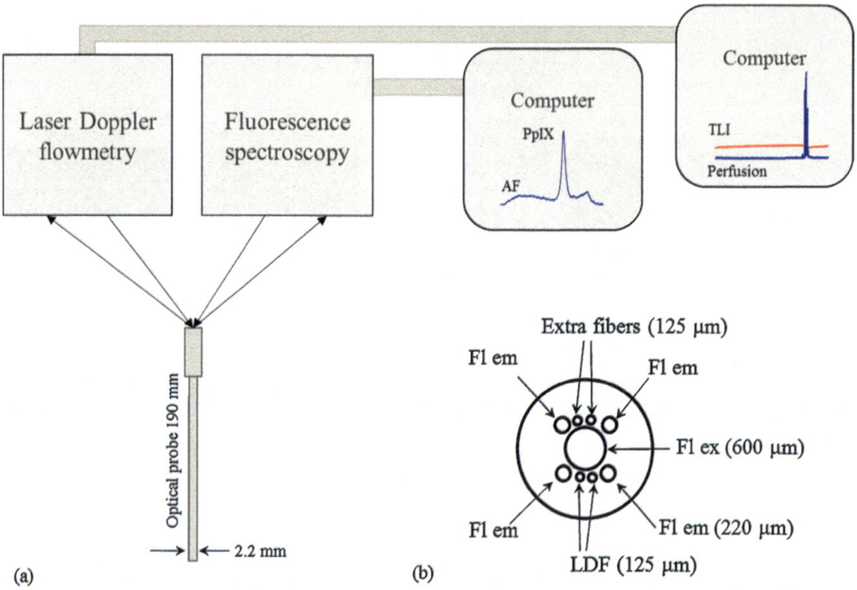

Fig. 11.16 (**a**) The system design concept and (**b**) configuration of the fibers in the probe. The system was composed of a fluorescence spectroscopy and a laser Doppler flowmetry system. The probe incorporated several optical fibers that were connected to the two systems. Fl: fluorescence, LDF: Laser Doppler flowmetry, em: emission, ex: excitation, AF: autofluorescence, TLI: total backscattered laser light intensity, PpIX: protoporphyrin IX (reprinted with permission from [57])

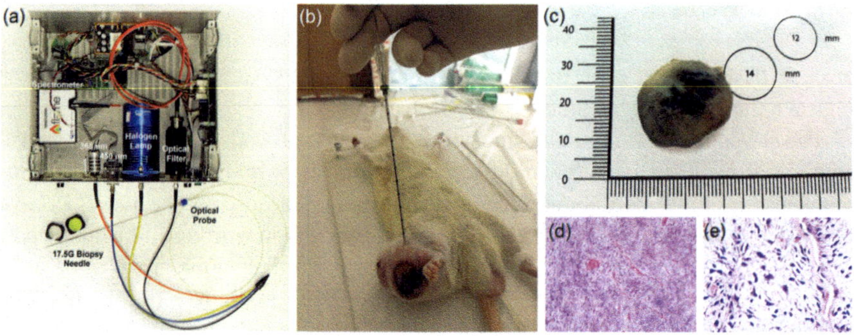

Fig. 11.17 (**a**) The multimodal FS/DRS fine-needle setup; (**b**) rat tumor experiment; (**c**) fragment of a tumor node up to 2.5 cm in a diameter; histological examination of tumor tissue with hematoxylin and eosin staining at (**d**) 100× and (**e**) 400× (reprinted with permission from [58])

tracking" tool to reduce the risk of hemorrhage due to rupture of vessels during needle insertion.

The optical needle for *in vivo* diagnosis of tumor (see Fig. 11.17) and the technique of its clinical application in the liver examination have recently been presented by the research group of Dunaev [58, 59]. A fine-needle fiber probe contains

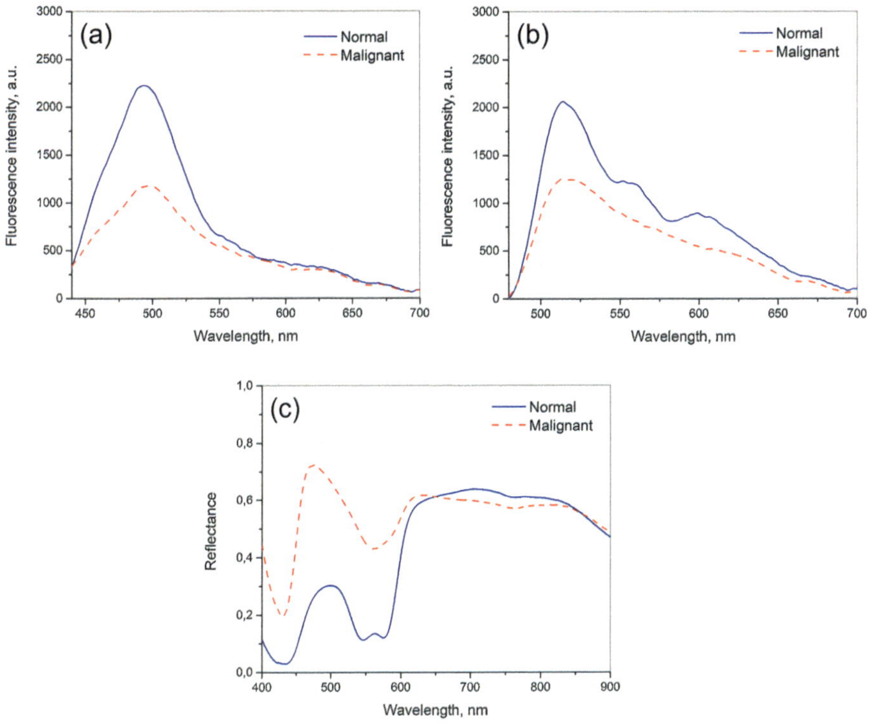

Fig. 11.18 Fluorescence spectra at excitation wavelengths (**a**) 365 nm, (**b**) 450 nm and (**c**) diffuse reflectance spectra of normal (blue line) and tumor (red line) rat tissue (reprinted with permission from [58])

emitting and collecting fibers for measurements of fluorescence intensity (with excitation at 365 nm and 450 nm) and diffuse reflectance (in the range 400–900 nm).

The optic probe has been developed to be compatible with the 17.5G biopsy needle standard and has a diameter of 1 mm. The probe has 10 optical fibers. The nine transmitting ones (100 μm each) include three fibers connected to a halogen light source, three fibers connected to a laser diode, and three fibers connected to a LED. The fibers are located around the central one (200 μm), which delivers the collected light to spectrometer. The quantity and orientation of optical fibers inside the fiber-optic probe provides uniform and bright illumination of the diagnostic volume and high signal-to-noise ratio. The probe has a bevel angle of 20°, which ensures reliable contact of the probe with dense tissues.

With the probe, *in vivo* test measurements of pathologically altered rat tissues (low-grade myxofibrosarcoma) were carried out. As one may see in Fig. 11.18a, b there are significant differences between fluorescence spectra of intact tissue and myxofibrosarcoma. The authors attribute this to changes in NADH, FAD, and collagen content. It was also observed significant differences between the diffuse reflectance spectra of normal and tumor tissue, notably in the wavelength range

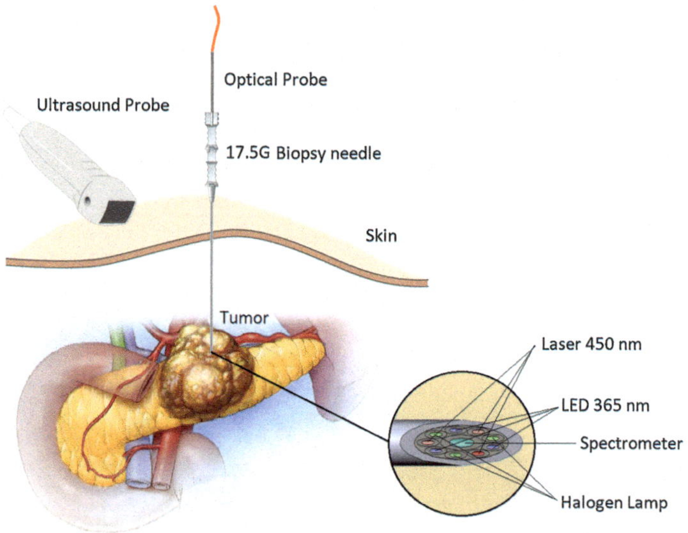

Fig. 11.19 The scheme of measurement during standard FNAB procedure (reprinted with permission from [59])

400–600 nm (see Fig. 11.18c). Reflectance measurements clearly show the pronounced reflectance peak in region 450–500 nm. There is the absence of peaks of oxyhemoglobin absorption on the tumor spectra, which indicates tissue ischemia.

In addition, the proposed methodology has been tested in clinical conditions. The methodology assumes the use of the setup during the standard FNAB procedure under the ultrasound control. The surgeon inserts a fiber-optic probe through the outer 17.5G needle into the tumor using the shortest trajectory (see Fig. 11.19). Before the probe reaches the suspicious neoplasm, measurements of fluorescence and diffuse reflectance are performed in healthy liver parenchyma. Then, the needle probe is introduced in tumor itself and the measurements are repeated. After the measurements, the surgeon obtains a tissue sample from the same area for histological and cytological analysis. The key point of this approach is comparison of optical and traditional biopsies results for more accurate interpretation.

This study involved 5 patients with adenocarcinoma. Figure 11.20 shows a typical example of fluorescence and diffuse reflectance measurements. The fluorescence intensity is higher in liver parenchyma for excitation at 365 nm and in tumor tissue for excitation at 450 nm. This may be caused by changes in the metabolic activity of cells in the tumor tissue, as well as a lower blood content that absorbs radiation, especially in the green region of the spectrum.

Diffuse reflectance spectra show significant changes in the presence of blood and oxy- and deoxyhemoglobin. The intensity of diffuse reflectance in tumors is higher than in parenchyma, which may indicate the presence of morphological changes in the tissue.

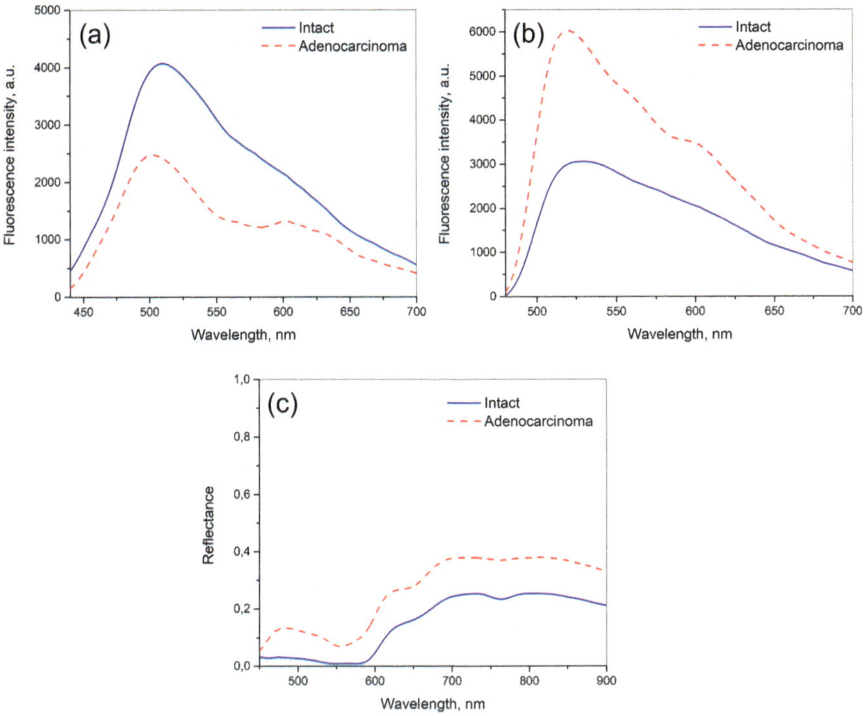

Fig. 11.20 Fluorescence spectra at excitation wavelengths (**a**) 365 nm, (**b**) 450 nm and (**c**) diffuse reflectance spectra of normal (blue line) and tumor (red line)

11.4 Conclusions

The high prevalence and mortality of cancer lead to a clinical desire for earlier detection of the disease. Great efforts are being made to achieve this goal by the implementation of optical spectroscopy and imaging techniques into minimally invasive technologies. Optical diagnostic techniques can provide additional information about the state of biological tissues with accuracy comparable to histological data and are structurally compatible with standard tools for MIS. Along with traditionally used optical tools such as endoscopes, colposcopes for cancer diagnosis and treatment, a new generation of tools is being developed that can acquire not only white light image, but also additional signals arising from cancer biomarkers, polarization and fluorescence imaging, narrow-band light reflection, and Raman scattering. Integration of several diagnostic methods in a single instrument significantly increases the effectiveness of determining the presence of pre-cancerous and malignant lesions at the early stages of their development.

Multimodal optical technologies can also be very useful for clinicians as a unique guide tool for biopsy. Optical spectroscopy and high-resolution imaging can provide alternative means of evaluating cell morphology in real time. Also, the introduction of optical diagnostic technologies can guide the doctor to select the most suitable sites, which helps to reduce the cost and improve the diagnostic effectiveness of biopsies and following histopathology analysis.

Thus, the analysis of the results of multimodal optical diagnostics can provide a great amount of potentially useful information during minimally invasive interventions. At the same time, multimodal optical technology has the advantages of simplicity of sample preparation, lack of violation of sample biochemistry, and the possibility of real-time diagnostics.

Acknowledgments This study was supported by the Russian Science Foundation under project No 18-15-00201. E. Zherebtsov also acknowledges the support of the Academy of Finland (grant No 318281) and V. Dremin also acknowledges the support of the Russian Science Foundation under project No 19-79-00082.

References

1. Wickham, J.E.A.: The new surgery. Br. Med. J. (Clin. Res. Ed.). **295**, 1581–1582 (1987). https://doi.org/10.1136/bmj.295.6613.1581
2. Siddaiah-Subramanya, M., Tiang, K., Nyandowe, M.: A new era of minimally invasive surgery: progress and development of major technical innovations in general surgery over the last decade. Surg. J. **03**, e163–e166 (2017). https://doi.org/10.1055/s-0037-1608651
3. Pazouki, A.: Minimally invasive surgical sciences: a new scientific opportunity for all scientists. J. Minim. Invasive Surg. Sci. **1**, 9–10 (2012). https://doi.org/10.5812/jmiss.2976
4. Breedveld, P., Stassen, H.G., Meijer, D.W., Jakimowicz, J.J.: Observation in laparoscopic surgery: overview of impeding effects and supporting aids. J. Laparoendosc. Adv. Surg. Tech. A. **10**, 231–241 (2000). https://doi.org/10.1089/lap.2000.10.231
5. Dankelman, J., Wentink, M., Stassen, H.G.: Human reliability and training in minimally invasive surgery. Minim. Invasive Ther. Allied Technol. **12**, 129–135 (2003). https://doi.org/10.1080/13645700310007689
6. Jones, D.B., Brewer, J.D., Soper, N.J.: The influence of three-dimensional video systems on laparoscopic task performance. Surg. Laparosc. Endosc. **6**, 191–197 (1996)
7. Lewin, J.S.: Future directions in minimally invasive intervention. Trans. Am. Clin. Climatol. Assoc. **128**, 346–352 (2017)
8. Wang, T.D., Van Dam, J.: Optical biopsy: a new frontier in endoscopic detection and diagnosis. Clin. Gastroenterol. Hepatol. Off. Clin. Pract. J. Am. Gastroenterol. Assoc. **2**, 744–753 (2004)
9. Kiesslich, R., Goetz, M., Hoffman, A., Galle, P.R.: New imaging techniques and opportunities in endoscopy. Nat. Rev. Gastroenterol. Hepatol. **8**, 547–553 (2011). https://doi.org/10.1038/nrgastro.2011.152
10. Hoffman, A., Manner, H., Rey, J.W., Kiesslich, R.: A guide to multimodal endoscopy imaging for gastrointestinal malignancy-an early indicator. Nat. Rev. Gastroenterol. Hepatol. **14**, 421–434 (2017). https://doi.org/10.1038/nrgastro.2017.46
11. Bedard, N., Pierce, M., El-Naggar, A., Anandasabapathy, S., Gillenwater, A., Richards-Kortum, R.: Emerging roles for multimodal optical imaging in early cancer detection: a global challenge. Technol. Cancer Res. Treat. **9**, 211–217 (2010). https://doi.org/10.1177/153303461000900210
12. Georgakoudi, I., Jacobson, B.C., Van Dam, J., Backman, V., Wallace, M.B., Müller, M.G., Zhang, Q., Badizadegan, K., Sun, D., Thomas, G.A., Perelman, L.T., Feld, M.S.: Fluorescence, reflectance, and light-scattering spectroscopy for evaluating dysplasia in patients with Barrett's esophagus. Gastroenterology. **120**, 1620–1629 (2001). https://doi.org/10.1053/gast.2001.24842

13. Kara, M.A., Bergman, J.J.: Autofluorescence imaging and narrow-band imaging for the detection of early neoplasia in patients with Barrett's esophagus. Endoscopy. **38**, 627–631 (2006). https://doi.org/10.1055/s-2006-925385
14. Curvers, W.L., Herrero, L.A., Wallace, M.B., Wong Kee Song, L.M., Ragunath, K., Wolfsen, H.C., Prasad, G.A., Wang, K.K., Subramanian, V., Weusten, B.L.A.M., Ten Kate, F.J., Bergman, J.J.G.H.M.: Endoscopic tri-modal imaging is more effective than standard endoscopy in identifying early-stage neoplasia in Barrett's esophagus. Gastroenterology. **139**, 1106–1114.e1 (2010). https://doi.org/10.1053/j.gastro.2010.06.045
15. von Holstein, C.S., Nilsson, A.M., Andersson-Engels, S., Willén, R., Walther, B., Svanberg, K.: Detection of adenocarcinoma in Barrett's oesophagus by means of laser induced fluorescence. Gut. **39**, 711–716 (1996). https://doi.org/10.1136/gut.39.5.711
16. Wang, K.K., Okoro, N., Prasad, G., WongKeeSong, M., Buttar, N.S., Tian, J.: Endoscopic evaluation and advanced imaging of Barrett's esophagus. Gastrointest. Endosc. Clin. N. Am. **21**, 39–51 (2011). https://doi.org/10.1016/j.giec.2010.09.013
17. Kato, M., Kaise, M., Yonezawa, J., Goda, K., Toyoizumi, H., Yoshimura, N., Yoshida, Y., Kawamura, M., Tajiri, H.: Trimodal imaging endoscopy may improve diagnostic accuracy of early gastric neoplasia: a feasibility study. Gastrointest. Endosc. **70**, 899–906 (2009). https://doi.org/10.1016/j.gie.2009.03.1171
18. van den Broek, F.J.C., Fockens, P., Van Eeden, S., Kara, M.A., Hardwick, J.C.H., Reitsma, J.B., Dekker, E.: Clinical evaluation of endoscopic trimodal imaging for the detection and differentiation of colonic polyps. Clin. Gastroenterol. Hepatol. **7**, 288–295 (2009). https://doi.org/10.1016/j.cgh.2008.10.025
19. Joshi, B.P., Pant, A., Duan, X., Prabhu, A., Wamsteker, E.J., Kwon, R.S., Elta, G.H., Owens, S.R., Appelman, H.D., Wang, T.D., Turgeon, D.K.: Multimodal video colonoscope for targeted wide-field detection of nonpolypoid colorectal neoplasia. Gastroenterology. **150**, 1084–1086 (2016). https://doi.org/10.1053/j.gastro.2016.02.075
20. Kim, Y.-I., Jeong, S., Jung, K.O., Song, M.G., Lee, C.-H., Chung, S.-J., Park, J.Y., Cha, M.G., Lee, S.G., Jun, B.-H., Lee, Y.-S., Hwang, D.W., Youn, H., Kang, K.W., Lee, Y.-S., Jeong, D.H., Lee, D.S.: Simultaneous detection of EGFR and VEGF in colorectal cancer using fluorescence-Raman endoscopy. Sci. Rep. **7**, 1035 (2017). https://doi.org/10.1038/s41598-017-01020-y
21. Jeong, S., Kim, Y.I., Kang, H., Kim, G., Cha, M.G., Chang, H., Jung, K.O., Kim, Y.H., Jun, B.H., Hwang, D.W., Lee, Y.S., Youn, H., Lee, Y.S., Kang, K.W., Lee, D.S., Jeong, D.H.: Fluorescence-Raman dual modal endoscopic system for multiplexed molecular diagnostics. Sci. Rep. **5**, 9455 (2015). https://doi.org/10.1038/srep09455
22. Miller, S.J., Lee, C.M., Joshi, B.P., Gaustad, A., Seibel, E.J., Wang, T.D.: Targeted detection of murine colonic dysplasia *in vivo* with flexible multispectral scanning fiber endoscopy. J. Biomed. Opt. **17**, 021103 (2012). https://doi.org/10.1117/1.JBO.17.2.021103
23. Tumlinson, A.R., Hariri, L.P., Utzinger, U., Barton, J.K.: Miniature endoscope for simultaneous optical coherence tomography and laser-induced fluorescence measurement. Appl. Opt. **43**, 113–121 (2004). https://doi.org/10.1364/AO.43.000113
24. Winkler, A.M., Rice, P.F.S., Weichsel, J., Watson, J.M., Backer, M.V., Backer, J.M., Barton, J.K.: *In vivo*, dual-modality OCT/LIF imaging using a novel VEGF receptor-targeted NIR fluorescent probe in the AOM-treated mouse model. Mol. Imaging Biol. **13**, 1173–1182 (2011). https://doi.org/10.1007/s11307-010-0450-6
25. Carbary-Ganz, J.L., Welge, W.A., Barton, J.K., Utzinger, U.: *In vivo* molecular imaging of colorectal cancer using quantum dots targeted to vascular endothelial growth factor receptor 2 and optical coherence tomography/laser-induced fluorescence dual-modality imaging. J. Biomed. Opt. **20**, 096015 (2015). https://doi.org/10.1117/1.JBO.20.9.096015
26. Li, Y., Jing, J., Yu, J., Zhang, B., Huo, T., Yang, Q., Chen, Z.: Multimodality endoscopic optical coherence tomography and fluorescence imaging technology for visualization of layered architecture and subsurface microvasculature. Opt. Lett. **43**, 2074–2077 (2018). https://doi.org/10.1364/ol.43.002074

27. Li, Y., Zhu, Z., Chen, J.J., Jing, J.C., Sun, C.-H., Kim, S., Chung, P.-S., Chen, Z.: Multimodal endoscopy for colorectal cancer detection by optical coherence tomography and near-infrared fluorescence imaging. Biomed. Opt. Express. **10**, 2419–2429 (2019). https://doi.org/10.1364/boe.10.002419
28. Charanya, T., York, T., Bloch, S., Sudlow, G., Liang, K., Garcia, M., Akers, W.J., Rubin, D., Gruev, V., Achilefu, S.: Trimodal color-fluorescence-polarization endoscopy aided by a tumor selective molecular probe accurately detects flat lesions in colitis-associated cancer. J. Biomed. Opt. **19**, 126002 (2014). https://doi.org/10.1117/1.jbo.19.12.126002
29. Skubleny, D., Dang, J.T., Skulsky, S., Switzer, N., Tian, C., Shi, X., de Gara, C., Birch, D.W., Karmali, S.: Diagnostic evaluation of sentinel lymph node biopsy using indocyanine green and infrared or fluorescent imaging in gastric cancer: a systematic review and meta-analysis. Surg. Endosc. **32**, 2620–2631 (2018). https://doi.org/10.1007/s00464-018-6100-9
30. Zheng, C., Lau, L.W., Cha, J.: Dual-display laparoscopic laser speckle contrast imaging for real-time surgical assistance. Biomed. Opt. Express. **9**, 5962–5981 (2018). https://doi.org/10.1364/boe.9.005962
31. Heeman, W., Dijkstra, K., Hoff, C., Koopal, S., Pierie, J.-P., Bouma, H., Boerma, E.C.: Application of laser speckle contrast imaging in laparoscopic surgery. Biomed. Opt. Express. **10**, 2010–2019 (2019). https://doi.org/10.1364/boe.10.002010
32. Hariri, L.P., Bonnema, G.T., Schmidt, K., Winkler, A.M., Korde, V., Hatch, K.D., Davis, J.R., Brewer, M.A., Barton, J.K.: Laparoscopic optical coherence tomography imaging of human ovarian cancer. Gynecol. Oncol. **114**, 188–194 (2009). https://doi.org/10.1016/j.ygyno.2009.05.014
33. Zuzak, K.J., Naik, S.C., Alexandrakis, G., Hawkins, D., Behbehani, K., Livingston, E.H.: Characterization of a near-infrared laparoscopic hyperspectral imaging system for minimally invasive surgery. Anal. Chem. **79**, 4709–4715 (2007). https://doi.org/10.1021/ac070367n
34. Baltussen, E.J.M., Kok, E.N.D., Brouwer de Koning, S.G., Sanders, J., Aalbers, A.G.J., Kok, N.F.M., Beets, G.L., Flohil, C.C., Bruin, S.C., Kuhlmann, K.F.D., Sterenborg, H.J.C.M., Ruers, T.J.M.: Hyperspectral imaging for tissue classification, a way toward smart laparoscopic colorectal surgery. J. Biomed. Opt. **24**, 016002 (2019). https://doi.org/10.1117/1.jbo.24.1.016002
35. Kikuchi, H., Kamiya, K., Hiramatsu, Y., Miyazaki, S., Yamamoto, M., Ohta, M., Baba, S., Konno, H.: Laparoscopic narrow-band imaging for the diagnosis of peritoneal metastasis in gastric cancer. Ann. Surg. Oncol. **21**, 3954–3962 (2014). https://doi.org/10.1245/s10434-014-3781-8
36. Schnelldorfer, T., Jenkins, R.L., Birkett, D.H., Wright, V.J., Price, L.L., Georgakoudi, I.: Laparoscopic narrow band imaging for detection of occult cancer metastases: a randomized feasibility trial. Surg. Endosc. **30**, 1656–1661 (2016). https://doi.org/10.1007/s00464-015-4401-9
37. Kandurova, K., Dremin, V., Zherebtsov, E., Potapova, E., Alyanov, A., Mamoshin, A., Ivanov, Y., Borsukov, A., Dunaev, A.: Fiber-optic system for intraoperative study of abdominal organs during minimally invasive surgical interventions. Appl. Sci. **9**, 217 (2019). https://doi.org/10.3390/app9020217
38. Becker, G.J.: 2000 RSNA annual oration in diagnostic radiology: the future of interventional radiology. Radiology. **220**, 281–292 (2001). https://doi.org/10.1148/radiology.220.2.r01au39281
39. Cope, C.: Percutaneous nonvascular abdominal interventions: reflections on the past and ideas for the future. J. Vasc. Interv. Radiol. **14**, 861–864 (2003). https://doi.org/10.1097/01.RVI.0000064854.87207.8
40. Lin, Q., Yang, R., Cai, K., Guan, P., Xiao, W., Wu, X.: Strategy for accurate liver intervention by an optical tracking system. Biomed. Opt. Express. **6**, 3287–3302 (2015). https://doi.org/10.1364/boe.6.003287
41. Watkinson, A.F., Adam, A. (eds.): Interventional Radiology: A Practical Guide. Radcliffe Medical Press, Oxford and New York (1996)

42. Bartolomé, F., Abramov, A.Y.: Measurement of mitochondrial NADH and FAD autofluorescence in live cells. Methods Mol. Biol. **1264**, 263–270 (2015). https://doi.org/10.1007/978-1-4939-2257-4_23
43. Kandurova, K., Dremin, V., Zherebtsov, E., Potapova, E., Filina, M., Dunaev, A., Mamoshin, A., Alyanov, A., Muradyan, V.: Optical diagnostics of bile duct tissues state with tumor compression. In: Proc. SPIE 11065, p. 1106508. SPIE-Intl Soc Optical Eng (2019)
44. Anidjar, M., Ettori, D., Cussenot, O., Meria, P., Desgrandchamps, F., Cortesse, A., Teillac, P., Le Duc, A., Avrillier, S.: Laser induced autofluorescence diagnosis of bladder tumors: dependence on the excitation wavelength. J. Urol. **156**, 1590–1596 (1996)
45. Bogomolov, A., Belikova, V., Zabarylo, U.J., Bibikova, O., Usenov, I., Sakharova, T., Krause, H., Minet, O., Feliksberger, E., Artyushenko, V.: Synergy effect of combining fluorescence and mid infrared fiber spectroscopy for kidney tumor diagnostics. Sensors. **17**, E2548 (2017). https://doi.org/10.3390/s17112548
46. Koenig, F., McGovern, F.J., Althausen, A.F., Deutsch, T.F., Schomacker, K.T.: Laser induced autofluorescence diagnosis of bladder cancer. J. Urol. **156**, 1597–1601 (1996)
47. Mayinger, B., Jordan, M., Horner, P., Gerlach, C., Muehldorfer, S., Bittorf, B.R., Matzel, K.E., Hohenberger, W., Hahn, E.G., Guenther, K.: Endoscopic light-induced autofluorescence spectroscopy for the diagnosis of colorectal cancer and adenoma. J. Photochem. Photobiol. B. **70**, 13–20 (2003)
48. Palmer, S., Litvinova, K., Dunaev, A., Yubo, J., McGloin, D., Nabi, G.: Optical redox ratio and endogenous porphyrins in the detection of urinary bladder cancer: a patient biopsy analysis. J. Biophotonics. **10**, 1062–1073 (2017). https://doi.org/10.1002/jbio.201600162
49. Tam, A.L., Lim, H.J., Wistuba, I.I., Tamrazi, A., Kuo, M.D., Ziv, E., Wong, S., Shih, A.J., Webster, R.J., Fischer, G.S., Nagrath, S., Davis, S.E., White, S.B., Ahrar, K.: Image-guided biopsy in the era of personalized cancer care: proceedings from the society of interventional radiology research consensus panel. J. Vasc. Interv. Radiol. **27**, 8–19 (2016). https://doi.org/10.1016/j.jvir.2015.10.019
50. Marshall, D., Laberge, J.M., Firetag, B., Miller, T., Kerlan, R.K.: The changing face of percutaneous image-guided biopsy: molecular profiling and genomic analysis in current practice. J. Vasc. Interv. Radiol. **24**, 1094–1103 (2013). https://doi.org/10.1016/j.jvir.2013.04.027
51. Francque, S.M., De Pauw, F.F., Van den Steen, G.H., Van Marck, E.A., Pelckmans, P.A., Michielsen, P.P.: Biopsy of focal liver lesions: guidelines, comparison of techniques and cost-analysis. Acta Gastroenterol. Belg. **66**, 160–165 (2003)
52. Choi, S.H., Han, K.H., Yoon, J.H., Moon, H.J., Son, E.J., Youk, J.H., Kim, E.-K., Kwak, J.Y.: Factors affecting inadequate sampling of ultrasound-guided fine-needle aspiration biopsy of thyroid nodules. Clin. Endocrinol. (Oxf.). **74**, 776–782 (2011). https://doi.org/10.1111/j.1365-2265.2011.04011.x
53. Gomez-Macías, G.S., Garza-Guajardo, R., Segura-Luna, J., Barboza-Quintana, O.: Inadequate fine needle aspiration biopsy samples: pathologists versus other specialists. Cytojournal. **6**, 9 (2009). https://doi.org/10.4103/1742-6413.52831
54. Lloyd, W.R., Wilson, R.H., Lee, S.Y., Chandra, M., McKenna, B., Simeone, D., Scheiman, J., Mycek, M.-A.: *In vivo* optical spectroscopy for improved detection of pancreatic adenocarcinoma: a feasibility study. Biomed. Opt. Express. **5**, 9–15 (2013). https://doi.org/10.1364/BOE.5.000009
55. Harris, K., Rohrbach, D.J., Attwood, K., Qiu, J., Sunar, U.: Optical imaging of tissue obtained by transbronchial biopsies of peripheral lung lesions. J. Thorac. Dis. **9**, 1386–1392 (2017). https://doi.org/10.21037/jtd.2017.03.113
56. Scolaro, L., Lorenser, D., Madore, W.-J., Kirk, R.W., Kramer, A.S., Yeoh, G.C., Godbout, N., Sampson, D.D., Boudoux, C., McLaughlin, R.A.: Molecular imaging needles: dual-modality optical coherence tomography and fluorescence imaging of labeled antibodies deep in tissue. Biomed. Opt. Express. **6**, 1767–1781 (2015). https://doi.org/10.1364/boe.6.001767

57. Haj-Hosseini, N., Richter, J.C.O., Milos, P., Hallbeck, M., Wårdell, K.: 5-ALA fluorescence and laser Doppler flowmetry for guidance in a stereotactic brain tumor biopsy. Biomed. Opt. Express. **9**, 2284–2296 (2018). https://doi.org/10.1364/BOE.9.002284
58. Dremin, V., Potapova, E., Zherebtsov, E., Kozlov, I., Seryogina, E., Kandurova, K., Alekseyev, A., Piavchenko, G., Kuznetsov, S., Mamoshin, A., Dunaev, A.: Optical fine-needle aspiration biopsy in a rat model. In: Proc. SPIE 10877, p. 108770K. SPIE-Intl Soc Optical Eng (2019)
59. Kandurova, K., Potapova, E., Shupletsov, V., Kozlov, I., Seryogina, E., Dremin, V., Zherebtsov, E., Alekseyev, A., Mamoshin, A., Dunaev, A.: Optical fine-needle biopsy approach for intra-operative multimodal diagnostics in minimally invasive abdominal surgery. In: Proc. SPIE 11079, p. 110791C. SPIE-Intl Soc Optical Eng (2019)

Chapter 12
Multimodal OCT for Malignancy Imaging

Grigory Gelikonov, Valentin Gelikonov, Alexander Moiseev, Pavel Shilyagin, Sergey Ksenofontov, Irina Kasatkina, Dmitriy Terpelov, Lev Matveev, Alexander Matveyev, Vladimir Zaitsev, Alexander Sovetsky, Natalia Gladkova, Elena V. Zagaynova, Marina Sirotkina, Ekaterina Gubarkova, Elena Kiseleva, Anton Plekhanov, Vadim Elagin, Konstantin Yashin, Dmitry Vorontsov, Elena Sedova, Anna Maslennikova, Sergey Kuznetsov, and Alex Vitkin

12.1 Introduction

Optical coherence tomography (OCT) is a minimally invasive imaging yielding cross-sectional and volumetric subsurface tissue images with spatial resolution of several micrometers, to a depth of 1–3 mm [1]. OCT is based on low-coherence interferometry in the near IR range of wavelengths (700–1300 nm). Since the first successful demonstrations of biological tissue microstructure imaging by OCT [2], in the further studies significant attention was focused not only on improvement of resolution and increasing the rate of acquiring conventional structural images but also on modifications of the very principle of OCT image formation and the development of new methods of OCT image processing aimed at visualization of new types of tissue parameters. Indeed, simultaneous mapping of several functionally different characteristics of biological tissues can significantly increase the information content and correspondingly improve the specificity of diagnostics based on OCT inspection. Such functionally different types of OCT imaging include polarization-sensitive images, elastographic images that characterize mechanical properties of tissues (stiffness), and OCT-based mapping of microcirculation of blood in the inspected region (i.e., angiographic imaging). Besides these, one can also combine structural OCT images with images of tissue fluorescence [3],

G. Gelikonov (✉) · V. Gelikonov · A. Moiseev · P. Shilyagin · S. Ksenofontov · I. Kasatkina
D. Terpelov · L. Matveev · A. Matveyev · V. Zaitsev · A. Sovetsky
Institute of Applied Physics RAS, Nizhny Novgorod, Russian Federation
e-mail: grgel@ipfran.ru

N. Gladkova · E. V. Zagaynova · M. Sirotkina · E. Gubarkova · E. Kiseleva · A. Plekhanov
V. Elagin · K. Yashin · D. Vorontsov · E. Sedova · A. Maslennikova · S. Kuznetsov
Privolzhsky Research Medical University, Nizhny Novgorod, Russian Federation

A. Vitkin
University of Toronto and University Health Network, Toronto, ON, Canada

V. V. Tuchin et al. (eds.), *Multimodal Optical Diagnostics of Cancer*,
https://doi.org/10.1007/978-3-030-44594-2_12

combining multispectral and spectral-domain OCT microscopy [4], performing OCT at significantly different spectral ranges [5–7], and so on.

12.1.1 Cross-Polarization OCT (CP OCT)

Polarization-sensitive methods extend biomedical imaging by increasing the information value of the OCT inspection via extraction of qualitatively new information about the tissue properties. For example, in conventional OCT the "regular" (spatially homogeneous) birefringence of tissue manifests itself in the form of modulation of the image brightness as a function of the depth (due to the phase delay between the normal optical modes propagating in the medium). In the more general case, the backscattered radiation acquires a cross-polarization component with the polarization orthogonal to that in the incident optical beam. It is known that the appearance of the cross-polarization signal can be related to such factors as regular birefringence, microscopic anisotropy of the scattering (for scatterers smaller than the optical wavelength), the influence of scattering particles shape and the spatial structure of the scattering medium [8]. In endoscopic inspection of the soft tissues, polarization-sensitive methods can significantly improve reliability of conclusions on the state of the tissue based on a qualitative consideration of polarization-sensitive images (however, their quantitative interpretation can be significantly complicated in view of strong deformability of surface tissue layers in some organs). By now, a number of methods for polarization-sensitive OCT imaging have been developed, and the corresponding biomedical experimental studies have begun to reveal the specificity of new information and estimate its diagnostic utility [9]. In most studies, the approach to improve the information value of polarization-sensitive imaging is based on the analysis of the relation between the regular birefringence and variation in the polarization state of the optical signal backscattered from different depths of the inspected tissue region [10–14]. The interferometric selection of the depth, from which the signal is backscattered, and measurement of the variation in its polarization with respect to the incident-wave polarization allow one to determine such properties as depolarization, birefringence, dichroism, and orientation of polarization axes [15, 16]. Such polarization characteristics provide information on the presence of certain ordered structures (e.g., concentration and type of collagen fibers, their local orientation in the near-surface layers [15, 17]), as well as on the tissue microstructural features [15]. The evolution of polarization as a function of the scattering depths is usually characterized using the formalism of Jones and Mueller matrices [11–13, 18–21].

Among various polarization-sensitive OCT methods, an important approach is when the backscattered radiation is analyzed in two orthogonal polarizations produced by different means [16]. The radiation with polarization coinciding with that of the initial signal is received in the so-called co-channel, whereas the signal with the orthogonal polarization is recorded in the cross-channel of the OCT scanner. In the cross-channel the backscattered radiation can appear due to the influence of both regular birefringence and scattering from individual heterogeneities in the tissue [6, 8, 17, 22–24]. The degree of backscattering in the orthogonal polarization depends on the size, structure, and anisotropy properties of the optical heterogeneities in the medium [8] and can characterize, for example, the state of collagen tissues, which is especially important for diagnostics of cancer pathologies [17, 25]. As experiments indicate [8], the effectiveness of cross-polarization scattering is especially high for relatively large-scale heterogeneities and is much weaker manifested for heterogeneities with sizes smaller than the optical wave length.

Under conditions of weak birefringence, the cross-polarization scattering can become the main mechanism of the appearance of signals with orthogonal polarization. Obtaining and comparing OCT scans visualizing the distribution of backscattering efficiency into the co-polarization component (the same polarization as in the initial wave) and into cross-polarization one (with orthogonal polarization) is the basis for one of the polarization-sensitive OCT methods, referred to as the cross-polarization OCT [6, 8, 17, 22–24]. The first results of application of the CP OCT [6, 8] showed that the images obtained via the cross-polarization channel (cross-polarization images) significantly differ from those in the co-polarization channel and provide new information about the tissue. For cross-polarization images, the signal is much weaker from the near-surface layers (where the photons propagate without efficient scattering into the cross polarization) than in the co-polarization images. This is favorable for reduction of over-illumination of the near-surface layers and allows for more detailed characterization of structures localized near the tissue boundary. Similarly, for deeper structures that were hardly distinguishable in the co-polarization images, the cross-polarization scans can yield significantly more contrast [6]. Relatively large-scale discrete scattering particles are often better distinguished in the cross-polarization images, and comparison of the latter with co-polarization ones can significantly improve the visualization contrast for certain types of structures in biological tissues. Additional detailed studies are required for determining the composition and morphology of biological structures that appear particularly bright in the CP OCT [6, 8].

Besides, cross-polarization images give very useful complementary structural information in comparison with co-polarization scans, which makes it possible to detect distortions caused by even a fairly weak regular birefringence. It is known, for example, that the level of the backscattered signal can be significantly reduced exclusively due to polarization effects, when the optical delay between the normal modes in the medium is small and does not exceed half of the wavelength [13, 15, 26].

It should be noted that an important issue in CP OCT is a rather complex dependence of the received signals on both polarization of the primary probing wave and orientation of anisotropy axes in the inspected tissue [9]. For example, if the vector of linear polarization of the probing wave occasionally coincides with the axis of the medium anisotropy, the presence of birefringence does not manifest itself in the obtained OCT image. This phenomenon can be efficiently eliminated by using modulation of the probing-wave polarization [27–29] or utilizing a pair of coherent [30, 31] or incoherent [16, 32] waves with orthogonal polarizations. Further, the signal in the cross-polarization channel depends on the polarization state of the primary probing wave, in contrast with the reception of only the co-polarized wave which is independent of it.

In the design of polarization-sensitive OCT, significant attention is paid to optimization of optical schemes and regimes of their operation in order to obtain the maximum information content and ensure feasibility in clinical practice. Presently, the best information value is reached in cross-polarization optical schemes of OCT based on bulk optical elements [10, 11, 13, 18, 33], as well as for diagnostic optical-fiber systems, in which the inspection does not require flexible variation of the optical paths [34, 35]. In the design of flexible endoscopic OCT systems, the transfer of the above-mentioned polarization-sensitive approaches to schemes based on optical fibers is challenging.

At a certain stage of the OCT development, an important role was played by optical-fiber systems based on anisotropic fibers that were able to maintain the interference signal under continuous deformation of the signal arm [6, 36, 37]. Then it was proposed to compensate for the influence of the optical-path deformation on the signal level by applying not only polarization-maintaining anisotropic fibers but also using the so-called common-path optical schemes based on isotropic fibers [30, 31, 38–41]. The main feature of such schemes is that both probing (sample) and reference beams propagate along the same path, because the reference beam is reflected back at the distal end of the optical-fiber probe. Due to this, the common-path schemes ensure reproducibility of flexible fiber-optic probes [31]. It can be noted that in the common-path interferometric optical schemes based on isotropic fibers, the probing-wave polarization can be arbitrary, whereas the optical-path difference between the reference and sample waves exceeds the coherence length. In view of this, to single out the interference signal one should use an auxiliary compensating interferometer. Such an optical scheme [39, 42] included a fiber-optic variant of the Fizeau interferometer which plays the role of a common optical part for the signal and reference waves [43], as well as an autocorrelator based on a Michelson interferometer with Faraday mirrors [44]. The first applications of $\pi/4$ Faraday cells for compensation for phase anisotropy in single-mode fibers were reported in [45, 46].

The usefulness of polarization-sensitive OCT is evident for inspection of birefringent tissues, such as muscles, cartilage, skin, coronary artery, outer ocular tissues, tissues of retina, and vocal cords. For soft biological tissues with weak birefringence (such as mucous and serous linings), polarization effects can also manifest themselves in OCT images. The modeling of polarization phenomena demonstrated that they can lead to the appearance of false layers in the images because of the influence of the above-mentioned modulation of the intensity due to the birefringence. Evidently, the appearance of such false layers is not acceptable in the interpretation of OCT images of layered structures [47]. Besides, because of the complexity of the near-surface structure of biological tissues, relatively simple algorithms that provide information on birefringence and orientation of optical axes can be developed only for the upper birefringent layer or for tissues with rather weak birefringence: the waves with orthogonal polarization should acquire a phase difference significantly smaller than 90° for all observation depths. Information about the direction of optical axes in deeper layers is distorted by birefringence in the upper layers, so that retrieving this information requires more complex algorithms based on information above the upper layer [20, 47]. To single out the regions of biological tissues with different directions of optical axes, color encoding can be used [20].

Evidently, for the complex multifragment structure of soft biological tissues with weak birefringence, exact quantitative information on the directions of the optical axes and the degree of birefringence in various tissue fragments plays a secondary role, especially for medical screening and clinical applications. In this context, it looks attractive to apply the capabilities of CP OCT for qualitative differentiation of various fragments by their polarization characteristics, and differences between OCT images obtained in co- and cross-channels.

The above-mentioned approaches to obtaining polarization-sensitive images can be used as a basis for designing improved variants of endoscopic OCT devices with faster image acquisition and simultaneous real-time presentation of complementary images formed via co- and cross-polarization channels. In particular, the common-path fiber-optic scheme ensures stable images in the co-polarization channel without individual tuning of the optical arms in changeable endoscopic probes in the course of OCT inspection of tissues. However, sensitivity of the cross-polarization OCT in the optical schemes based on single-mode fibers depends on the polarization of the incident interrogating beam. Elimination of unstable sensitivity of the cross-polarization channel should allow for both qualitative and quantitative comparison between the co- and cross-polarization images obtained under given conditions. Improvement of the cross-polarization channel in OCT imaging is important not only for obtaining complementary polarization-sensitive and conventional intensity imaging, but also for extracting information on the directions of optical axes. Integrated solution of such problems is important for improving reliability of endoscopic OCT-based diagnostics and quality of evaluation of biological tissue state in various clinical applications.

12.1.2 OCT-Based Angiography (OCA)

All functioning cells within the human body have to be within 100 microns of a viable vessel that carries oxygenated blood, to ensure adequate supply of oxygen. As such, blood hemodynamics and the human microvascular tree are an essential part of healthy biological tissues. Abnormalities in the vascular supply are associated with various diseases such as cancer and other pathologies. The process of microvascular alteration in tumors whereby new (but often chaotic and unhealthy) vessels are created is called neovascular angiogenesis. A variety of emerging cancer treatment strategies aim at neovasculature disruption ("anti-angiogenic therapies") rather than the more conventional approaches that target tumor cells directly. For these and other reasons, direct three-dimensional (3D) non-invasive *in vivo* imaging of tissue blood microcirculation is scientifically interesting, clinically important, and technologically surprisingly difficult [48].

Most volumetric medical imaging modalities have been adopted to tackle various aspects of this problem, including contrast-enhanced CT (computed tomography), Doppler ultrasound, and MR (magnetoresonance) angiography. As well, various optical approaches, including laser Doppler flowmetry, laser speckle contrast imaging, intravital fluorescence microscopy, diffuse correlation spectroscopy, photoacoustics, and polarization-based tissue viability imaging, are being developed to assess various aspects of blood flow in tissues. The literature on the subject is vast; an interested reader is referred to several selected key references [49–52].

Suffice it to say that each approach has its own strengths and weaknesses in terms of ease of implementation, invasiveness, tomographic (3D) imaging capability, sensitivity, penetration, contrast, resolution, contrast agents vs endogenous contrast, flow velocity vs microvascular tree mapping, direct microvascular visualization vs model-derived parametric maps, and so forth. Over the last 15 years, various research groups have also pursued OCT methods to visualize tissue blood flow, specifically in smaller blood vessels. Because OCT systems are widely used in ophthalmology, its application to blood flow visualization and measurement could make clinical use more practical. OCT techniques can be extended beyond their "standard" structural imaging capability to provide additional important information on tissue blood flow and microvasculature. This no-injection, dye-free contrast mechanism is very important in biology and medicine, in that it provides dynamic *in vivo* information on tissue functional status; this is implicated in development of various pathologies such as cancer, is often targeted in antivascular therapies, and so on. These techniques aim to contrast blood vessels from static tissue by assessing the change in the OCT signal caused by flowing blood cells. These intrinsic contrasts can be broadly classified as Doppler shift and speckle variance/decorrelation.

Volumetric angiography was not feasible until development of the two Fourier-domain OCT implementations: spectral-domain (SD-OCT) and swept-source. In 2006, Makita et al. used an 18.7-kHz SD-OCT system to perform volumetric angiography and visualization of retinal and choroidal vasculature [53]. As noted by Makita et al., the standard deviation/variance [54] or power [55] of the Doppler signal provided better results than the Doppler shift. Another approach called optical microangiography (OMAG) incorporated the amplitude of the OCT signal in addition to phase. An et al. [56] suggested that OMAG was better able to identify the microvasculature than previous methods utilizing only phase information.

With the continued improvement of OCT system speeds due to hardware advances, methods for OCA shifted from comparing between adjacent A-scans to between sequential cross-sectional B-scans. The increased time separation ensured that slower flow in the microvasculature would be detected. In 2009, Fingler et al. [57] used a 25-kHz SD-OCT system and a phase variance approach over 10 repeat B-scans at the same location to show microvasculature that was analogous to traditional fluorescein angiography in human eyes. In 2011, Kim et al. [58] used a 125-kHz SD-OCT system to image with a larger field of view. They used montaging/

stitching of 10 volumes to generate an OCT angiogram with coverage comparable to fluorescein angiography.

OCT-based angiography on amplitude or intensity was initially described in 2005 [59], when Barton et al. adapted laser speckle analysis for time-domain OCT. Speckle arises as a property of the interferometric nature of OCT, and speckle variation contains information regarding the motion of scatterers [60]. Specifically, the speckle pattern stays relatively constant over time for static objects while the pattern changes for objects in motion. Mariampillai et al. [61] extended the technique and presented speckle variance detection of microvasculature in a dorsal skinfold model using a swept-source OCT system in 2008. In their work, speckle variance was calculated as the variance of the OCT reflectance amplitude over three repeated B-scans at the same location. In optimizing the method, the authors in [62] noted that the B-scan rates for repeat scans needed to be fast enough such that bulk motion between B-scans was less than the OCT beam waist radius. Although "speckle variance" has been historically associated with amplitude-based OCA, fundamentally both amplitude and phase-based flow detection are based on variation in the speckle pattern and therefore provide largely equivalent information [63]. In addition to speckle variance, another intensity-based OCA approach was termed correlation mapping [64]. In correlation mapping OCA, cross-correlation of a grid on adjacent B-scans was performed to identify vasculature (weak correlation) versus static tissue (strong correlation).

OCA of retinal microvasculature in the human eye using methods based on amplitude or intensity was demonstrated in 2012. Motaghiannezam et al. [65] used logarithmic intensity variance and differential logarithmic intensity variance to capture the microvascular network near the fovea.

Because OCA is economical, non-invasive, and does not even require the use of bright visible light, it can be used more frequently than traditional angiography, which requires intravenous dye injection [66]. As a result, OCT *in vivo* microvascular imaging is a "hot topic" both in research and industrial laboratories, and in preclinical/clinical medical environments. The phase-resolved/Doppler based and temporal variance-based OCT techniques provide attractive options for detecting, visualizing, quantifying, and monitoring of tissue microcirculation. Significant technological challenges imposed by the microvascular detection difficulties are steadily met. While many outstanding problems remain to be solved, both on technological and clinical implementation fronts, the importance of the problem and the considerable progress to date bode well for microvascular OCT's eventual inclusion into the arsenal of widely used medical diagnostic technologies.

12.1.3 OCT-Based Elastography (OCE)

Elastographic mapping [67] is another direction in the development of OCT aimed at obtaining additional information to complement the conventional intensity (structural) images. The term "elastography" in OCT (OCE) is introduced by analogy with medical ultrasound and is usually understood as mapping of the spatial distribution of either shear elastic modulus G or the Young modulus E. It is well known that for most part of biological tissues (except for bones and cartilage) the shear modulus G is much less than the bulk elastic modulus K, so that the Poisson's ratio ν is very close to 0.5, the value typical of liquids (notice that the exact equality $\nu = 1/2$ corresponds to zero shear modulus $G = 0$). As a result, for almost all biological tissues, the shear and Young moduli are proportional to each other with the same coefficient, since in the isotropic approximation $E = 2(1 + \nu)G \approx 3G$ [68]. In view of this, the literature on elastography quite often does not indicate which particular modulus (G or E) is mapped, and the term "stiffness" that has no rigorous definition is used. For a given value of the bulk modulus, higher values of the shear modulus characterize the ability of a soft tissue to maintain its shape. Just by this reason, the procedure of conventional palpation gives elastographic information: fingers feel an inclusion with a higher shear modulus as a stiffer region surrounded by easier deformable soft tissue.

For medical purposes the mapping of stiffness distribution in soft (i.e., other than cartilage or bone) tissues is important, because variability of the shear modulus (stiffness) is much higher compared with variability of the bulk modulus. Namely, the stiffness of the same soft tissue in normal and pathological states may differ several times and even orders of magnitude in contrast to rather weak (on the order of a few per cent) complementary variations of the bulk modulus. On the other hand, it is the bulk modulus which determines the impedance of ultrasound waves, and just its weak variations are visualized in conventional ultrasonic medical scanners. Thus, eventual variations in the shear modulus are not directly visualized either in conventional ultrasound echography, or in OCT images, because they characterize the scattering properties of the tissue with respect to longitudinal ultrasonic waves and transverse optical waves.

Elastography with OCT was proposed and demonstrated in the landmark paper published by Schmitt in 1998 [67], in which many of the principles and issues of contemporary research were laid out. Schmitt used OCT to image a sample undergoing uniaxial compression and assessed the resulting local displacement field by examining the correlation between cross-sectional B-scans of the same location. The results were used to produce elastograms, i.e., cross-sectional maps of displacement, from which strain localized to a small region of the sample (i.e., the local relative change in length per unit length) could be estimated.

In many aspects, such studies have been based on the elastographic approaches used in ultrasound. However, unlike the medical ultrasonic scanners, OCT scanners capable of elastographic mapping are not yet commercially available. This fact is related, in particular, to certain features of OCT images which complicate direct transfer of elastographic processing principles from the medical ultrasound to OCT. Schmitt's work emphatically demonstrated that OCT could be used to gain information about a soft tissue's mechanical properties, and highlighted some of the opportunities and challenges that remain to this day. Firstly, mechanical image contrast is intrinsically different to optical contrast. To see an object in an OCT scan requires differences in scattering, produced by refractive index gradients, but mechanical contrast only requires sufficient scattering to obtain an OCT signal, and variations in the mechanical properties may be essentially unrelated to the optical parameters. Secondly, Schmitt measured displacements and only then inferred local strain. Strain is a relative quantity related to an absolute quantity, an elastic modulus or Young's modulus, but, in most practical scenarios, quantification of the elastic modulus is a non-trivial task. Thirdly, viscoelasticity implies hysteresis in the dynamic mechanical response, which can be a source of information or image contrast, or a source of complication in attempting to measure elastic parameters. Furthermore, even quasi-statically deformed tissues may exhibit pronouncedly nonlinear stress-strain dependences, including quasi-static hysteresis unrelated to viscoelasticity. All these features on the one hand may give additional diagnostic information, but on the other hand make interpretation rather non-trivial in comparison with simplest representation about the tissue as a linear elastic material.

After a slow start, the maturation of OCT technology in the early to mid-2000s has underpinned a recent acceleration in the field. Brett Bouma was the first to use the term "optical coherence elastography" in 2004 in [69]. Earlier that year, Rogowska et al. [70] had reported, in essence, a repeat of the Schmitt experiments on *in vitro* aorta samples, successively loaded with weights, instead of using an actuator. Later that year, Bouma and colleagues also began to explore the mechanical characterization of atherosclerotic plaque *in vivo* [69], an optical version of the then-emerging elastography based on intravascular ultrasound. They also demonstrated the first use of OCT to provide reference image data (in place of histology) for biomechanical modeling of arteries [71], but this has not become widely used.

Starting from 2003, OCT began the move towards Fourier- or spectral-domain [72], employing a static interferometer, with either spectrometer-based detection of broadband illumination or time-resolved detection of a swept-frequency optical source. These methods provided a gain in sensitivity approximately equal to the number of resolvable spectral channels, related to the reduction in shot noise per detection bin achieved without a reduction in overall signal [73, 74]. Wang et al. were the first to put this to work in OCE, demonstrating milliradian equivalent nanoscale displacements (with averaging) in a tissue phantom via comparison of A-scans [75] and B-scans [76]. Although no tissue results were presented, this direction proved to be key over the next decade.

OCE generally relies on an external or internal stimulation approach to load the tissue and an OCT-based detection method to measure the corresponding tissue response [77–81]. The early development of OCE features the static mechanical contact loading and the cross-correlation-based speckle tracking to capture the displacement [69–71]. The emerging of phase-resolved OCT detection utilizing the interferometric phase information from complex OCT signals, which reaches nanoscale and sub-nanoscale sensitivity to the tissue displacement [82, 83], enabled OCE techniques to assess and extract different parameters of the tissue deformation with high accuracy to reconstruct the tissue biomechanical properties [84–86]. Also, the development of OCE for different types of applications results in various loading approaches that have been proposed, such as the use of ultrasound for remote stimulation [84, 87, 88], the employment of air puff for noncontact excitation

[89–91], and applying magnetomotive nanoparticles as the internal transducers for vibration [85, 92, 93].

Reinforced from the high spatial resolution of OCT, the imaging or measurement resolution of OCE is spreading over the range from several microns to hundreds of microns depending on the methods used to reconstruct the biomechanical properties [94–98]. Generally, OCE maintains the same or even larger field of view compared with OCT at the millimeter level [99, 100].

OCE is characterized by its niche in intermediate spatial resolution (10s–100s μm) and degree of depth penetration and, by exploiting optical interferometry, its high sensitivity to small mechanical changes—at the microstrain level. Otherwise, it is following the evolution of elastography in the other imaging domains. In many ways, it forms an ideal bridge between the emerging understanding of the importance of mechanics in cell biology and the clinical application of this knowledge.

12.2 Description of an Optical Common-Path Scheme with Active Maintenance of the Circular Polarization of a Probing Wave for the Cross-Polarization Measurements

A custom-built Multimodal Spectral-Domain OCT device is used for presented research. The device is based on common-path optical scheme with precompensating Michelson type interferometer which forms probing and co- or cross- polarized reference wave pair. Optical radiation of 1325 nm central wavelength 100 nm FWHM spectrum and 10 mW optical power allows ~ 10 μm axial resolution in air. Optical system of the probe with ×3 magnification gives ~ 25 μm lateral resolution. 512-element InGaAs CCD sensor (Sensors Unlimited) provides scanning rate 20,000 A-scan per second. The device realizes: co- and cross-polarized reception of scattered radiation, active maintenance of circular polarization state of probing beam, and special lateral scanning regimes for angio- and elasto- mapping in real time. The device is compact (35 × 35 × 15 cm), has USB 2 interface to PC, and operates under custom built software.

The optical scheme of the system is shown in Fig. 12.1. The common-path optical scheme [16] with active maintenance of the circular polarization of a probing wave consists of five parts: superluminescent source of linearly polarized broadband radiation (I), fiber polarizer (P) with an extinction of about 30 dB [101, 102], and Lefevre polarization-control device (PC-1) supplies a linearly polarized wave to the input of the Michelson interferometer (II). The probing wave and the pre-delayed reference wave for the subsequent interference in the Fizeau interferometer are formed in the short and long arms of the Michelson interferometer, respectively. The long arm contains the electrically controlled Faraday cell (the polarization-rotation angle is 45°). When the Faraday cell is switched off (for odd B-scans) the polarization state of both reference and probe waves is the same and the system registers the so-called co-interference. When the Faraday cell is switched on the

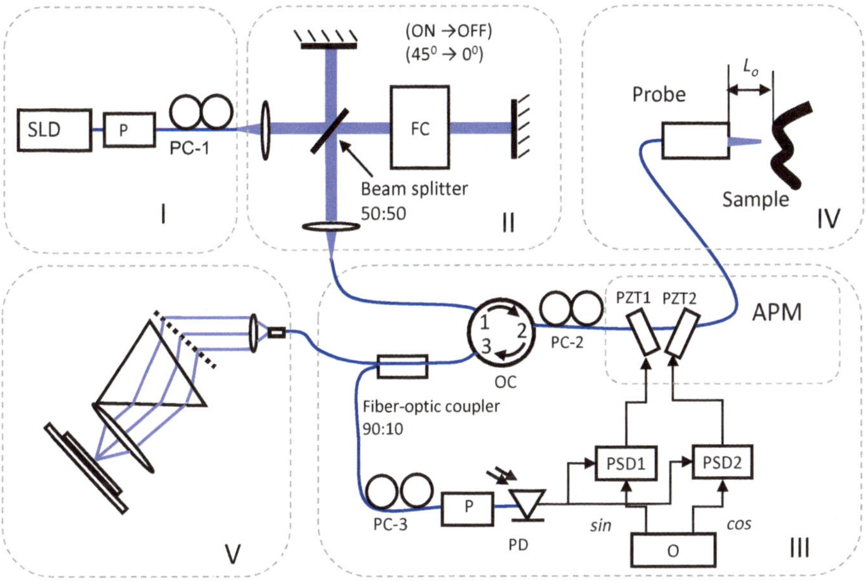

Fig. 12.1 Optical scheme of the multimodal SD-OCT system with automatic control of circular polarization of the probing wave. Here: I is the source of the linearly polarized broadband radiation; II is the precompensating Michelson interferometer with Faraday cell (FC) based reference wave polarization controller, which rotates the polarization to 45° when switched on; III is the system of circular polarization remote control, which is based on active polarization modulator (APM) comprised of two piezoelectric-crystal plates PZP1 and PZP2 with mutual 45° orientation which produce birefringence in optical fiber, and electronics; IV is the probe with the Fizeau interferometer; V is the spectrometer; SLD is the superluminescent diode; PC-1, PC-2, and PC-3 are the fiber polarization controllers; OC is the optical circulator; P denotes the fiber polarizers; PD is the photodiode; O is the oscillator, and PSD 1 and PSD 2 are the phase-sensitive detectors

polarization state of the reference wave rotates to the orthogonal state and the system registers cross-interference. "Cross" means portion of scattered light which changed polarization state in an object and has nonzero projection on orthogonal (relative to probing) polarization state. Unit (III) maintains the circular polarization of the probing wave, whose operation is described below. Probe (IV) with the measuring Fizeau interferometer at the output is the fourth part. In this interferometer, the reference wave reflected from the end of the probe fiber interferes with backscattered waves. The lateral scanning is performed by the mechanical transverse motion of the fiber end jointly with attached focusing lens. Spectrometer (V), which is linearized with respect to the wave number [103], records the interference spectrum of the reference and signal waves.

The method for active control of the polarization state is based on property of quarter-wave retarder to change the linear 45° oriented polarization to the circular one for forward pass and then to the linear polarization orthogonal to the input one on the back way [104–106]. In the considered case, the optical path of the polarization-control system including the polarization modulator (III) and the probe (IV) should be equivalent to quarter-wave plate.

The polarization controller PC-2 is used to form the linear input polarization state at 45° with respect to the axes of the first element of the active polarization modulator (APM). The minor part of light reflected back from the fiber end in the probe is directed via the 90:10 fiber-optic coupler and polarizer to the photodiode of the polarization-control system. The remaining major part of the light is supplied to the spectrometer. The control system of the probing-wave polarization mode is preliminary adjusted using the polarization controller (PC-3) to the minimum of the light transmission through the polarizer (P), while circular polarization at the probe output is maintained manually. In this case, the part of the optical scheme between the Michelson interferometer output and the probe output becomes equivalent to the part between the probe output and the polarizer from the viewpoint of the phase anisotropy and equal to quarter-wave retarder. In the active mode low level alternating voltage of 5 kHz is supplied from a common oscillator to the active phase modulators PZP1 and PZP2 with the quadrature phase shift by the method described in [107]. The error signals in both quadrature channels are formed by the value of the first harmonic at the output of the common photodiode, which detects the light, transmitted through the polarizer. The principle of operation of this scheme is justified in detail in the work [108]. The response time of the realized automatic-control system is about one-tenth of a second. This is enough to stably maintain the circular polarization state of the probing beam for all reasonable manipulations with the flexible probe. This allowed one to perform probing using manual positioning of the probe during both external and endoscopic studies of biological tissue.

12.3 Cross-Polarization Optical Coherence Tomography (CP OCT)

12.3.1 CP OCT in Collagen Assessment

Creation of applicable *in vivo* technologies for early diagnostics and monitoring of the disease progression and cancer invasion based on evaluation of microstructural changes has been remaining one of main trends in the development of optical imaging methods including intraoperative ones routine optical imaging [109]. The CP OCT method, described in this chapter, is a promising competitive method for aforementioned purposes due to the possibility of a separate analysis of anisotropic structures (for example, collagen), exhibiting cross-scattering and birefringence properties [110] along with the inherent for OCT technology high resolution (10–15 microns), 3D label-free imaging of tissue in volume of 2–3 mm^3 in a short time [111]. The possibility of registration of backscattered light in a cross-channel gives an opportunity to calculate relative (to co-channel) values of the OCT signal. CP OCT study of different pathological conditions of stratified tissues like mucosa

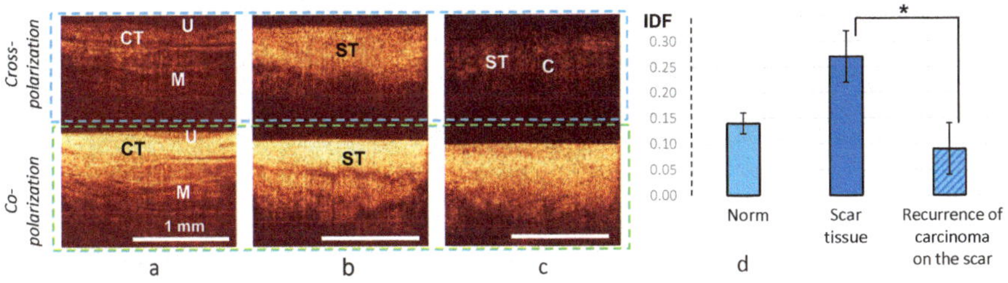

Fig. 12.2 CP OCT study of normal bladder mucosa (**a**), scar tissue (**b**) and recurrence of urothelial carcinoma in the scar region (**c**), and IDF values in these groups (**d**). *CT* connective tissue, *ST* scar tissue, *c* urothelial carcinoma

mostly was focused on epithelial-connective tissue interactions with evaluation of the epithelial thickness and a contrast between two layers [25, 112]. Later, it was shown that state of collagen—the main extracellular protein of the connective tissue matrix—reflects the diseases which occur not only in the connective tissue itself, but also in epithelium layer including cancer cells invasion.

An approach to quantitative, robust, and potentially automated assessment of CP OCT images reflecting the state of collagen fibers was developed [113]. Specifically, we propose to calculate an integral depolarization factor (IDF) that is a ratio of the OCT signal in orthogonal polarization to the analogous value calculated in the initial polarization, both averaged over the transverse coordinates (B-scan direction). It was demonstrated that IDF can be applied to *in vivo* detection of clinically relevant pathological states in urology [113, 114], and have a potential in cardiovascular application [115]. Figure 12.2 demonstrates effectiveness of IDF calculation (Fig. 12.2, d) in case of hardly distinguishable forms of cancer like recurrent carcinoma in the scar area (Fig. 12.2, c) in comparison to postoperative scar tissue (Fig. 12.2, b) and normal bladder mucosa (Fig. 12.2, a).

Thanks to ability of the myelinated nerve fibers for cross-scattering and birefringence like collagen [116], polarization-sensitive OCT can be successfully applied in neurosurgery for verity of different tasks. Visualization of fiber tracts [117] and brain tumors margins detection [118] are the most demanded (needy) one. In this chapter, our recent efforts to increase the accuracy of the tumorous and nontumorous tissues detection using CP OCT as well as identification of white matter location are presented.

12.3.2 *Optical Coefficients for Improvement of the Contrast of OCT Images*

Optical coefficients calculation and based on them color-coded maps construction is one of the methods for OCT contrast improving between different tissue types [119], especially if initial optical contrast is low and visual assessment of the OCT

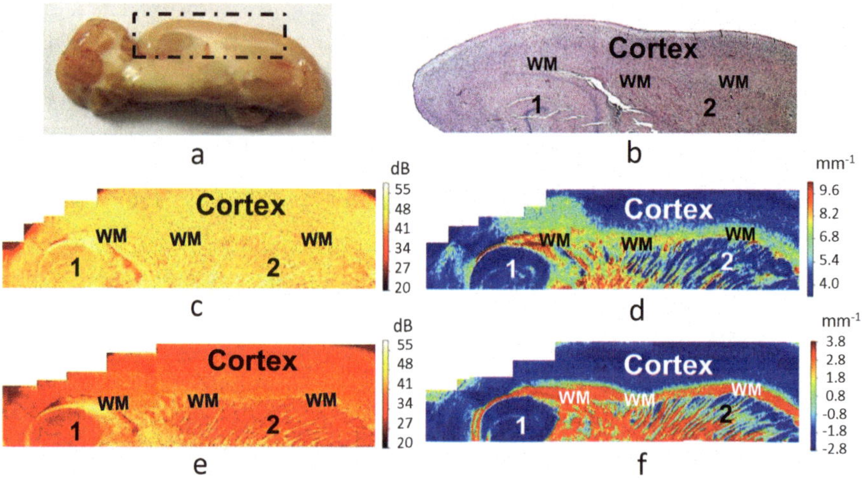

Fig. 12.3 Wide-field unprocessed CP OCT images and color-coded maps of the healthy rat brain (sagittal section): (**a**) photo of the brain. Rectangle indicates region of the CP OCT scanning. (**b**) corresponding histology, H&E staining. (**c**–**f**) wide-field brain images reconstruction from unprocessed images in co- (**c**) and cross-polarizations (**e**) and corresponding color-coded maps (**d**, **f**): attenuation coefficient (**d**), forward cross-scattering coefficient in log scale (**f**). Color-coded maps based on optical coefficients calculation provide better contrast for clearly visualization of the different brain structures: cerebral cortex, white matter, hippocampus (1), striatum (2)

image cause great doubts. It can occur in case of glial tumor margins detection, when high infiltrative and uneven nature of tumor growth creates difficulties in finding the true margins with normal brain tissue.

Using normal rat brains, a method for CP OCT image quantification was developed and tested. It is based on calculating the two optical coefficients (attenuation rate and forward cross-scattering (FCS)) and building en-face color-coded maps to show their 2D distribution. The details of the coefficients computations can be found in [120]. Figure 12.3 demonstrates an example of how optical coefficients can increase the contrast of white matter versus gray matter in normal rats brain.

12.3.3 Optical Coefficients That Improve the Contrast of OCT Images: A Patient with a Brain Glioma—Detection of a Clean Resection Edge

Compared with intensive *en face* CP OCT images, color-coded maps of glioma, and white matter based on different optical coefficients (Fig. 12.4. a3–f3, a4–f4) were more representative for visual assessment of the tissue. The color set of each map reflected the values of the corresponding optical coefficient; therefore, the areas of high attenuation or forward cross-scattering (myelin fibers, necrosis) appeared in

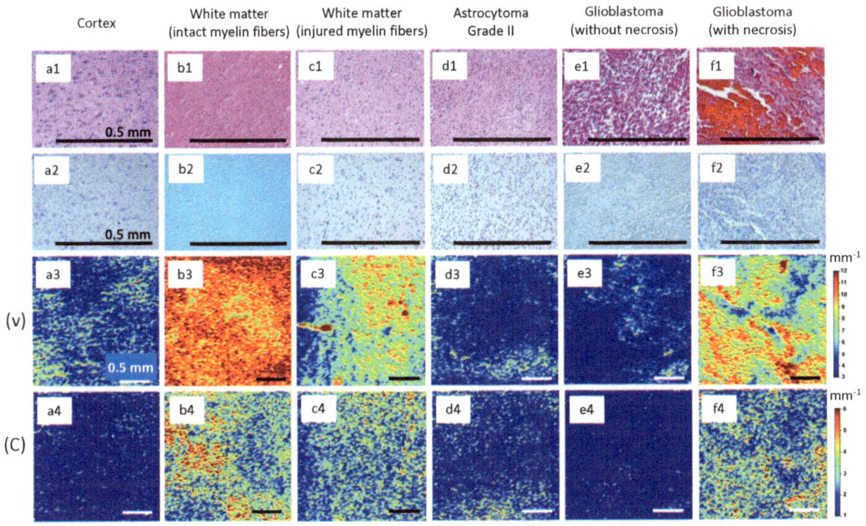

Fig. 12.4 Color-coded maps and corresponding histology of the tumorous and normal brain tissues. Based on attenuation (ν) (**a3**–**f3**) and forward cross-scattering (C) (**a4**–**f4**) coefficients *en-face* color-coded maps of cortex (**a1**–**a4**), white matter with intact (**b1**–**b4**) and injured (**c1**–**c4**) myelin fibers, astrocytoma Grade II (**d1**–**d4**), glioblastoma Grade IV with necrosis (**e1**–**e4**) and without it (**f1**–**f4**) corresponding to histology in H&E (**a1**–**f1**) and luxol blue (**a2**–**f2**) staining

light tones, and the areas of low attenuation or forward cross-scattering (cell clusters) appeared in darker tones.

Thus, on the attenuation map, white matter formed by densely packed myelin fibers appeared red and yellow (Fig. 12.4. b4), and tumors formed by cell clusters appeared dark or pale blue (Fig. 12.4. d3, e3). On the forward cross-scattering map, white matter appeared pale blue to red (Fig. 12.4 b4), whereas tumors appeared dark or pale blue (Fig. 12.4. d4, e4). These differences are more evident on color-coded maps of the margin between tumors and white matter. The dividing line was more clearly visualized by every coefficient (Fig. 12.4. c, d, dark green dotted line). Nevertheless, on two separate occasions, the white matter and tumor tissue looked similar on OCT images as follows: (1) white matter in the peritumoral area, particularly in high-grade gliomas, could be characterized by a decreased OCT signal because of myelin fiber destruction and/or intense edema (Fig. 12.4. c3, c4); in this area, tumor cells could likely persist and (2) tumor tissue with necrotic areas (e.g., glioblastoma core, tumor tissue after partial bipolar coagulation) had high attenuation and forward cross-scattering (Fig. 12.4. f3, f4).

The cortex comprises neurons and glial cells surrounded by uncompacted myelin fibers. For this reason, white matter on attenuation maps was represented by the whole color set, but on forward cross-scattering, it was mainly blue, similar to astrocytomas (Fig. 12.4. a3, a4). Thus, the identification of a well-marked margin could be difficult.

12.4 CP OCT for Breast Cancer Imaging

CP OCT system was used for assessment of the structural features of breast cancer tissue, permitting improvement in the contrast between benign and malignant tumorous tissues, based on a detailed evaluation of the degenerative changes of collagen associated with breast cancer invasion [121]. Figure 12.5 demonstrates CP OCT images (Fig. 12.5, c, d, i, j), color-coded maps (Fig. 12.5 e, f, k, l), and corresponding histological slice (Fig. 12.5, a, b, g, h) of the benign breast disease and malignant breast cancer features. It was shown that the use of optical coefficients increases the OCT contrast for breast cancer detection. In comparison to optical coefficients maps (Fig. 12.5 e, f), optical contrast based on the signal intensity (Fig. 12.5, c, d) difference is insufficient to distinguish breast tissue structures. For the purpose of malignant breast cancer identification, forward cross-scattering coefficient is more suitable. In the map of forward cross-scattering coefficient obtained for benign breast fibroadenoma (Fig. 12.5, upper box), regions of fibrous collagen fiber of stroma look as red areas. By contrast, invasive ductal carcinoma (Fig. 12.5, bottom box) demonstrated a low forward cross-scattering coefficient in the areas of tumor cells and appeared on color-coded map as blue areas with a higher coefficient from the rare collagen fibers of the tumor stroma (Fig. 12.5 j). Color-coded map of attenuation coefficient shows areas of high density of the scatterer's accumulation in large values (Fig. 12.5 k, yellow-red and bright-green colors).

12.5 Automated Segmentation of CP OCT Images

The possibility of the automated segmentation of CP OCT images was demonstrated on the pre-clinical murine ear model implanted with mouse colon carcinoma CT-26 [122]. Every pixel in the OCT cross-sectional image was represented as a set of decomposition coefficient of the local reflectivity profile into two orthogonal basics. Each basis was constructed from the set of profiles for normal and pathological OCT image regions manually segmented by an export using principal component analysis (PCA). A set of images manually segmented by comparison with histology was used to train the random forest trees classification algorithm. The outcome of the classification algorithm was a value in the range 0–1, which was interpreted as probability of a given pixel to represent a tumor. The example of the classification result is presented in Fig. 12.6 [122].

The algorithm was validated on the separate set of manually segmented images. The achieved sensitivity was equal to 0.83 and specificity equal to 0.82. For independent technique validation *en-face* borders of the tumors obtained from 3D volumetric classification results were superimposed with fluorescence image of CT-26 murine colon carcinoma expressed genetically encoded fluorescent protein KillerRed. The superimposed tumor boundaries imaged with fluorescence microscopy and classified from OCT volume datasets are presented in Fig. 12.7 [122].

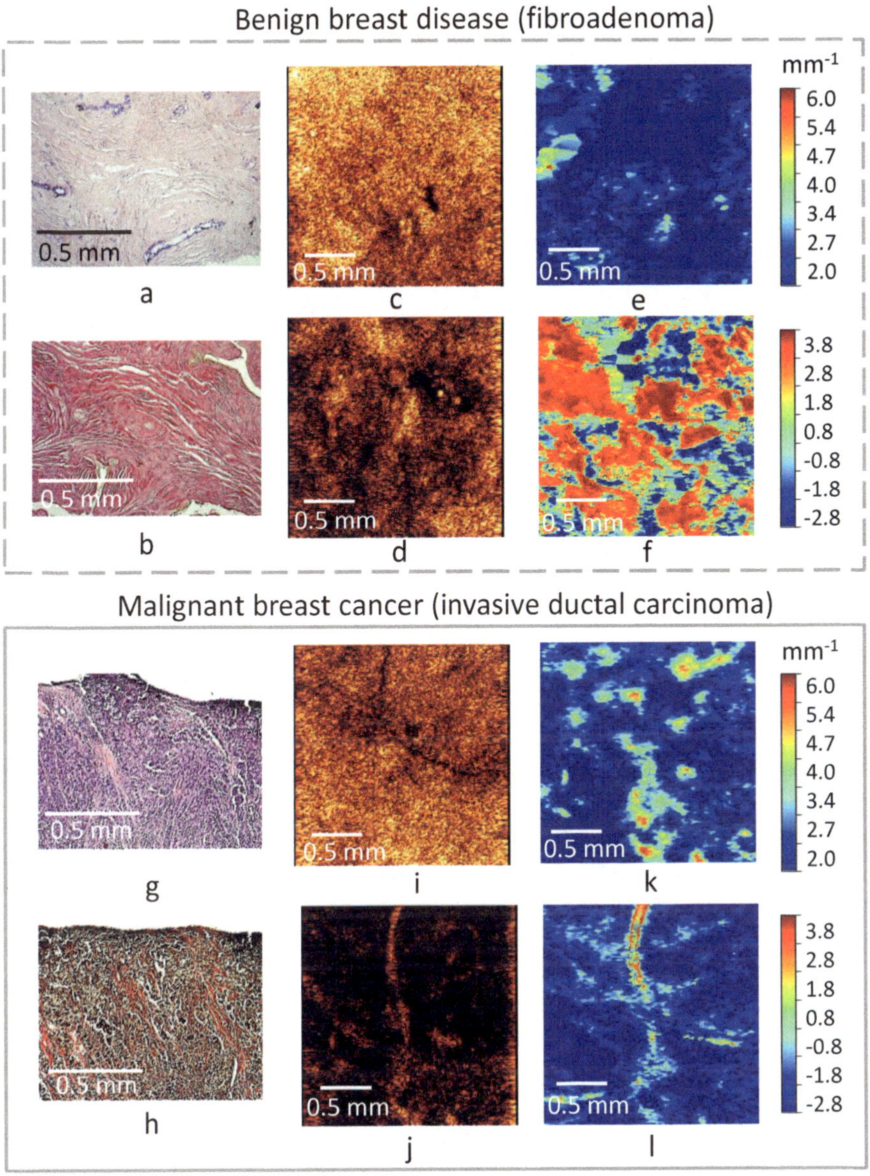

Fig. 12.5 CP OCT images (**c**–**d**, **i**–**j**), color-coded maps based on attenuation coefficients (**e**, **k**) and forward cross-scattering (**f**, **l**) coefficients of benign and malignant breast cancer. The corresponding to histology in hematoxylin and eosin (**a**, **g**) and Van-Gieson's (**b**, **h**) staining

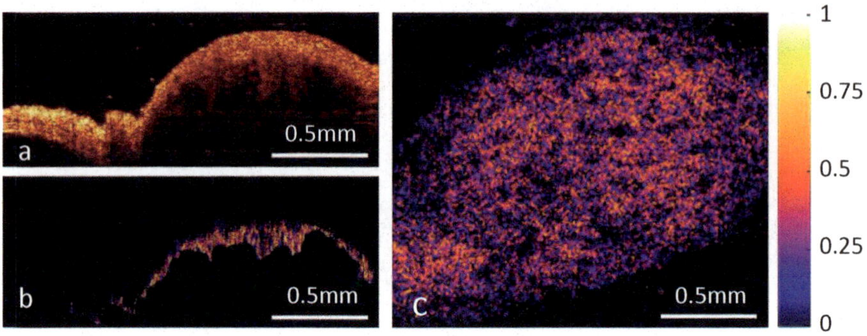

Fig. 12.6 Cross-sectional OCT images pixel classification process: (**a**) OCT image of a CT 26 on ear mice; (**b**) prediction of classification algorithm: map of tumor probability; (**c**) en face projection of the tumor region [122]

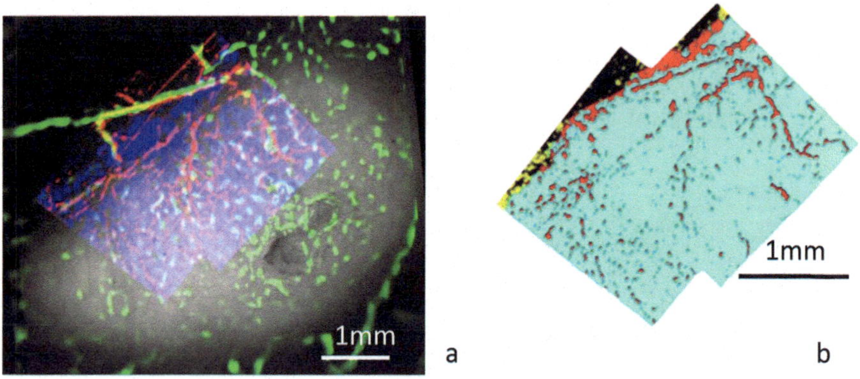

Fig. 12.7 (a) Alignment of fluorescence and angiographic OCT images by superimposing vessels meshworks (green—vessels visible in fluorescence, red—vessels obtained by OCT angiography processing). In purple en face projection of OCT classification results is presented, tumor visible in fluorescence channel is presented in gray. (**b**) Overlapping of tumor projections visible in fluorescence and obtained by the proposed OCT pixel classification method [122]

12.6 Optical Coherence Angiography (OCA)

12.6.1 Main Features of the Developed Real-Time OCA Realization

Vessels contrasting was made by high frequency filtration along the slow axis in the signal-domain [123]. One can expect that regions of the object, experienced motion during the scanning will induce higher frequency in the Fourier-domain:

$$V_{k,j+N/2,n} = \sum_{m=N-1}^{m=0} b_m B_{k,j-N/2+m,n}, \tag{12.1}$$

Here first index is a fast axis index, second index—slow axis index, and third is an axial index; $B_{k,j,n}$ is the jth B-scan, $V_{k,j,n}$ is the jth cross-section of the vasculature image, b_m are coefficients of high-pass filter impulse response.

Using such filtration one angiography B-scan from every 7 consecutive OCT B-scans can be obtained and every following angiography B-scan can be re-evaluated during the following B-scan acquisition.

Local phase correction preprocessing was applied to the OCT data to compensate for motion and compression artifacts which may occur during the data acquisition:

The corrected phase distribution can be written as:

$$\varphi_{k_0,n_0} = \arg\left(\sum_n B_{k,j,n} \cdot B^*_{k,j+1,n} \cdot \mathrm{rect}_M\left(n-n_0, k-k_0\right)\right), \tag{12.2}$$

where $\mathrm{rect}_M(n-n_0,k-k_0)$ is the rectangular window of size M, centered at the location (k_0,n_0).

To additionally contrast the signal from the vessels, amplitude of the OCT data was normalized prior to the filtration:

$$S_{k,j_0,n} = \frac{B_{k,j_0,n}}{\sqrt{\left(\frac{1}{N}\sum_{j=j_0-N/2}^{j=j_0+N/2} B_{k,j,n}\right)^2 + T^2}}, \tag{12.3}$$

where $S_{k,j,n}$ is the value substituted into Eq. (12.1) instead of $B_{k,j,k}$, T is a constant set to be equal to the mean amplitude in the dataset. The function in the denominator of Eq. (12.3) asymptotically approaches function y = x when signal's amplitude is high and constant T, when its amplitude is low.

12.6.2 Tumor Model for Optical Coherence Angiography Investigation

Ear tumor model is fairly well controllable model to study PDT vascular reaction. The specifics of the ear tumor model have been previously described [123, 124]. Typically, normal ear tissue vascularization is similar in different mice (Fig. 12.8 a). Visual comparison of OCA images of tumor and normal tissue reveals pronounced differences in vascular architecture and vascular density (Fig. 12.8 a, b). It is well known and can be seen in OCA images that tumor's vessels are abnormal with strongly increased tortuosity in comparison with normal vessels [125, 126]. Because of active neoangiogenesis, the tumor vasculature is significantly denser (Fig. 12.8

Fig. 12.8 OCA visualization of microvasculature network architecture in normal murine ear (**a**) and in murine colon carcinoma CT26 transplanted into the murine ear (**b**)

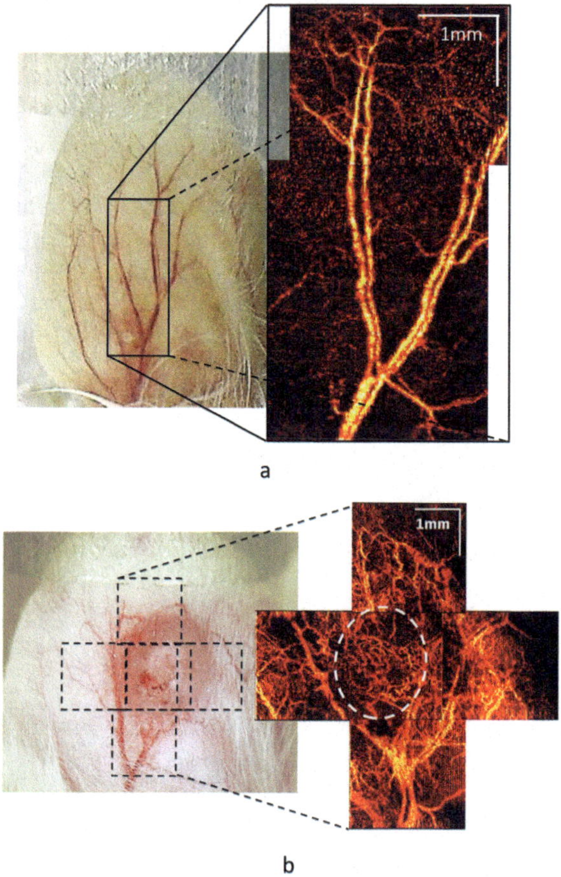

b). The small tumor depth and the small thickness of ear tissue make the model "transparent" for OCA, so that OCA can be used for characterizing the blood vessels [127].

12.6.3 OCA Monitoring of Tumor Response to Photodynamic Therapy (PDT)

On ear tumor model CT26, using OCA it was found that not all tumors responded to PDT in the same manner [128–130]. The tumors with severe microvascular damage (revealed by histology) displayed total disappearance of blood vessels on OCA images within the shorter time interval in 1 day (Fig. 12.9 upper row). Conversely, weak microvascular damage manifested itself as partial and reversible blood vessels disappearance (Fig. 12.9 middle row). Tumors after light exposure only as well as untreated tumors (Fig. 12.9 bottom row) demonstrated well-developed blood vessels network on OCA images throughout this observation time [128–130].

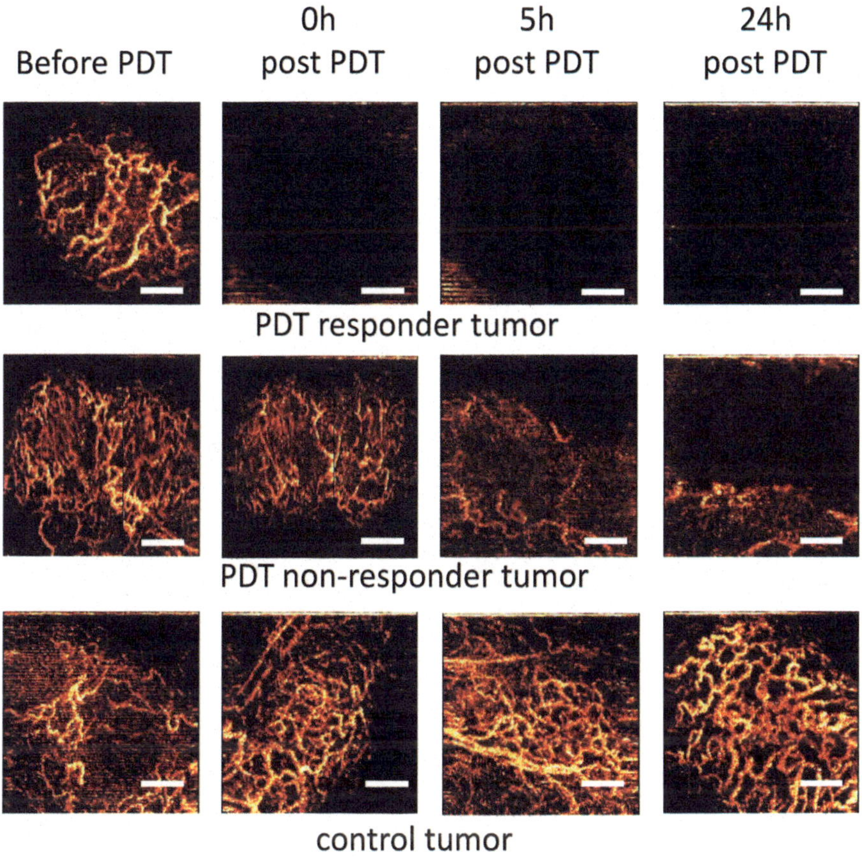

Fig. 12.9 OCA monitoring of tumor blood perfusion alteration dynamic before PDT, immediately post-PDT, 5 h and 24 h post-PDT. The upper row is an example of PDT responder; the middle row is an example of non-responder tumor; the bottom row is an example of control tumor (untreated). Scale bar size is 500 μm

To conclude, using the results for murine ear tumor model CT26, the following criterion of PDT success has been formulated: complete disappearance of blood vessels on OCA images in 24 h post PDT is associated with good clinical outcome.

12.6.4 OCA for Pre-Treatment Pathology Assessment and Treatment Monitoring Following PDT of Basal Cell Carcinoma Patients

PDT is emerging as a common and efficacious method for basal cell carcinoma (BCC) treatment, and new non-invasive imaging technology can further enhance it. OCA can accurately differentiate BCC from normal skin, assess BCC-related changes in the surrounding skin regions that appear normal (Fig. 12.10). Pre-PDT

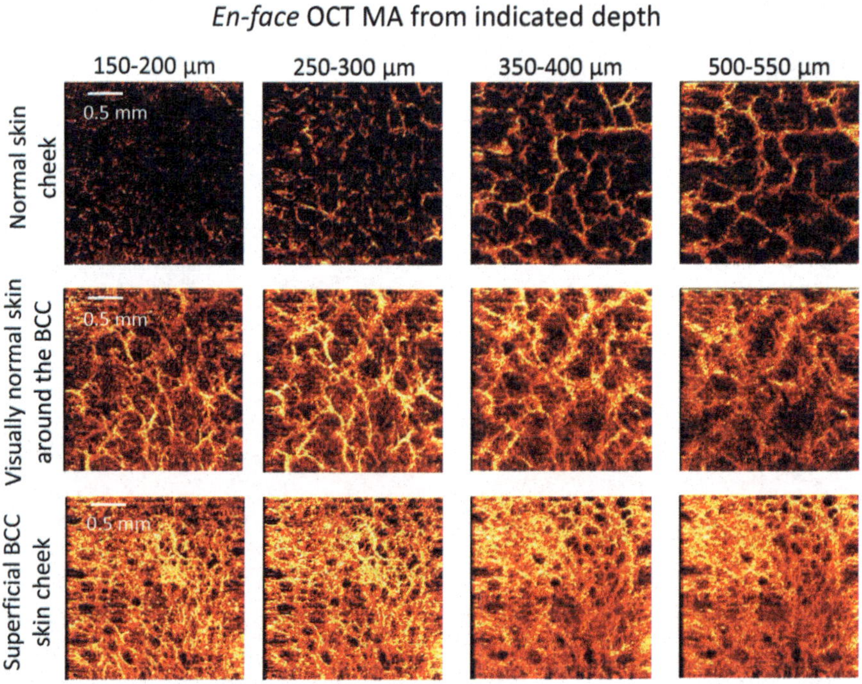

Fig. 12.10 *En-face* OCA images of normal skin (top row), compared with visually normal skin around the BCC (middle row), and superficial BCC lesion itself (bottom row). *En-face* OCA images are shown at four selected depths; the microvascular patterns are also seen to vary (thicker vessels, denser network) from normal skin to near-lesion regions to superficial BCC lesion core itself

BCC assessment *en-face* OCA images show microvascular networks at different indicated depths (Fig. 12.10, upper row), with average vessel diameter seen to increase with depth (from 150 to 550 μm). The OCA images of the visually normal skin around the BCC (Fig. 12.10, middle row) show to increased density of the microvascular networks, potentially indicating tissue inflammation around the visible lesion. OCA of the superficial BCC (Fig. 12.10, bottom row) exhibits a dense network of thin branching vessels at a depth of 250–300 μm. The deeper lesion vasculature is preserved, denser than much less dense network in normal skin, and is poorly visualized at depths exceeding ~500 μm [131].

The additional OCA imaging was used post-PDT to visualize treatment-induced microvascular alterations (immediately and 24 h after PDT) and healing dynamics (6 months after PDT) of BCC (Fig. 12.11). OCA analysis using vascular density demonstrated a complete (Fig. 12.11, upper row) and partial (Fig. 12.11, bottom row) vascular reaction at early stages post-PDT treatment of the different subtype BCC lesions. Immediately post-PDT, OCA shows a weak vascular reaction to the treatment; a small part of the vessels was well visualized in the tumor center. At 24 h post-PDT, there is an increase in vascular reaction. Complete disappearance of the

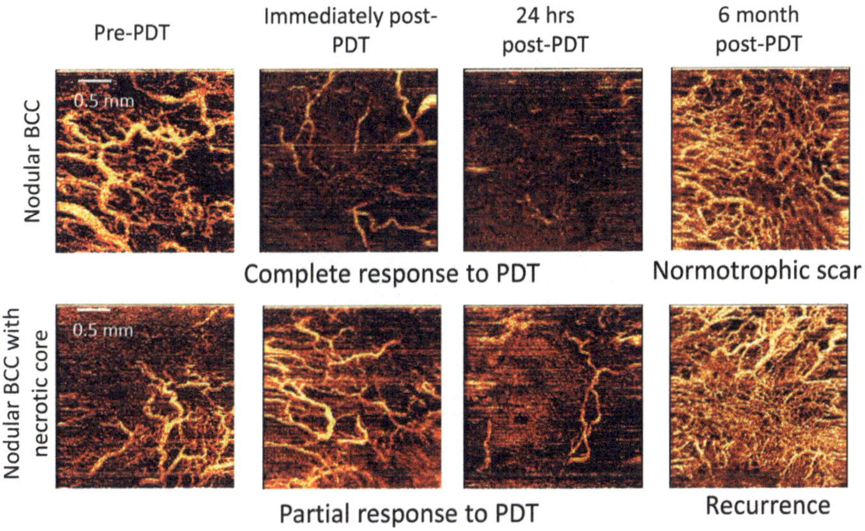

Fig. 12.11 OCA follow-up with two variants vascular responses to PDT in the different subtypes BCC before PDT, immediately and 24 h post-PDT. The demonstrates different clinical outcomes with complete (upper row) and partial (bottom row) responses 24 h post-PDT and scar formation or recurrence 6 month post-PDT

tumor vasculature at 24 h post-PDT could be due to direct PDT action. After at 6 months follow-up post-PDT OCA demonstrates high vascularization in the formed scar at its 6-months maturation stage in case of complete response (Fig. 12.11, upper row). In case of partial response OCA demonstrates an increased vascular density with slightly enlarged diameter blood vessels (Fig. 12.11, bottom row). Similar treatment response was observed for nodular BCC with necrosis, demonstrating partial microvascular damage within 24 h after PDT; preclinically, this correlated with long-term tumor recurrence.

Thus the demonstrated ability of OCA for BCC differentiation and subsequent PDT treatment response monitoring hold promise for improved managements of BCC patients [131].

12.6.5 OCA Mucositis Study During Radiation Therapy

Fifteen patients with stage II–IV squamous cell carcinoma of the oral cavity and pharynx were investigated. Patient study was approved by the Research Ethics Board of the Nizhny Novgorod State Medical Academy. Irradiation was performed using a linear accelerator (Varian Clinac 600) or Co60 system (Terabalt). Mucositis degree was scored by Radiation Therapy Oncology Group and European Organization for Research and Treatment of Cancer scale.

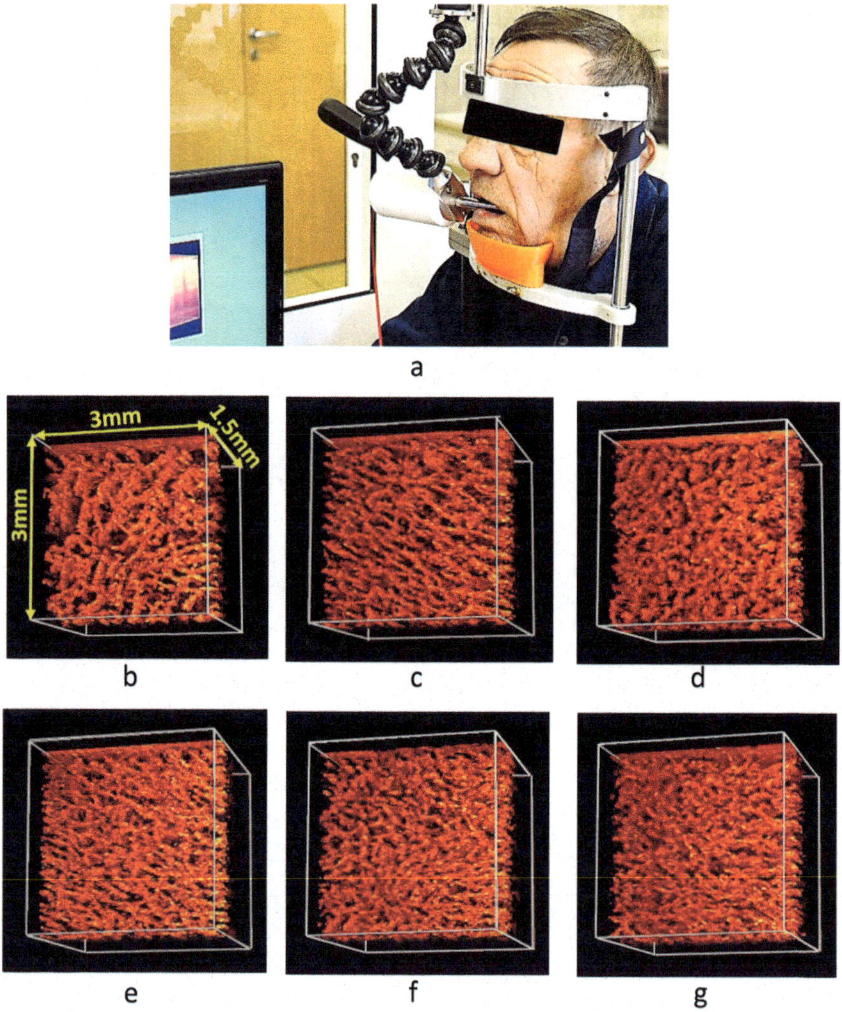

Fig. 12.12 (**a**) An example of OCA monitoring of buccal mucosa, (**b**) 3D OCA images before RT, (**c**) 3D OCA images before visual signs of mucositis appear (after 8 Gy), (**d**) 3D OCA images of grade 1 mucositis (10–12 Gy), (**e**) 3D OCA images of grade 2 mucositis (after 14 Gy), (**f**) 3D OCA images of grade 3 mucositis (after 20 Gy), (**g**) 3D OCA images after initiation of anti-mucositis therapy

OCT imaging was performed three times per week throughout the radiation therapy (RT) course on two symmetric sites on both cheeks (Fig. 12.12 a), along the centerline connecting the secretory duct of the salivary gland and the angle of the mouth. The probe was positioned on the mucosa with gentle contact [132].

OCA images of the normal buccal mucosa show dense, uniform microvascular networks (Fig. 12.12 b). At doses of 4–8 Gy, all patients exhibited an increased vessels density (Fig. 12.12 c). Continued dose accumulation caused an increase in the

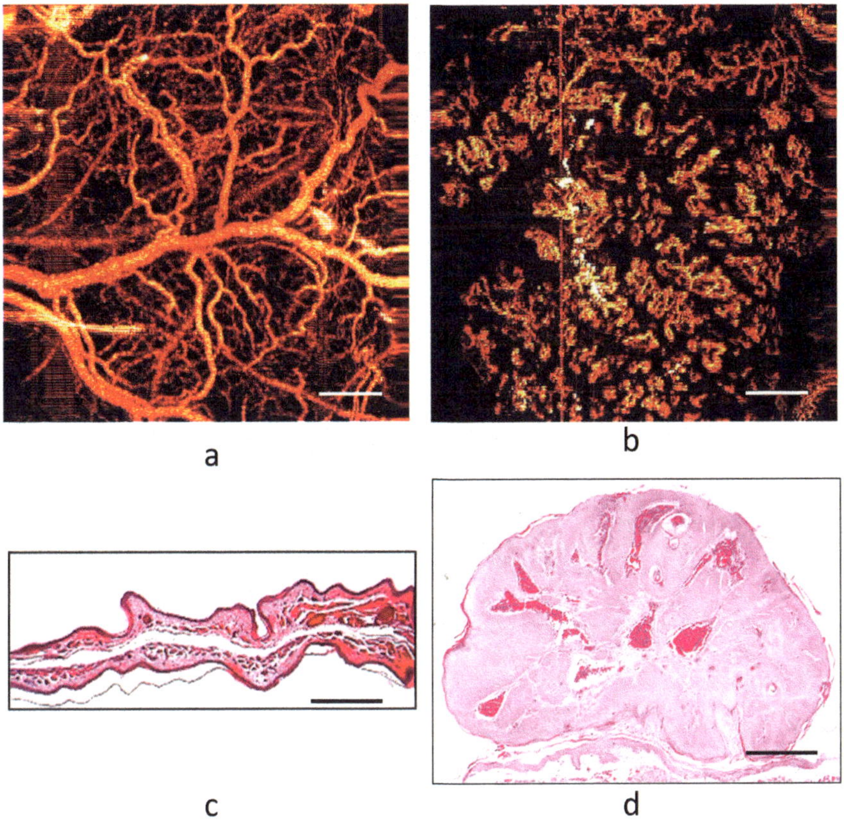

Fig. 12.13 (**a**) OCA images of normal hamster cheek pouch, (**b**) OCA images of chemical induced (7,12-Dimethylbenzanthracene DMBA) cancer on hamster cheek pouch, (**c**) Histological image of normal hamster cheek pouch, (**d**) Histological image of cancer induced by DMBA. Scale bar size is 500 μm

clinical manifestations of RT reaction, as also seen by the 3D OCA vascular density (Fig. 12.12 d–g). Statistically significant changes of vasculature microvascular parameters (compared to their initial pre-RT values) were detected when grade two and grade three mucositis developed (Fig. 12.12 e, f) [132].

The study showed that OCA monitoring in head and neck RT patients can be used for objective evaluation of radiation induced microvascular changes in the oral mucosa, for "shedding light" on the temporal sequence of early functional radiation toxicities, and potentially for design and effectiveness evaluation of anti-mucositis treatment and prophylaxis modalities.

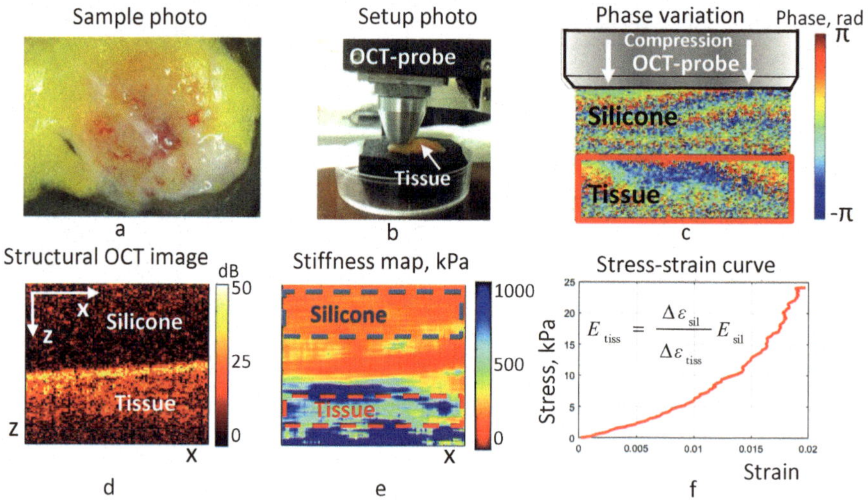

Fig. 12.14 Elucidation of experimental procedures for performing OCE imaging. An example of excised breast tissue sample (**a**), OCT probe pressing onto the studied sample (**b**), inter-frame phase variation (**c**), structural OCT B-scan of cancerous tissue under reference silicone (**d**), OCE image (**e**) and (**f**) is an example of stress-strain dependence for the tissue found with averaging over ROI windows shown as dashed-line rectangles in (**e**)

12.6.6 OCA for Studying Chemically Induced Cancer

Normal hamster cheek pouch is well vascularized. There are blood vessels with various diameter from 7 to 40 μm, and single vessels with diameter 100 μm (Fig. 12.13 a, c). The architecture of microvascular network contained long, thin vessels connected between each other (Fig. 12.13 a). OCA can detect the neovascularization in carcinoma. One of the criteria to distinguish squamous cell carcinoma is a large number of the short vessels looks like loops (Fig. 12.13 b) [133].

12.7 Optical Coherence Elastography (OCE)

12.7.1 Realization of Compressional OCE

A variant of compressional OCE described in [134–137] was used to visualize local inter-frame strains in the tissue induced by the tissue deformation. For estimating strains, local gradients of inter-frame phase-variations were calculated using an improved version of the phase-sensitive monitoring of displacements of scatterers in the tissue and a robust vector method for estimating phase gradients.

For quantification of tissue stiffness in compressional OCE, a calibration (silicone) layer with preliminary calibrated stiffness was used. The silicone used in the experimental examples described below had a Young's modulus in the range 50–100 kPa as the most suitable for studying tissue stiffness variations in the range from 20 kPa to 1000 kPa or even greater. A typical experimental configuration is shown in Fig. 12.14. In such a way, OCE examinations of both excised tissue samples and *in vivo* studies of experimental tumors inoculated on mice's ears were possible. The reference silicon layer with a known stiffness (used as stress sensor) was placed on the tissue surface, the probe was slightly pressed onto the studied tissue (Fig. 12.14 (b)), and the resultant strain distribution in the probe vicinity was reconstructed. Schematic Fig. 12.14 (c) shows a typical inter-B-scan phase variation in the studied tissue overlaid by the reference silicone layer. The latter is clearly seen in the structural B-scan OCT image (Fig. 12.14 (d)). Comparison of strain distribution within the pre-calibrated reference layer and examined tissue allowed for quantitative estimation of tissue stiffness.

The elastic Young's modulus of the tissue is defined as the ratio of strain increments in silicon and tissue multiplied by silicone stiffness. The stiffness B-scans derived from the analysis of inter-frame phase gradients could be represented as color-coded maps for the Young's modulus (kPa), usually in the range from several kPa to 1000–2000 kPa (Fig. 12.14 (e)) or can be analyzed in the form of stress-strain dependences obtained in a chosen ROI (Fig. 12.14 (f)).

A very important point in quantification of stiffness is that the Young moduli were determined for a pre-selected pressure exerted by the reference silicone layer onto the studied tissue. Indeed, our preliminary studies of breast cancer samples in agreement with previous data [138] indicated that the stress-strain relationship for such tissues may exhibit pronounced nonlinearity, so that for apparently very moderate straining (within a few per cent only), the Young modulus may vary several times. This fact should be taken into account when comparing different datasets. Therefore, to ensure the possibility of meaningful comparison between samples from different patients obtained in different days and/or using different calibration layers, the Young modulus was estimated for the same pre-selected pressure created within the calibration layer.

For feasibility of such measurements, the developed robust vector method [134, 136] of strain estimation was very important, because it ensured rather reliable estimation of inter-frame strains under aperiodic, approximately monotonic compression that was manually produced in the described below experiments. On the so-obtained stress-strain curves (obtaining of which was discussed in detail in [137]), narrow pre-selected ranges of pressure could be readily chosen to standardize the Young modulus quantification even for samples with different uneven thickness, potentially strongly variable stiffness, etc. For the quantitative estimates presented below, we used pressure ranges from 3 kPa to 5 kPa centered at 4 kPa. Reproducibility of such measurements, including the usage of calibration layers with different stiffness, was verified in a special series of experiments.

For quantitative characterization of stiffness in various breast tissue samples, ROI windows in OCE images could be chosen. The typical size of ROIs was 30 × 120 pixels (with pixel size ~ 4 microns in both directions). Examples of such windows in the silicone and tissue are shown in Fig. 12.14 (e). Using the averaging over such windows, plotting the strain in the silicone with pre-calibrated stiffness against the tissue strain yields stress-strain curves for the tissue [135, 137]. An example of such a curve is shown in Fig. 12.14 (f). Then one can find the ratio between the stress and strain increments for a standardized pre-selected pressure range as was explained above. In such a way different data obtained for different tissues in different experiments can be meaningfully compared. Examples of estimates of the Young modulus (stiffness) for several types of different tissues are presented in the next section. The results of OCE-based estimations of stiffness can be presented either in the form of spatially resolved color-coded maps or in the forms of dependences of stiffness on the pressure (stress) exerted onto the tissue.

Analysis of these OCE-based data can be used to develop procedures for differentiation of cancer and non-cancer regions (assessing the negative surgical margin of resection), differentiation of various cancer subtypes and monitoring of the tumor response to therapies.

12.7.2 OCE Monitoring of Tumor Response to Chemotherapy

The OCE monitoring of the elastic properties of the murine tumor model of breast cancer 4T1 (Fig. 12.15 a) made it possible to detect changes in the state of the cellular components of the tumor. The OCE monitoring (Fig. 12.15 b) was performed at different stages of tumor development in the control group and in the cytotoxic cisplatin multiple exposure group [139]. On day 5, the tumors in the treatment group (Fig. 12.15 f) were significantly softer than those in the control group (Fig. 12.15 c). This trend continued up to day 12 when the average stiffness of tumors in the therapeutic group (Fig. 12.15 g) was softer than those in the control group (Fig. 12.15 d). By the end of the observation, the tumors in both groups had low stiffness values, which were due to the developed necrosis of tumor cells. Pathological processes (irreversible changes in tumor cells and spontaneous or chemotherapy-induced necrosis of cells) found on histological examination (Fig. 12.15 e, h) lead to a decrease of stiffness on the tissue. The predominance of areas with pathological changes of cells over areas with viable cells in the tumors of the treatment group was shown both by histological examination and by OCE monitoring.

12.7.3 Optical Coherence Elastography for Diagnosing Breast Cancer and Detecting Breast Cancer Margins

Elastic properties of softbiological tissues may reflect their functional state and be a sensitive indicator of important pathological processes (including tumor) occurring in them [140]. Using compressional optical coherent elastography, significant differences were shown not only in the value of the Young modulus, but also in the character of elastic responses of nontumorous and tumorous breast tissues in *ex vivo* human tissues [141]. OCE is capable to reveal the difference between elastic

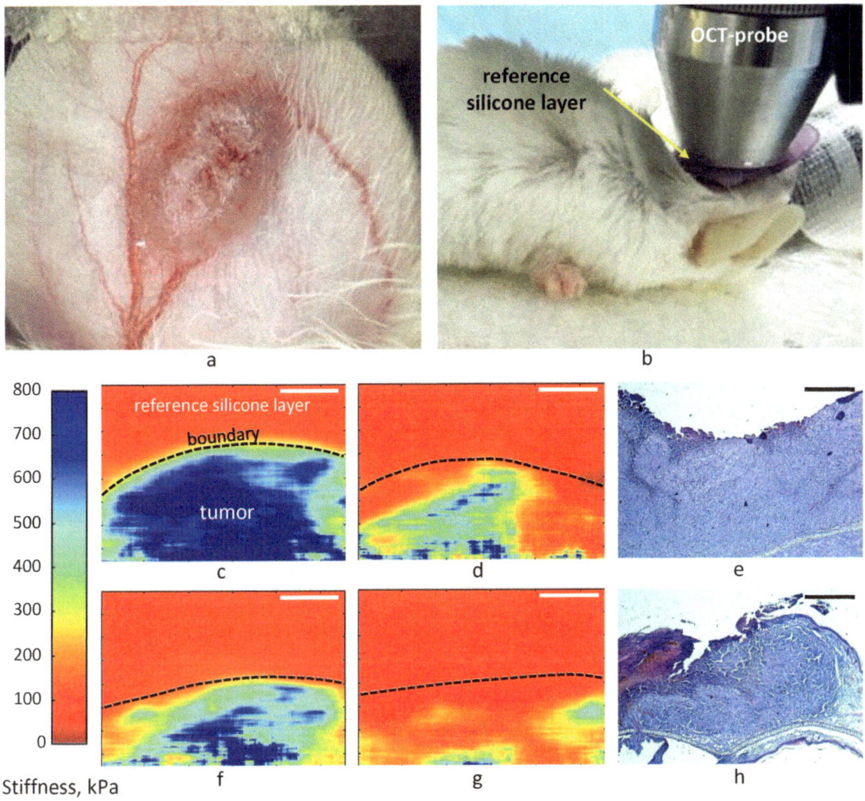

Fig. 12.15 (**a**) Tumor tissue on the mouse ear; (**b**) An example of OCE monitoring of mice tumor model; OCE images of stiffness distribution in the tumor tissues of mice in the control and therapeutic groups: changes in tumor stiffness in the control group as detected on days 5 (**c**) and 14 (**d**), changes in tumor stiffness in the therapeutic group as detected on days 5 (**f**) and 14 (**g**); on the left, the stiffness scale expressed in kPa for the Young's modulus. Panels (e) and (h) show the histological slides obtained on day 12 and demonstrating spontaneous and chemotherapy-induced necrosis of cells corresponiding to the elastographic images (d) and (g), respectively

properties of nontumorous (normal) tissue (Fig. 12.16 a, b); benign breast disease (fibroadenoma) (Fig. 12.16 c, d) and invasive ductal carcinoma (IDC) (Fig. 12.16 e, f). OCE shows clear correlation of the histologically revealed structural features (stromal and cellular components) of breast tissue with the character of elastic response and increase in stiffness with compression typical of tumor tissue. Figure 12.16 g represents the stiffness-strain dependencies for three samples and indicates the elastic behavior with rather low deformability of IDC (with stiffness ranged from 100 kPa for small deformations <1%, to about 800 kPa with further loading) compared to normal breast tissue and benign breast cancer with significantly smaller and fairly weakly varying stiffness.

Figure 12.16 f demonstrates that breast cancer is characterized by pronouncedly heterogeneous distribution of stiffness. Recently performed OCE examinations of several tens of breast cancer samples excised during surgical operations revealed

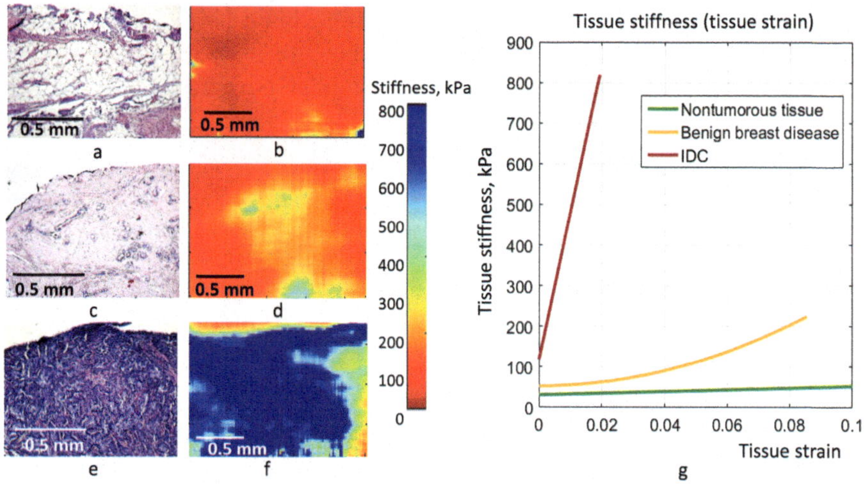

Fig. 12.16 Example of breast tissue samples in three states: nontumorous (normal) tissue (**a**, **b**); benign breast disease (fibroadenoma) (**c**, **d**) and invasive ductal carcinoma (IDC) (**e**, **f**). **a**, **c**, **e** —are the histological slides stained with H&E; **b**, **d**, **f**—are the corresponding OCE images; **g**—is the plot of the stiffness-strain dependence for the three samples

rather clear correlation between morphological tissue component and the corresponding range of its Young modulus. Thus it became possible to assess morphological composition of the studied tissue region by evaluating percentages of areas in OCE images falling into certain specific stiffness ranges. This analysis of the elastic modulus ranges in OCE images can be called "elasto-spectroscopy" by analogy with "mass-spectroscopy." The morphological structure of the tumor in its turn demonstrates pronounced correlation with molecular status of breast cancer. Thus obtaining OCE scans and their elasto-spectroscopic analysis opens promising prospects for express assessments of breast cancer subtypes [142].

Another application of OCE is delineation of tumor/non-tumor regions. An example of OCE visualization of such a transitional zone is shown in Fig. 12.17, where the border between non-tumor (normal) and tumorous breast tissue is clearly seen in the OCE image. Comparison between the OCE results (Fig. 12.17 bottom image) and histological data (Fig. 12.17 upper image) made it possible not only to distinguish between normal tissue and tumor (for determining resection boundary), but also to formulate OCE-based criteria allowing for differentiation between noninvasive (ductal carcinoma *in situ* (DCIS)) and invasive tumor (IDC). The OCE images very clearly show the cross sections of the ducts filled with tumor cells for DCIS as high-contrast zones with strongly increased stiffness, which will agree with the histological images. In the case of IDC is clearly seen via ever increasing proportion of stiff tumor cells over the scans and disappearance of highly localized regions of tumor cells embedded into surrounding much softer stromal tissue [142].

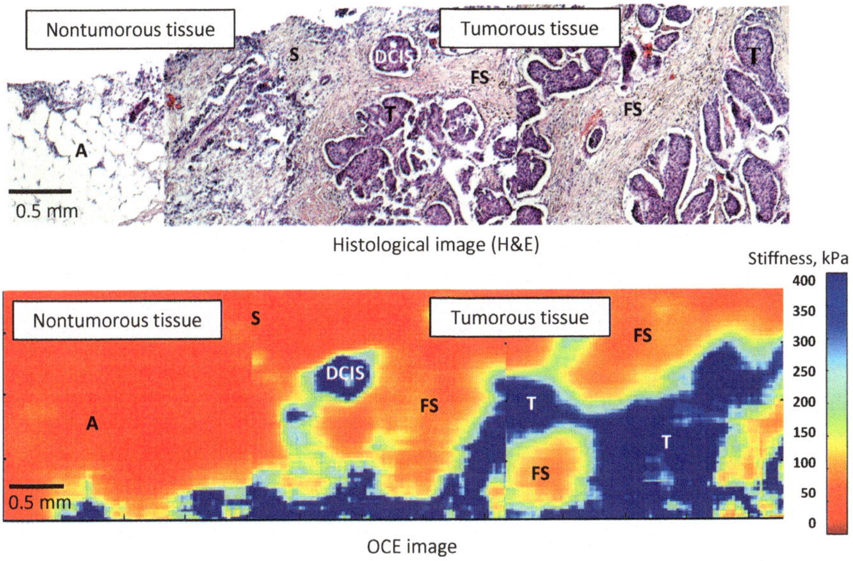

Fig. 12.17 Visualization of a transitional zone between peritumoral (nontumoral) breast tissue and tumorous tissue region using histological and OCE-based images. Tumorous tissue included both a non-invasive ductal carcinoma *in situ* (DCIS) and an invasive ductal carcinoma (IDC). Letters in the images show areas of adipose breast tissue (A), normal stroma (S), fibrous stroma (FS), and tumor cells (T) of IDC

12.8 Conclusion

The state-of-the-art multimodal OCT system can be used in multiple research and pre-clinical applications due to the various data processing approach applied to the data. The resulting multimodality of the device allows investigating tissue of interest from different points of view providing information about its physiological state as well as mechanical and optical characteristics. Combining the obtained information useful clinical insights about the morphological state of the tissue could be gathered. Collecting more data in multiple modalities can potentially give rise to the machine learning methods of analyzing such data, enabling faster proliferation of OCT technology into everyday clinical practice.

Acknowledgments Development of the pressure-normalization method for comparison of stiffness between different samples and areas with non-uniform straining was supported by RFBR grant 18-32-20056. The development of vector method of strain estimation and obtaining of stress-strain curves under aperiodic loading was supported by Russian Science Foundation (RSF) grant No.16-15-10274. The development of the multimodal OCT for detection of breast cancer was supported by RSF grant No.18-75-10068. The development of the CP OCT for detection of brain glioma was supported by RFBR grant 18-29-01049_mk.

References

1. Huang, D., Swanson, E., Lin, C., Schuman, J., Stinson, W., Chang, W., Hee, M., Flotte, T., Gregory, K., Puliafito, C.: Optical coherence tomography. Science. **254**(5035), 1178–1181 (1991). https://doi.org/10.1126/science.1957169
2. Swanson, E.A., Izatt, J.A., Hee, M.R., Huang, D., Lin, C.P., Schuman, J.S., Puliafito, C.A., Fujimoto, J.G.: *In vivo* retinal imaging by optical coherence tomography. Opt. Lett. **18**(21), 1864–1866 (1993). https://doi.org/10.1364/ol.18.001864
3. Pan, Y.T., Xie, T.Q., Du, C.W., Bastacky, S., Meyers, S., Zeidel, M.L.: Enhancing early bladder cancer detection with fluorescence-guided endoscopic optical coherent tomography. Opt. Lett. **28**(24), 2485–2487 (2003). https://doi.org/10.1364/ol.28.002485
4. Skala, M.C., Fontanella, A., Hendargo, H., Dewhirst, M.W., Izatt, J.A.: Combined hyperspectral and spectral domain optical coherence tomography microscope for noninvasive hemodynamic imaging. Opt. Lett. **34**(3), 289–291 (2009). https://doi.org/10.1364/ol.34.000289
5. Roper, S.N., Moores, M.D., Gelikonov, G.V., Feldchtein, F.I., Beach, N.M., King, M.A., Gelikonov, V.M. Sergeev, A.M., Reitze, D.H: *In vivo* detection of experimentally induced cortical dysgenesis in the adult rat neocortex using optical coherence tomography. J. of Neuroscience Methods, **80**(1), 91–98 (1998). https://doi.org/10.1016/s0165-0270(97)00202-1
6. Feldchtein, F.I., Gelikonov, G.V., Gelikonov, V.M., Iksanov, R.R., Kuranov, R.V., Sergeev, A.M., Gladkova, N.D., Ourutina, M.N., Warren, J.A., Reitze, D.H.: *In vivo* OCT imaging of hard and soft tissue of the oral cavity. Opt. Express. **3**(6), 239–250 (1998). https://doi.org/10.1364/oe.3.000239
7. Pan, Y., Farkas, D.L.: Noninvasive imaging of living human skin with dual-wavelength optical coherence tomography in two and three dimensions. J. Biomed. Opt. **3**(4), 446–455 (1998). https://doi.org/10.1117/1.429897
8. Schmitt, J.M., Xiang, S.H.: Cross-polarized backscatter in optical coherence tomography of biological tissue. Opt. Lett. **23**(13), 1060–1062 (1998). https://doi.org/10.1364/ol.23.001060
9. Drexler, W., Fujimoto, J.G.: Optical Coherence Tomography: Technology and Applications. Springer Science & Business Media (2008)
10. Hee, M.R., Huang, D., Swanson, E.A., Fujimoto, J.G.: Polarization-sensitive low-coherence reflectometer for birefringence characterization and ranging. J. Opt. Soc. Am. B. **9**(6), 903–908 (1992). https://doi.org/10.1364/josab.9.000903
11. de Boer, J.F., Milner, T.E., van Gemert, M.J.C., Nelson, J.S.: Two-dimensional birefringence imaging in biological tissue by polarization-sensitive optical coherence tomography. Opt. Lett. **22**(12), 934–936 (1997). https://doi.org/10.1364/ol.22.000934
12. de Boer, J.F., Srinivas, S.M., Malekafzali, A., Chen, Z., Stuart Nelson, J.: Imaging thermally damaged tissue by polarization sensitive optical coherence tomography. Opt. Express. **3**(6), 212–218 (1998). https://doi.org/10.1364/oe.3.000212
13. Everett, M.J., Schoenenberger, K., Colston, B.W., Da Silva, L.B.: Birefringence characterization of biological tissue by use of optical coherence tomography. Opt. Lett. **23**(3), 228–230 (1998). https://doi.org/10.1364/ol.23.000228
14. Booth, M.C., Di Giuseppe, G., Saleh, B.E.A., Sergienko, A.V., Teich, M.C.: Polarization-sensitive quantum-optical coherence tomography. Phys. Rev. A. **69**(4), 043815 (2004). https://doi.org/10.1103/PhysRevA.69.043815
15. Liu, B., Harman, M., Giattina, S., Stamper, D.L., Demakis, C., Chilek, M., Raby, S., Brezinski, M.E.: Characterizing of tissue microstructure with single-detector polarization-sensitive optical coherence tomography. Appl. Opt. **45**(18), 4464–4479 (2006). https://doi.org/10.1364/ao.45.004464
16. Kim, K.H., Park, B.H., Tu, Y., Hasan, T., Lee, B., Li, J., de Boer, J.F.: Polarization-sensitive optical frequency domain imaging based on unpolarized light. Opt. Express. **19**(2), 552–561 (2011). https://doi.org/10.1364/OE.19.000552
17. Kuranov, R.V., Sapozhnikova, V.V., Turchin, I.V., Zagainova, E.V., Gelikonov, V.M., Kamensky, V.A., Snopova, L.B., Prodanetz, N.N.: Complementary use of cross-polarization

and standard OCT for differential diagnosis of pathological tissues. Opt. Express. **10**(15), 707–713 (2002). https://doi.org/10.1364/oe.10.000707
18. de Boer, J.F., Milner, T.E., Nelson, J.S.: Determination of the depth-resolved stokes parameters of light backscattered from turbid media by use of polarization-sensitive optical coherence tomography. Opt. Lett. **24**(5), 300–302 (1999). https://doi.org/10.1364/ol.24.000300
19. Yao, G., Wang, L.V.: Two-dimensional depth-resolved Mueller matrix characterization of biological tissue by optical coherence tomography. Opt. Lett. **24**(8), 537–539 (1999). https://doi.org/10.1364/ol.24.000537
20. Hitzenberger, C.K., Götzinger, E., Sticker, M., Pircher, M., Fercher, A.F.: Measurement and imaging of birefringence and optic axis orientation by phase resolved polarization sensitive optical coherence tomography. Opt. Express. **9**(13), 780–790 (2001). https://doi.org/10.1364/oe.9.000780
21. Ghosh, N., Vitkin, I.A.: Tissue polarimetry: concepts, challenges, applications, and outlook. J. Biomed. Opt. **16**(11), 3652896 (2011). https://doi.org/10.1117/1.3652896
22. Fried, D., Xie, J., Shafi, S., Featherstone, J.D., Breunig, T.M., Le, C.: Imaging caries lesions and lesion progression with polarization sensitive optical coherence tomography. J. Biomed. Opt. **7**(4), 618–627 (2002). https://doi.org/10.1117/1.1509752
23. Sankaran, V., Walsh Jr., J.T., Maitland, D.J.: Comparative study of polarized light propagation in biologic tissues. J. Biomed. Opt. **7**(3), 300–306 (2002). https://doi.org/10.1117/1.1483318
24. Kuranov, R.V., Sapozhnikova, V.V., Shakhova, N.M., Gelikonov, V.M., Zagainova, E.V., Petrova, S.A.: Combined application of optical methods to increase the information content of optical coherent tomography in diagnostics of neoplastic processes. Quant. Electron. **32**(11), 993–998 (2002). https://doi.org/10.1070/qe2002v032n11abeh002334
25. Gladkova, N., Streltsova, O., Zagaynova, E., Kiseleva, E., Gelikonov, V., Gelikonov, G., Karabut, M., Yunusova, K., Evdokimova, O.: Cross-polarization optical coherence tomography for early bladder-cancer detection: statistical study. J. Biophotonics. **4**(7–8), 519–532 (2011). https://doi.org/10.1002/jbio.201000088
26. De Boer, J.F., Srinivas, S.M., Park, B.H., Pham, T.H., Zhongping, C., Milner, T.E., Nelson, J.S.: Polarization effects in optical coherence tomography of various biological tissues. IEEE J. Sel. Top. Quantum Electr. **5**(4), 1200–1204 (1999). https://doi.org/10.1109/2944.796347
27. Pierce, M.C., Hyle Park, B., Cense, B., de Boer, J.F.: Simultaneous intensity, birefringence, and flow measurements with high-speed fiber-based optical coherence tomography. Opt. Lett. **27**(17), 1534–1536 (2002). https://doi.org/10.1364/ol.27.001534
28. Oh, W.Y., Yun, S.H., Vakoc, B.J., Shishkov, M., Desjardins, A.E., Park, B.H., de Boer, J.F., Tearney, G.J., Bouma, B.E.: High-speed polarization sensitive optical frequency domain imaging with frequency multiplexing. Opt. Express. **16**(2), 1096–1103 (2008). https://doi.org/10.1364/oe.16.001096
29. Oh, W.Y., Vakoc, B.J., Yun, S.H., Tearney, G.J., Bouma, B.E.: Single-detector polarization-sensitive optical frequency domain imaging using high-speed intra A-line polarization modulation. Opt. Lett. **33**(12), 1330–1332 (2008). https://doi.org/10.1364/ol.33.001330
30. Gelikonov, V.M., Gelikonov, G.V.: New approach to cross-polarized optical coherence tomography based on orthogonal arbitrarily polarized modes. Laser Phys. Lett. **3**(9), 445–451 (2006). https://doi.org/10.1002/lapl.200610030
31. Gelikonov, V.M., Gelikonov, G.V.: Fibreoptic methods of cross-polarisation optical coherence tomography for endoscopic studies. Quantum Electr. **38**(7), 634–640 (2008). https://doi.org/10.1070/qe2008v038n07abeh013857
32. Jiao, S., Yu, W., Stoica, G., Wang, L.: Optical-fiber-based Mueller optical coherence tomography. Opt. Lett. **28**(14), 1206–1208 (2003). https://doi.org/10.1364/ol.28.001206
33. de Boer, J.F., Milner, T.E.: Review of polarization sensitive optical coherence tomography and stokes vector determination. J. Biomed. Opt. **7**(3), 359–371 (2002). https://doi.org/10.1117/1.1483879
34. Roth, J.E., Kozak, J.A., Yazdanfar, S., Rollins, A.M., Izatt, J.A.: Simplified method for polarization-sensitive optical coherence tomography. Opt. Lett. **26**(14), 1069–1071 (2001). https://doi.org/10.1364/ol.26.001069

35. Cense, B., Chen, T.C., Park, B.H., Pierce, M.C., de Boer, J.F.: *In vivo* depth-resolved birefringence measurements of the human retinal nerve fiber layer by polarization-sensitive optical coherence tomography. Opt. Lett. **27**(18), 1610–1612 (2002). https://doi.org/10.1364/ol.27.001610
36. Davé, D.P., Akkin, T., Milner, T.E.: Polarization-maintaining fiber-based optical low-coherence reflectometer for characterization and ranging of birefringence. Opt. Lett. **28**(19), 1775–1777 (2003). https://doi.org/10.1364/ol.28.001775
37. Zhang, J., Guo, S., Jung, W., Nelson, J.S., Chen, Z.: Determination of birefringence and absolute optic axis orientation using polarization-sensitive optical coherence tomography with PM fibers. Opt. Express. **11**(24), 3262–3270 (2003). https://doi.org/10.1364/oe.11.003262
38. Yoshino, T., Hashimoto, T., Nara, M., Kurosawa, K.: Common path heterodyne optical fiber sensors. J. Lightwave Technol. **10**(4), 503–513 (1992). https://doi.org/10.1109/50.134205
39. Bush, J., Feldchtein, F., Gelikonov, G., Gelikonov, V., Piyevsky, S.: Cost effective all-fiber autocorrelator for optical coherence tomography imaging. In: 17th International Conference on Optical Fibre Sensors, 2005. SPIE
40. Feldchtein, F., Bush, J., Gelikonov, G., Gelikonov, V., Piyevsky, S.: Cost-effective all-fiber autocorrelator-based 1300-nm OCT system. In: Coherence Domain Optical Methods and Optical Coherence Tomography in Biomedicine IX, 2005. Proc. SPIE. https://doi.org/10.1117/12.589502
41. Sharma, U., Fried, N.M., Kang, J.U.: All-fiber common-path optical coherence tomography: sensitivity optimization and system analysis. IEEE J. Sel. Top. Quantum Electr. **11**(4), 799–805 (2005). https://doi.org/10.1109/jstqe.2005.857380
42. Bush, I.J.: All fiber autocorrelator. US Patent 6,847,453, 2005
43. Drake, A.D., Leiner, D.C.: Fiber-optic interferometer for remote subangstrom vibration measurement. Rev. Sci. Instrum. **55**(2), 162–165 (1984). https://doi.org/10.1063/1.1137721
44. Bush J, Davis PG, Marcus, M.A.: All-fiber optic coherence domain interferometric techniques. In: Fiber Optic Sensor Technology II, 2001. Proc. SPIE. https://doi.org/10.1117/12.417430
45. Authorship certif. No. 1315797 A1 SSSR OfsiR, VM. Gelikonov, V. I. Leonov, and M.A. Novikov, Priority of 04.05.1984; Published 07.06.1987, Bullet. 21, p. 3
46. Gelikonov, V.M., Gusovskiy, D.D., Leonov, V.I.: Optical fiber sensor. Pis'ma v Zhurn Tekhn Fiz [in Russian]. **13**(13), 775 (1987)
47. Liu, B., Harman, M., Brezinski, M.E.: Variables affecting polarization-sensitive optical coherence tomography imaging examined through the modeling of birefringent phantoms. J. Opt. Soc. Am. A. **22**(2), 262–271 (2005). https://doi.org/10.1364/josaa.22.000262
48. Zaitsev, V.Y., Vitkin, I.A., Matveev, L.A., Gelikonov, V.M., Matveyev, A.L., Gelikonov, G.V.: Recent trends in multimodal optical coherence tomography. II. The correlation-stability approach in OCT elastography and methods for visualization of microcirculation. Radiophys. Quantum Electr. **57**(3), 210–225 (2014). https://doi.org/10.1007/s11141-014-9505-x
49. Yu, G., Durduran, T., Zhou, C., Wang, H.-W., Putt, M.E., Saunders, H.M., Sehgal, C.M., Glatstein, E., Yodh, A.G., Busch, T.M.: Noninvasive monitoring of murine tumor blood flow during and after photodynamic therapy provides early assessment of therapeutic efficacy. Clin. Cancer Res. **11**(9), 3543–3552 (2005). https://doi.org/10.1158/1078-0432.ccr-04-2582
50. Zhang, H.F., Maslov, K., Li, M.-L., Stoica, G., Wang, L.V.: *In vivo* volumetric imaging of subcutaneous microvasculature by photoacoustic microscopy. Opt. Express. **14**(20), 9317–9323 (2006). https://doi.org/10.1364/oe.14.009317
51. Leahy, M.J.: Microcirculation Imaging. Microcirculation Imaging. Wiley—Blackwell, New York (2012) https://doi.org/10.1002/9783527651238
52. Dunn, A.K., Leitgeb, R., Wang, R.K., Zhang, H.F.: Introduction: feature issue on *in vivo* microcirculation imaging. Biomed. Opt. Express. **2**(7), 1861–1863 (2011). https://doi.org/10.1364/boe.2.001861
53. Makita, S., Hong, Y., Yamanari, M., Yatagai, T., Yasuno, Y.: Optical coherence angiography. Opt. Express. **14**(17), 7821–7840 (2006). https://doi.org/10.1364/oe.14.007821

54. Zhao, Y., Chen, Z., Saxer, C., Shen, Q., Xiang, S., de Boer, J.F., Nelson, J.S.: Doppler standard deviation imaging for clinical monitoring of *in vivo* human skin blood flow. Opt. Lett. **25**(18), 1358–1360 (2000). https://doi.org/10.1364/ol.25.001358
55. Hyle Park, B., Pierce, M.C., Cense, B., Yun, S.-H., Mujat, M., Tearney, G.J., Bouma, B.E., De Boer, J.F.: Real-time fiber-based multi-functional spectral-domain optical coherence tomography at 1.3 μm. Opt. Express. **13**(11), 3931–3944 (2005). https://doi.org/10.1364/opex.13.003931
56. An, L., Wang, R.K.: *In vivo* volumetric imaging of vascular perfusion within human retina and choroids with optical micro-angiography. Opt. Express. **16**(15), 11438–11452 (2008). https://doi.org/10.1364/oe.16.011438
57. Fingler, J., Zawadzki, R.J., Werner, J.S., Schwartz, D., Fraser, S.E.: Volumetric microvascular imaging of human retina using optical coherence tomography with a novel motion contrast technique. Opt. Express. **17**(24), 22190–22200 (2009). https://doi.org/10.1364/oe.17.022190
58. Kim, D.Y., Fingler, J., Werner, J.S., Schwartz, D.M., Fraser, S.E., Zawadzki, R.J.: *In vivo* volumetric imaging of human retinal circulation with phase-variance optical coherence tomography. Biomed. Opt. Express. **2**(6), 1504–1513 (2011). https://doi.org/10.1364/boe.2.001504
59. Barton, J.K., Stromski, S.: Flow measurement without phase information in optical coherence tomography images. Opt. Express. **13**(14), 5234–5239 (2005). https://doi.org/10.1364/opex.13.005234
60. Aizu, Y., Asakura, T.: Bio-speckle phenomena and their application to the evaluation of blood flow. Opt. Laser Technol. **23**(4), 205–219 (1991). https://doi.org/10.1016/0030-3992(91)90085-3
61. Mariampillai, A., Standish, B.A., Moriyama, E.H., Khurana, M., Munce, N.R., Leung, M.K.K., Jiang, J., Cable, A., Wilson, B.C., Vitkin, I.A., Yang, V.X.D.: Speckle variance detection of microvasculature using swept-source optical coherence tomography. Opt. Lett. **33**(13), 1530–1532 (2008). https://doi.org/10.1364/ol.33.001530
62. Mariampillai, A., Leung, M.K.K., Jarvi, M., Standish, B.A., Lee, K., Wilson, B.C., Vitkin, A., Yang, V.X.D.: Optimized speckle variance OCT imaging of microvasculature. Opt. Lett. **35**(8), 1257–1259 (2010). https://doi.org/10.1364/ol.35.001257
63. Liu, G., Lin, A.J., Tromberg, B.J., Chen, Z.: A comparison of Doppler optical coherence tomography methods. Biomed. Opt. Express. **3**(10), 2669–2680 (2012). https://doi.org/10.1364/boe.3.002669
64. Enfield, J., Jonathan, E., Leahy, M.: *In vivo* imaging of the microcirculation of the volar forearm using correlation mapping optical coherence tomography (cmOCT). Biomed. Opt. Express. **2**(5), 1184–1193 (2011). https://doi.org/10.1364/boe.2.001184
65. Motaghiannezam, R., Fraser, S.: Logarithmic intensity and speckle-based motion contrast methods for human retinal vasculature visualization using swept source optical coherence tomography. Biomed. Opt. Express. **3**(3), 503–521 (2012). https://doi.org/10.1364/boe.3.000503
66. Gao, S.S., Jia, Y., Zhang, M., Su, J.P., Liu, G., Hwang, T.S., Bailey, S.T., Huang, D.: Optical coherence tomography angiography. Invest. Ophthalmol. Vis. Sci. **57**(9), OCT27–OCT36 (2016). https://doi.org/10.1167/iovs.15-19043
67. Schmitt, J.M.: OCT elastography: imaging microscopic deformation and strain of tissue. Opt. Express. **3**(6), 199–211 (1998). https://doi.org/10.1364/oe.3.000199
68. Lifshitz, E.M., Kosevich, A.M., Pitaevskii, L.P.: Chapter III - elastic waves. In: Lifshitz, E.M., Kosevich, A.M., Pitaevskii, L.P. (eds.) Theory of Elasticity, 3rd edn, pp. 87–107. Butterworth-Heinemann, Oxford (1986). https://doi.org/10.1016/B978-0-08-057069-3.50010-3
69. Chan, R.C., Chau, A.H., Karl, W.C., Nadkarni, S., Khalil, A.S., Iftimia, N., Shishkov, M., Tearney, G.J., Kaazempur-Mofrad, M.R., Bouma, B.E.: OCT-based arterial elastography: robust estimation exploiting tissue biomechanics. Opt. Express. **12**(19), 4558–4572 (2004). https://doi.org/10.1364/opex.12.004558
70. Rogowska, J., Patel, N.A., Fujimoto, J.G., Brezinski, M.E.: Optical coherence tomographic elastography technique for measuring deformation and strain of atherosclerotic tissues. Heart. **90**(5), 556–562 (2004). https://doi.org/10.1136/hrt.2003.016956

71. Chau, A.H., Chan, R.C., Shishkov, M., MacNeill, B., Iftimia, N., Tearney, G.J., Kamm, R.D., Bouma, B.E., Kaazempur-Mofrad, M.R.: Mechanical analysis of atherosclerotic plaques based on optical coherence tomography. Ann. Biomed. Eng. **32**(11), 1494–1503 (2004). https://doi.org/10.1114/B:ABME.0000049034.75368.4a
72. Leitgeb, R., Hitzenberger, C.K., Fercher, A.F.: Performance of Fourier domain vs. time domain optical coherence tomography. Opt. Express. **11**(8), 889–894 (2003). https://doi.org/10.1364/oe.11.000889
73. de Boer, J.F., Cense, B., Park, B.H., Pierce, M.C., Tearney, G.J., Bouma, B.E.: Improved signal-to-noise ratio in spectral-domain compared with time-domain optical coherence tomography. Opt. Lett. **28**(21), 2067–2069 (2003). https://doi.org/10.1364/ol.28.002067
74. Choma, M.A., Sarunic, M.V., Yang, C., Izatt, J.A.: Sensitivity advantage of swept source and Fourier domain optical coherence tomography. Opt. Express. **11**(18), 2183–2189 (2003). https://doi.org/10.1364/oe.11.002183
75. Wang, R.K., Ma, Z., Kirkpatrick, S.J.: Tissue Doppler optical coherence elastography for real time strain rate and strain mapping of soft tissue. Appl. Phys. Lett. **89**(14), 144103 (2006). https://doi.org/10.1063/1.2357854
76. Wang, R.K., Kirkpatrick, S., Hinds, M.: Phase-sensitive optical coherence elastography for mapping tissue microstrains in real time. Appl. Phys. Lett. **90**(16), 164105 (2007). https://doi.org/10.1063/1.2724920
77. Liang, X., Crecea, V., Boppart, S.A.: Dynamic optical coherence elastography: a review. J. Innov. Opt. Health Sci. **03**(04), 221–233 (2010). https://doi.org/10.1142/s1793545810001180
78. Sun, C., Standish, B.A., Yang, V.X.D.: Optical coherence elastography: current status and future applications. J. Biomed. Opt. **16**(4), 043001 (2011). https://doi.org/10.1117/1.3560294
79. Kennedy, B.F., Kennedy, K.M., Sampson, D.D.: A review of optical coherence elastography: fundamentals, techniques and prospects. IEEE J. Sel. Top. Quantum Electr. **20**(2), 272–288 (2014). https://doi.org/10.1109/jstqe.2013.2291445
80. Zaitsev, V.Y., Gelikonov, V.M., Matveev, L.A., Gelikonov, G.V., Matveyev, A.L., Shilyagin, P.A., Vitkin, I.A.: Recent trends in multimodal optical coherence tomography. I. Polarization-sensitive OCT and conventional approaches to OCT elastography. Radiophys. Quantum Electr. **57**(1), 52–66 (2014). https://doi.org/10.1007/s11141-014-9493-x
81. Larin, K.V., Sampson, D.D.: Optical coherence elastography - OCT at work in tissue biomechanics [invited]. Biomed. Opt. Express. **8**(2), 1172–1202 (2017). https://doi.org/10.1364/BOE.8.001172
82. Kirkpatrick, S.J., Wang, R.K., Duncan, D.D.: OCT-based elastography for large and small deformations. Opt. Express. **14**(24), 11585–11597 (2006). https://doi.org/10.1364/oe.14.011585
83. Wang, R.K., Nuttall, A.L.: Phase-sensitive optical coherence tomography imaging of the tissue motion within the organ of Corti at a subnanometer scale: a preliminary study. J. Biomed. Opt. **15**(5), 05600 (2010). https://doi.org/10.1117/1.3486543
84. Liang, X., Orescanin, M., Toohey, K.S., Insana, M.F., Boppart, S.A.: Acoustomotive optical coherence elastography for measuring material mechanical properties. Opt. Lett. **34**(19), 2894–2896 (2009). https://doi.org/10.1364/ol.34.002894
85. Crecea, V., Oldenburg, A.L., Liang, X., Ralston, T.S., Boppart, S.A.: Magnetomotive nanoparticle transducers for optical rheology of viscoelastic materials. Opt. Express. **17**(25), 23114–23122 (2009). https://doi.org/10.1364/oe.17.023114
86. Razani, M., Mariampillai, A., Sun, C., Luk, T.W.H., Yang, V.X.D., Kolios, M.C.: Feasibility of optical coherence elastography measurements of shear wave propagation in homogeneous tissue equivalent phantoms. Biomed. Opt. Express. **3**(5), 972–980 (2012). https://doi.org/10.1364/boe.3.000972
87. Qi, W., Li, R., Ma, T., Shung, K.K., Zhou, Q., Chen, Z.: Confocal acoustic radiation force optical coherence elastography using a ring ultrasonic transducer. Appl. Phys. Lett. **104**(12), 123702 (2014). https://doi.org/10.1063/1.4869562

88. Wang, S., Aglyamov, S., Karpiouk, A., Li, J., Emelianov, S., Manns, F., Larin, K.V.: Assessing the mechanical properties of tissue-mimicking phantoms at different depths as an approach to measure biomechanical gradient of crystalline lens. Biomed. Opt. Express. **4**(12), 2769–2780 (2013). https://doi.org/10.1364/boe.4.002769
89. Alonso-Caneiro, D., Karnowski, K., Kaluzny, B.J., Kowalczyk, A., Wojtkowski, M.: Assessment of corneal dynamics with high-speed swept source optical coherence tomography combined with an air puff system. Opt. Express. **19**(15), 14188–14199 (2011). https://doi.org/10.1364/oe.19.014188
90. Dorronsoro, C., Pascual, D., Pérez-Merino, P., Kling, S., Marcos, S.: Dynamic OCT measurement of corneal deformation by an air puff in normal and cross-linked corneas. Biomed. Opt. Express. **3**(3), 473–487 (2012). https://doi.org/10.1364/boe.3.000473
91. Wang, S., Larin, K.V., Li, J., Vantipalli, S., Manapuram, R.K., Aglyamov, S., Emelianov, S., Twa, M.D.: A focused air-pulse system for optical-coherence-tomography-based measurements of tissue elasticity. Laser Phys. Lett. **10**(7), 075605 (2013). https://doi.org/10.1088/1612-2011/10/7/075605
92. Crecea, V., Ahmad, A., Boppart, S.A.: Magnetomotive optical coherence elastography for microrheology of biological tissues. J. Biomed. Opt. **18**(12), 121504 (2013). https://doi.org/10.1117/1.JBO.18.12.121504
93. Ahmad, A., Kim, J., Sobh, N.A., Shemonski, N.D., Boppart, S.A.: Magnetomotive optical coherence elastography using magnetic particles to induce mechanical waves. Biomed. Opt. Express. **5**(7), 2349–2361 (2014). https://doi.org/10.1364/boe.5.002349
94. Nahas, A., Bauer, M., Roux, S., Boccara, A.C.: 3D static elastography at the micrometer scale using full field OCT. Biomed. Opt. Express. **4**(10), 2138–2149 (2013). https://doi.org/10.1364/boe.4.002138
95. Kennedy, K.M., Ford, C., Kennedy, B.F., Bush, M.B., Sampson, D.D.: Analysis of mechanical contrast in optical coherence elastography. J. Biomed. Opt. **18**(12), 121508 (2013). https://doi.org/10.1117/1.JBO.18.12.121508
96. Li, C., Guan, G., Cheng, X., Huang, Z., Wang, R.K.: Quantitative elastography provided by surface acoustic waves measured by phase-sensitive optical coherence tomography. Opt. Lett. **37**(4), 722–724 (2012). https://doi.org/10.1364/ol.37.000722
97. Adie, S.G., Liang, X., Kennedy, B.F., John, R., Sampson, D.D., Boppart, S.A.: Spectroscopic optical coherence elastography. Opt. Express. **18**(25), 25519–25534 (2010). https://doi.org/10.1364/oe.18.025519
98. Song, S., Huang, Z., Nguyen, T.M., Wong, E.Y., Arnal, B., O'Donnell, M., Wang, R.K.: Shear modulus imaging by direct visualization of propagating shear waves with phase-sensitive optical coherence tomography. J. Biomed. Opt. **18**(12), 121509 (2013). https://doi.org/10.1117/1.JBO.18.12.121509
99. Kennedy, B.F., Liang, X., Adie, S.G., Gerstmann, D.K., Quirk, B.C., Boppart, S.A., Sampson, D.D.: *In vivo* three-dimensional optical coherence elastography. Opt. Express. **19**(7), 6623–6634 (2011). https://doi.org/10.1364/oe.19.006623
100. Li, C., Guan, G., Reif, R., Huang, Z., Wang Ruikang, K.: Determining elastic properties of skin by measuring surface waves from an impulse mechanical stimulus using phase-sensitive optical coherence tomography. J. R. Soc. Interface. **9**(70), 831–841 (2012). https://doi.org/10.1098/rsif.2011.0583
101. Gelikonov, V.M., Gusovskii, D.D., Konoplev, Y.N., Leonov, V.I., Mamaev, Y.A., Turkin, A.A.: Investigation of a fiber-optic polarizer with a metal film and a dielectric buffer layer. Sov. J. Quantum Electron. **20**(1), 76 (1990). https://doi.org/10.1070/QE1990v020n01ABEH004800
102. Gelikonov, V.M., Konoplev, Y.N., Kucheva, M.N., Mamaev, Y.A., Turkin, A.A.: Effect of buffer layer on extinction coefficient of fiber-optic polarizer with metallic coating. Opt. Spectrosc. **71**(4), 397–398 (1991)

103. Gelikonov, V.M., Gelikonov, G.V., Shilyagin, P.A.: Linear-wavenumber spectrometer for high-speed spectral-domain optical coherence tomography. Opt. Spectrosc. **106**(3), 459–465 (2009). https://doi.org/10.1134/s0030400x09030242
104. Goto, K., Sueta, T., Makimoto, T.: Traveling-wave light-intensity modulators using the method of polarization-rotated reflection. IEEE J. Quantum Electron. **8**(6), 486–493 (1972). https://doi.org/10.1109/jqe.1972.1077092
105. Kuwahara, H.: Optical isolator for semiconductor lasers. Appl. Opt. **19**(2), 319–323 (1980). https://doi.org/10.1364/ao.19.000319
106. Vansteenkiste, N., Vignolo, P., Aspect, A.: Optical reversibility theorems for polarization: application to remote control of polarization. J. Opt. Soc. Am. A: Opt. Image Sci. Vision. **10**(10), 2240–2245 (1993). https://doi.org/10.1364/josaa.10.002240
107. Ezekiel, S., Arditty, H.J.: Fiber-Optic Rotation Sensor and Related Technologies. Springer, Berlin (1982)
108. Gelikonov, V.M., Romashov, V.N., Shabanov, D.V., Ksenofontov, S.Y., Terpelov, D.A., Shilyagin, P.A., Gelikonov, G.V., Vitkin, I.A.: Cross-polarization optical coherence tomography with active maintenance of the circular polarization of a sounding wave in a common path system. Radiophys. Quantum Electr. **60**(11), 897–911 (2018). https://doi.org/10.1007/s11141-018-9856-9
109. Wang, J., Xu, Y., Boppart, S.A.: Review of optical coherence tomography in oncology. J. Biomed. Opt. **22**(12), 121711 (2017). https://doi.org/10.1117/1.JBO.22.12.121711
110. Gubarkova, E.V., Dudenkova, V.V., Feldchtein, F.I., Timofeeva, L.B., Kiseleva, E.B., Kuznetsov, S.S., Shakhov, B.E., Moiseev, A.A., Gelikonov, V.M., Gelikonov, G.V., Vitkin, A., Gladkova, N.D.: Multi-modal optical imaging characterization of atherosclerotic plaques. J. Biophotonics. **9**(10), 1009–1020 (2016). https://doi.org/10.1002/jbio.201500223
111. Fujimoto, J., Swanson, E.: The development, commercialization, and impact of optical coherence tomography history of optical coherence tomography. Invest. Ophthalmol. Vis. Sci. **57**(9), OCT1–OCT13 (2016). https://doi.org/10.1167/iovs.16-19963
112. Gladkova, N., Maslennikova, A., Balalaeva, I., Feldchtein, F., Kiseleva, E., Karabut, M., Iksanov, R.: Application of optical coherence tomography in the diagnosis of mucositis in patients with head and neck cancer during a course of radio(chemo)therapy. Med. Laser Appl. **23**(4), 186–195 (2008). https://doi.org/10.1016/j.mla.2008.07.008
113. Kiseleva, E., Kirillin, M., Feldchtein, F., Vitkin, A., Sergeeva, E., Zagaynova, E., Streltzova, O., Shakhov, B., Gubarkova, E., Gladkova, N.: Differential diagnosis of human bladder mucosa pathologies *in vivo* with cross-polarization optical coherence tomography. Biomed. Opt. Express. **6**(4), 1464–1476 (2015). https://doi.org/10.1364/boe.6.001464
114. Kiseleva, E.B., Gladkova, N.D., Streltzova, O.S., Kirillin, M.Y., Maslennikova, A.V., Dudenkova, V.V., Yunusova, K.E., Sergeeva, E.A.: Cross-polarization OCT for *in vivo* diagnostics and prediction of bladder cancer. Chapter 3. In: Hammad Ather, M. (ed.) Bladder Cancer, vol. 43–61, p. 137. InTech (2017)
115. Gubarkova, E.V., Kirillin, M.Y., Dudenkova, V.V., Timashev, P.S., Kotova, S.L., Kiseleva, E.B., Timofeeva, L.B., Belkova, G.V., Solovieva, A.B., Moiseev, A.A., Gelikonov, G.V., Fiks, I.I., Feldchtein, F.I., Gladkova, N.D.: Quantitative evaluation of atherosclerotic plaques using cross-polarization optical coherence tomography, nonlinear, and atomic force microscopy. J. Biomed. Opt. **21**(12), 126010 (2016). https://doi.org/10.1117/1.JBO.21.12.126010
116. Wang, H., Akkin, T., Magnain, C., Wang, R., Dubb, J., Kostis, W.J., Yaseen, M.A., Cramer, A., Sakadžić, S., Boas, D.: Polarization sensitive optical coherence microscopy for brain imaging. Opt. Lett. **41**(10), 2213–2216 (2016). https://doi.org/10.1364/ol.41.002213
117. Wang, H., Black, A.J., Zhu, J., Stigen, T.W., Al-Qaisi, M.K., Netoff, T.I., Abosch, A., Akkin, T.: Reconstructing micrometer-scale fiber pathways in the brain: multi-contrast optical coherence tomography based tractography. NeuroImage. **58**(4), 984–992 (2011). https://doi.org/10.1016/j.neuroimage.2011.07.005
118. Yashin, K.S., Kiseleva, E.B., Moiseev, A.A., Kuznetsov, S.S., Timofeeva, L.B., Pavlova, N.P., Gelikonov, G.V., Medyanik, I.A., Kravets, L.Y., Zagaynova, E.V., Gladkova, N.D.:

Quantitative nontumorous and tumorous human brain tissue assessment using microstructural co- and cross-polarized optical coherence tomography. Sci. Rep. **9**(1), 2024 (2019). https://doi.org/10.1038/s41598-019-38493-y

119. Kut, C., Chaichana, K.L., Xi, J., Raza, S.M., Ye, X., McVeigh, E.R., Rodriguez, F.J., Quiñones-Hinojosa, A., Li, X.: Detection of human brain cancer infiltration *ex vivo* and *in vivo* using quantitative optical coherence tomography. Sci. Transl. Med. **7**(292), 292ra100 (2015). https://doi.org/10.1126/scitranslmed.3010611
120. Kiseleva, E.B., Yashin, K., Moiseev, A., Snopova, L.B., Gelikonov, G.V., Medyanik, I.A., Kravets, L.Y., Karyakin, N.N., Vitkin, I.A., Gladkova, N.: Quantitative cross-polarization optical coherence tomography detection of infiltrative tumor margin in a rat glioma model: a pilot study. Sovrem. Tehnol. Med. **10**(1), 6–13 (2018). https://doi.org/10.17691/stm2018.10.1.01
121. Gubarkova, E.V., Elagin, V.V., Moiseev, A.A., Sirotkina, M.A., Pavlova, N.P., Kuznetsov, S.S., Vorontsov, D.A., Vorontsov, A.Y., Gelikonov, G.V., Zagaynova, E.V., Gladkova, N.D.: Multimodal optical imaging as breast cancer margins assessment methods. In: Proc SPIE, Optical Coherence Tomography and Coherence Domain Optical Methods in Biomedicine XXIII 10867 (2019)
122. Moiseev, A., Snopova, L., Kuznetsov, S., Buyanova, N., Elagin, V., Sirotkina, M., Kiseleva, E., Matveev, L., Zaitsev, V., Feldchtein, F., Zagaynova, E., Gelikonov, V., Gladkova, N., Vitkin, A., Gelikonov, G.: Pixel classification method in optical coherence tomography for tumor segmentation and its complementary usage with OCT microangiography. J. Biophotonics. **11**(4), (2018). https://doi.org/10.1002/jbio.201700072
123. Moiseev, A., Ksenofontov, S., Sirotkina, M., Kiseleva, E., Gorozhantseva, M., Shakhova, N., Matveev, L., Zaitsev, V., Matveyev, A., Zagaynova, E., Gelikonov, V., Gladkova, N., Vitkin, A., Gelikonov, G.: Optical coherence tomography-based angiography device with real-time angiography B-scans visualization and hand-held probe for everyday clinical use. J. Biophotonics. **11**(10), (2018). https://doi.org/10.1002/jbio.201700292
124. Sirotkina, M.A., Buyanova, N.L., Kalganova Jr., T.I., Karabut, M.M., Elagin, V.V., Kuznetsov, S.S., Snopova, L.B., Gelikonov, G.V., Zaitsev, V.Y., Matveev, L.A., Zagaynova, E.V., Vitkin, A., Gladkova, N.D.: The development of the methodology of monitoring experimental tumors using multimodal optical coherence tomography: the choice of an optimal tumor model. Sovremennye Tehnologii v Med. **7**(2), 6–15 (2015). https://doi.org/10.17691/stm2015.7.2.01
125. Kuznetsov, S.S., Snopova, L.B., Karabut, M.M., Sirotkina, M.A., Buyanova, N.L., Kalganova, T.I., Elagin, V.V., Senina-Volzhskaya, I.V., Barbashova, L.N., Shumilova, A.V., Zagaynova, E.V., Vitkin, A., Gladkova, N.: Features of morphological changes in experimental CT-26 tumors growth. Sovremennye Tehnologii v Med. **7**(3), 32–39 (2015). https://doi.org/10.17691/stm2015.7.3.04
126. Hall, A.P.: The role of angiogenesis in cancer. Comp. Clin. Pathol. **13**(3), 95–99 (2005). https://doi.org/10.1007/s00580-004-0533-3
127. Hanahan, D., Weinberg, R.A.: Hallmarks of cancer: the next generation. Cell. **144**(5), 646–674 (2011). https://doi.org/10.1016/j.cell.2011.02.013
128. Sirotkina, M.A., Matveev, L.A., Shirmanova, M.V., Zaitsev, V.Y., Buyanova, N.L., Elagin, V.V., Gelikonov, G.V., Kuznetsov, S.S., Kiseleva, E.B., Moiseev, A.A., Gamayunov, S.V., Zagaynova, E.V., Feldchtein, F.I., Vitkin, A., Gladkova, N.D.: Photodynamic therapy monitoring with optical coherence angiography. Sci. Rep. **7**, 41506–41511 (2017). https://doi.org/10.1038/srep41506
129. Sirotkina, M.A., Gubarkova, E.V., Kiseleva, E.B., Zaitsev, V.Y., Kirillin, M.Y., Sovetsky, A.A., Matveyev, A.L., Matveev, L.A., Kuznetsov, S.S., Zagaynova, E.V., Vitkin, A., Gladkova, N.D.: Multimodal OCT for assessment of vasculature-targeted PDT success. In: SPIE BiOS, 2017. SPIE
130. Sirotkina, M.A., Kiseleva, E.B., Gubarkova, E., Buyanova, N.L., Elagin, V.V., Zaitsev, V.Y., Matveev, L., Matveev, A., Kirillin, M., Gelikonov, G.V., Gelikonov, V., Kuznetsov, S.S., Zagaynova, E.V., Gladkova, N.: Multimodal optical coherence tomography in the assessment

of cancer treatment efficacy. Bull. Russ. State Med. Univ. **4**, 19–26 (2016). https://doi.org/10.24075/brsmu.2016-04-03

131. Sirotkina, M.A., Gubarkova, E.V., Matveev, L.A., Zaitsev, V.Y., Moiseev, A.A., Feldchtein, F.I., Zagaynova, E.V., Vitkin, A., Gladkova, N.D.: Optical coherence angiography monitoring of tumor early response to PDT in experimental and clinical studies. In: Proc SPIE, PDT Delivery and Monitoring I (2019)
132. Maslennikova, A.V., Sirotkina, M.A., Moiseev, A.A., Finagina, E.S., Ksenofontov, S.Y., Gelikonov, G.V., Matveev, L.A., Kiseleva, E.B., Zaitsev, V.Y., Zagaynova, E.V., Feldchtein, F.I., Gladkova, N.D., Vitkin, A.: *In-vivo* longitudinal imaging of microvascular changes in irradiated oral mucosa of radiotherapy cancer patients using optical coherence tomography. Sci. Rep. **7**(1), 16505 (2017). https://doi.org/10.1038/s41598-017-16823-2
133. Elagin, V., Karabut, M., Kiseleva, E., Sirotkina, M., Dudenkova, V., Pavlova, N., Kuznetsov, S., Matveev, L., Moiseev, A., Shumilova, A., Feldchtein, F., Zagaynova, E., Vitkin, A., Gladkova, N.: Multiphoton tomography and multimodal OCT for *in vivo* visualization of oral malignancy in the hamster cheek pouch. In: SPIE Photonics Europe, 2018. SPIE
134. Zaitsev, V.Y., Matveyev, A.L., Matveev, L.A., Gelikonov, G.V., Sovetsky, A.A., Vitkin, A.: Optimized phase gradient measurements and phase-amplitude interplay in optical coherence elastography. J. Biomed. Opt. **21**(11), (2016). https://doi.org/10.1117/1.jbo.21.11.116005
135. Zaitsev, V.Y., Matveyev, A.L., Matveev, L.A., Gubarkova, E.V., Sovetsky, A.A., Sirotkina, M.A., Gelikonov, G.V., Zagaynova, E.V., Gladkova, N.D., Vitkin, A.: Practical obstacles and their mitigation strategies in compressional optical coherence elastography of biological tissues. J. Innov. Opt. Health Sci. **10**(6), 1742006–1742013 (2017). https://doi.org/10.1142/s1793545817420068
136. Matveyev, A.L., Matveev, L.A., Sovetsky, A.A., Gelikonov, G.V., Moiseev, A.A., Zaitsev, V.Y.: Vector method for strain estimation in phase-sensitive optical coherence elastography. Laser Phys. Lett. **15**(6), (2018). https://doi.org/10.1088/1612-202X/aab5e9
137. Sovetsky, A.A., Matveyev, A.L., Matveev, L.A., Shabanov, D.V., Zaitsev, V.Y.: Manually-operated compressional optical coherence elastography with effective aperiodic averaging: demonstrations for corneal and cartilaginous tissues. Laser Phys. Lett. **15**(8), (2018). https://doi.org/10.1088/1612-202X/aac879
138. Krouskop, T.A., Wheeler, T.M., Kallel, F., Garra, B.S., Hall, T.: Elastic moduli of breast and prostate tissues under compression. Ultrason. Imaging. **20**(4), 260–274 (1998). https://doi.org/10.1177/016173469802000403
139. Plekhanov, A.A., Gubarkova, E.V., Sovietsky, A.A., Zaitsev, V.Y., Matveev, L.A., Matveyev, A.L., Timofeeva, L.B., Kuznetsov, S.S., Zagaynova, E.V., Gladkova, N.D., Sirotkina, M.A.: Optical coherence elastography for non-invasive monitoring of tumor elasticity under chemotherapy: pilot study. Sovremennye Tehnologii v Med. **10**(3), 43–49 (2018). https://doi.org/10.17691/stm2018.10.3.5
140. Fung, Y.C.: Mechanical Properties of Living Tissues. Biomechanics. Springer, New York (1993). https://doi.org/10.1007/978-1-4757-2257-4
141. Gubarkova, E.V., Sovetsky, A.A., Matveev, L.A., Matveyev, A.L., Zaitsev, V.Y., Moiseev, A.A., Vorontsov, D.A., Vorontsov, A.Y., Kuznetsov, S.S., Gladkova, N.D., Sirotkina, M.A.: Multimodal OCT characterization of human breast cancer morphological types: preliminary study. In: Proc SPIE, Biophotonics: Photonic Solutions for Better Health Care VI 10685 (2018) https://doi.org/10.1117/12.2306450
142. Gubarkova, E.V., Sovetsky, A.A., Zaitsev, V.Y., Matveyev, A.L., Vorontsov, D.A., Sirotkina, M.A., Matveev, L.A., Plekhanov, A.A., Pavlova, N.P., Kuznetsov, S.S., Vorontsov, A.Y., Zagaynova, E.V., Gladkova, N.D.: OCT-elastography-based optical biopsy for breast cancer delineation and express assessment of morphological/molecular subtypes. Biomed. Opt. Express. **10**(5), 2244–2263 (2019). https://doi.org/10.1364/BOE.10.002244

Chapter 13
Texture Analysis in Skin Cancer Tumor Imaging

Oleg O. Myakinin, Alexander G. Khramov, Dmitry S. Raupov, Semyon G. Konovalov, Sergey V. Kozlov, and Alexander A. Moryatov

13.1 Introduction

Medical diagnostics (Greek: Diagnostikós—able to recognize) is a process of recognizing a disease and its identification using an approved medical technology. Diagnosis is often challenging, because many signs and symptoms are nonspecific. Thus, several possible explanations are required for comparison and contrast of different methods. This involves the correlation of various pieces of information followed by the recognition and differentiation of patterns. As for computer aided systems in medicine, one of the first (probably, even the first ever built) medical expert system MYCIN was developed in the early 1970s at Stanford University [1] for reconition of bacteriological infections. In addition, the system contained the decisive rules on the appointment of antibiotics [2]. The use of computers (hardware nodes) became absolutely necessary after the development of the first diagnostic imaging systems, which were associated with the emergence of promising optical diagnostic methods and significant progress in signal processing and pattern recognition. The first CT scanner for tissue visualization was developed in 1972 by Godfrey Hounsfield and Allan Cormac, who were awarded the Nobel Prize for this solution [3].

O. O. Myakinin (✉) · D. S. Raupov · S. G. Konovalov
Department of Laser and Biotechnical Systems, Samara National Research University, Samara, Russian Federation
e-mail: myakinin.oo@ssau.ru

A. G. Khramov
Department of Technical Cybernetics, Samara National Research University, Samara, Russian Federation

Image Processing Systems Institute of RAS - Branch of the FSRC "Crystallography and Photonics" RAS, Samara, Russian Federation
e-mail: khramov@smr.ru

S. V. Kozlov · A. A. Moryatov
Department of Oncology, Samara State Medical University, Samara, Russian Federation

V. V. Tuchin et al. (eds.), *Multimodal Optical Diagnostics of Cancer*,
https://doi.org/10.1007/978-3-030-44594-2_13

Since then, technology has gone far ahead, but the two diagnostic basics remain the same: “imaging” and “recognition.” As for “imaging,” everything is quite clear: a hardware-software complex visualizes a tissue under study, and this is the gold standard for diagnosing many diseases of internal organs, but in the case of “recognition,” everything is much more difficult. In Russia (as in many other countries), only a physician has the right to make a diagnostic decision, of course, using medical imaging (MRI, for instance), but the personal medical history, biochemistry analysis, and other factors are also taken into account. In this regard, the recognition software can be as an assistant only, which helps to make a preliminary diagnosis and reduce a rate of over-diagnosis by a general practitioner excluding false positive cancer diagnosis for patients with benign tumors, therefore minimizing psychological stress for these patients and a costly extra study in a specialized oncological hospital.

The problem of cancer diagnostics is the most acute problem facing the twenty-first century medicine. The number of annually recorded malignant neoplasms is strongly increasing [4]. Over the past 20–25 years, the number of newly detected malignant and benign tumors of skin, lung, gastrointestinal tract, and other internal organs has increased by almost one order of magnitude [5].

More than 550,000 cancer cases have been registered in Russia. In addition, according to statistics, about 13–15% among all fatal cases are associated with oncology [6]. Among men, cancer of the digestive system (37.2% of all fatal cases of cancer), lung and bronchus (30.2%), and breast cancer (17.1%) for women are the most dangerous forms [7]. In addition, malignant melanoma (MM) of skin is one of the most dangerous forms of cancer in terms of average survival and overall mortality as well. The overall mortality rate among all cancers of skin is more than 76% and the risk is steadily increasing with age of patients [8]. The situation is extremely negative due to late diagnosis: most cases are detected in the third or fourth stage that reduces all survival rates. The lacks of effectiveness of medical examinations, as well as reliable diagnostic tools for general practitioners, provoke an increase of mortality, which is almost twice as high as in the USA or the UK.

The diagnostics efficiency of malignant tumors varies with cancer type, but in general remain unsatisfactory due to the complexity of interpreting clinical differential features at an early stage of disease [9] and, ultimately, increases the risk of death. In this regard, the development of instrumental methods of diagnosis (including software) seems absolutely necessary. Optical Coherence Tomography (OCT) is successfully used to build high-precision spatial images of biological objects in real time [10]. Particularly great success has been achieved in ophthalmology [11], where accurate multidimensional OCT images provide comprehensive information on the morphological features of the tissue under study.

In contrast to such imaging technologies as ultrasound, CT, or MRI, OCT allows one studying the structure of a tumor with a spatial resolution of several microns [12], highlighting the area and the boundary of the tumor invasion at the cellular level. However, the strict differentiation of various tissues and the reliable determination of a type of pathology during OCT studies are possible only in some cases. In particular, OCT is a powerful tool for the diagnosis of Basal Cell Carcinoma (BCC) due to its specific spatial localization and topology [13]. Nevertheless, in most cases, OCT provides only detailed information about the morphological features of the tumor without the possibility of accurately determining of cancer type. As a rule, the sensitivity and specificity of OCT diagnosis does not exceed 75% for different types of skin cancer. Moreover, an increase in the sensitivity (up to 85%) is accompanied by a decrease in the specificity of the method (60–70%).

The surface localization of a tumor on the skin makes it possible the visual inspection of suspicious areas. Dermatoscopy (Greek: Derma—skin and scopeo—view) is one of the techniques that facilitate the recognition of skin lesions increasing the accuracy of clinical diagnosis [14] by magnified visualizing many details and structures, which stay invisible by the naked eye.

Dermatoscopy can be carried out with a conventional digital camera, as well as with a specialized device—dermatoscope. Currently, there is a large number of different models of dermotoscopes. In the simplest case, a specialized dermatoscope is a magnifier with a 10× zoom equipped with LED backlight (including LEDs with different wavelengths for shooting in different spectral ranges). A polarizing dermatoscope uses polarizing filters to increase contrast, reduce glare, and improve the quality of color skin images (multicomponent, multispectral). Another type of the dermatoscopy is epiluminescent (or contact) dermatoscopy [15, 16], when the lens of the dermatoscope is closely pressed to a neoplasm surface with filling the gap by immersion fluid. A design of dermatoscope can be either a self-sufficient medical device or an attachment for a camera, such as DermLite Carbon or DermLite Foto X (both 3Gen, USA). All devices operate in polarized light. The Foto X can operate in four modes at once: contact/contactless/polarized/unpolarized [17].

Multispectral dermatoscopy is another perspective approach, which uses a spectrophotometric intracutaneous analysis of skin scattering. The examples of widely used multispectral devices are SIAscope (MedX Health Corp, Mississauga, ON) and MelaFind (MELA Sciences, Irvington, NY). The idea of SIAscope was proposed by Simon Cotton in 1998 [18]. Implemented complex algorithms return high-resolution information regarding total melanin content of the epidermis, collagen, and hemoglobin content as well as the presence of melanin in the papillary dermis. SIAScope and MelaFind examine consequently in five and ten different spectral ranges from blue to infrared capturing up to 2.5 mm depth [19].

These two modalities (dermatoscope and OCT) are an excellent presentation of 2D and 3D skin imaging technologies, and we are going to discuss the application of texture analysis as a main part of a tumor recognition procedure for both of the modalities in this chapter.

Texture analysis is widely used in processing of various image types [20–22], despite the fact that currently a concept of texture is not well defined. Haralick noted [23] that there is no formal approach to texture description and definition, and methods of texture distinguishing are usually developing for each specific case separately. This is due to the fact that the nature of perception by the human visual system is unknown. It can be found at least three different approaches for texture analysis [23–25]: statistical, structural, and stochastic. We are going to describe all these approaches briefly in the following paragraphs.

This chapter does not include the results of neural networks analysis because, in our opinion, neural networks (especially convolutional networks) evaluate an average image of an object class (containing, for example, shape, general characteristic features) and lose details, which are not important for the class. A texture is a quasi-periodic image, usually of a small size (in comparison with the initial image), which describes just the details (for example, strokes, lines, and their periodicity). Thus, the ideology of the neural networks and the texture analysis complement each other. On the other hand, the neural networks can be used not only as a feature extraction tool but also as a decision-making one (VGG16/19 is an example of such a classic network combining both extractor and decision features [26]). However, in this chapter we have focused on texture analysis and have restricted ourselves by classical approaches, such as Fisher Linear Discriminant Analysis [27] and the Support Vector Machine (SVM).

The next sections describe software and hardware technologies for texture analysis of 2D (Dermatoscopy) and 3D (OCT) skin images, as well as the results of *ex vivo* and *in vivo* studies of different tumor types (classes) and statistical assessment of their separability. The discussion and comparative analysis may be found in the last section.

13.2 Quality Metrics

The main classification task is the recognition of cancerous tumor among all studied tumors. It is a binary classification problem, which divides all investigated tumors in two classes—Malignant class and Benign class. Or, if we take into account that malignant melanoma (MM) is the most dangerous skin cancer, the pattern recognition problem may be defined as MM vs other tumors. The sensitivity and specificity are usually used as quality metrics of the binary classification algorithm. The sensitivity (also called Recall Rec, or probability of detection) measures the ability of correct detection of Malignant tumor

$$\text{Sens} = \text{Rec} = \frac{tp}{tp + fn}, \tag{13.1}$$

where tp is the number of true positive and fn is the number of false negative predictions. Specificity measures the proportion of actual negatives that are correctly identified as such

$$\text{Spec} = \frac{tn}{tn + fp}, \tag{13.2}$$

where fp is the number of false positive and tn is the number of true negative.

However, each malignant or benign classes is compiled from several different types of tumor. Actually, the malignant class unites MM, BCC, Squamous Cell Carcinoma (SCC). The same for benign class, which unites nevi, keratosis, and other benign tumors. As a fact, the most complex classification problem is the recognition of the concrete tumor type or at least a cancerous type inside the complex Malignant class. It is a multi-class problem, when sensitivity/specificity metrics lose its meaning and we need to determine other accuracy metrics. In this case precision and recall may be used as quality metrics. *Recall* has the same formula as sensitivity. *Precision* (also called *Positive Precision Value,* PPV) is a fraction of true predictions inside all predictions by an algorithm:

$$\text{Prec} = \text{PPV} = \frac{tp}{tp + fp}. \tag{13.3}$$

It is also convenient to use F_1-measure (or F-score) [29], which combines precision and recall as equal-weighted average:

$$F_1 = \frac{2 * \text{Prec} * \text{Rec}}{\text{Prec} + \text{Rec}}. \tag{13.4}$$

13.3 Dermatoscopy Imaging for *In Vivo* Skin Cancer Diagnosis

13.3.1 Dermatoscopy Unit

Clinical study was performed using a home-made hand-held dermatoscope with a texture analysis software supporting several modes of imaging, including multispectral imaging with polarized light and skin autofluorescence analysis with UV excitation. A principle scheme of the device is presented in Fig. 13.1. For skin illumination, it was used four white LEDs FM-5630WDS-460W-R80 (with polarization coating and without it) and three groups of visible color LEDs (models CREE XPCReD-L1-0000-00301, CREE XRCGRN-L1-0000-00N01, CREE XREBLU-L1-0000-00K01) with peak spectral intensity in red (620 nm), green (530 nm), and blue (470 nm) ranges, respectively. The autofluorescence was excited by UV LEDs LEUVA77V20RV00 with peak intensity in 365 nm.

The digital camera was equipped with a long-pass filter (ThorLabs FGL435) for filtering autofluorescence images and a cross polarizer for deeper epidermis structure visualization. The typical examples of skin images for different imaging modes are shown in Fig. 13.2.

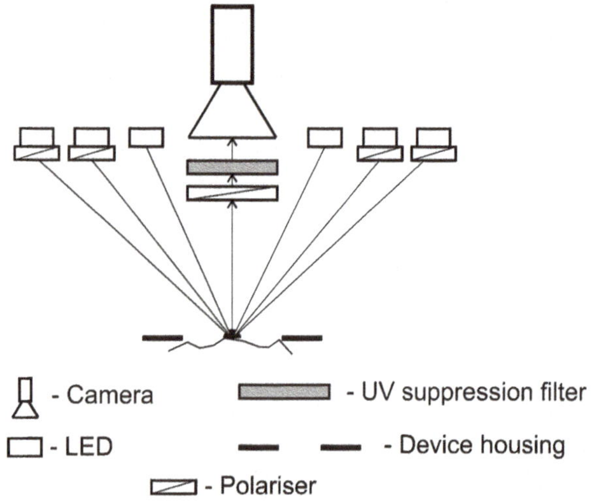

Fig. 13.1 Principle scheme of dermatoscopy hand-held unit

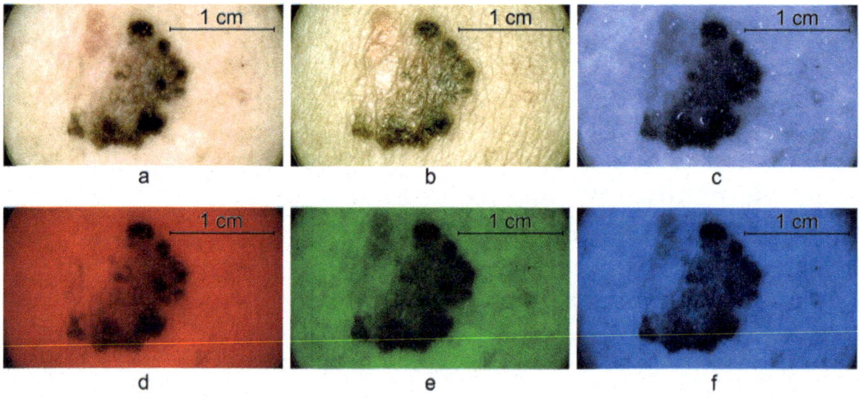

Fig. 13.2 Multispectral Images of Nonpigmented Melanoma: (**a**) White polarized light, (**b**) White non-polarized light, (**c**) Autofluorescence image, (**d**) Red light, (**e**) Green light, (**f**) Blue light

13.3.2 Algorithms of Texture Analysis in Tumor Recognition

In this paragraph, we describe tumor recognition algorithms based on an analysis of color and texture features evaluated from images acquired in polarized white light. This is the closest approach for conventional dermatoscopy.

Traditionally, physicians use some kind of checklist to assess the malignancy of a tumor. For example, the simplest and the most known of them, ABCD, is a set of *A*symmetry, *B*order irregularity, *C*olor variation, and *D*iameter [30]. This checklist does not require high resolution of images, because all features are integral, and images can be taken, for example, from a smartphone camera. Therefore, ABCD can be well suited for an automatic analysis of images taken only in macro mode. In other cases, an appropriate alternative is a 7-point checklist (7PCL) including the following properties [31]:

1. Atypical pigment network: hyperpigmented areas and irregular lines/dots patterns.
2. Gray-Blue areas: confluent, irregular gray-blue to whitish-blue diffuse pigmentation associated with pigment network melanophages and midreticular dermis melanocytes. Also referred as blue-whitish veil.
3. Atypical vascular pattern: linear irregular or dotted red structures outside regression zones and correlated to vascularized amelanotic nests.
4. Radial or irregular streaks: radially or asymmetrically arranged linear or bulbous structures at the edge of a tumor.
5. Irregular pigmentation: black, brown, or gray pigment areas with irregular shape or distribution.
6. Irregular dots and globules: black, brown, or gray round structures irregularly distributed within a lesion.
7. Regression pattern: white scar-like or blue-gray dots irregularly distributed inside a lesion.

On the one hand, it is just an extended ABCD method. However, Argenziano et al. [15] have showed that 7-point checklist may be effectively applied for visualization of a number of distinctive features of melanoma extracted from high quality epiluminescent microscopy skin images.

An alternative approach is supervised learning model with associated learning algorithm that analyzes and classifies images based on color and texture features (see Fig. 13.3). The texture-related recognition algorithm consists of the following steps [32]:

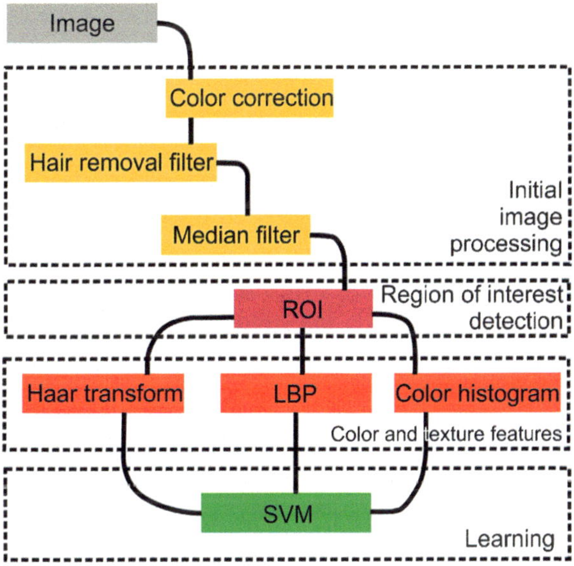

Fig. 13.3 Algorithm of automatic dermatoscopy lesion detection

- initial image pre-processing: color correction, hair removal, and median filtering;
- region of interest (ROI) detection (tumor area, tumor-healthy skin boundary, and healthy area);
- calculation of color (histogram) and texture features (Haar, Local Binary Patterns–LBP) for tumor and healthy skin areas;
- calculation of comparative (tumor/healthy) features;
- final diagnostic recommendation using support vector machine (SVM) classifier.

Initial image pre-processing consists of noise subtraction and hair removal also as histogram and color correction for image contrast increasing and for the highest available dynamic range.

Region of Interest (ROI) Detection The aim of this procedure is an image splitting on an area related to the lesion and another area that is associated with healthy skin (Fig. 13.4). K-means clustering algorithm [33] calculated on pixels brightness is applied to the image, which produces 40 (a predefined parameter) classes. 10% of classes with the lowest brightness are forcibly labeled as the lesion area, the rest of the classes are assigned to surrounding healthy skin. The image is spatially divided by 2D square blocks. The size of the block is determined as 1/20 of the smallest image dimension. Then, a rate of "lesion" pixels is calculated for each block based on the determined classes. If a block rate exceeds 5%, the block is marked as a lesion block. One "layer" of blocks (by 4 connections) around the lesion ROI is marked as transient and form the tumor boundary.

Color and Texture Features Evaluation Color and texture features are represented by a multidimensional composite vector, which combines together different

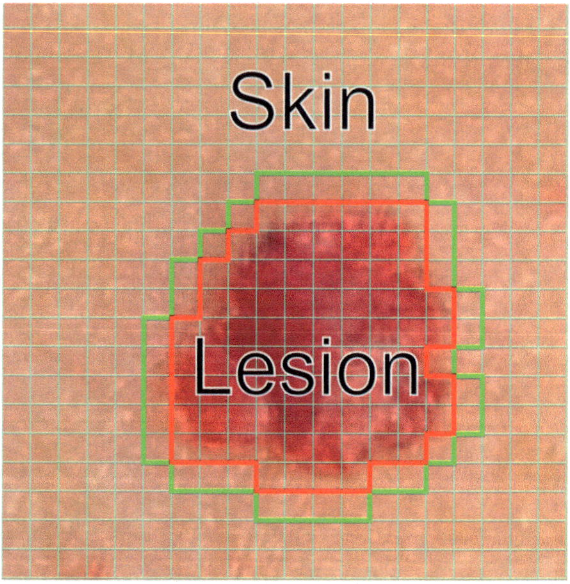

Fig. 13.4 Normal skin and lesion regions

image properties calculated by several approaches [32, 34]: Haar transform forms H_i vector; Local Binary Patterns (LBP) approach forms L_i vector; and color analysis defines C_i vector. These vectors are combined together in one vector later.

Haar Transform Discrete 3-level Haar transform [35] has been performed for each detected ROI block by applying two (high-pass and low-pass) filters using following algorithm:

- The filters have been applied to each row of the initial image resulting in two matrixes: one contains approximated part and another matrix contains fine details of the initial image.
- Then, this transform has been performed again for each column of these matrixes, resulting in four total matrixes, contained an approximated part, horizontal, vertical, and diagonal details of the image.

The result of 3-level Haar transform are 10 sub-images (see Fig. 13.5). The mean value and variance of sub-images are calculated and combined in a feature vector for every ROI block. Thereby, the number of feature vectors equals to the number of ROI blocks for every image. Finally, the histogram H_i has been calculated using the K-means clustering algorithm. This histogram represents a distribution of the image feature vectors.

Local Binary Patterns (LBP) The main idea of LBP [36] is a p-bit number evaluation for every image pixel by comparing its brightness with the brightness of p surrounding pixels using the following algorithm (see Fig. 13.6):

- the brightness of the initial pixel (green) is used as a threshold;
- the surrounding pixels (red) are located on a circle with the radius r;
- if a surrounding pixel brightness value exceeds the threshold, the corresponding bit in the "number" is set to 1, and zero otherwise.

The tissue analysis software uses two LBP algorithms with following parameters: $p = 16$, $r = 2$ and $p = 24$, $r = 3$. The results are combined into a single 40-bit number. The resulting histogram is normalized and forms the vector L_i.

Color Analysis Color histograms have been used for classification as a part of the feature vector. Brightness range of every color channel has been divided on P fixed intervals and a pixel brightness distribution has been calculated for every channel in RGB, HSV, and LAB color spaces in terms of these intervals. Then, the histograms standard deviation and entropy values are calculated and added to the histogram data to define the vector C_i.

Comparative Features A comparative analysis has been performed in order to improve classification by taken into account the personal characteristics of skin. All features are evaluated separately for the lesion and for the surrounding healthy skin. Then, the difference metric between these areas is calculated: Euclidean distance d and cosine measure θ are defined as

Fig. 13.5 (**a**) Single-level Haar transform and (**b**) 3-level Haar transform

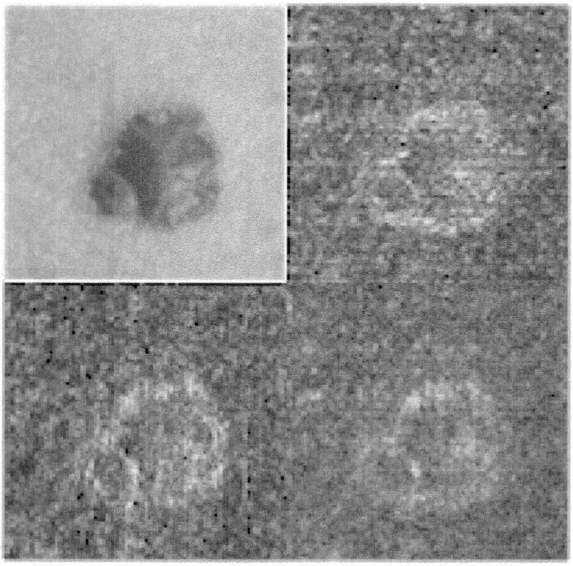

a

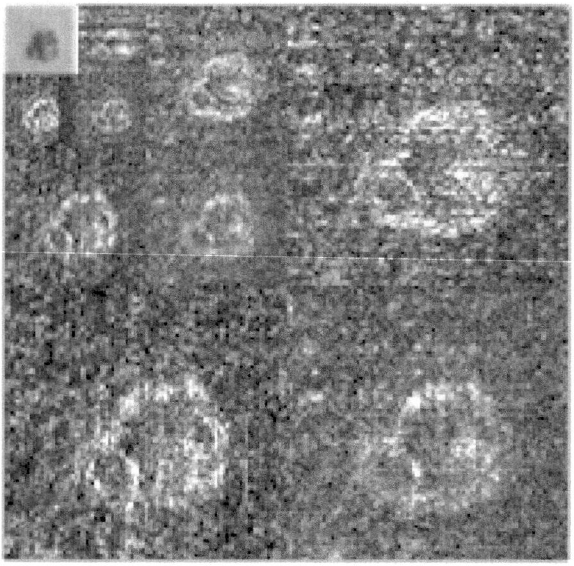

b

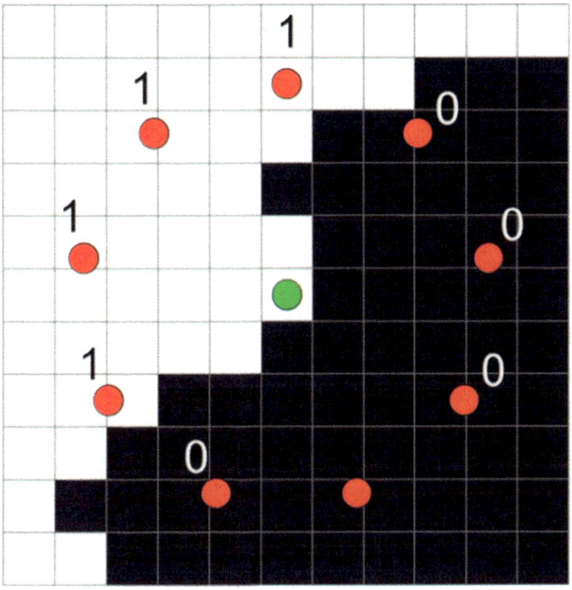

Fig. 13.6 LBP algorithm: an example of surrounding pixels (bilinear interpolation) selection and the central pixel as a threshold

$$d(\overline{x},\overline{y}) = \| \overline{x} - \overline{y} \|, \tag{13.5}$$

$$\theta(\overline{x},\overline{y}) = \frac{\overline{x}^T \overline{y}}{\| \overline{x} \| \| \overline{y} \|}, \tag{13.6}$$

where ||■|| means Euclidean norm, the magnitude of the vector; $\overline{x}, \overline{y}$ —feature vectors in the correspondent areas.

Classification Resulting feature vectors from every approach (H_i, L_i, and C_i) for the lesion area are combined in a single vector for every image. In the case of using comparative features (the so-called, personified mode), the final single vector is being also supplemented by Euclidean distance d and cosine measure θ. The linear SVM method is used for tumor classification [28].

13.3.3 Biological Samples

Three series of experiments were carried out to determine the cancer recognition quality of the proposed algorithm of the texture analysis of the dermatoscopy images. The first two series (70 images in first experiment and 130 images in second experiment) were performed with a manual immersion dermatoscope connected to the Canon EOS 1100D Kit camera (10× zoom). The resulting image dataset (JPEG RGB, 8 bit/channel image) includes 30% of melanomas and 70% of benign tumors. The third experiment was performed with the specialized dermatoscope described above. The third dataset includes 106 images: 53 Melanomas, 53 other types of tumors (22 Pigmented Nevi, 4 Papillomas, 1 Dermatofibroma, 25 Keratomas, and 1 Bowen's disease) in TIFF format (RGB, 12 bit/channel). As the software module for calculating color texture features supports only 24-bit images, the reduction 12-bit down to 8-bit was performed before the following texture analysis. ROI detection was switched on to the automatic mode.

The protocol of *in vivo* tissue diagnostics has been approved by the ethical committee of Samara State Medical University. All patients were at least 18 years old. Informed consents were acquired from all patients before the *in vivo* study. In all experiments images were made by a physician, and the final diagnosis has been histologically proven after resection of a suspicious tumor.

13.3.4 Classification Results

For the first two experimental datasets, the classification task has been defined as melanoma versus pigmented benign tumors. It was used two modes of ROI detection: automatic and semi-automatic when additional user correction of the tumor boundary was performed. The results of classification are collected in Tables 13.1 and 13.2. The proposed texture algorithm has been tested for parameters and features defined in the tumor area only (universal mode) as well as for the comparative features (personified mode), which includes personal features of skin. As it may be seen from the results of classification the user-defined correction of ROI detection may enhance the sensitivity of MM detection on several percentages only. At the same time the personalization of skin properties by comparison of tumor and healthy areas gives much more for the classification.

As a result, the third series of experiments was performed only in the personified mode. In experiments #3 the cross-validation was carried out according to the following principle: a sample set is divided randomly into two (training and test) sets in the ratio of 80/20%, 70/30%, 60/40%, and 50/50%. The qualifier is retrained on the new set of training samples and, then, tested on the test one. As a result of cross-validation, the average error of test samples for all cases is presented in Table 13.3.

Table 13.1 Classification results for the automatic ROI detection mode

	Experiment #1 70 samples		Experiment #2 130 samples	
Mode	Sensitivity (%)	Specificity (%)	Sensitivity (%)	Specificity (%)
Universal mode	74	62	63	80
Personified mode	94	94	90	90

Table 13.2 Classification results for the semi-automatic ROI detection mode

	Experiment #1 70 samples		Experiment #2 130 samples	
Mode	Sensitivity (%)	Specificity (%)	Sensitivity (%)	Specificity (%)
Universal mode	77	91	75	83
Personified mode	94	94	86	89

Table 13.3 Classification results for experiment #3 by using the automatic ROI detection mode

Training set		Cross-validation	
Sensitivity (%)	Specificity (%)	Sensitivity (%)	Specificity (%)
96	89	90	86

13.4 Optical Coherence Tomography for *Ex Vivo* Skin Cancer Detection

13.4.1 OCT Setup Scheme

The OCT system (Fig. 13.7) includes a broadband superluminescent laser diode (840 ± 45 nm wavelength range, 20 mW output power) and Michelson interferometer with 50/50 split ratio to the sample and reference arms, and a spectrometer as the detector. The spectrometer comprises of a diffraction grating (1200 grooves/mm) and a CCD line scan camera (4096 pixel resolution, 29.3 kHz line rate). The interference signal from the sample and the reference arms of the Michelson interferometer is detected by the spectrometer and digitized by an image acquisition card (NI-IMAQ PCI-1428). A depth profile (A-line) is obtained by converting the interference signal detected by the IMAQ into linear k-space.

13.4.2 Data Collection of Biological Samples

Altogether, we investigated 1008 OCT images from 33 *ex vivo* skin tissue samples (Caucasian, I and II skin phenotype). These OCT images have been obtained from regular patients of Samara Regional Clinical Oncology Dispensary. This set includes: basal cell carcinoma (BCC) (272 B-scans (2D), 11 samples (3D)); healthy skin (229 B-scans (2D), 8 samples (3D)); MM (254 B-scans (2D), 10 samples (3D)); nevus (253 B-scans (2D), 4 samples (3D)). The protocol of *ex vivo* tissue diagnostics has been approved by the ethical committee of Samara State Medical University. All patients were at least 18 years old. Informed consents were acquired from all patients before the *ex vivo* study.

Initially, a physician examines a patient for preliminary diagnosis. In the case of suspected malignant melanoma or another cancer (excluding BCC), the patient was subjected to resect this suspicious tissue. After this resection, the sample was

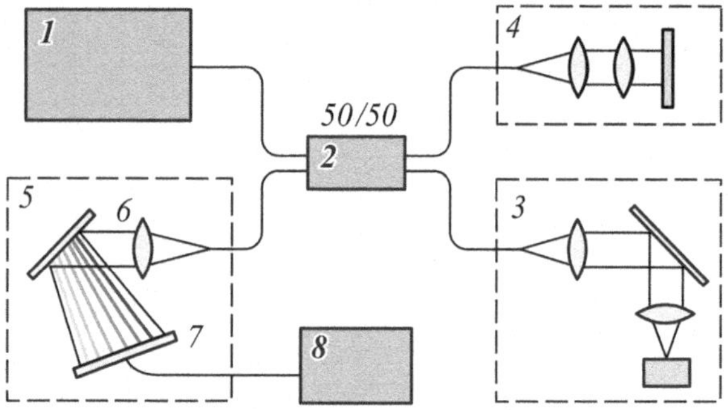

Fig. 13.7 Spectral domain OCT scheme: 1—broadband superluminescent source, 2—50/50 beamsplitter, 3—sample arm, 4—reference arm, 5—spectrometer with grating 6 and CCD camera 7, 8—computer with IMAQ [37]

carefully divided into two parts so that each part contains approximately half of the neoplasm area. One part was stored in a sterile box at +4 ± 2 °C and, then, delivered within 4 h to the laboratory for OCT investigation. The second part was histologically studied using standard protocols in Samara Regional Clinical Oncology Dispensary. The skin resection samples were approximately 2 × 2 × 1 cm. OCT scanning was performed as for tumor area as for apparently healthy tissue at the maximum possible distance from the outer border of the tumor.

Histopathological sections and OCT images of studied samples of malignant (MM and BCC) and benign (Pigmented Nevi) tumors are shown in Fig. 13.8. OCT image (B-scan) of healthy skin is presented in Fig. 13.9.

It can be noted that there are topologically and texturally similar structures on the histological and corresponding OCT images of tumor samples. One can clearly see the painted layers of tumor cells of round or elliptical shape in a histological section of BCC (Fig. 13.8c). Basal cell carcinoma cells on the periphery of the tumor layer are typically in the form of a lance-like row. This formation of malignant cells defines changes in optical density of basal cell carcinoma and normal tissue, so the basal cell carcinoma area has a darker color of circle form on the OCT image (Fig. 13.8d).

In OCT, normal epidermis (including the stratum corneum) is seen as a bright stripe on the tissue surface in malignant melanoma image (Fig. 13.8b). Diffusely scattering cells containing melanin complexes and small undifferentiated cells without pigment are located under the epidermis layer. This determines the optical properties of the tumor. Both OCT (Fig. 13.8b) and histopathology (Fig. 13.8a) reveal destruction of specific layered structure of normal skin. Randomly located multiform objects of different optical density are visualized on the OCT image instead of normal layered structure. Melanoma cells may have an excess amount of pigment or may include nonpigmented elements, which appear on the OCT image in form of dark and bright areas, respectively.

Normal tissue (Fig. 13.9) and Nevi (Fig. 13.8f) are apparently similar and represent structures without a clear texture pattern. Histologically, they are distinguishable due to the presence of a higher pigment concentration in the Nevi (Fig. 13.8e), which local change is not so strong and contrasted as in melanoma. On the OCT

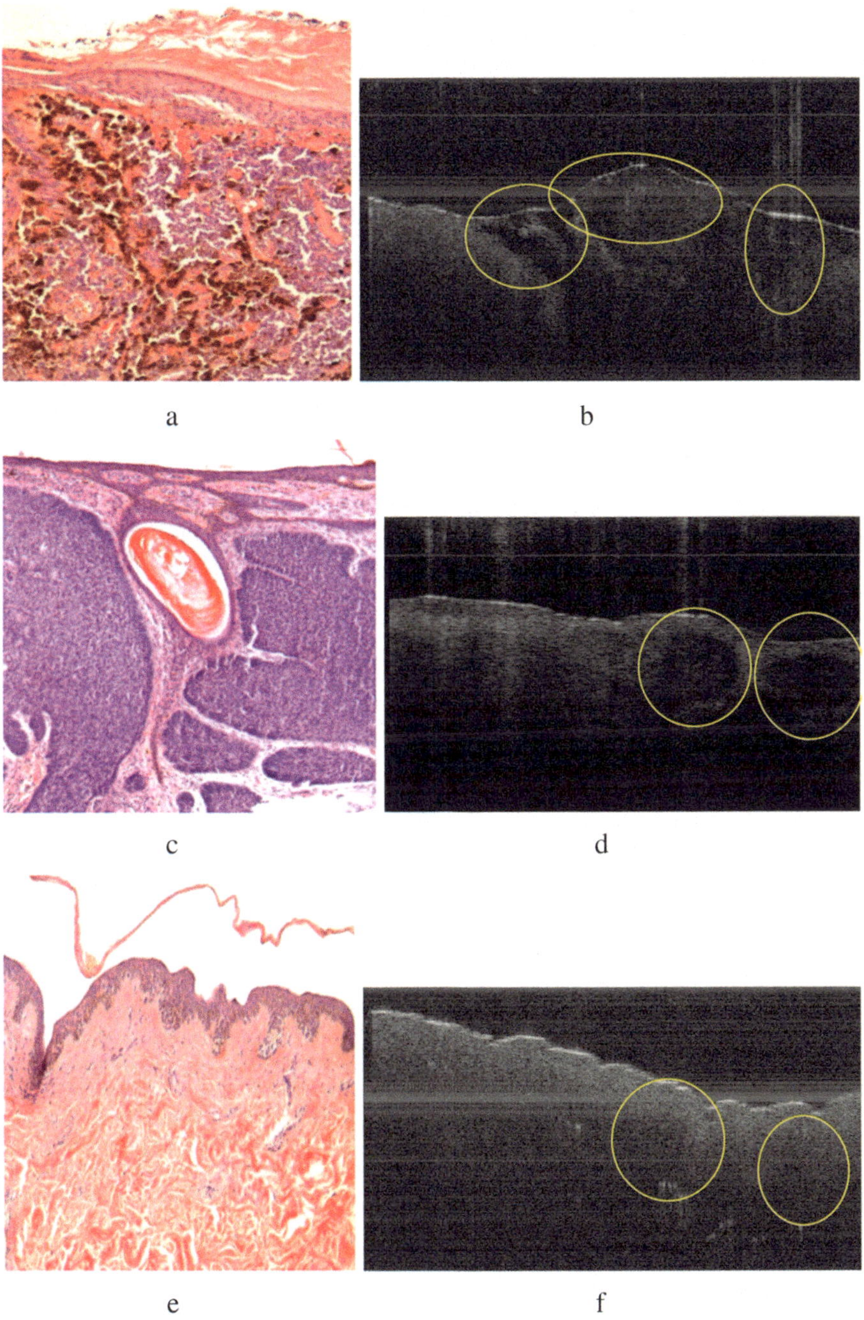

Fig. 13.8 Histopathological sections (100× magnification) and OCT images (B-scans) of studied skin samples: (**a**, **b**) Melanoma; (**c**, **d**) Basal Cell Carcinoma; (**e**, **f**) Pigmented Nevi [38]. Yellow ellipses mark tissues inhomogeneities

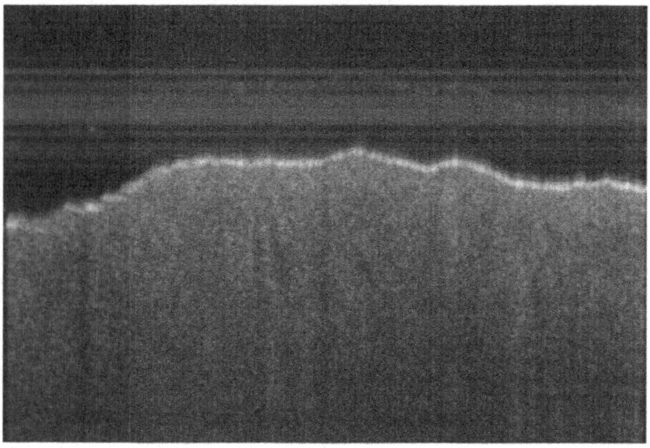

Fig. 13.9 OCT image (B-scan) of normal skin tissue

images, one can note the inhomogeneity of the dermis optical properties (Fig. 13.8f), notably a different depth of signal attenuation, which indicates differences in melanin distribution along the image spatial coordinate. In the normal tissue, the layer of the dermis is visualized uniformly in depth along the horizontal axis (Fig. 13.9).

13.4.3 Feature Extraction Algorithm

13.4.3.1 Denoising and ROI Selection

The interval type II fuzzy anisotropic diffusion filter was applied for all OCT data for images denoising [39]. The utilization of this filter effectively removes speckle noise and enhances the sharp regions (e.g., tumor boundaries) in OCT images [38]. All texture features described below have been calculated inside a ROI, which is manually selected by a rectangle covered the image inhomogeneity associated with the tumor. In this case, the ROI contains healthy tissue textures, but its volume is less than 1% and, as a fact, its influence is quite limited. Additional segmentation for tumor border areas inside the ROI rectangle has not been processed to simplify a calculation workflow.

13.4.3.2 Features Overview

The informative features of the algorithm of OCT image texture analysis are collected in Table 13.4. It must be mention that these feature categories have strong associations with such tumor features as asymmetry, borders irregularity, color, diameter, which are used in ABCD and 7-point algorithms in dermatoscopy. However, they are applied for 3D images and include an information about the tumor invasion and its spatial properties. In opposite, dermatoscopy images include

Table 13.4 Diagnostic features of OCT skin tumors

Group/method	Feature/variable name	Feature description
Haralick [40]	Correlation_0/45/90/135	Haralick correlation in 0/45/90/135 degrees direction over GLCM matrix
	Homogeneity_0/45/90/135	Haralick homogeneity in 0/45/90/135 degrees direction over GLCM matrix
	Contrast_0/45/90/135	Haralick contrast in 0/45/90/135 degrees direction over GLCM matrix
	Energy_0/45/90/135	Haralick energy in 0/45/90/135 degrees direction over GLCM matrix
Gabor [41]	Gabor_0/1/2/…/16	17 Gabor features, evaluated in ROI 0, 1,…,16
Tamura [42]	F_{cnt}	Contrast (Tamura)
	F_{dir}	Directionality
	F_{crs}	Coarseness
Fractal dimension	$D_{1\text{f}}$	1D-box counting fractal dimension [43]
	S_{df}	Standard deviation of 1D-box counting fractal dimension
	D_{ps}	2D-power spectrum fractal dimension [44]
	$D_{2\text{f}}$	2D-differential box counting fractal dimension [45]
Complex directional field (CDF) [46]	CDF_mean/variance, WF_mean/variance	Mean/variance of CDF phase and weight function (WF)
Markov random field (MRF) [47]	MeanR	Mean of autocorrelation function
	VarR	Variance of autocorrelation function

integrated by depth scattered radiation and, as a result, represent a depth-averaged image of the tumor.

13.4.3.3 Haralick Features

Weszka et al. [48] have shown that statistical descriptors are able to evaluate a texture better than features associated with spatial frequency. The statistical descriptors may be defined with the help of the Gray Level Co-Occurrence Matrix (GLCM) [23, 40], which characterizes the spatial relationship between the values of a brightness function in a certain local area of an image. The calculation of the matrixes is considered in detail elsewhere [23]. The matrixes depend on a distance between a pair of neighboring pixels with a specific gray level and an angular direction determined by those pair. We use the values in two adjacent points for calculation of GLCM as in a base method [40]. However, the method may be generalized to a case of three neighbors.

Technically, in our case, GLCM $P(i, j)$ is reflected the frequency of occurrence of a pixel with gray level value (intensity) i horizontally (vertically, diagonally) connected directly to a pixel with value *j*. For correct mathematical expression, $P(i, j)$ additionally has to be normalized and defined as $p(i, j)$. Thus, each element $p(i, j)$ in the normalized GLCM specifies a probability that a pixel with value *i* occurred horizontally (vertically, diagonally) adjacent to a pixel with value *j*.

These normalized matrixes $L \times L$ size are used for definition of texture features and image classification. We used four Haralick features most correlated (in our private opinion) with dermatoscopy features: homogeneity, correlation, contrast, and energy. More detailed information about the calculation procedure see, for example, in [49]

Second angular moment (squared energy) measures a local uniformity of the gray levels. Similar pixels maximize angular second moment.

$$f_E = \sum_{i=0}^{L-1} \sum_{j=0}^{L-1} p^2(i,j). \tag{13.7}$$

Contrast (variance) is a measure of intensity or gray level variations between two pixels. Larger contrast reflects larger differences in GLCM

$$f_{\text{Contr}} = \sum_{n=0}^{L-1} n^2 \sum_{i=0}^{L-1} \sum_{j:|i-j|=n} p(i,j). \tag{13.8}$$

Correlation of gray levels shows the linear dependency of these values in the GLCM

$$f_{\text{Cor}} = \frac{1}{\sigma_x \sigma_y} \left(\sum_{i=0}^{L-1} \sum_{j=0}^{L-1} ijp^2(\text{i,j}) - \mu_x \mu_y \right), \tag{13.9}$$

where μ_x, μ_y, σ_x, σ_y are mean values and standard deviations of pixel brightness by appropriate probabilities: $p_x(i) = \sum_{j=0}^{L-1} p(i,j)$, $p_y(j) = \sum_{i=0}^{L-1} p(i,j)$.

Inverse difference moment (homogeneity) measures how close a distribution of elements in a GLCM to the diagonal of the GLCM. Behavior of homogeneity is typically opposite to contrast

$$f_{\text{Hom}} = \sum_{i=0}^{L-1} \sum_{j=0}^{L-1} \frac{p(i,j)}{1 + (i-j)^2}. \tag{13.10}$$

The Haralick features (Eqs. (13.7)–(13.10)) are calculated across each of 4 directions (horizontal—0°, vertical—90°, and two diagonal directions—45° and 135°). Totally 16 Haralick features have calculated (Table 13.4). The final values of second angular moment, contrast, correlation, and homogeneity are evaluated by averaging over the angles of each feature, respectively.

13.4.3.4 Gabor Features

Gabor filters are a well-known methodology for texture analysis that has been used in various applications [41, 50, 51]. The image frequency component is very important for texture analysis; however, Fourier transform does not produce any information about its spatial distribution. One can use a windowed Fourier transform to extract this spatial information. The Gabor transform is obtained by implementation of Gaussian functions as windows, which have good spatial-frequency localization.

Gabor coefficients are defined as follow:

$$G_{f,\varphi}(n_0,n_1) = \mathrm{FT}^{-1}\left[\hat{I}(u_0,u_1)\cdot M_{f,\varphi}(u_0,u_1)\right], \tag{13.11}$$

$$M_{f,\varphi}(u_0,u_1) = \exp\left[-2\pi^2 a^2\left(\left(u_0' - f\right)^2 \lambda^2 + u_1'\right)\right], \tag{13.12}$$

where $\hat{I}(u_0,u_1)$ is a Fourier image of $I(n_0, n_1)$, λ is an anisotropy coefficient, $f = \sqrt{u_0^2 + u_1^2}$ and $(u_0',u_1') = (u_0 \cos\varphi + u_1 \sin\varphi, -u_0 \sin\varphi + u_1 \cos\varphi)$.

To describe an image $I(n_0, n_1)$ via Gabor functions, a new image has to be created:

$$\tilde{I}(n_0,n_1) = \begin{cases} I\left(n_0 + n_0^0, n_1 + n_1^0\right), \left(n_0 + n_0^0, n_1 + n_1^0\right) \in D_I. \\ 0, \text{ elsewhere.} \end{cases} \tag{13.13}$$

Here $n_0^0 = \min\{n_0 \in \mathbb{Z} \mid \exists n_1 : (n_0,n_1) \in D_I\}$ is a left boundary of set D_I, $n_1^0 = \min\{n_1 \in \mathbb{Z} \mid \exists n_0 : (n_0,n_1) \in D_I\}$ is a bottom boundary of set D_I.

A discrete Fourier transform is applied on this image:

$$\hat{I}(m_0,m_1) = \sum_{n_0=0}^{M-1} \sum_{n_1=0}^{N-1} I(n_0,n_1) \exp\left\{-2\pi i\left(\frac{m_0 n_0}{M} + \frac{m_1 n_1}{N}\right)\right\}. \tag{13.14}$$

Next, a resulting image is multiplied on various sampled Gaussian windows of the form

$$W_k(m_0,m_1) = \exp\left(\frac{\left(m_0 - m_0^0(k)\right)^2}{2\sigma_0^2(k)} + \frac{\left(m_1 - m_1^0(k)\right)^2}{2\sigma_1^2(k)}\right). \tag{13.15}$$

All window centers $\left(m_0^0(k), m_1^0(k)\right)$ are determined in advance. The resulting set of complex images of the original image spectra has a form

$$\hat{y}_k(m_0,m_1) = \hat{I}(m_0,m_1)\left(W_k(m_0,m_1) + \overline{W}_k(m_0,m_1)\right), \tag{13.16}$$

where a window $\overline{W}_k(m_0,m_1)$ has a center $\left(-m_0^0(k), -m_1^0(k)\right)$. Each complex image allocates a certain frequency range. The spectrum should remain symmetrical and match the real image.

The inverse Fourier transform over the images $\hat{y}(m_0,m_1)$ should produce a set of images $y(m_0, m_1)$, similar to the initial one, but each of them contains only certain textures. The average energies of these images may be used as Gabor features in the classification task. According to the Parseval theorem, one can obtain these features without application of the inverse Fourier transform:

$$g_k = \sum_{m_0=-\frac{M-1}{2}}^{\frac{M-1}{2}} \sum_{m_1=-\frac{N-1}{2}}^{\frac{N-1}{2}} \hat{y}_k(m_0,m_1)\hat{y}_k^*(m_0,m_1). \tag{13.17}$$

Finally, we have to determine the number of features, and also select for each of them the center of a Gaussian window $\left(m_0^0(k), m_1^0(k)\right)$ and standard deviations. For example, one can use a frequency domain splitting shown in Fig. 13.10, which allows one carefully separate high-frequency textures.

Thus, we have 17 Gabor features g_k (Gabor _ k), $k = 0, 1, 2, \ldots, 16$. The center of the k-th Gaussian window is located in the center of the k-th rectangular area shown

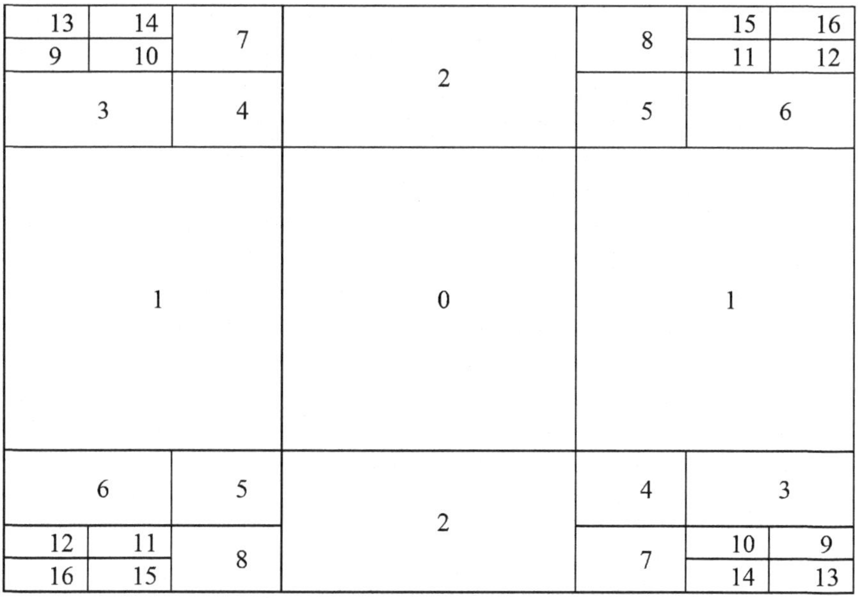

Fig. 13.10 Frequency domain splitting and Gaussian window centers selection

in the figure, and for each window, except the zero one, there is a paired one, the center of which is symmetric with respect to the axis origin.

Standard deviations $\sigma_0(k)$ and $\sigma_1(k)$ for the k-th window are selected based on a size of the k-th rectangular area, so the values of the spectrum inside the area are not too strongly extinguished by the Gabor filter, and the values outside of the area have been significantly reduced. We assume that $\sigma_0(k) = \frac{1}{2}w(k), \sigma_1(k) = \frac{1}{2}h(k)$, where $w(k)$ and $h(k)$ are the linear dimensions of the k-th region. In this case, the value of the window function $W_k(m_0, m_1)$ in the corner of this region is obviously equal to $e^{-1} \approx 0.37$, and the largest of the values on the border of the rectangular region is $e^{-\frac{1}{2}} \approx 0.61$. Taking into account that Gaussian function decreases much quickly, such a case suits us perfectly, and in all practical cases we use these values as standard deviations.

13.4.3.5 Tamura Features

Tamura et al. [42] proposed six textural attributes corresponding to the human perception: coarseness, contrast, directionality, line-likeness, regularity, and roughness. It was found that the first three features are the most important, because they are most strongly associated with visual perception. Three other features are highly correlated with the first ones and do not add much for texture analysis and classification.

Coarseness is a feature associated with the distance between noticeable spatial variations in gray levels, that is, with the size of primitive elements (textiles) that form a texture. As an image contains textures of several sizes, the coarseness aims to identify the larger size at which a texture exists. This feature may be calculated by the following algorithm. For each image pixel (supplemented with zeros outside the domain of definition) the average value $\overline{I}_k(m,n)$ is calculated in the six different windows with sizes $(2^k + 1) \times (2^k + 1)$ pixels, $k = 0, 1, \ldots 5$. For each image pixel and for each size k, the difference modulus is calculated with the nearest average of a non-overlapping window in each of the four directions θ:

$$E_{k\theta}(m,n) = \begin{cases} \left|\overline{I}_k(m,n) - \overline{I}_k(m+2^k+1,n)\right|, \text{ if } \theta = 0^\circ; \\ \left|\overline{I}_k(m,n) - \overline{I}_k(m,n+2^k+1)\right|, \text{ if } \theta = 45^\circ; \\ \left|\overline{I}_k(m,n) - \overline{I}_k(m-2^k-1,n)\right|, \text{ if } \theta = 90^\circ; \\ \left|\overline{I}_k(m,n) - \overline{I}_k(m,n-2^k-1)\right|, \text{ if } \theta = 135^\circ. \end{cases} \quad (13.18)$$

For each point in each direction θ, the window size k is calculated that maximizes the difference $E_{k\theta}(m, n)$ and determines the best size $S_\theta(m,n) = 2^{\underset{k}{\operatorname{argmax}} E_{k\theta}(m,n)}$.

Coarseness is calculated by averaging $S_\theta(m, n)$ over the entire image and directions θ

$$F_{\text{crs}} = \frac{1}{4|D_I|} \sum_{\theta=1}^{4} \sum_{(i,j)\in D_I} S_\theta(m,n), \tag{13.19}$$

where D_I is a set of all image points (i, j) and $|D_I|$ is a scalar defining the image area value.

Tamura *contrast* measures how dramatically the brightness varies in the image together with the polarization of the distribution of black and white. This feature may be calculated as follows:

$$F_{\text{cnt}} = \frac{\sigma}{\alpha_4^r}. \tag{13.20}$$

Here σ^2 is variance of the image brightness, $\alpha_4 = \frac{\mu_4}{\sigma^4}$ is kurtosis or normalized fourth central moment, $r = 0.25$ is an experimentally selected coefficient.

Directionality is a feature measured using a histogram of local contour directions. The feature measures not the orientation itself but the total degree of directionality in the image. First, the edge strength $e(m, n)$ and angular orientation $a(m, n)$ of the contour are measured:

$$e(m,n) = \frac{1}{2}\left(|\Delta_x(m,n)| + |\Delta_y(m,n)|\right); \tag{13.21}$$

$$a(m,n) = \operatorname{arctg}\frac{\Delta_y(m,n)}{\Delta_x(m,n)} + \frac{\pi}{2}, \tag{13.22}$$

where the horizontal and the vertical brightness differences between the neighboring pixels $\Delta_x(m, n)$ and $\Delta_y(m, n)$ can be obtained, for example, by processing the original image with the discrete differentiation Prewitt operator computing an approximation of gradient of the image intensity function.

The histogram $h_a(k)$ of quantized angular directions $a(m, n)$ is constructed by counting the edge strength $e(m, n)$ exceeding a certain preselected brightness threshold. This histogram reflects the degree of directionality. In order to measure the *directionality* one needs to calculate the number n_{ext} and relative positions a_p of the peaks (local maxima). The final equation of *directionality* is defined as follows:

$$F_{\text{dir}} = 1 - R \cdot n_{\text{ext}} \sum_{p=1}^{n_{\text{ext}}} \sum_{a\in w_p} \left(a - a_p\right)^2 \cdot h_a(a), \tag{13.23}$$

where R denotes a normalizing factor related to quantizing levels of the angles a and w_p is a range between valleys surrounding the corresponding peak p.

13.4.3.6 Fractal Analysis

The fractal analysis has been used to examine the structural changes in OCT images of biological tissue. The most popular algorithm for computing the fractal dimension of one-dimensional and two-dimensional data is the box counting method originally developed by Voss [43]. It was effectively applied for computing the fractal dimension characterizing arterial tissue [52] and the breast carcinoma [53]. Gao et al. [54] have applied the power spectrum method to carry out the fractal analysis on the layered retinal tissue for diagnosing the diabetic retinopathy.

In the box counting method, the fractal surface (volume) is covered with a grid of n-dimensional boxes (hyper-cubes) with side length ε, and the number of boxes that contains a part of the fractal $N(\varepsilon)$ are calculated. As for images, the grid consists of cubes. In our case, 1D fractal dimensions across A-scans have been calculated and, then, averaged. Therefore, the fractal surface is covered with squares of recursively different sizes, and the fractal dimension is defined as follows [55]:

$$D_{1f} = \frac{\log N(\varepsilon)}{\log 1/\varepsilon}. \tag{13.24}$$

As for OCT B-scans, we also use differential box counting method [45]. Let us consider the OCT scan image of size $M \times M$ pixels covered with a grid of a size $s \times s$, where s is an integer from the interval [1, $M/2$]. Let us determine the digital intensity unit measure $g = r\,G$, where G is the maximum intensity on the OCT image and $r = {}^{s}/_{M}$. Therefore, each (i, j)th element of the grid may be characterized by intensity distribution inside an $s \times s$ element measured in digital units g. In this case the difference in maximum ($g_{\max}$) and minimum ($g_{\min}$) digital intensities inside the (i, j)th grid element would determine the local contribution number

$$n_r(i,j) = g_{\max} - g_{\min} + 1. \tag{13.25}$$

Taking contributions from the whole grids, we have the total number

$$N(r) = \sum_{ij} n_r(i,j). \tag{13.26}$$

Calculating (13.26) for different grid sizes s the fractal dimension D_{2f} may bedefined similarly to Eq. (13.24) as the least square linear fit of $\log N(r)$ against $\log\left(\frac{1}{r}\right)$.

We also use the power spectrum method for counting fractal dimension [44, 56], which relates to the Fast Fourier Transform (FFT) of 2D gray-scale image $I(i, j)$:

$$P(k_{xi}, k_{yj}) - cf^{-\beta}, \tag{13.27}$$

where c is a constant, f is a spatial frequency $f = \sqrt{k_{xi}^2 + k_{yj}^2}$.

β may be examined by fitting the function in Eq. (13.27) to the calculated 2D power spectrum. In the practice, β is calculated as being a slope of the curve $\ln P \times \ln f$. The slope could be approximated by using the least square linear fit. Then, the 2D Fourier fractal dimension D_{ps} may be easily estimated through the following equation [44]:

$$D_{ps} = \frac{8-\beta}{2}. \tag{13.28}$$

13.4.3.7 Complex Directional Field

The development of a tumor significantly changes optical parameters of the tissue [57, 58], which allows it to distinguish them from healthy tissue in OCT images. Thus, pattern analyzing of tumor boundaries (pathological heterogeneities) may be helpful for evaluation of the degree of malignancy and a type of the tumor.

One of the ways to describe such complex structures as borders is the direction field that in the case of OCT image represents a field of angles $\varphi(x, y)$ characterizing a predominant direction of the boundary of heterogeneity in a local neighborhood of the point (x, y). The complex directional field (CDF) is defined as a brightness gradient of the image at the point (x, y)

$$\nabla I(x,y) = G_{max} \cdot w(x,y) \cdot \exp(i2\varphi(x,y)), \tag{13.29}$$

where $G_{max} = \max(|\nabla I(x, y)|)$, $w(x, y) = |\nabla I(x, y)|/G_{max}$ is the weight function, and a phase is determined on the interval $[0, \pi]$ as follows:

$$\tan\varphi(x,y) = -\frac{f_x}{f_y}, \tag{13.30}$$

$$(f_x, f_y) = \left(\frac{\partial I(x,y)}{\partial x}, \frac{\partial I(x,y)}{\partial y}\right). \tag{13.31}$$

The CDF method does not require image preparation, border selection or segmentation, and preliminary manual labeling; it can be applied directly to a gray-scale image that allows one to analyze any complex heterogeneous structures in OCT images.

Obviously, the weight function $w(x, y)$ must have a maximum value in areas with sharp changes of the brightness, and a value close to zero should be observed for image areas with almost constant brightness. This determines one of the main disadvantages of the directional field method— the high sensitivity to noise level, which as a rule characterized by edgy alteration of the brightness. In order to overcome this drawback, we use the local gradients calculation procedure, which is based on the fact that gradient of a function at any point is perpendicular to a tangent to the contour of the brightness level at the given point.

The local gradients counting procedure is based on a calculation of the function of brightness gradient ($f_x^{k,l}$, $f_y^{k,l}$) at various positions of the local mask inside the symmetric rectangular external window W of $M \times N$ size, that is scanning the image [46], where $1 \le k \le N$, $1 \le l \le M$.

The local gradients ($f_x^{k,l}$, $f_y^{k,l}$) define the local angles:

$$\varphi_{k,l} = -\tan^{-1} \frac{f_x^{k,l}}{f_y^{k,l}}. \tag{13.32}$$

Then direction of trace in the center of outer window W may be evaluated by averaging of local angles field:

$$\dot{\varphi} = \frac{1}{2} \arg \sum_{\mathrm{k}=1}^{\mathrm{N}} \sum_{\mathrm{l}=1}^{\mathrm{M}} \exp\left(\mathrm{i}2\varphi_{\mathrm{k,l}}\right), \tag{13.33}$$

and a value of the weight function:

$$\dot{w} = \left| \frac{1}{M \cdot N} \sum_{k=1}^{N} \sum_{l=1}^{M} \exp\left(i2\varphi_{kl}\right) \right|. \tag{13.34}$$

Equations (13.33) and (13.34) are used as texture descriptors.

13.4.3.8 Markov Random Field

According to the stochastic approach [24], a tumor texture inside an image $I(x, y)$ may be formalized through a random process (a random field). In this way, each pixel $I(x_k, y_l)$ has a probability function on all its neighbor pixels. Obviously, this neighbor set is limited by a region of the neoplasm texture, and the field parameters are being different for different types of tumors, thereby describing stochastic parameters of tissue heterogeneity, absorption, and scattering. This approach requires preliminary and accurate segmentation of the original OCT images.

This model may be simplified by mandatory reducing the neighbor set, for example, down to 3 × 3 pixels square window. Suggesting that this area is much smaller than a size of the neoplasm texture, we can conclude that accurate ROI extraction is not required in this case, and the image (texture) model may be described through Markov Random Field (MRF) [59]. This approach has been already applied for texture modeling, texture classification, and texture segmentation in image analysis [60].

To evaluate MRF parameters of the original image (inside the ROI) $I(x, y)$, an autocorrelation function [46] $R(x, y)$ over 3 × 3 area is calculated as follows:

$$R(x,y)=\sum_{1}^{i=-1}\sum_{1}^{j=-1} I(x,y)I(x+i,y+j),\ (i,j)\neq(0,0). \tag{13.35}$$

The mean and variance of the autocorrelation function Eq. (13.35) are used as textural features.

13.4.4 Classification Results

We have performed several classification experiments with biological sample collection described in Sect. 13.4.2. In first experiment, four methods have been used for texture feature classification: Haralick, Markov Random Fields and fractal features for analysis of OCT B-scans and complex directional field (CDF) for analysis of C-scans (see Table 13.4). The dataset included 488 OCT images: 320 C-scans (MM—80, BCC—80, nevus—80, healthy skin—80) and 168 B-scans (MM—42, BCC—42, nevus—42, healthy skin—42).

Fig. 13.11 shows the scatter plots for selected texture features pairs, illustrating interclass separability. A linear classifier has been used as a separator. Table 13.5 summarizes the results of the linear binary classification for MM vs healthy skin, MM vs nevus, MM vs BCC, BCC vs nevus, and nevus vs healthy skin.

The results of these binary classifications from Table 13.5 are highly contradictory and do not allow us to unambiguously choose one or another pair or group of texture features as the main one. The highest accuracy was achieved for MRF features—more than 92% for sensitivity and 95% for specificity for classification of almost all malignant tumors. But MRF fails in a case of nevus vs healthy skin, which is characterized by extremely low sensitivity. It may be explained by the fact that Nevus is often represents a hyperpigmented area of a skin only without a significant change in the pattern. In addition, there are more samples in the case of nevus vs healthy skin than, for example, in MM vs BCC. Fractal analysis does not show satisfactory results (sensitivity and specificity are always lower than 70%) in all cases that may be indirectly justified by instability of fractal dimension. Haralick features show slightly better results, except for MM vs nevus case, where specificity drops below 60% for all pairs of features. A confident classification with Tamura features was achieved only in the case of BCC vs nevus. Apart from this, it may be notes, that CDF, unlike all other features, is calculated using C-scans, and this is the group that showed excellent results in all considered cases. As for the B-scans analysis, multicomponent classifiers with multi-texture features have to be compiled.

Therefore, each texture group has advantages in binary classification only of a specific tumor type vs healthy skin. However, these results are illustrative only because there is absolutely no problem to distinguish melanoma from healthy skin for a physician if this is known in advance that both healthy tissue and melanoma samples are presented in the dataset. In a case of binary classification containing healthy skin, we have the same difficulty for any other tumor types. On the other

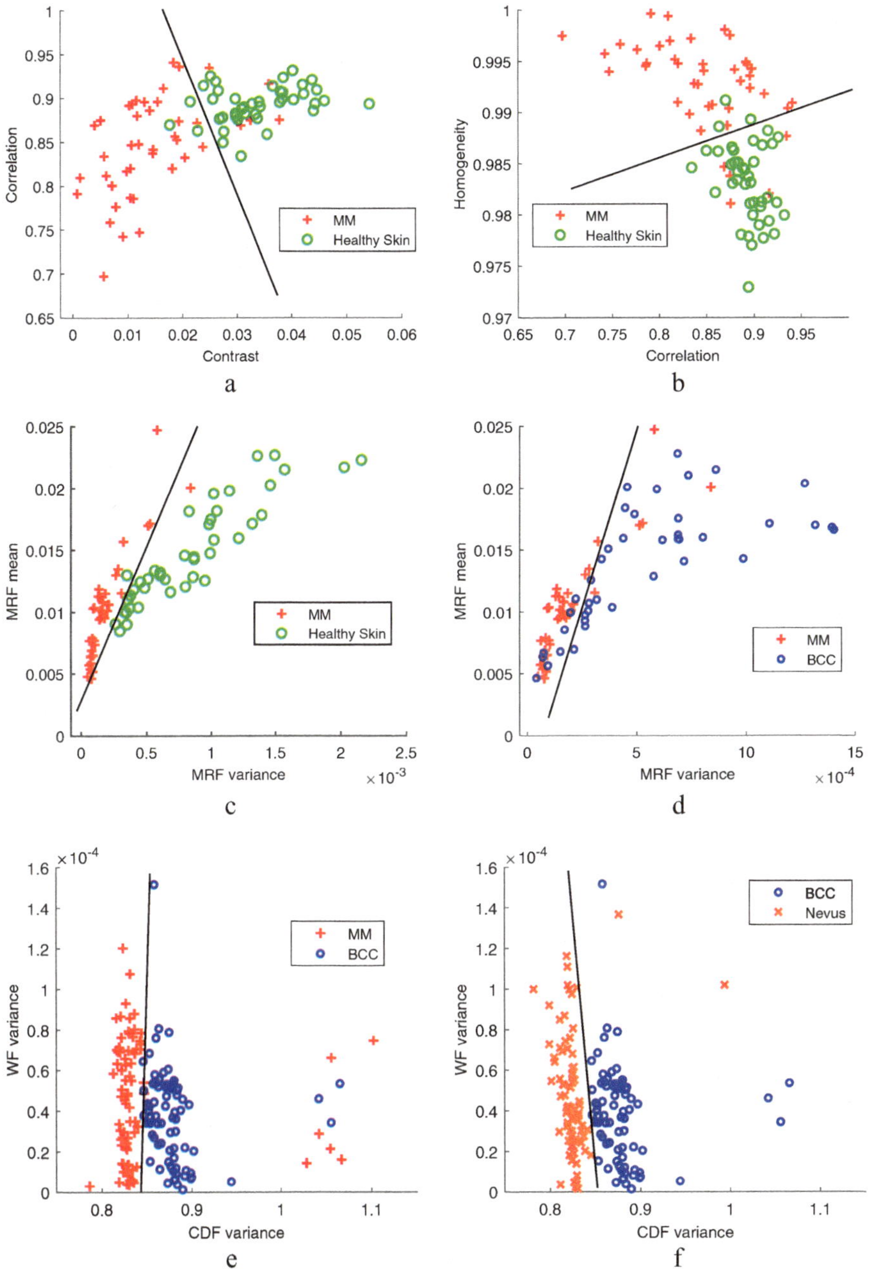

Fig. 13.11 Scatter plots for the binary classifier MM vs healthy skin by (**a**) Haralick contrast–correlation, (**b**) Haralick correlation–homogeneity, (**c**) MRF variance–MRF mean; MM vs BCC by (**d**) MRF variance–MRF mean; (**e**) CDF variance–WF variance; and BCC vs nevus by (**f**) CDF variance–WF variance [61]

Table 13.5 Statistical characteristics of tumor tissues texture binary classification [61]

Feature class	Features	Sensitivity-specificity	Samples	Scan type
		MM vs healthy skin		
Haralick	Contrast–Correlation	88%-92.8%	84 (42/42)	B
	Correlation–Homogeneity	88%-95.2%	84 (42/42)	B
	Contrast–Energy	79%-14%	492 (239/253)	B
MRF	MRF variance–MRF mean	92.8%-95.2%	84 (42/42)	B
Fractal analysis	$S_{df} - D_{1f}$	66%-69%	105(21/63)	B
	$D_{2f} - D_{ps}$	20%-99%	492 (239/253)	B
Tamura	Contrast–Coarseness	93%-8%	492 (239/253)	B
	Coarseness–Directionality	97%-7%	492 (239/253)	B
		MM vs nevus		
Haralick	Contrast–correlation	78.5%-59.5%	84 (42/42)	B
	Correlation–homogeneity	83.3%-57.1%	84 (42/42)	B
	Contrast–energy	78.5%-59.1%	84 (42/42)	B
	Correlation–energy	81%-50%	84 (42/42)	B
	Energy–homogeneity	83.3%-50%	84 (42/42)	B
MRF	MRF variance–MRF mean	97.6%-73.8%	84 (42/42)	B
Fractal analysis	$S_{df} - D_{1f}$	64.2%-66.7%	84 (42/42)	B
	$D_{2f} - D_{ps}$	93.6%-62%	84 (42/42)	B
Tamura	Contrast–coarseness	90%-68%	490 (239/251)	B
	Coarseness–directionality	94%-2%	490 (239/251)	B
		MM vs BCC		
Haralick	Contrast–energy	87%-100%	42 (21/21)	B
	Homogeneity–energy	66%-92%	42 (21/21)	B
	Contrast–correlation	93%-7%	510 (239/271)	B
	Correlation–Homogeneity	99%-1%	510 (239/271)	B
MRF	MRF variance–MRF mean	90.4%-83.3%	84 (42/42)	B
Fractal Analysis	$D_{2f} - D_{ps}$	85%-6%	510 (239/271)	B
Tamura	Contrast–coarseness	94%-28%	510 (239/271)	B
	Coarseness–directionality	94%-4%	510 (239/271)	B
CDF	WF variance–CDF variance	91.5%-100%	160 (80/80)	C
		Nevus vs healthy skin		
CDF	WF variance–CDF variance	97.5%-83.7%	160 (80/80)	C
Haralick	Contrast–correlation	12%-99%	504 (251/253)	B
	Correlation–homogeneity	18%-99%	504 (251/253)	B
	Contrast–energy	28%-72%	504 (251/253)	B
	Energy–homogeneity	95%-87%	504 (251/253)	B
MRF	MRF variance–MRF mean	24%-100%	504 (251/253)	B
Fractal Analysis	$D_{2f} - D_{ps}$	89%-25%	504 (251/253)	B
Tamura	Contrast–coarseness	70%-87%	504 (251/253)	B
	Coarseness–directionality	97%-8%	504 (251/253)	B

(continued)

Table 13.5 (continued)

Feature class	Features	Sensitivity-specificity	Samples	Scan type
		BCC vs nevus		
Haralick	Energy–homogeneity	93%-94%	526 (273/253)	B
	Contrast–correlation	98%-12%	526 (273/253)	B
	Correlation–homogeneity	95%-10%	526 (273/253)	B
	Contrast–energy	99%-24%	526 (273/253)	B
MRF	MRF variance–MRF mean	95%-96%	526 (273/253)	B
Fractal Analysis	$D_{2f} - D_{ps}$	21%-99%	526 (273/253)	B
Tamura	Contrast–coarseness	88%-91%	526 (273/253)	B
	Coarseness–directionality	96%-2%	526 (273/253)	B
CDF	WF variance–CDF variance	100%-97.5%	160 (80/80)	C

hand, the good preliminary binary classification results prove principal applicability of Haralick and MRF features for melanoma mapping and accurate detection of tumor boundaries with a resolution much higher than in the case of dermatoscopy [62], Raman tumor mapping [63], and laser-induced breakdown spectroscopy of melanoma [64].

The results of binary classification of melanoma versus benign tumors are much more interesting. One of the most typical clinical problems is distinguishing of malignant melanoma versus pigmented nevus, which have many similar visual properties. As one can see from Table 13.5, the accuracy of the texture classification is slightly reduced in the case in comparison with healthy skin but it still remains quite high for most Haralick features, and extremely high for MRF variance-mean: the sensitivity exceeds 97% with specificity of 73.8%. It must be mentioned that fractal analysis gives good results for 2D-differential box counting and 2D spectral power fractal dimensions only.

For the second experiment, the complete samples collection (see Sect. 13.4.2) have been examined using all declared texture features. Decision trees have been used for classification (depth = 5).

In contrast to the previous classification approach (Fig. 13.11), where a separation of classes occurs due to the plane split into two half-planes using a separation line, decision tree is a nonlinear method for making decision. First of all, a decision tree is a binary one, each node of which is a simple binary condition (classifier) of one of the variables (features). The next level contains two nodes for "True" and "False" values of the condition necessarily. Such node can be either a leaf describing a region of space with objects of one class only, or a sub-tree if the node covers objects of different classes. The latter is splitting recursively according to the same rule. The maximum depth of the tree (in our case depth = 5) is set up as a partition parameter. The deepest layer can contain only leaves (even if they cover objects from different classes). Thus, the decision tree "cuts" a phase plane (Fig. 13.12) into rectangles (leaves), each of which has a certain constant classifier of a class, in other words, can recognize one class only. In general, the decision tree, certainly, seeks to minimize the classification error.

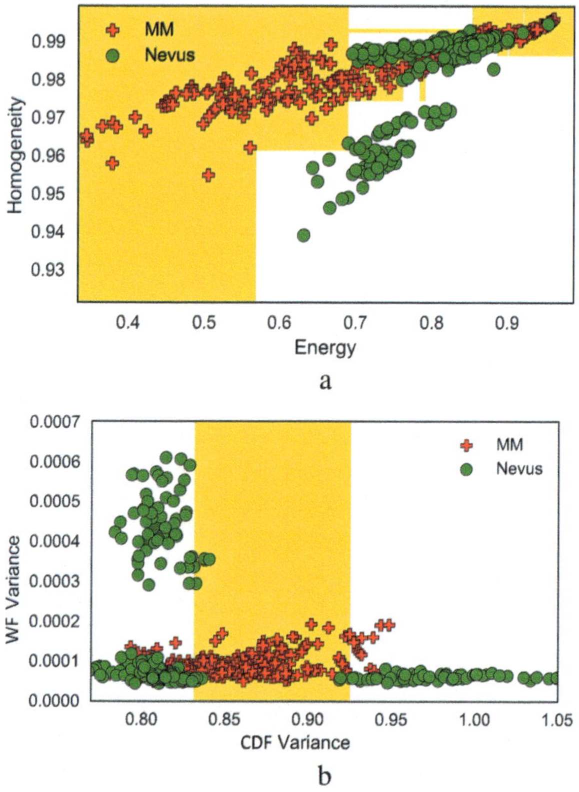

Fig. 13.12 Scatter plots for binary classifier MM vs nevus (**a**) Haralick: energy-homogeneity and (**b**) CDF: CDF variance–WF variance [65]

The results are presented using delayed sampling. The samples set was randomly divided for training and test sets in proportion 80%/20%. The k-fold cross-validation has been used with a random separation of the all samples into five groups (folds) for assessing the quality of the training. In this case, onefold (20%) plays a role of a test set and four others (80%) combine a training set each time. The results of classification are summarized in Table 13.6.

Scattered plots for the binary classifier MM vs nevus are shown in Fig. 13.12 for Haralick: energy-homogeneity for B-scans (a) and for CDF: CDF variance—WF variance for C-scans (b). As one can see, the classifier is characterized by fragmented phase surface for each class as for B-scans as for C-scans and consists of a set of detached areas after cross-validation (Fig. 13.12).

This is interesting to note that the best accuracy in identifying melanoma versus nevus was achieved with Tamura texture features, but the more stable classification results with almost equivalent values of specificity and sensitivity for different classes (MM vs nevus, BCC vs nevus) were obtained for Haralick: energy–homogeneity. At the same time, the accuracy of BCC versus benign tumor identification by Tamura (contrast–coarseness) is 10% less than the accuracy of MM identification. At the same time, the exact determination of the malignancy type is very

Table 13.6 Statistical characteristics (after cross-validation) of the tissues separation [65]

Feature class	Features	Tissue 1	Tissue 2	Sensitivity-specificity	Class tissue fraction (total number)
Haralick	Energy–homogeneity	BCC	Nevus	93%-94%	273/253 (526)
Haralick	Energy–homogeneity	MM	Nevus	92%-93%	254/253 (507)
Haralick	Energy–homogeneity	Nevus	Healthy skin	95%-87%	253/229 (482)
CDF	WF variance–CDF variance	MM	Nevus	83%-96%	254/253 (507)
CDF	WF variance–CDF variance	BCC	Nevus	77%-97%	273/229 (502)
MRF	MRF variance–MRF mean	BCC	Nevus	95%-96%	273/253 (526)
MRF	MRF variance–MRF mean	MM	Nevus	81%-89%	254/253 (507)
Tamura	Contrast–coarseness	BCC	Nevus	88%-91%	273/253 (526)
Tamura	Contrast–coarseness	MM	Nevus	96%-93%	254/253 (507)

Table 13.7 Binary SVM classifiers

	Sensitivity (%)	Specificity (%)	Total number of images
Malignancy vs all	90	89	1011
MM vs tumor	96	96	490

important because a treatment option depends on the type of the detected malignant tumor: surgery resection for MM and laser ablation for BCC.

Considering the standard diagnostic workflow, this is usually multi-stage and includes sequential determination of a malignant neoplasm (usually by a general practitioner) and, after that only, the determination and confirmation of the type of the malignancy. Therefore, a "classifier of malignancy" is very popular in medicine, its main aim is to distinguish a "malignant" tumor from a "benign" one. This gives a physician a "direction" of what to do next to determine a tumor type by another way, if it is a malignant neoplasm. But, in the same time, the physician, first of all, wants to know whether the detected tumor under study is melanoma due its dangerous features, as it is mentioned above. In this case, a classifier MM vs other tumors is looking very appropriative. It is well known that several features together may reach much better results than each of these features by one. Therefore, the two-stage classification was performed using support vector machine (SVM), where the vector is defined by all declared texture features. The results of classification are presented in Table 13.7: the accuracy of malignancy identification exceeds 90% with following possibility of melanoma determination with sensitivity and specificity about 96%.

Table 13.8 Three-class SVM classifier (748 images)

	Precision (%)	Recall (%)	F_1-score (%)	Total number of images
BCC	96	98	97	268
MM	97	95	96	239
Nevus	99	99	99	251

Table 13.9 Four-class SVM classifier (1011 images)

	Precision (%)	Recall (%)	F_1-score (%)	Total number of images
BCC	91	93	92	268
MM	93	90	91	239
Healthy skin	88	89	88	253
Nevus	99	99	99	251

Then, the SVM kernel was generalized for multi-class classifiers. Tables 13.8 and 13.9 show these classifiers obtained by using SVM for the considered sets of images. In these tables, each row can be interpreted as a binary classifier of the corresponding class versus all other examples in the dataset. For instance, there are BCC vs MM + nevus in Table 13.8 or BCC vs MM + nevus + healthy skin in Table 13.9. Precision, recall, and F_1-score (see Sect. 13.2 for details) have been calculated for all possible binary cases by using a linear binary SVM and all texture features for B-scans.

The results are very promising. Precision does not drop below 88% for normal skin, i.e., the classifier initially focuses more towards accurate diagnosis of neoplasms, both precision and specificity (Recall) of which do not drop below 89% for the four-class and 95% for the three-class classifiers.

13.5 Discussion and Conclusions

We analyzed quite large datasets for various tumors using two different technologies—dermatoscopy and OCT. Naturally, the fundamental difference in the collected data lies not in the image resolution only, but in the physical principles of acquiring of visual information about the tumor. If in the case of a dermatoscope, we integrate the scattered radiation from all the skin layers and get a 2D image of the tumor, in the case of OCT, we have a 3D image of higher resolution, which may be analyzed layer by layer both in depth direction (B-Scan) and in the layers parallel to the skin surface (C-Scan). Therefore, the texture properties of dermatoscopic 2D images have averaged optically by depth already. Apparently, C-scans for OCT and 2D images obtained by dermatoscope should have similar texture features. However, a more accurate cancer recognition by OCT is achieved with the texture analysis of B-scans. The latter is due to the fact that a malignant neoplasm is characterized by invasion of the tumor in depth, which is detected by the analysis of B-scans implicitly and has more explicit texture features. In this regard, the texture analysis of dermatoscopic and OCT images complement each other.

Despite the fact, that the algorithms of analysis for both imaging modalities are based on textural features and SVM classification, there are a number of differences between them. For OCT, the vector of classifying descriptors is formed by classical basic textural features (Haralick, Gabor, Tamura, Markov random field), which are commonly used for digital image processing. It may be noted the similarity between the Gabor and Haar features. Both features evaluate a texture by its spatial-frequency distribution. In terms of algorithm performance, Window Fourier Transform is faster than Discrete Wavelet Transform. But linearizing the Haar transformation by applying simple linear filters significantly reduces the complexity of calculating descriptors, which makes the procedure much more faster.

In dermatoscopy, the features transformation and optimization (Haar Transform, LBP, and color histograms) have been used, since the algorithm is intended originally to be implemented as a software for the dermatoscopic device. Another fundamental feature of the dermatoscopic images is the color texture analysis, which implicitly includes the distribution of natural skin fluorophores with different absorption and fluorescence bands. The intensive growth of a malignant neoplasm induces the development of the microvasculature, morphologic, and biochemical changes specific for different cancer types that appears in alteration of the color of texture compared to the healthy skin. At the same time, a large variability in the color characteristics of human skin leads to a large variance of these features and requires their correction in texture analysis. It was carried out by calculating comparative textural descriptors, which include the personal properties of the healthy skin (Personified Mode). All analyzed sets of dermatoscopic images (70, 130, and 106 images) have showed comparable classification results. At the same time, personalized diagnostics (evaluation of comparative textural descriptors of tumor and surrounding healthy tissue) have given a significant increase (15–20p.p) compared with the general mode (only tumor descriptors calculation).

The model is quite stable as evidenced by the results of cross-validation. The decrease of sensitivity and specificity was 6p.p. after cross-validation only and 3p.p., respectively. The obtained results are much better than the classification results described in [15, 16, 31, 34]: sensitivity 96% against 93% and specificity 89% against 80%. This is an application consequence of the complex texture features and the personalized mode in the texture analysis algorithm. The method of dermatoscopy is well suited for screening when high sensitivity indices are required (at the level of 95–97%), that ensures by the proposed algorithm, while the high specificity (89%) minimizes the number of false positive cases. In addition, the low weight and mobility of the device allow one conducting screening procedures outside a hospital without the need for sophisticated laboratory equipment.

Separately, it is possible to highlight studies about the dermatoscopic classification by using convolutional neural networks (CNNs), but CNNs are very demanding on computing resources, including memory. For example, a quite popular VGG16 contains up to 144 million parameters [26]. CNN has a problem with on-line training also, i.e., quick retraining of the classifier/network after adding a new sample. In

[74], the authors used Google Inception v.3 CNN, trained on approximately 129,000 images and 2032 various diseases. This network contains much less parameters, but their number still has the order of millions. The authors obtained accuracy of about 70%, but without total validation by biopsy/histopathology.

A number of disadvantages of the proposed algorithm should be mentioned. All medical computer diagnostic methods are extremely dependent on the source data. In this case, the dataset of 100–300 samples is not quite enough, and additional model training on a large dataset is required for obtaining exact statistical results. At the moment, the dependence of the result on the color correction stage has not been studied in details. Due to the different sensitivity of chromatic components inside a camera matrix, a RAW image looks greenish. Color correction allows us to correct the color temperature, making the images more "pleasant" for an operator. However, it is more important to make the image more contrast for dermatoscopy analysis, emphasizing certain dermatoscopic elements. For example, there are the increase of brightness/contrast of the white-blue veil by correction of the blue contribution, or enhancement of the contrast of melanin spots and atypical pigment networks by red one. The use of the additional color backlighting in the dermatoscope would allow us to obtain brighter and more contrast chromatic components on the same camera. Combining with the original RGB color image, one could increase the brightness of selected chromophores (the choice of chromophores depends on the switched color LEDs), simplifying their detection by texture features.

The better classification results were achieved for texture analysis of OCT images. It was demonstrated that Precision and Recall exceeds 97% in the multi-class recognition procedure due to the implementation of multi-texture analysis when used texture features complement each other. It makes possible the recognition of various tumors (as malignant as benign) with high-score identification of a tumor type in real clinical conditions.

Quite few successful applications of texture analysis of OCT images are known. For example, Sawyer et al. [66] have achieved 78% accuracy for ovarian cancer detection on a mouse model by using texture features including Haralick ones. The mouse model was also used in texture studies of lungs [67] and cervical cancer [68]. A study of the human intraretinal cystoid regions may be found in [69], where the authors used Gabor texture features, fractal dimension, and a number of various features based on intensity and gradient properties of the image. In principle, we use the similar approach being that, for instance, Haralick features describe indirectly the intensity properties and their distribution on the image. The complex directional field method allows us to work with boundaries and brightness differences similarly to gradients.

As for skin cancer research, Marvdashti with co-authors [70] have received the accuracy 95% in detecting BCC in human skin samples collected from 42 patients using polarization sensitive optical coherence tomography. The authors used a more sophisticated algorithm for ROI detection and extracted a number of various features for A- and B-scans, including intensity and phase retardation statistics (range, standard deviation, mean, median), histogram statistics, Haralick texture features, and morphological ones. It must be noted that our OCT system is equipped by a shorter wavelength light source (840 nm vs 1325 nm swept-source laser [70]), which characterized by higher noise level. But this fact does not greatly affect the classification result, same as our ROI selection algorithm, which captures a small part of healthy tissue along with tumor. We have obtained slightly lower BCC recognition result (precision 91% and sensitivity-recall 93%—see Table 13.9), but this is for four-class classifier, versus 95% sensitivity for a binary classifier (BCC vs healthy skin) in [70]. However, for the 3-class classifier, we have achieved higher sensitivity 98% and F_1-score 97%. In both cases, we have determined contemporaneously different types of tumors in comparison with binary classifier in [70].

In another study [71], the authors were able to distinguish BCC and squamous cell carcinoma (SCC) from healthy tissue using quadratic and linear SVM with accuracy of 87% and 81%, respectively. The authors used Haralick texture features only and a number of high-order statistical features. According to comparison of different texture methods (see Table 13.5), MRF autocorrelation function and Tamura Contrast–Coarseness texture features are characterized by higher sensitivity-specificity of cancer recognition than Haralick texture analysis. In addition, the disadvantage of high-order statistical features usage is their robust dependence on noise influence.

Melanoma, as a class of OCT images, is quite rare in studies [72] due to the lower prevalence of melanoma compared to BCC and its large absorption. For example, Boone et al. [73] used the relative attenuation factor only of A-scan signal and have obtained high sensitivity and specificity rates of 93% and 97%, respectively, in differentiation of MM vs non-malignant lesions on 15 MM examples. However, the BCC inclusion in that dataset should significantly reduce these rates since non-melanoma tumors do not have such a strong dependence on absorption.

It is interesting to estimate the significance of different texture methods for different neoplasm identification and cancer differentiation. Figure 13.13a shows the rates of different texture features contribution to the recognition result. Figure 13.13b demonstrates the significance of different groups, which combined by the contribution of the features of the same texture method. Each texture feature in a group has been normalized by total fraction of correspondent texture group, thus it make clear visible the order of texture features in the recognition procedure.

As one may see from Fig. 13.13a the Top-8 rates include all MRF and Tamura features and three from four fractal features. The contribution of this Top-8 features in cancer recognition is 69.7%. All other 34 features give 30.3% only. On the other hand, each texture method (group) contains different number of features, which means an average contribution of each feature, for example, from Haralick group (16) is less than from MRF group (2), then, Haralick significance is "blurred" between a large number of the same group features. But Haralick Correlation (first four features) has an advantage over the rest. Thus, if we compare the contribution of different methods (groups), we can see that Haralick group gives 22.1% (see Fig. 13.13b), which is more than Tamura or fractal groups and this is comparable to the MRF group with the maximum 33.5% contribution to the probability of tumor recognition. At the same time, the calculation of 16 Haralick features is quite fast procedure due to the fact that all features are processed from GLCM evaluation.

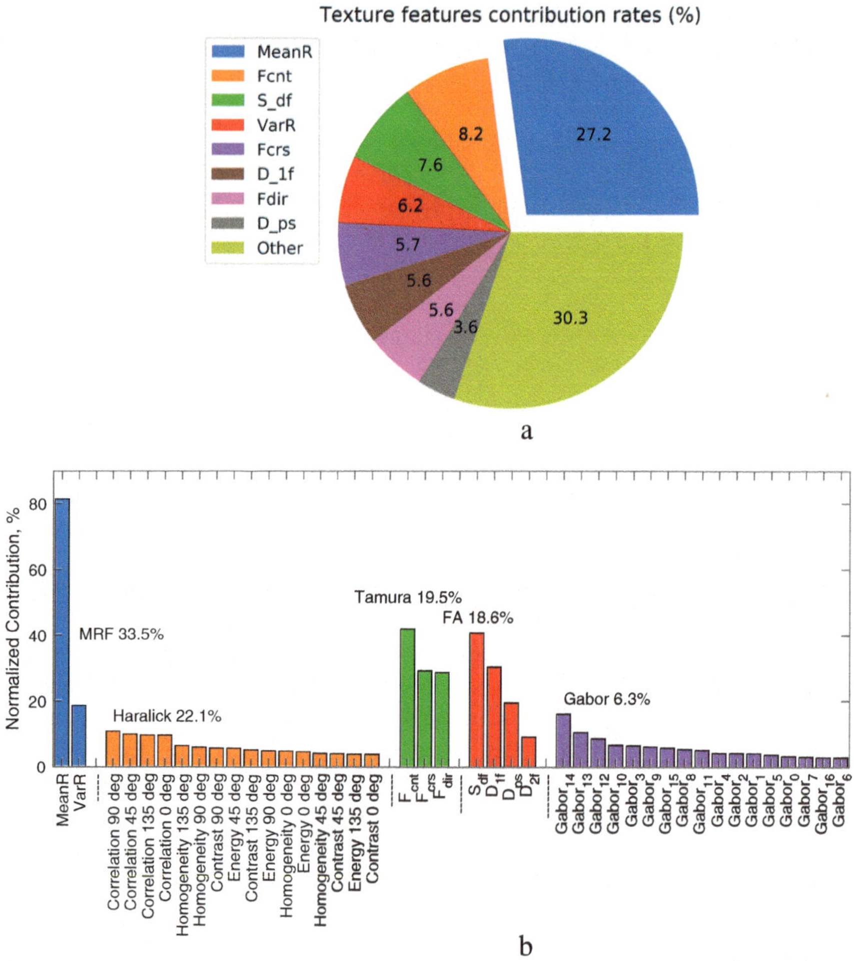

Fig. 13.13 The significance of different texture features (**a**) and groups (**b**) in tumor recognition. Each texture feature in a group (**b**) has been normalized by total fraction of correspondent texture group

Gabor features turned out to be the most uninformative in this study, their total contribution does not exceed 6.3%. This fact confirms the advantage of stochastic as well as statistical texture descriptors over frequency ones [48]. The features set may be optimized for enhance performance by exclusion of Gabor group with total significance of all other groups of 93.6%.

Therefore, the texture analysis of 2D dermatoscopic and 3D OCT images looks like a highly promising approach for automated classification and recognition of various malignant and benign neoplasms of the skin. It may be used as an interactive deep learning kernel interface for novel OCT devices.

Acknowledgments This research was supported by the grant # 19-52-18001 Bolg_a of the Russian Foundation of Basic Research. We are very thankful to Dr. Wei Gao from Ningbo University of Technology, China for Matlab scripts for denoising and fractal dimension calculating, as well as not a long but productive work together in Samara National Research University.

References

1. Shortliffe, E.H., Buchanan, B.G.: A model of inexact reasoning in medicine. Math. Biosci. **23**, 351–379 (1975). https://doi.org/10.1016/0025-5564(75)90047-4
2. Buchanan, B.G., Shortliffe, E.H.: Rule-Based Expert Systems: the MYCIN Experiments of the Stanford Heuristic Programming Project. Addison Wesley, Reading (1984)
3. Philbin, T.: The 100 Greatest Inventions of All Time: A Ranking Past and Present. Citadel Press, New York (2003)
4. Mathers, C.D., Loncar, D.: Projections of global mortality and burden of disease from 2002 to 2030. PLoS Med. **3**(11), e442 (2006). https://doi.org/10.1371/journal.pmed.0030442
5. Islami, F., Miller, K.D., Jemal, A.: Cancer burden in the United States – a review. Ann. Cancer Epidemiol. **2**(1), 39 (2018). https://doi.org/10.21037/ace.2018.08.02
6. Goss, P.E., Strasser-Weippl, K., Lee-Bychkovsky, B.L., Fan, L., Li, J., Chavarri-Guerra, Y., Chen, Z.: Challenges to effective cancer control in China, India, and Russia. Lancet Oncol. **15**(5), 489–538 (2014). https://doi.org/10.1016/S1470-2045(14)70029-4
7. Davydov, M.I., Aksel, E.M. (eds.): Statistika zlokachestvennykh novoobrazovaniy v Rossii i stranakh SNG v 2012 g. (Statistics of Malignant Neoplasms in Russia and the CIS Countries in 2012). Izdatelskaya gruppa RONTS, Moscow (2014)
8. Boyle, P., Levin, B. (eds.): World Cancer Report 2008. IARC Press, Geneva (2008)
9. Friedman, R.J., Gutkowicz-Krusin, D., Farber, M.J., Warycha, M., Schneider-Kels, L., Papastathis, N., Kopf, A.W.: The diagnostic performance of expert dermoscopists vs a computer-vision system on small-diameter melanomas. Arch. Dermatol. **144**(4), 476–482 (2008). https://doi.org/10.1001/archderm.144.4.476
10. Drexler, W., Fujimoto, J.G. (eds.): Optical Coherence Tomography: Technology and Applications. Springer, New York (2008)
11. Drexler, W., Fujimoto, J.G.: State-of-the-art retinal optical coherence tomography. Prog. Retin. Eye Res. **27**(1), 45–88 (2008). https://doi.org/10.1016/j.preteyeres.2007.07.005
12. Mogensen, M., Thrane, L., Jørgensen, T.M., Andersen, P.E., Jemec, G.B.: OCT imaging of skin cancer and other dermatological diseases. J. Biophotonics. **2**(6-7), 442–451 (2009). https://doi.org/10.1002/jbio.200910020
13. Mogensen, M., Nürnberg, B.M., Forman, J.L., Thomsen, J.B., Thrane, L., Jemec, G.B.E.: *In vivo* thickness measurement of basal cell carcinoma and actinic keratosis with optical coherence tomography and 20-MHz ultrasound. Br. J. Dermatol. **160**(5), 1026–1033 (2009). https://doi.org/10.1111/j.1365-2133.2008.09003.x
14. Massone, C., Di Stefani, A., Soyer, H.P.: Dermoscopy for skin cancer detection. Curr. Opin. Oncol. **17**(2), 147–153 (2005). https://doi.org/10.1097/01.cco.0000152627.36243.26
15. Argenziano, G., Fabbrocini, G., Carli, P., De Giorgi, V., Sammarco, E., Delfino, M.: Epiluminescence microscopy for the diagnosis of doubtful melanocytic skin lesions: comparison of the ABCD rule of dermatoscopy and a new 7-point checklist based on pattern analysis. Arch. Dermatol. **134**(12), 1563–1570 (1998). https://doi.org/10.1001/archderm.134.12.1563
16. Benvenuto-Andrade, C., Dusza, S.W., Agero, A.L.C., Scope, A., Rajadhyaksha, M., Halpern, A.C., Marghoob, A.A.: Differences between polarized light dermoscopy and immersion contact dermoscopy for the evaluation of skin lesions. Arch. Dermatol. **143**(3), 329–338 (2007). https://doi.org/10.1001/archderm.143.3.329

17. Kaliyadan, F.: The scope of the dermoscope. Indian Dermatol. Online J. **7**, 359–363 (2016). https://doi.org/10.4103/2229-5178.190496
18. Moncrieff, M., Cotto, S., Claridge, E., Hall, P.: Spectrophotometric intracutaneous analysis: a new technique for imaging pigmented skin lesions. Br. J. Dermatol. **146**(3), 448–457 (2002). https://doi.org/10.1046/j.1365-2133.2002.04569.x
19. Monheit, G., Cognetta, A.B., Ferris, L., Rabinovitz, H., Gross, K., Martini, M., King, R.: The performance of MelaFind: a prospective multicenter study. Arch. Dermatol. **147**(2), 188–194 (2011). https://doi.org/10.1001/archdermatol.2010.302
20. Mirmehdi, M., Xie, X., Suri, J. (eds.): Handbook of Texture Analysis. Imperial College Press, London (2008)
21. Petrou, M.: Image Processing: Dealing with Texture, vol. 1. Wiley, Chichester (2006)
22. Pietikäinen, M.K. (ed.): Texture Analysis in Machine Vision, pp. 197–206. World Scientific, Singapore (2000)
23. Haralick, R.M.: Statistical and structural approaches to texture. Proc. IEEE. **67**(5), 786–804 (1979)
24. Dubes, R.C., Jain, A.K.: Random field models in image analysis. J. Appl. Stat. **20**(5-6), 121–154 (1993). https://doi.org/10.1080/02664769300000062
25. Ahuja, N., Rosenfeld, A.: Mosaic models for textures. IEEE Trans. Pattern Anal. Mach. Intell. **1**, 1–11 (1981). https://doi.org/10.1109/TPAMI.1981.4767045
26. Simonyan, K., Zisserman, A.: Very deep convolutional networks for large-scale image recognition (arXiv, 2014). https://arxiv.org/abs/1409.1556. Accessed 14 June 2019
27. Fisher, R.A.: The use of multiple measurements in taxonomic problems. Ann. Eugenics. **7**(2), 179–188 (1936). https://doi.org/10.1111/j.1469-1809.1936.tb02137.x
28. Cristianini, N., Shawe-Taylor, J.: An Introduction to Support Vector Machines and Other Kernel-Based Learning Methods. Cambridge University Press, Cambridge (2000)
29. Olson, D.L., Delen, D.: Advanced Data Mining Techniques. Springer, New York (2008)
30. Abbasi, N.R., Shaw, H.M., Rigel, D.S., Friedman, R.J., McCarthy, W.H., Osman, I., Polsky, D.: Early diagnosis of cutaneous melanoma: revisiting the ABCD criteria. JAMA. **292**(22), 2771–2776 (2004). https://doi.org/10.1001/jama.292.22.2771
31. Walter, F.M., Prevost, A.T., Vasconcelos, J., Hall, P.N., Burrows, N.P., Morris, H.C., Emery, J.D.: Using the 7-point checklist as a diagnostic aid for pigmented skin lesions in general practice: a diagnostic validation study. Br. J. Gen. Pract. **63**(610), e345–e353 (2013). https://doi.org/10.3399/bjgp13X667213
32. Myakinin, O.O., Zakharov, V.P., Bratchenko, I.A., Artemyev, D.N., Neretin, E.Y., Kozlov, S.V.: Proc. SPIE. **9599**, 95992B (2015). https://doi.org/10.1117/12.2188165
33. Tou, J.T., Gonzalez, R.C.: Pattern Recognition Principles, 2nd edn. Addison-Wesley Pub. Co., Boston (1977)
34. Wadhawan, T., Situ, N., Rui, H., Lancaster, K., Yuan, X., Zouridakis, G.: 2011 Annual International Conference of the IEEE Engineering in Medicine and Biology Society, pp. 3180–3183 (2011). https://doi.org/10.1109/IEMBS.2011.6090866
35. Gonzalez, R.C., Woods, R.E.: Digital Image Processing, 3rd edn. Pearson, London (2007)
36. Wang, L., He, D.C.: Texture classification using texture spectrum. Pattern Recogn. **23**(8), 905–910 (1990). https://doi.org/10.1016/0031-3203(90)90135-8
37. Raupov, D.S., Myakinin, O.O., Bratchenko, I.A., Zakharov, V.P., Khramov, A.G.: Multimodal texture analysis of OCT images as a diagnostic application for skin tumors. J. Biomed. Photon. Eng. **3**(1), 010307 (2017). https://doi.org/10.18287/JBPE17.03.010307
38. Gao, W., Zakharov, V.P., Myakinin, O.O., Bratchenko, I.A., Artemyev, D.N., Kornilin, D.V.: Medical images classification for skin cancer using quantitative image features with optical coherence tomography. J. Innovative Opt. Health Sci. **9**(2), 1650003 (2016). https://doi.org/10.1142/S1793545816500036
39. Puvanathasan, P., Bizheva, K.: Interval type-II fuzzy anisotropic diffusion algorithm for speckle noise reduction in optical coherence tomography images. Opt. Express. **17**(2), 733–746 (2009). https://doi.org/10.1364/OE.17.000733

40. Haralick, R.M., Shanmugam, K.: Textural features for image classification. IEEE Trans. Syst. Man Cybern. **3**(6), 610–621 (1973). https://doi.org/10.1109/TSMC.1973.4309314
41. Fogel, I., Sagi, D.: Gabor filters as texture discriminator. Biol. Cybern. **61**(2), 103–113 (1989). https://doi.org/10.1007/BF00204594
42. Tamura, H., Mori, S., Yamawaki, T.: Textural features corresponding to visual perception. IEEE Trans. Syst. Man Cybern. **8**(6), 460–473 (1978). https://doi.org/10.1109/TSMC.1978.4309999
43. Voss, R.F.: Fundamental algorithms for computer graphics. In: Earnshaw, R.A. (ed.) Random Fractal Forgeries, pp. 805–835. Springer, Berlin (1985)
44. Ahammer, H.: Higuchi dimension of digital images. PLoS One. **6**(9), e24796 (2011). https://doi.org/10.1371/journal.pone.0024796
45. Sarkar, N., Chaudhuri, B.B.: An efficient differential box-counting approach to compute fractal dimension of image. IEEE Trans. Syst. Man Cybern. **24**(1), 115–120 (1994). https://doi.org/10.1109/21.259692
46. Ilyasova, N.U., Ustinov, A.V., Khramov, A.G.: Comput. Opt. **18**, 150–164 (1998)
47. Plastinin, A.I., Kupriyanov, A.V.: A model of Markov random field in texture image synthesis and analysis. Proc. Samara State Aerosp. Univ. **2**, 252–257 (2008)
48. Weszka, J.S., Dyer, C.R., Rosenfeld, A.: A comparative study of texture measures for terrain classification. IEEE Trans. Syst. Man Cybern. **4**, 269–285 (1976). https://doi.org/10.1109/TSMC.1976.5408777
49. Zayed, N., Elnemr, H.A.: Statistical analysis of Haralick texture features to discriminate lung abnormalities. J. Biomed. Imaging. **2015**, 267807 (2015). https://doi.org/10.1155/2015/267807
50. Park, M., Jin, J.S., Wilson, L.S.: Fifth IEEE Southwest Symposium on Image Analysis and Interpretation, pp. 178–182 (2002). https://doi.org/10.1109/IAI.2002.999914
51. Palm, C., Keysers, D., Lehmann, T., Spitzer, K.: Gabor filtering of complex hue/saturation images for color texture classification. Proc. JCIS. **2000**, 45–49 (2000)
52. Flueraru, C., Popescu, D.P., Mao, Y., Chang, S., Sowa, M.G.: Added soft tissue contrast using signal attenuation and the fractal dimension for optical coherence tomography images of porcine arterial tissue. Phys. Med. Biol. **55**(8), 2317 (2010). https://doi.org/10.1088/0031-9155/55/8/013
53. Sullivan, A.C., Hunt, J.P., Oldenburg, A.L.: Fractal analysis for classification of breast carcinoma in optical coherence tomography. J. Biomed. Opt. **16**(6), 066010 (2011). https://doi.org/10.1117/1.3590746
54. Gao, W.. PhD thesis, University of Miami (2012)
55. Annadhason, A.: Methods of fractal dimension computation. IRACST. **2**(1), 166–169 (2012)
56. Florindo, J.B., Martinez Bruno, O.: Fractal descriptors in the Fourier domain applied to color texture analysis. Chaos. **21**(4), 043112 (2011). https://doi.org/10.1063/1.3650233
57. Salomatina, E.V., Jiang, B., Novak, J., Yaroslavsky, A.N.: Optical properties of normal and cancerous human skin in the visible and near-infrared spectral range. J. Biomed. Opt. **11**(6), 064026 (2006). https://doi.org/10.1117/1.2398928
58. Yamashita, T., Kuwahara, T., Gonzalez, S., Takahashi, M.: Non-invasive visualization of melanin and melanocytes by reflectance-mode confocal microscopy. J. Investig. Dermatol. **124**(1), 235–240 (2005). https://doi.org/10.1111/j.0022-202X.2004.23562.x
59. Winkler, G.: Image Analysis, Random Fields and Dynamic Monte Carlo Methods. Springer, New York (1995)
60. Li, S.Z.: Markov Random Field Modeling in Image Analysis. Springer, New York (2009). https://doi.org/10.1007/978-1-84800-279-1
61. Raupov, D.S., Myakinin, O.O., Bratchenko, I.A., Zakharov, V.P., Khramov, A.G.: Skin cancer texture analysis of OCT images based on Haralick, fractal dimension, Markov random field features, and the complex directional field features. Proc. SPIE. **10024**, 100244I (2016). https://doi.org/10.1117/12.2246446
62. Celebi, M.E., Wen, Q., Iyatomi, H., Shimizu, K., Zhou, H., Schaefer, G.: A state-of-the-art survey on lesion border detection in dermoscopy images. Dermosc. Image Anal. **2015**, 97–129 (2015). https://doi.org/10.1201/b19107-5

63. Zakharov, V.P., Bratchenko, I.A., Myakinin, O.O., Artemyev, D.N., Kornilin, D.V., Kozlov, S.V., Moryatov, A.A.: Multimodal diagnosis and visualisation of oncologic pathologies. Quantum Electron. **44**(8), 726–731 (2014). https://doi.org/10.1070/QE2014v044n08ABEH015545
64. Moon, Y., Han, J.H., Choi, J.H., Shin, S., Kim, Y.C., Jeong, S.: Mapping of cutaneous melanoma by femtosecond laser-induced breakdown spectroscopy. J. Biomed. Opt. **24**(3), 031011 (2018). https://doi.org/10.1117/1.JBO.24.3.031011
65. Raupov, D.S., Myakinin, O.O., Bratchenko, I.A., Zakharov, V.P., Khramov, A.G.: Analysis of 3D OCT images for diagnosis of skin tumors. Proc. SPIE. **10716**, 1071608 (2018). https://doi.org/10.1117/12.2305405
66. Sawyer, T.W., Chandra, S., Rice, P.F., Koevary, J.W., Barton, J.K.: Three-dimensional texture analysis of optical coherence tomography images of ovarian tissue. Phys. Med. Biol. **63**(23), 235020 (2018). https://doi.org/10.1088/1361-6560/aaefd2
67. Gossage, K.W., Tkaczyk, T.S., Rodriguez, J.J., Barton, J.K.: Texture analysis of optical coherence tomography images: feasibility for tissue classification. J. Biomed. Opt. **8**(3), 570–576 (2003). https://doi.org/10.1117/1.1577575
68. Lindenmaier, A.A., Conroy, L., Farhat, G., DaCosta, R.S., Flueraru, C., Vitkin, I.A.: Texture analysis of optical coherence tomography speckle for characterizing biological tissues *in vivo*. Opt. Lett. **38**(8), 1280 (2013). https://doi.org/10.1364/ol.38.001280
69. de Moura, J., Vidal, P.L., Novo, J., Rouco, J., Ortega, M.: Proc. Comput. Sci. **112**, 1369–1377 (2017). https://doi.org/10.1016/j.procs.2017.08.043
70. Marvdashti, T., Duan, L., Aasi, S.Z., Tang, J.Y., Bowden, A.K.E.: Classification of basal cell carcinoma in human skin using machine learning and quantitative features captured by polarization sensitive optical coherence tomography. Biomed. Opt. Express. **7**(9), 3721–3735 (2016). https://doi.org/10.1364/BOE.7.003721
71. Adabi, S., Hosseinzadeh, M., Noei, S., Conforto, S., Daveluy, S., Clayton, A., Nasiriavanaki, M.: Universal *in vivo* textural model for human skin based on optical coherence tomograms. Sci. Rep. **7**(1), 17912 (2017). https://doi.org/10.1038/s41598-017-17398-8
72. Xiong, Y.-Q., Mo, Y., Wen, Y.-Q., Cheng, M.-J., Huo, S.-T., Chen, X.-J., Chen, Q.: Optical coherence tomography for the diagnosis of malignant skin tumors: a meta-analysis. J. Biomed. Opt. **23**(2), 020902 (2018). https://doi.org/10.1117/1.JBO.23.2.020902s
73. Boone, M.A.L.M., Suppa, M., Dhaenens, F., Miyamoto, M., Marneffe, A., Jemec, G.B.E., Del Marmol, V., Nebosis, R.: *In vivo* assessment of optical properties of melanocytic skin lesions and differentiation of melanoma from non-malignant lesions by high-definition optical coherence tomography. Arch. Dermatol. Res. **308**(1), 7–20 (2016). https://doi.org/10.1007/s00403-015-1608-5
74. Esteva, A., Kuprel, B., Novoa, R.A., Ko, J., Swetter, S.M., Blau, H.M., Thrun, S.: Dermatologist-level classification of skin cancer with deep neural networks. Nature. **542**(7639), 115–118 (2017). https://doi.org/10.1038/nature21056

Chapter 14
Application of Acousto-Optical Hyperspectral Imaging for Skin Cancer Diagnostics

Vitold E. Pozhar, Alexander S. Machikhin, Oleg O. Myakinin, and Ivan A. Bratchenko

14.1 Introduction

In recent decades, optical biopsy has been widely used for the investigation of almost all types of cancer. These studies include two major objectives: firstly, the precise definition of the boundaries and topology of the tumor, and, secondly, the exact identification of the type of cancer. Spectroscopic response from key native molecules reveals the differences between cancerous and normal skin areas due to morphological and molecular changes in the tissue. The key methods in optical biopsy are fluorescence, Stokes shift, elastic backscattering reflectance, and time-resolved spectroscopy [1]. In majority of applications, the spectroscopic signal is captured from a single point in the tissue or averaged over the macroscopic volume of the tissue. However, there is an opportunity to collect a signal from the vast tissue area, and this approach may be implemented with hyperspectral imaging (HSI) [2].

The hyperspectral image is a three-dimensional array of $M \times N \times K$ data consisting of a sequence of K images of $M \times N$ size, and each of these images corresponds to an intensity of optical signal in the spectral band $\lambda_K \pm \Delta\lambda$ at each point of the tested object surface. Thus, native spectral anomalies in absorption, reflection, and fluorescent signal of different chemical components of tissue pathology may be used for contrast imaging of cancer lesion [3]. Thereby different applications of HSI systems have been utilized in cancer study.

Spectral imaging based on acousto-optical tunable filters (AOTF) is a promising rapidly developing technique for tissue analysis and diagnostics due to their ability to obtain the entire spatial information (x-axis and y-axis) of the studying object for

V. E. Pozhar (✉) · A. S. Machikhin
Scientific and Technological Center of Unique Instrumentation of the Russian Academy of Sciences, Moscow, Russian Federation
e-mail: vitold@ntcup.ru

O. O. Myakinin · I. A. Bratchenko
Samara National Research University, Samara, Russian Federation

V. V. Tuchin et al. (eds.), *Multimodal Optical Diagnostics of Cancer*,
https://doi.org/10.1007/978-3-030-44594-2_14

a certain wavelength at one time, performing a scanning in the spectral dimension (λ-axis). The tunable optical filters exploit ultrasonic acoustic waves for generation of the diffraction grating and selection of spectral components. Correspondently, acousto-optical (AO) instruments can achieve high spatial resolutions and fast acquisition times. Such realization of spectral instruments makes the filter wavelength tuning easier and more rapid and, thus, the spectral features detection becomes more convenient and effective. High spectral resolution allows one to study in detail the optical properties of the tissues, and high spatial resolution permits the analysis of samples by combining the spectral and the morphological properties of the tissue.

In this chapter, the basic principles of AOTF are presented and most important results of AOTF-based HSI implementation for skin cancer detection are discussed.

Section 14.1 describes the physical principles that govern AOTF operations and analyzes methods and techniques for correcting image distortions. Section 14.2 collects data about AOTF-based spectral imaging devices for different endoscopy, microscopy, and remote sensing applications. Section 14.3 discusses the results of the AO HSI device testing for skin cancer detection and classification in clinical trials.

14.2 AOTF-Based Spectral Imaging

14.2.1 Operation Principle of Acousto-Optical Hyperspectral Imager

For spectral analysis and spectral selection one needs an optical tunable monochromator with following features: relatively narrow bandwidth, wide spectral range, minimal tuning time, maximal transmission coefficient, high level of out-of-band light suppression, the ability of image spectral filtration without information losses or at least without significant image distortion. Such spectral imaging device with fast tuning can be developed using acousto-optical (AO) technology based on Bragg light diffraction on acoustic waves. Below, the physical base of AO filtration is presented.

AOTF [4] is a spectral element, which comprises AO crystal cell and usually a pair of polarizers. It allows selecting of narrow bandpass, which localization in the spectral axis is determined by supplied ultrasound frequency. In AO cell, the light beam diffracts on the Bragg grating, which is induced in the crystal by propagating acoustic wave due to the resonance elasto-optic effect. The matching (resonant) wavelength λ_f of selecting spectral component is determined by the grating period d, which is inversely proportional to the ultrasound frequency f:

$$\lambda_f \sim \frac{V_{\mathrm{ac}}}{f}. \tag{14.1}$$

Here V_{ac} is the speed of acoustic wave. Such dependence determines the method of wavelength selection—by choosing the appropriative electric signal frequency of radiofrequency (RF) generator, which is transformed by piezotransducer into the ultrasonic wave in AO cell. The bandwidth of the spectral optical window $\Delta\lambda$ is defined by the number N_{us} of ultrasound diffraction grating periods, which are crossed by the light beam [5]:

$$\Delta\lambda \approx \frac{\lambda}{N_{us}} \tag{14.2}$$

The bandwidths can be chosen in the wide interval, as usually $N_{us} \sim 10^2/10^3$. Therefore, for visible spectral range typical bandwidth is $\Delta\lambda \sim 0.5/5$ nm. As the spectral range of AOTFs usually does not exceed an octave $\lambda_{max}/\lambda_{min} < 2$ (typically, 0.45/0.8 μm, 0.9/1.7 μm, 2/4 μm, or 0.25/0.4 μm), the total number of resolvable spectral channels are approximately equal to the effective period number of ultrasound grating $N_{AOTF} \sim N_{us}$.

Another notable feature of the resonant interaction is the change in the direction of propagation of the diffracted light beam and the alternation of the polarization plane. Therefore, the diffracted beam may be filtered either by output polarizer (polarization selection, see Fig. 14.1) or by the diaphragm (angular selection).

The maximum value of the diffraction coefficient $T(\lambda)$ depends on the power density of ultrasound signal P_{ac} and the size of the grating in the direction of the incident light beam L:

$$T_{max} \equiv T(\lambda_f) = \sin^2(\Gamma L), \quad \Gamma \propto \frac{\sqrt{M_2 P_{ac}}}{\lambda_f} \tag{14.3}$$

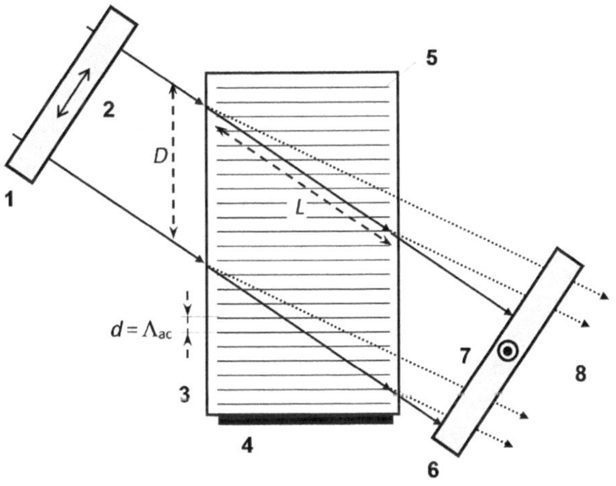

Fig. 14.1 Principal parameters of Bragg diffraction in AO cell: 1—input polarizer (arrow shows the direction of polarization); 2, 7, 8—incident, non-diffracted, and diffracted light beams; 3—acousto-optic crystal cell; 4—acoustic wave emitter (piezotransducer); 5—Bragg grating generated by propagating acoustic wave; 6—output polarizer. Beams intersection region is approximately $L \times D$

The diffraction coefficient can reach 100% at some level of ultrasound power depending on the crystal acousto-optic quality factor $M_2 \approx p^2n^6(V_{ac})^{-3}/\rho$, which in turn is defined by the material characteristics: elasto-optic constant p, refractive index n, and density ρ. Thus, theoretically all the energy of the selected spectral component can be completely transmitted through the filter. The required power is $P_{ac} \sim 0.1/10$ W.

The minimum time interval for switching from one to another wavelength is determined by the ultrasound transit time across the light beam:

$$\tau = \frac{D}{V_{ac}} \tag{14.4}$$

As the typical size of light and sound waves interaction region $D \sim 1$–10 cm, the switching time is usually $\tau \sim 1/10$ μs.

For imaging, the very important characteristic is the angular field-of-view (FoV). In contrast with ordinary diffraction gratings, which are capable of one-dimensional spatial resolution, AOTFs exhibit almost equal FoV in each direction. This value is not high as it is determined by the number of ultrasound grating periods $\Delta\theta \approx \sqrt{N_{us}}$ [5]. Though the angle is rather moderate $\Delta\theta \sim 1/3$, the total number of resolvable points in the image can reach 10^3 for light divergence $\lambda/D \sim 10^{-4}$.

The most important property providing the high quality spectral imaging is the ability to transfer images through AOTF without distortion. The diffraction on spatially distributed structure of the complex AO cell with significant number of auxiliary elements (polarizers, prisms) forms the impression that it is extremely difficult to satisfy this condition. But, since the early days of AO, it was demonstrated imaging capabilities of AOTF with rather high spatial resolution (over 100 pairs of lines per millimeter) [6, 7]. However, for a long time it was assumed that AO filter image quality is quite moderate and inferior to liquid crystal tunable filters [8]. Later, the theoretical studies [9, 10] proved that linear, non-linear, and chromatic aberrations (elongations, distortions, drift, etc.) can be nearly vanished, so that the number of resolvable spatial points can approximately reach 10^3.

The situation changed when double AOTF monochromators with compensation of image deformation were developed. It was shown experimentally [11] and theoretically [12] that the second identical AOTF, being placed inversely, corrects image deformations of the primary AOTF. The detail theory of double AOTF image transformation may be found elsewhere [10, 13]. Later on this basis, a software module was developed for the Zemax environment [14], which paves the way for optimizing optical systems comprising both classical optical components (lenses, plates, etc.) and AOTFs. Such an approximating AOTF module is the first step towards to solving the fundamental problem of precise rays-tracing modeling the heterogeneous systems containing two types of components: classical refractive optical elements and AO diffractive units, i.e. combining elements of geometrical and wave optics.

Below, the general theoretical relations of light beam transmission through AOTF are presented, different optical schemes for acousto-optical hyperspectral

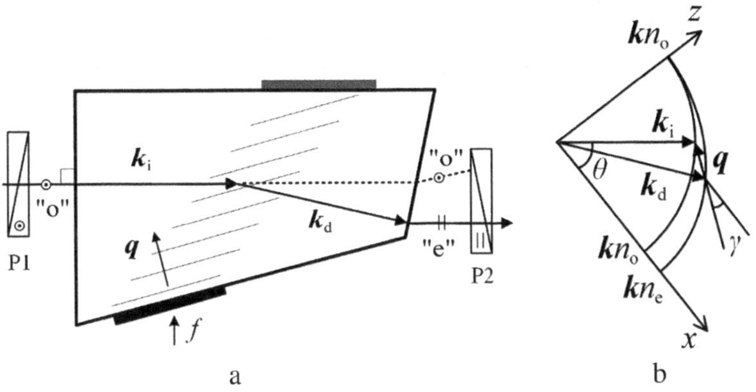

Fig. 14.2 Acousto-optical tunable filter: optical scheme (**a**) and wave vector diagram (**b**)

imagers (AO HSI) are compared and the basic types of registered aberrations are described [14].

Figure 14.2 represents the arrangement of a classical non-collinear AOTF [15], comprising an AO cell and a pair of crossed polarizers. An ordinary (o) polarized beam is directed at the Bragg angle to the acoustic beam and, consequently, it diffracts on the acoustic grating, deflecting at a certain angle and changing the polarization into extraordinary (e) mode (diffraction process $o \rightarrow e$). Only narrow spectral interval of light $\Delta\lambda$ near central wavelength λ_f corresponding to Bragg condition is diffracted in that way, while the output polarizer eliminates the rest of radiation. Usually, the input AO crystal facet is oriented normally to the direction of input light beam $\boldsymbol{k}_i$, while the output facet is cut to compensate the spectral drift of the output beam and minimize the axial chromatic shift of the image [16].

AO filter is characterized by high operational speed, reliability, accuracy, and stability as it is free from any movable elements (mirrors, prisms, etc.) due to electrical radiofrequency tuning principle.

A very important AO feature is the capability to operate not only in parallel rays or in collimated-beam scheme (Fig. 14.3a), but also in confocal or converging schemes (Fig. 14.3b, c) [14]. Such schematic variability facilitates the optical scheme optimization, improving the image quality and helps to increase the transmission coefficient with simultaneous minimization of device dimensions.

As determined experimentally, the relative image deformations caused by AOTF can reach 3–5% [17, 18]. They are inhomogeneous both in spectrum and in field-of-view that leads to spectral drift, asymmetric scaling, and non-linear distortions.

Examples of images detected using the AO spectral imager (Fig. 14.4) demonstrate fairly good image quality at each wavelength, but unsatisfactory overlay results. Spectral drift and other deformations cause blur and distort the spectrum in each pixel.

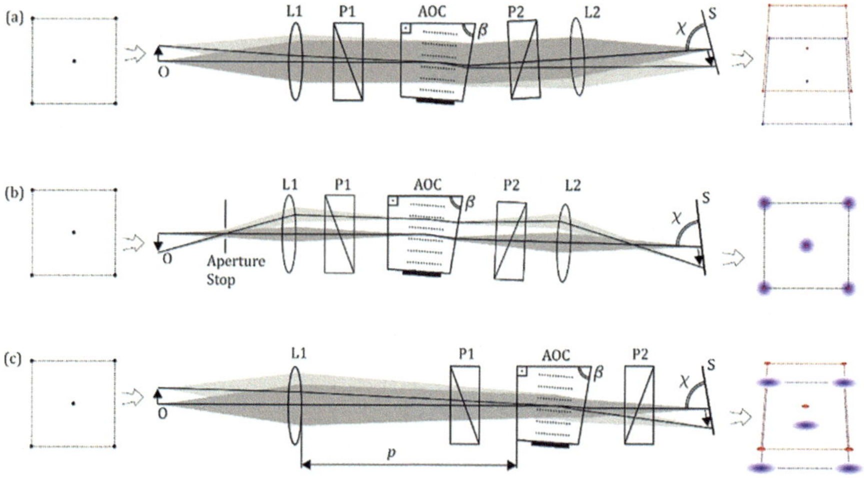

Fig. 14.3 Various optical schemes for imaging AOTF-based spectrometer: collimated (**a**), confocal (**b**), and converging (**c**) beam. *O* object; *L1, L2* input and output lenses, *AOC* AO cell, *P1, P2* input and output crossed polarizers, *S* spectral image. Optimum angles β and χ are selected in each scheme

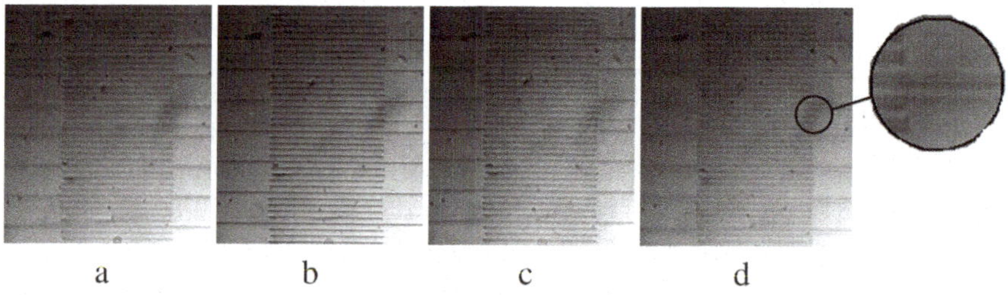

Fig. 14.4 Spectral images of optical standard pattern: (**a**) 480 nm; (**b**) 580 nm; (**c**) 720 nm; (**d**) their superposition

14.2.2 Double Acousto-Optical Monochromatization

Cascade filtration by two AO cells is used for increasing spectral contrast, spectral resolution enhancement, and for simultaneous filtration of both polarization components of light [19–26]. In hyperspectral imaging, double monochromatization is also a technique for correction of spectral-spatial distortions in images caused by a single AO cell [11, 27, 28].

The basic principles of cascade diffraction in double AOTF are presented in Fig. 14.5. The inspected object 1 is placed in the focus of the input optics 2. The collimated beam comes through the double (tandem) monochromator composed of two identical AO cells 4, 6 and three polarizers 3, 5, 7. The wavelength to be selected by AO monochromator is specified by the computer 10 according to the program.

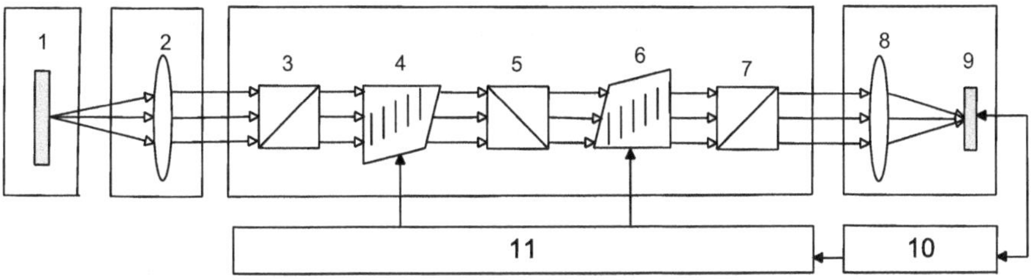

Fig. 14.5 Double AOTF configuration of HIS: 1—object; 2—input optics; 3, 5, 7—polarizers; 4, 6—AO cells; 8—output optics; 9—photodetector array; 10—computer; 11—control unit

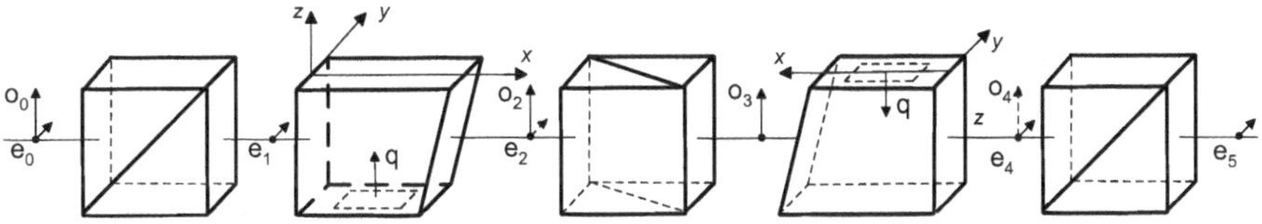

Fig. 14.6 Aberration-compensated double imaging monochromator. Optical scheme. (Refraction at the oblique facets is not depicted. The shape of AO cells is schematic)

Controller 11 contains radiofrequency synthesizer and amplifier, which generate driving signal delivered to each AOTF. The light transmitted through the monochromator is focused by the lenses 8 at camera detector 9.

Both AOTFs must be identical to avoid divergence of their transmission windows due to mismatch of their spectral tuning characteristics $\lambda(f)$ as they are controlled by the single RF generator. Some technological tricks are used for AOTF pair manufacturing.

To eliminate image deformations two AO cells must be oriented as shown in Fig. 14.6. The second AO cell should be turned over at 180° to transmit the light beam in the opposite direction compared with the first AO cell. In this way, most distortions caused by diffraction in the first AO cell are compensated by diffraction in the second AO cell. Therefore, this configuration is preferably used for double AO monochromators designed for imaging applications [11, 29–32].

To illustrate the idea of distortion compensation let us consider wave vector diagrams (Fig. 14.7). For the central ray of the divergent beam the Bragg condition $\boldsymbol{k}_d = \boldsymbol{k}_i + \boldsymbol{q}$ connecting wave vectors of diffracted, incident, and acoustic wave is represented as closed rectangular. As can be seen the diagrams in the first and second AO cells are represented with the same triangular so the distortions in the second AO cell compensate those of the first one.

The modeling based on these diagrams and the exact formulas [10] was performed for single and double AOTF monochromators made of the same AO cells (Fig. 14.8). The standard test pattern in the form of square grid demonstrates the significant compensations of distortions in the double monochromator.

In general, to analyze distortions in real devices one needs to account non-identity of AO cells, imperfections of their fabrication, and inaccuracy in orientation

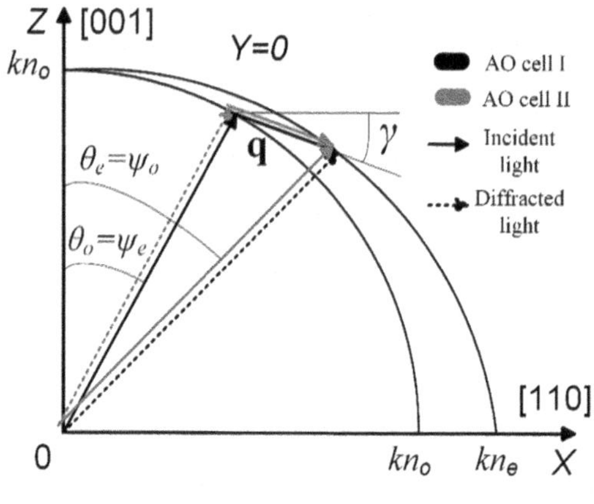

Fig. 14.7 Wave vector diagrams representing light diffraction in AO cells of double imaging monochromator

of their AO crystals. In linear approximation the deformations of images caused by an AOTF are following: spectral drift in the direction of optical axis, unequal scaling in diffraction and in the orthogonal direction. These deformations can be compensated with classical optical elements. Yet, the effects of higher orders (non-linear deformations) cannot be removed in a simple way.

14.3 Acousto-Optical Hyperspectral Imagers

Nowadays, there is a variety of applications, in which AO imaging spectrometers and spectral imagers are applied: biomedicine [2, 33, 34], remote sensing [35, 36], agriculture [37, 38]. These instruments facilitate and speed up investigations. The key characteristic of these devices is image quality.

A series of AOTF-based instruments for spectral endoscopy [31], spectral microscopy [15], spectral and hyperspectral imaging [30], stereospectroscopy [39, 40] were developed and this family is rather plentiful and diverse [41]. Their key characteristic is the extent of image deformation, which can prevent the correct comparison of images detected at different wavelengths and also prevents direct spectral functions determination. One of the solutions to this problem is the instrument calibration [42, 43]. The other approach is an instrumental compensation of aberrations. The advance feature of such devices is the capability to reconstruct spectral characteristics by means of simple stacking the spectral images.

Below, different AO devices for spectral imaging are described, including AO hyperspectral imager based on compensated double AOTF monochromator, which was used in clinical neoplasia studies.

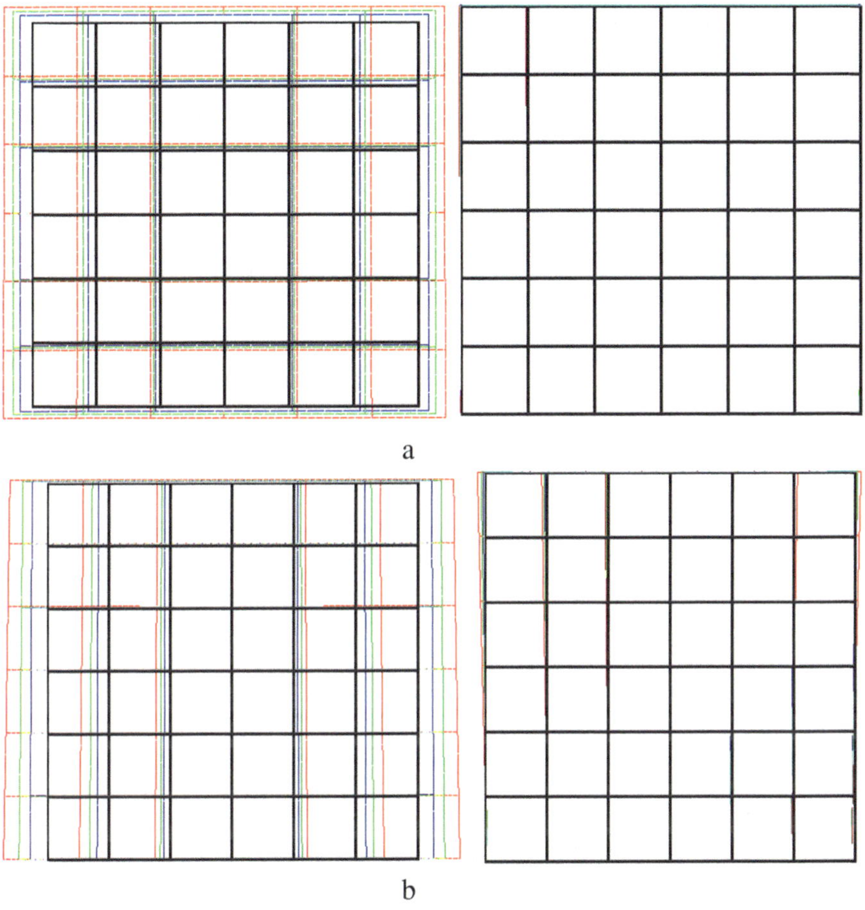

Fig. 14.8 Image deformations modeling for identical single (*left*) and double (*right*) AOTF monochromators. Test object is square-cells grid (*black*). Calculated images for different wavelengths are depicted with corresponding colors: red—0.7 μm, green—0.55 μm, blue—0.38 μm. Calculations were performed for extraordinary polarized input radiation for the following diffraction geometries: (**a**) TeO_2, $\theta_1 = 32.6°$, $\gamma_1 = -18.9°$, $\theta_2 = \gamma_2 = 0$, $\Delta\theta_i = 3°$; (**b**) TeO_2, $\theta_1 = 73.6°$, $\gamma_1 = -7.1°$, $\theta_2 = \gamma_2 = 0$, $\Delta\theta_i = 3°$. (*The distortions are significantly enlarged for visibility*)

14.3.1 Endoscopic Hyperspectral Imagers

Spectral investigations are more and more often used for diagnostics and therapy of tumors, predominantly for cancer [2, 44–46] as well as for the study of hardly accessible technical objects for non-destructive testing of cavities in industrial production (engines of rockets and aviation, pipes, steam generators, etc.) with industrial endoscopes [47, 48] That is why the exclusive attention is now devoted to the analytical modules attachable to endoscopes [49–54]. Such a spectral attachment can be designed in the form of transocular module (Fig. 14.9, outlined) for standard optical

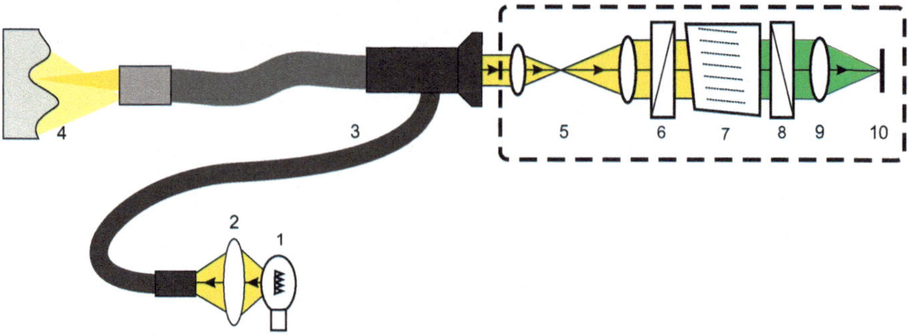

Fig. 14.9 AOTF-based hyperspectral imaging module attached to an endoscope: 1—incandescent lamp, 2, 5—optical conjugation systems, 3—fiber-optic light guide, 4—object, 6, 8—crossed polarizers, 7—AO cell, 9—lens, 10—photodetector array

endoscope. The type of endoscope, rigid or flexible, does not matter, while the output eyepiece should be standard.

However, it is a big problem to ensure effective optical conjugation of endoscopic eyepiece and AOTF due to highly differing optical characteristics. Below, the basic problems of endoscopic AO hyperspectral imagers development are considered.

The existing AO endoscopic spectrometric devices are used as transocular modules for rigid and flexible fiber-optic endoscopes are intended for photoluminescence study of internal organs and contain single AOTF monochromators [55–58]. They really provide spectral imaging, but hardly guarantee the precision measurement of the spectra in every point of the image. It is caused by the spectral-dependent deformations of the image, which prevent direct identification of the image points. Thus, for the exact spectra determination one needs to apply some additional procedure for deformations correction, which requires the preliminary calibration of the system and results in rather complicated, laborious, time-consuming detection procedure.

Another problem is the limited "energy" efficiency of direct AOTF coupling with endoscopic optics. That is because AO monochromator possesses rather narrow FoV $2\varpi_{ao} < 3°$ while the input pupil is quite large $D_{ao} \approx 10$ mm (which is limited by the crystal size). The FoV parameter is determined by the effective number of Bragg diffraction grating periods $2\varpi_{ao} \approx N^{1/2}$ so that for typical value of spectral resolving force $\lambda/\delta\lambda \approx N \sim 10^3$, the angular interval of synchronism $2\varpi_{ao} \approx 2°$. Another important characteristic is the defection angle between the diffracted and transmitted beams. Usually this angle is only 7–15°. For comparison, standard eyepieces for a visual device have the following characteristics: $D'_{oc} = 2$–6 mm, $2\varpi'_{oc} \approx 20$–$80°$. Due to inequality $\varpi'_{oc} >> \varpi_{AO}$, placement of AOTF just after the ocular is not reasonable due to FoV cutting down (Fig. 14.10). Therefore, the problem can be resolved only by means of adequate optical conjugation.

The conjugating system was modeled in optical design program Zemax [59] with use of specially developed module imitating light beams transmission through

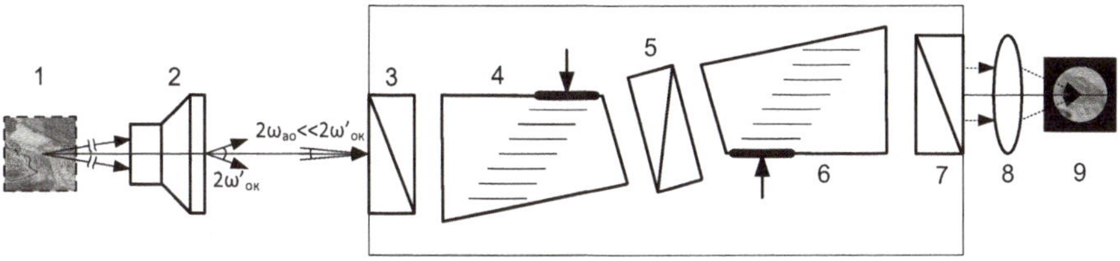

Fig. 14.10 Direct coupling of endoscope with an AOTF monochromator: 1—object, 2—eyepiece of visual device, 3, 5, 7—polarizers, 4, 6—AO cells, 8—output lens, 9—spectral image

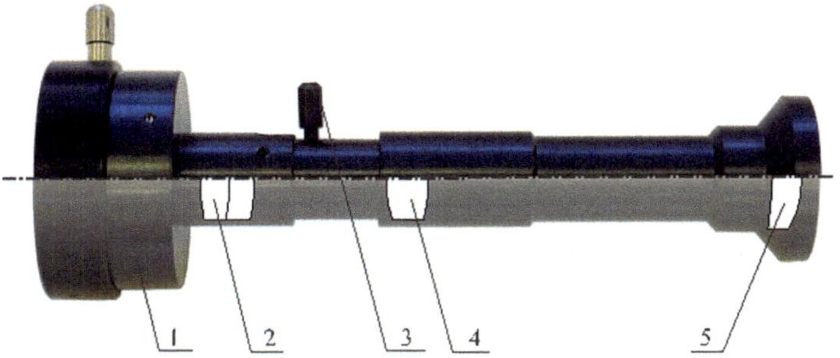

Fig. 14.11 Optical system for conjugation of AOTF with endoscope eyepieces: 1—attachment port to ocular, 2—two-lens glued assembly, 4—collective lens, 3—check screw for variable focusing, 5—output lens

AOTF. The designed optical system was implemented in form of adjustable optical tube (Fig. 14.11) [60] with the following characteristics: magnification $\Gamma = 0.4\times$, output pupil diameter—6 mm, output FoV—4°, total length—100 mm.

The developed conjugating optical system was tested for coupling the double AOTF monochromator to visual devices: a rigid lens-type endoscope, a flexible fiber-optic endoscope, and a telescope [60]. Despite the fundamental differences between their optical schemes and optical characteristics, the conjugated system significantly increases FoV and raises the relative brightness of the out-of-center points. This system was used in endoscopic add-on module (Fig. 14.12a)

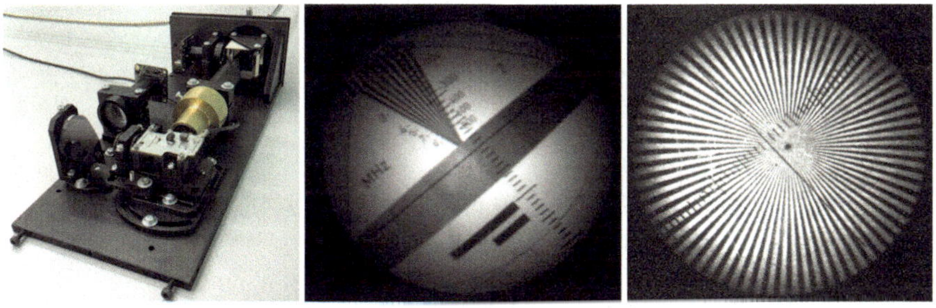

Fig. 14.12 Testing of endoscopic HSI module based on the single AOTF: (**a**) internal view; (**b**, **c**) standard pattern images detected with a rigid endoscope at 555 nm wavelength

The developed module was tested with the use of standard patterns (Fig. 14.12b, c). At the distance of 25 mm it provides resolution 8–10 lines/mm across the total FoV. Due to the use of compensated double monochromator, the spectral shift does not exceed 1% in the range 400–700 nm [61–63].

14.3.2 Hyperspectral Imaging Microscopic Module

Detection of spectral images of microscopic objects is one of the main applications of AOTFs [8, 64–70]. They are used for contrast visualization, fluorescent and Raman spectroscopy, border selection, etc. Often, for this purpose, a specialized or even unique installations are assembled which require complex adjustment and are not subjected to any transportation. Unfortunately, hyperspectral imagers are not adapted to classical microscopes. Therefore, an actual problem is the development of HSI module, which would be entirely compatible with modern commercial microscopes, would not require any constructive upgrading of them, and would be able to provide spectral images simultaneously with classical RGB recording with no special readjustment. Some pioneer technical solutions to this problem such as add-on module for inverted microscope "Leitz DM-IRBE" are known [55]. However, the problems of optical conjugation and ergonomics issues are not considered.

As a rule, for spectral imaging, AOTF is located in the camera port of the modern trinocular microscope. This is the most convenient location as it does not require any modification of the standard microscope and binocular vision function is kept unchanged. The standard non-inverted microscope construction permits placement of an axially symmetric assembly consisting of AOTF, lenses, and camera. However, this solution is not applicable for synchronous two-channel detection of the object: (1) in white light with color (RGB) camera and (2) in specified spectral interval with a monochrome camera. Then the concept of recording and joint processing of the pair images cannot be realized in this approach, as image is caught in the same time from the same object location with near identical observation conditions (FoV, focus distance).

Another concept was implemented for add-on hyperspectral module (Fig. 14.13), which is located between objective lenses and binoculars [71]. In this case, all the constituent parts of the microscope system remain unchanged, including RGB camera, and there is enough space for an additional optical system and even for a long-size HSI based on double AOTF monochromator.

This module operates in the following way. Wide-band radiation from the source 1 is focused on the object with Köhler illumination system, comprising lens elements 2 and 5, field diaphragm 3, and aperture stop 4. After successive transmission through the inspected object and the microobjective 6, the light radiation is partly reflected in the AO hyperspectral module with the help of beam-splitting cube 12, while the rest of light flow propagates directly upward. The last is distributed by the

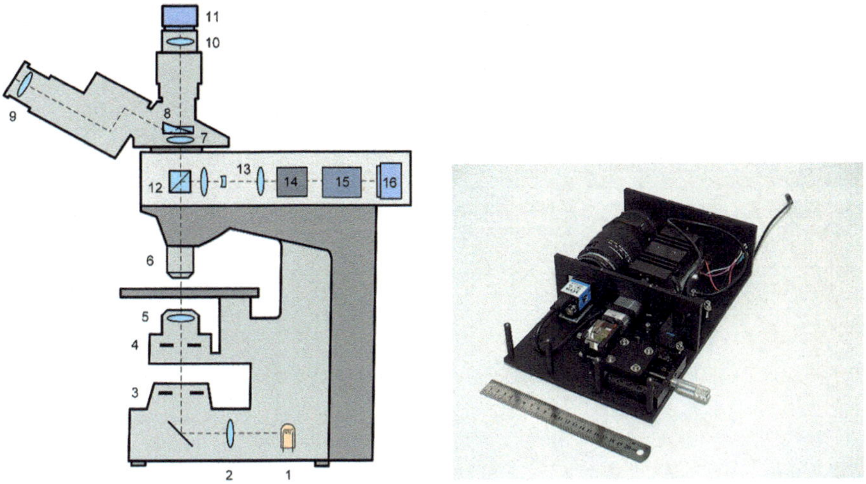

a b

Fig. 14.13 Transmission microscope with AO HSI module: (**a**) structure, (**b**) internal view of AO imaging module. 1—white light source, 2, 5—condenser, 3—field diaphragm, 4—aperture diaphragm, 6, 15—lens, 7—tube lens, 9—eyepiece, 8, 12—beam splitter, 10—optical adapter, 13—conjugating optical system, 14—AOTF, 11—RGB camera, 16—monochrome camera

beam splitter element 8 into two channels: first—to the binoculars 9 for visual control, second—to RGB camera 11 through the optical adapter 10, which provides necessary magnification.

The spectral imaging module (Fig. 14.13b) is placed between the basement and trinocular head of the microscope. It contains the beam splitter cube 12, conjugating optical system 13, AOTF 14, lens 15, and monochrome camera 16. As the beam splitter is placed between microobjective lens 6 and tube lens 7, the collimated radiation enters AO HSI module and then transmits the conjugating system 13. When spectral images are not required, the beam splitter 12 can be simply shifted aside to avoid light losses.

A varifocal lens 15 with microobjective lens 6 ensures variability of magnifying power. Acousto-optical tunable filter consists of a pair of opposite oriented identical wide-angle AO cells made of TeO2 (cut-off angle $\gamma = 6°$) and three polarizers. Spectral range covers 450–750 nm, spectral resolution is approximately 1.5 nm (at $\lambda = 630$ nm), and the number of resolvable elements is 600 × 500. The device is controlled via USB2.0 port.

The optical system was modeled in Zemax program [14, 59]. The detailed analysis including account of Bragg diffraction was performed to optimize the conjugating system for ensuring high quality imaging for microobjectives 4–20×. The parameters of calculation and other details are presented elsewhere [71].

For testing, AO HSI module was inserted into a trinocular optical transmission microscope, like in Fig. 14.13. In the upper row of Fig. 14.14, the images of a standard pattern are presented, which were obtained with 10× magnification, while in

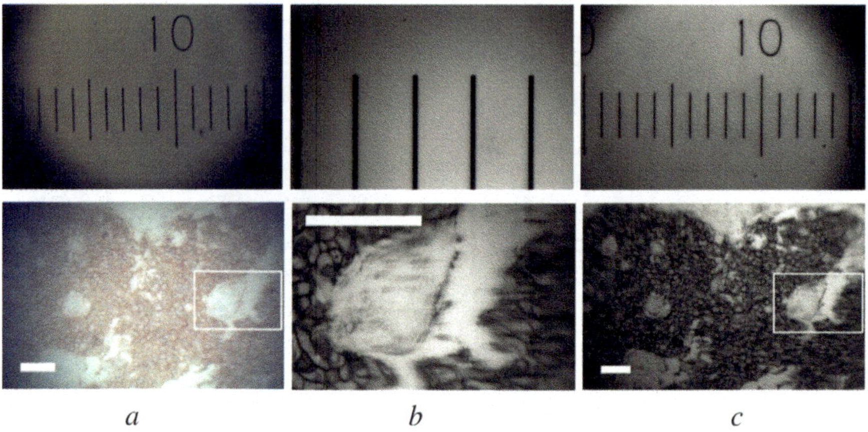

Fig. 14.14 Test objects images: *upper row*—standard pattern (scale bar), *lower row*—breast cancer tumor; (**a**) color (RGB) images, (**b**) direct spectral images obtained with no conjugation, (**c**) spectral images obtained with conjugating optical system (λ = 530 nm). The scale division size is 100 μm as well as the scale bar length below

the lower row images of the local area of some biomedical object are shown with 20× objective lens.

These images demonstrate that the conjugating optical system operates properly and expands the field-of-view of monochrome camera to that of RGB one, as well as increases brightness of the spectral image to the level comparable to that of white-light channel. Thus, double monochromatization and use of the conjugating optical system make it possible the cooperative processing of images from two channels (spectral and visible to human eye) without additional step for image correction. The developed spectral imaging module is the unit completed both physically and functionally. It can be used to diversify functions of standard commercial microscopes without their modifications.

14.3.3 Hyperspectral Instrument for Remote Spectral Imaging

Among various applications of hyperspectral imagers, temperature remote detection is one of the oldest. The structure of AO device for detection of spatial distribution of spectral features is presented in Fig. 14.15 [72].

From thermal radiation flow illuminated by the heated object 1 the input objective I forms collimated beam, which is directed at AOTF. The last comprises a pair of identical wide-angle AO cells 3, 5 placed in opposite manner and three polarizers: co-directed input 2 and output 6, crossed to the intermediate 4 one. In AO cell 3, the radiation spectral component with wavelength corresponding to acoustic Bragg grating period $\lambda \sim 1/f$ is diffracted ($o \to e$). The extraordinary polarized diffracted light beam is selected by the polarizer 4. The diffraction process is repeated

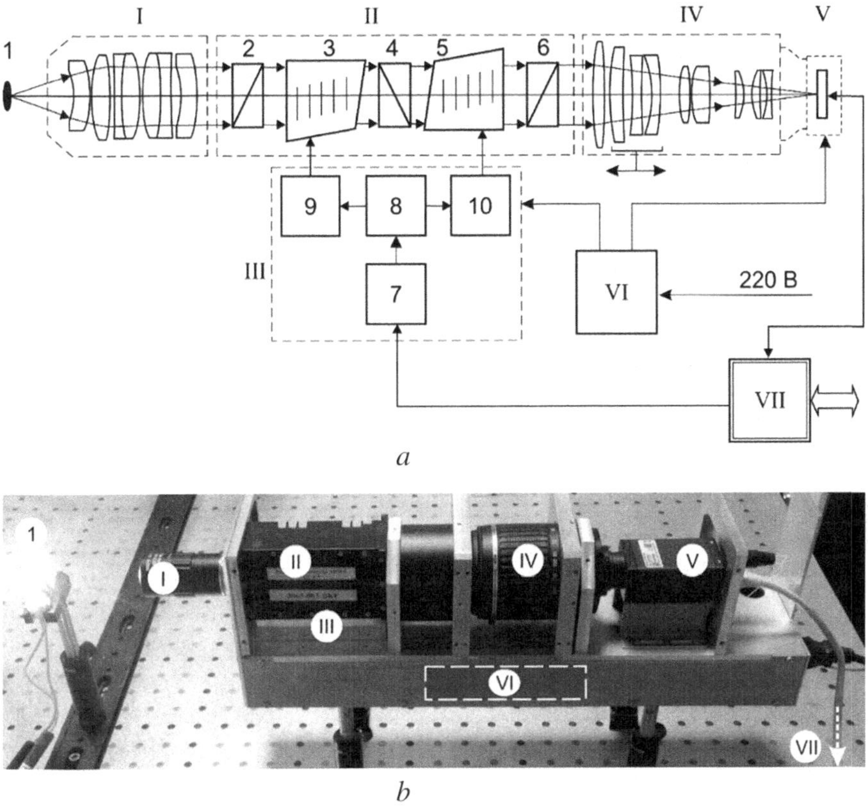

Fig. 14.15 AO hyperspectral temperature imager: principle diagram (**a**) and internal view at the optical table (**b**) I—microobjective, II—AOTF, III—control unit, IV—intermediate objective, V—monochrome camera, VI—power supply, VII—computer, 1—object; 2, 4, 6—polarizers; 3, 5—AO cells, 7—controller, 8—RF generator, 9, 10—RF amplifiers

in the next AO cell 5 that results in the additional contrast of the selected spectral component. Then with use of the lens IV, the selected radiation with specified wavelength λ is focused at the camera photodetector V.

With changing acoustic wave frequency *f* in the range 43–82 MHz by means of variation of RF electric signal 8 supplied at the piezotransducers in AO cells 3 and 5, one can tune the period of the diffraction grating and get the image of the object 1 at any wavelength λ in the range 650–1110 nm with rather high spectral resolution ($\delta\lambda$ = 1.5 nm at λ = 780 nm) and spatial resolution (500 × 500 elements). Synchronization of camera V and spectral tuning of AO cells 3 and 5 permits to obtain a set of spectral images of the object 1 at arbitrary wavelengths in the inspected spectral range.

Associated spatial and spectral distortions caused by AO cells constitute a principal problem for correct comparison of detected spectral images and spectra reconstructions in different points of the image. These deformations, are almost completely compensated with the use of a pair of identical AO cells placed symmetrically, configuring for filtration the radiation of different polarization [13]. The developed AO HSI has the following characteristics: FoV $2\varpi_{AO} = 3°$, input opening $D_{AO} = 8$ mm, the number of spatially resolved elements approximately 500 × 500, and, hence, the angular resolution $\psi_{AO} \approx 20''$.

Replaceable microobjective 1 ensures variability of the magnification and the working distance can be chosen in correspondence with the task. Varifocal lens IV is used for adjusting the image size for any microobjective 1. Video camera Manta G-419 NIR with CMOS array of 2048 × 2048 elements of 6.5 × 4.8 mm^2 area is used for image acquisition. The frame rate is up to 58 Hz for resolution 1000 × 1000 pixels.

To reach the highest characteristics of HSI, one needs to optimize its elements. Light flux maximization and aberration minimization of the device were performed in Zemax program [14, 73].

The spectral images of the optical standard pattern detected with the developed AO HSI are shown in Fig. 14.16. The device provides high quality images in the total operation spectral range with no additional focusing.

The developed AO HSI permits detection of spectral images of high spectral and spatial resolution and makes possible advancing techniques of remote and non-contact sensing.

In contrast to multi-channel diffraction spectrometers, which instantly detect the spectrum with one-dimensional spatial resolution, the developed AO HSI is capable of synchronous detection of spectral characteristics of all the image elements and, thus, can detect the temperature distribution over space that is necessary in many technical and physical problems.

It is highly important that due to aberration minimization the developed AO HSI provides correct detection of spectra in every point of the image. This property allows the device application for measuring the temperature of heated (~1000 K) objects [72].

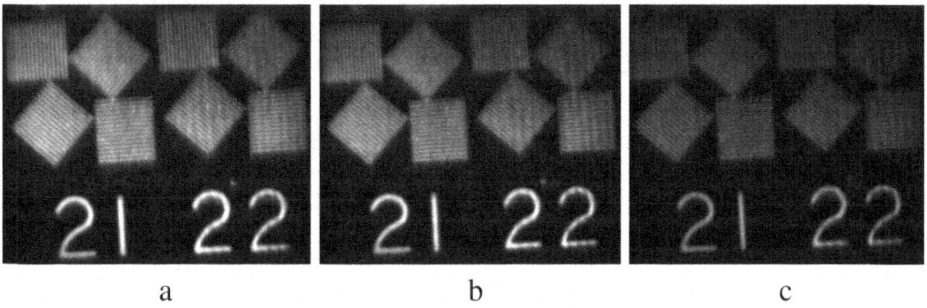

Fig. 14.16 Comparison of spectral images of the same standard pattern recorded by once adjusted AO HSI: (**a**) 750 nm, (**b**) 800 nm, (**c**) 850 nm

14.3.4 *Hyperspectral Instrument for Skin Pathologies Assessment*

One of the most promising optical techniques for skin pathology study is hyperspectral imaging. This technique provides studying both the tumor shape and its composite. Therefore, it became implementable in medical technologies and has already found some applications [74, 75]. For example, it was shown [76] that the sensitivity of HSI in melanoma detection is approximately 90%, while specificity is 84%. Thus, the development and implementation of HSI instruments and analytical techniques is an important task for *in vivo* diagnostics of malignant neoplasms.

The developed AO HSI apparatus based on compensated double monochromator is presented in Fig. 14.17.

The radiation from light-emitting diode (LED) source 2, reflected and scattered by the inspected object 1, is transformed with teleobjective 3 into collimated light beam and directed at the AO tunable monochromator. The last comprises a pair of wide-angle AO cells 5, 7 and 3 polarizers: input 4, intermediate 6, and output 8. The direction of polarization for the middle polarizer is orthogonal to that for polarizers 4 and 8. In AO cell 5, the optical radiation with wavelength λ corresponding to synchronism condition is diffracted at Bragg grating induced by the propagating acoustic wave. The diffracted radiation changes the linear polarization ($o \rightarrow e$) and is selected by the polarizer 6, while the non-diffracted beam is eliminated. The next-step diffraction takes place in the second AO cell 7 for higher contrast of selected radiation. Then the lens 9 focuses the filtered radiation at the monochrome photodetector array 10.

The device operates as follows. Digital commands from the computer 15 are sent to the controller 14, which initiates the synthesis of the corresponding RF signal by the synthesizer 13. This signal is amplified by the RF amplifiers 11, 12 and then is supplied to piezotransducers of AO cells 5, 7. They generate the ultrasonic acoustic waves, which due to elasto-optic effect change periodically the refraction indices of

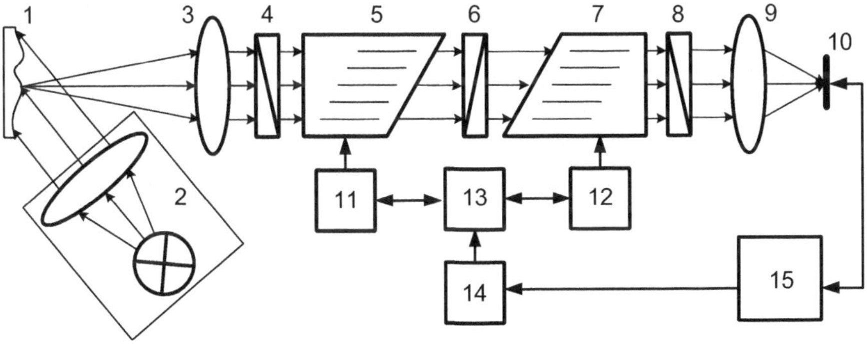

Fig. 14.17 AO HSI apparatus for skin lesions study. 1—object (patient), 2—LEDs-based optical source, 3—input teleobjective, 5, 7—AO cells, 4, 6, 8—polarizers, 9—output lens, 10—monochrome camera, 11, 12—RF amplifiers, 13—RF synthesizer, 14—controller plate, 15—computer

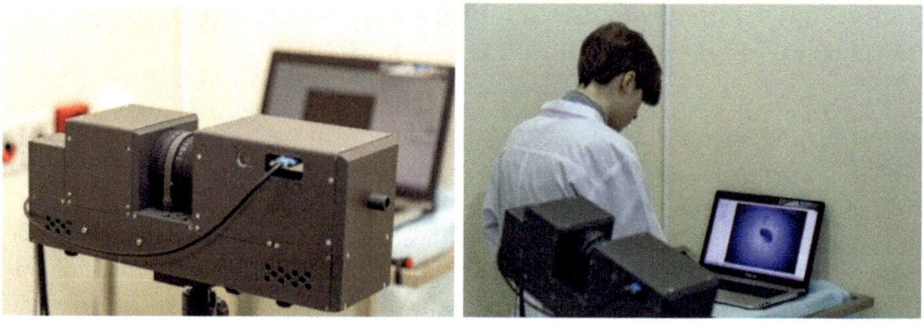

Fig. 14.18 External view of AO HSI and experimental environment

the medium resulting in the creation of dynamic Bragg gratings. By tuning the frequency of the acoustic wave, one can specify the wavelength of the diffracted light beam. So different spectral components can be selected and recorded by the camera.

The developed device exhibits the following characteristics: spectral range 440–750 nm, spectral resolution $\delta\lambda = 2.5$ nm at $\lambda = 633$ nm, spatial resolution 500 × 500 elements. The HSI system was constructed in two assumptions: the distance to the object is 1 m and the inspected area is 7 × 7 cm^2. The aberration and "energetic" optimization were performed in the same manner as it was described above.

The developed HSI device (Fig. 14.18) was used for differentiation of malignant tumors implementing data about skin reflection in the visible spectral range. The preliminary processing procedure is targeted at the artifacts compensation in biomedical images recording *in vivo* caused by the object movements. Another aim of the image processing is an estimation of the object brightness distribution [77].

The analysis of the skin tissues was based on the estimation of skin chromophores. Average concentrations of basic chromophores can be determined with account of spectral features of the tested tissue. As a principal diagnostic quantitative factor, the normalized integral index of optical density was used. The local changes of optical indexes concentration in comparison with surrounding healthy skin indicate the high risk of tissue malignization. The results obtained with the developed AO HSI complemented with data extracted with other techniques (dermatoscopy, Raman scattering, etc.) open the way to effective differentiation of benign and malignant skin neoplasms [30, 77, 78]. The next section highlights various applications of the developed instrument in skin tumor diagnostics.

14.4 Experimental Study of Skin Using AOTF-Based Imagers

The most convenient approach of HSI implementation in cancer screening is the examination of superficial tissues, and many studies were conducted for skin and oral cancer detection. For example, Johansen et al. [79] demonstrated a number of hyperspectral cameras suitable for skin cancer detection with an analysis of back-scattered and autofluorescence signals. Different HSI approaches find sensitivities and specificities of melanoma detection in wide range from 100% to 23%. However, critical remarks made by Johansen et al. point out some limitations of most recent HSI skin cancer studies, especially with small datasets, and further investigations are required for correct estimation of HSI approach skin cancer detection accuracy. In other study, Fei et al. [80] utilized HSI for oral and thyroid cancer margins detection with backscattered light. The authors demonstrated an average accuracy of 90% ± 8%, sensitivity of 89% ± 9%, and specificity of 91% ± 6% for oral tissue discrimination. For tissue specimens from the thyroid, the method achieved an average accuracy of 94% ± 6%, sensitivity of 94% ± 6%, and specificity of 95% ± 6%.

For the integral organs malignancy inspection HSI was used for colorectal cancer with fluorescence registration [81], for brain cancer [82], prostate cancer [83], and backscattered light, and many other applications [1, 2]. For instance, Fabelo et al. [82] registered 300,000 spectra of brain tissue images and reached more than 97% accuracy in brain cancer detection with backscattered light analysis. Such studies clearly prove the applicability of HSI techniques based on backscattered and fluorescent modalities application in cancer screening.

More weak optical signals, such as Raman, also may be used in HSI applications. The *in vivo* applications are complicated due to an extremely low Raman signal, and most efforts of HSI implementations are focused on such implementations as coherent anti-Stokes Raman scattering, stimulated Raman scattering, and surface-enhanced Raman scattering. In turn, application of nonconventional Raman techniques is mainly focused on cell cultures and performed for prior known chemical components [84–86]. However, the authors clearly demonstrate applicability of Raman techniques for future cancer studies.

In addition, promising applications of HSI in oncology include image-guided surgeries, including robotic-assisted applications. Bravo et al. [87] demonstrated the application of HSI in *in vivo* case examples from clinical neurosurgeries revealed changes to the localization and contrast of protoporphyrin IX maps, making concentrations accessible that were not visually apparent. The adoption of these methods has the potential to maintain sensitive and accurate visualization of different chemicals contrast over the course of surgery. Besides *in vivo* studying of cancers, HSI was used for the differentiation of tissue histological samples and biopsies. Berisha et al. [88] demonstrated the application of Fourier transform infrared spectroscopy for the quantitative study of breast tissue histology sections. Kho et al. [89] showed the application of near infrared HSI to determine borders of breast cancer with 84–93% accuracy. Akbari et al. [90] demonstrated HSI study of gastric cancer in 1000–2500 nm area with more than 90% sensitivity and specificity and ability to detect the tumors with a size not exceeding 0.5 mm.

Some other examples of HSI implementation in cancer study can be found in recent reviews [91, 92]. Experimental results show that the use of spatial information in addition to the spectral information brings significant improvements in the cancer classifier performance and allows classification of cellular subtypes. However, for the moment showed examples are limited due to the small datasets, and the preliminary results warrant further research across all organ systems to determine if HSI has a place in the operating theater or in the clinics [91].

All reviewed examples of HSI implementation in cancer screening use different instrumentation and different approaches of optical signal acquisition. Different systems are adopted for the registration of a certain optical modality from certain tissue type. This makes every HSI system unique and applicable in strict conditions. In our recent studies [93, 94] we demonstrated *in vivo* implementation of the AO HSI device in skin cancer detection and differentiation.

14.4.1 Skin Samples

There were two studied cohorts, one included 45 patients with skin neoplasms (19 basal cell carcinomas (BCC), 10 benign tumors (BN), and 16 malignant melanomas (MM)) and the other included 91 patients (19 BCC, 22 MM, and 50 BN neoplasms, including 16 nevi, 5 papilomas, 4 hemangiomas, 7 different inflammatory diseases, 12 keratomas, 1 fibroma, 1 Bowen disease, 4 kerato papillomas). Every hyperspectral tumor study was accompanied by histological analysis to make a final diagnosis. The protocols of *in vivo* tissue diagnostics were approved by the ethical committee of Samara State Medical University. The Institutional Review Board of Samara National Research University approved the study protocols. This research adhered to the tenets set forth in the Declaration of Helsinki. Informed consent of each subject was obtained.

14.4.2 HSI Data Analysis

Generally, to obtain quantitative estimations of various human skin tissues backscattered spectra we used data about spectral optical density (OD) of skin tissues. The diffuse backscattered reflectance from skin in spatial point (x, y) may be characterized by OD on the selected wavelength λ as:

$$\mathrm{OD}(x,y,\lambda) = \lg \frac{I_o(x,y,\lambda)}{I(x,y,\lambda)} \tag{14.5}$$

where $I_0(x, y, \lambda)$ is the backscatter intensity from the background of light source, $I(x, y, \lambda)$ is the backscattered intensity from the tissue sample. While the exact determination of tumor type may be performed with OD analysis based on the above spectral bands:

$$\mathrm{OD}_i(x,y) = \frac{1}{\lambda_2 - \lambda_1} \int_{\lambda_1}^{\lambda_2} \mathrm{OD}(x,y,\lambda)\, d\lambda \tag{14.6}$$

where hemoglobin-related optical density OD_H defined for (λ_1 = 530 nm, λ_2 = 570 nm) and melanin-related OD_M—for (λ_1 = 600 nm, λ_2 = 670 nm). Further, we calculated mean OD_H and OD_M values for all studied tumors and healthy skin. Allocation of tumor area was performed on the basis of the approach presented in our previous studies and was performed based on image segmentation by Otsu method [95]. Healthy skin region was chosen near the tumor area with 1.5–2 cm indentation from the tumor and included approximately 1 × 1 cm area of healthy skin. Typical normalized spectra and tumor image presented in Fig. 14.19. Details of hyperspectral image with tumor and healthy skin areas allocation as well as distribution of OD_M coefficient may be found elsewhere [93].

Discrimination of tumors was performed based on two principles. First principle took into account only raw values of OD_H and OD_M coefficients, while second principle used data for the normal skin of the same patient for OD^N_H and OD^N_M coefficient as follows:

$$\mathrm{OD}_i^N = \frac{\mathrm{OD}_i^T}{\mathrm{OD}_i^H} \tag{14.7}$$

where OD^T_i and OD^H_i are OD_i coefficients calculated for one patient for tumor (index ^{T}) and healthy skin (index ^{H}) areas, respectively.

Finally, we analyzed integral properties of skin for allocated tumor area and healthy skin area with an unbiased dispersion of OD_i and OD^N_i coefficients inside the area of interest:

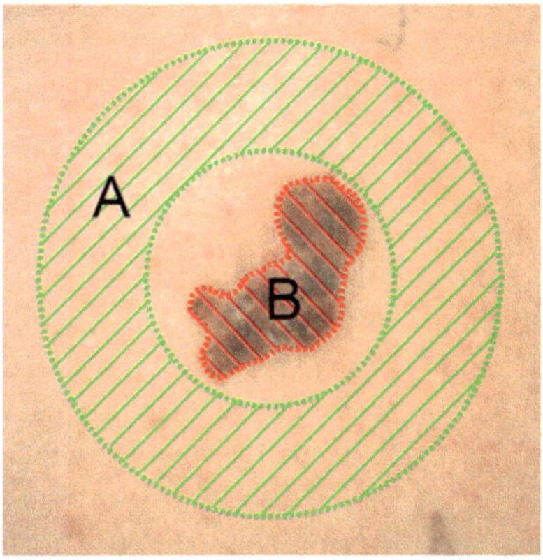

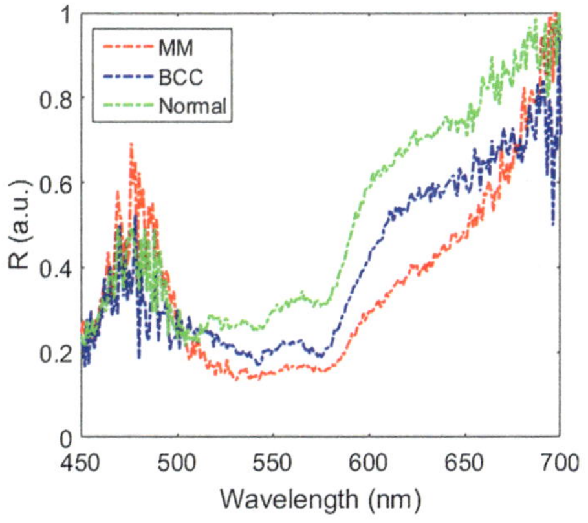

Fig. 14.19 Tumor image (5 × 5 cm) (**a**) and normalized spectral diffuse reflection (**b**) for MM, BCC, and normal skin. A—normal skin area, B—tumor area

$$\overline{\mathrm{OD}}_i = \frac{1}{N-1}\sum_{n=1}^{N}\left(\mathrm{OD}_n - \bar{M}\right)^2 \tag{14.8}$$

where N is the number of pixels in the area of interest, OD_n is the OD (or normalized OD) in the nth pixel of area of interest, $\bar{M}$ is the mean value of OD_n inside the area of interest.

In skin tissue classification we utilized $\overline{\mathrm{OD}}_i$ and OD^N_i coefficients. The choice of these coefficients was based on coefficients utility. While OD^N_i coefficients provide information about changes of chromophores content in the tumor in comparison to healthy skin of the same person, $\overline{\mathrm{OD}}_i$ coefficients provide information about mean values of chromophores concentration in the tumor area. Thus, these two approaches use different features of chromophores distributions and both may be useful in skin tissue classification.

$\overline{\mathrm{OD}}_i$ and OD^N_i coefficients were used for tissue classification by discriminant analysis (DA). DA can separate two or more classes based on different statistical parameters of Gaussian distributions. The efficiency of the proposed approach is characterized by sensitivity and specificity and ability to select defined classes in different areas of phase plane. The analysis of skin tissue data allocation was performed using quadratic DA classifiers [96]. In addition, data were analyzed using SPSS software version 23 (SPSS Inc., Chicago, IL) with logistic regression, and a P-value of <0.01 was considered to indicate statistically significant differences. Then, the plot of sensitivity versus 1-specifity (called receiver operating characteristic (ROC) curve) is built, and the area under the curve (AUC) was used as an effective criterion of skin tissue discrimination accuracy. This curve plays a central role in evaluating diagnostic ability of tests to discriminate the true state of subjects, finding the optimal cut-off values, and comparing alternative diagnostic tasks when each task is performed on the same subject [97].

14.4.3 Skin Cancer Differentiation Based on Hyperspectral Imager Features

Typical *in vivo* images are presented in Fig. 14.20 for BCC (top) and MM (bottom). Here (a, d) are digital images, (b, e) and (c, f) are distribution of ODH and ODM correspondently. Significant pigmentation of all tumor surface characterizes MM, unlike BCC sample. It is clearly visible that BCC and MM have great differences in distribution of hemoglobin and melanin concentrations (OD_H and OD_M maps), which make possible visual identification of cancer type and define tumor boundary [95].

As can be seen from Fig. 14.21a (corresponding to the first cohort) HSI determines skin cancer with very high sensitivity and specificity of 84 and 87%, respectively. Almost the same value (86%) characterizes the specificity of MM differentiation from other cancer types, but the sensitivity decreases up to 56% (see

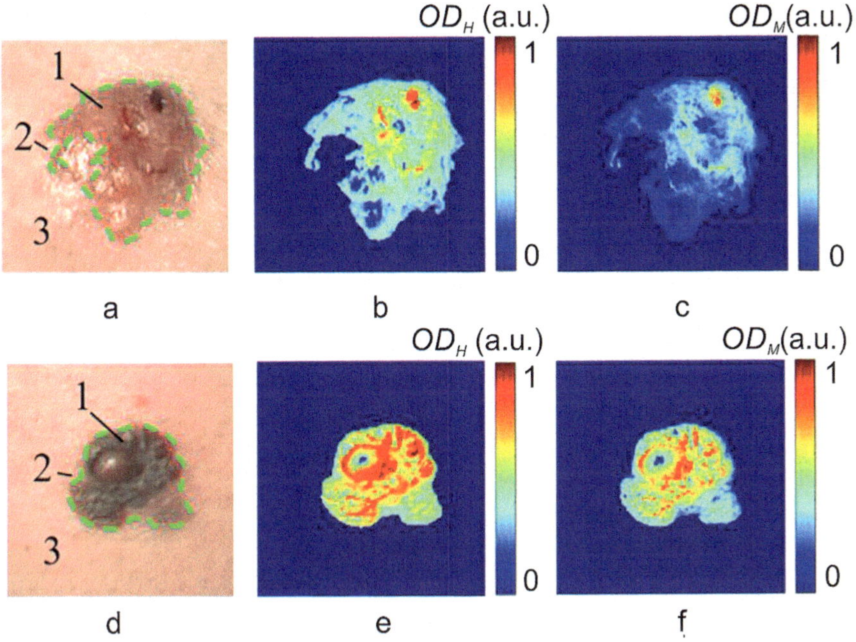

Fig. 14.20 *In vivo* digital images (**a**, **d**), OD_H (**b**, **e**), and OD_M (**c**, **f**) maps of BCC (top) and MM (bottom): image sizes are 3 × 3 cm; 1—pathology area, 2—tumor boundary, 3—healthy skin

Fig. 14.21b). MM have higher values of OD_H and OD_M due to the high concentration of the melanocytes in epidermal cells of melanoma unlike BCC or BN. Most MM samples classified as positive are pigmented melanoma, while three MM samples of I–II stages classified as negative belong to epithelioid melanoma and atypical melanoma, containing significantly less melanin. The presence of the negative results in both groups can also explained by skin redness in the study area (for MM of III–IV stages) and features of skin phenotype.

The results of the second studied cohort analysis are presented in Fig. 14.22 and Table 14.1. Application of single $\overline{OD}_i$ or OD^N_i coefficient for melanin content leads to 60–70% accuracy, and these coefficients for hemoglobin content provide a slightly lower accuracy of 55–65%. It is important to note that accounting of normal healthy skin helps to improve accuracy for $\overline{OD}_H$ and decreases accuracy for $\overline{OD}_M$. These facts may be caused by inhomogeneity of melanin and hemoglobin distribution in skin. While neoplasms have heterogeneous distribution of hemoglobin and capillary network, normal skin has more homogeneous one, and, thus, comparison of $\overline{OD}_H$ dispersion in neoplasm and normal tissue provides additional information for tissue discrimination.

Melanin content of neoplasms frequently may be increased in comparison to healthy skin, but as every person have their own skin phenotype and unique level of melanin [98] it may be complicated to estimate differences in melanin content of neoplasm and healthy skin. That is why most likely in our study application of OD^N_M

Fig. 14.21 Discriminant analysis in $OD_H - OD_M$ plane: (**a**) skin cancer (MM, BCC, and BN; red cycles) vs healthy skin (green triangles); (**b**) MM (red cycles) vs other cancer types (BCC and BN; blue squares); (**c**) MM of I–II stage (red cycles) vs MM of III–IV stage (blue stars)

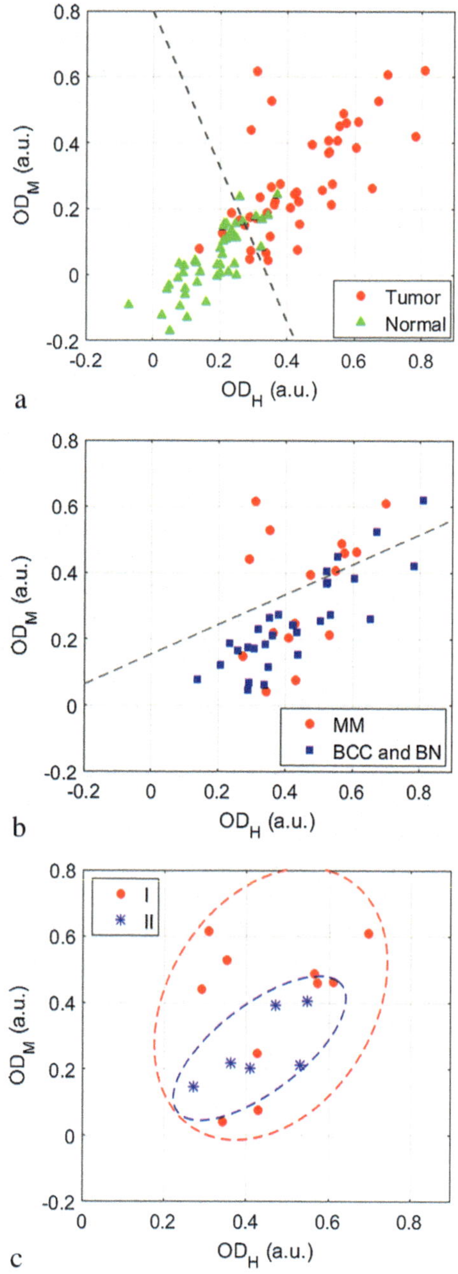

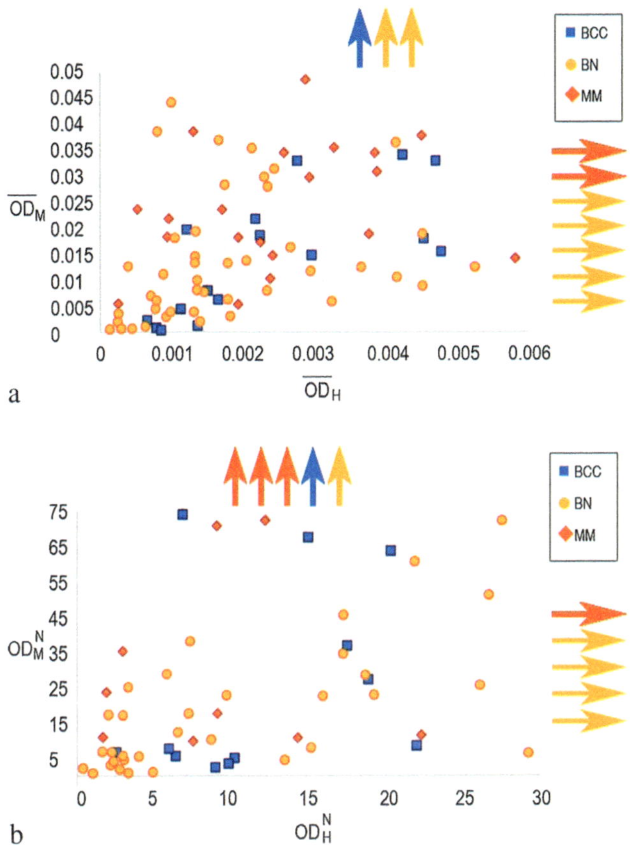

Fig. 14.22 Distribution of skin neoplasms on phase planes with $\overline{OD}_i$ (**a**) and OD^N_i (**b**) axes

Table 14.1 Accuracy (%) of neoplasm classification

Classes	$\overline{OD}_H$	$\overline{OD}_M$	OD^N_H	OD^N_M	$\overline{OD}_H + \overline{OD}_M$	$OD^N_H + OD^N_M$	Physician
MM + BCC vs BN	61	71	65	60	76	78	51
MM vs BCC	54	59	66	59	68	79	56
MM vs BCC + BN	60	69	68	63	75	75	51

coefficient showed rather low accuracy of skin neoplasm classification in comparison to $\overline{OD}_M$.

Application of coefficients combination ($\overline{OD}_H + \overline{OD}_M$ and $OD^N_H + OD^N_M$) helps to improve accuracy of tissue classification by 5–15% in comparison with single coefficient implementation. $OD^N_H + OD^N_M$ combination in comparison with $\overline{OD}_H + \overline{OD}_M$ pair shows almost the same rates of accuracy for cases of MM vs other neoplasms and malignant neoplasms vs BN classification, but demonstrates 11% higher accuracy in case of MM vs BCC classification. That may be explained by OD^N_H coefficient usage as it shows high rates of tissues classification in comparison with unnormalized on healthy skin $\overline{OD}_H$ coefficient. Thus, we may try to combine normalized and unnormalized coefficients to achieve the highest performance of tissues classification.

In order to do this we performed ROC analysis for combination of proposed coefficients. Figure 14.23 demonstrates ROC curves for different pairs of coefficients. One may see that the best AUC values show a combination of normalized and unnormalized coefficients. This fact proves that the best performance of HSI skin neoplasms study may be implemented with joint analysis of pathology area and healthy skin. Combination of $OD^N_H + \overline{OD}_M$ and $\overline{OD}_H + \overline{OD}_M$ coefficients shows 0.693 and 0.686 AUC values, respectively, while $OD^N_H + OD^N_M$ coefficients pair shows only 0.608 values. As higher AUC value the higher final accuracy of classification for unnormalized and combined approaches of neoplasm, hemoglobin and melanin dispersion in clinical practice. Moreover, comparison of HSI analysis and physician survey accuracy presented in Table 14.1 shows that automatic analysis may significantly improve accuracy of skin neoplasm detection. HSI analysis may

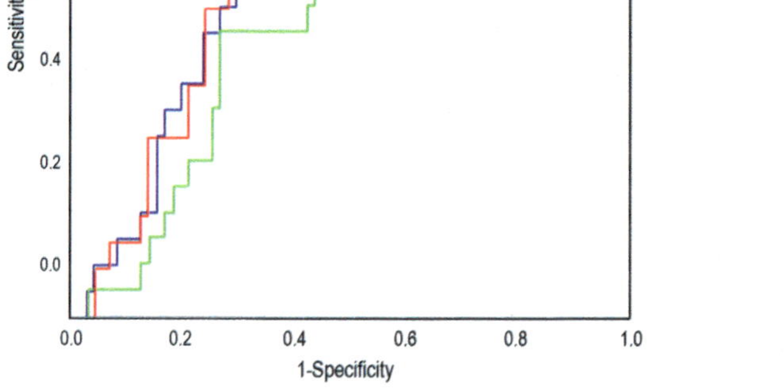

Fig. 14.23 ROC curves of MM vs BCC + BN classification with logistical regression for $OD^N_H + \overline{OD}_M$, $OD^N_H + OD^N_M$ and $\overline{OD}_H + \overline{OD}_M$ coefficients

be used during the first hand examinations, which suffer from very low specificity of malignant tumors detection [99].

Considering alternative optical methods of the proposed approach, one may say that the most frequently used method today is multispectral analysis with additional spectral ranges or native fluorescence analysis. For example, it was proposed to use the spectral region up to 1000 nm and analyze the content of NADH/NAD+ [100, 101]. The total sensitivity and specificity for separation of MM from nevus and BCC were 97% and 96%, respectively. However, this approach requires an additional light source (either a separate source for the near-IR or a white diode with a near-IR coating up to 1000 nm), as well as a camera with a higher sensitivity (in comparison with CCD) in the 1000–1100 nm range, which significantly complicates the design of the multispectral camera. In a study by Nagaoka et al. [76] an index is proposed that describes the heterogeneity of the tumor, which is the analog of the color index in dermatoscopy. The sensitivity of the determination of melanoma was 90%, and specificity was 84%. However, it is important to note that the above-described results [76, 100, 101] were obtained on a small number of samples (only five–ten melanomas were included in the studied cohort).

Evolution of the proposed technique may be based on the approach of spatial evaluation of diagnostic dermatoscopic signs, and, for example, may be implemented with convolutional neural networks analysis [102]. The authors [102] claimed maximum sensitivity and specificity of melanoma diagnosis at 90% and 85%, respectively. Therefore, image analysis approaches may be combined with proposed technique to improve the overall accuracy in HSI skin cancer imaging.

14.5 Conclusions

The proposed approaches to skin tissue classification showed 68–78% accuracy. These values of accuracy are not quite high in comparison to 90% and higher accuracies of such optical approaches as Raman or fluorescent spectroscopy [103]. However, integral information about skin chromophores may significantly increase accuracy of visual inspection by dermatologist, as proposed technique demonstrated 15–20% higher accuracy in skin tumors classification. Wherein application of both OD_i and $OD^{N}i$ coefficients and their combination leads to the comparable accuracy in tumors classification. This fact proves that choice of the visible range area (450–750 nm) for the analysis of the chemical composition of skin may be useful in oncodermatological applications.

It should be noted that in the proposed approach the processing of hyperspectral data is carried out in real time and does not require operator participation. In general, the simplicity of the equipment and the speed of processing results may allow for adapting the proposed method for tasks of mass screening of skin tumors. Moreover, in future works on the development of the proposed approach, we may increase the number of examined skin neoplasms samples and combine the proposed technique with image pattern analysis methods.

Acknowledgments This study was supported by the Russian Federation President grant for state support of young scientists (project MK-1888.2019.2).

References

1. Alfano, R., Pu, Y.: Laser for Medical Applications. Woodhead Publishing Limited, Philadelphia (2013). https://doi.org/10.1533/9780857097545.3.325
2. Lu, G., Fei, B.: Medical hyperspectral imaging: a review. J. Biomed. Opt. **19**(1), 010901 (2014). https://doi.org/10.1117/1.jbo.19.1.010901
3. Zakharov, V.P., Bratchenko, I.A., Artemyev, D.N., Myakinin, O.O., Kozlov, S.V., Moryatov, A.A., Orlov, A.E.: Neurophotonics and Biomedical Spectroscopy, pp. 449–476. Elsevier, Förlag (2018)
4. Yariv, A., Yeh, P.: Optical Waves in Crystals. Wiley, New York (1984)
5. Pozhar, V.E., Mazur, M.M., Machikhin, A.S.: Comparison of two modes of Bragg diffraction on broadband ultrasound. Proc. SPIE. **10815**, 108151I (2018). https://doi.org/10.1117/12.2502414
6. Chang, I.C.: Acousto-optic tunable filters. Opt. Eng. **20**, 824–829 (1981)
7. Belikov, I.B., Ya, G., Voloshinov, V.B., Magdich, L.I., Mit'kin, M.I., Parygin, V.N.: Acousto-optic images filtering. Tech. Phys. Lett. **10**(10), 1225–1229 (1984)
8. Levin, I.W., Lewis, I.N.: Near-infrared acousto-optic spectroscopic microscopy: a solid state approach to chemical imaging. Appl. Spectrosc. **46**, 553–558 (1992)
9. Pustovoit, V., Pozhar, V.: Collinear diffraction of light by sound waves in crystals: devices, applications, new ideas. Photon. Optoelectron. **2**(2), 53–69 (1994)
10. Pozhar, V., Machihin, A.: Image aberrations caused by light diffraction on ultrasonic waves in uniaxial crystals. Appl. Opt. **51**(20), 4513–4519 (2012)
11. Pustovoit, V.I., Pozhar, V.E., Mazur, M.M., Shorin, V.N., Kutuza, I.B., Perchik, A.V.: Double-AOTF spectral imaging system. Proc. SPIE. **5953**, 59530P (2005)
12. Machihin, A.S., Pozhar, V.E.: Image distortions caused by transmission through double acousto-optical monochromator. Electromagn. Waves Electron. Syst. **14**(11), 63–68 (2009)
13. Machihin, A.S., Pozhar, V.E.: Spatial and spectral image distortions caused by diffraction of an ordinary polarised light beam by an ultrasonic wave. Quantum Electron. **45**(2), 161–165 (2015)
14. Machikhin, A., Batshev, V., Pozhar, V.: Aberration analysis of AOTF-based spectral imaging systems. J. Opt. Soc. Am. A. **34**(7), 1109–1113 (2017)
15. Machikhin, A.S., Polschikova, O.V., Ramazanova, A.G., Pozhar, V.E.: Multi-spectral quantitative phase imaging based on filtration of light via ultrasonic wave. J. Opt. **19**(7), 075301 (2017)
16. Epikhin, V.M., Kalinnikov, Y.K.: Elimination of spectral drift of diffraction angle in noncollinear acousto-optical filter. Tech. Phys. **59**(2), 160–163 (1989)
17. Pozhar, V.E., Machihin, A.S.: Proceedings of X International Conference for Young Researchers, WECONF-2010, p. 23. Media Publisher, St. Petersburg (2010)
18. Batshev, V.I., Machikhin, A.S., Pozhar, V.E., Burmak, L.I.: Proceedings of International Conference on Coherent and Nonlinear Optics, Conference on Lasers, Applications, and Technologies '2016, St. Petersburg, pp. 17–18 (2016)
19. Pozhar, V.E., Pustovoit, V.I.: Consecutive collinear diffraction of light in several acoustooptic cells. Sov. J. Quantum Electron. **15**(10), 1438–1439 (1985)
20. Mazur, M.M., Shorin, V.N., Chizhikov, S.I., Leonov, S.N.: Double acousto-optic monochromator made of $CaMoO_4$. Opt. Spectrosc. **67**, 736 (1989)
21. Zakharov, V.G., Parygin, V.N.: Acousto-optic diffraction in two crystals. Sov. J. Commun. Technol. Electron. **35**, 175 (1990)

22. Mazur, M.M., Shorin, V.N., Abramov, A.Y., Magometov, Z.A., Mazur, L.I.: Spectrometer with double acousto-optical monochromator. Opt. Spectrosc. **81**(3), 475–477 (1996)
23. Epikhin, V.M., Rokos, I.A.: A double acousto-optic monochromator for the UV region with improved optical characteristics based on a KDP single crystal. Opt. Spectrosc. **96**(3), 465–469 (2004)
24. Yushkov, K., Dupont, S., Kastelik, J., Voloshinov, V.: Polarization-independent imaging with an acousto-optic tandem system. Opt. Lett. **35**, 1416–1418 (2010)
25. Kastelik, J., Yushkov, K., Dupont, S., Voloshinov, V.: Cascaded acousto-optical system for the modulation of unpolarized light. Opt. Express. **17**(15), 12767–12776 (2009)
26. Zhang, C., Zhang, Z., Wang, H., Yang, Y.: Spectral resolution enhancement of acousto-optic tunable filter by double-filtering. Opt. Express. **16**(14), 10234–10239 (2008)
27. Mazur, M.M., Suddenok, Y.A., Shorin, V.N.: Double acousto-optic monochromator of images with tunable width of the transmission function. Tech. Phys. Lett. **4**, 167–169 (2014)
28. Mazur, M.M., Pozhar, V.E.: Spectrometers based on acousto-optical filters. Optophys. Meas. **58**(9), 982–988 (2015)
29. Mazur, M.M., Mazur, L.I., Pustovoit, V.I., Suddenok, Y.A., Shorin, V.N.: High-transmission two-crystal acousto-optic monochromator. Tech. Phys. **87**(9), 1407–1410 (2017). https://doi.org/10.1134/S1063784217090183
30. Zherdeva, L., Bratchenko, I., Alonova, M., Myakinin, O., Artemyev, D., Moryatov, A., Kozlov, S., Zakharov, V.: Hyperspectral imaging of skin and lung cancers. Proc. SPIE. **9887**, 98870S (2016)
31. Machikhin, A., Pozhar, V., Batshev, V.: Double-AOTF-based aberration-free spectral imaging endoscopic system for biomedical applications. J. Innovative Opt. Health Sci. **8**(3), 1541009 (2015)
32. Machikhin, A., Zinin, P., Shurygin, A., Khokhlov, D.: Imaging system based on a tandem acousto-optical tunable filter for *in-situ* measurements of the high temperature distribution. Opt. Lett. **41**(5), 901–904 (2016)
33. Mudry, K., Plonsey, R., Bronzino, J.: Biomedical Imaging. CRC Press, New York (2003)
34. Vo-Dihn, T., Kasili, P., Cullum, B.: Multispectral imaging for medical diagnostics. Proc. SPIE. **4615**, 13–19 (2002)
35. Shaw, G.A., Burke, H.K.: Spectral imaging for remote sensing. Lincoln Lab. J. **14**(1), 3–26 (2003)
36. Chang, C.: Hyperspectral Imaging: Techniques for Spectral Detection and Classification. Springer, New York (2003)
37. Sun, D.-W.: Hyperspectral Imaging for Food Quality Analysis and Control. Elsevier, Amsterdam (2009)
38. Renfu, L., Bosoon, P.: Hyperspectral Imaging Technology in Food and Agriculture. Springer, New York (2015)
39. Machikhin, A.S., Pozhar, V.E.: Obtaining spectral stereo images with electron spectral tuning and polarization separation. Tech. Phys. Lett. **40**(9), 823–826 (2014)
40. Machikhin, A.S., Pozhar, V.E.: Single-AOTF-based stereoscopic 3-dimensional spectral imaging systems. J. Phys. Conf. Ser. **661**, 012041 (2015)
41. Pozhar, V.E., Pustovoit, V.I.: Acousto-optical spectral technologies. Bull. Russ. Acad. Sci. Phys. **79**(10), 1221–1226 (2015)
42. Machikhin, A.S., Shurygin, A.V., Pozhar, V.E.: Spatial and spectral calibration of an endoscopic acousto-optical spectrometer. Instrum. Exp. Tech. **59**(5), 692–697 (2016)
43. Fadeyev, A., Pozhar, V.: Optimization of measuring and calibration procedures for gas analyser based on acousto-optical tunable filters. Proc. SPIE. **8**, 808242 (2011)
44. Tuchin, V.V. (ed.): Optical Biomedical Diagnostics. SPIE Press, Bellingham (2016)
45. Mycek, M., Pogue, B.W.: Handbook of Biomedical Fluorescence. CRC Press, New York (2003)
46. Cherry, S.R., Badawi, R.D., Qi, J.: Essentials of *In Vivo* Biomedical Imaging. CRC Press, New York (2015)

47. Czichos, H.: Handbook of Technical Diagnostics. Fundamentals and Application to Structures and Systems. Springer, New York (2013)
48. Machikhin, A.S., Pozhar, V.E.: Application of acousto-optical light filtration in non-destructive testing. Vestn. Mosk. Energ. Inst. **6**, 120–127 (2018). https://doi.org/10.24160/1993-6982-2018-6-120-127
49. Liu, J.J., Droller, M.J., Liao, J.C.: New optical imaging technologies for bladder cancer: considerations and perspectives. J. Urol. **188**(2), 361–368 (2012)
50. Lim, H., Murukeshan, V.M.: A four-dimensional snapshot hyperspectral video-endoscope for bio-imaging applications. Sci. Rep. **6**, 24044 (2016)
51. Sharma, N., Takeshita, N., Ho, K.Y.: Raman spectroscopy for the endoscopic diagnosis of esophageal, gastric, and colonic diseases. Clin. Endosc. **49**(5), 404–407 (2016)
52. Fedeli, P., Gasbarrini, A., Cammarota, G.: Spectral endoscopic imaging: the multiband system for enhancing the endoscopic surface visualization. J. Clin. Gastroenterol. **45**(1), 6–15 (2011)
53. Yang, C.: PhD thesis, University of Washington (2014)
54. Ratone, J., Bories, E., Caillol, F., Pesenti, C., Godat, S., Sellier, F., Hoibian, S., Landon, M., Servajean, C., De Cassan, C., Lestelle, V., Casanova, J., Poizat, F., Giovannini, M.: Impact of full spectrum endoscopy (fuse, endochoice) on adenoma detection: a prospective French pilot study. Ann. Gastroenterol. **30**(5), 512–517 (2017). https://doi.org/10.20524/aog.2017.0176
55. Bouhifd, M., Whelan, M., Aprahamian, M.: Fluorescence imaging spectroscopy utilizing acousto-optic tunable filters. Proc. SPIE. **5826**, 185–193 (2005)
56. Leitner, R., DeBiasio, M., Arnold, T., Dinh, C., Loog, M., Duin, R.: Multi-spectral video endoscopy system for the detection of cancerous tissue. Pattern Recogn. Lett. **34**(1), 85–93 (2013)
57. Bouhifd, M., Whelan, M., Aprahamian, M.: Use of acousto-optic tunable filter in fluorescence imaging endoscopy. Proc. SPIE. **5143**, 305–314 (2003)
58. Martin, M., Wabuyele, M., Panjehpour, M., Phan, M., Overholt, B., DeNovo, R., Moyers, T., Song, S., Vo-Dinh, T.: Dual modality fluorescence and reflectance hyperspectral imaging: principle and applications. Proc. SPIE. **5692**, 133–139 (2005). https://doi.org/10.1117/12.604445
59. Zemax. www.zemax.com
60. Machikhin, A.S., Batshev, V.I.: An optical system for coupling double acousto-optic monochromators and eyepieces of visual optical instruments. Instrum. Exp. Tech. **57**(6), 736–741 (2014). https://doi.org/10.1134/S0020441214060086
61. Machikhin, A., Pozhar, V., Batshev, V.: Acousto-optic video spectrometer module for medical endoscopic studies. J. Opt. Technol. **80**(7), 439–443 (2013). https://doi.org/10.1364/JOT.80.000439
62. Machikhin, A.S., Pozhar, V.E., Batshev, V.I.: An acousto-optic endoscopic imaging spectrometer. Instrum. Exp. Tech. **56**(4), 477–481 (2013). https://doi.org/10.1134/S0020441213030202
63. Pozhar, V., Machihin, A., Batshev, V.: Compact AOTF-based videospectrometer for endoscopic medical research. Photon. Lasers Med. **2**(2), 153–157 (2013)
64. Morris, H., Hoyt, C., Treado, P.: Imaging spectrometers for fluorescence and Raman microscopy: acousto-optic and liquid crystal tunable filters. Appl. Spectrosc. **48**(7), 857–865 (1994)
65. Park, B., Lee, S., Yoon, S.-C., Sundaram, J., Windham, W.R., Hinton, A., Lawrence, K.: AOTF hyperspectral microscopic imaging for foodborne pathogenic bacteria detection. Proc. SPIE. **8027**, 802707 (2011)
66. Rajwa, B., Ahmed, W., Venkatapathi, M., Gregoria, G., Jinb, F., Soosb, J., Trivedib, S., Robinsona, J.: AOTF-based system for image cytometry. Proc. SPIE. **5694**, 591115 (2005). https://doi.org/10.1117/12.591115
67. Levenson, R.H., Hoyt, C.C.: Spectral imaging and microscopy. Am. Lab. **32**(22), 26–33 (2000)

68. Yushkov, K., Molchanov, V., Belousov, P., Abrosimov, A.: Contrast enhancement in microscopy of human thyroid tumors by means of acousto-optic adaptive spatial filtering. J. Biomed. Opt. **21**(1), 016003 (2016)
69. Li, Q., He, X., Wang, Y., Liu, H., Xu, D., Guo, F.: Review of spectral imaging technology in biomedical engineering: achievements and challenges. J. Biomed. Opt. **18**(10), 100901 (2013)
70. Wachman, E., Niu, W., Farkas, D.: AOTF microscope for imaging with increased speed and spectral versatility. Biophys. J. **73**, 1215–1222 (1997)
71. Polschikova, O., Machikhin, A., Batshev, V., Ramazanova, A., Belov, A., Pozhar, V.: AOTF-based optical system of a microscope module for multispectral imaging techniques. Proc. SPIE. **10592**, 105920H (2017)
72. Machikhin, A.S., Batshev, V.I., Zinin, P.V., Shurygin, A.V., Khokhlov, D.D., Pozhar, V.E., Martyanov, P.S., Bykov, A.A., Boritko, S.V., Troyan, I.A., Kazakov, V.A.: An imaging spectroradiometry system for measuring spatial temperature distributions in microscopic objects. Instrum. Exp. Tech. **60**(3), 401–406 (2017). https://doi.org/10.1134/S0020441217020105
73. Batshev, V.I., Machikhin, A.S., Pozhar, V.E.: Determining the aberration characteristics of optical systems containing acousto-optical diffraction elements. Tech. Phys. Lett. **43**(2), 216–219 (2017)
74. Dicker, D.T., Lerner, J., Van Belle, P., Guerry, D., Herlyn, M., Elder, D.E., El-Deiry, W.S.: Differentiation of normal skin and melanoma using high resolution hyperspectral imaging. Cancer Biol. Ther. **5**(8), 1033–1038 (2006)
75. Koprowski, R., Wilczyński, S., Wróbel, Z., Kasperczyk, S., Błońska-Fajfrowska, B.: Automatic method for the dermatological diagnosis of selected hand skin features in hyperspectral imaging. Biomed. Eng. Online. **13**, 47 (2014)
76. Nagaoka, T., Nakamura, A., Okutani, H., Kiyohara, Y., Koga, H., Saida, T., Sota, T.: Hyperspectroscopic screening of melanoma on acral volar skin. Skin Res. Technol. **19**(1), e290–e296 (2012). https://doi.org/10.1111/j.1600-0846.2012.00642.x
77. Bratchenko, I.A., Alonova, M.V., Myakinin, O.O., et al.: Hyperspectral visualization of skin pathologies in visible region. Comput. Opt. **40**(2), 240–248 (2016)
78. Zherdeva, L., Bratchenko, I., Myakinin, O., Moryatov, A., Kozlov, S., Zakharov, V.: *In vivo* hyperspectral imaging and differentiation of skin cancer. Proc. SPIE. **24**, 100244G (2016)
79. Johansen, T.H., Møllersen, K., Ortega, S., Fabelo, H., Garcia, A., Callico, G.M., Godtliebsen, F.: Recent advances in hyperspectral imaging for melanoma detection. WIREs Comput. Stat. **12**, e1465 (2019). https://doi.org/10.1002/wics.1465
80. Fei, B., Lu, G., Wang, X., Zhang, H., Little, J.V., Patel, M.R., Griffith, C.C., El-Diery, M.W., Chen, A.Y.: Label-free reflectance hyperspectral imaging for tumor margin assessment: a pilot study on surgical specimens of cancer patients. J. Biomed. Opt. **22**(8), 086009 (2017)
81. Leavesley, S.J., Deal, J., Martin, W.A., Lall, M., Lopez, C., Rich, T.C., Boudreaux, C.W., Rider, P.F., Hill, S.: Optical Biopsy XVI: Toward Real-Time Spectroscopic Imaging and Diagnosis, vol. 10489, p. 19. SPIE, Washington, DC (2018)
82. Fabelo, H., Ortega, S., Lazcano, R., et al.: An intraoperative visualization system using hyperspectral imaging to aid in brain tumor delineation. Sensors. **18**(2), 430 (2018)
83. Akbari, H., Halig, L.V., Schuster, D.M., Osunkoya, A., Master, V., Nieh, P.T., Chen, G.Z., Fei, B.: Hyperspectral imaging and quantitative analysis for prostate cancer detection. J. Biomed. Opt. **17**(7), 076005 (2012)
84. Hartsuiker, L., Petersen, W., Rayavarapu, R.G., et al.: Raman and fluorescence spectral imaging of live breast cancer cells incubated with PEGylated gold nanorods. Appl. Spectrosc. **66**(1), 66–74 (2012)
85. Masia, F., Karuna, A., Borri, P., Langbein, W.: Hyperspectral image analysis for CARS, SRS, and Raman data. J. Raman Spectrosc. **46**(8), 727–734 (2015)
86. Karuna, A., Masia, F., Wiltshire, M., Errington, R., Borri, P., Langbein, W.: Label-free volumetric quantitative imaging of the human somatic cell division by hyperspectral coherent

anti-stokes Raman scattering. Anal. Chem. **91**(4), 04706 (2019). https://doi.org/10.1021/acs.analchem.8b04706
87. Bravo, J.J., Olson, J.D., Davis, S.C., Roberts, D.W., Paulsen, K.D., Kanick, S.C.: Hyperspectral data processing improves PpIX contrast during fluorescence guided surgery of human brain tumors. Sci. Rep. **7**, 9455 (2017)
88. Berisha, S., Lotfollahi, M., Jahanipour, J., et al.: Deep learning for FTIR histology: leveraging spatial and spectral features with convolutional neural networks. Analyst. **144**(5), 1642–1653 (2018). https://doi.org/10.1039/c8an01495g
89. Kho, E., de Boer, L.L., Van de Vijver, K.K., et al.: Hyperspectral imaging for resection margin assessment during cancer surgery. Clin. Cancer Res. **25**(12), 3572–3580 (2019). https://doi.org/10.1158/1078-0432.CCR-18-2089
90. Akbari, H., Uto, K., Kosugi, Y., et al.: Cancer detection using infrared hyperspectral imaging. Cancer Sci. **102**(4), 852–857 (2011)
91. Halicek, M., Fabelo, H., Ortega, S., et al.: *In-vivo* and *ex-vivo* tissue analysis through hyperspectral imaging techniques: revealing the invisible features of cancer. Cancer. **11**, 756 (2019). https://doi.org/10.3390/cancers11060756
92. Ortega, S., Fabelo, H., Iakovidis, D.K., et al.: Use of hyperspectral/multispectral imaging in gastroenterology. Shedding some–different–light into the dark. J. Clin. Med. **8**, 36 (2019). https://doi.org/10.3390/jcm8010036
93. Zherdeva, L.A., Bratchenko, I.A., Myakinin, O.O., et al.: *In vivo* hyperspectral imaging and differentiation of skin cancer. Proc. SPIE. **10024**, 100244G (2016)
94. Bratchenko, I.A., Sherendak, V.P., Myakinin, O.O., et al.: *In vivo* hyperspectral analysis of skin hemoglobin and melanin content for neoplasia detection. J. Biomed. Photon. Eng. **4**(4), 040301 (2018)
95. Zherdeva, L.A., Bratchenko, I.A., Myakinin, O.O., Moryatov, A.A., Kozlov, S.V., Zakharov, V.P.: *In vivo* hyperspectral imaging and differentiation of skin cancer. Proc. SPIE. **10024**, 100244G (2015)
96. Duda, R.O., Hart, P.E., Stork, D.G.: Pattern Classification. Wiley, New York (2001)
97. Hajian-Tilaki, K.: Receiver operating characteristic (ROC) curve analysis for medical diagnostic test evaluation. Caspian J. Intern. Med. **4**(2), 627–635 (2013)
98. Brenner, M., Hearing, V.J.: The protective role of melanin against UV damage in human skin. Photochem. Photobiol. **84**(3), 539–549 (2008)
99. Eikje, N.S., Aizawa, K., Ozaki, Y.: Vibrational spectroscopy for molecular characterisation and diagnosis of benign, premalignant and malignant skin tumours. Biotechnol. Annu. Rev. **11**, 191–225 (2005)
100. Lihachev, A., Derjabo, A., Ferulova, I., et al.: Autofluorescence imaging of basal cell carcinoma by smartphone RGB camera. J. Biomed. Opt. **20**(12), 120502 (2015). https://doi.org/10.1117/1.JBO.20.12.120502
101. Lihacova, L., Bolocko, K., Lihachev, A.: Semi-automated non-invasive diagnostics method for melanoma differentiation from nevi and pigmented basal cell carcinomas. Proc. SPIE. **10592**, 1059206 (2017). https://doi.org/10.1117/12.2295773
102. Esteva, A., Kuprel, B., Novoa, R.A., Ko, J., Swetter, S.M., Blau, H.M., Thrun, S.: Dermatologist-level classification of skin cancer with deep neural networks. Nature. **542**, 115–118 (2017). https://doi.org/10.1038/nature21056
103. Bratchenko, I.A., Artemyev, D.N., Myakinin, O.O., Khristoforova, Y.A., et al.: Combined Raman and autofluorescence *ex vivo* diagnostics of skin cancer in near-infrared and visible regions. J. Biomed. Opt. **22**(2), 027005 (2017). https://doi.org/10.1117/1.JBO.22.2.027005

Chapter 15
Multimodal Imaging at Depth Using Innovations in Raman Spectroscopy and Optical Coherence Tomography

Mingzhou Chen and Kishan Dholakia

15.1 Introduction

The interest and scope for application of optical imaging in early diagnosis of disease have gained exceptional attention. Optical approaches may offer a highly specific and sensitive approach potentially coupled with an inexpensive implementation. A number of varied optical technologies are at our disposal for use. The choice of method is dictated by a number of factors, primarily by the type of information required, resolution required, and depth information needed. Microscopy in the form of confocal or multiphoton imaging in an epi-geometry can be very useful for fluorescent samples. Light sheet microscopy avoids the epi-geometry and instead uses orthogonal illumination and detection paths to deliver wide field, rapid imaging of fluorescent samples at high speed, with low photodamage as this particular geometry avoids unnecessary irradiation of parts of the sample not under investigation. Such geometries can be used with non-fluorescent samples, for example, capturing autofluorescence or molecular signatures. In addition other modalities may be needed if we wish to capture morphological information on the sample. This chapter will focus on combinations of Raman spectroscopy and optical coherence tomography for generating label-free three-dimensional images.

M. Chen (✉) · K. Dholakia
SUPA, School of Physics and Astronomy, University of St Andrew, St Andrews, Great Britain
e-mail: mingzhou.chen@st-andrews.ac.uk; kd1@st-andrews.ac.uk

V. V. Tuchin et al. (eds.), *Multimodal Optical Diagnostics of Cancer*,
https://doi.org/10.1007/978-3-030-44594-2_15

15.2 Basic Principles of Raman Spectroscopy and Optical Coherence Tomography

The light–matter interaction is at the heart of understanding Raman spectroscopy. This well-known and powerful method dates back to around 1928 with Raman receiving the Nobel Prize for his discovery in 1930. The Raman signature refers to the inelastic scattering component when monochromatic light is incident upon a sample. The scattering of light may occur due to molecular modes primarily associated with vibration and rotation. Specifically, at a sub-atomic scale, Raman scattering is based upon the electron cloud from a given sample interacting with the external electric field of the incident optical field. This results in an induced dipole moment within the molecule that is correlated with its polarizability. In turn this can result in a re-emission of light with energy lower or higher than that associated with the incident field, as shown in Fig. 15.1. Typically we record the Stokes lines where the molecule of interest initially resides in its ground state and is left in an excited vibrational state by the scattering process, emitting a photon of lower energy than the incident field. This inelastic scattering process is very weak (typically one photon in a million or less is inelastically scattered). Molecules offer a very small Raman cross section (around 10^{-30} cm^2 per molecule). As a consequence,

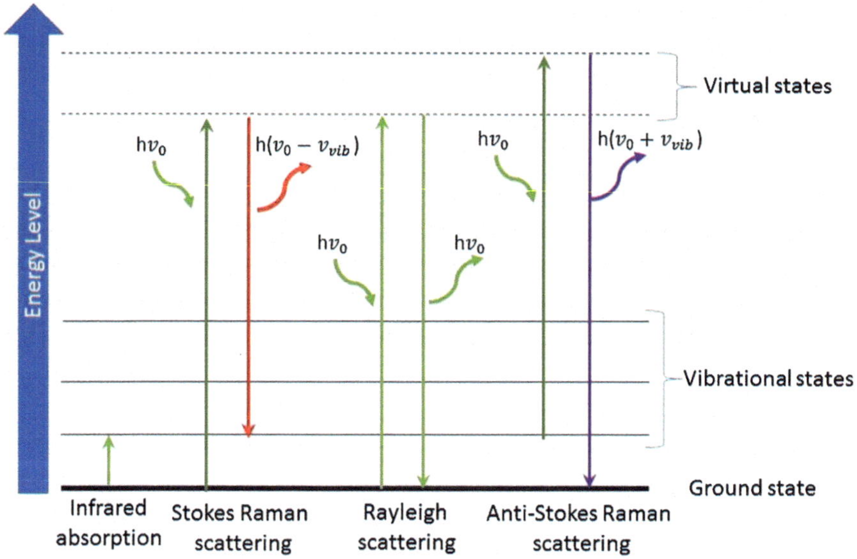

Fig. 15.1 The energy transitions for scattering in the light–matter interaction. In Rayleigh scattering, the molecule can relax back to the ground state and emit a photon with same energy as the incident photon. In Stokes Raman scattering, the molecule can relax to a vibrational state and emit a photon with lower energy than the incident photon. Alternatively in Anti-Stokes Raman scattering, a molecule, already in an excited vibrational state, can relax down to the ground state, emitting a photon with more energy than the incident photon

spontaneous Raman scattering requires a long acquisition time, which can be the order of minutes or hours especially when considering wide field of view recordings. This may be an issue for biomedical applications. Also, it is difficult to obtain signals from depth. A large body of Raman studies use the well-known therapeutic window for research where work is performed at near infra-red wavelengths: an excitation wavelength of 785 nm has proved very popular in this regard [1–4].

As an optical technology based on inelastic light scattering, Raman spectroscopy can analyze the molecular composition of tissue. Therefore, it provides a potential modality to identify the regions of the tumor *in vitro* or intraoperatively [5]. For example, a handheld probe for contact Raman spectroscopy identified gliomas in the brain [6]. Intraoperative use of the probe allowed the team to differentiate normal brain cells from dense cancer and normal brain invaded by cancer cells, with both a sensitivity and a specificity in excess of 90%. This Raman-based approach allowed brain cancer cells detection for patients with gliomas graded from 2 to 4. Furthermore, the same group progressed to show highly accurate detection of cancer *in situ* can be achieved by Raman spectroscopy combined with intrinsic fluorescence spectroscopy and diffuse reflectance spectroscopy [7]. More broadly in the cancer area, studies have been performed on cancers in the gastrointestinal tract [8], skin cancer [9], breast [10], and lung tissue [11]. In particular some good degrees of success in classifying normal versus malignant tissue have been seen using fiber probes that can integrate signals over larger areas versus acquisition of Raman signatures from single cells.

Recording high resolution label-free information on the morphology of biomedical samples is also invaluable to numerous areas of fundamental bioscience and early diagnosis of disease. Obviating the need for labeling means we require a source of natural contrast that is strong enough to allow us to distinguish boundaries between cell layers. Optical coherence tomography has emerged over the last 25 years as a powerful, rapid tool to address this very goal [12]. It relies on backscattered light from cell layers where the scattering results from subtle but key differences in refractive index between these layers, leading to intrinsic contrast and thus morphological information. By gating the scattered light through interference with a reference signal, an image can be acquired to show contrast and image the microstructures in the sample.

In the time domain OCT embodies the principle of a Michelson interferometer. The motion of a mirror in a reference arm results in interference when the path length matches that of the sample arm which is reflected from tissue for the case of a very low coherence source (e.g. a superluminescent diode). In this situation excellent discrimination of a given tissue plane is achieved versus the background. By translating the reference mirror a three-dimensional image of the tissue can be captured. Time domain OCT is powerful and useful but is typically very slow.

A more rapid, practical alternative is to move from the time to the frequency domain with these physical principles. We have two forms of OCT in the Fourier domain. In spectral domain OCT, a broadband light source illuminates a sample and we use a grating for dispersion to reflect the whole backscattered signal onto a spectrometer. Swept source OCT goes further by using a very fast tunable light source and appropriate detector to capture the whole OCT signal [13, 14].

OCT offers some key advantages which explains its rapid rise to prominence in the last few decades. Firstly the optical fluence level for image acquisition is low so OCT causes very little photodamage to the issue, making it useful for analyses of organs such as the eye [15–17]. Further, OCT may be free space, but an exceptional effort has resulted in a suite of fiber-based probes for implementing this imaging mode. These have allowed access to organs within the body such as the gastrointestinal (GI) tract [18–20] or arteries near the heart to explore cardiovascular disease [21]. OCT can achieve high sensitivity and specificity in distinguishing between cancer core and normal white matter in brain tissue based on optical attenuation values. In this way, it provides direct visual cues for cancer versus noncancer areas for brain cancer detection [22]. Comparable with histopathology, high resolution, cross-sectional OCT images of tissue provide information about depth of tumor growth and therefore enhance the efficiency and sensitivity of early cancer diagnosis [23]. Though OCT in its simplest form can yield morphological images and depth-resolved structural images for tissue, more sophisticated implementations can deliver more information. This may be in the form of motional information of blood flow (e.g., Doppler OCT) [24, 25], more detail on the structural composition of tissue (birefringence OCT) [26, 27], and details of the distribution of specific contrast agents (molecular contrast OCT) [28–30].

15.3 Combination of Raman Spectroscopy and OCT for Label-Free Multimodal Imaging

Raman spectroscopy is a label-free imaging method that can reveal molecular markers of disease with high specificity. OCT is also label-free and offers morphological data, and can unravel the microstructures of tissue with high sensitivity. A key challenge is to acquire both this molecular and morphological information simultaneously from the specimen under investigation in order to gain complementary information and ultimately analyze the sample accurately [31–34]. Previous studies have indicated that obtaining information from both Raman spectroscopy and OCT in tandem may be used as an effective multimodal tool for an efficient non-invasive optical biopsy in disease diagnosis [3, 35].

OCT can penetrate several millimeters deep into the samples while scanning over a large transverse area (normally several mm^2). However, conventional Raman spectroscopy can only deliver superficial information (100 μm) as Raman signals from the deeper layers are very weak and photons experience stronger diffusion to and from the deeper areas. This discrepancy in penetration depth between OCT and Raman spectroscopy prohibits a useful combination of these two modalities to obtain both OCT and Raman images at large depths and co-register data from images. Spatially offset Raman spectroscopy (SORS) is one possible way to increase the Raman penetration depth into a sample by simply shifting the collection point away from the excitation point [36–38]. With a careful consideration of the geometry for both OCT and SORS, a combination of these two modalities can be fulfilled and ultimately lead to an enhancement of sensitivity and specificity for identification [39–41].

15.4 Imaging at Depth with Wavelength Modulated SORS and OCT with Co-Registered Information

Here, we describe a combination of wavelength modulated spatially offset Raman spectroscopy (WM-SORS) and optical coherence tomography (OCT) that overcomes some key roadblocks. Then we show how these two modalities can be integrated in one hybrid system. The performance of such a hybrid system is then characterized by using a phantom.

15.4.1 Raman at Depth with WM-SORS

In a conventional confocal Raman system, the excitation laser beam is focused into the turbid sample and scattered Raman photons will be collected by the same optics. Therefore, collection is at the same point of excitation, which maintains highest efficiency of laser excitation and collection. However, the main drawback of this geometry is the Raman signals from depth decay exponentially due to the scattering and attenuation. Therefore, it can only acquire superficial Raman signals a few hundreds micrometers of the surface of sample [42]. In order to recover biological Raman signals from depth, the collection point is deliberately shifted away from the laser point for a small offset, *s*, in standard SORS, as shown in Fig. 15.2. Usually

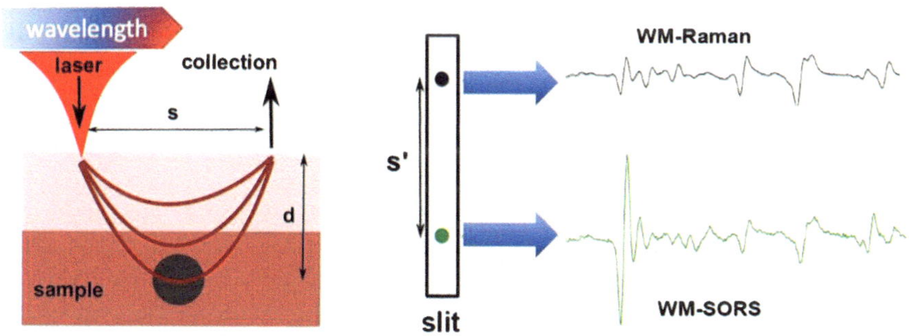

Fig. 15.2 Principle of wavelength modulated spatially offset Raman spectroscopy. Left shows the excitation laser (which can be wavelength-tuned over 1.5 nm range with central wavelength at 785 nm) along with the turbid sample, the target deep in the sample and the spatially offset collection point. Right shows the images of laser focal point and spatially offset collection point at the slit of spectrometer, and their corresponding wavelength modulated Raman/SORS spectra using a grating. Figure was adapted with permission from Chen et al. [43]

different spatial offsets can be achieved by using movable collection optics, such as a piece of fiber or a fiber bundle. However, the relationship between Raman signals from the target at different depths d and the spatially offset s needs to be calibrated for different samples.

In WM-SORS, there is no need to use any movable optics. The optical setup is the same as a standard confocal Raman system [44] except the collection fiber is replaced by a telescope in order to image the laser spot and the collection point onto the slit of the spectrometer. The light at each point along the slit will be dispersed to form a spectrum by the same single grating in the spectrometer. The standard Raman spectrum can be acquired from the rows of binned pixels corresponding to the image of laser spot, while other rows of binned pixels on the CCD with a spatial offset s_0 away from the image of laser spot form a SOR spectrum. The physical spatial offsets s on the sample plane can be calculated from the s_0 on the slit with the known magnification factor (2.78 in our setup) of the imaging system between the sample plane and the slit. Ideally, each row of pixels corresponds to a Raman spectrum with a spatial offset s. In this way, by simply extracting the spectra from different rows of pixels, we can make full use of the two dimensions of the CCD camera to acquire both standard Raman spectra and SOR spectra with different offsets, with one single exposure. By fully using the rows of CCD pixels, one standard Raman spectrum and eight SOR spectra at different offsets (maximum 0.7 mm) can be extracted from one single CCD frame.

Similar to our previous wavelength modulated Raman spectroscopy (WMRS) work [43–48], five spectra need to be acquired when the laser wavelength is tuning over 1 nm with a step of 0.2 nm. These can be reconstructed into one single WMR spectrum using principal component analysis (PCA). Therefore, one WMR spectrum and eight WM-SORS spectra can be calculated simultaneously from the acquired five frames of the CCD camera of the spectrometer using the analyzing method described above. We will use the WM-SORS spectra for imaging because the background fluorescence can be suppressed [45] while signals from depth are more pronounced than for standard SOR spectra.

15.4.2 Morphological Information at Depth: OCT

Optical coherence tomography (OCT) is a powerful way to acquire morphological information from sample at millimeter depth levels. Our approach uses the standard Fourier domain OCT based on a Michelson interferometer with which it is easy to couple a Raman excitation laser onto the sample as the simple geometry used. In practice, one can use any other format of OCT with sufficient accessibility for the Raman system. Sharing the optical axis is the key factor when one design the multimodal system as images from both modalities need to be co-localized for final analysis.

15.4.3 Multimodal Optical System with WM-SORS and OCT Co-Registered at Depth

In order to obtain both morphological and molecular imaging at depth, we integrate OCT and WM-SORS into a hybrid system as shown in Fig. 15.3. In the modality of WM-SORS, a tunable 785 nm laser source (Spectra-Physics, 1 W, Ti:Sapphire 3900s) is used as an excitation laser source for WM-SORS with a wavelength tuning range of 1 nm, which is very similar with the one we used for our previous WMRS system [44, 46, 48]. The collection path is in free space as described in the Sect. 15.4.1. In the modality of OCT, a broadband SLD (Superlum, S850, central wavelength 850 nm, $\Delta\lambda = 30$ nm) is used as an OCT light source and it works based on a Michelson interferometer as a Fourier domain OCT. In this setup, a low NA objective (Thorlabs, LSM02-BB) and a Galvo mirror are used for both OCT and WM-SORS. To avoid any signal loss, a flip mirror (F_1) is used to switch between OCT mode and WM-SORS mode.

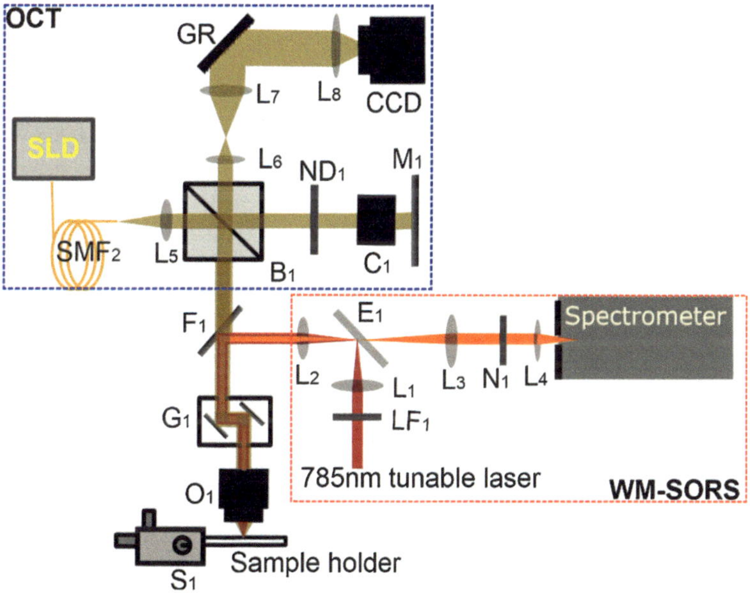

Fig. 15.3 Multimodal optical hybrid system with WM-SORS and OCT. Main parts of these two modalities, OCT and WM-SORS, are represented by the blue and red dash-lined rectangles, respectively. Both OCT and WM-SORS share the remaining optics, Galvo mirros G1 and objective O1 before the sample holder. F1 is a flip mirror to switch between OCT and WM-SORS. A superluminescent diode (SLD) laser is used for OCT illumination (Superlum, Cork, Ireland, S850). S1 is a X-Y-Z stage to hold the sample. SMF2 is a single mode fiber. L1 to L8 are lenses. E1 is an edge filter. LF1 is a laser line filter. N1 is a notch filter. GR is the grating. ND1 is a neutral density filter B1 is a 50:50 beam splitter. C1 is a dispersion compensating block. M1 is a reference mirror. A line CCD camera (AViiVA EM1) and a spectrometer (Andor SR303i)) are used. Figure was adapted with permission from Chen et al. [43]

Normally the relationship between the spatial offset *s* and the depth information *d* for different samples must be calibrated due to the unknown optical refractive index of the sample. With the common optical path used in our system, WM-SOR spectra or Raman images from hidden objects deep in the sample can be acquired with the depth information *d* co-registered using OCT images. In this way, the complicated calibration procedure of *s* and *d* can be avoided, although a low NA objective for WM-SORS may reduce the performance of Raman spectroscopy.

15.4.4 Characterization of Optical System

In order to characterize the WM-SORS, layered samples can be used with the layer thickness being measured by OCT. As an example, a block of lard is used as a phantom with a piece of polystyrene sheet (about 1.5 mm thickness) being inserted inside diagonally. The excitation laser is focused on the surface of the lard. Due to the depth in the lard and the high scattering property of lard, the standard WM Raman signals from polystyrene dramatically decrease when the polystyrene sheet penetrates deeper into the lard with respect to the laser focus at the surface. This is revealed by an exponential decay in the intensity of the typical polystyrene Raman peak at 1001.4 cm^{-1}. When a spatial offset is used in WM-SORS, the Raman peak intensity ratio between the polystyrene (1001.4 cm^{-1}) and the top layer of lard (1430 cm^{-1}) increases more rapidly, indicating that an enhancement of Raman signal from hidden polystyrene can be achieved. As shown in Fig. 15.4, the enhancement is related to the depth of the polystyrene in the lard and the offset used if the properties of the top layer are fixed. Here, a maximum 14-fold enhancement can be achieved in this sample. Note that the enhancement is related to the scattering property of the sample itself. Therefore, different samples may need to use a different optimized offset in order to extract information from a specific depth.

15.5 Multimodal Deep Imaging with WM-SORS and OCT

15.5.1 Imaging Pharmaceutical Particles Deep Through Lard

By discriminating the slightly different refractive index of the targets and surrounding medium, OCT can easily locate the position of targets and show the shapes, volumes or margins. However, this is not enough to tell the differences between the targets or know the compositions of the targets. Then the biochemical information acquired from WM-SORS can provide a strong ability to detect the difference in these targets. Here we show an example of how this hybrid system can locate and identify different pharmaceutical particles hidden deep in the lard block.

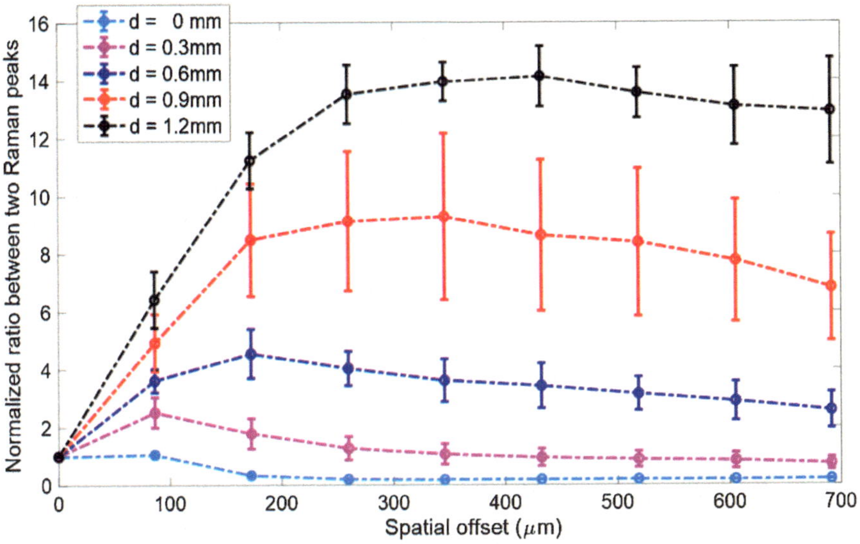

Fig. 15.4 Normalized ratios between the polystyrene Raman peak at 1001.4 cm^{-1} and the lard Raman peak at 1430 cm^{-1} as a function of spatial offset s. The thickness d of lard on top of polystyrene is measured from the co-registered OCT images. Standard deviations were obtained from six measurements on the samples for each condition. Figure was adapted with permission from Chen et al. [43]

OCT slice images at different depths are shown in Fig. 15.5 (a–d), where one can find that there are two triangular particles located around 1 mm deep in the lard block. However, it is difficult to discriminate which pharmaceutical particle we are observing. With OCT images only, it is impossible to identify which is aspirin and which is ibuprofen.

In order to identify these different compounds, one can use several characteristic Raman peaks, for example, lines at 840 cm^{-1} of ibuprofen, 1188 cm^{-1} of aspirin, and 1430 cm^{-1} of lard, which can be used for identifying these three compounds clearly. The reconstructed Raman images from these characteristic Raman peaks using different spatial offsets are shown in Fig. 15.5 (e–p). As both OCT and WM-SORS share the same optical axis, these 42 × 42-pixel Raman images can be co-registered with the OCT images. With non-offset $s = 0$ μm used, the Raman signals are dominated by the top lard layer except for the big and shallowly-buried aspirin particle (Fig. 15.5 (e)) in the lard. However, the Raman signals from the smaller ibuprofen particle which is deeper in the lard are completely in the noise level (Fig. 15.5 (i)). Stronger Raman signals can be obtained using spatially offsets s, which can be revealed in higher image contrast in Fig. 15.5 (f–h). WM-SORS images in Fig. 15.5 (j–l)) also show the signals from ibuprofen particles. Due to distinct Raman peaks for aspirin and ibuprofen used, there is no cross talk in WM-SORS images for both compounds. Therefore, these two different compounds can be easily discriminated. Furthermore, multi-spectral images can provide plenty of information. As shown in

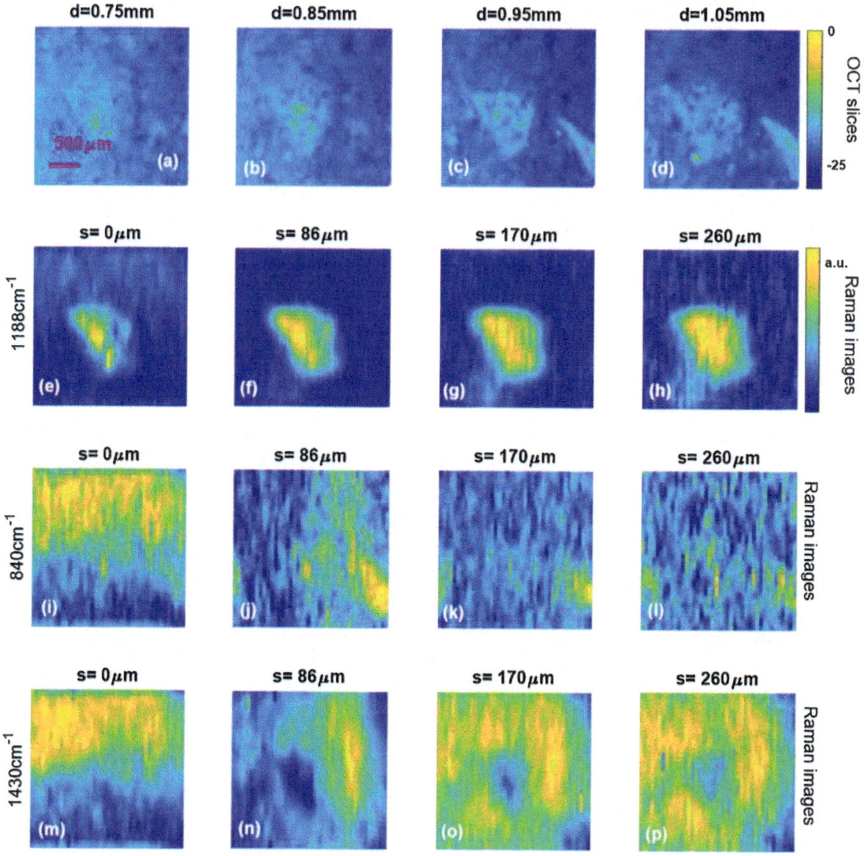

Fig. 15.5 Imaging through lard for pharmaceutical particles using WM-SORS and OCT. The top row (**a–d**) shows the OCT slices at a different depth of d. Aspirin Raman images (**e**–**h**), ibuprofen Raman images (**i**–**l**), and lard Raman images(**m**–**p**), using different spatially offsets s, are reconstructed by using Raman peaks at 1188 cm^{-1} of aspirin, 840 cm^{-1} of ibuprofen, and 1430 cm^{-1} of lard, respectively. Color legends indicate the intensity in dB for OCT images and Raman peak intensity in arbitrary units for Raman images. Scale bar indicates 500 μm. Figure was adapted with permission from Chen et al. [43]

reconstructed lard WM-SORS images in Fig. 15.5 (n–p), the shadows indicate the location of aspirin particle as a lack of lard.

15.5.2 Imaging Layered Rat Brain Tissue

In most tumor detection applications, one can do diagnosis based on the changes of optical properties between tumor and normal tissue. However, with this one piece of information, the sensitivity and specificity of detection will be much lower [3].

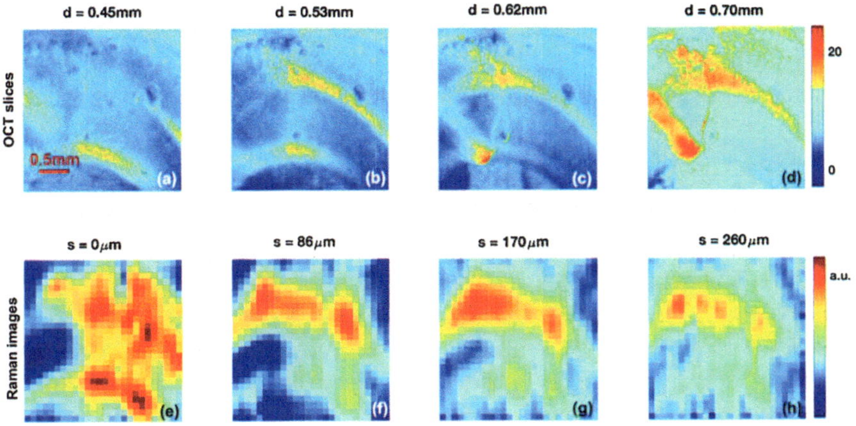

Fig. 15.6 Brain tissue imaging using WM-SORS and OCT. (**a**)–(**d**) show OCT images of brain tissue at different depths. Note that OCT images are averaged over 10 slices around the depth in order to increase the contrast. (**e**)–(**h**) show Raman images (at 1430 cm^{-1}) of the same brain tissue at different offsets. Color legends indicate the intensity in dB for OCT images and Raman peak intensity in arbitrary units for Raman images. Figure was adapted with permission from Chen et al. [43]

Specific Raman signals from a tumor would improve the detection ability, but is only suitable for the tumor at the tissue surface. Therefore, WM-SORS and OCT may provide a multimodal imaging method to detect the hidden tumor deep in the tissue more accurately.

Here we only show the ability of this hybrid system by detecting the white matter in the rat brain tissue. The white matter in the rat brain tissue has slightly different optical properties comparing with the surrounding tissue and therefore its OCT intensity is higher. In Fig. 15.6 (a–d), OCT images can reveal these white tissues which are located at different depths. However, there are some artifacts shown in Fig. 15.6 (c, d) which may be incorrectly identified as white matter in the analysis step.

The white matter in the rat brain tissue actually have a very distinct Raman peak around 1430 cm^{-1}. Using the Raman intensity of this characteristic peak, Raman images can be reconstructed as shown in Fig. 15.6 (e–h). As shown in Fig. 15.6 (e), without any spatial offset ($s = 0$ μm), the Raman image is blur due to the mixture of contributions from white matter at the surface and the depth. However, in Fig. 15.6 (f–h), the image contrast is substantially enhanced because Raman signals from deep white matter are enhanced due to the spatial offsets used. As the specific biomedical information is acquired from the white matter only, the artifacts in the OCT images, as shown in Fig. 15.6 (c, d), do not exist in these WM-SORS images. This further improve the detection ability of this novel hybrid system.

15.6 Conclusion

Multimodal imaging by acquiring both morphological and molecular information in deep tissue or through turbid media would greatly advance the diagnosis of early stage diseases, such as tumor detection. Our hybrid WM-SORS/OCT system can acquire both WMR spectra and WM-SOR spectra at different spatial offsets simultaneously on one single CCD camera shot without using any movable optics and additionally removes fluorescence. With the co-alignment of optical axis in the OCT and WM-SORS, our system can probe 1.2 mm deep into strong scattering media (lard) to acquire molecular information from hidden targets. Therefore, it provides a deep tissue characterization technique acquiring molecular vibrational spectra co-located with microstructure information, with potential increases in sensitivity and specificity for cancer detection. Raman spectroscopy highlights differences in signals from DNA, proteins, amides, and lipids among other molecules. Such features are associated with neoplastic progression with genomic and metabolic changes with the onset of cancer. OCT can highlight morphological differences in cancer and OCT can provide information on the cell nucleus, i.e. the size and shape, the ratio of the nucleus to cytoplasm, and information on the organization of structures such as glands. Furthermore, it also offers aspects of functional data such as blood flow or tissue birefringence. The combination of Raman and OCT, particularly recording co-localized three-dimensional information as we have described, could lead to a new form of diagnostic or screening tool for cancer studies. Indeed such a combination in a fiber implementation may be of relevance to guide interventional therapy and record responses in a clinical setting. We contend this is an area of great future promise for cancer research.

Acknowledgments The authors thank the UK Engineering and Physical Sciences Research Council (EPSRC: EP/P030017/1 and EP/M000869/1) for funding. We also thank Graham Bruce for useful comments on the manuscript.

References

1. Krafft, C., Knetschke, T., Siegner, A., Funk, R.H., Salzer, R.: Vib. Spectrosc. **32**(1), 75 (2003)
2. Petry, R., Schmitt, M., Popp, J.: ChemPhysChem. **4**(1), 14 (2003)
3. Ashok, P.C., Praveen, B.B., Bellini, N., Riches, A., Dholakia, K., Herrington, C.S.: Biomed. Opt. Express. **4**(10), 2179 (2013)
4. Canetta, E., Riches, A., Borger, E., Herrington, S., Dholakia, K., Adya, A.K.: Acta Biomater. **10**(5), 2043 (2014)
5. Jermyn, M., Mok, K., Mercier, J., Desroches, J., Pichette, J., Saint-Arnaud, K., Bernstein, L., Guiot, M.C., Petrecca, K., Leblond, F.: Sci. Transl. Med. **7**(274), 274ra19 (2015)
6. Kast, R., Auner, G., Yurgelevic, S., Broadbent, B., Raghunathan, A., Poisson, L.M., Mikkelsen, T., Rosenblum, M.L., Kalkanis, S.N.: J. Neuro-Oncol. **125**(2), 287 (2015)
7. Jermyn, M., Mercier, J., Aubertin, K., Desroches, J., Urmey, K., Karamchandiani, J., Marple, E., Guiot, M.C., Leblond, F., Petrecca, K.: Cancer Res. **77**(14), 3942 (2017)

8. Hsu, C.W., Huang, C.C., Sheu, J.H., Lin, C.W., Lin, L.F., Jin, J.S., Chau, L.K., Chen, W.: PLoS One. **11**(7), e0159829 (2016)
9. Lui, H., Zhao, J., McLean, D., Zeng, H.: Cancer Res. **72**(10), 2491 (2012)
10. Haka, A.S., Shafer-Peltier, K.E., Fitzmaurice, M., Crowe, J., Dasari, R.R., Feld, M.S.: Cancer Res. **62**(18), 5375 (2002)
11. Jess, P.R.T., Mazilu, M., Dholakia, K., Riches, A.C., Herrington, C.S.: Int. J. Cancer. **124**(2), 376 (2009)
12. Fujimoto, J.G., Brezinski, M.E., Tearney, G.J., Boppart, S.A., Bouma, B., Hee, M.R., Southern, J.F., Swanson, E.A.: Nat. Med. **1**(9), 970 (1995)
13. Chinn, S.R., Swanson, E.A., Fujimoto, J.G.: Opt. Lett. **22**(5), 340 (1997)
14. Choma, M., Sarunic, M., Yang, C., Izatt, J.: Opt. Express. **11**(18), 2183 (2003)
15. Drexler, W., Morgner, U., Ghanta, R.K., Kartner, F.X., Schuman, J.S., Fujimoto, J.G.: Nat. Med. **7**(4), 502 (2001)
16. Wojtkowski, M., Leitgeb, R., Kowalczyk, A., Bajraszewski, T., Fercher, A.F.: J. Biomed. Opt. **7**(3), 457 (2002)
17. Fujimoto, J.G.: Nat. Biotechnol. **21**(11), 1361 (2003)
18. Sivak, M.V., Kobayashi, K., Izatt, J.A., Rollins, A.M., Ung-runyawee, R., Chak, A., Wong, R.C., Isenberg, G.A., Willis, J.: Gastrointest. Endosc. **51**(4), 474 (2000)
19. Shen, B., Zuccaro, G.: Endoscopy. **32**(10), 796 (2000)
20. Tsai, T.H., Fujimoto, J., Mashimo, H.: Diagnostics. **4**(2), 57 (2014)
21. Liu, L., Gardecki, J.A., Nadkarni, S.K., Toussaint, J.D., Yagi, Y., Bouma, B.E., Tearney, G.J.: Nat. Med. **17**(8), 1010 (2011)
22. Kut, C., Chaichana, K.L., Xi, J., Raza, S.M., Ye, X., McVeigh, E.R., Rodriguez, F.J., Quinones-Hinojosa, A., Li, X.: Sci. Transl. Med. **7**(292), 292ra100 (2015)
23. Karl, A., Stepp, H., Willmann, E., Buchner, A., Hocaoglu, Y., Stief, C., Tritschler, S.: Eur. J. Med. Res. **15**(3), 131 (2010)
24. Barton, J.K., Stromski, S.: Opt. Express. **13**(14), 5234 (2005)
25. Vakoc, B.J., Yun, S.H., de Boer, J.F., Tearney, G.J., Bouma, B.E.: Opt. Express. **13**(14), 5483 (2005)
26. Everett, M.J., Schoenenberger, K., Colston, B.W., Da Silva, L.B.: Opt. Lett. **23**(3), 228 (1998)
27. De Boer, J., Srinivas, S., Malekafzali, A., Chen, Z., Nelson, J.: Opt. Express. **3**(6), 212 (1998)
28. Rao, K.D., Choma, M.A., Yazdanfar, S., Rollins, A.M., Izatt, J.A.: Opt. Lett. **28**(5), 340 (2003)
29. John, R., Rezaeipoor, R., Adie, S.G., Chaney, E.J., Oldenburg, A.L., Marjanovic, M., Haldar, J.P., Sutton, B.P., Boppart, S.A.: Proc. Natl. Acad. Sci. **107**(18), 8085 (2010)
30. de la Zerda, A., Prabhulkar, S., Perez, V.L., Ruggeri, M., Paranjape, A.S., Habte, F., Gambhir, S.S., Awdeh, R.M.: Clin. Exp. Ophthalmol. **43**(4), 358 (2015)
31. Sudheendran, N., Qi, J., Young, E.D., Lazar, A.J.: Laser Phys. Lett. **11**(10), 105602 (2014)
32. Patil, C.A., Kalkman, J., Faber, D.J., Nyman, J.S., van Leeuwen, T.G., Mahadevan-Jansen, A.: J. Biomed. Opt. **16**(1), 011007 (2011)
33. Liu, C.H., Qi, J., Lu, J., Wang, S., Wu, C., Shih, W.C., Larin, K.V.: J. Innov. Opt. Health Sci. **08**(04), 1550006 (2015)
34. Gupta, R.K., Chen, M., Malcolm, G.P.A., Hempler, N., Dholakia, K., Powis, S.J.: Opt. Express. **27**(10), 13706 (2019)
35. Patil, C.A., Bosschaart, N., Keller, M.D., van Leeuwen, T.G., Mahadevan-Jansen, A.: Opt. Lett. **33**(10), 1135 (2008)
36. Matousek, P., Morris, M.D., Everall, N., Clark, I.P., Towrie, M., Draper, E., Goodship, A., Parker, A.W.: Appl. Spectrosc. **59**(12), 1485 (2005)
37. Keller, M.D., Majumder, S.K., Mahadevan-jansen, A.: Opt. Lett. **34**(7), 926 (2009)
38. Keller, M.D., Vargis, E., de Matos Granja, N., Wilson, R.H., Mycek, M., Kelley, M.C., Mahadevan-Jansen, A.: J. Biomed. Opt. **16**(7), 077006 (2011)
39. Khan, K.M., Krishna, H., Majumder, S.K., Rao, K.D., Gupta, P.K.: J. Biophotonics. **7**(1–2), 77 (2014)

40. Maher, J.R., Chuchuen, O., Henderson, M.H., Kim, S., Rinehart, M.T., Kashuba, A.D.M., Wax, A., Katz, D.F.: Biomed. Opt. Express. **6**(6), 2022 (2015)
41. Wang, J., Zheng, W., Lin, K., Huang, Z.: Opt. Lett. **41**(13), 3045 (2016)
42. Stone, N., Baker, R., Rogers, K., Parker, A.W., Matousek, P.: Analyst. **132**(9), 899 (2007)
43. Chen, M., Mas, J., Forbes, L.H., Andrews, M.R., Dholakia, K.: J. Biophotonics. **11**(1), e201700129 (2018)
44. Chen, M., McReynolds, N., Campbell, E.C., Mazilu, M., Barbosa, J., Dholakia, K., Powis, S.J.: PLoS One. **10**(5), e0125158 (2015)
45. De Luca, A.C., Mazilu, M., Riches, A., Herrington, C.S., Dholakia, K.: Anal. Chem. **82**(2), 738 (2010)./>
46. McReynolds, N., Cooke, F.G.M., Chen, M., Powis, S.J., Dholakia, K.: Sci. Rep. **7**(1), 43631 (2017)
47. Baron, V.O., Chen, M., Clark, S.O., Williams, A., Hammond, R.J.H., Dholakia, K., Gillespie, S.H.: Sci. Rep. **7**(1), 9844 (2017)
48. Woolford, L., Chen, M., Dholakia, K., Herrington, C.S.: J. Biophotonics. **11**(4), e201700244 (2018)

Chapter 16
Terahertz Spectroscopy and Imaging of Brain Tumors

Kirill I. Zaytsev, Irina N. Dolganova, Valery E. Karasik, Vladimir N. Kurlov, Igor V. Reshetov, Valery V. Tuchin, Sheyh-Islyam T. Beshplav, and Alexander A. Potapov

16.1 A Problem of Intraoperative Diagnosis of Human Brain Tumors

Intraoperative diagnosis of human brain tumors remains an important problem of medicine, applied physics, and engineering sciences [1]. Among all tumors of the brain, gliomas form the most common class of primary tumors and represent about

K. I. Zaytsev (✉)
Prokhorov General Physics Institute of the Russian Academy of Sciences, Moscow, Russian Federation

Bauman Moscow State Technical University, Moscow, Russian Federation

I. N. Dolganova
Bauman Moscow State Technical University, Moscow, Russian Federation

Institute of Solid State Physics of the Russian Academy of Sciences, Moscow, Russian Federation

V. E. Karasik
Bauman Moscow State Technical University, Moscow, Russian Federation

V. N. Kurlov
Institute of Solid State Physics of the Russian Academy of Sciences, Moscow, Russian Federation

I. V. Reshetov
First Moscow State Medical University (Sechenov University), Moscow, Russian Federation

V. V. Tuchin
Saratov State University, Saratov, Russian Federation

National Research Tomsk State University, Tomsk, Russian Federation

Institute of Precision Mechanics and Control of the Russian Academy of Sciences, Saratov, Russian Federation

S.-I. T. Beshplav · A. A. Potapov
Burdenko Neurosurgery Institute, Moscow, Russian Federation

V. V. Tuchin et al. (eds.), *Multimodal Optical Diagnostics of Cancer*,
https://doi.org/10.1007/978-3-030-44594-2_16

26% of all primary tumors and about 81% of malignant primary tumors [2]. Modern paradigms of their treatment rely on multidisciplinary approaches, merging the expertise of neurosurgeons, neurooncologists, radiation oncologists, neuropathologists, and radiologists and bringing the latest advantages in the area of tissue imaging and exposure to a clinical practice. Malignant gliomas, especially glioblastoma (or glioma of grade IV according to the World Health Organization (WHO) classification [3]) have a poor prognosis and a bad patients' outcome, even when applying an aggressive therapy [4, 5]. Among the prognostic factor of glioma treatment, its gross-total resection seems to be a crucial one for reduction of the tumor recurrence probability and improvement of the patients' survival [6]. At the same time, gliomas usually possess unclear margins, making achievement of their gross-total resection challenging, even for a highly experienced surgeon [7].

Nowadays, novel brain imaging modalities are rapidly developed or have been already applied in a clinical practice in order to mitigate the challenging problems posed by the intraoperative delineation of tumor margins [8, 9]. Among them, the intraoperative magnetic resonance imaging (MRI) [10, 11] demonstrates high efficiency, but remains time-consuming and very expensive, since it required adaptation of the neurosurgical workflow for work with high magnetic fields, including the use of non-magnetic materials [12–14]. In turn, fluorescence spectroscopy and imaging, relying either on 5-aminolevulinic acid (5-ALA)-induced fluorescence of protoporphyrin IX (ppIX) [15–17] or on fluorescein sodium [18], are widely employed in intraoperative diagnosis of human brain tumors, being rather convenient and inexpensive. They require administration of exogenous markers into the body and provide high sensitivity and specificity for high-grade gliomas, while their performance for low-grade gliomas is rather poor. Capabilities of such promising techniques, as Raman spectroscopy and imaging [19–22], photoacoustic imaging [22, 23], multiphoton microscopy, and optical coherence tomography (OCT) [9, 24–28], are vigorously explored. However, they are still considered as laboratory research tools, and it would take a certain time to transfer them to a neurosurgical practice. All the aforementioned justify an importance of further development of novel instruments for the intraoperative diagnosis of brain tumors.

16.2 Terahertz Biophotonics

During the past decades, terahertz (THz) spectroscopy and imaging have been abundantly developed with a rapid progress in femtosecond laser technologies and with an appearance of novel effective approaches for the THz-wave generation and detection [29]. THz spectroscopy exploit electromagnetic waves in the frequency range of 0.1–3.0 THz, or in the wavelength range of about 3 mm–100 μm (see Fig. 16.1). Since the end of the twentieth century, the THz technology has been considered as a promising instrument of medical diagnosis [30], [31], including the non-invasive, least-invasive, and intraoperative label-free diagnosis of malignances of the skin [32–37], mucosa [38, 39], colon [40–42], liver [43], gastric [44, 45], breast [46–48], etc. Most recently, a potential of THz instruments in the

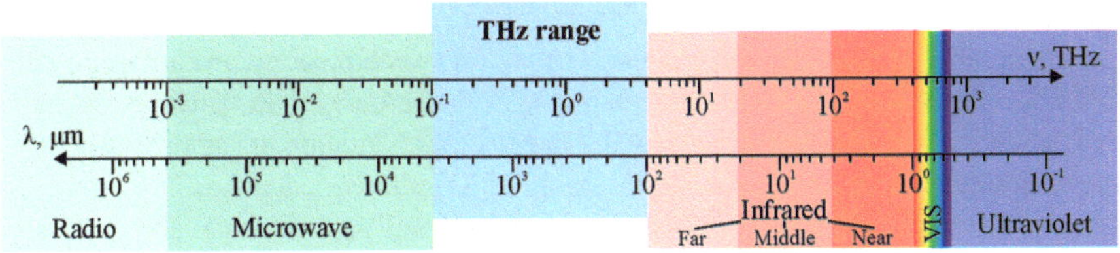

Fig. 16.1 THz range of electromagnetic spectrum. Courtesy of K.I. Zaytsev

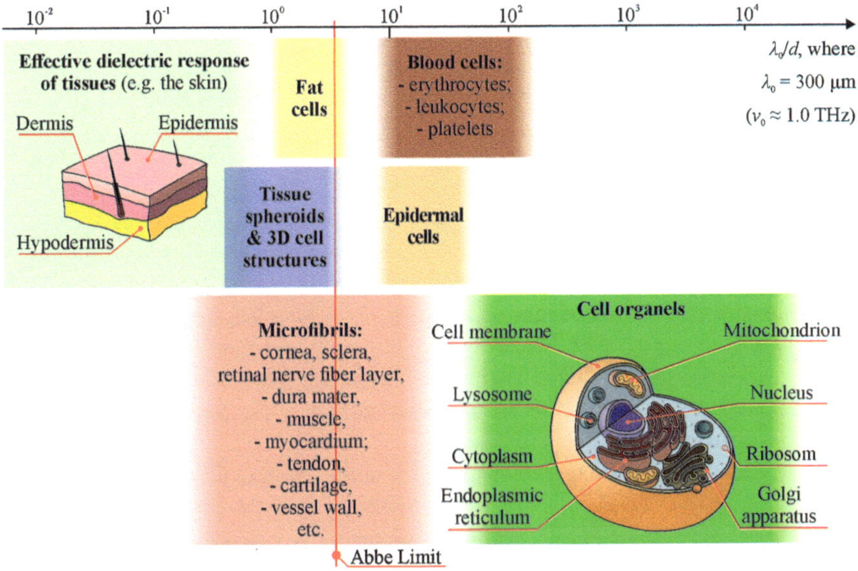

Fig. 16.2 A scheme illustrating structural inhomogeneities of tissues at the scale posed by THz wavelengths, where the dimensions of the considered tissues components are normalized by the THz wavelength of $\lambda_0 = 300$ μm, for convenience. Here, the $\lambda/2$ Abbe diffraction limit of the spatial resolution is shown by the vertical red line. Courtesy of K.I. Zaytsev

intraoperative neurodiagnosis has been highlighted with a strong emphasis to differentiation between intact (healthy) tissues and tumors of the brain [49]. Before proceeding to the in-depth analysis of modern results in THz diagnosis of human brain tumors, in this section, some general peculiarities of THz-wave–tissue interactions, as well as main principles of THz pulsed spectroscopy are briefly discussed.

16.2.1 *Terahertz-Wave–Tissue Interaction*

As shown in Fig. 16.2, dimensions of the structural elements in tissues (such as microfibrils, separate cells and their agglomerates, cell organelles, etc.) are usually negligibly small at the scale posed by THz wavelengths. On the one side, such small elements cannot be resolved using conventional THz imaging modalities, exploiting the diffraction-limited optical systems. Thus, THz spectroscopy and imaging usually allow for studying an effective response of tissues, which is spatially averaged within the area of THz beam spot [30]. On the other hand, small dimensions of tissue components lead to the Rayleigh regime of THz-wave scattering. Therefore, an assumption of homogeneous character of tissues usually takes place, and the effective medium theory is applied for describing the THz-wave–tissue interactions [30]. In this case, electrodynamic properties of tissues at THz frequencies are described completely by an effective complex dielectric permittivity

$$\tilde{\varepsilon} = \varepsilon' - i\varepsilon'', \tag{16.1}$$

with its real ε' and imaginary ε'' parts, or by a complex refractive index

$$\tilde{n} = n' - in'' \equiv n - i\frac{c}{2\pi\upsilon}\alpha \equiv \sqrt{\tilde{\varepsilon}}, \tag{16.2}$$

with its real $n \equiv n'$ and imaginary n'' parts, where $c = 3\times10^8$ m/s stands for the speed of light in a free space, υ is an electromagnetic wave frequency in [Hz], and α is an amplitude absorption coefficient in [cm^{-1}].

Dielectric response of both healthy and pathological tissues (*in vivo* or *ex vivo* freshly excised/non-dehydrated specimens) does not possess any sharp resonant features in the THz range and is primary determined by the content and state (free or bound) of water [30]. Similarly to the THz dielectric permittivity of water, the THz response of tissues has a relaxation dynamics and could be described by the Debye, Cole–Cole, Davidson-Cole, or Havriliak–Negami models [50–55]. All these models describe electro-dipole excitation of matter, for which the real part of a complex dielectric permittivity monotonically decays with frequency, while the imaginary part features a broad absorption band, centered at the inverse relaxation time [50, 51]. The double-Debye model (i.e., the one, comprised of the two Debye relaxation terms) is the most widely applied one for describing the THz dielectric properties of water, biological tissues and liquids [30, 54, 55]

$$\tilde{\varepsilon} = \varepsilon_\infty + \frac{\Delta\varepsilon_1}{1 + i2\pi\upsilon\tau_1} + \frac{\Delta\varepsilon_2}{1 + i2\pi\upsilon\tau_2}, \tag{16.3}$$

where ε_∞ stands for the dielectric permittivity at infinitely high frequencies (as compared to the considered spectral range); $\Delta\varepsilon_1$ and $\Delta\varepsilon_2$ regulate a contribution of the Debye kernels to the resultant dielectric permittivity; τ_1 and τ_2 are relaxation times in [ps], corresponding to the "fast" and "slow" relaxations [30]. The double-Debye model forms a conventional approach for parametrizing the THz response of tissues using only 5 independent coefficients—ε_∞, $\Delta\varepsilon_1$, $\Delta\varepsilon_2$, τ_1, τ_2. At the same time, one should notice that this model is not physically rigorous, since it provides a fit of the experimental data with two broad absorption band, the maxima of which are centered beyond (or at the edge of) the considered THz spectral range; they are located at the inverse relaxation times $\upsilon_1 = \tau_1^{-1} \sim 10^{-1}$ and $\upsilon_2 = \tau_2^{-1} \sim 10^{1}$ THz. In this case, the double-Debye model provides an extrapolation of the experimental data. Nevertheless, many papers report parameters of the double-Debye model for various tissues, relying the data of THz pulsed spectroscopy; for example, see Refs. [54–61].

As an evidence of strong impact of water on the THz response of tissues, one could refer to a difference between the THz optical properties of freshly excised fibrous corrective tissues and adipose tissues of the breast *ex vivo* [47]. The connective tissues possess much higher water content, as compared to that in fat cells; thus, their THz refractive index and absorption coefficient are much higher, being closer to that of water. High water content in tissues leads to strong THz-wave absorption and limits the depth of THz-wave penetration into tissues by hundreds, or even tens of microns, depending on the tissue type and the electromagnetic wave frequency. Therefore, only the reflection-mode measurements are reliable for the THz spectroscopy and imaging of both tissues *in vivo* and freshly excised tissues *ex vivo*, and only superficial properties of tissues could be probed by THz waves [62].

The label-free contrast between healthy and malignant tissues in the THz range reportedly originates from difference in water content and structural variation in tissues. As a result of abundant vascularity and edema, malignant tissues can content more water and possess higher THz refractive index and absorption coefficient [49, 63, 64]. Among other factors, microscopic variations of tissue properties (variational tissue microenvironments [65], deteriorative cellular morphology and necrotic debris [66], presence of mutative biomolecule [67]) and large-scale fluctuations of the volumetric and surface structure of tissues [36, 68] might either form an additional source of contrast or alter a quality of the detected THz signals, decreasing the data reproducibility.

Knowledge of the complex dielectric permittivity (Eq. 16.1) or the complex refractive index (Eq. 16.2) allows for complete description of the THz-wave–tissue interactions in the framework of classical electrodynamics; particularly, it yields:

- analytical description of the THz-wave propagation in layered media using plane-wave approximation along with the Fresnel formulas and the Bouguer–Lambert–Beer law; this is of crucial importance for spectroscopic applications [49];
- numerical simulation of the THz–wave interaction with biological objects of a complex shape using methods of computational electrodynamics [54, 69, 70] or statistical Monte Carlo approaches [68];
- differentiation between healthy and pathological tissues using either absolute values of dielectric constants [36, 49] or parameters of the double-Debye models of a complex dielectric permittivity, as it was demonstrated in Ref. [60].

Despite the attractiveness of the effective medium theory in the THz biophotonics, tissues might possess structural inhomogeneities, the dimensions of which are comparable to the THz wavelengths [69, 70]. For example, recent results of THz imaging of tissues using the sub-wavelength-resolution solid immersion microscopy revealed their mesoscale structural elements, such as separate fat cells and their agglomerates embedded into connective fibrous tissues of the breast, lactiferous ducts of the breast, muscle fibers of the tongue, etc. [71, 72, 73]. These elements might lead to the Mie scattering of THz waves in the tissues [74]. Furthermore, the polarization-sensitive THz imaging provide useful information for differentiation of healthy and cancerous tissues [35, 41], which cannot be predicted by the effective medium theory. It emphasizes the importance of taking into account the dispersion, absorption, and scattering properties of the tissues. The novel models could be constructed relying on the radiation transfer theory [74], which is widely applied in visible and infrared optics [75]. However, this demanding problem of THz biophotonics is still to be addressed.

16.2.2 Terahertz Pulsed Spectroscopy and Imaging

Majority of THz systems for biomedical applications relies on the principles of THz pulsed spectroscopy (TPS) and THz pulsed imaging; the latter implies point-by-point scanning of the sample surface with a focused beam of the TPS system, acquisition and processing of 3D data set [29, 62]. One of the main reasons for such an extensive use of TPS in biophotonics is associated with an ability for simultaneous detection of both frequency-dependent amplitude and phase of the THz waveform in a broad spectral range as a result of a single measurement. Furthermore, TPS yields analysis of a physical response of matter either in time domain (pulse response) or in frequency domain (complex dielectric permittivity or complex refractive index). The existing component base of TPS is well developed; it yields production of portable, handheld and ergonomic devices, capable for operation not only in laboratory conditions, but also in a clinical environment [76, 77]. Modern measurement procedures and methods for processing the data and solving the inverse ill-posed problems in TPS and related imaging modalities are well developed and yield accurate characterization of tissues [78–82].

TPS implies probing a sample with THz pulses of sub-picosecond duration, which feature a broad spectral operation range, typically spanning the frequencies between 0.1 and 4.5 THz [29]. This instrument appeared as a result of Auston's research on optoelectronic switching (photoconductivity) and gating in semiconductors, pumped by ultrashort optical laser pulses [83]. Since then, many approaches have been proposed for generation and coherent detection of THz pulsed using ultrashort optical excitation [29]; nevertheless, the THz photoconductive antennas (PCA) still represent the most prevalent type of THz pulsed emitter and detector in TPS systems due to simplicity, flexibility, and reliability of their design and technical characteristics [84]. Particularly, during the past few years, a dramatic enhancement in PCA performance was demonstrated relying on effects of electromagnetic field confinement behind plasmonic and dielectric nanostructures incorporated into a photoconductive gap [85, 86].

A number of TPS schemes are now available; but we would demonstrate its basic principle in a context of reflection-mode measurements of a tissue using the scheme presented in Fig. 16.3. In this system, the laser beam is divided by the beamsplitter in pump and probe beams, which proceed to PCA-emitter and PCA-detector of THz waves, respectively, with the difference in optical path, obtained by adding a time-delay in the probing path. THz detection is implemented by mixing of THz and laser pulses, commonly PCA are used for this purpose. The generated photocurrent in PCA, proportional to the THz electric field, provides the detected TDS signal. The changing temporal delay results in registered temporal waveform of THz response from the sample.

The reflection-mode scheme with an oblique THz-beam incidence angle and a reference quartz window, atop of which the object of interest is handled, is demonstrated in Fig. 16.4a [49]. A sample holder can be equipped with the possibility of lateral scanning, in order to record a 2D distribution of sample reflectivity, transmittance or dielectric properties. In this case, TDS becomes a powerful imaging tool, with significant capabilities in medical imaging [30]. As shown in Fig. 16.4b and c, the received THz waveforms, reflected from a sample, are comprised of a sequence of pulses, the interference of which leads to appearance of the Fabry–Perot resonances in the Fourier domain. Extraction of the sample optical properties is not a straightforward procedure, since one should perform apodization (window filtering) of the time-domain waveforms with considering possible multiple THz pulse reflections in a reference window, the uncertainties of sample thickness, illumination conditions and noises [30]. The mentioned errors and uncertainties reduce accuracy of

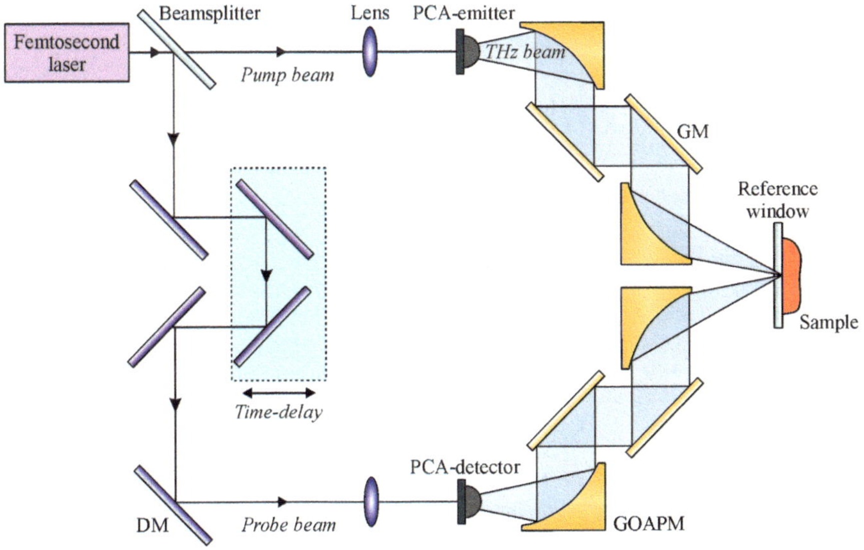

Fig. 16.3 A typical scheme of TPS based on THz-wave generation and detection in PCAs, where DM stands for the near-infrared dielectric mirror, GM stands for the THz gold-coated mirror, GOAPM stands for THz gold-coated off-axial parabolic mirror. Courtesy of I.N. Dolganova

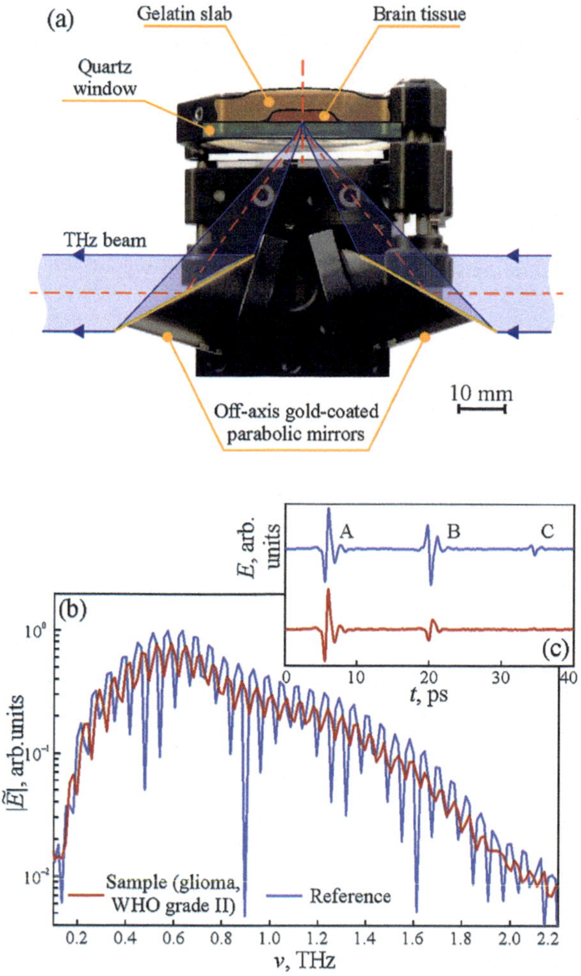

Fig. 16.4 Experimental characterization of biological tissues *ex vivo* using reflection-mode TPS measurements: (**a**) a unit for reflection-mode measurements based on a pair of GOAPMs and a reference quartz window; (**b**), (**c**) reference and sample waveforms, represented in Fourier domain and time domains, respectively. Here, the reference waveform corresponds to the THz wave reflection from the empty quartz window, while the sample waveform corresponds to the THz wave reflection from the quartz window with a tissue sample (human brain glioma) handled behind. Adapted from Ref. [49], published by SPIE under a Creative Commons (CC BY) license

measurements and restrict spectral operation range. For the considered geometry of measurements (see Figs. 16.3 and 16.4), the procedure of sample optical properties reconstruction is described in details in Ref. [49].

A typical spectral operation range corresponds to 0.1–4.5 THz for a common TPS system, while the frequency-domain dynamic range is usually 60 dB near the maximum of the TPS spectrum (0.5–1.0 THz). Meanwhile, unique TPS setups produce signal above 10 THz and even up to 200 THz, when extremely short laser pulses (less than 10 fs) are used for the THz-wave generation in air plasma; but there is still a problem of obtaining smooth and stable spectrum over 10 THz [87–89]. At the same time, majority of THz measurements of tissues involving TPS are performed in a low-frequency part of the THz range (below 1.5–2.0 THz), due to high signal-to-noise ratio and low impact of water vapors along the THz-beam path on the measured data.

16.3 Terahertz Spectroscopy and Imaging of Brain Tumors

An initial study of snap-frozen human brain tissues *ex vivo* using TPS system in conjunction with closed-cycle cryostat was performed in [90], where the authors demonstrated significant contrast between healthy and diseased tissues, which were neuropatologically classified as containing abnormally high number of protein plagues, inherent to the Alzheimer's disease. The contrast between healthy and abnormal tissues was observed for all samples taken from three distinct regions of cerebral cortex: superior frontal gyrus, inferior frontal gyrus, and cingulate gyrus. Since an impact of water on THz dielectric response of tissues was suppresses by their freezing [38] and there were no resonant features in the THz absorption coefficients of tissues, the observed contrast was attributed to a collective response of a variety of abnormal proteins, that are accumulated in brain tissues with the Alzheimer disease. The described research work was followed by [91], where a mouse model of the Alzheimer's disease was studied using THz spectroscopy [92]. Particularly, a transmission-mode TPS was applied to measure the optical properties of dried tissue sliced *ex vivo*, which allows for confirming the contrast between normal and diseased tissues of the brain.

Only in 2014, a considerable attention was paid to studying an ability of THz technology use in diagnosis of human brain tumors, started with two papers on THz spectroscopy of freshly excised and paraffin-embedded models of gliomas from mice and rats [66, 93].

16.3.1 Study of Glioma Model in Mice and Rats

Oh et al. [66] applied TPS equipped with a raster-scan imaging unit in order to study an orthotopic glioma model from a rat. For performing this model, the 9L/lacZ rat glioma cells were implanted into male 9-week-old rats [94], and the tumors were allowed to grow for about 3–4 weeks, until it reached about 1 cm^3 in size. Unlike a common human brain glioma, this model provides a distinct boundary between normal and malignant tissues, which makes it possible to visually determine the tumor margins in freshly excised brain tissues. All rat brains were excised and divided into equal parts with a knife. Then the THz images of the freshly excised whole brain sample were collected by scanning of their surface with the focused THz beam. The observed results of the freshly excised *ex vivo* brain imaging are shown in Fig. 16.5 [66]. Here, the THz images were formed by extracting peak-to-peak values from the TPS data at each point of the sample surface [66]. One could notice that tumor margins in THz images agree with those in MRI and visible data; thus, all these methods yield clear determination of the tumor margins. The results of this study indicate that THz spectroscopy and imaging offer reliable diagnostic tumor images and have a potential to become useful tools to determine the tumor margins during the neurosurgery.

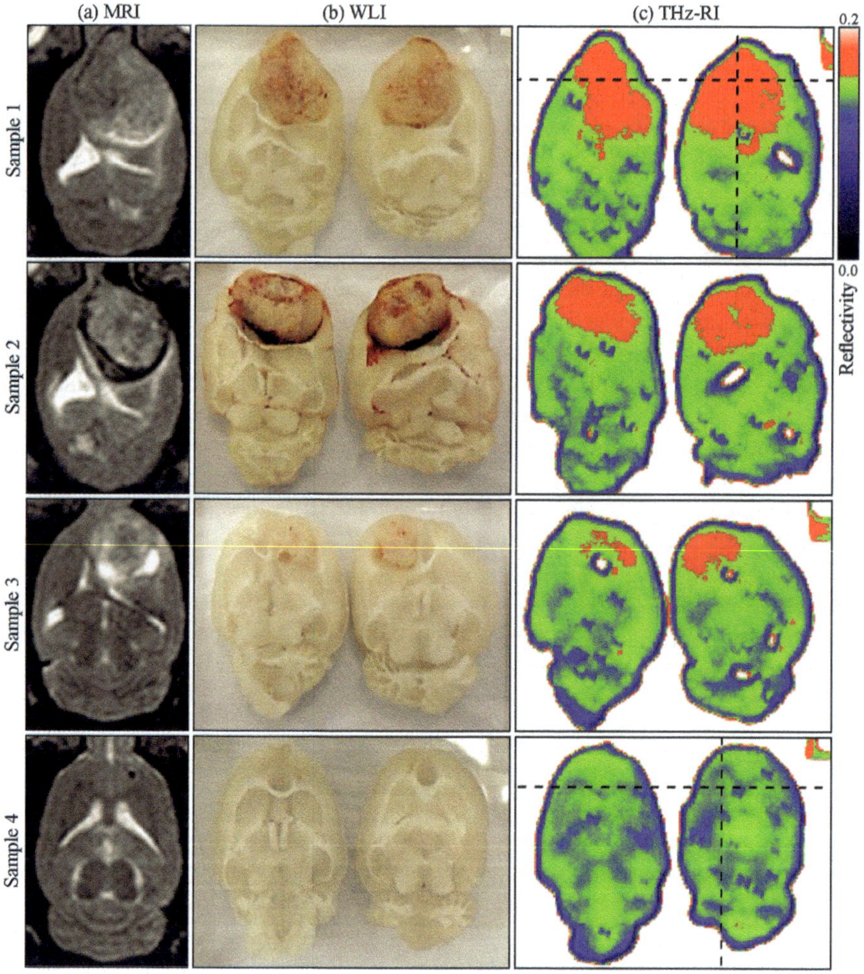

Fig. 16.5 A comparison of (**a**) magnetic resonance images (MRI), (**b**) white light images (WLI), and (**c**) THz reflection-mode images (THz-RI) of the freshly excised whole rat brain tissues with (rows 1–3) and without (row 4) tumors—i.e. an orthotopic glioma model for a rat with a clear tumor margins is used. Adapted from Ref. [66] with the permission of OSA

In order to study an origin of the observed contrast between healthy and malignant tissues of the rat brains, aside from water, Oh et al. [66] investigated the fresh brain samples dehydrated using a paraffin-embedding technique. The blocks with dehydrated tissues were further examined using both THz imaging and the hematoxylin and eosin (H&E)-stained histology. The authors observed the difference between the paraffin-embedded healthy and pathological tissues, but this difference was several time smaller than that observed in studies of freshly excised tissues (see Fig. 16.5); it does not allow for clear detection of the tumor margins. Using H&E-stained histology, the authors found that cell density near the tumor boundaries was higher than that in the tumor center and in the intact tissues, leading to increased THz reflectivity of paraffin-embedded tissues near the tumor boundary. Thus, when the effects of water on the THz dielectric response of tissues are eliminated, the observed contrast might originate from different densities of cells. Furthermore the necrotic debris in the central part of glioma was reported to impact the THz response of tissues [66].

A significant contrast between white matter and gray matter was observed in THz images of the whole rat brain *ex vivo*, which is due to the high content of myelin in a white matter [66]. As myelin is mainly formed by lipids and concentrates around axons, the white matter features a visible white color and lower THz optical constants, as compared to that of gray matter and tumor. Nevertheless, the difference in the THz response of white and gray matter does not affect capabilities of THz imaging for delineation of the tumor margins [66]. Here, we should notice that significant impact of myelin content in brain tissues on their THz dielectric response was further investigated in [95], where a potential of THz technology for label-free diagnosis of myelin deficit in paraffin-embedded tissues from mice and rhesus monkeys was demonstrated experimentally using reflection-mode TPS system.

The contrast between paraffin-embedded intact tissues and gliomas was confirmed using glioma models from mice [93]. For this aim, the authors used GL261 gliomas cells line, implanted into male 6-week-old mice. The mice were maintained for 20 days in order to let the tumor growth. Then, the whole mice brains were excised and embedded in paraffin blocks. The results of THz dielectric spectroscopy of the paraffin-embedded tissues demonstrate statistical differences between the THz response of healthy tissues and gliomas. It allows for selecting the optimal spectral bands and features for differentiation between normal tissues and tumors in paraffin blocks.

Finally, a potential of the THz reflectometry and imaging in intraoperative neurodiagnosis have been highlighted in several later papers using TPS, along with different measurement conditions and glioma types [96, 97]. Yamaguchi et al. [96] applied reflection-mode THz dielectric spectroscopy for investigating of the *ex vivo* freshly excised glioma model, in which C6 gliomas cells line were implanted into rats. Ji et al. [97] used reflection-mode for THz imaging of *ex vivo* freshly excised and *in vivo* glioma model, in which human glioblastoma tumorspheres (transfected with the enhanced GFP) were implanted into mice. Nevertheless, a comprehensive study of intact tissues and gliomas of the human brain was still required in order to objectively uncover strengths and weaknesses of THz technology in intraoperative diagnosis of gliomas in humans.

16.3.2 Study of Gliomas of the Human Brain

In Ref. [97], authors expand their studies to human samples and demonstrate results of THz reflectometry of *ex vivo* human brain gliomas of WHO grades II (6 samples), III (4 samples), and IV (4 samples). Unfortunately, the images were performed only for single frequency 0.5 THz, however, highlighting the potential of THz reflection imaging for human brain tissues. Nevertheless, THz dielectric spectroscopy of glioma samples in a broad spectral range could open new perspectives of THz radiation and further studies.

For instance, in Ref. [98], reflection-mode THz dielectric spectroscopy of two *ex vivo* human brain gliomas of WHO grade IV (glioblastoma) was considered. For both gliomas, the statistical differences in the THz spectra were observed, but only one of them shown the ALA-5 induced ppIX fluorescence during surgery. The more detailed study of human brain spectroscopic features was presented in Ref. [49]. In this work, the reflection-mode TPS was used to study a set of human brain glioma samples *ex vivo*. The tissue samples were surgically excised, according to the preliminary medical diagnosis and guided by the ALA-5-induced ppIX fluorescence imaging. These samples were classified by WHO grade; among them: 2, 9, 4, and 11 gliomas of grade I, II, III, and IV, respectively. In order to fix tissues for the THz measurements, they were placed on a reference substrate and covered with a gelatin slab; such gelatin embedding allows for preserving tissues from hydration/dehydration during transportation and THz measurements and, thus, sustains the THz response of tissues unaltered for several hours after the surgery, as compared to that of freshly excised tissues [99]. Along with the tumor, the samples contained perifocal regions, which were comprised of intact and edematous tissues. After the THz measurements, the gelatin slabs were removed, the tissue samples were fixed in a formalin and transferred back to the medical institution, where routine histopathological analysis, including H&E-stained histology, was applied in order to confirm the diagnosis.

The TPS setup with a unit for reflection-mode measurements (see Fig. 16.4a) was used for the THz dielectric spectroscopy of brain tissues. The experimental setup was covered by the plastic housing and purged by the nitrogen gas, in order to decrease humidity along the THz beam path and, thus, suppress an impact of water vapors on the measured data. The reference quartz window serves as a part of the housing; namely, it is in contact with a nitrogen gas from the bottom side, and with an air (or gelatin-embedded tissues) from the top side. Examples of reference and sample THz waveforms are presented in Fig. 16.4b and c, along with their Fourier spectra, which were detected using the above-mentioned system. The details of THz optical constants reconstruction based on these TPS waveforms are described elsewhere [49].

The THz refractive index and absorption coefficients of human brain gliomas of different WHO grades, as well as representative data of the H&E-stained histology are shown in Fig. 16.6 [49]. The THz response of perifocal tissues splits into two distinct classes, which correspond to the intact and edematous tissues. Therefore, along with the THz optical properties of grade I–IV gliomas, we show the response of intact and edematous tissues. In Fig. 16.6, the error bars represent the $\pm 2\sigma$ (or 95%) confidential interval of measurements, where σ stands for a standard deviation and accounts for fluctuations of the optical properties both within each tissue sample and within each tissue class.

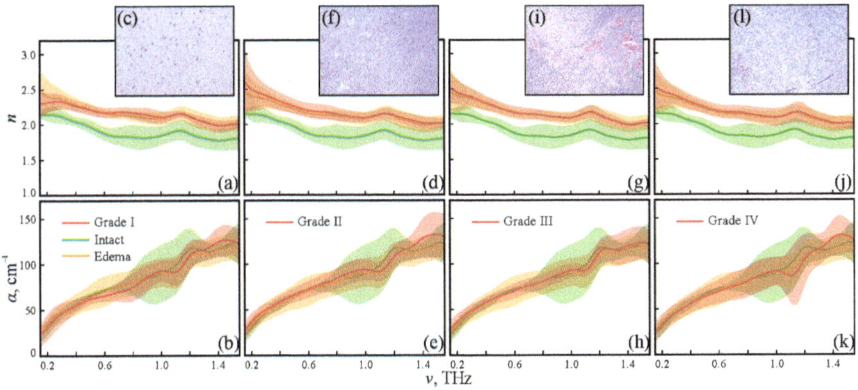

Fig. 16.6 Refractive index n, amplitude absorption coefficient α, and H&E-stained histology of gelatin-embedded human brain gliomas of different WHO grades *ex vivo*: (**a**)–(**c**) grade I; (**d**)–(**f**) grade II; (**g**)–(**i**) grade III; and (**j**)–(**l**) grade IV. The THz dielectric response of gliomas is compared to that of intact (healthy) and edematous tissues. The error bars represent the 95% confidential interval of measurements. Adapted from Ref. [49], published by SPIE under a Creative Commons (CC BY) license

In complete accordance with the results of preliminary studies involving the use of glioma models from rats and mice [66, 93, 96, 97], the statistical difference between the THz response of intact tissues and gliomas of the human brain is observed (Fig. 16.6) [49]. The refractive index of gliomas of all WHO grades is higher than that of healthy tissues, which might lead to the label-free contrast of several percent as demonstrated in Ref. [97]. In turn, the results of Fig. 16.6 do not allow to discriminate between the edematous tissues and tumors, as well as between different grades of gliomas using the THz spectroscopy. Significant dispersion of the THz optical properties within each class of tissues observed in Fig. 16.6 is usual for most of the label-free modalities of spectroscopy and imaging. Finally, in this study, white and gray matter were not differentiated in distinct classes (they form a single class of intact tissues); therefore, it might be a reason of high dispersion in the response of intact tissues.

Thus, an ability for using the THz spectroscopy and imaging in diagnosis of human brain gliomas of different WHO grades has been revealed in Ref. [49]. This confirmed a potential of THz technology in the intraoperative diagnosis of human brain gliomas.

16.3.3 Study of Meningiomas of the Human Brain

Although most research papers were dedicated to the measurements of glioma models from mice and rats, or human brain gliomas, the THz technologies might have a potential in diagnosis of other types of primary and secondary tumors of the human brain. For example, in Ref. [98], an ability for differentiation between human brain meningioma of WHO grade I and intact tissues of the brain has been demonstrated.

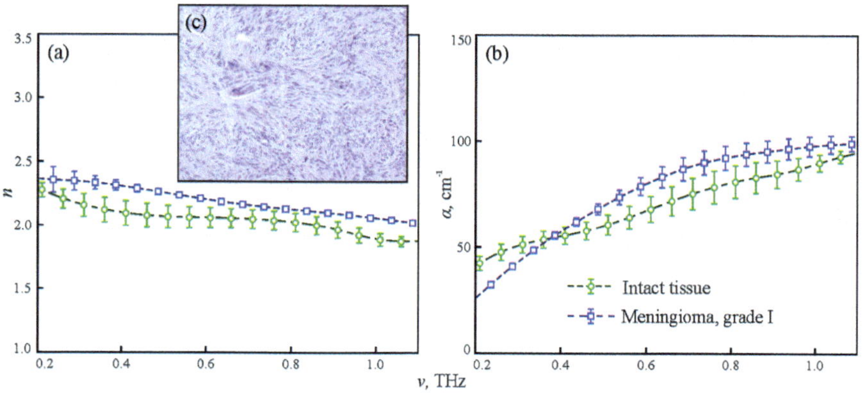

Fig. 16.7 THz dielectric spectroscopy of gelatin-embedded human brain meningioma, WHO grade I: (**a**) refractive index *n*, (**b**) absorption coefficient *α*, (**c**) H&E-stained histology. Courtesy of K.I. Zaytsev

In Fig. 16.7, we present the THz refractive index, absorption coefficient and H&E-stained histology of gelatin-embedded grade I human brain meningioma *ex vivo*, as compared to that of intact tissues. However, an attractiveness of THz technology in diagnosis of other types of human brain tumors should be further confirmed, since it would take significant amount of research and engineering efforts to collect the THz optical properties of tumors, as well as to analyze systematically an ability for their intraoperative diagnosis using THz spectroscopy and imaging.

16.3.4 Capabilities of Multimodal Diagnosis of Brain Tumors

Similarly to other instruments of label-free diagnosis of malignancies, a significant variability of measured data is inherent to THz spectroscopy and imaging [30], which might lead to the low sensitivity and specificity of THz diagnosis. A prospective way for improving the performance of THz diagnosis, while sustaining its non-invasive character, might be associated with combination of several modalities of label-free tissue imaging. As an example, a synergetic effect polarization-sensitive imaging in THz and optical domains has been demonstrated in Ref. [44]. A beautiful illustration of multimodal imaging of glioma models has been demonstrated in Fig. 16.8 [97], where the performance of preoperative MRI, WLI, GFP, H&E, OCT, THz, and 5-ALA-induced ppIX fluorescence imaging modalities are compared using an equal set of samples—the *ex vivo* freshly excised glioma models, based on implantation of human glioblastoma tumor spheres (transfected with the enhanced GFP) into mice. By combining different physical effects of electromagnetic wave–tissues interaction, standing behind the considered imaging modalities, one could achieve an improved performance of intraoperative diagnosis of brain tumors.

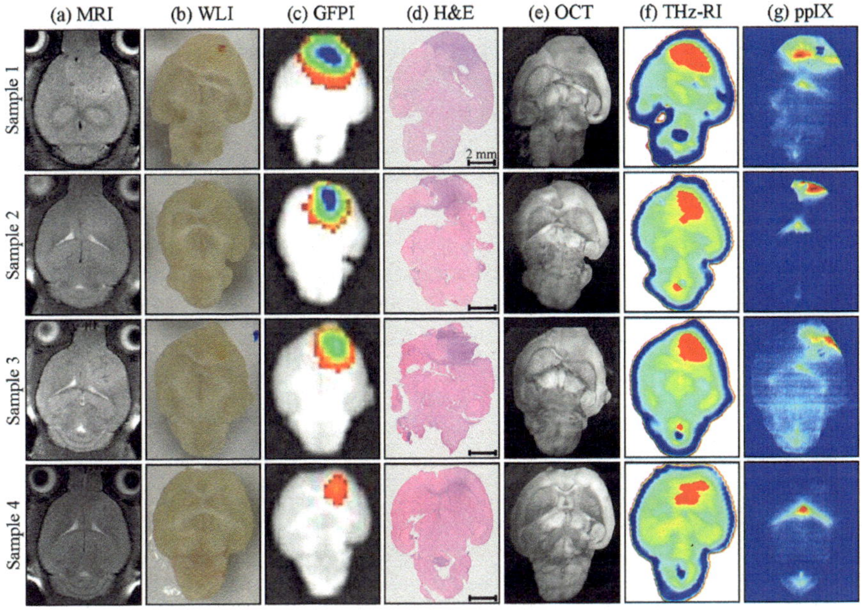

Fig. 16.8 Differentiation between healthy brain tissues and glioma model from 4 samples of whole mouse brain using various imaging modalities: (**a**) preoperative magnetic resonance imaging (MRI), (**b**) white light imaging (WLI) of the excised brain; (**c**) green-fluorescent protein imaging (GFP); (**d**) hematoxylin and eosin-stained histology (H&E); (**e**) optical coherence tomography (OCT); (**f**) THz reflection-mode images (THz-RI); (**g**) 5-ALA-induced protoporphyrin IX fluorescence imaging (ppIX). Adapted from Ref. [97], published by Springer Nature under a Creative Commons (CC BY) license

Indeed, such multimodal imaging of tumors represents a prospective direction for further research.

16.4 Pilot Terahertz Measurements of Traumatic Brain Injuries

Besides application of THz imaging and spectroscopy for investigation and diagnosis of brain tumors, the non-neoplastic diseases, caused by traumatic brain injuries (TBI) [100] and often accompanied by cerebral edema and pericontusional penumbra, also demonstrate specific features in THz range [101, 102].

Recent research with rat models demonstrates possibility of finding injury area by means of THz spectroscopy and continuous-wave imaging [101]. In these pilot experiments, four cases were considered: normal brain and mild, moderate, and severe TBI tissue samples. TBIs were obtained after opening a scalp of previously anesthetized and shaved rats, without laceration of the dura, by dropping a 30-g stainless steel rod on the dura from a certain height, i.e. 15, 25, 35 cm in cases of making mild, moderate, and severe TBI, respectively. A continuous-wave THz imaging (at the frequency of 2.52 THz) of 40-μm thick brain tissue slices covered by oleic acid in order to prevent their dehydration demonstrate significant differences between averaged transmittance of normal and injured samples (see Fig. 16.9): 23, 20, 19% for normal, mild, moderate, and severe TBI, respectively. Significantly, that these results are in good correlation with increased water content of samples from normal to severe TBIs [101]. Thus, water concentration, which is a well-known label of tissue malignancies, indicates the TBI as well. This should be carefully accounted and kept in mind, since traumatic conditions and presence of edematous regions of brain tissue could possibly yield interpretation errors in THz spectroscopic measurements and imaging of tumor margins.

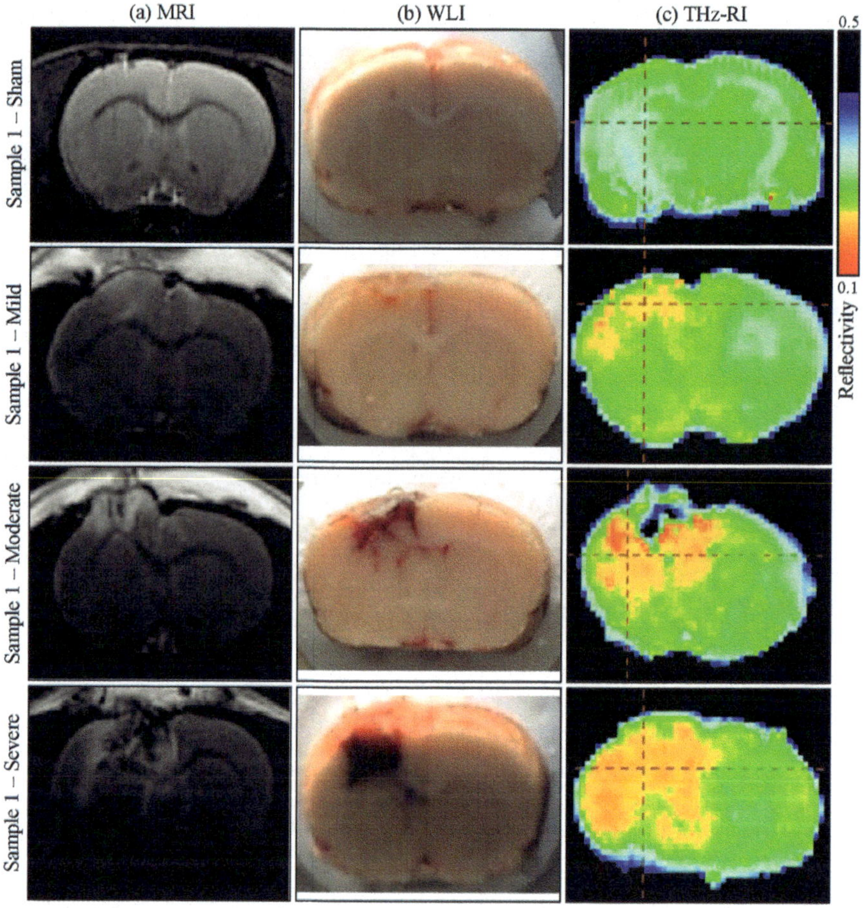

Fig. 16.9 A data of (**a**) magnetic resonance imaging (MRI), (**b**) white light imaging (WLI), and (**c**) THz reflection-mode images of freshly excised brain tissues without (row 1) and with (rows 2–4) different degrees of TBI . Adapted from Ref. [101], published by SPIE under a Creative Commons (CC BY) license

Along with water content, structural changes in injured tissues also impact their THz response, as it was demonstrated by spectroscopic measurements of dehydrated paraffin-embedded brain tissue slices [101]. Increasing in the severity of injury possesses the increasing in averaged THz absorption coefficient in a wide spectral range. The authors suggested that the possible explanation is the cell density decreasing with increasing severity, which was approved by H&E-stained microscopy of studied samples. Nevertheless, the exact origin of THz spectral differences between normal and various types of TBI needs further exploration.

To use THz imaging for distinguish between different degrees of TBI, machine learning approach can be applied. Different algorithms of classification were compared: support vector machine, k-nearest neighbor, random forest classifiers, and specific feature of transmittance distribution in spatial domain and statistical distribution features in normalized gray histogram [102]. Though random forest provided better accuracy in [102], the optimal algorithm remains to be established.

THz imaging of TBI, described in [101], suffers from low spatial resolution, which is clear from comparison of MRI and THz imaging of TBI margins. The latter approach provides insufficiently larger detected size of injury. The promising techniques, such as THz solid immersion microscopy [72, 74], could probably solve such a problem.

16.5 Conclusions

In this chapter, we considered recent research results, highlighting a prospect of THz technology in the intraoperative diagnosis of brain tumors in label-free detection of tumor margins during surgery in order to ensure its gross-total resection. Despite the attractiveness of THz technology for solving this demanding and socially important problem, there are still a number of factors, which restrain transfer of THz spectroscopy and imaging from laboratory environment to a clinical practice. Particularly, significant research and engineering efforts are required for development of rather low-cost, ergonomic, fast and effective THz instruments, which might be integrated into the modern neurosurgical workflows and existing methods of the intraoperative neurodiagnosis or express-histology of excised tissues. Nevertheless, delineation of brain tumor margins during surgery remains very important task, which gives hope that all above-mentioned challenging problems of THz diagnosis would be addressed in the nearest future.

Acknowledgments The work was supported by the grant # 17-79-20346 (Sect. 16.1) and grant # 18-12-00328 (Sects. 16.2 and 16.3) of the Russian Science Foundation. The authors are thankful to Nikita V. Chernomyrdin (from the Bauman Moscow State Technical University), Alexandra V. Kosyrkova and Pavel V. Nikitin (both from the Burdenko Neurosurgery Institute), Igor E. Spektor and Gennady A. Komandin (both from the Prokhorov General Physics Institute of the Russian Academy of Sciences), and Marina A. Schcedrina (from the Sechenov First Moscow State Medical University) for fruitful discussions.

References

1. Gunel, J.M., Piepmeier, J.M., Baehring, J.M.: Malignant Brain Tumors. State-of-the-Art Treatment, p. 297. Springer International Publishing AG, Switzerland (2017). https://doi.org/10.1007/978-3-319-49864-5
2. Ostrom, Q., Gittleman, H., Truitt, G., Boscia, A., Kruchko, C., Barnholtz-Sloan, J.: CBTRUS statistical report: primary brain and other central nervous system tumors diagnosed in the United States in 2011–2015. Neuro Oncol. **20**(Suppl 4), iv1–iv86 (2018). https://doi.org/10.1093/neuonc/noy131
3. Louis, D.N., Perry, A., Reifenberger, G., von Deimling, A., Figarella-Branger, D., Cavenee, W.K., Ohgaki, H., Wiestler, O.D., Kleihues, P., Ellison, D.W.: The 2016 World Health Organization classification of tumors of the central nervous system: a summary. Acta Neuropathol. **131**(6), 803–820 (2016). https://doi.org/10.1007/s00401-016-1545-1
4. Stupp, R., Mason, W., van den Bent, M., Weller, M., Fisher, B., Taphoorn, M., Belanger, K., Brandes, A., Marosi, C., Bogdahn, U., Curschmann, J., Janzer, R., Ludwin, S., Gorlia, T., Allgeier, A., Lacombe, D., Cairncross, J., Eisenhauer, E., Mirimano, R.: Radiotherapy plus concomitant and adjuvant temozolomide for glioblastoma. N. Engl. J. Med. **352**(10), 987–996 (2005). https://doi.org/10.1056/NEJMoa043330
5. Comprehensive, integrative genomic analysis of diffuse lower-grade gliomas. N. Engl. J. Med. **372**(26), 2481–2498 (2015). https://doi.org/10.1056/NEJMoa1402121
6. Chang, E.F., Clark, A., Jensen, R.L., Bernstein, M., Guha, A., Carrabba, G., Mukhopadhyay, D., Kim, W., Liau, L.M., Chang, S.M., Smith, J.S., Berger, M.S., McDermott, M.W.: Multiinstitutional validation of the University of California at San Francisco low-grade glioma prognostic scoring system. J. Neurosurg. **111**(2), 203–210 (2009). https://doi.org/10.3171/2009.2.JNS081101
7. Hefti, M., Mehdorn, H., Albert, I., Dorner, L.: Fluorescence-guided surgery for malignant glioma: a review on aminolevulinic acid induced protoporphyrin ix photodynamic diagnostic in brain tumors. Curr. Med. Imag. Rev. **6**(4), 254–258 (2010). https://doi.org/10.2174/157340510793205503
8. Garzon-Muvdi, T., Kut, C., Li, X., Chaichana, K.: Intraoperative imaging techniques for glioma surgery. Future Oncol. **13**(19), 1731–1745 (2017). https://doi.org/10.2217/fon-2017-0092
9. Vase, F., MacKinnon, N., Farkas, D., Kateb, B.: Review of the potential of optical technologies for cancer diagnosis in neurosurgery: a step toward intraoperative neurophotonics. Neurophotonics. **4**(1), 011010 (2016). https://doi.org/10.1117/1.NPh.4.1.011010
10. Mehdorn, H., Schwartz, F., Dawirs, S., Hedderich, J., Dorner, L., Nabavi, A.: High-field iMRI in glioblastoma surgery: improvement of resection radicality and survival for the patient? In: Pamir, M., Seifert, V., Kiris, T. (eds.) Intraoperative Imaging, pp. 103–106. Springer Vienna, Vienna (2011). https://doi.org/10.1007/978-3-211-99651-5_16
11. Schatlo, B., Fandino, J., Smoll, N., Wetzel, O., Remonda, L., Marbacher, S., Perrig, W., Landolt, H., Fathi, A.-R.: Outcomes after combined use of intraoperative MRI and 5-aminolevulinic acid in high-grade glioma surgery. Neuro Oncol. **17**(12), 1560 (2015). https://doi.org/10.1093/neuonc/nov049
12. Shikunova, I.A., Stryukova, D.O., Rossolenko, S.N., Kiselev, A.M., Kurlov, V.N.: Neurosurgery contact handheld probe based on sapphire shaped crystal. J. Cryst. Growth. **457**, 265–269 (2017). https://doi.org/10.1016/j.jcrysgro.2016.08.062
13. Shikunova, I.A., Zaytsev, K.I., Stryukov, D.O., Dubyanskaya, E.N., Kurlov, V.N.: Neurosurgical sapphire handheld probe for intraoperative optical diagnostics, laser coagulation and aspiration of malignant brain tissue. Proc. SPIE. **10411**, 104110Q (2017). https://doi.org/10.1117/12.2286124
14. Katyba, G.M., Zaytsev, K.I., Dolganova, I.N., Shikunova, I.A., Chernomyrdin, N.V., Yurchenko, S.O., Komandin, G.A., Reshetov, I.V., Nesvizhevsky, V.V., Kurlov, V.N.: Sapphire

shaped crystals for waveguiding, sensing and exposure applications. Prog. Cryst. Growth Charact. Mater. **64**(4), 133–151 (2018). https://doi.org/10.1016/j.pcrysgrow.2018.10.002
15. Stummer, W., Pichlmeier, U., Meinel, T., Wiestler, O.D., Zanella, F., Reulen, H.-J.: Fluorescence-guided surgery with 5-aminolevulinic acid for resection of malignant glioma: a randomised controlled multicentre phase III trial. Lancet Oncol. **7**(5), 392–401 (2006). https://doi.org/10.1016/S1470-2045(06)70665-9
16. Pustogarov, N., Panteleev, D., Goryaynov, S., Ryabova, A., Rybalkina, E., Revishchin, A., Potapov, A., Pavlova, G.: Hiding in the shadows: CPOX expression and 5-ALA induced fluorescence in human glioma cells. Mol. Neurobiol. **54**(7), 5699–5708 (2017). https://doi.org/10.1007/s12035-016-0109-7
17. Potapov, A., Goryaynov, S., Okhlopkov, V., Shishkina, L., Loschenov, V., Savelieva, T., Golbin, D., Chumakova, A., Goldberg, M., Varyukhina, M., Spallone, A.: Laser biospectroscopy and 5-ALA fluorescence navigation as a helpful tool in the meningioma resection. Neurosurg. Rev. **39**(3), 437–447 (2016). https://doi.org/10.1007/s10143-015-0697-0
18. Chen, B., Wang, H., Ge, P., Zhao, J., Li, W., Gu, H., Wang, G., Luo, Y., Chen, D.: Gross total resection of glioma with the intraoperative fluorescence-guidance of fluorescein sodium. Int. J. Med. Sci. **9**(8), 708–714 (2012). https://doi.org/10.7150/ijms.4843
19. Jermyn, M., Mok, K., Mercier, J., Desroches, J., Pichette, J., Saint-Arnaud, K., Bernstein, L., Guiot, M.-C., Petrecca, K., Leblond, F.: Intraoperative brain cancer detection with Raman spectroscopy in humans. Sci. Transl. Med. **7**(274), 274ra19 (2015). https://doi.org/10.1126/scitranslmed.aaa2384
20. Orringer, D., Pandian, B., Niknafs, Y., Hollon, T., Boyle, J., Lewis, S., Garrard, M., Hervey-Jumper, S., Garton, H., Maher, C., Heth, J., Sagher, O., Wilkinson, D., Snuderl, M., Venneti, S., Ramkissoon, S., McFadden, K., Fisher-Hubbard, A., Lieberman, A., Johnson, T., Xie, X., Trautman, J., Freudiger, C., Camelo-Piragua, S.: Rapid intraoperative histology of unprocessed surgical specimens via fibre-laser-based stimulated Raman scattering microscopy. Nat. Biomed. Eng. **1**, 0027 (2017). https://doi.org/10.1038/s41551-016-0027
21. Jermyn, M., Desroches, J., Mercier, J., St-Arnaud, K., Guiot, M.-C., Leblond, F., Petrecca, K.: Raman spectroscopy detects distant invasive brain cancer cells centimeters beyond MRI capability in humans. Biomed. Opt. Express. **7**(12), 5129–5137 (2016). https://doi.org/10.1364/BOE.7.005129
22. Kircher, M., de la Zerda, A., Jokerst, J., Zavaleta, C., Kempen, P., Mittra, E., Pitter, K., Huang, R., Campos, C., Habte, F., Sinclair, R., Brennan, C., Mellingho, I., Holland, E., Gambhir, S.: A brain tumor molecular imaging strategy using a new triple-modality MRI-photoacoustic-Raman nanoparticle. Nat. Med. **18**, 829–834 (2012). https://doi.org/10.1038/nm.2721
23. Liang, B., Liu, W., Zhan, Q., Li, M., Zhuang, M., Huo, Q., Yao, L.J.: Impacts of the murine skull on high-frequency transcranial photoacoustic brain imaging. J. Biophotonics. (2010). https://doi.org/10.1002/jbio.201800466
24. Yashin, K.S., Kiseleva, E.B., Gubarkova, E.V., Moiseev, A.A., Kuznetsov, S.S., Shilyagin, P.A., Gelikonov, G.V., Medyanik, I.A., Kravets, L.Y., Potapov, A.A., Zagaynova, E.V., Gladkova, N.D.: Cross-polarization optical coherence tomography for brain tumor imaging. Front. Oncol. **9**, 201 (2019). https://doi.org/10.3389/fonc.2019.00201
25. Yashin, K.S., Kiseleva, E.B., Moiseev, A.A., Kuznetsov, S.S., Timofeeva, L.B., Pavlova, N.P., Gelikonov, G.V., Medyanik, I.A., Kravets, L.Y., Zagaynova, E.V., Gladkova, N.D.: Quantitative nontumorous and tumorous human brain tissue assessment using microstructural co- and cross-polarized optical coherence tomography. Sci. Rep. **9**, 2024 (2019). https://doi.org/10.1038/s41598-019-38493-y
26. Kiseleva, E.B., Yashin, K., Moiseev, A.A., Sirotkina, M.A., Timofeeva, L.B., Fedoseeva, V.V., Alekseeva, A.I., Medyanik, I.A., Karyakin, N.N., Kravets, L.Y., Gladkova, N.: Cross-polarization optical coherence tomography in comparative *in vivo* and *ex vivo* studies of the optical properties of normal and tumorous brain tissues. Sovremennye Tehnologii v Medicine. **9**(4), 177–187 (2017). https://doi.org/10.17691/stm2017.9.4.22

27. Dolganova, I., Chernomyrdin, N., Aleksandrova, P., Beshplav, S.-I., Potapov, A., Reshetov, I., Kurlov, V., Tuchin, V., Zaytsev, K.: Nanoparticle-enabled experimentally trained wavelet-domain denoising method for optical coherence tomography. J. Biomed. Opt. **23**(9), 091406 (2018). https://doi.org/10.1117/1.JBO.23.9.091406
28. Dolganova, I., Aleksandrova, P., Beshplav, S.-I., Chernomyrdin, N., Dubyanskaya, E., Goryaynov, S., Kurlov, V., Reshetov, I., Potapov, A., Tuchin, V., Zaytsev, K.: Wavelet-domain de-noising of OCT images of human brain malignant glioma. Proc. SPIE. **10717**, 107171X (2018). https://doi.org/10.1117/12.2314727
29. Lee, Y.-S.: Principles of Terahertz Science and Technology, p. 340. Springer (2009)
30. Smolyanskaya, O.A., Chernomyrdin, N.V., Konovko, A.A., Zaytsev, K.I., Ozheredov, I.A., Cherkasova, O.P., Nazarov, M.M., Guillet, J.-P., Kozlov, S.A., Kistenev, Y.V., Coutaz, J.-L., Mounaix, P., Vaks, V.L., Son, J.-H., Cheon, H., Wallace, V.P., Feldman, Y., Popov, I., Tuchin, V.V.: Terahertz biophotonics as a tool for studies of dielectric and spectral properties of biological tissues and liquids. Prog. Quant. Electron. **62**, 1–77 (2018). https://doi.org/10.1016/j.pquantelec.2018.10.001
31. K.I. Zaytsev, I.N. Dolganova, N.V. Chernomyrdin, G.M. Katyba, A.A. Gavdush, O.P. Cherkasova, G.A. Komandin, M.A. Shchedrina, A.N. Khodan, D.S. Ponomarev, I.V. Reshetov, V.E. Karasik, M. Skorobogatiy, V.N. Kurlov, and V.V. Tuchin, "The progress and perspectives of terahertz technology for diagnosis of neoplasms: a review," Journal of Optic 22(1), 013001 (2020), https://doi.org/10.1088/2040-8986/ab4dc3
32. Woodward, R., Wallace, V., Pye, R., Cole, B., Arnone, D., Lineld, E., Pepper, M.: Terahertz pulse imaging of *ex vivo* basal cell carcinoma. J. Investig. Dermatol. **120**(1), 72–78 (2003). https://doi.org/10.1046/j.1523-1747.2003.12013.x
33. Wallace, V., Fitzgerald, A., Shankar, S., Flanagan, N., Pye, R., Cluff, J., Arnone, D.: Terahertz pulsed imaging of basal cell carcinoma *ex vivo* and *in vivo*. Br. J. Dermatol. **151**(2), 424–432 (2004). https://doi.org/10.1111/j.1365-2133.2004.06129.x
34. Wallace, V., Fitzgerald, A., Pickwell, E., Pye, R., Taday, P., Flanagan, N., Ha, T.: Terahertz pulsed spectroscopy of human basal cell carcinoma. Appl. Spectrosc. **60**(10), 1127–1133 (2006). https://doi.org/10.1366/000370206778664635
35. Joseph, C., Patel, R., Neel, V., Giles, R., Yaroslavsky, A.: Imaging of *ex vivo* nonmelanoma skin cancers in the optical and terahertz spectral regions. Optical and terahertz skin cancers imaging. J. Biophotonics. **7**(5), 395–303 (2014). https://doi.org/10.1002/jbio.201200111
36. Zaytsev, K., Kudrin, K., Karasik, V., Reshetov, I., Yurchenko, S.: *In vivo* terahertz spectroscopy of pigmentary skin nevi: pilot study of non-invasive early diagnosis of dysplasia. Appl. Phys. Lett. **106**(5), 053702 (2015). https://doi.org/10.1063/1.4907350
37. Zaytsev, K., Chernomyrdin, N., Kudrin, K., Reshetov, I., Yurchenko, S.: Terahertz spectroscopy of pigmentary skin nevi *in vivo*. Opt. Spectrosc. **119**(3), 404–410 (2015). https://doi.org/10.1134/S0030400X1509026X
38. Sim, Y., Park, J., Ahn, K.-M., Park, C., Son, J.-H.: Terahertz imaging of excised oral cancer at frozen temperature. Biomed. Opt. Express. **4**(8), 1413–1421 (2013). https://doi.org/10.1364/BOE.4.001413
39. Sim, Y., Ahn, K., Park, J., Park, C., Son, J.: Temperature-dependent terahertz imaging of excised oral malignant melanoma. IEEE J. Biomed. Health Inform. **17**(4), 779–784 (2013). https://doi.org/10.1109/JBHI.2013.2252357
40. Reid, C., Fitzgerald, A., Reese, G., Goldin, R., Tekkis, P., O'Kelly, P., Pickwell-MacPherson, E., Gibson, A., Wallace, V.: Terahertz pulsed imaging of freshly excised human colonic tissues. Phys. Med. Biol. **56**(14), 4333–4353 (2011). https://doi.org/10.1088/0031-9155/56/14/008
41. Doradla, P., Alavi, K., Joseph, C., Giles, R.: Detection of colon cancer by continuous-wave terahertz polarization imaging technique. J. Biomed. Opt. **18**(9), 090504 (2013). https://doi.org/10.1117/1.JBO.18.9.090504
42. Wahaia, F., Kasalynas, I., Venckevicius, R., Seliuta, D., Valusis, G., Urbanowicz, A., Molis, G., Carneiro, F., Carvalho Silva, C., Granja, P.: Terahertz absorption and reflection imaging of

carcinoma-affected colon tissues embedded in paraffin. J. Mol. Struct. **1107**, 214–219 (2016). https://doi.org/10.1016/j.molstruc.2015.11.048
43. Chen, H., Ma, S.-H., Yan, W.-X., Wu, X.-M., Wang, X.-Z.: The diagnosis of human liver cancer by using THz fiber-scanning near-field imaging. Chin. Phys. Lett. **30**(3), 030702 (2013). https://doi.org/10.1088/0256-307x/30/3/030702
44. Hou, D., Li, X., Cai, J., Ma, Y., Kang, X., Huang, P., Zhang, G.: Terahertz spectroscopic investigation of human gastric normal and tumor tissues. Phys. Med. Biol. **59**(18), 5423–5440 (2014). https://doi.org/10.1088/0031-9155/59/18/5423
45. Ji, Y., Park, C., Kim, H., Kim, S.-H., Lee, G., Noh, S., Jeon, T.-I., Son, J.-H., Huh, Y.-M., Haam, S., Oh, S., Lee, S., Suh, J.-S.: Feasibility of terahertz reflectometry for discrimination of human early gastric cancers. Biomed. Opt. Express. **6**(4), 1398–1406 (2015). https://doi.org/10.1364/BOE.6.001398
46. Fitzgerald, A., Wallace, V., Jimenez-Linan, M., Bobrow, L., Pye, R., Purushotham, A., Arnone, D.: Terahertz pulsed imaging of human breast tumors. Radiology. **239**(2), 533–540 (2006). https://doi.org/10.1148/radiol.2392041315
47. Ashworth, P., Pickwell-MacPherson, E., Provenzano, E., Pinder, S., Purushotham, A., Pepper, M., Wallace, V.: Terahertz pulsed spectroscopy of freshly excised human breast cancer. Opt. Express. **17**(15), 12444–12454 (2009). https://doi.org/10.1364/OE.17.012444
48. Fitzgerald, A., Wallace, V., Pinder, S., Purushotham, A., O'Kelly, P., Ashworth, P.: Classification of terahertz-pulsed imaging data from excised breast tissue. J. Biomed. Opt. **17**(1), 016005 (2012). https://doi.org/10.1117/1.JBO.17.1.016005
49. Gavdush, A., Chernomyrdin, N., Malakhov, K., Beshplav, S.-I., Dolganova, I., Kosyrkova, A., Nikitin, P., Musina, G., Katyba, G., Reshetov, I., Cherkasova, O., Komandin, G., Karasik, V., Potapov, A., Tuchin, V., Zaytsev, K.: Terahertz spectroscopy of gelatin-embedded human brain gliomas of different grades: a road toward intraoperative THz diagnosis. J. Biomed. Opt. **24**(2), 027001 (2019). https://doi.org/10.1117/1.JBO.24.2.027001
50. Cole, K., Cole, R.: Dispersion and absorption in dielectrics I. Alternating current characteristics. J. Chem. Phys. **9**(4), 341–351 (1941). https://doi.org/10.1063/1.1750906
51. Cole, K., Cole, R.: Dispersion and absorption in dielectrics II. Direct current characteristics. J. Chem. Phys. **10**(2), 98–105 (1942). https://doi.org/10.1063/1.1723677
52. Davidson, D.: Dielectric relaxation in liquids: I. The representation of relaxation behavior. Can. J. Chem. **39**(3), 571–594 (1961). https://doi.org/10.1139/v61-069
53. Havriliak, S., Negami, S.: A complex plane analysis of dispersions in some polymer systems. J. Polym. Sci. C: Polym. Symp. **14**(1), 99–117 (1966). https://doi.org/10.1002/polc.5070140111
54. Pickwell, E., Cole, B.E., Fitzgerald, A.J., Wallace, V.P., Pepper, M.: Simulation of terahertz pulse propagation in biological systems. Appl. Phys. Lett. **12**, 2190–2192 (2004). https://doi.org/10.1063/1.1688448
55. Pickwell, E., Fitzgerald, A.J., Cole, B.E., Taday, P.F., Pye, R.J., Ha, T., Pepper, M., Wallace, V.P.: Simulating the response of terahertz radiation to basal cell carcinoma using *ex vivo* spectroscopy measurements. J. Biomed. Opt. **10**(6), 064021 (2005). https://doi.org/10.1117/1.2137667
56. Walker, G.C., Berry, E., Smye, S.W., Zinov'ev, N.N., Fitzgerald, A.J., Miles, R.E., Chamberlain, M., Smith, M.A.: Modelling the propagation of terahertz radiation through a tissue simulating phantom. Phys. Med. Biol. **49**(10), 1853–1864 (2004). https://doi.org/10.1088/0031-9155/49/10/002
57. Pickwell, E., Cole, B.E., Fitzgerald, A.J., Pepper, M., Wallace, V.P.: *In vivo* study of human skin using pulsed terahertz radiation. Phys. Med. Biol. **49**(9), 1595–1607 (2004). https://doi.org/10.1088/0031-9155/49/9/001
58. Ney, M., Abdulhalim, I.: Modeling of reflectometric and ellipsometric spectra from the skin in the terahertz and submillimeter waves region. J. Biomed. Opt. **16**(6), 067006 (2011). https://doi.org/10.1117/1.3592779

59. Fitzgerald, A.J., Pickwell-MacPherson, E., Wallace, V.P.: Use of finite difference time domain simulations and Debye theory for modelling the terahertz reflection response of normal and tumour breast tissue. PLoS One. **9**, e99291 (2014). https://doi.org/10.1371/journal.pone.0099291
60. Truong, B.C.Q., Tuan, H.D., Wallace, V.P., Fitzgerald, A.J., Nguyen, H.T.: The potential of the double Debye parameters to discriminate between basal cell carcinoma and normal skin. IEEE Trans. Terahertz Sci. Technol. **5**(6), 990–998 (2015). https://doi.org/10.1109/TTHZ.2015.2485208
61. Truong, B.C.Q., Tuan, H.D., Fitzgerald, A.J., Wallace, V.P., Nguyen, H.T.: A dielectric model of human breast tissue in terahertz regime. IEEE Trans. Biomed. Eng. **62**(2), 699–707 (2015). https://doi.org/10.1109/TBME.2014.2364025
62. Parrott, E.P., Sy, S.M.Y., Blu, T., Wallace, V.P., Pickwell-MacPherson, E.: Terahertz pulsed imaging *in vivo*: measurements and processing methods. J. Biomed. Opt. **16**(10), 106010 (2011). https://doi.org/10.1117/1.3642002
63. Chen, H., Chen, T.-H., Tseng, T.-F., Lu, J.-T., Kuo, C.-C., Fu, S.-C., Lee, W.-J., Tsai, Y.-F., Huang, Y.-Y., Chuang, E.Y., Hwang, Y.-J., Sun, C.-K.: High-sensitivity *in vivo* THz transmission imaging of early human breast cancer in a subcutaneous xenograft mouse model. Opt. Express. **19**(22), 21552–21562 (2011). https://doi.org/10.1364/OE.19.021552
64. Yamaguchi, S., Fukushi, Y., Kubota, O., Itsuji, T., Ouchi, T., Yamamoto, S.: Origin and quantification of differences between normal and tumor tissues observed by terahertz spectroscopy. Phys. Med. Biol. **61**(18), 6808–6820 (2016). https://doi.org/10.1088/0031-9155/61/18/6808
65. Sy, S., Huang, S., Wang, Y.-X.J., Yu, J., Ahuja, A.T., Zhang, Y.-T., Pickwell-MacPherson, E.: Terahertz spectroscopy of liver cirrhosis: investigating the origin of contrast. Phys. Med. Biol. **55**(24), 7587–7596 (2010). https://doi.org/10.1088/0031-9155/55/24/013
66. Oh, S.J., Kim, S.-H., Ji, Y.B., Jeong, K., Park, Y., Yang, J., Park, D.W., Noh, S.K., Kang, S.-G., Huh, Y.-M., Son, J.-H., Suh, J.-S.: Study of freshly excised brain tissues using terahertz imaging. Biomed. Opt. Express. **5**(8), 2837–2842 (2014). https://doi.org/10.1364/BOE.5.002837
67. Joseph, C.S., Yaroslavsky, A.N., Al-Arashi, M., Goyette, T.M., Dickinson, J.C., Gatesman, A.J., Soper, B.W., Forgione, C.M., Horgan, T.M., Ehasz, E.J., Giles, R.H., Nixon, W.E.: Terahertz spectroscopy of intrinsic biomarkers for non-melanoma skin cancer. Proc. SPIE. **7215**, 72150I (2009). https://doi.org/10.1117/12.809402
68. Ney, M., Abdulhalim, I.: Comprehensive Monte-Carlo simulator for optimization of imaging parameters for high sensitivity detection of skin cancer at the THz. Proc. SPIE. **9721**, 97210W (2016). https://doi.org/10.1117/12.2209486
69. Vilagosh, Z., Lajevardipour, A., Wood, A.: Modelling terahertz radiation absorption and reflection with computational phantoms of skin and associated appendages. Proc. SPIE. **10456**, 104560M (2018). https://doi.org/10.1117/12.2283206
70. Vilagosh, Z., Lajevardipour, A., Wood, A.W.: Computational phantom study of frozen melanoma imaging at 0.45 terahertz. Bioelectromagnetics. **40**(2), 118–127 (2019). https://doi.org/10.1002/bem.22169
71. Chernomyrdin, N.V., Kucheryavenko, A.S., Kolontaeva, G.S., Katyba, G.M., Dolganova, I.N., Karalkin, P.A., Ponomarev, D.S., Kurlov, V.N., Reshetov, I.V., Skorobogatiy, M., Tuchin, V.V., Zaytsev, K.I.: Reflection-mode continuous-wave 0.15λ-resolution terahertz solid immersion microscopy of soft biological tissues. Appl. Phys. Lett. **113**(11), 111102 (2018). https://doi.org/10.1063/1.5045480
72. Chernomyrdin, N.V., Kucheryavenko, A.S., Rimskaya, E.N., Dolganova, I.N., Zhelnov, V.A., Karalkin, P.A., Gryadunova, A.A., Reshetov, I.V., Lavrukhin, D.V., Ponomarev, D.S., Karasik, V.E., Zaytsev, K.I.: Terahertz microscope based on solid immersion effect for imaging of biological tissues. Opt. Spectrosc. **126**(5), 560–567 (2019). https://doi.org/10.1134/S0030400X19050059
73. N.V. Chernomyrdin, V.A. Zhelnov, A.S. Kucheryavenko, I.N. Dolganova, G.M. Katyba, V.E. Karasik, I.V. Reshetov, and K.I. Zaytsev, "Numerical analysis and experimental study

of terahertz solid immersion microscopy," Optical Engineering 59(6), 061605 (2019), https://doi.org/10.1117/1.OE.59.6.061605 .
74. Ishimaru, A.: Electromagnetic Wave Propagation, Radiation, and Scattering: From Fundamentals to Applications, p. 946. Wiley–IEEE Press, Piscataway
75. Tuchin, V.: Tissue Optics: Light Scattering Methods and Instruments for Medical Diagnostics, 3rd edn, p. 1000. SPIE Press, Bellingham
76. Wilmink, G.J., Ibey, B.L., Rivest, B.D., Grundt, J.E., Roach, W.P., Tongue, T.D., Schulkin, B.J., Laman, N., Peralta, X.G., Roth, C.C., Cerna, C.Z.: Development of a compact terahertz time-domain spectrometer for the measurement of the optical properties of biological tissues. J. Biomed. Opt. **16**(4), 047006 (2011). https://doi.org/10.1117/1.3570648
77. Grootendorst, M.R., Fitzgerald, A.J., Brouwer de Koning, S.G., Santaolalla, A., Portieri, A., van Hemelrijck, M., Young, M.R., Owen, J., Cariati, M., Pepper, M., Wallace, V.P., Pinder, S.E., Purushotham, A.: Use of a handheld terahertz pulsed imaging device to differentiate benign and malignant breast tissue. Biomed. Opt. Express. **8**(6), 2932–2945 (2017). https://doi.org/10.1364/BOE.8.002932
78. Reid, C.B., Pickwell-MacPherson, E., Laufer, J.G., Gibson, A.P., Hebden, J.C., Wallace, V.P.: Accuracy and resolution of THz reflection spectroscopy for medical imaging. Phys. Med. Biol. **55**(16), 4825–4837 (2010). https://doi.org/10.1088/0031-9155/55/16/013
79. Huang, S., Ashworth, P.C., Kan, K.W.C., Chen, Y., Wallace, V.P., Zhang, Y.-T., Pickwell-MacPherson, E.: Improved sample characterization in terahertz reflection imaging and spectroscopy: erratum. Opt. Express. **19**(5), 24782–24782 (2011). https://doi.org/10.1364/OE.17.003848
80. Fan, S., Parrott, E.P.J., Ung, B.S.Y., Pickwell-MacPherson, E.: Calibration method to improve the accuracy of THz imaging and spectroscopy in reflection geometry. Photon. Res. **4**(3), A29–A35 (2016). https://doi.org/10.1364/PRJ.4.000A29
81. Chen, X., Parrott, E.P.J., Ung, B.S.Y., Pickwell-MacPherson, E.: A robust baseline and reference modification and acquisition algorithm for accurate THz imaging. IEEE Trans. Terahertz Sci. Technol. **7**(5), 493–501 (2017). https://doi.org/10.1109/TTHZ.2017.2722981
82. Zhang, R., He, Y., Liu, K., Zhang, L., Zhang, S., Pickwell-MacPherson, E., Zhao, Y., Zhang, C.: Composite multiscale entropy analysis of reflective terahertz signals for biological tissues. Opt. Express. **25**(20), 23669–23676 (2017). https://doi.org/10.1364/OE.25.023669
83. Auston, D.H.: Picosecond optoelectronic switching and gating in silicon. Appl. Phys. Lett. **26**(3), 101 (1975). https://doi.org/10.1063/1.88079
84. Lavrukhin, D.V., Yachmenev, A.E., Pavlov, A.Y., Khabibullin, R.A., Goncharov, Y.G., Spektor, I.E., Komandin, G.A., Yurchenko, S.O., Chernomyrdin, N.V., Zaytsev, K.I., Ponomarev, D.S.: Shaping the spectrum of terahertz photoconductive antenna by frequency-dependent impedance modulation. Semicond. Sci. Technol. **34**(3), 034005 (2019). https://doi.org/10.1088/1361-6641/aaff31
85. Lepeshov, S., Gorodetsky, A., Krasnok, A., Rafailov, E., Belov, P.: Enhancement of terahertz photoconductive antenna operation by optical nanoantennas. Laser Photon. Rev. **11**(1), 1600199 (2017). https://doi.org/10.1002/lpor.201600199
86. Lavrukhin, D.V., Yachmenev, A.E., Glinskiy, I.A., Khabibullin, R.A., Goncharov, Y.G., Ryzhii, M., Otsuji, T., Spector, I.E., Shur, M., Skorobogatiy, M., Zaytsev, K.I., Ponomarev, D.S.: Terahertz photoconductive emitter with dielectric-embedded high-aspect-ratio plasmonic grating for operation with low-power optical pumps. AIP Adv. **9**(1), 015112 (2019). https://doi.org/10.1063/1.5081119
87. Matsubara, E., Nagai, M., Ashida, M.: Ultrabroadband coherent electric field from far infrared to 200 THz using air plasma induced by 10 fs pulses. Appl. Phys. Lett. **101**(1), 011105 (2012). https://doi.org/10.1063/1.4732524
88. Dai, J., Xie, X., Zhang, X.-C.: Detection of broadband terahertz waves with a laser-induced plasma in gases. Phys. Rev. Lett. **97**(10), 103903 (2006). https://doi.org/10.1103/PhysRevLett.97.103903

89. Andreeva, V.A., Kosareva, O.G., Panov, N.A., Shipilo, D.E., Solyankin, P.M., Esaulkov, M.N., Gonzalez de Alaiza Martinez, P., Shkurinov, A.P., Makarov, V.A., Berge, L., Chin, S.L.: Ultrabroad terahertz spectrum generation from an air-based filament plasma. Phys. Rev. Lett. **116**(6), 063902 (2016). https://doi.org/10.1103/PhysRevLett.116.063902
90. Png, G.M., Flook, R., Ng, B.W.-H., Abbott, D.: Terahertz spectroscopy of snap-frozen human brain tissue: an initial study. Electron. Lett. **45**(7), 343–345 (2009). https://doi.org/10.1049/el.2009.3413
91. Shi, L., Shumyatsky, P., Rodríguez-Contreras, A., Alfano, R.: Terahertz spectroscopy of brain tissue from a mouse model of Alzheimer's disease. J. Biomed. Opt. **21**(1), 015014 (2016). https://doi.org/10.1117/1.JBO.21.1.015014
92. Oddo, S., Caccamo, A., Shepherd, J.D., Murphy, M.P., Golde, T.E., Kayed, R., Metherate, R., Mattson, M.P., Akbari, Y., Ferla, F.M.L.: Triple-transgenic model of Alzheimer's disease with plaques and tangles: intracellular Aβ and synaptic dysfunction. Neuron. **39**(3), 409–421 (2003). https://doi.org/10.1016/S0896-6273(03)00434-3
93. Meng, K., Chen, T.-N., Chen, T., Zhu, L.-G., Liu, Q., Li, Z., Li, F., Zhong, S.-C., Li, Z.-R., Feng, H., Zhao, J.-H.: Terahertz pulsed spectroscopy of paraffin-embedded brain glioma. J. Biomed. Opt. **19**(7), 077001 (2014). https://doi.org/10.1117/1.JBO.19.7.077001
94. Doblas, S., He, T., Saunders, D., Pearson, J., Hoyle, J., Smith, N., Lerner, M., Towner, R.A.: Glioma morphology and tumor-induced vascular alterations revealed in seven rodent glioma models by *in vivo* magnetic resonance imaging and angiography. J. Magn. Reson. Imaging. **32**(2), 267–275 (2010). https://doi.org/10.1002/jmri.22263
95. Zou, Y., Li, J., Cui, Y., Tang, P., Du, L., Chen, T., Meng, K., Liu, Q., Feng, H., Zhao, J., Chen, M., Zhu, L.-G.: Terahertz spectroscopic diagnosis of myelin deficit brain in mice and rhesus monkey with chemometric techniques. Sci. Rep. **7**, 5176 (2017). https://doi.org/10.1038/s41598-017-05554-zID
96. Yamaguchi, S., Fukushi, Y., Kubota, O., Itsuji, T., Ouchi, T., Yamamoto, S.: Brain tumor imaging of rat fresh tissue using terahertz spectroscopy. Sci. Rep. **6**, 30124 (2016). https://doi.org/10.1038/srep30124
97. Ji, Y.B., Oh, S.J., Kang, S.-G., Heo, J., Kim, S.-H., Choi, Y., Song, S., Son, H.Y., Kim, S.H., Lee, J.H., Haam, S.J., Huh, Y.M., Chang, J.H., Joo, C., Suh, J.-S.: Terahertz reflectometry imaging for low and high grade gliomas. Sci. Rep. **6**, 36040 (2016). https://doi.org/10.1038/srep36040
98. Chernomyrdin, N.V., Gavdush, A.A., Beshplav, S.-I.T., Malakhov, K.M., Kucheryavenko, A.S., Katyba, G.M., Dolganova, I.N., Goryaynov, S.A., Karasik, V.E., Spektor, I.E., Kurlov, V.N., Yurchenko, S.O., Komandin, G.A., Potapov, A.A., Tuchin, V.V., Zaytsev, K.I.: *In vitro* terahertz spectroscopy of gelatin-embedded human brain tumors: a pilot study. Proc. SPIE. **10716**, 107160S (2018). https://doi.org/10.1117/12.2316302
99. Fan, S., Ung, B., Parrott, E.P.J., Pickwell-MacPherson, E.: Gelatin embedding: a novel way to preserve biological samples for terahertz imaging and spectroscopy. Phys. Med. Biol. **60**(7), 2703 (2015). https://doi.org/10.1088/0031-9155/60/7/2703
100. Zakharova, N., Kornienko, V., Potapov, A., Pronin, I.: Neuroimaging of Traumatic Brain Injury, p. 154. Springer, USA (2014)
101. Zhao, H., Wang, Y., Chen, L., Shi, J., Ma, K., Tang, L., Xu, D., Yao, J., Feng, H., Chen, T.: High-sensitivity terahertz imaging of traumatic brain injury in a rat model. J. Biomed. Opt. **23**(3), 036015 (2018). https://doi.org/10.1117/1.JBO.23.3.036015
102. Shi, J., Wang, Y., Chen, T., Xu, D., Zhao, H., Chen, L., Yan, C., Tang, L., He, Y., Feng, H., Yao, J.: Automatic evaluation of traumatic brain injury based on terahertz imaging with machine learning. Opt. Express. **26**, 6371–6381 (2018). https://doi.org/10.1364/OE.26.006371

Index

V. V. Tuchin et al. (eds.), *Multimodal Optical Diagnostics of Cancer*,
https://doi.org/10.1007/978-3-030-44594-2

E

Printed by Printforce, the Netherlands